7 DAY

University of Plymouth
Charles Seale Hayne Library
Subject to status this item may be renewed
via your Voyager account

http://voyager.plymouth.ac.uk
Tel: (01752) 232323

Dietary Sugars
Chemistry, Analysis, Function and Effects

Food and Nutritional Components in Focus

Series Editor:
Professor Victor R Preedy, *School of Medicine, King's College London, UK*

Titles in the Series:
1: Vitamin A and Carotenoids: Chemistry, Analysis, Function and Effects
2: Caffeine: Chemistry, Analysis, Function and Effects
3: Dietary Sugars: Chemistry, Analysis, Function and Effects

How to obtain future titles on publication:
A standing order plan is available for this series. A standing order will bring delivery of each new volume immediately on publication.

For further information please contact:
Book Sales Department, Royal Society of Chemistry, Thomas Graham House, Science Park, Milton Road, Cambridge, CB4 0WF, UK
Telephone: +44 (0)1223 420066, Fax: +44 (0)1223 420247
Email: booksales@rsc.org
Visit our website at http://www.rsc.org/Shop/Books/

Dietary Sugars
Chemistry, Analysis, Function and Effects

Edited by

Victor R Preedy
School of Medicine, King's College London, UK

RSCPublishing

Food and Nutritional Components in Focus No. 3

ISBN: 978-1-84973-370-0
ISSN: 2045-1695

A catalogue record for this book is available from the British Library

Published by The Royal Society of Chemistry,
Thomas Graham House, Science Park, Milton Road,
Cambridge CB4 0WF, UK

Registered Charity Number 207890

For further information see our web site at www.rsc.org

Printed and bound in Great Britain by CPI Group (UK) Ltd, Croydon, CR0 4YY, UK

Preface

In the past three decades there have been major advances in our understanding of the chemistry and function of nutritional components. This has
been enhanced by rapid developments in analytical techniques and instrumentation. Chemists, food scientists and nutritionists are, however, separated
by divergent skills, and professional disciplines. Hitherto, this transdisci-
plinary divide has been difficult to bridge.

The series **Food and Nutritional Components in Focus** aims to cover in a single
volume the chemistry, analysis, function and effects of single components in the
diet or its food matrix. Its aim is to embrace scientific disciplines so that
information becomes more meaningful and applicable to health in general.

The series **Food and Nutritional Components in Focus** covers the latest
knowledge base and has a structured format.

Dietary Sugars has 4 major sections, namely:

- Caffeine in Context;
- Chemistry and Biochemistry;
- Analysis; and
- Function and Effects.

Coverage includes sugars in the context of honey, dental caries, whole-body
metabolism and glycemic load. Thereafter, there are sections on the chemistry
of glucose, galactose, maltose, fructose, sucrose, lactose. Other sugars are also
described throughout the book. Methodical aspects include characterization
and assays of sugars in seeds, urine, blood, human milk, vegetables, dairy
produce and other foods. The techniques cover gas, ion and thin layer
chromatography, UV spectrophotometry, electrochemical detection, mass
spectrometry, biosensors, enzymatic reactions, high-temperature liquid
chromatography, Raman spectroscopy and other methodology. In terms of

Food and Nutritional Components in Focus No. 3
Dietary Sugars: Chemistry, Analysis, Function and Effects
Edited by Victor R Preedy
© The Royal Society of Chemistry 2012
Published by the Royal Society of Chemistry, www.rsc.org

function and effects on health and disease there are chapters on obesity and childhood BMI, total parenteral nutrition, intestinal transport, tracers in metabolic studies, modelling neurological aging, sugar preference, beer, metabolic syndrome, nonalcoholic fatty liver disease, antioxidant defence, young children, lactose intolerance, industry, technology and biotechnology.

Each chapter transcends the intellectual divide with a novel cohort of features, namely by containing:

- Key Facts (areas of focus explained for the lay person);
- Definitions of Words and Terms; and
- Summary Points.

The book is designed for chemists, food scientist and nutritionists, as well as health care workers and research scientists. Contributions are from leading national and international experts, including contributions from world renowned institutions.

Professor Victor R Preedy,
King's College London

Contents

Food and Nutritional Components in Focus No. 3
Dietary Sugars: Chemistry, Analysis, Function and Effects
Edited by Victor R Preedy
© The Royal Society of Chemistry 2012
Published by the Royal Society of Chemistry, www.rsc.org

Chemistry and Biochemistry

Chapter 9 Sucrose Chemistry **138**

Leonardo M. Moreira, Juliana P. Lyon, Patrícia Lima, Vanessa J. S. V. Santos and Fabio V. Santos

Analysis

Function and Effects

Chapter 35 Dextrose in Total Parenteral Nutrition **619**
Karen C. McCowen

Chapter 36 The Interstinal Transport of Galactase **635**
María Jesús Rodríguez Yoldi

Chapter 47 Technology and Biotechnology of Lactose Contained in Raw Food Materials
Magdalini Soupioni, Maria Kanellaki and Loulouda A. Bosnea

Dietary Sugars in Context

CHAPTER 1
Sugars in Honey

SEVGI KOLAYLI,[a] LAÏD BOUKRAÂ,*[b]
HÜSEYIN ŞAHIN[c] AND FATIHA ABDELLAH[d]

[a] Karadeniz Technical University, Department of Chemistry, Faculty of Sciences, 61080 Trabzon, Turkey; [b] Institute of Veterinary Sciences, Ibn-Khaldun University of Tiaret, 14000 Tiaret, Algeria; [c] Karadeniz Technical University, Department of Chemistry, Faculty of Sciences, 61080 Trabzon, Turkey; [d] Institute of Veterinary Sciences, Ibn-Khaldun University of Tiaret, 14000 Tiaret, Algeria
*Email: l_boukraa@mail.univ-tiaret.dz

1.1 Introduction

Honey is produced by honey bees from the nectar of plants, as well as from dew. For a long period of human history, honey was an important source of carbohydrates and the only widely available sweetener, until the production of industrial sugar began to replace it after 1800. The natural product mainly consists of carbohydrates and a small amount of other compounds such as phenolics, proteins, amino acids, minerals, vitamins, pigments and organic acids (Figure 1.1). Honey has been used in folk medicine since the early ages of human history and in more recent times their role in the treatment of burns, gastrointestinal disorders, asthma, infected wounds and skin ulcers has been reinvestigated. The composition, nutritional value, appearance and sensory properties of honey differ in relation to its botanical origin and the geographical area where bee hives are located.

Sugars are the main constituents of honey containing about 95% of honey dry weight. In general, honey sugars contain 70% monosaccharide and 10–15% disaccharides and oligosaccharides composed of glucose and fructose

Food and Nutritional Components in Focus No. 3
Dietary Sugars: Chemistry, Analysis, Function and Effects
Edited by Victor R Preedy
© The Royal Society of Chemistry 2012
Published by the Royal Society of Chemistry, www.rsc.org

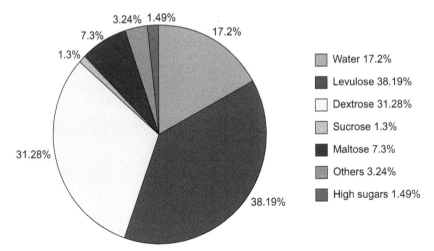

Figure 1.1 Pie-chart of honey composition indicating the percentage share of various sugars, water and other minor constituents.

(Ouchemoukh *et al.* 2010). Honey sugars are responsible for many of the physicochemical properties of honey, such as sweetness, viscosity, hygroscopy, granulation and energy value (Cavia *et al.* 2002). Sugar composition has also been used to distinguish honey samples by botanical origin or geographical origin. Sugar's composition is affected by contributions of the plant and environmental conditions. Sugars contained in nectars are mainly fructose, glucose and sucrose, but their relative proportions are usually variable, however, they are quite consistent for certain botanical families. On the other hand, the relative amount of the two major monosaccharaides (fructose and glucose) is useful for the classification of monofloral honeys, as well as the fructose : glucose and the glucose : water ratios. The minor sugars have a relatively low value for the determination of botanical origin. For example, honeydew honey contains a higher amount of di- and trisaccharides, especially melezitose, which are present in blossom honeys. The numerous di- and trisaccharides in honey are produced by microbial activity and enzymatic reactions in the intestinal tract of the aphids and during honey ripening. The small differences in the sugars' spectra of blossom honeys are explained by the fact, that di- and tri-saccharides are mainly produced through trans-glycosylation or enzymatic reversion by the alpha-glycosidase in honey (Ruoff 2006). Honey sugar profile was used to detect adulteration with cheap invert sugars.

1.2   Main Sugars

1.2.1   Monosaccharides

The main sugars are the hexose monosaccharides fructose and glucose, which are produced by the hydrolysis of the disaccharide of sucrose (Table 1.1).

Table 1.1 The most common sugar types in honeys.

Monosaccharides	Disaccharides	Trisaccharides
Glucose	Sucrose (max. %5)	Melezitose
Fructose	Maltose (0.5–3.5)	Isomaltotriose
Arabinose	Isomaltose	Raffinose
	Turanose	Theanderose
	Trehalose	Isopanose
	Neotrehalose	Erlose
	Melibiose	Panose
	Maltulose	Maltotriose
	Kojibiose	Laminaritriose
	Gentiobiose	Kestose
	Palatinose	Neokestose (Fructose Trisaccharide)
	Nigerose	Cellobiose
	Laminaribiose	
	Difructose Anhydride	

Fructose is the main monosaccharide that is found in three main forms in the diet: as free fructose (in fruits and honey), as a constituent of the disaccharides sucrose or as fructans, a polymer of fructose usually in oligosaccharide form (in some vegetables and wheat) (Shepherd and Gibson 2006). Fructose, a natural sugar found in honey and many fruits, is consumed in significant amounts in Western diets. Fructose is approximately two times sweeter than sucrose, glucose is less sweet and maltose even less sweet and therefore fructose is responsible for honey sweetness and is commonly used as a bulk sweetener (Rizkalla 2010). In nearly all honey types, fructose prodominates and only a few honeys, such as rape (*Brassica napus*), dandelion (*Taraxacum officinale*) and blues curls (*Trihostema lanceolatum*). Rape appears to contain more glucose than fructose (Kaškoniene *et al.* 2010). The relative amount of the two monosaccharides fructose and glucose is frequently useful for the classification of unifloral honeys (Bogdanov 2009).

Glucose is the second main sugar in honey and varies from 25 g to 42 g per 100 g. The actual proportion of glucose and fructose in any particular honey depends largely on the source of nectar (Anklam 1998). It is reported that a F : G ratio of 1.14 or less would indicate fast granulation, while values over 1.58 are associated with no tendency to granulation (Tosi *et al.* 2004; White *et al.* 1975). The F : G ratio found in chestnut honeys is higher and, thus, these honeys are not prone to crystallization (Tezcan *et al.* 2010). It is also suggested that the ratio of fructose to glucose could be used to typify honey samples from different origins, moreover it may indicate the tendency of honey to crystallize (Mendes *et al.* 1998).

1.2.2 Disaccharides

Sucrose, which is composed of fructose and glucose linked together, is a disaccharide. It comprises a little over 1% of the composition of honey. Honey contains other disaccharides which make up over 7% of its composition.

Some of the disaccharides in honey are sucrose, maltose, kojibiose, turanose, isomaltose and maltulose. Sucrose content is important to detect heavy sugars feeding of the bees or adulteration by direct addition of saccharose. According to some studies, the amount of sucrose has been used to distinguish the adulteration of honey samples by sugar syrups (Özcan *et al.* 2006). For example, supplementary feeding of honey bees with saccharose syrup caused a higher saccharose level in honey (Özcan *et al.* 2006). Other studies reported that sucrose content of honey was not an effective property for distinguishing pure blossom honey from adulterated (with sucrose syrup) honey (Guler *et al.* 2007).

1.2.3   Oligosaccharides

Generally, honeydew honey compared with blossom honey contains higher amounts of di-, tri-, and higher oligosaccarides, non-reducing sugars, such as melezitose, maltotriose and raffinose, which are usually not found in blossom honeys (Diez *et al.* 2004). However, Weston and Brocklebank (1999) had studied oligosaccharide fraction in samples of manuka, heather, clover and honeydew honey from New Zealand by high-performance anion-exchange chromatography with pulsed amperometric detection. The monosaccaharide amount was the lowest in honeydew honey among these honeys, while oligosaccharide amount was the highest in the study. In addition, isomaltose (maltulose), kojibiose, turanose (gentiobiose), nigerose and maltose were the major oligosaccharide component, and isomaltose was the major component in heather honey.

Honey sugars are mainly formed by the action of several enzymes on nectar sucrose. Although adult bees can use glucose, fructose, sucrose, trehalose, maltose and melezitose, they cannot use rhamnose, xylose, arabinose, galactose, mannose, lactose, raffinose, dextrin or inulin. The differences in carbohydrate utilization between larvae and adults may be due to the absence of appropriate enzymes. Food enters the alimentary canal by way of the mouth of the bee and passes through the esophagus to the honey stomach. Invertase, in the honey stomach, breaks down the sucrose of nectar to the samples monosaccharides, glucose and fructose present in honey (Özcan *et al.* 2006).

Oligosaccharide profiling could also be used to check the authenticity of honey types and adulteration. For example, theanderose, a glucosyl-sucrose, was detected in all cane sugar samples examined and has not been detected in any of the beet sugar samples. It was proposed that the presence of theanderose is a better indicator for distinguishing cane and beet white sugars (Morel du Boil 1996).

Maltose, a reducing sugar, is a disaccharide, produced by breakdown of starch by enzymatic digestion, and a partial hydrolysis. It is present in germination grain, in a small proportion, in corn syrup.

Raffinose is a trisaccharide composed of galactose, fructose and glucose. It was reported to be present in heterofloral honey (0.3–0.4 mg/g). The origin of raffinose in floral honey is not clear; it is suggested that rafinose could be a nectar constituent or could be present through honeydew contamination (Kaškonienè *et al.* 2010). Clarck *et al.* (1992) reported that raffinose amount in the range of 300–1200 mg/kg is indicative of beet sugars, since no cane sugar tested contained observable raffinose. It is reported that raffinose and melezitose were observed only in chestnut and fir honey samples among the 469 monofloral honeys, such as the rape, sunflower, acacia, lavender and cinder heather (Table 1.2). Melezitose was found in fir honey and chestnut honey, 2.22 ± 0.48 and 0.42 ± 0.31 g per 100 g of honey, respectively, and was not detected in cinder heather, lavender, acacia, rape and sunflower.

Erlose is an intermediate trisaccharide in the metabolism of nectar sugars by honey bees. It is made from sucrose by transglucosylation of the α-D-glucosyl group of a molecule of sucrose to the fourth position of the glucose moiety of another sucrose.

Table 1.2 The detected maximum amount of sucrose concentration according to the honey types (Arvanitoyannis 2005).

Sugar composition criteria for honey according to Council Directive 2001/110/EC of 20 December 2001, and CL 1998/12-S of the Codex Alimentarius 2001	Limit
Total content of fructose and glucose	
-blossom honey	Not less than 60 g/100 g
-honeydew honey, blends of honeydew honey with blossom honey	Not less than 45 g/100 g
Sucrose content	
-in general	Not more than 5 g/100 g
-avacado (*Persea americano* Mill.)	
-chestnut (*Castenae sativa*)	
-erica (*Heather*)	
-linden (*Tilia argentea*)	
-leatherwood (*Eucryphia lucida*, eucriphia milliganii)	
-phacelia (*Phacelia tancetifolia*)	
-rosemary (*Rosmarinus officinals*)	
-rhododendron (*Rhododendron ponticum*)	
-sunflower (*Helianthus annus*)	
-taraxacum (dandelion)	
-thyme (Thymus)	
-honeydew	
-false acacia (*Robinia pseudoacacia*)	Not more than 10 g/100 g
-alfalfa (*Medicoga sativa*)	
-citruss spp.	
-french honeysuckle (*Hedysarum*)	
-Mensies Banksia (*Banksia menziesil*)	
-red gum (*Eucalyptus camadulensis*)	
-trifolium (Trifolium spp.)	
-borage (*Borage officinalis*)	Not more than 15 g/100 g
-calothamnus san.	
-eucalyptus scab.	
-Banksia gr.	
-lavander (*Laavandula* spp.)	
-Xantharrhoea pr.	

Difructose anhydrides (DFAs) are pseudodisaccharides produced by condensation of two fructose molecules by means of a caramelization reaction which takes place during heating of sugars (Montilla *et al.* 2006). The concentration of DFAs in honey samples is recommended as an indicator of honey adulteration, with high fructose corn syrup or invert syrup, allowing the detection of values as low as 5% (w/w).

1.3   Adulteration of Honey

For centuries, the purity and naturalness of commercialized honey has been questioned. Because honey composition is highly variable, adulteration is very easy to do by overfeeding with inexpensive sweeteners such as saccharose syrups, corn syrups, high fructose syrups, invert syrups and saccharide variants. Overfeeding bees with saccharide or invert saccharide derivatives is practiced by beekeepers to increase honey production.

Sucrose analysis has been frequently used to determine the adulteration, but the test is not adequate, because worker bees convert saccharose to glucose and fructose by digestive enzymes (Ruiz-Matute *et al.* 2010). The ratio of fructose and glucose indicates honey adulteration. However, some researchers have reported that saccharose, fructose and glucose amount can be used to distinguish pure honey from adulterated honey. Maltose is usually present in honey in low quantities (30 mg/g) and is suggested as a marker of natural honey. Higher amounts of maltose concentration may indicate adulteration of honey by sugar syrup or starch hydrolysate. Maltose and isomaltose are anomers and honeydew honeys showed significanlty higher isomaltose concentrations than flora honeys.

It was suggested that high maltose : isomaltose ratio may indicate adulteration of honey by starch hydrolysate; on the contrary, low maltose : isomaltose values may indicate the use of high fructose-containing syrup (Kaškonienè *et al.* 2010). In recent studies, some of the carbohydrates such as inulin, inulobiose, inulotriose, were used as adulteration markers. For example, inulobiose is not present in honeys, but is detected in adulterated honeys.

1.4   Crystallization of Honey

Crystallization of honey is an undesirable process because it affects its textural properties, making it less appealing to the consumer, and in many cases, it results in increased moisture of the liquid phase, which can allow naturally occurring yeast cells to multiply, causing fermentation of the honey. Glucose may crystallize as a α-D-glucose monohydrate with the stable crystalline form below 50 °C, as an α-D-glucose anhydrous, stable form between 50 °C and 80 °C and β-anhydrous form, stable above 80 °C (Young 1957). In solutions saturated with fructose, the transition temperature from glucose monohydrate to anhydrous glucose has found to be below 30 °C (Cavia *et al.* 2002). Glucose solubility is lower than fructose; a high glucose ratio may facilitate the crystallization of honeys. This natural phenomenon happens when glucose, one of the three main sugars in honey, spontaneously precipitates out of the super-saturated honey solution. The glucose loses water (becoming glucose mono-hydrate) and takes the form of a crystal (a solid body with a precise and orderly structure) (Assil *et al.* 1991). The crystals form a lattice that immobilizes other components of honey in a suspension, thus creating a semi-solid state (McGee 1984). The tendency of honey to crystallize depends primarily on its glucose content and moisture level. The overall composition of honey, which includes sugars other than glucose and more than 180 identified substances such as minerals, acids and proteins, also influences crystallization. Additionally, crystallization can be stimulated by any small particles – dust, pollen, bits of wax or propolis, air bubbles – that are present in the honey. These factors are related to the type of honey and are influenced by how the honey is handled and processed. Storage conditions (temperature, relative humidity and type of container) may also influence the tendency of honey to crystallize.

1.5 Is Honey Sweeter than Sugar?

Fructose is slightly sweeter than sucrose; glucose is less sweet, and maltose even less sweet. In most honeys, fructose predominates and tends to make honey taste slightly sweeter than sugar. Some honeys that are very rich in fructose tend to taste very sweet, but there are a few types of honeys which contain more glucose than fructose. On the average honey is 1 to 1.5 times sweeter (on a dry weight basis) than sugar. Liquid honey is approximately as sweet as sugar, yet it contains only 82.4 g carbohydrates/100 g (vs. 100 g for sucrose) and provides only 304 Kcal/100 g (*vs.* 400 Kcal for sucrose) (National Honey Board 2011). Honey and sugar have a lot in common, but there are some differences that do appear to give honey an edge. The question is, are those differences significant? First, let's examine the similarities. Both honey and sugar are calorie-dense sweeteners and both are primarily made up of a combination of fructose and glucose, though their chemical structure differs. In sugar, the fructose and glucose are bound together. The combination is called sucrose. In honey, fructose and glucose are primarily independent of each other, although honey also contains a bit of sucrose. Honey generally contains more fructose than glucose, and since fructose tastes sweeter than glucose, you might find yourself using less honey than you would sugar. At the same time, honey is denser, so it weighs more than sugar. Because of this, a tablespoon of honey has 64 calories, compared with 45 calories for the same amount of table sugar. So, even if you use less honey, the calorie intake might be higher (Martha 2011).

1.6 Honey Sugars and Health

1.6.1 Honey and Blood Glucose Level

The glycemic index of honey and sugar are roughly the same. The glycemic index is the ranking of carbohydrates according to their effect on blood glucose levels. A piece of bread's glycemic index score is around 100, while sugar's is 58 and honey's can range from 30–58. Low glycemic index foods may help people lose weight, improve diabetes control and help keep people full for longer. Just because sugar and honey are made up of the same ingredients does not mean they have the same amount of nutrients. In a comparison of the two side by side, weight for weight, it may look like honey is better in calories. However, when looking at the amount of tablespoons that they equal, sugar may look better, because 1 tablespoon of sugar would be 46 calories (Ischayek and Kern 2006). Some honeys, *e.g.* acacia and yellow box, with relatively high concentration of fructose, have a lower GI than other honey types (Table 1.3).

1.6.2   Honey as an Anti-infective Agent

The intrinsic properties of honey affect the growth and survival of microorganisms; in particular, the low pH and high sugar content of undiluted honeys prevent the growth of many species of microorganisms (Iurlina and Fritz 2005). Honey has potent antibacterial activity and is effective in preventing and clearing wound infections (Allen 2000). Topical honey was shown to be effective in treating postoperative skin wounds in neonates that had failed to respond to antibiotic therapy (Vardi *et al.* 1998). It has been demonstrated in many studies that the antibacterial effects of honey are attributed to its high osmolarity, low pH, hydrogen peroxide content and presence of other uncharacterized compounds (Molan 1995).

1.6.3   Honey as a Prebiotic

It has been found that the activity of certain species of *Bifidobacterium* in the colon can be stimulated by fructo-oligosaccharides (Shin *et al.* 2000) and the oligosaccharides from soya beans (Liu 1997). Honey is also effective

Table 1.3   Glycemic index (GI) and glycemic load (GL) for a serving (25 g) of honey (Foster-Powell *et al.* 2002).

	Honey Origin	*Fructose g/100 g*	*GI*	*AC g/serving*	*GL (per serving)*
Acacia (black lockust)	Romania	43	32	21	7
Yellow box	Australia	46	35 ± 4	18	6
Stringy bark	Australia	52	44 ± 4	21	9
Red gum	Australia	35	46 ± 3	18	8
Iron bark	Australia	34	48 ± 3	15	7
Yapunya	Australia	42	52 ± 5	17	9
Pure Australia	Australia		58 ± 6	21	12
Commercial blend	Australia	38	62 ± 3	18	11
Salvation June	Australia	32	64 ± 5	15	10
Commercial blend	Australia	28	72 ± 6	13	9
Honey of unspecified origin	Canada		87 ± 8	21	18
Average		55	55 ± 5	18	10
Glucose			100		
Fructose			19		

AC = available carbohydrate.

in stimulating the growth of bifidobacteria (Kajiwara *et al.* 2002), and again it may be the oligosaccharides that are the active components (Weston and Brocklebank 1999). The importance of these oligosaccharides lies in the fact that the α – galactosidic linkages present cannot be digested by humans (Gopal *et al.* 2001) but, on entering the colon, they can be metabolised by *Bifidobacteria*. Although the stimulatory role of oligosaccharides on the gut flora(s) has received the most attention, it has been speculated that the same components in honey could inhibit the development of pathogens like *Helicobacter pylori* or *Staphylococcus aureus* in the body. More specifically, it has been proposed that the oligosaccharides could become attached to the cell walls of the bacteria and prevent adhesion to human tissues (Somal *et al.* 1994).

In conclusions, honey is a mixed viscous natural product; it is difficult to evaluate exactly its sugar composition and other properties. Fructose and glucose are the main sugars and it contains several oligosaccharides in small amounts. Although several works have focused on the study of mono- and disaccharides, knowledge of tri- and tetrasaccharides is still limited. Up to now, there were 25–30 individual sugars detected and evaluated. The profiles and ratio of the sugars are used to determine the botanical properties and adulteration of honey.

Summary Points

- Honey is one of the most complex natural mixtures and mainly consists of carbohydrates.
- The sugar composition is variable and depends on the type of flower used by the bees.
- Fructose, glucose and sucrose are the main types of sugars in honey.
- Honey sugars are responsible for many of the physicochemical properties of honey, such as sweetness, viscosity and energy value.
- Honeydew honey contains a higher amount of di- and trisaccharides, especially melezitose, which is present in blossom honeys.
- Honey sugar profile is used to detect adulteration with cheap invert sugars.
- Crystallization happens when glucose, one of three main sugars in honey, spontaneously precipitates out of the supersaturated honey solution.
- Fructose is slightly sweeter than sucrose; glucose is less sweet, and maltose even less sweet.
- Both honey and sugar are calorie-dense sweeteners.
- Both honey and sugar are primarily made up of a combination of fructose and glucose, though their chemical structures differ.
- The glycemic index of honey and sugar are relatively the same.
- Some honeys, *e.g.* acacia and yellow box, which have a relatively high concentration of fructose, have a lower GI than other honey types.
- Honey has potent antibacterial activity and is effective in preventing and clearing wound infections.
- Honey is also effective in stimulating the growth of *Bifidobacteria* and acts as a prebiotic.

Key Facts

1. As it is highly concentrated in carbohydrates, honey is a good source of energy.
2. For a long period of human history, honey was an important source of carbohydrates and was the only widely available sweetener. The sweet taste is mainly due to the amount of fructose.
3. Honey seems to have the potential to clear infection as well as to be an effective prophylactic agent that may contribute to reducing the risks of cross-infection. This is due, in a great part, to its high concentration in honey.
4. In diabetics, consuming honey may have the same effect as sugar, however, honey with a high concentration of fructose has a low glycemic index.
5. Honey contains a wide variety of substances that can function as pre-biotics, which are substances that increase the growth and activity of good bacteria to help with the GI tract. Researches have shown that adding honey to dairy products such as yogurt or milk may enhance the growth, activity and viability of the prebiotics.

Definition of Words and Terms

Adulteration: means to make impure or inferior by adding foreign substances to a product.

Antioxidant: An antioxidant is a molecule capable of inhibiting the oxidation of another molecule by removing free radical intermediates, so antioxidants are often reducing agents such as thiols, ascorbic acid or polyphenols.

Bifidobacteria: they are one of the major genera of bacteria that make up the colon flora. They aid digestion and are associated with a lower incidence of allergies. They also prevent some forms of cancer.

Honeydew honey: is a classification that refers to honey produced by honey bees collecting nectar that is exuded from another insect, such as an aphid or scale insect.

Glycemic index: or GI is a measure of the effects of carbohydrates on blood sugar levels. Carbohydrates that break down quickly during digestion and release glucose rapidly into the bloodstream have a high GI

Hygroscopy: it is the ability of a substance to attract and hold water molecules from the surrounding environment which leads to an increase in volume, stickiness, or other physical characteristic of the material.

Monofloral: Monofloral honey is a type of honey which has a high value in the marketplace because it has a distinctive flavor or other attribute due to its being predominantly from the nectar of one plant species.

Prebiotic: Prebiotics are non-digestible food ingredients that stimulate the growth and/or activity of bacteria in the digestive system in ways claimed to be beneficial to health.

List of Abbreviations

F   Fructose
G   Glucose
IHC   The Institute for Human Continuity
mg   Milligram
g   Gram
w   Weigh
DFAs Di-Fructose Anhydrides
max   Maximum
spp   Is Used To Denote All Species Of A Higher Taxon
AOAC Association of Official Analytical Chemists International
HPLC High Performance Liquid Chromatography
GC   Gas Chromatography
MS   Mass Spectrometry

References

Allen, K.L. The potential for using honey to treat wounds infected with MRSA and VRE. First World Wound Healing Congress, Melbourne, Australia. September 12–13, 2000.

Arvanitoyannis, I.S., 2005. Novel quality control methods in conjuction with chemometrics (multivariate analysis) for detecting honey authenticity. *Critical Reviews in Food Science and Nutrition*. 45: 193–203.

Assil, H.I., Sterling, R. and Sporns, P., 1991. Crystal control in processed liquid honey. *Journal of Food Science*. 56 (4): 1034.

Bogdanov, S., 2009. Honey composition. Bee Products Science, Book of Honey, Chap. 5. www.bee-hexagon.net

Cavia, M.M., Fernandez-Muino, M.A., Gomez-Alonso, E., Montes-Pérez, M.J., Huidobro, J.F., and Sancho, M.T., 2002. Evolution of fructose and glucose in honey over one year: Influence of induced granulation. *Food Chemistry*. 78: 157–161.

Codex Alimentarius, 2001. *Organically Produced Foods*. Food and agriculture organization of the United Nations world health organization, Rome.

Diez, M.J., Andrés, C., and Terrab, A., 2004. Physicochemical parameters and pollen analysis of Moroccan honeydew honeys. *International Journal of Food Sciences and Technomogy*. 39: 167–176.

Foster-Powell, K., Holt, S.H.A. and Brand-Miller, J.C., 2002. International table of glycemic index and glycemic load values: 2002. *The American Journal of Clinical Nutrition*. 76: 5–56.

Gopal, P.K., Sullivan, P.A. and Smart, J.B., 2001. Utilisation of galacto-oligosaccharides as selective substrates for growth by lactic acid bacteria including Bifidobacterium lactis and Lactobacillus rhamnosus. *International Dairy Journal*. 1: 19–25.

Ischayek, J.I. and Kern, M., 2006. US honeys varying in glucose and fructose content elicit similar glycemic indexes. *The Journal of the American Dental Association.* 106 (8): 1260–1262.

Iurlina, M.O. and Fritz, R., 2005. Characterization of microorganisms in Argentinean honeys from different sources. *International Journal of Food Microbiology.* 105: 297–305.

Kaškonienè, V., Venskutonis, P.R. and Ceksterytè, V., 2010. Carbohydrate composition and electrical conductivity of different origin honeys from Lithuania. *LWT- Food Science and Technology.* 43: 801–807.

Liu, K., 1997. *Soybeans: Chemistry, Technology and Utilisation,* Chapman & Hall, London.

McGee, H., 1984. *On Food and Cooking: The science and lore of the kitchen.* Macmillan Publishing Company, New York.

Mendes, E., Proenca, M.E.B., Ferreira, I.M.P.L.V.O. and Ferreira, M.A., 1998. Quality evaluation of Portuquese honey. *Carbohyrate Poylmers.* 37: 219–223.

Molan P.C., 1995. The antibacterial properties of honey. *Chemistry in New Zealand.* 59: 10–14.

Montilla, A., Ruiz-Matute, A.I., Sanz, M.L., Martíez-Castro, I. and del Castillo, M.D., 2006. Difructose anhydrides as quality markers of honey and coffee. *Food Research International.* 39: 801–806.

Morel du Boil, P.G., 1996. Theanderose-A characteristic of cane sugar crystals. *Proc S Afr Sud Technol Ass.* 70: 141–144.

National Honey Board, 2011: Carbohydrates and the sweetness of honey. 303: 776–2337. www.honey.com

Ouchemoukh, S., Schweitzer, P., Bachir Bey, M., Djoudad-Kadji, H. and Louaileche, H., 2010. HPLC sugar profiles of Algerian honeys. *Food Chemistry.* 121: 561–568.

Özcan, M., Arslan, D. and Durmuş, A.C., 2006. Effect of inverted saccharose on some properties of honey. *Food Chemistry.* 99: 24–29.

Rizkalla, S.W., 2010. Health implications of fructose consumption: A review of recent data. *Nutrition and Metabolism.* 7: 82.

Ruiz-Matute, A.I., Rodrìguez-Sànchez, S., Sanz, M.L. and Matìnez-Castro, I., 2010. Detection of adulteration of honey with high fructose syrups from inulin by GC analysis. *Journal of Food Composition and Analysis.* 23: 273–276.

Ruoff, K., 2006. Doctor tesis/Authentication of the botanical origin of honey. Helsinki University, Diss. Eth.no.16850.

Shepherd, S.J. and Gibson, P.R., 2006. Fructose malabsorption and symptoms of irritable bowel syndrome: guidelines for effective dietary management. *Journal of the American Dietetic Association.* 106: 1631–1639.

Shin, H.S., Lee, J.H., Pestka, J.J. and Ustunol, Z. 2000. Growth and viability of commercial Bi.dobacterium spp. in skim-milk containing oligosaccharides and inulin. *Journal of Food Science,* Vol. 65 No. 5, pp. 884–7.

Somal, N.A., Coley, K.E., Molan, P.C. and Hancock, B.M. 1994. Susceptibility of Helicobacter pylori to the antibacterial activity of manuka honey. *Journal Royal Society of Medicine.* Vol. 87, pp. 9–12.

Tezcan, F., Kolaylı, S., Sahin, H., Ulusoy, E. and Erim, B.F., 2010. Evaluation of organic acid, sacchraride composition and antioxidant properties of some authentic Turkish honeys. *Journal of Food and Nutrition Research.* 50: 33–40.

Tosi, E.A., Rè, E., Lucero, H. and Bulacio, L., 2004. Effect of honey high temperature short-time heating on parameters related to quality, crystallization phenomena and fungal inibition. *Lebensmittel Wissenschaft und Technologie.* 37: 669–678.

Vardi A., Barzilay Z. and Linder N., 1998. Local application of honey for treatment of neonatal postoperative wound infection. *Acta Paediatrica.* 87: 429–432.

Weston, R. and Brocklebank, L.K., 1999. The oligosaccharide composition of some New Zealand honeys. *Food Chemistry.* 64: 33–37.

White, J.W., Willson, R.B., Maurizio, A. and Smith, F.G., 1975. *Honey. A Comprehensive Survey.* London: Heinemann, pp. 608.

Young, F.E., 1957. D-glucose-water phase diagram. *Journal of Physical Chemistry.* 61: 616–619.

CHAPTER 2
Sugars and Dental Caries

ANNA HAUKIOJA*[a] AND MERJA LAINE[b]

[a] Institute of Dentistry, Faculty of Medicine, University of Turku, Lemminkäisenkatu 2, FI-20520 Turku, Finland; [b] Docent, Specialist in Clinical Dentistry, Institute of Dentistry, Faculty of Medicine, University of Turku, Lemminkäisenkatu 2, FI-20520 Turku, Finland
*Email: anna.haukioja@utu.fi

2.1 Dental Caries

2.1.1 Dental Caries and Erosion

The tooth is composed of three hard tissues called enamel, dentin, cementum and a soft tissue called dental pulp (Figure 2.1). When healthy, there is a balance between remineralisation and demineralisation of dental hard tissues (Equation 1).

$$\text{Hydroxyapatite: } Ca_5(PO_4)_3(OH)_2(s) \leftrightarrow 5Ca^{2+}(aq) + 3PO_4{}^{3-}(aq) + 2OH^-(aq)$$

$$(1)$$

Caries is a process in which this balance is disturbed: acids produced by bacteria on the tooth surface shift the equilibrium towards demineralisation. If the acidic challenges are too strong, too long or too frequent, the softening of the enamel surface allows the microbes to invade deeper into the underlying tooth tissues and cause tooth decay. In particular, newly erupted teeth are at caries risk. In adults, pH 5.5 is often considered the critical pH for enamel demineralisation. On the other hand, clinical studies show that the caries process can be arrested and even reversed.

Food and Nutritional Components in Focus No. 3
Dietary Sugars: Chemistry, Analysis, Function and Effects
Edited by Victor R Preedy
© The Royal Society of Chemistry 2012
Published by the Royal Society of Chemistry, www.rsc.org

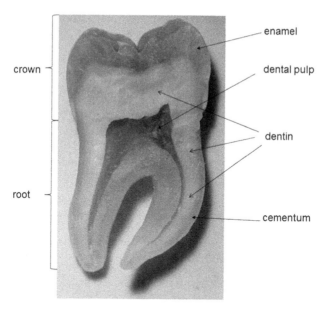

Figure 2.1 The structure of a tooth. A tooth consists of a crown and a root. The three hard tissues of a tooth are enamel, dentin and cementum, and a soft tissue called the dental pulp. Enamel is the hardest tissue in the human body; the structure of dentin and cementum resembles that of bone. The innermost tissue is the dental pulp. It contains nerves, lymph and blood vessels and keeps the tooth alive by nourishing and protecting the tooth. (Unpublished).

Dental erosion is the irreversible loss of tooth structure due to dissolution of minerals caused by acids of non-bacterial, *e.g.* dietary or gastrointestinal, origin. The prevalence and severity of erosion has increased in younger age groups during the last decades due to increased consumption of acidic soft drinks and juices (Tahmassebi *et al.* 2006).

2.1.2 Dental Plaque in Caries

Dental plaque is a biofilm: an organised community of bacteria attached to tooth surface and covered by an extracellular polymer matrix. According to the ecological plaque hypothesis, oral diseases with microbial aetiology are not caused by specific microbes but by a disturbance in the microbial balance of the plaque (Marsh 2006; Figure 2.2). With respect to caries, there is a shift towards dominance by acidogenic and acid-tolerant gram-positive species (Marsh 2006). These species benefit from the acid challenge. While other bacteria are inhibited in acidic conditions, acid-tolerant species are able to live and even grow in acidic conditions. Bacterial species associated with dental caries include mutans streptococci, *i.e. Streptococcus mutans* and *Streptococcus sobrinus* as well as certain species of lactobacilli, bifidobacteria and *Actinomyces* (Beighton 2005).

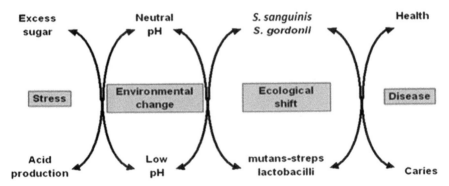

Figure 2.2 The ecological plaque hypothesis. Acidogenic bacteria in dental plaque
produce acids from dietary sugars making their environment acidic. This
environmental change favours the proliferation of acid tolerant species
such as mutans streptococci and lactobacilli and shifts the balance on
tooth surface towards demineralisation (*cf.* Equation 1). Dental caries can
not be prevented by only targeting the putative pathogens. Interference
with the factors influencing the environment and dental plaque ecology are
also needed.
(Slightly modified from Marsh 2006, with permission).

2.1.3 Saliva Protects Oral Tissues

Saliva is a natural mouthwash, having an important role in the protection of all
oral tissues. With respect to dental hard tissues, it has an important role in
maintaining the balance between the demineralisation and remineralisation
processes. Salivary calcium and phosphate ions are in equilibrium with the solid
tooth tissues and when the pH is high enough the balance is in the side of
precipitation (*cf.* Equation 1). The salivary pH is normally close to neutral.
Saliva resists pH changes but there are individual differences in the buffer
capacity. Good buffer capacity is related to high salivary flow rate and favours
remineralisation. In addition, saliva coats the enamel with a protein layer called
pellicle, which protects the enamel against demineralisation. Furthermore,
saliva flushes the oral cavity and dilutes harmful substances, *e.g.* acids from
beverages or produced by oral microbes. Finally, saliva contains several innate
defence factors, which can – at least *in vitro* – inhibit the acid production as well
as the attachment of cariogenic bacteria onto dental surfaces. The defence
factors also have bacteriostatic properties (Tenovuo 1998).

2.2 The Influence of Sugars in Diet on Caries Risk

Caries develops when frequent acid attacks soften the dental enamel. Thus,
fermentable carbohydrates are the key dietary components in the caries
process. However, when the cariogenicity of a particular food is evaluated
other properties as well as the sugar content need to be taken account. These

include *e.g.* the presence of protective factors, such as calcium in milk, frequency of consumption, and so-called food factors such as pH, food consistency and retention in the mouth, as well as factors influencing the oral microbiota (Zero 2004). Sucrose is called "the arch criminal" in dental caries, and with good reason. The evidence from human, animal and *in vitro* studies confirm the role of dietary sugars in dental caries, and the role of sucrose, in particular, is evident (Moynihan and Petersen 2004). Still, the role of an individual food component is never as important as the diet as a whole (Table 2.1).

The frequency of meals containing fermentable carbohydrate is the most important dietary factor influencing the caries risk. A direct and consistent

Table 2.1 Dietary and behavioural factors that influence the caries risk (unpublished).

Increase the Risk of Caries	*Examples*
Snacks instead of regular meals	Sweeteners • sugars
Frequent intake or use of large amounts of sugars, in particular sucrose	• syrup • honey
Use of cariogenic foods • high fermentable carbohydrate content • induce acid attack • sticky consistency • acidic	Drinks • soft drinks with sugars • sports drinks • energy drinks • sweetened fruit juices Candy Snacks • potato crisps • cream crackers Some dried fruits • dates • figs • raisins
Protective Measures	*Examples*
Use of topical fluoride	Non-caloric sweeteners Xylitol, sorbitol
Good oral hygiene	Milk products • cheese
Use of low cariogenic or protective foods • low carbohydrate content • high pH • contains calcium and phosphorus • stimulates salivary flow	• milk Peanuts Eggs Meat Vegetables
Use of cheese or sugar-free chewing gum after meals Use of water as thirst-quencher	Dietary fibre Tea without added sugar or honey

relationship has been observed between caries frequency and snacking (Newbrun 1982). Snacks are often rich in sucrose or other refined carbohydrates, which makes them harmful. Furthermore, the injuriousness of sucrose seems to be more strongly related to the frequency of consumption than to the amount (Zero 2004). On the other hand, sucrose-containing foods are not very harmful when consumed as a part of regular meals.

The role of the amount of consumed sugar is hard to separate from the frequency by which the sugars are consumed. Still, there seems to be an S-shaped relationship between the sugar intake and dental caries incidence: with low amounts of sugars, there is no correlation, but when the intake exceeds the safe amount, there is a strong positive correlation between the sugar intake and dental caries (Zero 2004). The safe amount of sugar seems to be approximately 50 g/person/day, which is well in line with the current population nutrient intake goals of WHO for preventing diet-related chronic diseases (Joint WHO/FAO Expert Consultation 2003, Moynihan and Petersen 2004). Fluoride intake shifts the S-shape curve, meaning that higher amounts of sugars can be consumed before caries occurs. In individuals with good oral hygiene and frequent fluoride exposure, even higher levels of sugar consumption may be tolerated (Zero 2004).

2.3 The Relation between Different Sugars and Dental Health

Although caries is a multifactorial disease and no specific sugar can explain the caries risk in contemporary industrialised countries (Marshall *et al.* 2007), there are differences between different sugars in respect to dental caries.

2.3.1 Sugar Polymers

2.3.1.1 Starches

Starches are glucose polymers which vary both in length and branching. They can be considered as cariogenic for several reasons. Firstly, the hydrolysis of starch by salivary or bacterial amylases results in maltose, maltotriose and low molecular weight dextrins, which can be fermented by oral acidogenic bacteria (Clarkson *et al.* 1987) and used as acceptors during glucan synthesis by streptococcal glycosyltransferases (Vacca-Smith *et al.* 1996). In addition, starch may make food stickier, thus its oral clearance is slower, in particular if the salivary flow rate is low at the same time. Finally, starch may influence the cariogenity of oral biofilms by affecting the structure of biofilms and the metabolism of oral bacteria. When comparing sucrose and the combination of sucrose and starch, the latter may increase the number of lactobacilli in dental plaque, enhance the acidogenity, increase the biofilm mass and expression of biofilm formation-associated genes of *Streptococcus mutans* (Duarte *et al.* 2008; Ribeiro *et al.* 2005). However, not all of these findings have been confirmed *in vivo* (Aires *et al.* 2008).

To summarise, starch can be considered as cariogenic, in particular in the presence of sucrose. In addition, soft products containing cooked or hydrolysed starch differ from whole grain products and fruits. Many snacks such as potato crisps and cream crackers are not sweet but cariogenic due to the starch they contain. On the other hand, the influence of starch on dental caries is most likely minimal, if it is consumed as part of a diet low in sucrose and limited eating frequency (Zero *et al.* 2008). Furthermore, substituting fructose or xylitol for sucrose with no difference in starch consumption, resulted in a significant reduction in caries (Scheinin *et al.* 1974). Finally, less refined starchy foods such as whole grain products have properties which protect the teeth (Moynihan and Petersen 2004).

2.3.1.2   Glucose Syrups and Maltodextrins

Glucose syrups and maltodextrins are glucose polymers, which are added to soft drinks, infant food, sports drinks, some pharmaceuticals *etc.* Just as starch, they can be cleaved by amylases to shorter units. If the length is short enough (up to 3 glucose units), they are fermentable by oral microbes (Grenby and Mistry 2000). In addition, they seem to be even more cariogenic than mono- and disaccharides (including sucrose) in rats (Grenby and Mistry 2000). Thus, these glucose polymers are potentially cariogenic, although, they are generally not considered as cariogenic as sucrose in humans (Zero *et al.* 2008).

2.3.1.3   Novel Oligosaccharides and Structural Isomers of Sucrose

Novel oligosaccharides and structural isomers of sucrose are used as sweeteners and prebiotics, for example. Trehalulose and leucrose are examples of sucrose isomers that are used as sweeteners but which are not cariogenic according to animal studies (Zero *et al.* 2008). Prebiotics are typically oligosaccharides, which are not digested by humans but enhance the growth of one or a selected number of bacterial species which are considered beneficial. Isomalto- and fructooligosaccharides are typical examples of prebiotics. Research on dental caries and prebiotics is sparse, but *in vitro* evidence and animal studies suggest that they could be cariogenic; isomaltooligosaccharides less than sucrose but fructooligosaccharides even as cariogenic as sucrose (Moynihan 1998a). However, prebiotics are used in such small amounts that their clinical relevance in caries is most probably minor.

2.3.2   Disaccharides

Disaccharides in a normal diet include sucrose, lactose, maltose and trehalose. The role of sucrose and lactose is discussed in detail below, the role of maltose was discussed in connection with starch and other glucose polymers. Like maltose, trehalose is also a glucose dimer; it can be found in small amounts in mushrooms, bread, beer, soybeans and shrimp. Most probably, the role of trehalose in caries is minor; yet, it has been suggested as a tooth-friendly sugar substitute (Neta *et al.* 2000).

2.3.2.1 Sucrose

Sucrose, the common sugar, is considered as the most cariogenic carbohydrate. Sucrose is the most common form of added sugar; it is widely used as a sweetener, a spice and a preservative. In addition, sucrose is cheap, easy to handle and good-tasting.

Sucrose is acidogenic; already the classical study of Stephan (1940) demonstrated that ingestion of sucrose results in a rapid pH fall in dental plaque. In acid production the phosphorylated sucrose is cleaved by bacterial invertase (sucrose-6-phosphate hydrolase) to glucose-6-phosphate and fructose, which are used in the metabolism of acidogenic bacteria (Figure 2.3).

The special role of sucrose in caries is not only due to its acidogenic potential but also its involvement as a substrate in the synthesis of exopolysaccharides,

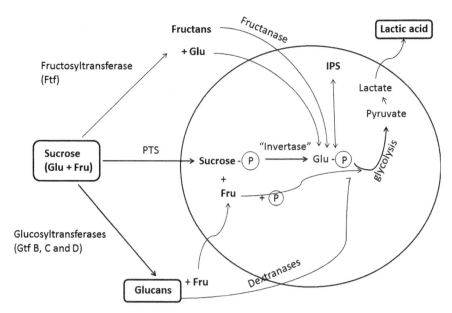

Figure 2.3 Utilisation of sucrose by *Streptococcus mutans*. *Streptococcus mutans* is cariogenic because of the efficient production of acids (lactic acid) and ability to adhere on tooth surface as well as form biofilm. *Streptococcus mutans* make both extracellular and intracellular polysaccharides from sucrose: glucans are important in adherence and biofilm formation, fructans are more important as an energy source, although they both act in both roles. Intracellular polysaccharides serve as an energy storage. The diagram is simplified, not all metabolic pathways or details are shown. Glu = glucose, Fru = fructose, PTS = phosphoenolpyruvate:sugar phosphotransferase system, "Invertase" = sucrose-6-phosphate hydrolase, IPS = intracellular polysaccharide.
(Modified from Hamada *et al.* 1984).

extracellular glucans and fructans (Figure 2.3). Sucrose enhances the exopolysaccharide production also by inducing the expression of biofilm related genes of *Streptococcus mutans* (Shemesh *et al.* 2006). Extracellular glucans are synthesised by bacterial glucosyltransferases and they can form the major part of the extracellular matrix of dental plaque. Sucrose increases the exopolysaccharide mass of *in situ* formed biofilms (Tenuta *et al.* 2006). Glucans enhance the accumulation of mutans streptococci on teeth by mediating interactions between bacteria and tooth surface and between the bacteria themselves. Furthermore, it has been suggested that glucans increase the plaque porosity, which allows deeper penetration of dietary sugars and greater acid production near to the tooth surface (Zero 2004). Finally, sucrose seems to increase the cariogenicity of other sugars and starch (Firestone *et al.* 1982).

2.3.2.2 Lactose

Lactose is fermented by several oral bacteria, including species belonging to the genera that are often associated with dental caries. Thus, lactose could be considered as cariogenic. However, lactose does not result in as deep a pH fall in dental plaque as other dietary sugars and the view of milk products as cariogenic is not supported by epidemiologic evidence (Johansson *et al.* 2010; Stephan 1940). In contrast, milk seems to have a protective effect against dental caries in children with high caries risk (Petti *et al.* 1997). Furthermore, milk with lactose and lactose-free milk do differ in their cariogenicity, when studied in experimental animals (Bowen *et al.* 1991). This could be explained by other components in milk and milk products, which protect the teeth. These include *e.g.* calcium, phosphate, casein and other protein and lipid components. Finally, one can speculate that milk products are often consumed as part of meals, thus they do not increase the frequency of acid challenges. In fact, consumption of cheese after meals seems to protect the tooth by stopping the acid attack (Herod 1991). Naturally, if milk or a milk product contains added sugar (sucrose) and the product is used frequently between meals, some of them *e.g.* sweetened yogurt, may be cariogenic.

Lactose is commonly used in pharmaceuticals. For example, some medicines for treatment of asthma bronciale are administered *via* dry powder inhalers containing lactose-monohydrate. Because these medicines are used frequently and they are acidic by themselves, the sugars in the products may in combination with acidity and possible drug-induced dry mouth increase the risk of dental caries (Thomas *et al.* 2010). Lactulose, on the other hand, is a synthetic isomer of lactose, and thus, also a disaccharide of fructose and galactose. Lactulose is used in laxative products and as a prebiotic. It is fermented by some oral acidogenic bacteria, but at such slow rate that the fermentation does not have any significance *in vivo* (Moynihan *et al.* 1998b).

Breast milk contains high amounts of lactose, and therefore there has been a concern that breast feeding could be a risk factor for early childhood caries. However, several studies indicate that breast feeding *per se*, even when continued for a long time or during the night, cannot be connected with increased risk of early childhood caries (*e.g.* Alaluusua *et al.* 1990; Kramer *et al.* 2007), although contradictory findings exist as well (Folayan *et al.* 2010). Furthermore, according to a systematic literature review, there is no such evidence from which one could determine a period after which breast feeding would increase the risk of dental caries (Valaitis *et al.* 2000). In combination with sucrose consumption and high mutans streptococcus counts, breast feeding increases the caries risk, if it continues long after tooth eruption. Therefore, it is important that parents and guardians are well-adviced about their own and the child's oral hygiene procedures, not to give candies to small children, and to avoid sugar-containing and acidic drinks in nursing bottles and the transmission of cariogenic microbes from themselves to their babies. The health benefits of breast feeding are evident. WHO recommends continuing breast feeding together with complementary food after six months for up to two years of age or beyond (WHO, www-link).

2.3.3   Monosaccharides

2.3.3.1   *Glucose and Fructose*

The most common dietary monosacharides are glucose, the grape sugar or dextrose, and fructose, the fruit sugar. They are found in high amounts in fruits and honey. They are fermented by oral acidogenic bacteria, and can, thus, be considered as cariogenic. There is only a minor difference between the acidogenic potential of fructose, glucose and sucrose (Zero 2004). Although, the cariogenicity of glucose and fructose seems to be dependent on whether they are intrinsic or extrinsic. Consumption of fresh fruit (intrinsic sugar) does not increase the risk of dental caries (Moynihan and Petersen 2004), but fruit juices and some dried fruits such as raisins can be considered as cariogenic. However, according to the Turku sugar studies, substituting fructose for sucrose resulted in a significant decrease in caries (Scheinin *et al.* 1974). In addition, people with hereditary fructose intolerance seem to have less caries-associated microbes and also less caries than controls (Saxén *et al.* 1989).

2.3.3.2   *Galactose*

Free galactose is found in low amounts in milk products, some vegetables and soy beverages (Kim *et al.* 2007). An exception is the low lactose products in which the lactose is hydrolysed to glucose and galactose. Research on galactose and dental caries is sparse, but galactose is likely to have no or only minor influence on dental health. Still, galactose results in an acid attack when pure solution is used (Salako and Kleinberg 1992).

2.3.4   Sugar Alcohols

Sugar alcohols or polyols and non-caloric sweeteners are commonly used in the production of "tooth-friendly" products. Non-caloric sweeteners such as aspartame and sucralose are not fermented by oral bacteria. Thus they can not be cariogenic. Sugar alcohols are polyhydric alcohols, which do not have more than one hydroxy group attached to each carbon atom. Commonly used sugar alcohols include 12-carbon maltitol, six-carbon sorbitol (D –glucitol) and mannitol, the five-carbon xylitol and four-carbon erythritol.

Maltitol and the six-carbon polyols sorbitol and mannitol are to some extent fermented by some oral bacteria, including mutans streptococci and some lactobacilli (Zero *et al.* 2008). However, they are fermented at a slow rate and the degradation yields also other end-products than lactate. With respect to sorbitol, the enzymes needed for metabolism are inducible, thus it is questionable whether sorbitol is fermented in significant amounts *in vivo*. In fact, the pH drop of dental plaque after rinsing with sorbitol is minor when compared to sucrose and does not reach the critical pH needed for dentin demineralisation (Zero *et al.* 2008). Xylitol and erythritol are not fermented in the oral cavity. In addition to being non-fermentable, sweet sugar alcohols stimulate saliva flow and thereby promote remineralisation (Söderling *et al.* 1976). Furthermore, xylitol inhibits the production of insoluble extracellular polysaccharides (Mäkinen 2011), whereas sorbitol and mannitol seem to upregulate the expression of genes encoding enzymes that are associated with biofilm formation *in vitro* (Shemesh *et al.* 2006).

Clinical studies on xylitol have been conducted for over 30 years and the evidence for xylitol being not only non-cariogenic but protecting against dental caries is convincing. The effectiveness of xylitol is based on both chemical and biological properties, such as the pentitol structure, complex formation with calcium, interactions with *Streptococcus mutans* and overall plaque metabolism. Habitual xylitol consumption reduces dental caries (Mäkinen 2011). In addition, maternal use of xylitol can prevent the transmission of mutans streptococci to their children and thereby prevent dental caries of their children although they have not received xylitol (Söderling *et al.* 2001).

Summary Points

- This chapter focuses on the relation between sugars and dental caries.
- Dental caries, tooth decay, is a localised destruction of dental hard tissue by acids, which are produced by bacteria in dental plaque.
- Sugars, in particular sucrose, have an aetiological role in dental caries.
- Sucrose is cariogenic because it is fermented by acidogenic bacteria and it serves as a substrate for bacterial glucosyltransferases.
- Foods containing hydrolysed or cooked starch can be cariogenic.
- In the mouth, starch is cleaved to smaller units which are fermented by oral bacteria and starch may also make the food sticky, which slows down the oral clearance.
- Lactose, the milk sugar, is less cariogenic than many other sugars. In addition, milk and milk products contain substances that are protective against dental caries.

- The use of chewing gum sweetened with sugar alcohols, xylitol or sorbitol, is recommended.
- Modest amounts of sugars can be tolerated, if good oral hygiene habits are practised and topical fluoride is in use.
- Although sugars and foods differ in their cariogenicity, the role of an individual food component is never as important as the diet as a whole.

Key Facts

- In Western world, the prevalence and severity of dental caries has declined during the last decades, but caries treatment is still one of the most common procedures in dental practice.
- The prevalence of caries is highly polarised among children and young adults: caries-free dentitions are found more and more often, but a small minority has a severe caries problem.
- Root caries is a major dental problem in older people.
- Globally, differences in caries experience are related to environmental factors such as use of fluoride, behavioural factors and diet.
- The use of topical fluorides, dietary counselling, and good oral hygiene are the primary prevention methods of dental caries.

Definitions of Words and Terms

Acid attack: The fall in pH after consumption of food or drink below the critical pH for demineralisation of tooth surface.

Acidogenic bacteria: Bacteria, which produce acids in their metabolism.

Acid tolerant bacteria: Bacteria, which are able to grow in an acidic environment.

Amylase: An enzyme that hydrolyses dietary starch into smaller units. The amylase found in human saliva is called alpha-amylase.

Biofilm: An organised community of bacteria attached to a surface interface and covered by extracellular polymeric matrix. The biofilm on tooth surface is usually called the dental plaque.

Dental caries: Tooth decay.

Demineralisation: Removal of mineral ions (mainly calcium and phosphate) from the tooth surface.

Exopolysaccharides: High-molecular-weight sugar polymers, which locate outside the cell. Exopolysaccharides are produced by microbial enzymes and they are important in bacterial attachment and biofilm formation. Glucans and fructans are examples of exopolysaccharides formed of glucose and fructose units, respectively.

Extrinsic sugars: Sugars in food or drinks, which locate outside the cells. They can be initially free, like milk sugar, released from food components, such as fruits, or added to food during preparation.

Fructosyltransferase: Extracellular enzyme which synthesize fructans (see exopolysaccharides).

Glucosyl transferases: Extracellular enzymes which synthesize glucans (see exopolysaccharides).

Intrinsic sugars: Sugars within the cellular structure of fruits and vegetables (compare extrinsic sugars).

Remineralisation: Restoring of mineral ions (mainly calcium and phosphate) to the enamel.

Sugar clearance: Removal of sugars from the oral cavity. Saliva dilutes and removes sugars or sugar-containig foods from the oral cavity. Both the salivary flow rate and the consistence of food are important in this respect. The faster is the clearance (short clearance time) the better for dental health.

References

Aires, C.P., Del Bel Cury, A.A., Tenuta, L.M., Klein, M.I., Koo, H., Duarte, S., and Cury, J.A., 2008. Effect of starch and sucrose on dental biofilm formation and on root dentine demineralization. *Caries Res.* 42: 380–386.

Alaluusua, S., Myllärniemi, S., Kallio, M., Salmenperä, L., and Tainio V.M., 1990. Prevalence of caries and salivary levels of mutans streptococci in 5-year-old children in relation to duration of breast feeding. *Scand J Dent Res.* 98: 193–196.

Beighton, D., 2005. The complex oral microflora of high-risk individuals and groups and its role in the caries process. *Community Dent Oral Epidemiol.* 33: 248–255.

Bowen, W.H., Pearson, S.K., VanWuyckhuyse, B.C., and Tabak, L.A., 1991. Influence of milk, lactose-reduced milk, and lactose on caries in desalivated rats. *Caries Res.* 25: 283–286.

Clarkson, B.H., Krell, D., Wefel, J.S., Crall, J., and Feagin, F.F., 1987. In vitro caries-like lesion production by *Streptococcus mutans* and *Actinomyces viscosus* using sucrose and starch. *J Dent Res.* 66: 795–798.

Duarte, S., Klein, M.I., Aires, C.P., Cury, J.A., Bowen, W.H., and Koo, H., 2008. Influences of starch and sucrose on *Streptococcus mutans* biofilms. *Oral Microbiol Immunol.* 23: 206–212.

Firestone, A.R., Schmid, R., and Mühlemann, H.R., 1982. Cariogenic effects of cooked wheat starch alone or with sucrose and frequency-controlled feedings in rats. *Arch Oral Biol.* 27: 759–763.

Folayan, M.O., Sowole, C.A., Owotade, F.J., and Sote, E., 2010. Impact of infant feeding practices on caries experience of preschool children. *J Clin Pediatr Dent.* 34: 297–301.

Grenby, T.H., and Mistry, M., 2000. Properties of maltodextrins and glucose syrups in experiments in vitro and in the diets of laboratory animals, relating to dental health. *Br J Nutr.* 84: 565–574.

Hamada, S., Koga, T., and Ooshima T., 1984. Virulence factors of *Streptococcus mutans* and dental caries prevention.

Herod, E.L., 1991. The effect of cheese on dental caries: a review of the literature. *Aust Dent J*. 36: 120–125.

Johansson, I., Holgerson, P.L., Kressin, N.R., Nunn, M.E., and Tanner, A.C., 2010. Snacking habits and caries in young children. *Caries Res*. 44: 421–430.

Joint WHO/FAO Expert Consultation, 2003. Diet, nutrition and the prevention of chronic diseases. http://whqlibdoc.who.int/trs/who_trs_916.pdf

Kim, H.O., Hartnett, C., and Scaman, C.H., 2007. Free galactose content in selected fresh fruits and vegetables and soy beverages. *J Agric Food Chem*. 55: 8133–8137.

Kramer, M.S., Vanilovich, I., Matush, L., Bogdanovich, N., Zhang, X., Shishko, G., Muller-Bolla, M., and Platt, R.W., 2007. The effect of prolonged and exclusive breast-feeding on dental caries in early school-age children. New evidence from a large randomized trial. *Caries Res*. 41: 484–488.

Mäkinen, K., 2011. Sugar alcohol sweeteners as alternatives to sugar with special consideration of xylitol. *Med Princ Pract*. 20: 303–320.

Marsh, P.D., 2006. Dental plaque as a biofilm and a microbial community – implications for health and disease. *BMC Oral Health*. 6: S14.

Marshall, T.A., Eichenberger-Gilmore, J.M., Larson, M.A., Warren, J.J., and Levy, S.M., 2007. Comparison of the intakes of sugars by young children with and without dental caries experience. *J Am Dent Assoc*. 138: 39–46.

Moynihan, P., and Petersen, P.E., 2004. Diet, nutrition and the prevention of dental diseases. *Public Health Nutr*. 7: 201–226.

Moynihan, P.J., 1998a. Update on the nomenclature of carbohydrates and their dental effects. *J Dent*. 26: 209–218.

Moynihan, P.J., Ferrier, S., Blomley, S., Wright, W.G., and Russell, R.R., 1998b. Acid production from lactulose by dental plaque bacteria. *Lett Appl Microbiol*. 27: 173–177.

Neta, T., Takada, K., and Hirasawa, M., 2000. Low-cariogenicity of trehalose as a substrate. *J Dent* 28: 571–576.

Newbrun, E., 1982. Sugar and dental caries: a review of human studies. *Science*. 217: 418–423.

Petti, S., Simonetti, R., and Simonetti D'Arca, A., 1997. The effect of milk and sucrose consumption on caries in 6-to-11-year-old Italian schoolchildren. *Eur J Epidemiol*. 13: 659–664.

Ribeiro, C.C., Tabchoury, C.P., Del Bel Cury, A.A., Tenuta, L.M., Rosalen, P.L., and Cury, J.A., 2005. Effect of starch on the cariogenic potential of sucrose. *Br J Nutr*. 94: 44–50.

Salako, N.O., and Kleinberg, I., 1992. Comparison of the effects of galactose and glucose on the pH responses of human dental plaque, salivary sediment and pure cultures of oral bacteria. *Arch Oral Biol*. 37: 821–829.

Saxén, L., Jousimies-Somer, H., Kaisla, A., Kanervo, A., Summanen, P., and Sipilä, I., 1989. Subgingival microflora, dental and periodontal conditions in patients with hereditary fructose intolerance. *Scand J Dent Res*. 97: 150–158.

Scheinin, A., Mäkinen, K.K., and Ylitalo, K., 1974. Turku sugar studies. I. An intermediate report on the effect of sucrose, fructose and xylitol diets on the caries incidence in man. *Acta Odontol Scand*. 32: 383–412.

Shemesh, M., Tam, A., Feldman, M., and Steinberg, D., 2006. Differential expression profiles of Streptococcus mutans ftf, gtf and vicR genes in the presence of dietary carbohydrates at early and late exponential growth phases. *Carbohydr Res.* 341: 2090–2097.

Söderling, E., Rekola, M., Mäkinen, K.K., and Scheinin, A., 1976. Turku sugar studies XXI. Xylitol, sorbitol-, fructose- and sucrose-induced physico-chemical changes in saliva. *Acta Odontol Scand.* 34: 397–403.

Söderling, E., Isokangas, P., Pienihäkkinen, K., Tenovuo, J., and Alanen, P., 2001. Influence of maternal xylitol consumption on mother-child transmission of mutans streptococci: 6-year follow-up. *Caries Res.* 35: 173–177.

Stephan, R.M., 1940. Changes in Hydrogen-Ion Concentration on Tooth Surfaces and in Carious Lesions. *JADA.* 27: 718–723.

Tahmassebi, J.F., Duggal, M.S., Malik-Kotru, G., and Curzon, M.E., 2006. Soft drinks and dental health: A review of the current literature. *J Dent.* 34: 2–11.

Tenovuo, J., 1998. Antimicrobial function of human saliva--how important is it for oral health? *Acta Odontol Scand.* 56: 250–256.

Tenuta, L.M., Ricomini Filho, A.P., Del Bel Cury A.A., and Cury, J.A., 2006. Effect of sucrose on the selection of mutans streptococci and lactobacilli in dental biofilm formed in situ. *Caries Res.* 40: 546–549.

Thomas, M.S., Parolia, A., Kundabala, M., and Vikram, M., 2010. Asthma and oral health: a review. *Aust Dent J.* 55: 128–133.

Vacca-Smith, A.M., Venkitaraman, A.R., Quivey, R.G. Jr and Bowen, W.H., 1996. Interactions of streptococcal glucosyltransferases with alpha-amylase and starch on the surface of saliva-coated hydroxyapatite. *Arch Oral Biol.* 41: 291–298.

Valaitis, R. Hesch, R., Passarelli, C., Sheehan, D., and Sinton, J., 2000. A systematic review of the relationship between breastfeeding and early childhood caries. *Can J Public Health.* 91: 411–417.

WHO: http://www.who.int/nutrition/topics/exclusive_breastfeeding/en/

Zero, D.T., 2004. Sugars - the arch criminal? *Caries Res.* 38: 277–285.

Zero, D.T., Moynihan, P., Lingström, P., and Birkhed, D., 2008. The role of dietary control. In: Fejerskov O., and Kidd E. (ed.) *Dental caries. The disease and its clinical management.* Second edition. Blackwell Munksgaard Ltd, Oxford, UK, pp. 329–352.

CHAPTER 3

Whole Body Glucose Metabolism

JØRGEN JENSEN*[a] AND JESPER FRANCH[b]

[a] Department of Physical Performance, Norwegian School of Sport Sciences, P.O. Box 4014 Ullevål Stadion, 0806 Oslo, Norway; [b] Department of Health Science and Technology, Aalborg University, Fredrik Bajers Vej 7D, 9220 Aalborg, Denmark
*Email: jorgen.jensen@nih.no

3.1 Introduction

Carbohydrate is the major energy component in bread, pasta, rice, potatoes, and most people get the majority of their energy from carbohydrates. Various diets, however, differ dramatically, from some people who live mainly on carbohydrates, to others who live mainly on fat and protein. Indeed, the human body needs protein and fat, and has the ability to use them as energy components. The body has a marked ability to adapt, and even with a large difference in carbohydrate intake, blood glucose is maintained relatively stably around 5 mM. In total, humans have ~4 g of glucose in the blood (Wasserman 2009).

Glucose is a water-soluble monosaccharide which is transported in the blood at high concentrations and delivers energy to cells. Glucose is transported across cell membranes by glucose transporters (GLUTs) and metabolised *via* glycolysis for energy production (Jensen 2009). Glycolysis describes the ten reactions in which glucose is transformed into pyruvate. Glycolysis is similar in all cells, but several key enzymes have different isoforms with different

Food and Nutritional Components in Focus No. 3
Dietary Sugars: Chemistry, Analysis, Function and Effects
Edited by Victor R Preedy
© The Royal Society of Chemistry 2012
Published by the Royal Society of Chemistry, www.rsc.org

regulation (Jensen 2009). Such isoform-specific expression of glycolytic enzymes is important for regulation of whole body glucose metabolism. Glycolysis in the liver occurs *e.g.* only in the postprandial phase, whereas glycolytic flux is high in skeletal muscles during high intensity exercise.

Excess intake of dietary carbohydrates can be stored as glycogen in skeletal muscles and in the liver (Jensen and Lai 2009). Skeletal muscle glycogen synthesis plays an important role for insulin-mediated regulation of blood glucose (Shulman *et al.* 1990). At rest, glucose oxidation is low and accounts $\sim 100 \, \text{mg} \cdot \text{min}^{-1}$. During high-intensity exercise, carbohydrate oxidation increases above $3 \, \text{g} \cdot \text{min}^{-1}$, but the glucose concentration in the blood remains stable. However, as the glycogen content in skeletal muscles and liver decreases during prolonged exercise ($<2 \, \text{h}$), a reduction in blood glucose occurs, which contributes to fatigue (Coyle *et al.* 1986). Low glucose concentration ($<2 \, \text{mM}$) may cause acute death, but such low concentrations does not occur in healthy people even during severe exercise or prolonged fasting.

Glucose is also a substrate for lipid synthesis and high intake of carbohydrate causes obesity. Obesity and inactivity are the main risk factors for development of type 2 diabetes. Diabetes is by definition a disease with elevated blood glucose concentration. Acute elevation of blood glucose (*e.g.* doubling, as may occur after intake of a large amount of carbohydrates) is harmless, but prolonged elevation causes diabetic complications like cardiovascular disease, nephropathy, neuropathy and blindness. Exercise prevents development of type 2 diabetes (Knowler *et al.* 2002), and the increased glucose utilization during exercise most likely contributes to the improved regulation of glucose metabolism.

3.2 Glucose Transport

All carbohydrates have to be broken down to monosaccharides in the intestine for uptake and glucose is transported *via* the portal vein to the liver and enter the circulation (Frayn 2010). Glucose is the "currency" of carbohydrate in the blood ($\sim 5 \, \text{mM}$) and the only sugar which can be transported into all cells (Figure 3.1). Sucrose is a disaccharide built by a glucose molecule and a fructose molecule; lactose is a disaccharide built by a glucose molecule and a galactose molecule. Fructose and galactose are mainly metabolised in the liver; fructose being an excellent substrate for lipid formation, and galactose is converted to glucose. Part of the glucose is immediately taken up by the liver, but the majority continues in the circulation, becoming available for the brain and other cells.

The major tissues involved in regulation of blood glucose concentration are the liver, skeletal muscles, adipose tissue, the brain and pancreas. The brain utilizes mainly glucose as substrate and oxidises daily about 120 g of glucose. Skeletal muscles have the capacity to metabolize large quantities of glucose during exercise. See Keith Frayn's excellent book *Metabolic regulation. A human perspective* for further reading (Frayn 2010).

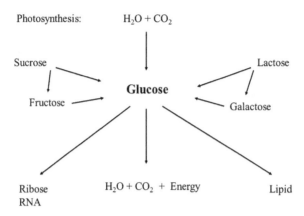

Figure 3.1   Glucose is the centre of carbohydrate metabolism in humans. Glucose is produced by photosynthesis in plants. Carbohydrate molecules like fructose and galactose are converted to glucose (mainly in the liver). Glucose can be used as energy or converted to ribose for RNA synthesis or stored as lipid (triacylglycerol).

3.2.1   Glucose Transporters

Glucose is transported into various cell types by a family of glucose transporter proteins, (GLUTs), which are expressed differently in various tissues (Frayn 2010). In total, there exist 14 different GLUTs and they have different kinetic properties and regulation (Thorens and Mueckler 2010). Some cell types (*e.g.* liver and pancreatic β-cell) expresses GLUT2 which is present in the cell membrane continuously. GLUT2 has a high K_m for glucose and glucose transport is therefore determined by concentration of extracellular glucose (Thorens and Mueckler 2010).

GLUT4 has particular importance in the regulation of whole body glucose metabolism because this is the transporter that is regulated by insulin; defective regulation of GLUT4 causes insulin resistance (Thorens and Mueckler 2010). GLUT4 is expressed in skeletal muscle and adipocytes and insulin regulates glucose uptake by translocation of GLUT4 from intracellular vesicles to the cell membrane, where they transport glucose into the cell (Figure 3.2). Rate of glucose transport in muscles is determined by translocation of GLUT4 to the membrane (Larance *et al.* 2008) and insulin-mediated translocation of GLUT4 is impaired in insulin resistant muscles (Etgen *et al.* 1996). GLUT4 is also translocated to the membrane during muscle contraction (Etgen *et al.* 1996).

Glucose metabolism is also regulated by adrenaline (Figure 3.2). Adrenaline does not impair basal or insulin-stimulated glucose transport in skeletal muscles, but inhibits glucose metabolism *via* accumulation of glucose 6-phosphate and inhibition of hexokinase activity (Aslesen and Jensen 1998). Adrenaline also activates glycogen phosphorylase and stimulates glycogen breakdown and inhibits glycogen synthesis (Jensen and Lai 2009).

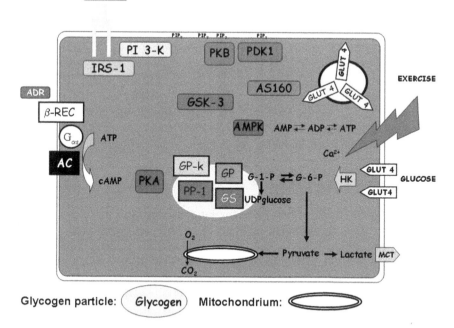

Figure 3.2 Regulation of glucose uptake in skeletal muscles. Insulin and contraction stimulates translocation of GLUT4 to the cell membrane. Insulin stimulates translocation *via* tyrosine phosphorylation of IRS-1 and activation of PI 3-kinase and PKB; activated PKB phosphorylates AS160 and GLUT4 vesicles translocates and fuses with the cell membrane. During insulin stimulation, the major part of glucose taken up is incorporated into glycogen by glycogen synthase (GS) activated via PI 3-kinase, PKB and inhibition of GSK3 activity. Muscle contraction increases energy consumption and activation of AMPK which mediates phosphorylation of AS160 and translocation of GLUT4. During muscle contraction, the glucose taken up is metabolized immediately. Adrenaline stimulates glycogen breakdown *via* PKA mediated phosphorylation of glycogen phosphorylase kinase (GP-K) which phosphorylates and activates glycogen phosphorylase (GP). Protein phosphatase-1 (PP-1) dephosphorylates GP and GS.

Abbreviations: ADR: adrenaline; AC: Adenylate cyclase, β-REC: β-Adrenergic receptor; Gαs: G-protein stimulating AC activity; IRS-1: Insulin receptor substrate-1; PI-3-K: Phosphatidylinositol 3-kinase; PKB: Protein kinase B (also called AKT), PDK-1: Phosphoinositide dependent protein kinase 1; PIP$_2$: Phosphatidylinositol 4,5bisphosphate; PIP$_3$: Phosphatidylinositol 3,4,5triphosphate; GSK-3: glycogen synthase kinase-3; AS160: Akt substrate of 160 kDa; AMPK: AMP-activated protein kinase; HK: Hexokinase; PKA: cAMP dependent kinase; GP-k: Glycogen phosphorylase kinase; GP: Glycogen phosphorylase; PP-1: Protein phosphatase-1; GS: Glycogen synthase; MCT: Mono carboxylate transporter.

3.3 Metabolism of Glucose

Glucose is an important energy substrate metabolised by most, if not all, cell types. Fat is the other major energy substrate and the carbon skeleton of proteins also provide energy, fuelling the Krebs cycle with substrate to generate NADH and $FADH_2$ for the electron transport chain and ATP synthesis. Most cells metabolize both glucose and fat simultaneously. However, glucose (and glycogen) has two major advantages: first, glucose can be metabolised anaerobically; and, second, oxidation of carbohydrates can generate energy at a much higher rate than oxidation of fat. The advantages of fat are that the high energy density, and subcutaneous fat can isolate and reduce heath loss.

3.3.1 Glycogen Synthesis

Glucose can be stored as glycogen for later utilization. Glycogen is mainly stored in the liver and skeletal muscles. The liver is relatively small (~ 1.5 kg) and contains about ~ 100 g of glycogen (Table 3.1). Skeletal muscle is the largest tissue in the body and accounts for $\sim 40\%$ of the bodyweight of healthy young men, and skeletal muscles store ~ 500 g glycogen (Jensen and Lai 2009).

Glycogen is synthesized at a protein called glycogenin, and glycogen has a fractal structure and can contain $\sim 50,000$ glucose molecules in skeletal muscles (Meléndez *et al.* 1999). Glycogen synthase is the enzyme that incorporates glucose into the glycogen particle with UDG-glucose as substrate (Jensen and Lai 2009). For glycogen synthesis, glucose has to be transported into cells, phosphorylated to glucose 6-fosfate, isomerised to glucose 1-phosphate, converted to UDP-glucose before glycogen synthase incorporate the glucosyl unit into the glycogen particle (Figure 3.3).

The function of liver glycogen is to maintain blood glucose constant in the period where glucose is not absorbed. Liver glycogen is broken down to glucose 1-phosphate by glycogen phosphorylase, converted to glucose 6-phosphate and dephosphorylated by glucose 6-phosphatase and glucose can be released into the blood (Frayn 2010). Skeletal muscles do not express glucose 6-phosphate and cannot release glucose.

Skeletal muscle glycogen is the main energy substrate during high intensity exercise, and depletion of skeletal muscles glycogen causes fatigue (Hermansen *et al.* 1967). Skeletal muscle glycogen has also an important role in whole body regulation of blood glucose after a meal. During insulin stimulation, the major part of the disposed blood glucose is incorporated into skeletal muscle glycogen

Table 3.1 Major carbohydrate stores in humans.

	Blood	*Muscles*	*Liver*
Concentration	5 mM	100–120 mmol/kg	350–400 mmol/kg
Total content	4–5 g	~ 500 g	~ 100 g

Hypothetical data for a 70 kg person with a muscles mass corresponding to $\sim 40\%$ of body weight.

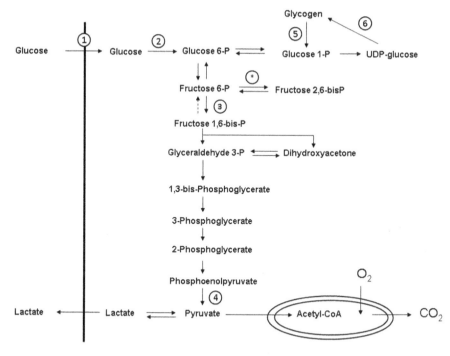

Figure 3.3 Schematic overview of glucose metabolism. The regulated steps are marked with numbers. (1) Glucose transport; (2) hexokinase; (3) 6-phosphofructokinase (PFK-1); (4) pyruvate kinase; (5) glycogen phosphorylase; (6) glycogen synthase. *Denotes the bifunctional enzyme 6-phosphofructo-2-kinase/fructose-2,6-bisphosphatase (PFK-2) which regulates concentration of fructose 2,6-bisphosphate (the strongest allosteric activator of PFK-1). For further reading, see Jensen (2009).

(Shulman *et al.* 1990). It is therefore obvious that appropriate insulin sensitivity in skeletal muscles is required for normal regulation of blood glucose concentration (Jensen, 2009; Shulman *et al.* 1990).

The glycogen content in skeletal muscles is limited and high glycogen content decreases glycogen synthase activity (Jensen *et al.* 2006; Jensen and Lai 2009). In muscles with high glycogen content, rate of glycogen synthesis is reduced and glycolytic flux increased (Jensen *et al.* 2006). We believe such increased glycolytic flux in skeletal muscle during insulin stimulation is unhealthy and will increase *de novo* lipid synthesis. Although the major part of *de novo* lipid synthesis occurs in the liver, hyperglycemia can stimulate *de novo* lipogenesis in skeletal muscle cells (Aas *et al.* 2004).

3.3.2 Glycolysis

Glycolysis describes the chain of reactions where one glucose molecule is broken down to two pyruvate molecules *via* 10 enzymatic reactions (Figure 3.3) (Newsholme and Leech 2009). Most cells contain only a small amount of free glucose, and glycolysis is fed by glucose transported into the cells or from degradation of glycogen (Jensen 2009). Physiologically, glycolysis has two substrates (glycogen and glucose) and regulation of the initial steps differ under these conditions. Glycolysis from glucose has the following rate-limiting steps: (1) glucose transport, (2) phosphorylation (hexokinase), (3) 6-phospho-fructokinase (PFK-1) and (4) pyruvate kinase (Figure 3.2). Glycogen phos-phorylase catalyses the breakdown of glycogen to glucose 1-phosphate, and the rate-limiting steps of glycogen metabolism *via* glycolysis are: (1) glycogen phosphorylase, (2) 6- phosphofructokinase and (3) pyruvate kinase.

Gluconeogenesis is the metabolic pathway where glucose is formed, *e.g.* from pyruvate, lactate or amino acids. Glycolysis and gluconeogenesis have opposite purposes (degradation versus formation of glucose) and many of the enzymatic reactions are the same (Newsholme and Leech 2009). Indeed, only three of the reactions in glycolysis are not reversible (hexokinase, PFK-1 and pyruvate kinase). Glycolysis and gluconeogenesis shall not occur simultaneously, and it is therefore important to block glycolysis to obtain flux through the gluco-neogenic pathway.

In most physiological conditions, 6-phosphofructokinase (PFK-1) regulates glycolytic flux (Connett and Sahlin 1996; Jensen 2009). Three isoforms of PFK-1, coded from different genes, allow that glycolysis is differently regulated in various cell types. Such different regulation is crucial for whole body reg-ulation of glucose metabolism. The regulation of PFK-1 is complex and incomplete understood. The PFK-1 activity is regulated by allosteric regulators (many metabolites), pH, phosphorylation, complex formation and reversible binding to the cytoskeleton (Connett and Sahlin 1996; Jensen 2009; Kemp and Foe 1983). The most important allosteric activators are fructose 2,6-bispho-sphate, fructose 6-phosphate, glucose 1,6-bisphosphate, AMP, P_i and the most important inhibitors are ATP, citrate and CrP (Kemp and Foe 1983; Newsholme and Leech 2009). The tetrameric form of PFK-1 is active, but PFK-1 can dissociates to dimers and monomer (Kemp and Foe 1983) and allosteric regulators influence complex formation.

The concentration of fructose 2,6-bisphosphate, the strongest activator of PFK-1, is regulated by the enzyme 6-phosphofructo-2-kinase/fructose-2,6-bisphosphatase (PFK-2). This bifunctional enzyme catalyse the inter-conver-sion of fructose 6-phosphate and fructose 2,6-phosphate (Rider *et al.* 2004). This enzyme is also expressed tissue specific and regulation of PKF-2 activity regulates fructose 2,6-bisphosphate concentration and glycolytic flux (Rider *et al.* 2004).

The last reaction in glycolysis (pyruvate kinase) is also tightly regulated in the liver where gluconeogenesis is important (Jensen 2009). Pyruvate kinase cata-lyses conversion of phosphoenolpyruvate to pyruvate with the formation of ATP. Activity of pyruvate kinase is inhibited in cells when gluconeogenesis occurs (liver and kidney). Glucagon stimulates gluconeogenesis in liver, because PKA phosphorylates the liver isoform of pyruvate kinase which inhibit activity and therefore glycolysis. PKA also phosphorylates PFK-2 and acti-vates fructose-2,6-bisphosphatase activity (Rider *et al.* 2004).

The metabolism of pyruvate varies and depends on many factors. Pyruvate can be reduced to lactate under anaerobic conditions or oxidised in the mitochondria under aerobic conditions. In skeletal, a substantial amount of pyruvate is transaminated to alanine and transported to the liver where nitrogen disposal occurs *via* the urea cycle.

3.3.3 Alternative Routes of Glucose Metabolism

The majority of glucose is metabolised *via* glycolysis, but small amounts enter the pentose phosphate pathway and the hexosamine biosynthetic pathway. The pentose pathway has several functions and provides ribose for RNA synthesis, deoxyribose for DNA synthesis and supply NADPH for *de novo* lipogenesis. Glucose also regulates gene expression *via* carbohydrate response element binding protein (ChREBP). The pentose phosphate pathway intermediate xylulose 5-phosphate activates PP2A and the dephorphorylated ChREBP is translocated to the nucleus (Ilzuka and Horikawa 2008). ChREBP-mediated gene regulation is extremely important in the liver and regulates expression of genes involved in lipid synthesis. ChREBP-mediated gene expression has not been described in skeletal muscles.

Regulation of glycolysis also channels glucose into the hexosamine biosynthetic pathway, which has been shown to cause insulin resistance. The hexosamine pathway produces UDP-glucosamine, which is the substrate for reversal O-linked glycosylation of proteins (Copeland *et al.* 2008). Reversible O-linked protein glycosylation at serines and threonines, often the same amino acids that become phosphorylated, has shown to be an important signalling mechanism (Copeland *et al.* 2008). Glucose is not only an energy substrate, but also a signalling molecule communication between tissues and regulates gene expression specifically in many cells.

3.4 Glucose and the Mitochondria

The pyruvate produced in glycolysis can be metabolised *via* several pathways. During inadequate oxygen supply, pyruvate can be reduced to lactate by lactate dehydrogenase with regeneration of NAD^+, which is required in glycolysis. For complete oxidation, pyruvate enters the mitochondria for degradation in the Krebs cycle yielding energy-rich electrons to reduction of NAD^+ and FAD providing energy to the electron transport chain for ATP synthesis (Newsholme and Leech 2009). Rate of glucose metabolism in the mitochondria is determined by energy requirement, and increases 100-fold during exercise. The rate of ATP synthesis in the electron transport chain in skeletal muscles is limited by O_2 delivery.

3.4.1 The Krebs Cycle

Glycolysis occurs in the cytosol, and pyruvate is transported into the mitochondria for complete degradation. Pyruvate is degraded to acetyl-CoA by pyruvate dehydrogenase (PDH). Fat, in the form of long chain acyl-CoA, is also degraded to acetyl-CoA in the mitochondria and provide substrate for the Krebs cycle (Frayn 2010). Metabolism of glucose and fat occurs simultaneously, and excess of glucose inhibits fat metabolism and *vice versa* (Hue and Taegtmeyer 2009).

The glucose-fatty acid cycle explain how FFA decreases glucose utilization. The glucose-fatty acid cycle states that high rate of β-oxidation increases the concentration of acetyl-CoA and citrate. Acetyl-CoA inhibits PDH and therefore pyruvate degradation while citrate inhibits PFK-1 and glycolysis (Newsholme and Leech 2009). Inhibition of PFK-1 will favour accumulation of glucose 6-phosphate and inhibition of hexokinase activity and therefore glucose phosphorylation. Randle and colleagues did the majority of experiments on the heart to describe the glucose-fatty acid cycle and glycolysis is differently regulated heart and muscle (Depre *et al.* 1998). It has been difficult to show that the glucose-fatty acid cycle operates in skeletal muscles during exercise.

Ingestion of carbohydrates increases glucose oxidation and decreases fat oxidation rather the other way around (Flatt 1995). Excess glucose stimulates accumulation of malonyl-CoA and inhibition of lipid oxidation (Ruderman *et al.* 1999). Malonyl-CoA is produced by acetyl-CoA carboxylase (ACC) and malonyl-CoA is the substrate for lipid synthesis (Frayn 2010). However, malonyl-CoA also inhibits CPT-1 activity and therefore transport of long chain acyl-CoA into the mitochondria for β-oxidation (Hue and Taegtmeyer 2009; Ruderman *et al.* 1999).

3.4.2 *De Novo* Lipid Synthesis

The human body has limited capacity to store carbohydrates and when the glycogen stores are filled, excess carbohydrates have to be used or converted to lipids. *De novo* lipid synthesis from glucose requires that glucose pass through glycolysis (Jensen 2009). The majority of *de novo* lipid synthesis occurs in the liver and the synthesised lipid is secreted from the liver as triacylglycerol packet into VLDL. Lipid synthesis, and glycolytic glucose metabolism in the liver occurs only after carbohydrate rich meals.

De novo lipid synthesis has also been demonstrated in skeletal muscle cells (Aas *et al.* 2004), but the physiological importance of lipid synthesis in skeletal muscles has not been clarified. However, it is clear that accumulation of triacylglycerol and long chain acyl-CoA occurs in skeletal muscles (Franch *et al.* 2002). It seems likely that skeletal muscles metabolise a large part of the glucose *via* glycolysis and release lactate for lipid synthesis in liver when the glycogen content is high (Jensen *et al.* 2006).

3.5   Regulation of Whole Body Glucose Metabolism

The glucose concentration in the blood has to be maintained within narrow limits for homeostasis. Glucose is not only an energy substrate required for survival, but glucose also contributes to its own regulation as a signalling molecule. High concentration of glucose stimulates secretion of insulin from the β-cells in the pancreas; low concentration of glucose stimulates secretion of glucagon from the α-cells in pancreas. Glucagon stimulates glucose release from the liver to maintain glucose homeostasis (Frayn 2010).

3.5.1 Glucose Metabolism after Meals

Most people get more than 50% of their energy from carbohydrates, which have to be broken down to monosaccharides for uptake in intestines and transport across cell membranes. Frayn describes carefully the degradation and metabolism of carbohydrates, fat and proteins in *Metabolic regulation. A human perspective*. In brief, ingestion of carbohydrate will increase the blood glucose concentration which stimulates insulin secretion. Insulin stimulates glucose uptake in skeletal muscles and adipocytes, which will reduce blood glucose concentration. Insulin will also inhibit glucose release from the liver and restore a glucose concentration of ~ 5 mM.

In insulin-resistant people and people with type 2 diabetes, insulin is unable to stimulate glucose disposal and the glucose concentration remains elevated (Frayn 2010). Insulin sensitivity can be tested by ingestion of a standardized amount of glucose (75 g) and blood glucose and insulin concentrations are followed. Type 2 diabetic subjects have still elevated glucose 2 h after ingestion of a glucose load of 75 g.

3.5.2 Glucose Metabolism in the Post-absorptive Phase

In the absence of intestinal glucose uptake, the liver supplies blood with glucose. Liver glycogen content initially supplies glucose release, but gluconeogenesis increases gradually during fasting. Glucose release is regulated by glucagon and insulin. Fasting increases plasma glucagon which stimulates glycogen breakdown and gluconeogenesis in the liver (Frayn, 2010). During fasting, insulin concentration decreases and the insulin-mediated inhibition of glucose release is removed. Fasting for more days will increase the concentration of ketone bodies to provide an alternative for glucose as substrate for the brain (Frayn, 2010).

3.5.3 Glucose Metabolism During Exercise

During exercise, glucose metabolism is increased. Skeletal muscles are the motors for movement where chemical energy in glucose (and fat) is transformed to mechanical work. Oxidation of one glucose molecule requires 6 molecules of O_2, and energy production can be calculated from oxygen uptake and respiratory exchange ratio (RER) (Frayn 1983). At rest, oxygen uptake is about 0.2–0.3 L $\cdot$ min^{-1} and glucose oxidation is about 0.1 g $\cdot$ min^{-1}.

During exercise the main carbohydrate source is skeletal muscles glycogen, but glucose is also extracted from the blood at high rate when skeletal muscle glycogen content decreases and carbohydrate is supplied (Coyle *et al.* 1986). The three major factors determining the relative contribution of glycogen/ glucose to whole body energy metabolism are: (1) exercise intensity, (2) duration of the exercise and (3) fitness level of the subjects. At rest and during moderate intensity exercise (<60% of VO_{2max}) glucose/glycogen, fat and protein will be oxidised in concert and reflect the relative composition of the diet. With increasing exercise intensities glycogen/glucose will contribute more to energy production (van Loon *et al.* 2001) and at 90–120% of VO_{2max}, glycogen/glucose will cover nearly 100% of the energy expenditure needed for muscle contraction (Gollnick 1985).

During low-intensity endurance exercise, glycogen content decreases gradually and fat oxidation increases (Christensen and Hansen 1939); at exhaustion after 2–3 hrs of exercise only $\sim 30\%$ of the energy is derived from carbohydrates. An improvement in fitness level by raising VO_{2max} will most often result in less carbohydrate oxidation (and a higher reliance for fat oxidation) when exercise is performed at the same absolute intensity and even at the same relative intensity (Figure 3.4). However, trained subjects also have a high capacity for glucose oxidation and well-trained young men can oxidise more than $3 \, g \cdot min^{-1}$ of glucose for 60–90 min (Hermansen *et al.* 1967; van Loon *et al.* 2001).

During exercise, muscle glycogen is metabolized together with blood glucose. Exercise activates glycogen phosphorylase via Ca^{2+}-mediated activation of phosphorylase kinase. Exercise also stimulates translocation of GLUT4 to the cell membrane (Figure 3.2). The intracellular signalling mechanism stimulating GLUT4 translocation remains to be determined, but the most likely signalling pathway is via 5′-AMP-activated protein kinase (Hardie and Sakamoto 2006).

3.5.4 Type 2 Diabetes

Insulin stimulates glucose uptake in skeletal muscles and increases whole body glucose oxidation in healthy subjects (Højlund *et al.* 2006). In type 2 diabetic subjects, insulin is unable to increase glucose oxidation and insulin-stimulated glycogen synthesis in skeletal muscles is impaired (Højlund *et al.* 2006) and blood glucose concentration remains elevated. Type 2 diabetes is an "energy over-supply syndrome" where excess glucose storage is impaired (Jensen 2009).

Humans have high ability to convert glucose into lipid when the glycogen stores are filled and a high carbohydrate diet is consumed (Figure 3.5). Acheson *et al.* reported that healthy humans were able to convert $\sim 475 \, g$ of glucose per day to lipid ($\sim 150 \, g$) when fed a high amount of carbohydrates for several days (Acheson *et al.* 1988). Acute elevation of glycogen content in skeletal muscles impairs insulin-stimulated glycogen synthesis without impairing insulin signalling or insulin-stimulated glucose uptake (Jensen *et al.* 2006). However,

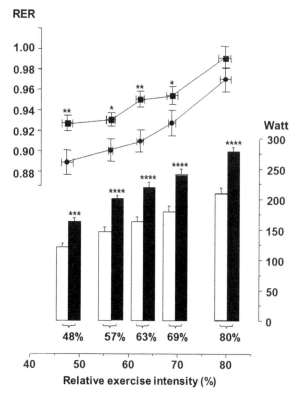

Figure 3.4 Substrate oxidation at different exercise intensities in trained (n = 10) and untrained (n = 10) male subjects. RER is lower in trained subjects (circles) compared to untrained subjects (squares) at same relative intensities at and below 69% of maximal oxygen uptake (VO$_{2max}$). The lower RER in trained subjects expresses that a higher percentage of energy is derived from lipid oxidation. Untrained subjects was characterized by a VO$_{2max}$ of 3.62 ± 0.14 L · min^{-1} (48.8 ± 1.52 ml · kg^{-1} · min^{-1}) trained by 4.63 ± 0.13 L · min^{-1} (63.1 ± 1.17 ml · kg^{-1} · min^{-1}). The two groups had similar body weight and fat percentage. The trained group had trained regularly for more than 6 years (unpublished results Franch *et al.*).

Abbreviation: RER: Respiratory exchange ratio.

insulin-stimulated glucose uptake and insulin signalling is impaired in muscles from insulin-resistant rats (Etgen *et al.* 1996; Ruzzin *et al.* 2005), and insulin-resistant skeletal muscles are characterized by a high content of triacylglycerol rather than elevated glycogen content (Franch *et al.* 2002; Ruzzin *et al.* 2005). These data suggest that high glycogen content in muscles prevents the normal storage of glucose as muscle glycogen and increases glycolysis and lactate release. When the glycogen stores are filled, glucose may be metabolised *via* glycolysis and enter the Cori cycle or remain in the blood until used as energy or converted to lipid. Under such metabolic condition, insulin resistance will gradually arise and increase the risk for developing type 2 diabetes (Figure 3.5).

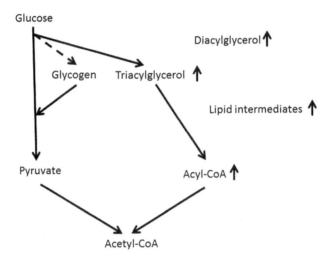

Figure 3.5   Glucose can be stored as glycogen or converted to triacylglycerol. Accumulation of triacylglycerol will increase the concentration of diacylglycerol, acyl-CoA and other lipid intermediates causing insulin resistance. Exercise increases oxidation of acyl-CoA and increases insulin sensitivity.

Indeed, low aerobic fitness and oxidative capacity is a hallmark of type 2 diabetes, and endurance training increases mitochondrial function and insulin sensitivity. Indeed, increased oxidative capacity may contribute to the beneficial effect of training on insulin sensitivity. However, perhaps the most important beneficial effect of endurance training is that muscle glycogen is utilized. Muscles with low glycogen efficiently stores glucose as glycogen (Jensen *et al.* 2006), which improves whole body glucose regulation. We believe that healthy storage of glucose is as glycogen.

Summary Points

- Carbohydrates are the major energy source for most people.
- Carbohydrates are broken down to glucose for transport in the blood.
- A family of facilitative glucose transporters (GLUTs) transport glucose across cell membranes.
- Glucose is an energy-rich molecule but glucose is also a signalling molecule controlling secretion of hormones and regulating gene expression.
- Glucose is stored as glycogen; a glycogen particle can contain 50 000 glucose molecules.
- Glycogen is most abundant in skeletal muscles and in the liver.
- Glycolysis metabolises one glucose molecule into two molecules of pyruvate in ten enzymatic reactions.

- Pyruvate can be transported into mitochondria for complete oxidation and ATP production.
- Glycolysis can occur in the absence of oxygen; pyruvate is converted to lactate under anaerobic conditions.
- Glucose oxidation at rest is about $0.1 \, g \cdot min^{-1}$ and can increase above $3 \, g \cdot min^{-1}$ in well-trained individuals during exercise.
- Glucose can be converted to lipid and high glucose intake causes obesity.
- Type 2 diabetes is a disease where regulation of blood glucose is insufficient due to insulin resistance in muscle and liver.
- Exercise improves insulin sensitivity and prevents development of type 2 diabetes.

Key Facts

1. Glucose transport and basic metabolic pathways

Carbohydrate is a major component in the food, and carbohydrates have to be broken down to monosaccharides for transport across cell membranes. Glucose concentration in blood has to be maintained relatively stably to ensure energy for cells and avoid the harmful effects of high concentrations. Carbohydrate is stored as glycogen in skeletal muscles and the liver; liver glycogen is important to supply glucose to the blood when intestinal glucose uptake does not occur.

- Carbohydrates are broken down to monosaccharides for transport across cell membranes.
- Glucose is the form of carbohydrate transported in the blood.
- 14 different isoform of glucose transporters regulate glucose transport over cell membranes.
- Glycolysis is the metabolic conversion of glucose into pyruvate for anaerobic ATP production.
- Pyruvate is further oxidised in the mitochondria (Krebs cycle) or reduced to lactate depending on oxygen supply.
- Glucose has to be metabolised *via* glycolysis for *de novo* lipogenesis.
- Glycolysis is similar in all cell types but key enzymes have different isoforms which are expressed tissue specific and regulated differently.
- Carbohydrates are stored as glycogen in the liver ($\sim 100 \, g$) and skeletal muscles ($\sim 500 \, g$).

2. Glucose metabolism during rest and exercise

Glucose metabolism is low at rest but exercise can increase glucose metabolism to $3 \, g \cdot min^{-1}$ during exercise in healthy, well-trained young men. Glycogen depletion in skeletal muscles occurs after $\sim 90 \, min$ of high intensity exercise and glycogen depletion causes fatigue.

- Glucose metabolism at rest is low but the brain (and some other cell) uses glucose continuously.
- Many tissues *e.g.* liver, skeletal muscles and heart adapt glucose utilization to glucose delivery.

- Glucose metabolism can increase to $3\,\text{g}\cdot\text{min}^{-1}$ during exercise in healthy young people.
- Diet, exercise intensity and training status determine the rate of glucose oxidation during exercise.
- Endurance trained individuals have increased muscle glycogen stores compared to untrained individuals.
- Glycogen depletion in active muscles causes fatigue.

3. Glucose, lipogenesis and type 2 diabetes

Diabetes is defined from elevated blood glucose concentration in fasting condition. Type 2 diabetes occurs mainly in obese people and results from insulin resistance in skeletal muscles and insufficient insulin production. Exercise improves insulin sensitivity and prevents development of type 2 diabetes.

- Insulin resistance occurs often in obese and inactive elderly people.
- Type 2 diabetes occurs because skeletal muscles are unable to remove glucose from the blood during insulin stimulation.
- Glucose is converted to lipid when glycogen content in skeletal muscle is high.
- Lipid synthesis from glucose occurs mainly in the liver but skeletal muscles can also synthesize lipid from glucose.
- Accumulation of fat in skeletal muscles causes insulin resistance and abolishes appropriate glucose regulation.
- Exercise decreases glycogen content in skeletal muscles and improves insulin sensitivity.

Definitions of Words and Terms

ATP (Adenosine triphosphate): Molecule with the ability to transfer energy from its phosphate bonds to energy demanding processes in the cell.

Glycolysis: The ten enzymatic reaction converting one glucose molecule into two pyruvate molecules.

Glycogen: The polymer form of glucose in humans and other mammalian.

Glycogenin: Protein/enzyme responsible for initial glucose incorporation into glycogen.

GLUT (Glucose transporters): Proteins that transport glucose across cell membranes (bi-directional).

GLUT2 (Glucose transporter 2): Mainly expressed in liver and pancreatic β-cells. Present in the cell membrane continuously.

GLUT4 (Glucose transporter 4): Mainly expressed in skeletal muscle cells. Translocates to the cell membrane when activated by insulin or muscle contraction.

Insulin resistance: When insulin, within the normal blood concentration range, fail to exert standard response in skeletal muscle. Insulin resistance is a hallmark of type 2 diabetes.

Insulin signalling: Intracellular processes that transmit information from the insulin receptor to physiological responses (*e.g.* translocation of GLUT4).

Lactate: The anion produced when pyruvate is converted into lactic acid. The reaction occurs when the glycolytic flux exceed the aerobic capacity of the mitochondria.

PFK-1 (6-Phosphofructokinase): The most important regulator of glycolytic flux.

Maximal oxygen uptake (VO_{2max}): The highest rate of oxygen consumption which occurs during strenuous exercise.

Respiratory exchange ratio (RER): Ratio between CO_2 expiration and O_2 uptake (VCO_2 divided by VO_2). RER reflects the ratio between carbohydrate and fat oxidation.

De novo lipogenesis: Synthesis of lipids from glucose or other nor-lipid carbon sources.

VCO_2: Whole body CO_2 expiration, measured in $L \cdot min^{-1}$.

VO_2: Whole body O_2 uptake, measured in $L \cdot min^{-1}$.

VO_{2max}: The maximal whole body oxygen uptake (consumption) rate. Higher values indicate higher fitness levels ($ml\ O_2 \cdot kg^{-1} \cdot min^{-1}$).

Acknowledgements

JJ's research is supported by grants from the Novo Nordisk Foundation and *via* participation in COST Action BM0602 (Network supported by European).

References

Aas, V., Kase, E.T., Solberg, R., Jensen, J., and Rustan, A.C., 2004. Chronic hyperglycaemia promotes lipogenesis and triacylglycerol accumulation in human skeletal muscle cells. *Diabetologia.* 47: 1452–1461.

Acheson, K.J., Schutz, Y., Bessard, T., Anantharaman, K., Flatt, J.P., and Jequier, E., 1988. Glycogen storage capacity and de novo lipogenesis during massive carbohydrate overfeeding in man. *Am J Clin Nutr.* 48: 240–247.

Aslesen, R. and Jensen, J., 1998. Effects of epinephrine on glucose metabolism in contracting rat skeletal muscle. *Am J Physiol.* 275: E448–E456.

Christensen, E.H. and Hansen, O., 1939. Arbeitsfähigkeit und Ernährung. *Skand Archiv Physiol.* 81: 160–171.

Connett, R.D. and Sahlin, K., 1996. Control of glycolysis and glycogen metabolism. In *Handbook of Physiology. Integration of Motor, Circulatory, Respiratory and Metabolic Control during Exercise.*, eds. Rowell, L.B. and Shepherd, J.T., pp. 870–911. American Physiological Society.

Copeland, R.J., Bullen, J.W., and Hart, G.W., 2008. Cross-talk between GlcNAcylation and phosphorylation: roles in insulin resistance and glucose toxicity. *Am J Physiol Endocrinol Metab.* 295: E17–E28.

Coyle, E.F., Coggan, A.R., Hemmert, M.K., and Ivy, J.L., 1986. Muscle glycogen utilization during prolonged strenuous exercise when fed carbohydrate. *J Appl Physiol.* 61: 165–172.

Depre, C., Rider, M.H., and Hue, L., 1998. Mechanisms of control of heart glycolysis. *Eur J Biochem.* 258: 277–290.

Etgen, G.J., Wilson, C.M., Jensen, J., Cushman, S.W., and Ivy, J.L., 1996. Glucose transport and cell surface GLUT-4 protein in skeletal muscle of the obese Zucker rat. *Am J Physiol.* 271: E294–E301.

Flatt, J.P., 1995. Use and storage of carbohydrate and fat. *Am J Clin Nutr.* 61: S952–S959.

Franch, J., Knudsen, J, Ellis, B.A., Pedersen, P.K., Cooney, G.J., and Jensen, J., 2002. Acyl-CoA binding protein expression is fibre type specific and elevated in muscles from obese insulin-resistant Zucker rat. *Diabetes.* 51: 449–454.

Frayn, K.N., 1983. Calculation of substrate oxidation rates in vivo from gaseous exchange. *J Appl Physiol.* 55: 628–634.

Frayn, K.N., 2010. *Metabolic regulation. A human perspective*, 3rd ed., pp. 1–371. Wiley-Blackwell, Chichester.

Gollnick, P.D., 1985. Metabolism of substrates: energy substrate metabolism during exercise and as modified by training. *Fed Proc.* 44: 353–357.

Hardie, D.G. and Sakamoto, K., 2006. AMPK: a key sensor of fuel and energy status in skeletal muscle. *Physiology (Bethesda).* 21: 48–60.

Hermansen, L., Hultman, E., and Saltin, B., 1967. Muscle glycogen during prolonged severe exercise. *Acta Physiol Scand.* 71: 129–139.

Højlund, K., Frystyk, J., Levin, K., Flyvbjerg, A., Wojtaszewski, J.F., and Beck-Nielsen, H., 2006. Reduced plasma adiponectin concentrations may contribute to impaired insulin activation of glycogen synthase in skeletal muscle of patients with type 2 diabetes. *Diabetologia.* 49: 1283–1291.

Hue, L. and Taegtmeyer, H., 2009. The Randle cycle revisited: a new head for an old hat. *Am J Physiol Endocrinol Metab.* 297: E578–E591.

Ilzuka, K. and Horikawa, Y., 2008. ChREBP: a glucose-activated transcription factor involved in the development of metabolic syndrome. *Endocr J.* 55: 617–624.

Jensen, J., 2009. The role of skeletal muscle glycolysis in whole body metabolic regulation and type 2 diabetes. In: *Glycolysis: Regulation, Processes and Diseases*, ed. Lithaw H, pp. 65–83. Nova Science Publishers, Inc., New York.

Jensen, J., Jebens, E., Brennesvik, E.O., Ruzzin, J., Soos, M.A., Engebretsen, E.M., O'Rahilly, S., and Whitehead, J.P., 2006. Muscle glycogen inharmoniously regulates glycogen synthase activity, glucose uptake, and proximal insulin signaling. *Am J Physiol Endocrinol Metab.* 290: E154–E162.

Jensen, J. and Lai, Y.C., 2009. Regulation of muscle glycogen synthase phosphorylation and kinetic properties by insulin, exercise, adrenaline and role in insulin resistance. *Arch Physiol Biochem.* 115: 13–21.

Kemp, R.G. and Foe, L.G., 1983. Allosteric regulatory properties of muscle phosphofructokinase. *Mol Cell Biochem.* 57: 147–154.

Knowler, W., Barret-Connor, E., Fowler, S., Hamman, R., Lachin, J.M., Walker, E., and Nathan, D., 2002. Reduction in the incidence of type 2 diabetes with lifestyle intervention or metformin. *The New England Journal of Medicine.* 346: 393–403.

Larance, M., Ramm, G., and James, D.E., 2008. The GLUT4 code. *Mol Endocrinol.* 22: 226–233.

Meléndez, R., Meléndez-Hevia, E., and Canela, E.I., 1999. The fractal structure of glycogen: A clever solution to optimize cell metabolism. *Biophys J.* 77: 1327–1332.

Newsholme, E. and Leech, T., 2009. *Functional Biochemistry in Health and Disease*, pp. 1–543. Wiley-Blackwell, Chichester.

Rider, M.H., Bertrand, L., Vertommen, D., Michels, P.A., Rousseau, G.G., and Hue, L., 2004. 6-phosphofructo-2-kinase/fructose-2,6-bisphosphatase: head-to-head with a bifunctional enzyme that controls glycolysis. *Biochem J.* 381: 561–579.

Ruderman, N.B., Saha, A.K., Vavvas, D., and Witters, L.A., 1999. Malonyl-CoA, fuel sensing, and insulin resistance. *Am J Physiol.* 276: E1–18.

Ruzzin, J., Wagman, A.S., and Jensen, J., 2005. Glucocorticoids-induced insulin resistance: defects in insulin signalling and the effects of a selective glycogen synthase kinase-3 inhibitor. *Diabetologia.* 48: 2119–2130.

Shulman, G.I., Rothman, D.L., Jue, T., Stein, P., DeFronzo, R.A., and Shulman, R.G., 1990. Quantification of muscle glycogen synthesis in normal subjects and subjects with non-insulin-dependent diabetes by 13C nuclear magnetic resonance spectroscopy. *New Engl J Med.* 322: 223–228.

Thorens, B. and Mueckler, M., 2010. Glucose transporters in the 21st Century. *Am J Physiol Endocrinol Metab.* 298: E141–E145.

van Loon, L.J., Greenhaff, P.L., Constantin-Teodosiu, D., Saris, W.H., and Wagenmakers, A.J., 2001. The effects of increasing exercise intensity on muscle fuel utilisation in humans. *J Physiol.* 536: 295–304.

Wasserman, D.H., 2009. Four grams of glucose. *Am J Physiol Endocrinol Metab.* 296: E11–E21.

CHAPTER 4

Medical Implications of Dietary Simple Sugars and Complex Carbohydrates, Glycemic Index and Glycemic Load

MARY DOWNES GASTRICH*[a] AND MICHELLE WIEN[b]

[a] UMDNJ-Robert Wood Johnson Medical School, Department of Obstetrics, Gynecology and Reproductive Sciences, Clinical Administration Building (CAB), Rm. 2104, 125 Paterson St., New Brunswick, NJ08901-1977; [b] Loma Linda University, School of Public Health, Department of Nutrition, Nichol Hall, Rm. 1107, Loma Linda, CA 92350
*Email: gastrimd@umdnj.edu

4.1 Introduction

The concepts of GI and GL have been useful in facilitating the understanding of the relationship of dietary CHO and various health/medical outcomes. In general, it has been stated that GI has proven to be a more useful nutritional concept than the type of CHO (Foster-Powell *et al.* 2002). GI is an index which is used as a reliable classification of the amount of CHO in foods according to their postprandial glycemic effect (Foster-Powell *et al.* 2002). The relationship of GI and GL is based on the assumption that the quantity and quality of CHO influence the glycemic response (Foster-Powell *et al.* 2002). Foster-Powell *et al.* (2002) indicated that GI compares "equal quantities of CHO and provides a

Food and Nutritional Components in Focus No. 3
Dietary Sugars: Chemistry, Analysis, Function and Effects
Edited by Victor R Preedy

measure of carbohydrate quality but not quantity". The general concept is that the higher the GL, the greater the elevation in blood glucose level, and, if elevated over the long-term it can lead to increased risk of developing CVD and T2DM (Foster-Powell *et al.* 2002).

Previously, Gastrich *et al.* 2007 reviewed studies that assessed the impact of additive sugar in the diet. The studies generally had small sample sizes or included both the pediatric and the adult populations. Results of the Gastrich *et al.* (2008) review of randomized control trials of the effects of CHO and low GI/GL on levels of specific metabolic syndrome/CVD risk factors in adults are summarized below; however, these studies were either very small (N < 50) or relatively small (N ≥ 100 but N ≤ 200).

- The medical implication of ingesting simple sugars (*e.g.* fructose) appears to promote hypertriglyceridemia in subjects more than complex CHO (not including starches with a high GI) (Fried and Rao 2003).
- The type of (nonstarchy) CHO in foods appears to have medical implications in lowering TG levels in subjects; diets high in CHO accompanied by a low dietary GI tend to decrease TG levels (De Lorgeril *et al.* 1999; Esposito *et al.* 2004; Vincent-Baudry *et al.* 2005).
- The US CARDIA study and other studies have indicated that diets low in GI are associated with a reduced LDL-C, which may in turn have the beneficial effect of risk factors for IHD (Sloth *et al.* 2004); a diet that has a low daily GI or GL may decrease levels of specific metabolic syndrome/ CVD risk factors; a high-polysaccharide, high-fiber diet was recommended by several studies to achieve a reduction of more than 10% of metabolic syndrome factors (TC and LDL-C) (Poppitt *et al.* 2002).
- In a few studies, there appear to be gender differences in the medical implications of simple and complex CHO on TG levels; in men, a high intake of fructose and glucose increases TG levels, whereas in women a high sucrose intake elevates fasting plasma TG levels (Raben *et al.* 2001).
- A low-fat, high-polysaccharide diet (especially one high in dietary fiber) was shown to lower serum cholesterol in overweight subjects (Hellerstein 2002).

4.2   Purpose of this Review

This review provides a comprehensive summary of the medical implications of dietary CHO, GI and GL from predominately 20 large-scale studies. The following criteria and levels of evidence were used to select the studies:

- Studies conducted since 2000 in English, with a large sample size (N ≥ 1000), plus two medium-sized studies (N ≥ 100); all studies had appropriate SA meeting the level of evidence criteria described below.
- Study references were available in online journals, PubMed or at the authors' medical library; no studies in journals with a 12-month delay.

- Studies conducted on relatively healthy adult men and/or women (18 yrs or older); children were excluded.
- Studies were designed to assess dietary CHO without a mixed diet or mixed fat-diet approach; undertaken at multicenters, institutions, major research or medical centers.
- Reviews were excluded.
- Studies were ranked according to the highest grades of evidence, based partly on Evans' hierarchy of ranked evidence as follows:
  - **Level 1: Highest levels of evidence**
    Randomized clinical trials (RCTs) including adequately sampled studies and controlled crossover clinical trials with a $p \leq 0.05$ CI.
  - **Level 2: Medium levels of evidence**
    Matched and/or case control studies, or randomized crossover design involving several health centers with a $p \leq 0.05$ CI.
  - **Level 3: Adequate and/or lowest levels of evidence**
    Prospective, cohort or comparative studies (studies in which the subjects are identified and then followed forward in time) and cross-sectional observational studies (those that draw inferences about the possible effect of a treatment on subjects where the assignment of subjects into a treated group versus a control group is outside the control of the investigator).

4.3   Results

A total of 20 studies met the criteria above and most had a sample size greater than 1000. Tables 4.1–4.3 summarize the study results; Table 4.1 provides an overall summary and detail of the studies of the effects of dietary CHO, GI and GL and various medical conditions, which are discussed under specific headings in the Results section below; Table 4.2 tabulates results and conclusions of these studies in relation to the differing medical implications; and Table 4.3 categorizes the significant results in terms of gender differences.

4.4   The Relationship Between the Quality and Quantity of Dietary CHO and Differing Medical Implications

4.4.1   Dietary CHO and Metabolic Syndrome/CVD Risk Factors

This category included five studies (including three Level 1 and two Level 3 studies). It is significant that all these large-scale studies, including those of differing levels of evidence, showed a significant positive relationship between the quality and/or quantity of dietary CHO and CVD risk.In terms of CHO quality, Esposito *et al.* (2004) indicated that a fiber-rich Mediterranean diet

Table 4.1 Large-scale studies from 2000 to 2011 showing the relationship between medical implications and additive dietary sugars and complex carbohydrates (CHOs), GI and GL. Studies are arranged by medical categories and further stratified by differing levels of evidence (as defined in the text). The results show the negative and/or no significant impact of dietary sugar and various outcomes.

Medical Condition and Name of Study	Number of Subjects, Experimental Design and Methods and SA	Results and Conclusions
1. Level 1 (highest level of evidence) Randomized clinical trials (RCTs) including adequately sampled studies and single randomized clinical trials and/or controlled crossover clinical trial. Studies are listed in chronological order by largest sample size first.		
1.1. Dietary CHO and T2DM/Insulin Resistance		
1.1.1. Lau *et al.* 2005 Dietary Glycemic Index, Glycemic Load,Fiber, Simple Sugars, and Insulin Resistance	N = 5675; a non-pharmacological intervention study (used data from the Danish Inter99 intervention study on diet, physical activity, and smoking and effects of CVD); study used food frequency questionnaires to estimate dietary intake of men and women. SA: Linear and logistic regression, HOMA model	Intake of dietary fiber was inversely associated with the probability of having insulin resistance; insulin resistance was not associated with diets high or low in GI or simple sugars.
1.2. CVD and Dietary Sugars		
1.2.1. Esposito *et al.* 2004 Effect of a Mediterranean-Style Diet on Endothelial Dysfunction and Markers of Vascular Inflammation in the Metabolic Syndrome	N = 180 (90 in the intervention group l and 90 controls); randomized, single-blind trial conducted in Italy on men and women from 2001–2004; experimental (intervention) group given a Mediterranean diet while control group a prudent diet SA: Wilcoxon test and HOMA model	Patients in the experimental group had significantly reduced serum concentrations of *hs*-CRP (P = 0.01), IL-6 (P = 0.04), IL-7 (P = 0.4), and IL-18 (P = 0.3), as well as decreased insulin resistance (P < 0.001) and significantly less features of the metabolic syndrome than controls (P < 0.001).
1.2.2. Yancy *et al.* 2004 Assessment of Low-Carbohydrate, Ketogenic Diet versus a Low-Fat Diet To Treat Obesity and Hyperlipidemia	N = 120 overweight, hyperlipidemic men/women; a low CHO diet was compared to CVD measurements (body weight, body composition, fasting serum lipid levels, and tolerability). SA: chi-square test or Fisher exact test and linear mixed-effects models.	Yancy *et al.* reported that subjects on the low-CHO diet had greater decreases in serum triglyceride levels (change, and greater increases in high-density lipoprotein cholesterol levels but there were not significant for LDL-C levels among groups.

Table 4.1 *(Continued)*

Medical Condition and Name of Study	Number of Subjects, Experimental Design and Methods and SA	Results and Conclusions
2. Level 2 Studies: Matched, case control, and randomized cross over studies		
2.1. Diabetes and Dietary Sugars		
2.1.1. Krishnan *et al.* 2007 Glycemic Index, Glycemic Load, and Cereal Fiber Intake and Risk of Type 2 Diabetes in US Black Women	N = 59,000; women subjects were from the US Black Women Study; a prospective cohort study that evaluated biennial National Cancer Institute (NCI)-Block food frequency questionnaire (FFQ) to assess the association of glycemic load, glycemic index, and cereal fiber with risk of type 2 diabetes. Cereal fiber content for each ingredient for all-grain containing foods was obtained from the US Department of Agriculture. *Statistical Analysis:* Cox proportional hazards models were used to calculate incidence rate ratios (IRRs), also known as hazard ratios, and 95% confidence intervals (CIs while controlling for lifestyle and dietary factors).	Krishnan *et al.* concluded that GI was associated with the risk of diabetes: the IRR for the highest quintile relative to the lowest was 1.23 (95% CI, 1.05–1.44). Cereal (a CHO) fiber intake was inversely associated with risk of diabetes, with an IRR of 0.82 (95% CI, 0.70–0.96) for the highest *vs.* lowest quintiles of intake. Krishnan *et al.* concluded that Increased cereal fiber in the diet may be preventative in reducing the risk of type 2 diabetes in black women.
2.1.2. Schulze *et al.* 2004 Sugar-Sweetened Beverages, Weight Gain, and Incidence of Type 2 Diabetes in Young and Middle-Aged Women	N = 91,249 women free of diabetes and other major chronic diseases. This was a prospective cohort analyses conducted from 1991 to 1999 among women in the Nurses' Health Study (a large sized prospective cohort study of 116,671 female US nurses aged 24 to 44 years at study initiation in 1989) in the US. The study used a questionnaire that included a 133-item semi quantitative FFQ. An assessment of	Schultz *et al.* concluded that higher consumption of sugar sweetened beverages was associated with larger weight gain in middle aged women and this may be associated with an increased risk of type 2 DM in women.

non-dietary exposures (BMI, DM, physical activity level, smoking, etc.) was conducted. SA: calculation of mean weight changes, the estimation of the relative risk (RR) of diabetes for each category of intake compared with the lowest category using Cox proportional hazards analysis stratified by 5-year age categories and 2-year intervals.

2.2. Other Medical Complications

2.2.1. Shaw et al. 2003

Neural tube defects (NTDs) associated with maternal periconceptional dietary intake of simple sugars and glycemic index

$N = 1224$ (613 eligible women experimentals and 611 women controls without NTDs); population-based case-control study; a 100-item FFQ was administered to the subjects to assess nutrient intakes from the diet. An estimation of GI was conducted using the FFQ. The study assessed the maternal periconceptionaldietary intakes of sucrose, glucose, fructose, and foods with higher glycemic index values to determine the risk of having NTD-affected pregnancies.
SA: OD ratios at 95% CIs was used to calculate risks of NTD

Shaw et al. found that higher intakes of sucrose and foods with higher GI values were significantly associated with elevated risks of (NTDs) $\approx$2-fold but not glucose or fructose. Higher GI values increased the risk of NTD, especially spina bifida, for women whose BMI values were >29. They concluded that BMI is predictive of elevated glucose concentrations in non-diabetic women and in diabetic women, it is associated with an increased risks of NTD-affected pregnancies
In addition to an association between hyperinsulinemia and an increased risk of delivering infants with NTDs.

3. Level 3 Studies

3.1 Dyslipidmia and Dietary Sugars

3.1.1. Welsh et al. 2010

Caloric Sweetener (= added sugars) Consumptionand Dyslipidemia Among US Adults

$N = 6113$; Cross-sectional study among US adults. Subjects were part of the National Health and Nutrition Examination Survey (NHANES) 1999–2006 study. Subjects were grouped by intake of added sugars using limits specified in dietary recommendations

Welsh et al. reported that the adjusted mean HDL-C, geometric mean TGs, and mean LDL-C levels and adjusted odds ratios of dyslipidemia, including low HDL-C levels (<40 mg/dL for men; 50 mg/dL for women), high triglyceride levels (≤150 mg/dL), high

Table 4.1 (*Continued*)

Medical Condition and Name of Study	Number of Subjects, Experimental Design and Methods and SA	Results and Conclusions
	(<5% [reference group], 5%–<10%, 10%–<17.5%, 17.5%–<25%, and ≤25% of total calories. Subjects were interviewed using a 24-hour dietary recall to assess dietary intake. SA: Linear and logistical regression	LDL-C levels (≤130 mg/dL), or high ratio of triglycerides to HDL-C (<3.8). An important conclusion of Welsh *et al.* was that there was a statistically significant correlation between dietary added sugars and blood lipid levels among US adults. Further, increased added sugars were associated with critical CVD risk factors such as lower HDL-C levels, higher triglyceride levels, and higher ratios of triglycerides to HDL-C. Welsh *et al.* recommended further studies are needed to examine dietary sugars in terms of CVD risk – especially since dietary fats may have decreased in diets with concomitant increases in CHO levels.
3.2. Cancer and Dietary Sugars/CHOs 3.2.1. Franceschi *et al.* 2001 Dietary glycemic load and colorectal cancer risk	N = 6107 Subjects in Italy; (a total of 1953 subjects in the experimental group consisted of 1125 men and 828 women with histologically confirmed incident cancer of the colon or rectum). The controls (N = 4154) consisted of 2073 men and 2081 women) were hospitalized for acute conditions and were in-patients with no history of cancer, a case-control study on colorectal cancer conducted in Italy between January 1992	GI was positively correlated to GL (Pearson correlation coefficient, *r* = 0.59), and colorectal cancer risk increased with an increase in dietary GI. Franceschi *et al.* reported that they found a direct association of colorectal cancer risk with GI (odds ratio (OR) in highest *vs.* lowest quintile = 1.7; 95% confidence interval (CI): 1.4–2.0) and GL (OR = 1.8; 95% CI: 1.5–2.2).

and June 1996. A validated FFQ was conducted on all subjects. The average daily dietary GI and glycemic load (GL), and fiber intake was calculated.

SA: Pearson correlation coefficient.

3.2.2. Higginbotham *et al.* 2004.
Dietary Glycemic Load and Risk of Colorectal Cancer in the Women's Health Study

N = 38,451 US women. Prospective cohort study cancer study that used baseline dietary intake measurements, assessed with a semi quantitative FFQ, to examine the associations of dietary GI, overall dietary GI, CHO, fiber, nonfiber CHO, sucrose, and fructose with the subsequent development of colorectal cancer.

SA: Cox proportional hazardsmodels were used to estimate relative risks (RRs).

3.2.3. Howarth *et al.* 2008
The association of glycemic load and carbohydrate intake with colorectal cancer risk in the Multiethnic Cohort Study

N = 191,004. The Multiethnic Cohort (MEC) was conducted in 1993 and consisted of subjects residing in Hawaii or Southern California and predominantly of 5 ethnic groups: African American, white, Latino, Native Hawaiian, or Japanese American. A baseline quantitative FFQ data was used to assess usual dietary intake.

SA: Cox regression, was used to calculate adjusted RRs and 95% CIs.

3.2.4. Larsson *et al.* 2006
Dietary Carbohydrate, Glycemic Index, and Glycemic Load in Relation to Riskof Colorectal Cancer in Women

N = 61,433, prospective study in Sweden to assess the associations of dietary CHO, GI and GL on the incidence of colorectal cancer among Swedish women (40 to 76 y) who were free of cancer in 1987–1990; a 67-item FFQ was administered.

They concluded that the positive associations of GI and GL with colorectal cancer strongly suggests the contributory role of dietary refined simple sugars.

Higginbotham *et al.* found that dietary GL was statistically significantly associated with an increased risk of colorectal cancer (adjusted RR = 2.85, 95% confidence interval [CI] = 1.40 to 5.80, comparing extreme quintiles of dietary glycemic load ($P_{trend} = 0.004$). Increased risk of colorectal cancer was associated with increase dietary fructose, high levels of CHO, and non-fiber CHO. The authors concluded that a high dietary glycemic load may increase the risk of colorectal cancer in women.

Howarth *et al.* found differing results among men and women. The RRs, as well as an inverse association with GL and colorectal cancer was found in women of all ethnic groups but not in men. White men had a positive association with increasing GL but not the other ethnic groups.

Larsson *et al.* reported that results of the hazard age-adjusted ratio indicated that GI, but not CHO intake or GL, was associated with an increased risk of colorectal cancer but this relationship was not found in the multivariate model analysis where a high

Table 4.1 (*Continued*)

Medical Condition and Name of Study	Number of Subjects, Experimental Design and Methods and SA	Results and Conclusions
3.2.5. Larsson *et al.* 2009 Glycemic load, glycemic index and breast cancer risk in a prospective cohortof Swedish women	SA: statistical procedures were implemented with SAS software, version 9.1; all p values are two sided (at 95% CI). N = 61,433 Swedish women; prospective study assessed he associations of CHO intake, GI and GL with risk of hormone receptor-defined breast cancer using a FFQ during 1987–1990. SA: Cox proportional hazards models were used to estimate RRs and 95% CIs.	CHO intake, a high GI, and a high GL did not increase the risk of colorectal cancer. Larsson *et al.* found that GL but not CHO intake or GI was weakly positively associated with overall breast cancer risk (p for trend = 0.05). In the study, the analyses was stratified by estrogen receptor (ER) and progesterone receptor (PR). In these cases, the status of the breast tumors were observed to be statistically significant positive associations of CHO intake, GI and GL with risk of ER1/PR2 breast cancer.
3.2.6. Nielsen *et al.* 2005 Dietary Carbohydrate Intake Is Not Associated with the Breast Cancer Incidence Rate Ratio in Postmenopausal Danish Women1	N = 23,870 Danish women aged 50–65 y, who participated in the "Diet, Cancer and Health" cohort, which evaluated the associations between dietary CHO intake and breast cancer incidence. SA: Cox's proportional hazard model stratified according to age at entry (1-y intervals) and Cox regression models.	Nielsen *et al.* found no significant association for intake of various simple CHOs (*e.g.* glucose, fructose, sucrose, maltose, lactose) or starch and breast cancer incidence rate, furthermore, they found no association for GI or GL. They concluded dietary intake of various carbohydrates was not associated with breast cancer incidence rates for either estrogen receptor positive (ER+) or (ER−) breast cancer. There was a slight significant positive association between glycemic index and (ER−) breast cancer was observed (*P* = 0.05). In conclusion, Nielsen *et al.* found no significant or clear associations of dietary CHO, GI or GL with breast cancer.

3.2.7. Nothlings *et al.* 2008
Dietary glycemic load, added sugars, and carbohydrates as risk factors for pancreatic cancer: the Multiethnic Cohort Study

N = 162,150; Subjects were part of the Multiethnic Cohort Study in Hawaii and Los Angeles (1993–1996) which assessed diet, in relation to cancer outcomes. The purpose of the study was to assess associations between GL, dietary CHO and simple sugars such as sucrose, fructose, total sugars, and added sugars in relation to the risk of pancreatic cancer; a quantitative FFQ was used.
SA: Cox proportional hazards models using age as the time metric were calculated to derive relative risks (RRs).

While Nothlings *et al.* found an increased risk between higher intakes of total sugars, fructose, and sucrose and pancreatic cancer. The association with fructose was significant when the highest and lowest quartiles were compared (relative risk: 1.35; 95% CI: 1.02, 1.80; *P* for trend = 0.046).
Nothlings *et al.* conjectured that a high dietary intake of fructose and sucrose intakes may play a role in pancreatic cancer but other medical conditions (overweight or obesity concomitant with some insulin resistance) may play a role. More research is needed on this.

3.2.8. Simon *et al.* 2010
Glycemic index, glycemic load, and the risk of pancreatic cancer among postmenopausal women in the women's health initiative observational study and clinical trial

N = 161,809 postmenopausal women of ages 50–79 from the Women's Health Initiative Study (WHI). WHI data was used to assess the relationship between dietary factors that are associated with increased postprandial blood glucose levels are also associated with an increased risk of pancreatic cancer.
FFQs were used to assess GI, GL, CHO, and simple sugars (fructose and sucrose).
SA: Chi-square tests and Cuzick's trend test was used for ordinal variables and t-tests for continuous variables

Simon *et al.* results indicated that pancreatic cancer was not associated with dietary GI, GL or simple CHOs and sugars

3.2.9. Terry *et al.* 2003
Glycemic Load, Carbohydrate Intake, and Risk of Colorectal Cancer in Women: A Prospective Cohort Study

N = 49,124 women participating in a randomized, controlled trial of screening for breast cancer in Canada.
SA: quintiles reported at 95% CI

Terry *et al.* data did not support the hypothesis that high dietary GL CHO or or sugar increased colorectal cancer risk in women.

Table 4.1 (*Continued*)

Medical Condition and Name of Study	Number of Subjects, Experimental Design and Methods and SA	Results and Conclusions
3.3. CVD and Dietary Sugars 3.3.1. Liu *et al.* 2000 A prospective study of dietary glycemic load, carbohydrate intake, and risk of coronary heart disease in US women	N = 75,521 healthy women cohort aged 38–63 y; use of FFQ to calculate GL. SA: multivariate analysis.	Liu *et al.* results indicated that after a 10 y follow-up, 761 cases of CHD were reported that dietary GL was directly associated with risk of CHD
3.4. Metabolic Syndrome and Dietary Sugar 3.4.1. Brown *et al.* 2010 SSBs, and BP	N = 2,696 people (40 to 59 years of age) were randomly recruited from US and UK (subset of the Intermap Study (cross sectional study using FFQs); metabolic measurements (BP) were analyzed and compared to SSBs, and simple sugars data from the FFQs. SA: Pearson correlation, multiple regression analysis	Brown *et al.* results indicated that dietary SSBs were directly related to BP (*P* values of 0.005 to <0.001 (systolic BP) and 0.14 to <0.001 (diastolic BP) as well as the direct association of fructose and glucose to BP. Brown *et al.* concluded that dietary intake of SSBs, sugars, and salt be substantially reduced to prevent an increase in BP.
3.5. Other Medical Disease and Dietary Sugar 3.5.1. Gopinath *et al.* 2011 Carbohydrate Nutrition Is Associated with the 5-Year Incidence of Chronic Kidney Disease	N = 3,508 participants (49 y or older) during 1992–1994; community based cohort study that assessed the association between CHO nutrition, including mean dietary GI/GL, and the dietary intakes of CHO, sugar, starch and cereal, vegetable and fruit fiber, and both the prevalence and incidence of moderate CKD.	Gopinath *et al.* results showed that subjects in the highest quartile of mean dietary GI intake compared with those in the first quartile (reference) had a 55% increased likelihood of having eGFR (measure of kidney function), 60 mL $\times$ min^{-1} $\times$ 1.73 m^{-2} [multivariable-adjusted OR = 1.55 (95% CI = 1.07–2.26). There was no significant

SA: SAS statistical software (SAS Institute) version 9.1 was used for analyses including t tests, x2 tests, and logistic regression.

3.5.2. Tsai *et al.* 2005
Dietary carbohydrates and glycemic load and the incidence of symptomatic gallstone disease in men

N = 51,529 US male health professionals (40–75 years of age) in 1986 and part of the Health Professionals Follow up Study in Kentucky were administered a semi-quantitative FFQ and followed up until 1998 to assess the relationship between the relation between a high carbohydrate intake and the risk of symptomatic gall stones
SA: Mantel-Haenszel summary estimator and multivariate analyses.

relation among mean dietary GL, carbohydrate or sugar intakes, and an eGFR, $60\ L \times min^{-1} \times 1.73\ m^{-2}$. After a five-year follow-up with subjects, the authors reported that a high consumption of energy-dense, nutrient-poor (simple sugars such as soft drinks, cordials, cookies, cakes, buns, scones, pastries, confectionary, sugar, honey, jams, and syrups) had an adverse influence on renal function 5 y later. Tsai *et al.* results indicated that the estimated relative risk for men in the highest quintile compared with those in the lowest quintile of energy adjusted dietary GL was 1.28 (95% CI 1.10, 1.49; p for trend = 0.02). For CHO intake, the RR for the highest compared with the lowest was 1.59 (95% confidence interval (CI) 1.25, 2.02; p for trend = 0.002). For dietary GI (95% CI 1.01, 1.39; p for trend = 0.04), there was a positive relationship as well as for dietary intakes of starch, sucrose, and fructose. Tsai *et al.* concluded that that a high intake of CHO, GL and GI increases the risk of symptomatic gallstone disease in men.

Table 4.2 Tabulation of results of large-scale studies assessing the relationship between differing medical implications and dietary CHOs. Studies are grouped by medical implication, and significant outcomes in relation to Dietary CHO, GI and GL. √ = indicates a significant association.

Medical Condition or Implication/ Name of Study and Level of Evidence	Medical Implication	Significant Positive Relationship between Dietary CHOs and Medical Implication	Significant Inverse Relationship between dietary CHO and medical implication	No Significant Relationship
1. Metabolic Syndrome and Cardiovascular Disease (6 studies)				
Esposito et al. 2004 (Level 1)	Metabolic Syndrome in men/women	√ Mediterranean-style diet appears to be effective in reducing risk of metabolic syndrome and its associated cardiovascular risk		
Yancy et al. 2004 (Level 1)	Hyperlipidemia men/women	√ the low-CHO diet had greater decreases in serum triglyceride levels		
Welsh et al. 2010 (Level 2)	Blood Lipid Levels men/women	√ significant correlation between dietary added sugars and blood lipid levels; increased added sugars were associated with critical CVD risk factors such as lower HDL-C levels, higher triglyceride levels, and higher ratios of triglycerides to HDL-C.		
Liu et al. 2000 (Level 1)	Coronary heart disease (CHD) in US women	√ dietary GL was directly associated with risk of CHD in women		
Brown et al. 2010 (Level 3)	Blood pressure men/women	√ for SSBs, fructose and glucose		

2. Insulin Resistance and Type 2 Diabetes Mellitus

Reference	Outcome			
Lau et al. 2005 (Level 1)	Insulin Resistance in men/women			√ Dietary GI, GL and high CHO were not significantly associated with the risk of insulin resistance
Krishnan et al. 2007 (Level 2)	Risk of T2DM in women		√ Dietary fiber was inversely associated with the probability of having insulin resistance.	
Schulze et al. 2004 (Level 2)	Risk of T2DM in women	√ Increased dietary SSBs is associated with an increased risk for development T2DM	√ Increasing dietary fiber may decrease T2DM	

3. Risk of Cancer (5 studies)

Reference	Outcome			
Franceschi et al. 2001	Colorectal cancer risk in men and women	√ GI and GL and colorectal cancer		
Higginbotham et al. 2004	Colorectal cancer in women	√ GL and colorectal cancer		
Howarth et al. 2008	Colorectal cancer risk in men/women	√ for GL, CHO and sucrose white men only but not other ethnic groups	√ for GL, CHO and sucrose in women of all ethnic groups	
Larsson et al. 2006	Colorectal cancer risk in women	√ Dietary GI		
Terry et al. 2003	Colorectal cancer in women	√ Dietary GL		
Larsson et al. 2009 Nielsen et al. 2005	Breast cancer in women Breast cancer in women	√ Dietary GL		√ GL CHO or sugar and colorectal cancer risk √ Dietary CHOs for CHOs, simple sugars, GL and GI
3.2.10. Nothlings et al. 2008	Pancreatic cancer in men/women	√ Total sugars, fructose, and sucrose and pancreatic cancer		
Simon et al. 2010	Pancreatic cancer in women			√ GI, GL, CHO or sugars and pancreatic cancer

Table 4.2 *(Continued)*

Medical Condition or Implication/ Name of Study and Level of Evidence	Medical Implication	Significant Positive Relationship between Dietary CHOs and Medical Implication	Significant Inverse Relationship between dietary CHO and medical implication	No Significant Relationship
4. Other Medical Implications/Conditions				
Shaw et al. 2003 (Level 2)	NTD women	√ Sucrose and foods with higher GI values		
Gopinath et al. 2011	Chronic Kidney Disease (CKD) in men/women		√ High cereal fiber intake and reduced incidence of moderate CKD	
Tsai et al. 2005	Gallstone disease in men	√ high intake of CHO, GL and GI increases the risk of symptomatic gall stone disease in men.		

Abbreviations: See text above.

Table 4.3 Comparison of gender differences in medical implications and results of dietary CHOs in all studies reviewed. Men and women are compared by health implications to significant results of the relationship of dietary CHOs to medical implications. (+) = positive relationship; (−) = negative relationship; (0) = no significant relationship.

Medical Condition	Men and Women Studies			Men only Studies			Women only Studies		
	+	−	0	+	−	0	+	−	0
Total Studies: 22	+								0
Metabolic Syndrome and Cardiovascular Disease Total Studies: 7	Esposito et al. 2004; Yancy et al. 2004; Welsh et al. 2010						Liu et al. 2000; Brown et al. 2010;		
Insulin Resistance and T2DM Total Studies: 3		Lau et al. 2005 (fiber)	Lau et al. 2005 (GI, GL, CHO;				Krishnan et al. 2007; Schulze et al. 2004		
Cancer Total Studies: 10									
Colorectal Cancer Subtotal: 6	Franceschi et al. 2001 (GI and GL); Higginbotham et al. 2004 (GL);			Howarth et al. 2008 (white men only had a positive association with increasing GL			Larsson et al. 2006 (GI)	Howarth et al. 2008 (GL, CHO and sucrose)	Terry et al. 2003 (GL) CHO or sugars
Pancreatic Cancer Subtotal: 2	Nothlings et al. 2008 (for total sugars, fructose, and sucrose)								Simon et al. 2010 (GI, GL, CHO or sugars)
Breast Cancer Subtotal: 2							Larsson et al. 2009 (GL)		Larsson et al. 2009 (CHOs); Nielsen et al. 2005

Table 4.3 (Continued)

Medical Condition	Men and Women Studies			Men only Studies			Women only Studies		
	+	0	−	+	0	−	+	−	0
Total Studies: 22									
Other Medical Conditions Total Studies: 3									
NTD							Shaw et al. 2003 (sucrose and foods with higher GI values		
CKD			Gopinath et al. 2011 (high cereal fiber intake and reduced incidence of moderate CKD)						
Gallstone Disease	Tsai et al. 2005 (CHO, GL and GI)								

Abbreviations: See text above.

seemed to be effective in reducing the risk of metabolic syndrome. Welsh *et al.* (2010) found a significant correlation between dietary added sugars and blood lipid levels, and, increased dietary sugars were associated with critical CVD risk factors such as lower HDL-C levels, higher TG levels, and a higher ratio of TG to HDL-C. The increased risk of CHD is directly related to increased daily GL in women (Liu *et al.* 2000). Liu *et al.* (2002) found that dietary GL is significantly and positively associated with plasma *hs*-CRP, a metabolic marker, in his cohort of healthy middle-aged women. Finally, SSB, fructose and glucose were directly related to elevated BP in both genders (Brown *et al.* 2010). Thus, these four studies reflect evidence of a clear and unequivocal risk of CVD for the specific quality of dietary CHO. In the context of dietary quantity, Yancy *et al.* (2004) showed that a low-CHO diet yielded greater decreases in serum TG levels.

4.4.2   Dietary CHO and T2DM/Insulin Resistance

It is significant that both studies assessing dietary fiber (CHOs) demonstrated that increasing dietary fiber reduced the risk of T2DM in both men and women (Lau *et al.* 2005; Krishnan *et al.* 2007). SSB were significantly positively related to the risk of developing T2DM – probably because of weight gain (Schulze *et al.* 2004). However, dietary GI and GL were not significantly associated with the risk of T2DM (Lau *et al.* 2005). These null findings may be the result of the self-administered FFQ used in the study, which may not be able to capture the factors that contribute to the dietary GI and GL, *i.e.* individual food composition, food preparation methods and overall meal composition.

4.4.3   Dietary CHO and Cancer Risks

4.4.3.1   Colorectal Cancer

Of the five large-scale studies, four show a positive relationship between dietary GI, GL and CHO (in women and white men) and colorectal cancer risk (Franceschi *et al.* 2001; Higginbotham *et al.* 2004; Howarth *et al.* 2008; Larsson *et al.* 2006). However, one study found no significant relationship with GL, CHO or sugar (Terry *et al.* 2003). Howarth *et al.* (2008) found an inverse relationship with GL, CHO and sucrose in women of all ethnic groups, and, for white men, but not other ethnic groups. It appears that ethnicity and gender may play a role in these mixed findings.

4.4.3.2   Breast Cancer

There are inconsistent results in this category and the two large-scale studies show inconsistent findings. While Larsson *et al.* (2009) found a significant positive relationship with dietary GL and breast cancer, no significant relationship with CHO intake was found. Nielsen *et al.* (2005) found no significant relationships with CHO intake, simple sugars, GI or GL. These mixed results may be due to the aforementioned challenges associated with the use of FFQ.

Additionally, dietary GL is capable of capturing both the quantity and quality of the CHO (simple *vs.* complex), whereas a high CHO diet may contain predominantly low GI foods, thus diminishing the overall insulin demand of the diet and reducing breast cancer risk.

4.4.3.3   Pancreatic Cancer

The two large-scale studies indicated no significant relationship of dietary GI, GL, CHO, or simple sugars with pancreatic cancer (Nothlings *et al.* 2008, Simon *et al.* 2010).

4.4.4   Dietary CHOs and Other Medical Conditions

One study of NTD in women indicated a positive association with sucrose (simple sugar) and dietary GI (Shaw *et al.* 2003). Gopinath *et al.* (2011) found an inverse association with dietary fiber and reduced incidence of moderate CKD. For gallstone disease in men, Tsai *et al.* (2005) found a positive relationship with dietary CHO, GL and GI in the context of increased risk of symptomatic CKD.

In conclusion, the strongest trend in large-scale studies on gender differences is that all studies demonstrated a significant positive relationship between simple CHO and CVD risk or risk of metabolic syndrome. A similar trend can be seen for SSB and risk of T2DM. Dietary fiber appears important in reducing the risk of T2DM and CKD. Because of mixed results in large-scale studies utilizing FFQ to measure dietary intake, the interpretation of the relationship among dietary GI, GL, CHO, and simple sugars is challenging. In terms of dietary CHO and cancer risk, the strongest indication of a positive association with GI, GL and CHO and colorectal cancer exists among women and white men.

4.5   Gender Differences in Trials of the Effects of Dietary CHO and Various Medical Implications

4.5.1   Trends in Gender Differences in Dietary CHO and Metabolic Syndrome and CVD

All large-scale studies conducted on men and women (Esposito *et al.* 2004, Yancy *et al.* 2004, Welsh *et al.* 2010, Brown *et al.* 2011), including one on women (Liu *et al.* 2000), resulted in a significantly positive association between dietary CHO intake and risk of CVD (metabolic syndrome, hyperlipidemia, TG, BP, CHD, MI and IHD).

4.5.2   Trends in Gender Differences in Dietary CHO and Insulin Resistance/T2DM

In terms of gender, it is important that both of the large-scale studies indicated that dietary fiber may be important in decreasing the incidence of T2DM in both men and women (Lau *et al.* 2005) and in women within an all-black female cohort (Krishnan *et al.* 2007).

Regarding other types of CHO, there were some potential ethnic differences for CHO and T2DM that require further investigation. While there was no significant relationship between dietary GI, GL and CHO and T2DM in a large-scale study of men and women in a Danish population-based Inter99 study (Lau *et al.* 2005), GI was significantly and positively associated with T2DM in a large-scale black female cohort participating in the US Black Women's Health Study (Krishnan *et al.* 2007). These results appear to indicate the need for further investigations to assess the relationship between the quality and quantity of dietary CHO and T2DM across multi-ethnic populations.

Lastly, with regard to SSB, Schulze *et al.* (2004) indicated that they were positively associated with development of T2DM in women, which is consistent with similar findings in adolescents.

4.5.3   Trends in Gender Differences in Dietary CHO and Cancers

Of five studies, three (two studies on men/women, and one in women) found that GL was significantly associated with the incidence of colorectal cancer (Franceschi *et al.* 2001; Higginbotham *et al.* 2004; Howarth *et al.* 2008). In white men only, the RR for colorectal cancer was associated with GL only (Howarth *et al.* 2008). Howarth *et al.* (2008) further concluded that in their large, older, multi-ethnic population, higher dietary GL and CHO intake were protective against colorectal cancer in women but not men. One study found an association of GI and cancer (Larsson *et al.* 2006) in women within a female cohort while others either found an inverse relationship (Howarth *et al.* 2008) or no significant relationship of dietary GL, CHO and sucrose (Larsson *et al.* 2006 and Terry *et al.* 2003). Among the six studies of colorectal cancer, only one was conducted in the US; other studies were conducted in Italy, Canada and Sweden.

For the two pancreatic cancer studies conducted in the US, Nothlings *et al.* (2008) found a significantly positive relationship for men and women for the intake of total sugars, fructose and sucrose. Simon *et al.* (2010) found no relationship with dietary GI, GL, CHO or sugars in women within a female cohort.Clearly, further studies are needed to determine the role of dietary factors in the context of pancreatic cancer risk.

In terms of breast cancer, Larsson *et al.* (2009) found a significant positive relationship with dietary GL and breast cancer risk but no relationship with dietary CHO – the latter finding being consistent with the Nielsen *et al.* (2005) findings.

4.5.4   Trends in Gender Differences in Dietary CHO and Other Medical Conditions

Shaw *et al.* (2003) found a significant relationship between sucrose and high GI foods and NTD, Gopinath *et al.* (2011) found an inverse relationship between dietary fiber and CKD in both genders, and, Tsai *et al.* (2005) found a positive relationship between high dietary intake of CHO, GL and GI and increased risk of symptomatic gall stone disease in men within an all male cohort. However, these were single studies on each topic and further research is needed on these individual medical conditions.

4.6   Conclusions

In conclusion, all large-scale studies conducted in men and women, including one study on women indicated a significantly positive association between the quantity and/or quality of dietary CHO and risk of CVD.In addition, studies found that high dietary fiber had a significant relationship in reducing T2DM risks in both men and women as well as moderate CKD. In terms of dietary GL, there were mixed results.However, three large-scale studies found a significant relationship between dietary GL and colorectal cancer in men and women. There are too few large-scale studies assessing the relationship of dietary CHO and pancreatic and breast cancer to draw any firm conclusions.

The relationship between high dietary CHO, GI and GL is complex with respect to medical implications. The inconsistencies in the findings found across studies and genders may be due to the varying proportions and types of CHO (rice *vs.* bread *vs.* potatoes) being consumed within a specific cohort. Howarth *et al.* (2008) speculated that a rice-based diet may not provide as robust a measure or physiologic response to GL as do potatoes and bread, and, GL may not be a good measure of insulin response in women. The variety of rice, cooking method and cooking times can also affect GL calculations.

Cohort study results are also affected by systematic error due to the under-reporting of intake within specific population subgroups.A limitation of the cohort studies described in this review included the use of a self-administered FFQ, which leads to measurement error and attenuated risk estimates. Misclassification of CHO intake due to recall errors also exists with the use of FFQ to measure diet.Hence, estimates of CHO, GI and GL derived from the FFQ may not accurately reflect the physiologic glycemic and insulinemic effects of intake and metabolism of mixed meals and prepared foods.

4.7   Discussion

The results of this review include two primary conclusions: (1) dietary fiber appears to attenuate CVD and T2DM risks; and (2) high intakes of simple dietary CHO is positively associated with CVD and metabolic syndrome risks. This former result is also consistent with the results of the large-scale Framingham Offspring Study (N = 2941) that showed that increased intakes of whole grains, which are good sources of fiber, attenuate CVD, metabolic syndrome and T2DM risks. (McKeown *et al.* 2002). In addition, a low GI diet was associated with a reduced CVD risk in a medium-scale randomized control study (N = 129) which assessed mixed fat and CHO diets in overweight and obese young adults (McMillan-Price *et al.* 2006).

The quality of dietary CHO appears to be important in terms of gender differences and risk of disease The medium-scale Vestfold Heartcare Study Group study (N = 197) indicated that simple sugars such as fructose and glucose increase TG levels in men whereas dietary sucrose increases TG in women.

Because of mixed results, additional large-scale studies are needed to further explore the relationship of dietary CHO, GI and GL in the context of T2DM and cancer research.

Summary Points

- There are differing effects of dietary CHO, GI and GL in different medical conditions.
- Gender differences appear to be important in terms of dietary CHO, GI, GL, in relation to specific medical conditions.
- In terms of CVD risk:
  - Large-scale studies seem to show that gender and ethnic differences in populations may influence the effect of dietary CHO, GI and GL on CVD risks and/or risks of T2DM; small- to medium-scale studies show more variation in results.
  - Studies demonstrated a significant positive relationship between the quality and/or quantity of dietary CHO and CVD risks or risk of metabolic syndrome.
  - There is an increased risk of CHD in response to an increased dietary GL in women.
  - SSB, fructose and glucose were directly related to elevated BP in both genders.
  - Simple additive sugars were associated with the development of critical CVD risk factors such as lower HDL-C levels, higher TG levels, and a higher ratio of TG to HDL-C.
  - The Mediterranean diet seems to be effective in reducing the risk of developing metabolic syndrome.
- In terms of T2DM risk:
  - Dietary fiber appears to be preventative in both men and women with regard to attenuating CVD and T2DM risks.
  - SSB are significantly positively related to the risk of developing T2DM in women.
  - GI was significantly and positively associated with T2DM in a large-scale study of women participating in the US Black Women's Health Study.
- In terms of cancer risk:
  - Four of five large-scale studies indicated a positive relationship with dietary GI, GL and CHO (in women and white men) and colorectal cancer risk.
  - In a larger, older, multi-ethnic population, higher GL and higher CHO intake may be protective against colorectal cancer in women but not men.

- ○ Large-scale studies show no significant relationship of GI, GL CHO, or simple sugars with pancreatic cancer risk.
- ○ Half the large-scale studies found that GL was significantly associated with the incidence of colorectal cancer.
- ○ Half the large-scale studies found that GL was significantly associated with the incidence of colorectal cancer.
- Other Medical Conditions.
  - ○ There is a positive relationship between a higher GI and sucrose-containing diet with NTD in women.
  - ○ Increasing dietary fiber, which is positive in reducing CVD and T2DM risk may also be beneficial in reducing the incidence of moderate CKD.
  - ○ High dietary CHO, GL and GI increases the risk of gallstone disease in men.

Key Facts

- Metabolic Syndrome is a concept that includes 5 risk factors that are combined to include atherogenic dyslipidemia, elevated blood pressure (BP), elevated glucose level, a prothrombotic state, and a proinflammatory state.
- While GI is a useful tool, some caution is needed in using the GI because GI was arbitrarily assigned to represent the relative glycemic potency of glycemic carbohydrate in food but not in the total food itself.
- The quality of quantity of CHO is important; complex CHO, versus simple CHO (*e.g.* sugars) appear more beneficial to reducing the risks of CVD and other medical implications.
  - ○ A higher dietary GL is a reflection of either greater CHO consumption, an intake of high GI foods, or both.
  - ○ The CHO content of foods vary according to season and country, which contributes to imprecise derived estimates of dietary GI and GL intakes across studies.
  - ○ The use of FFQ in large-scale observational studies fails to capture individual food composition, preparation methods, and the composition of the total meal. Thus, the estimated derived dietary CHO, GI and GL may not be capable of detecting the physiological responses of glucose metabolism demonstrated in tightly controlled metabolic feeding studies.

Definitions

Glycemic index = the effect of glycemic carbohydrate in a food on blood glucose as a percentage of the effect of an equal amount of glucose (Monro 2003).

Glycemic Load = GI x dietary carbohydrate content (Foster-Powell 2002).

List of Abbreviations

ANOVA	Analysis of Variance
BMI	body mass index
BP	blood pressure
CAD	coronary artery disease
CHD	coronary heart disease
CHO	carbohydrates
CKD	chronic kidney disease
CI	confidence intervals
CVD	cardiovascular disease
DM	diabetes mellitus
GFR	glomerular filtration rate
FFQ	food frequency questionnaires
GI	glycemic index
GL	glycemic load
HDL-C	high density lipoprotein cholesterol
HOMA	homeostatic model assessment
hs-CRP	high sensitive C-reactive protein
IHD	ischemic heart disease
IRR	incidence rate ratios
IHD	ischemic heart disease
LDL-C	low density lipoprotein cholesterol
MI	myocardial infarction
NTD	neural tube defects
N	number (sample size)
Omni-Heart	Optimal Macronutrient Intake Trial to Prevent Heart Disease
OR	Odds Ratio
RR	relative risks
SA	statistical analysis
SSB	sugar-sweetened beverages
T2DM	type 2 diabetes Mellitus
TC	total cholesterol
TG	triglycerides
US Cardia Study	Coronary Artery Risk Development in Young Adults

References

Brown, I.J., Stamler, J., Van Horn, L., Robertson, C.E., Chan, Q., Dyer, A.R., Huang, C.C., Rodriguez, B.L., Zhao, L., Daviglus, M.L., Ueshima, H., and Elliott, P., 2011, for the International Study of Macro/Micronutrients and Blood Pressure Research Group. Sugar-Sweetened Beverage, Sugar Intake of Individuals, and Their Blood Pressure: International Study of Macro/Micronutrients and Blood Pressure. *Hypertension*. 57(4): 695–701.

De Lorgeril, M., Salen, P., Martin, J.L., Monjaud, I., Delaye, J., and Mamelle, N., 1999. Mediterranean diet, traditional risk factors, and the rate of cardiovascular complications after myocardial infarction: final report of the Lyon diet heart study. *Circulation*. 99: 779–785.

Esposito, K., Marfella, R., Ciotola, M., DiPalo, C., Giugliano, F., Guigliano, G., D'Armiento, M., D'Andrea F., and Giugliano D., 2004. Effect of a Mediterranean-Style Diet on Endothelial Dysfunction and Markers of Vascular Inflammation in the Metabolic Syndrome A Randomized Trial. *JAMA*. 292: 1440–1446.

Foster-Powell, K., Holt, S.H.A, and BrandMiller, J.C., 2002. International table of glycemic index and glycemic load values. *Am J Clin Nutr*. 76(1): 556.

Fried, S.K., and Rao, S.P.. 2003. Sugars, hypertriglyceridemia, and cardiovascular disease. *Am J Clin Nutr*. 78(4): 873S–880S.

Franceschi, S., Dal Maso, L., Augustin, L., Negri, E., Parpinel, M., Boyle, P., Jenkins, D.J.A., and LaVecchia, C., 2001. Dietary glycemic load and colorectal cancer risk. *Annals of Oncology*. 12: 173–178.

Gastrich, M.D., Lasser, N.L., Wien, M., and Bachmann G., 2008. Dietary complex carbohydrates and low glycemic index/load decrease levels of specific metabolic syndrome/CVD risk factors. *Topics in Clinical Nutrition*. 23(1): 76–96.

Gastrich, M.D., Bachmann, G., and Wien, M., 2007. A Review of Recent Studies Assessing the Impact of Additive Sugar in the Diet. *Topics in Clinical Nutrition*. 22(2): 137–155.

Gopinath, B., Harris, D.C., Flood, V.M., Burlutsky, G., Brand-Miller, J., and Mitchell, P., 2011. Carbohydrate nutrition is associated with the 5-year incidence of chronic kidney disease. *J Nutr*. 141(3): 433–439. Epub 2011 Jan 12.

Hellerstein, M.K., 2002. Carbohydrate-induced hypertriglyceridemia: modifying factors and implications for cardiovascular risk. *Curr Opin Lipidol*. 13: 33–40.

Higginbotham, S., Zhang S.F., Lee, I.M., Cook, N.R., Giovannucci, E., Buring, J.E., and Liu, S., Dietary, 2004. Glycemic Load and Risk of Colorectal Cancer in the Women's Health Study. *J Natl Cancer Inst*. 96: 229–233.

Howarth, N.C., Murphy, S.P., Wilkens, L.R., and Henderson, B.E., Kolonel 2008. The association of glycemic load and carbohydrate intake with colorectal cancer risk in the Multiethnic Cohort Study. *Am J Clin Nutr*. 88(4): 1074–1082.

Krishnan, S., Rosenberg, L., Singer, M., Hu, F.B., Djoussé, L., Cupples, L.A., and Palmer, J.R., 2007. Glycemic index, glycemic load, and cereal fiber intake and risk of type 2 diabetes in US black women. *Arch Intern Med*. 167(21): 2304–2309.

Larsson, S.C., Bergkvist, L., and Wolk, A., 2009. Glycemic load, glycemic index and breast cancer risk in a prospective cohort of Swedish women. *Int J Cancer*. 125(1): 153–157.

Larsson, S.C., Giovannucci, E., and Wolk, A., 2006. Dietary Carbohydrate, Glycemic Index, and Glycemic Load in Relation to Risk of Colorectal Cancer in Women. *Am J Epidemiol*. 165: 256–261.

Lau, C., Færch, K., Glümer, C., Tetens, I., Pedersen, O., Carstensen, B., Jørgensen, T., and Borch-Johnsen, K., 2005. Dietary Glycemic Index, Glycemic Load, Fiber, Simple Sugars, and Insulin Resistance. The Inter99 study. *Diabetes Care.* 28: 1397–1403.

Liu, S., Willett, W.C., Stampfer, M.J., Hu, F.B., Franz, M., Sampson, L., Hennekens, C.H., and Manson, J.E., 2000. A prospective study of dietary glycemic load, carbohydrate intake, and risk of coronary heart disease in US women. *Am J Clin Nutr.* 71(6): 1455–1461.

Liu, S., Manson, J.E., Buring, J.E., Stampfer, M.J., Willett, W.C., and Ridker, P.M., 2002. Relation between a diet with a high glycemic load and plasma concentrations of high-sensitivity (*hs*) C-reactive protein in middle-aged women. *Am J Clin Nutr.* 75: 492–498.

McKeown, N.M., Meigs, J.B., Liu, S., Wilson, P.W.F., and Jacques, P.F., 2002. Whole-grain intake is favorably associated with metabolic risk factors for type 2 diabetes and cardiovascular disease in the Framingham Offspring Study. *Am J Clin Nutr.* 76(2): 390–398.

McMillan-Price, J., Petocz, P., Atkinson, F., O'neill, K., Samman, S., Steinbeck, K., Caterson, I., and Brand-Miller, J., 2006. Comparison of 4 diets of varying glycemic load on weight loss and cardiovascular risk reduction in overweight and obese young adults: a randomized controlled trial. *Arch Intern Med.* 166(14): 1466–1475.

Monro, J., 2003. Redefining the glycemic index for dietary management of postprandial glycemia. *J Nutr.* 133: 4256–4258.

Nielsen, T.G., Olsen, A., Christensen, J., Overvad, K., and Tjønneland A., 2005. Dietary carbohydrate intake is not associated with the breast cancer incidence rate ratio in postmenopausal Danish women. *J Nutr.* 135(1): 124–128.

Nothlings, U., Murphy, S.P., Wilkens, L.R., Henderson, B.E., and Kolonel, L.N., 2007. Dietary glycemic load, added sugars, and carbohydrates as risk factors for pancreatic cancer: the Multiethnic Cohort Study. *Am J Clin Nutr.* 86: 1495–1501.

Poppitt, S.D., Keogh, G.F., Prentice, A.M., Williams, D.E., Sonnemans, H.M., Valk, E.E., Robinson, E., and Wareham, N.J., 2002. Long-term effects of ad libitum low-fat, high carbohydrate diets on body weight and serum lipids in overweight subjects with metabolic syndrome. *Am J Clin Nutr.* 75: 11–20.

Raben, A., Holst, J.J., Madsen, J., and Astrup, A., 2001. Diurnal metabolic profiles after 14 d of an ad libitum highstarch, high-sucrose, or high fat diet in normal-weight never-obese and postobese women. *Am J Clin Nutr.* 73: 177–189.

Schulze, M.B., Manson, J.E., Ludwig, D.S., Colditz, G.A., Stampfer, M.J., Willett, W.C., and Hu, F.B., 2004. Sugar-Sweetened Beverages, Weight Gain, and Incidence of Type 2 Diabetes in Young and Middle-Aged Women. *JAMA.* 292: 927–934.

Shaw, G.M., Quach, T., Nelson, V., Carmichael, S.L., Schaffer, D.M., Selvin, S., and Yang, W., 2003. Neural tube defects associated with maternal periconceptional dietary intake of simple sugars and glycemic index. *Am J Clin Nutr.* 78(5): 972–978.

Simon, M.S., Shikany, J.M., Neuhouser, M.L., Rohan, T., Nirmal, K., Cui, Y., and Abrams, J., Women's Health Initiative, 2010. Glycemic index, glycemic load, and the risk of pancreatic cancer among postmenopausal women in the women's health initiative observational study and clinical trial. *Cancer Causes Control.* 21(12): 2129–2136.

Sloth, B., Krog-Mikkelsen, I., Flint, A., Tetens, I., Björck, I., Vinoy, S., Elmståhl, H., Astrup, A., Lang, V., and Raben, A., 2004. No difference in body weight decrease between a lowglycemic-index and a high-glycemic-index diet but reduced LDL cholesterol after 10-wk ad libitum intake of the low-glycemic-index diet. *Am J Clin Nutr.* 80: 337–347.

Terry, P.D., Jain, M., Miller, A.B., Howe, G.R., and Rohan, T.E., 2003. Glycemic Load, Carbohydrate Intake, and Risk of Colorectal Cancer in Women: A Prospective Cohort Study. *J Natl Cancer Inst.* 95: 914–916.

Welsh, J.A., Sharma, A., Abramson, J.L., Vaccarino, V., Gillespie, C., and Vos, M.B., Caloric sweetener consumption and dyslipidemia among US adults. *JAMA.* 303(15): 1490–1497.

Tsai, C.J., Leitzmann, M.F., Willett, W.C., and Giovannucci, E.L., 2005. Dietary carbohydrates and glycaemic load and the incidence of symptomatic gall stone disease in men. *Gut.* 54(6): 823–828.

Vincent-Baudry, S., Defoort, C., Gerber, M., Bernard, M.C., Verger, P., Helal, O., Portugal, H., Planells, R., Grolier, P., Amiot-Carlin, M.J., Vague, P., and Lairon, D., 2005 The Medi-RIVAGE study: reduction of cardiovascular disease risk factors after a 3-mo intervention with a Mediterranean-type diet or a low-fat diet. *Am J Clin Nutri.* 82: 964–971.

Yancy, W.S. Jr., Olsen, M.K., Guyton, J.R., Bakst, R.P., and Westman, E.C., 2004. A Low-Carbohydrate, Ketogenic Diet versus a Low-Fat Diet To Treat Obesity and Hyperlipidemia. *Ann Intern Med.* 140: 769–777.

Chemistry and Biochemistry

CHAPTER 5

Glucose Chemistry

BILAL AHMAD MALIK AND
MOHAMMED BENAISSA*

Electronic and Electrical Engineering Department, The University of
Sheffield, Mappin Street, Sheffield, S1 3JD, United Kingdom
*Email: m.benaissa@sheffield.ac.uk

5.1 Introduction

Carbohydrates are one of the most abundant bio-molecules on Earth. More
than 100 billion metric tonnes of CO_2 and H_2O are converted into cellulose and
other plant products each year. Certain carbohydrates (sugar and starch) are an
essential part of food in most places of the world. Carbohydrates are aldehydes,
ketones with many hydroxyl groups. They act as energy stores, fuels and
metabolic intermediates and are part of the structural framework of RNA and
DNA. Carbohydrates consist of carbon, oxygen and hydrogen, where the
proportion of hydrogen and oxygen is (H:O = 2:1). Carbohydrates have the
empirical formula of $(CH_2O)_n$ and are classified as monosaccharides, oligosaccharides and polysaccharides (the origin of "saccharide" is the Greek word
sakcharon, meaning "sugar"). Monosaccharides are composed of a single
polyhydroxy aldehyde or ketone unit. Figure 5.1 below shows the structure of
monosaccharides, disaccharide and polysaccharide (Holdsworth *et al.* 1973).

As shown in Figure 5.2 a monosaccharide, a disaccharide and a polysaccharide have one, two and many rings with hemiacetal, acetal and many
acetal functional groups, respectively.

Monosaccharides are colourless crystalline solids, insoluble in non-polar
solvents, but soluble in water. The taste of most of the monosaccharides is

Food and Nutritional Components in Focus No. 3
Dietary Sugars: Chemistry, Analysis, Function and Effects
Edited by Victor R Preedy
© The Royal Society of Chemistry 2012
Published by the Royal Society of Chemistry, www.rsc.org

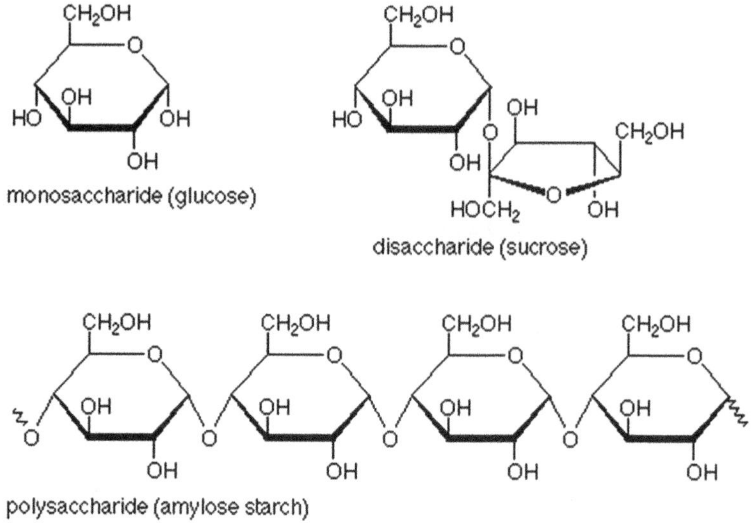

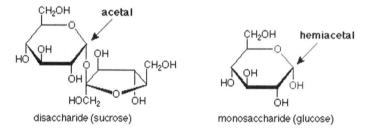

Figure 5.1   Structure of monosaccharides, disaccharide and polysaccharide. This figure shows the structure of a monosaccharide, a disaccharide and a polysaccharide.

Figure 5.2   Ring shape of monosaccharide (glucose) and disaccharide (sucrose). This figure shows that a monosaccharide (glucose), a disaccharide (sucrose) have one and two with hemiacetal, acetal functional groups, respectively.

sweet. The backbone of common monosaccharide molecules is unbranched carbon chains where the carbon atoms are linked by single bonds. The open-chain form has a carbonyl and a hydroxyl group. The monosaccharides are classified as either aldoses or ketoses depending on the position of the carbonyl group. The carbonyl group is at the end of the carbon chain in case of an aldose and at any other position in case of a **ketose**. Tetroses, pentoses, hexoses and heptoses have four, five, six and seven carbon atoms respectively. The most common monosaccharides in nature are D-glucose and D-fructose. This chapter presents the 6-C monosaccharide sugar "glucose" widely distributed among animal and plant tissues and cells (Elliott *et al.* 1997).

5.2 Structure

Glucose ($C_6H_{12}O_6$) is an aldohexose, since it contains an aldehyde group and six carbon atoms; it exists in two forms: open-chain (acyclic) and ring (cyclic) form. In aqueous solution, both forms are in equilibrium, but the cyclic form is predominant at a pH value of 7. The cyclic form of glucose is also known as glucopyranose, as its structure is similar to that of pyran (Brown and McClarin 1981). All carbon atoms are linked to a hydroxyl group except the fifth atom, which forms CH_2OH group by linking to a sixth carbon atom outside the ring as shown in Figure 5.3. Table 5.1 lists the key chemical properties of glucose.

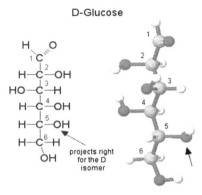

Figure 5.3 Glucose Structure. This figure shows two forms of glucose, open-chain (acyclic) and ring (cyclic) form. The cyclic form of glucose is also known as glucopyranose, as its structure is similar to that of pyran.

Table 5.1 Properties of Glucose. This table lists the abbreviation, other names, and chemical properties of glucose.

Glucose	
Chemical formula	$C_6H_{12}O_6$
IUPAC name	D-glucose
Chemical name	6-(hydroxymethyl)oxane-2,3,4,5-tetrol
Synonym for D-glucose	dextrose
Varieties of D-glucose	α-D-glucose; β-D-glucose
According to position of the hydroxyl group	
Abbreviations	Glc
Molecular mass	180.16 g/mol
Exact Mass	180.063388
Melting Point	α-D-glucose: 146 °C
	β-D-glucose: 150 °C
Density	1.54 g/cm^3
Solubility in water	91 g/100 mL

5.3   Chemical Reactions with Glucose

Glucose is reduced with concentrated hydriodic acid in the presence of red phosphorus to form a trace amount of n-hexane. This indicates an unbranched chain of 6-carbon atoms which is formed by the six carbon atoms in the glucose (Nelson *et al.* 2008).

$$C_6H_{12}O_6 \xrightarrow{HI/p} CH_3-CH_2-CH_2-CH_2-CH_2-CH_3$$
Glucose   **n-hexane**

Glucose quickly dissolves in water, to form a neutral solution. This reaction indicates absence of a carboxyl group (COOH) in glucose molecule.

Glucose reacts with hydroxylamine (NH_2OH) to form glucose oxime.

$$C_6H_{12}O_6 \xrightarrow{NH_2OH} HC=NOH-(CHOH)_4-CH_2OH$$
Glucose   **Glucose Oxine**

Glucose reacts with hydrogen cyanide (HCN) to form cyanohydrins.

$$C_6H_{12}O_6 \xrightarrow{HCN} CN-(CHOH)_5-CH_2OH$$
Glucose   **Glucose cyanohydrin**

Glucose on oxidation with bromine water (Br_2/H_2O) is converted to gluconic acid. The gluconic acid is converted to glucaric acid on further oxidation with dilute nitric acid (HNO_3). This indicates a primary alcoholic group is present.

$$C_6H_{12}O_6 \xrightarrow{Br_2/H_2O} COOH-(CHOH)_4-CH_2OH \xrightarrow{HNO_3} COOH-(CHOH)_4-COOH$$
Glucose   **Gluconic acid**   **Glucaric acid**

Glucose reduces ammoniacal solution of silver nitrate, which is known as Tollen's reagent, to metallic silver (Ag).

Glucose forms pentacetate with acetic anhydride in the presence of pyridine. This reaction infers that five hydroxyl groups are present in a glucose molecule. These evidences lead us to the conclusion that glucose is a pentahydroxyhexanal (an aldohexose).

Glucose in methanol reacts with the small amount of gaseous HCl to form anomeric methyl acetal as shown in Figure 5.4.

Glucose when it reacts with concentrated hydrochloric acid forms laevulinic acid.

$$C_6H_{12}O_6 \xrightarrow{Conc.HCl} CH_3CO-CH_2-CH_2-COOH + HCOOH + H_2O$$
Glucose   **Laevulinic acid**

Glucose undergoes dehydration with concentrated sulphuric acid to form hydroxy methyl furfural as shown in Figure 5.5.

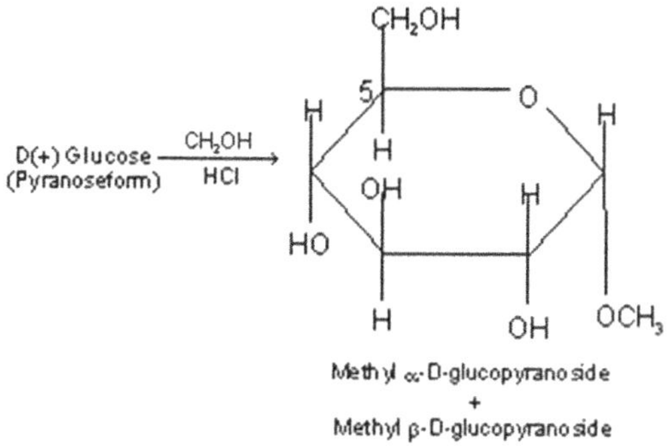

Figure 5.4 Reaction of glucose in methanol with small amount of gaseous HCl. This figure shows that glucose in methanol reacts with the small amount of gaseous HCl to form anomeric methyl acetal.

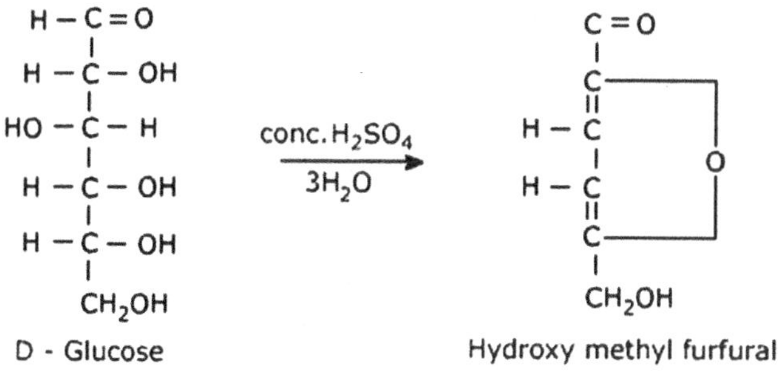

Figure 5.5 Reaction of glucose with concentrated sulphuric acid. This figure shows that glucose undergoes dehydration with concentrated sulphuric acid and forms hydroxy methyl furfural.

$$D-glucose \rightleftharpoons D-maltose \rightleftharpoons D-fructose$$

Figure 5.6 Reaction of glucose with sodium hydroxide. This figure shows what is known as Glucose Lobery de Bruyn-van Ekenstein rearrangement. Glucose when heated with concentrated sodium hydroxide (NaOH), forms a mixture of D-glucose, D-fructose and D-maltose as it exhibits a reversible isomerisation.

Glucose, when heated with concentrated sodium hydroxide (NaOH), changes the colour. It first changes to yellow and then to a brown colour and finally resinifies. However, when treated with dilute NaOH solution, it forms a mixture of D-glucose, D-fructose and D-maltose as it exhibits a reversible isomerisation. This reaction of glucose is also known as Lobery de Bruyn-van Ekenstein rearrangement as illustrated in Figure 5.6.

5.4   The Configuration of Glucose

In order to understand the configuration of Glucose, we first need to understand the stereoisomers and Chirality (McGraw-Hill 2004).

The molecular formula of stereoisomers is the same. The order of attachment of atoms in their molecules is also the same but they differ in the three-dimensional orientations of their atoms in space. Stereoisomers are divided into two groups: enantiomers and diastereomers.

Chirality is encountered in three-dimensional objects. Chiral (pronounced Ki-ral meaning "hand" derived from the Greek word "cheir") molecules show optical activity. Some of them rotate polarized light to the left (counter-clockwise) and are said to be levorotatory, whereas others rotate the polarized light to the right (clockwise) and are said to be dextrorotatory. By convention, rotation to the left is given a minus sign ($-$) and rotation to the right is given a plus sign ($+$).

Emil Fischer, a German scientist, was a pioneer in the study of Stereoisomers of monosaccharides. Fischer was able to assign relative configurations to each of the stereocentres in glucose, by using the phenylhydrazine reaction and reaction sequences by which monosaccharide chains be extended by one carbon at a time. Fischer helped to unravel the complexity of carbohydrate structure and established the validity of the tetrahedral carbon atom, by establishing the relative configurations of stereocentres of glucose. Fischer was awarded the Nobel Prize for his work on the chemistry of carbohydrates and purines.

The work of Fischer concluded that glucose exists in two stereoisomeric forms. The form, which is derived from the "right-handed form" of glucose, is known as D-glucose. D-glucose is often referred to as dextrose. Polarized light is rotated to the right by solutions of dextrose. Polymers, which are derived from the dehydration of D-glucose, are starch and cellulose. L-glucose, the other stereoisomeric form of the glucose, is scarcely found in nature. However, D-glucose is found in abundance.

The open isomer D-glucose forms four different cyclic isomers: α-D-gluco-pyranose, α-D-glucofuranose, β-D-glucopyranose, and β-D-glucofuranose. All of these isomers exhibit chirality as shown in Figure 5.7.

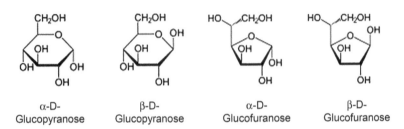

| α-D- | β-D- | α-D- | β-D- |
| Glucopyranose | Glucopyranose | Glucofuranose | Glucofuranose |

Figure 5.7   Four different cyclic isomers of D-Glucose. This figure shows cyclic isomers of glucose: α-D-glucopyranose, α-D-glucofuranose, β-D-glucopyranose, and β-D-glucofuranose. All of these isomers exhibit Chirality.

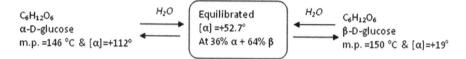

Figure 5.8 Mutarotation of glucose. This figure shows Pure α-D-glucopyranose has a specific rotation of + 112.2° and a melting point of 146 °C, and Pure β-D-glucopyranose has a specific rotation of + 18.7° and a melting point of 148 to 155 °C. Due to change in optical rotation, the slow interconversions of the pure anomers give a 37:63 equilibrium mixture.

5.5 Mutarotation of Glucose

Mutarotation is defined as the change in optical activity that occurs when α or β form of a carbohydrate is converted to an equilibrium mixture of the two forms (McMurry *et al.* 1996). Both anomers of D-glucopyranose *i.e.* α-D-glucopyranose and β-D-glucopyranose can be crystallized and purified. Pure α-D-glucopyranose has a specific rotation of + 112.2° and a melting point of 146 °C, and Pure β-D-glucopyranose has a specific rotation of + 18.7° and a melting point of 148 to 155 °C. A solution prepared by dissolving crystalline α-D-glucopyranose in water shows an initial rotation of + 112.2°. This slowly decreases to an equilibrium value of + 52.7°, as α-D-glucose reaches an equilibrium with β-D-glucose. A solution of β-D-glucose also undergoes mutarotation, during which, the specific rotation changes from an initial value of + 18.7° to the same equilibrium value of + 52.7°. This change in optical rotation is due to the slow interconversion of the pure anomers to give a 37:63 equilibrium mixture. Mutarotation occurs by a reversible ring-opening of each anomer to the open-chain aldehyde, followed by reclosure. Although the equilibrium is slow at neutral pH, it is catalyzed by both acid and base as illustrated in Figure 5.8.

Summary Points

- Glucose is readily soluble in water to give a neutral solution, which indicates the absence of a -COOH group.
- Glucose is reduced with concentrated hydriodic acid in the presence of red phosphorus to form trace amount of n-hexane.
- Glucose reacts with hydroxylamine (NH_2OH) to produce Glucose Oxime.
- Glucose reacts with hydrogen cyanide (HCN) to form cyanohydrins.
- Glucose on oxidation with bromine water (Br_2/H_2O) is converted to gluconic acid. The gluconic acid is converted to glucaric acid on further oxidation with dilute nitric acid.
- Glucose reduces Tollen's reagent (ammoniacal solution of silver nitrate) to metallic silver (Ag) and forms pentacetate with acetic anhydride in the presence of pyridine.
- Glucose in methanol reacts with the small amount of gaseous HCl to form anomeric methyl acetal.

- Glucose, when it reacts with concentrated hydrochloric acid forms laevulinic acid.
- Glucose undergoes dehydration with concentrated sulphuric acid and forms hydroxy methyl furfural.
- Glucose, when heated with concentrated sodium hydroxide (NaOH), turns yellow, then brown and eventually forms resin. However, with dilute NaOH, glucose exhibits a reversible isomerisation and forms a mixture of D-glucose, D-fructose and D-maltose. This reaction is known as Lobry de Bruyn-van Ekenstein transformation.

Definitions of Words and Terms

Monosaccharide: A single sugar.

Aldose: A polyhydroxy aldehyde, *i.e.*, a carbohydrate containing an aldehyde functional group.

Ketone: an organic compound that contains two alkyl groups attached to a carbonyl group.

Ketose: A carbohydrate which contains a ketone.

Furanose: A monosaccharide of five-member closed chain form consists of four carbon atoms and one oxygen atom.

Pyranose: A six-member cyclic form of a monosaccharide.

Mutarotation: A change in optical activity due to chemical change.

Fischer Projection: Fisher projection is a two dimensional representation of a three dimensional organic molecule. It communicates valuable stereochemical information without drawing 3-D structural representation.

Haworth Projection: It is a simple 3-D view of cyclic structure of monosaccharides.

Chair Conformation: It is the most accurate resprestation of orientation of cyclohexane rings in space.

Anomeric carbon: A carbon in glucose connected to two oxygen atoms by single bonds.

Concentration: The amount of substance within a second substance, usually a solution.

List of Abbreviations

ATP	Adenosine triphosphate
ADP	Adenosine diphosphate
°C	Degree Celsius
CAS	Chemical Abstracts Service
cm3	Cubic Centimeter
DNA	Deoxyribonucleic acid
g/mol	gram per mole
Glc	Glucose
IUPAC	International Union of Pure and Applied Chemistry

mL   milliliter
mg   milligram
mmol/L   millimole per litre
mg/dL   milligram per decilitre
nm   nanometer
RNA   Ribonucleic acid
UV   UltraViolet

References

Brown, W.H. and McClarin, J.A., 1981. *Introduction to Organic and Biochemistry*. Willard Grant Press: chem-online.org http://www.chem-online.org/carbohydrate/glucose.htm (accessed 19/11).

Elliott, W.H., Elliott, D.C., Jefferson, J.R. and Wheldrake, J., 1997. *Biochemistry and Molecular Biology*. Oxford University Press.

Holdsworth, D.K., Papua, U.O. and Chemistry, N.G.D.O., 1973. *Introduction to Organic Chemistry*. Dept. of Chemistry, University of Papua and New Guinea.

McGraw-Hill, 2004. *McGraw-Hill Concise Encyclopedia of Chemistry*. McGraw-Hill Professional.

McMurry, J., Castellion, M.E. and Ballantine, D.S., 1996. *Fundamentals of General, Organic, and Biological Chemistry*. Prentice Hall.

Nelson, D.L., Cox, M.M. and Lehninger, A.L., 2008. *Principles of Biochemistry*. Freeman.

CHAPTER 6
Galactose Chemistry

FABIO VIEIRA DOS SANTOS,*[a] VANESSA JAQUELINE
DA SILVA VIEIRA DOS SANTOS,[b] JULIANA PEREIRA
LYON[c] AND LEONARDO MARMO MOREIRA[d]

[a] Centro de Ciências da Saúde, Universidade Federal de São João del Rei,
Rua Sebastião Gonçalves Coelho, 400 Divinópolis, MG, Brazil; [b] Centro de
Ciências da Saúde, Universidade Federal de São João del Rei, Rua Sebastião
Gonçalves Coelho, 400 Divinópolis, MG, Brazil; [c] Departamento de Ciências
Naturais, Universidade Federal de São João del Rei, Praça Dom Helvécio,
74, Fábricas, São João del Rei, MG, Brazil; [d] Universidade Federal de São
João del Rei, Praça Dom Helvécio, 74, Fábricas, São João del Rei, MG,
Brazil
*Email: santos_fv@yahoo.com.br

6.1 Introduction

Carbohydrates or saccharides (Greek: sakcharon, suggar) are important
structural and functional components in all living organisms and represent
some of the more abundant biological molecules. This molecular group can be
defined as polyhydroxy aldehydes and polyhydroxy ketones or as substances
that can release these compounds when they are hydrolysed. Their general
chemical formula is $[C(H_2O)]_n$, where $n \geq 3$. However, even though the majority
of the saccharides present this empirical formula, some carbohydrates can
include nitrogen, phosphorus or sulphur in their structures.

The simplest carbohydrates, which cannot be hydrolysed into smaller subunits, are known as monosaccharides. Monosaccharides can be one unit
polyhydroxy aldehydes or polyhydroxy ketones. D-glucose is the most

Food and Nutritional Components in Focus No. 3
Dietary Sugars: Chemistry, Analysis, Function and Effects
Edited by Victor R Preedy
© The Royal Society of Chemistry 2012
Published by the Royal Society of Chemistry, www.rsc.org

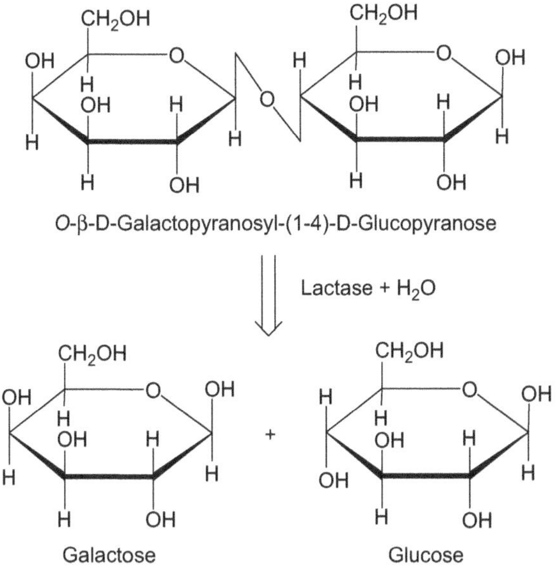

O-β-D-Galactopyranosyl-(1-4)-D-Glucopyranose

Lactase + H$_2$O

Galactose   Glucose

Figure 6.1 Lactose hydrolysis. The disaccharide lactose [*O*-β-D-galactopyranosyl-(1-4)-D-glucopyranose] is hydrolysed in the human intestine through the action of the enzyme lactase, which belongs to the family of β-D-galactosidases, and releases the monosaccharides glucose and galactose, which can be used as energy sources for the energetic metabolism of the cells.

abundant of the monosaccharides and has a fundamental role in the energetic metabolism of living organisms. Another important monosaccharide is D-galactose. This sugar contains six carbons (hexose) and is frequently observed associated with glucose in the disaccharide lactose (Figure 6.1), which is present in the milk of mammals.

D-galactose, like D-glucose, is an energetic source in cell metabolism. Information on galactose metabolism will be discussed in other chapters of this book. Briefly, in this biochemical process, D-galactose is converted into galactose-6-phosphate by the action of the enzyme galactokinase (GALK) through the consumption of ATP. After this reaction, UDP-galactose is obtained through the action of the enzyme galactose-1-phosphate uridyl-transferase (GALT). The enzyme UDP-galactose 4-epimerase catalyses the regeneration of UDP-glucose in the final step of normal galactose metabolism (Holden *et al.* 2003). After these reactions, glucose-6-phosphate is obtained through the action of a transferase and the enzyme phosphoglucomutase. Glucose-6-phosphate is then able to enter into the glycolysis pathway.

Individuals that have mutations in genes that encode the enzymes that participate in galactose metabolism can present disturbances due to the accumulation of galactose or its derivatives, which results from chemical and/or enzymatic reactions. In this chapter, the chemical properties and different chemical reactions related to D-galactose will be presented and discussed.

6.2 Chemical Classification of D-Galactose

As previously mentioned, D-galactose is a carbohydrate containing six carbons with the formula ($C_6H_{12}O_6$). The molecular weight of this sugar is 180.156 g mol^{-1}, and its carbon skeleton contains a carbonyl group (C=O) at its extremity, which characterises D-galactose as an aldose (aldohexose).In its pure form, this monosaccharide is a white powder and odorless, with a density of 1.732 g cm^{-3}. Table 1 provides the principal physical and chemical properties of D-galactose.

Monosaccharides are designated as "D" or "L", based on the configuration around their chiral centre, which is more distant from the carbonyl group. If the hydroxyl group in this chiral centre has the same configuration as (+)-glyceraldehyde, the sugar is designated "D". However, if the configuration is similar to that observed in (−)- glyceraldehyde, the designation is "L". In the case of D-galactose, the more distant chiral centre is at carbon 5, and the hydroxyl group of this carbon is on right, which is the configuration observed in (+)-glyceraldehyde (Figure 6.2). The majority of the hexoses present in living organisms are D-isomers. Thus, D-galactose is characterised as a D-glucose epimer because these two sugars have a stereochemical difference only at C-4 (Figure 6.3).

Monosaccharides containing five or six carbons are generally observed in aqueous solutions as ring structures. The cycling process is due to the establishment of a covalent bond between the carbonyl group of the sugar with a hydroxyl along the chain. Thus, the cyclic forms of D-glucose and D-galactose, for instance, are hemiacetal and formed by an intramolecular reaction between the hydroxyl group present at C-5 and the carbonyl group (Figure 6.4). This process generates a new chiral centre, which is observed at C-1, and

Table 6.1 Chemical information and physicochemical properties of D-Galactose. Principal information about the physical and chemical properties of the monosaccharide galactose. CAS – Chemical Abstracts Service; IUPAC – International Union of Pure and Applied Chemistry.

Formula	$C_6H_{12}O_6$
IUPAC Name	(3R,4S,5R,6R)-6-(Hydroxymethyl)oxane-2,3,4,5-tetrol
CAS	59-23-4
Appearance	White powder
Odor	Odorless
Molar Mass	180.156 g mol^{-1}
Density	1.732 g cm^{-3}
Molar Refractivity	37.254 cm^3
Melting Point	163 °C – 169 °C
Boiling Point	410.797 °C at 760 mmHg
Solubility in Water	680 g L^{-1}
Heat of Vaporization	92.2 kJ mol^{-1}
Heat of Combustion	−2792 kJ mol^{-1}

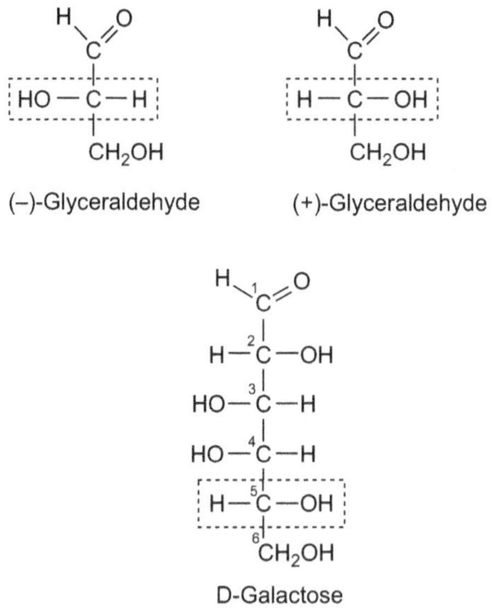

(–)-Glyceraldehyde   (+)-Glyceraldehyde

D-Galactose

Figure 6.2 D-Galactose. The configuration around the chiral centre more distant from the carbonyl group permits the designation of a monosaccharide as "D" or "L". In D-galactose, the hydroxyl group at the chiral centre at C-5 is right like that of (+)-glyceraldehyde (see highlighted sections in the structures).

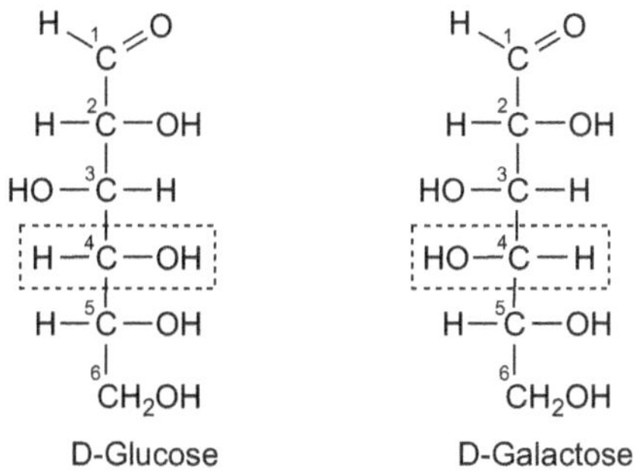

D-Glucose   D-Galactose

Figure 6.3 Similarities between D-glucose and D-galactose. D-glucose and D-galactose are epimers at C-4. These carbohydrates differ only in their configurations at their chiral centres, which are highlighted in the structures. Another epimer of D-glucose is D-mannose, which differs only in its configuration at C-2.

consequently, two cyclic forms are possible (diastereoisomers). These alternative forms are anomers, and the hemiacetal carbon is known as an anomeric carbon. As in other hexoses, D-galactose presents the anomeric α or β forms based on the position of the -OH group at the C-1 (Figure 6.5). In the α anomer, the C-1 hydroxyl group is trans to the –CH$_2$OH group, and in the

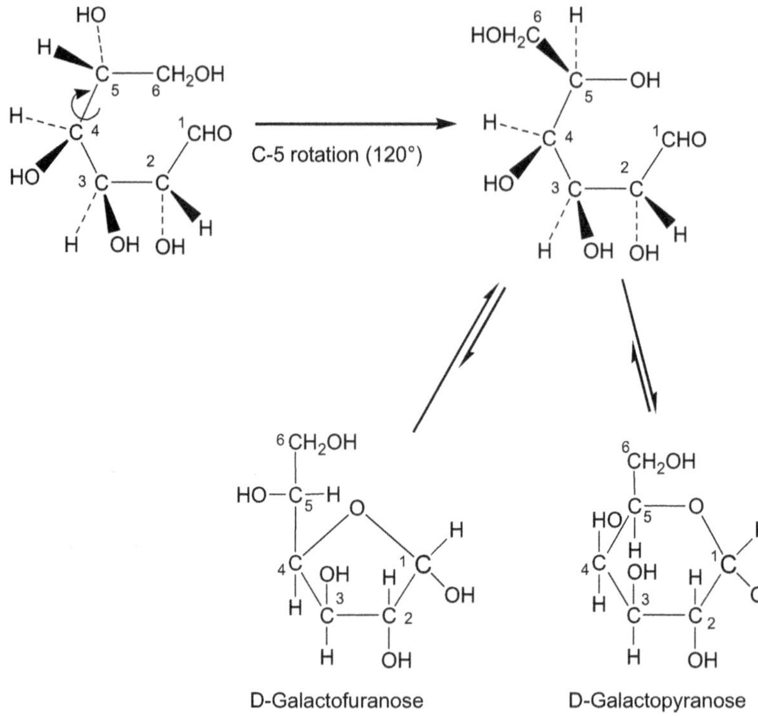

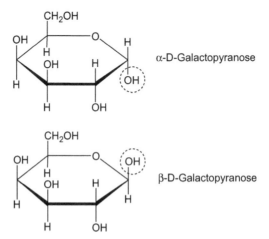

Figure 6.4   Formation of the ring structures of D-galactose. D-galactose is more frequently observed as a ring in an aqueous solution. D-galactopyranose and D-galactofuranose can be formed after a rotation of C-5 (120°) and the interaction between the hydroxyl group present in this carbon with the carbonyl group (C-1).

Figure 6.5   Alfa (α) and beta (β) anomers of D-galactose. As observed, the config-uration of the hydroxyl group in C-1 (anomeric carbon) is trans to the -CH$_2$OH in α-D-galactopyranose and cis in β-D-galactopyranose.

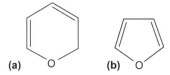

Figure 6.6 Pyran (a) and furan (b). Monosaccharides that exhibit rings with six atoms
are termed pyranoses because of their similarity with heterocyclic pyran.
Furanoses are sugars that present five atoms in their rings, like furans.

β anomer, the configuration is cis. Calorimetric studies performed by
Takahashi and Ono in 1973 indicate that β-D-galactose is energetically more
stable at 25 °C compared to the α anomer of the monosaccharide.

When a monosaccharide contains a ring structure with six atoms, as is
observed in pyran (a cyclic ether), it is known as a pyranose (see explanation in
Figure 6.6). If the sugar contains a ring with five atoms, as observed in furan, it
is known as a furanose. D-glucose is observed in solution only as α-D-glucopyranose or as β-D-glucopyranose. In an equilibrated solution, the proportions
of α and β anomers of glucose are 36% and 64%, respectively. Moreover, D-
galactose can be identified in an equilibrated aqueous solutions as pyranose or
furanose. A study performed by Pazourek (2010) using HPLC (high performance liquid chromatography) measurements demonstrated that aqueous
solutions of D-galactose contain α-D-galactopyranose (33.3%), β-D-galactopyranose (62.5%), α-D-galactofuranose (1.4%), and β-D-galactofuranose
(2.85%). These results are in agreement with previous NMR (nuclear magnetic
resonance) studies performed by Zhu *et al.* (2001). Acyclic forms of D-galactose
are observed in equilibrated solutions but at very low concentrations.

The presence of various forms of D-galactose in solutions is explained by the
process known as mutarotation. Each hexose form exhibits a specific rotation.
In aqueous solutions, these specific rotations are constantly altered until an
equilibrium state is achieved. The variations in specific rotations are a reflection
of the interconversion between the different forms (cyclic and linear) of the
monosaccharide. When the equilibrium in this interconversion is reached, the
rotation value of the solution is observed. An equilibrated aqueous solution of
D-galactose exhibits a rotation of $+80.2°$ ($[\alpha]^{25}_{D} = +80.2°$).

6.3   Chemical Reactions with D-Galactose

6.3.1   Galactoside Formation

The glycosides are compounds that can be chemically defined as substances
formed by the union of a sugar with a non-carbohydrate moiety, which is
normally a small organic molecule. A galactoside is a molecule where the
hydrogen from the hydroxyl group in the C-1 is substituted for an organic
moiety. A simple galactoside can be obtained by the treatment of D-galactose
with methanol (CH_3OH) under acidic conditions (in the presence of HCl for

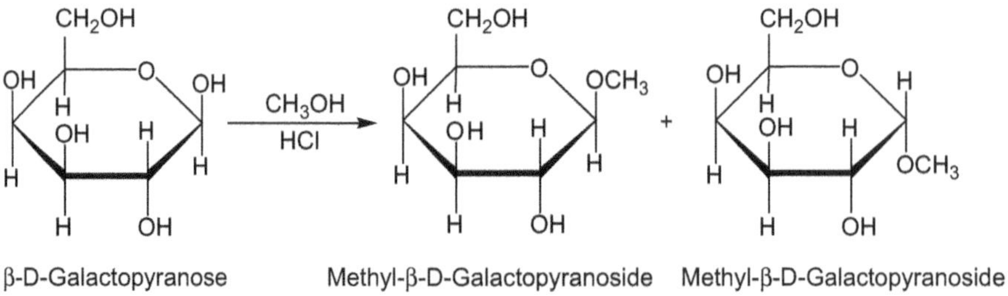

β-D-Galactopyranose   Methyl-β-D-Galactopyranoside   Methyl-β-D-Galactopyranoside

Figure 6.7   Galactoside. In the presence of methanol (CH_3OH) and under acidic conditions (hydrochloric acid – HCl), a methyl group (CH_3) substitutes for the hydrogen of the hydroxyl (OH) group of the anomeric carbon with the establishment of a glycosidic bond.

example). During this process, methyl-α-D-galactopyranoside and methyl-ß-D-galactopyranoside can be formed, which is in accordance with the configuration of the methyl group at C-1 (Figure 6.7). The hydrolysis of glycosides in living organisms is catalysed by enzymes known as galactosidases.

6.3.2   Ether Formation

If a monosaccharide such as D-galactose is diluted in a basic solution in the presence of good alkyl donors, such as dimethyl sulphate ($(CH_3O)_2SO_2$), methyl iodide (CH_3I) or silver oxide (Ag_2O), all of their hydroxyl groups can accept methyl groups. This process can result in the formation of ether groups at all these positions. At the anomeric carbon, the linkage with the methyl group is labile under acidic conditions, and this radical can easily be removed.

This chemical reaction is an important part of an approach to determine the ring size of a monosaccharide. Under acidic conditions, an open chain tetramethylated monosaccharide reacts with nitric acid (HNO_3), and a keto group is formed at C-5 (if the initial heterocyclic ring contains six atoms) or at C-4 (if the ring contains five atoms). Further oxidation caused by HNO_3 cleaves the carbon chain at both bonds surrounding the ketone group. This oxidation process results in a mixture of two dicarboxylic acids, which differ in length by one carbon atom. Once these dicarboxylic acids have been identified, the location of the ketone is established, and its initial ring size (furanose or pyranose) can be determined.

6.3.3   Ester Formation

As described above, the hydroxyl groups of a monosaccharide are acceptors of methyl groups, and they form ether groups at these positions. Thus, these hydroxyl groups can be acetylated with ester group formation at these sites. To perform this reaction, the compound that acts as the acyl donor is usually acetic anhydride ($(CH_3CO)_2O$). During this process, at a low temperature and in the presence of a weak base, all OH groups are acetylated, which include the anomeric group present at C-1, and the α anomer is generally more stable in an aqueous solution.

6.3.4 Cyclic Ketal Formation

In the presence of propanone (acetone – $(CH_3)_2CO$) and sulfuric acid (H_2SO_4), ketal groups can be selectively formed at the cis-vicinal hydroxyl of an α-D-galactose molecule.

6.3.5 Oxidation

Sugar reactions can be divided into two groups: reactions that occur with an open molecule and reactions that happen with the ring form of a molecule. In reactions of isomers with an open chain and a closed chain, competitiveness exists but the reactions of aldoses with oxidising agents do not occur with the hemiacetal function of cyclic isomers. However, this reaction will occur with the aldehyde of an open chain.

In Figure 6.8, the detection of D-galactose, a reducing sugar, is demonstrated through the employment of the Fehling test. Polyfunctional monosaccharides with open chains can typically react based on the functional groups present in their structures. For instance, the oxidation of formyl that happens with aldoses such as D-galactose results in a positive detection in standard oxidising tests. Thus, D-galactose is a reducing sugar that, like other aldoses, forms aldonic acid when oxidised. Galactonic acid, which is formed during the Fehling reaction with D-galactose, is also detected in the red blood cells of patients with galactosaemia, a genetic disturbance generally caused by a deficiency in galactose-1-phosphate uridyltransferase (GALT) (Yager *et al.* 2003). In mammals, galactonic acid is formed in an alternative D-galactose metabolism pathway by the action of the enzyme galactose dehydrogenase, and recent

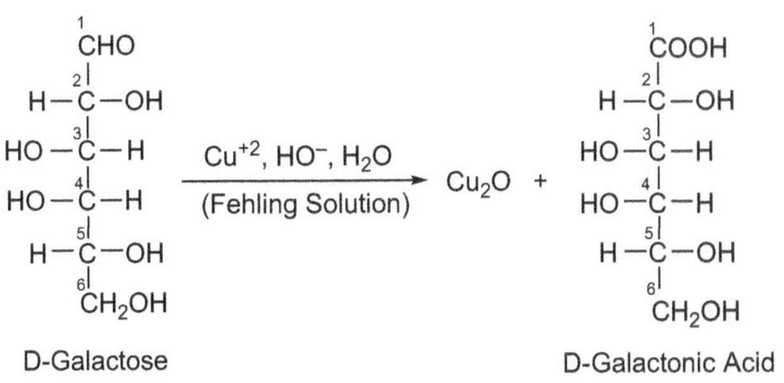

Figure 6.8 Fehling reaction with D-galactose, a reducing sugar. The oxidation of the formyl group in the presence of Fehling's solution results in the formation of Cu_2O (cupric oxide), which precipitates and presents a brick red color, and D-galactonic acid.

studies have shown that this metabolite is detected in the plasma of healthy humans after the consumption of beverages containing galactose (Bruce *et al.* 2010).

Heated, diluted nitric acid can be used to cause vigorous oxidations in aldoses because it leads to the attack of primary hydroxyls as well as formyl groups. When performed with D-galactose, this process results in the formation of galactaric acid (Figure 6.9). Galactaric acid ($C_6H_{10}O_8$) is also known as mucic acid and is insoluble in water and alcohol. This compound is used as a chelator and in skin care products. It was formerly used as a leavening agent in self-raising flour (Mojzita *et al.* 2010).

The specific oxidation of the primary alcohol groups of aldoses is another relevant event and results in the formation of uronic acids. With D-galactose, this process results in the formation of D-galacturonic acid (Figure 6.10). This acid sugar is a component of different types of pectin. The pectins are fundamental polysaccharides observed in the extracellular matrix of plant cells and as constituents of the cell walls of these organisms. Polygalacturonic acid or pectic acid is a common type of pectin and is composed of D-galacturonic acids linked by α (1-4) bonds with occasional rhamnosyl residues, which results in torsion in

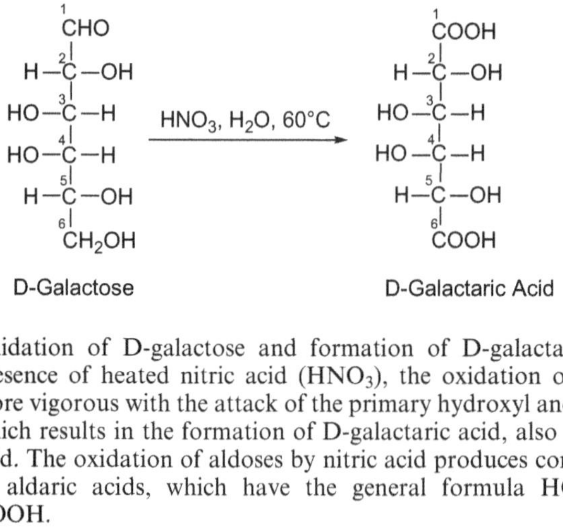

Figure 6.9 Oxidation of D-galactose and formation of D-galactaric acid. In the presence of heated nitric acid (HNO_3), the oxidation of D-galactose is more vigorous with the attack of the primary hydroxyl and formyl groups, which results in the formation of D-galactaric acid, also known as mucic acid. The oxidation of aldoses by nitric acid produces compounds known as aldaric acids, which have the general formula $HOOC\text{-}(CHOH)_n\text{-}COOH$.

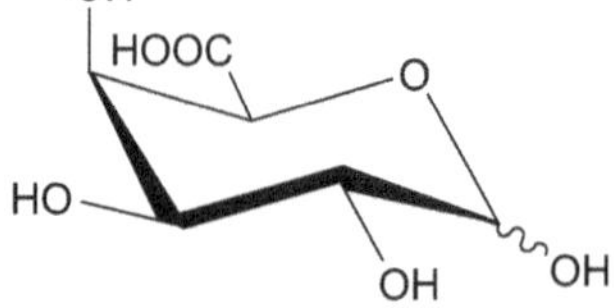

Figure 6.10 The galacturonic acid. This compound results from the oxidation of the primary alcohol group of D-galactose and is an important component of the cell wall of terrestrial plant species as a constituent of pectins.

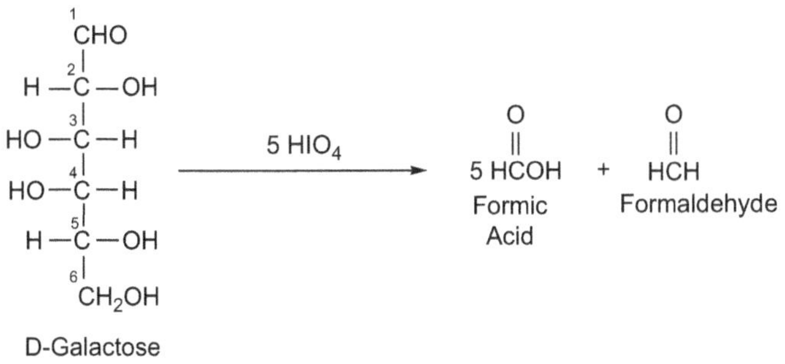

$$
\begin{array}{c}
\overset{1}{C}HO \\
H\overset{2}{-}\overset{|}{C}-OH \\
HO-\overset{3}{\underset{|}{C}}-H \\
HO-\overset{4}{\underset{|}{C}}-H \\
H-\overset{5}{\underset{|}{C}}-OH \\
\overset{6}{C}H_2OH
\end{array}
\quad\xrightarrow{\text{5 HIO}_4}\quad
\begin{array}{c}
O \\
\parallel \\
\text{5 HCOH} \\
\text{Formic} \\
\text{Acid}
\end{array}
\;+\;
\begin{array}{c}
O \\
\parallel \\
\text{HCH} \\
\text{Formaldehyde}
\end{array}
$$

D-Galactose

Figure 6.11 Oxidative degradation of D-galactose with periodic acid. Each cleavage process requires 1 mol of HIO_4 (periodic acid). Thus, to cleave 1 mol of D-galactose, 5 moles of HIO_4 are required and result in 5 moles of formic acid and 1 mol of formaldehyde.

the molecule. Rhamnogalacturonan I is a large and heterogeneous pectin whose basic structure is derived from an alternation of α(1-4) D-galacturonic acid and α(1-2) D-rhamnose. This pectin contains lateral chains that are normally formed by arabinans, galactans and arabinogalactans.

The oxidative degradation of vicinal diols in carbonyl compounds is possible through a reaction with periodic acid (HIO_4). The complete degradation of its carbon chain to compounds with one carbon atom is possible. After being oxidised with periodic acid, the majority of sugars generally form complex mixtures because they contain several pairs of vicinal diols. In Figure 6.11, the degradation of D-galactose in the presence of HIO_4 and its resultant products can be observed.

6.3.6 Reduction

Another important chemical reaction that can be performed using monosaccharides is the conversion of aldehydes in polyhydroxylated compounds known as alditols. For example, this reduction reaction can be performed by the treatment of D-galactose with sodium borohydride ($NaBH_4$) to produce D-galactitol (Figure 6.12). This compound is formed in living organisms because the enzyme aldose reductase can reduce the aldose group of D-galactose. Galactitol cannot be further metabolised and is predominantly excreted in urine (Segal and Berry 1995). Nevertheless, persons with galactosaemia can accumulate galactose in the lens of their eyes and, as a consequence, galactitol will be formed through the action of aldose reductase (Figure 6.12). This situation represents a risk for the development of cataracts because galactitol is osmotically active, which results in water diffusing into the lens of the eye. However, it should be highlighted that the high concentration of glucose in diabetics can result in a similar situation because of the production and accumulation of glucitol (sorbitol) due to the action of aldose reductase.

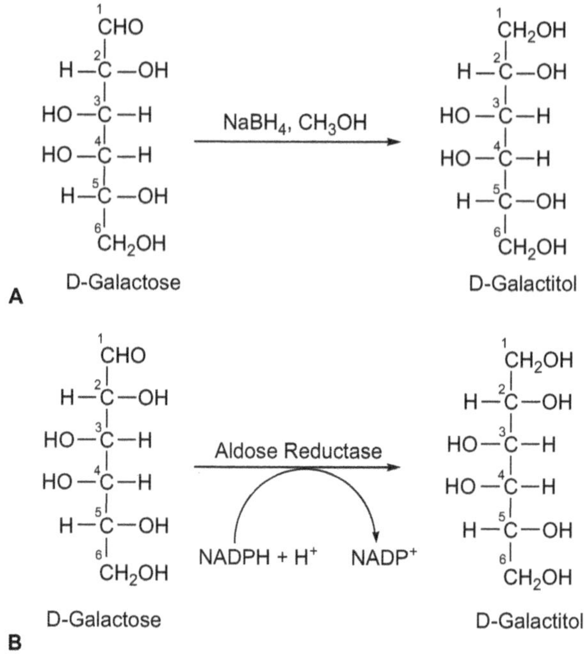

CHO
H—C—OH
HO—C—H
HO—C—H
H—C—OH
CH₂OH
D-Galactose

NaBH₄, CH₃OH →

CH₂OH
H—C—OH
HO—C—H
HO—C—H
H—C—OH
CH₂OH
D-Galactitol

A

CHO
H—C—OH
HO—C—H
HO—C—H
H—C—OH
CH₂OH
D-Galactose

Aldose Reductase →
NADPH + H⁺   NADP⁺

CH₂OH
H—C—OH
HO—C—H
HO—C—H
H—C—OH
CH₂OH
D-Galactitol

B

Figure 6.12   Reduction of D-galactose and production of D-galactitol. (A) Reduction reaction of D-galactose with sodium borohydride (NaBH₄), which results in the formation of D-galactitol. (B) The production of this compound by the action of aldose reductase.

6.4   Lactose: Disaccharide Formation and Hydrolysis

Lactose is a disaccharide formed by two monosaccharides joined covalently together by an *O*-glycosidic bond. This bond is established when the anomeric carbon (C1) from D-galactose reacts with the hydroxyl group present at the C4 of D-glucose. This process can be explained as the formation of an acetal from a hemiacetal (the anomeric carbon from D-galactose) and an alcohol (the hydroxyl present at the C4 of D-glucose).

The formal name for lactose is *O*-β-D-galactopyranosyl-(1-4)-D-glucopyranose (Figure 6.1). This name explicitly indicates the monosaccharide types present in the structure and the bond type (the C1 from the galactose residue to the C4 from the glucose residue) formed. The "pyranosyl" and "pyranose" terms indicate that both monosaccharides contain ring forms with six atoms. The "*O*" is present to indicate that the glycosidic bond involves an oxygen atom.

Lactose is naturally found only in milk. In this disaccharide, the anomeric carbon from the glucose residue is available and can be oxidised by ferric and cupric ions (similar to the formerly described reaction of Fehling with D-galactose). Thus, lactose is a reducing sugar like other common disaccharides.

However, the disaccharide sucrose (D-glucose + D-frutose) is not a reducing sugar because the anomeric carbons of the glucose and fructose residues are involved in the glycosid bond and not available for oxidation.

The sugar-sugar link defined by the glycosidic bond can be hydrolysed by heat under acidic conditions. However, this bond is resistant to alkaline conditions. In humans, lactose is hydrolysed by lactase, a β-galactosidase enzyme. This book presents other chapters that explore the disaccharide lactose and its metabolism.

Summary Points

- Galactose is an important monosaccharide and component of the disaccharide lactose, which is present in milk.
- D-galactose contains six carbons in its structure, and it is an epimer of D-glucose, as these sugars are only stereochemically different at carbon 4 (C-4).
- The most common ring form of D-galactose contains six atoms, and this sugar is normally observed as a pyranose.
- Based on the position of the hydroxyl group in C-1, the α and β anomeric forms of D-galactose are possible. However, studies indicate that β-D-galactose is energetically more stable.
- D-galactose is a reducing sugar because its hydroxyl group contains an anomeric carbon that can be oxidised by ferric and cupric ions, for example.
- Oxidation processes involving D-galactose can result in the production of D-galactonic acid, D-galactaric acid, and D-galacturonic acid, among other products.
- The reduction of D-galactose can result in the production of D-galactitol.
- The link between a D-galactose and a D-glucose by a glycosidic bond results in the formation of lactose (*O*-β-D-galactopyranosyl-(1-4)-D-glucopyranose).

Key Facts about Galactose and its Importance

- Lactose is a disaccharide, *i.e.*, a carbohydrate with two subunits, that occurs naturally only in the milk of the majority of mammals, including cows, goats and humans.
- Absorption of lactose in the intestinal mucosa requires that this disaccharide be hydrolysed into its components, D-glucose and D-galactose, which are linked by a β-glycosidic bond.
- Some human diseases are related to abnormalities in galactose metabolism and absorption, such as the genetic disease galactosaemia.
- Individuals affected by galactosaemia have impaired activities for the enzymes galactokinase (GALK), galactose-1-phosphate uridyltransferase (GALT) or UDP-galactose 4′-epimerase.

- An absence of the function of one of these enzymes results in the accumulation of galactose in the blood and, consequently, in the formation of subproducts that can be toxic to the organism.
- For example, the enzyme aldose reductase produces D-galactitol from D-galactose, and the accumulation of this product in the lens of the eye can result in the development of cataracts.
- Galactose is not exclusively produced by animals and has functions other than those related to energetic metabolism.
- Galactose and its derivatives are important structural components of the cell wall of plants.

Definitions of Words and Terms

Aldehyde. In organic chemistry, an aldehyde is a molecule characterised by the presence of a H–C=O group, also known as a formyl group. Aldehydes contain a sp^2-hybridised planar carbon centre that is linked by a double bond to an oxygen and a single bond to a hydrogen. The simplest aldehyde is formaldehyde.

Aldohexose. An aldohexose is a monosaccharide with six carbons in its structure and that has the general formula $C_6H_{12}O_6$. In this sugar type, the carbonyl group at its extremity is an aldehyde (carbon 1), and the other carbons are linked to hydroxyl groups.

Epimers. Epimers are molecules that differ only in the configuration of one asymmetric centre in their stereoisomers, which present more than two asymmetric centres.

Extracellular Matrix. A structure of polysaccharides and proteins secreted by cells. The extracellular matrix performs structural functions in animal and plant tissues and is fundamental in several physiological and developmental processes in multicellular organisms.

Fehling's Solution. Fehling's solution is used to test carbohydrates and identify aldehyde functional groups in water soluble compounds. The procedure was developed in 1849 by the German chemist Hermann von Fehling. Fehling's solution is composed of a blue aqueous solution of copper(II) sulphate (Fehling's solution "A") and a clear solution of aqueous potassium sodium tartrate and sodium hydroxide (Fehling's solution "B"). After heating the solution, the test substance is added, and the presence of an aldehyde is detected by the oxidation of this group with the concomitant reduction of Cu^{+2} to Cu^{+1}, which forms a red precipitate.

Galactose-1-Phosphate Uridyltransferase (GALT). This enzyme converts ingested galactose into glucose at the second step of the Leloir pathway. The absence of the expression of this protein characterises classical galactosaemia.

High Performance/Pressure Liquid Chromatography (HPLC). HPLC is a chromatographic methodology used in different areas of science such as biochemistry and analytical chemistry to identify, purify and quantify substances that are present in a mixture. The mobile phase in this method is under high pressure. The stationary phase presents particles with diameters of around five micrometers while increasing the superficial area and, consequently, the efficiency in the separation of the components.

Ketones. The organic function that is a carbon linked by a double bond to oxygen and two other organic radicals and that has the general structure R–C(=O)–R′ is termed a ketone. If R and R' are the same, the ketone is symmetric. However, if the ketone is characterised as asymmetric, R and R' are different.

Nuclear Magnetic Resonance Spectroscopy (NMR). The use of the magnetic properties of certain atomic nuclei to identify and characterise atoms or molecules is the basis of nuclear magnetic resonance spectroscopy. This methodology permits the determination of the structure, reaction state, dynamics and chemical environment of a molecule.

Pectins. Pectins constitute a family of structural heteropolysaccharides present in the cell wall of terrestrial plant species. These polysaccharides are normally ramified and can consist of galacturonic acid, rhamnose, arabinose, or galactose subunits among other components.

List of Abbreviations

°C	Degree Celsius
CAS	Chemical Abstracts Service
cm^3	Cubic Centimeter
GALK	Galactokinase
GALT	Galactose-1-Phosphate Uridyltransferase
HPLC	High Performance/Pressure Liquid Chromatography
IUPAC	International Union of Pure and Applied Chemistry
kJ	Kilojoule
L	Liter
mmHg	Millimeters of Mercury
NMR	Nuclear Magnetic Resonance
UDP	Uridine Diphosphate

References

Bruce, S.J., Breton, I., Decombaz, J., Boesch, C., Scheurer, E., Montoliu, I., Rezzi, S., Kochhar, S. and Guy, P.A., 2010. A plasma global metabolic profiling approach applied to an exercise study monitoring the effects of glucose, galactose and fructose drinks during post-exercise recovery. *Journal of Chromatography B*. 878: 3015–3023.

Holden, H.M., Rayment, I. and Thoden, J.B., 2003. Structure and Function of enzymes of the Leloir pathway for galactose metabolism. *The Journal of Biological Chemistry*. 278: 43885–43888.

Mojzita, D., Wiebe, M., Hilditch, S., Boer, H., Penttila, M. and Richard, P., 2010. Metabolic engineering of fungal strains for conversion of D-galacturonate to meso-galactarate. *Applied and Environmental Microbiology.* 76: 169–175.

Pazourek, J., 2010. Monitoring of mutarotation of monosaccharides by hydrophilic interaction chromatography. *Journal of Separation Science.* 33: 974–981.

Segal, S. and Berry, G.T., 1995. Disorders of galactose metabolism. In: Scriver, B.A., Sly, W. and Valle, D. (ed.) *The Metabolic Basis of Inherited Diseases.* McGraw-Hill, New York, USA, pp. 967–1000.

Takahashi, K. and Ono, S., 1973. Calorimetric studies on the mutarotation of D-galactose and D-mannose. *The Journal of Biochemistry.* 73: 763–770.

Yager, C.T. Chen, J., Reynolds, R. and Segal, S., 2003. Galactitol and galactonate in red blood cells of galactosemic patients. *Molecular Genetics and Metabolism.* 80: 283–289.

Zhu, Y., Zajicek, J. and Serianni, A.S., 2001. Acyclic forms of aldohexoses in aqueous solution: quantitation by 13C NMR and deuterium isotope effects on tautomeric equilibria. *Journal of Organic Chemistry.* 66: 6244–6251.

CHAPTER 7

Maltose Chemistry and Biochemistry

ROBERT R. CROW, SANATH KUMAR AND
MANUEL F. VARELA*

Department of Biology, Eastern New Mexico University, Station
33 Portales, NM, 88130, USA
*Email: Manuel.Varela@enmu.edu

7.1 Introduction

The sugar D-(+)-maltose is a glucoside consisting of two glucose monomers that are connected by an α-1,4 glycosidic bond to form maltose, 4-O-α-D-glucopyranosyl-D-glucose ($C_{12} H_{22} O_{11}$), of molecular weight of 342.3 Daltons (Figure 7.1). This chapter focuses on the biochemistry and transport of maltose. Much of the metabolic and transport machinery for maltose are conserved, from bacteria to humans (Henderson, *et al.* 1993). Hence, studies of maltose chemistry and biochemistry in lower organisms have been used as model systems for study in higher organisms, including humans. Metabolic disorders involving the malabsorption of maltose have been documented and discussed elsewhere (Raben, *et al.* 2002).

7.2 The Metabolism of Maltose

The digestion of starch in animals starts in the mouth by α-amylase present in the saliva and subsequently in the small intestine. However, amylases are

Food and Nutritional Components in Focus No. 3
Dietary Sugars: Chemistry, Analysis, Function and Effects
Edited by Victor R Preedy

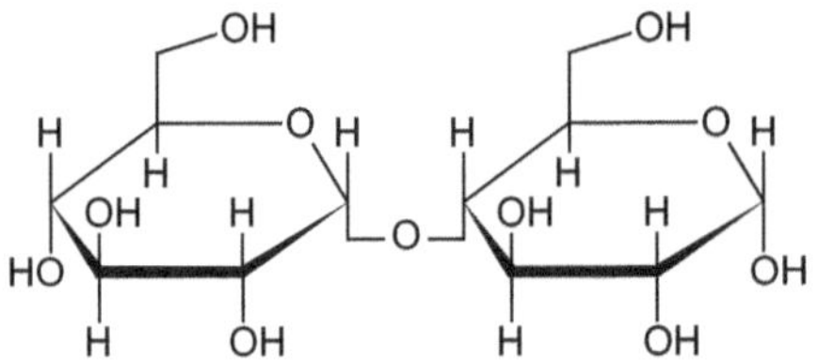

Figure 7.1   The structure of maltose.

Table 7.1   Key enzymes involved in the breakdown of starch in microorganisms.

Enzyme	E.C. number	Function
α-Amylase	3.2.1.1	Endohydrolase. Cleaves α, 1-4 glycosidic bonds.
β-Amylase	3.2.1.2	Exohydrolase. Cleaves α, 1-4 glycosidic linkages from the non-reducing end of the starch.
Pullulanase type I	3.2.1.41	Debranching enzyme. Hydrolyzes α, 1-6 glycosidic linkages of amylopectin. Does not hydrolyze α, 1-4 linkage.
Pullulanase type II		Hydrolyzes both α, 1-4 and α, 1-6 glycosidic linkages.
Glucoamylase	3.2.1.3	Exohydrolase. Cleaves glucose residues from the non-reducing end of starch producing β-D-glucose.
Isoamylase	3.2.1.68	Debranching enzyme. Hydrolyzes α, 1-6 glycosidic linkages.
α-D-glucosidase (Maltase)	3.2.1.20	Cleaves terminal 1-4 glycosidic linkages from the non-reducing end releasing α-D-glucose.

unable to completely degrade the starch molecule, since they cannot hydrolyze the α-1,6-glycosidic bonds of the amylopectin. α-Amylase hydrolyzes starch into limit dextrins, maltose and a small amount of glucose. Limit dextrins are short dextrins with many 1,6-glycosidic bonds produced by incomplete digestion of amylopectins by α-amylase. Thus, the complete digestion of starch requires debranching enzymes called isomaltases or α-1,6 glucosidase (Nigam and Singh 1995). Maltose is further hydrolyzed by maltase to yield two molecules of glucose. Glucose is absorbed into the blood stream and transported to different tissues. A brief overview of the various enzymes participating in starch hydrolysis is given in Table 7.1.

Starch is the most common storage polysaccharide for maltose in plants and a key carbohydrate in many important biotechnological applications. Cereals such as rice, wheat, maize, barley and tubers such as potato and cassava are well-known rich sources of starch. Starch is stored in the form of large water-insoluble granules of 0.1–10 μm size in the plastids of photosynthetic plants. Starch is composed of two large molecular weight components: amylose and amylopectin, and both of these macromolecules are made up of α-D-glucose monomers. Amylose is a linear polymer of α-D-glucose linked by α-1,4-glycosidic bonds and amylose constitutes 15–25% of the starch polymer.

Amylose folds itself into a helical conformation consisting of 6 glucose residues per helical turn. The amylose helix has high binding affinity for iodine which gives dark blue color with starch, a commonly used test to detect starch. Amylopectin, on the other hand, constitutes 75–85% of the starch polymer, is made up of α-1,4-linked glucose monomers branched in the α-1,6 position. On average, amylopectin has 4% of α-1,6 branch points in the polymer. The structural complexity of starch requires a combination of enzymes to completely hydrolyze the polymer into oligosaccharides, disaccharides such as maltose, and the monomer glucose. A number of starch-degrading enzymes known from plants and microorganisms are collectively called amylases, which are of two types: α- and β-amylases. α-Amylase is an endoglycosidase which has the ability to cleave α-glycosidic bonds anywhere along the starch polymer, while the β-amylase can do so only from the non-reducing end of the starch polymer (Ray and Nanda 1996).

The incompletely absorbed dietary carbohydrates such as the starch make an important source of nutrition for colonic microflora. Bacterial amylases are the predominant means of starch hydrolysis in the colon, produced by a number of starch-degrading resident flora in the large intestine of humans such as *Bifidobacterium, Bacteroides, Fusobacterium,* etc., while *Clostridium butyricum* is one of the predominant starch-degrading bacteria in the animal intestine (Wang, *et al.* 1999).

7.2.1 Maltose and Glycolysis

Metabolic systems for maltose have been thought to be conserved in many taxa of organisms, including those ranging from bacteria to humans. Thus, the study of maltose chemistry and biochemistry in lower organisms produces useful model systems for study in higher organisms, including humans. Maltose results from starch metabolism in plants and bacteria. In plants, the breakdown of transitory starch has been extensively studied, owing to its importance in plant response to environmental stresses (Lu and Sharkey 2006). The breakdown of transitory starch takes place in chloroplasts by debranching enzymes into glucose and maltose (Lu and Sharkey 2006). In bacteria, the utilization of polymeric organic compounds such as the starch is accomplished by exoenzymes that are exported across the cytoplasmic membrane to the outside of the cell. The products of exoenzymes action such as the dextrins and disaccharides are transported into the cell by specific transporters, where further metabolism takes place and which involves the enzymes amylomaltase (MalQ), maltodextrin phosphorylase (MalP) and glucokinase (Boos and Schuman 1998). Together these result in the formation of G1P and G6P from maltose and maltodextrins which undergo glycolysis (Figure 7.2).

7.2.2 Enzymes Involved in Maltose Metabolism

Three cytoplasmic enzymes are involved in the conversion of maltose and maltodextrins into glucose and glucose-1-phosphate namely MalQ

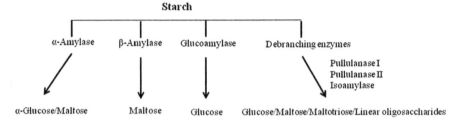

Figure 7.2 The action of various enzymes on starch and their end products.

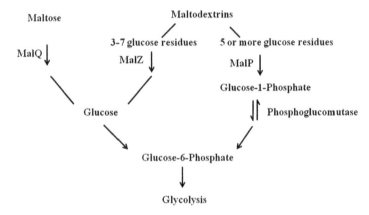

Figure 7.3 Enzymes involved in the metabolism of maltose and their products.

(amylomaltase), MalP (maltodextrin phosphorylase) and MalZ (maltodextrin glucosidase), as shown in Figure 7.3 (Boos and Schuman 1998). When maltose is being transported to inside of the cell, the enzyme MalQ transfers dextrinyl residues onto it forming glucose. MalP does not act on maltodextrins shorter than a maltopentaose. The action of this enzyme on maltodextrins results in the formation of a product dextrin shorter by one glucose residue and an α-glucose-1-phosphate, which is termed as phosphorolysis. MalZ acts on maltodextrins with 3–7 glucose residues (maltoheptose and smaller maltodextrins), releasing single glucose residues from the reducing end. Maltose is not a substrate for MalZ and the smallest maltodextrin hydrolyzed by it is a maltotriose. The end products of maltose metabolism, *i.e.*, glucose and α-glucose-1-phosphate, require further conversion to enter the glycolytic cycle. The enzyme glucokinase (GCK) phosphorylates glucose into glucose-6-phosphate, while the enzyme phosphoglucomutase (PGM) converts α-glucose-1-phosphate into glucose-6-phosphate.

7.3 Maltose Transport Systems

7.3.1 Passive Transport and Channels for Maltose

Passive transport systems allow translocation of small solute molecules such as maltose, across the membrane down their solute concentration in which the solute concentrations on both sides of the membrane reach an equilibrium such that the net solute movement is zero. A key passive transport system for maltose is the extensively studied LamB channel, known also as maltoporin (Klebba 2002; Ranquin and Van Gelder 2004). This transport system was initially discovered in the bacterium *Escherichia coli* as an outer membrane component serving also as a receptor for phage λ. The *lamB* gene from *E. coli* was cloned and found to be a component of the so-called maltose regulon, MalT. The nucleotide sequence of *lamB* was determined, and its gene product physiologically characterized in terms of both transport kinetics and downhill transport properties. In addition to harboring a binding site for maltose, LamB binds maltodextrin derivatives and other structurally unrelated small molecules. LamB binds to, but does not transport, starch.The LamB protein was purified to homogeneity and a high resolution crystal structure (3.1 Å) was determined in 1995 (Schirmer and Cowan 1995). The LamB crystal structure showed a trimeric arrangement (three identical monomers, called β-barrels), in which each β-barrel monomer had 18 transmembrane domains composed of anti-parallel β-strands (Schirmer *et al.* 1993) and where the transmembrane β-strands were linked continuously by extra-membranous loops. Recent work showed that down-regulation of LamB expression resulted in the enhancement of bacterial resistance to antimicrobial agents (Lin *et al.* 2010). Presumably, in such host cells lacking LamB, cellular entry cannot occur for small-molecule antimicrobial agents.

7.3.2 Primary Active Transporters for Maltose

The principal mode of energy for primary active transporters is the hydrolysis of ATP. Such ATP-driven solute transport systems are known as ATP-binding cassette (ABC) transporters and are homologous across many taxa, from bacteria to man (Jones *et al.* 2009). Such ABC transport systems, or pumps, have two distinctive functional domains: two ATPase enzyme domains and two transmembrane domains (TMD) which function as the solute transporter. The ATPase domain is called the nucleotide binding domain (NBD) and can often reversibly catalyze the hydrolysis of ATP into ADP and phosphate. The transporter domain, composed of the TMD, often has a precise solute specificity profile, and can range widely from amino acids, ions, nucleotides, antimicrobial agents, Krebs cycle intermediates, to a variety of structurally distinct sugars, including maltose.The TMDs of the ABC transporters can vary in their number of α-helices (*e.g.*, between 10 and 20 α-helices) that make up the integral membrane domains, depending on whether they are of the Type I (10-14 α-helices) or the Type II (20 or less α-helices) variety. A well-known and extensively studied ABC transporter for maltose, a Type I system, is the maltose/maltodextrin transport system (MalFGK$_2$E) from *E. coli* and from *Salmonella enterica*, both of which are Gram-negative bacteria. This MalFGK$_2$E transport system contains a maltose-binding protein, MalE, which resides in the periplasmic space of Gram-negative bacteria, two pore-forming subunits, called MalF and MalG, which constitute the component that translocates maltose across the membrane, and two molecules of MalK (MalK$_2$), which harbor the NBD and hydrolyzes ATP during maltose transport (Jones *et al.* 2009). In summary, extracellular maltose will enter a Gram-negative bacterium through its outer membrane (OM) *via* a maltoporin channel (*e.g.* LamB) that resides in this OM; then a maltose-binding periplasmic protein (*e.g.* MalE) will shuttle the maltose across the periplasm to the inner membrane (IM), where entry of maltose into the cytoplasm is conferred by MalFG; during this maltose transport across the IM, ATP is hydrolyzed by MalK$_2$. Intracellular maltose would subsequently be metabolized for cellular growth and homeostasis.

Although less well-known, another maltose primary active transport system, MAL, has been found in the fungus, a eukaryote, called *Saccharomyces cerevisiae* (Charron *et al.* 1986). This MAL locus confers the ability to utilize maltose in yeast and harbors genes for a maltose transporter, *MALX1*, a maltase enzyme, *MALX2,* and a transcriptional activator, *MALX3*. In this system, addition of maltose allows its binding to MALX3, which in turn results in the induction of MALX1 and MALX2 expressions in order to confer maltose transport across the membrane and its subsequent metabolic utilization inside the cell (Naumov *et al.* 1994).

7.3.3   Secondary Active Transporters for Maltose

Secondary active transporters utilize cation gradients as the principle mode of energy for active transport of solutes. Dietary maltose in the gut will be hydrolyzed by the enzyme maltase to yield 2 molecules of glucose, which then enter the gut epithelial cells via a sodium/glucose symporter, such as SLGT1 (Nigam and Singh 1995). A putative maltose secondary transporter, MaltP, has been found in the bacterium *Bacillus subtilis* (Tangney *et al.* 1992). Transport of maltose, however, has not yet been demonstrated for this particular protein. More recent work from our laboratory found that the sucrose permease from *E. coli*, CscB, also transports maltose (Peng *et al.* 2009). Prior to this finding, it was generally thought that the solute specificity of CscB was extremely limited to, namely, the sugar sucrose. To our knowledge, therefore, transport of maltose by CscB represents the first bacterial secondary active transporter that has been demonstrated to actively transport and accumulate maltose across the membrane against its maltose concentration gradient (Peng *et al.* 2009). Future work will focus on conducting structure-functional analyses in order to determine whether the sugar binding sites of CscB are functionally and structurally conserved in its closely-related homologs, the lactose permease of *E. coli*, LacY, and in the melibiose permease of *E. cloacae*, MelY.

7.3.4 Group Translocation Transport Systems for Maltose

Also known as the phosphoenolpyruvate-dependent phosphotransferase system (PTS), the group translocation system for maltose catalyzes the phosphorylation of maltose as it enters the cell (Postma *et al.* 1993). This system is unlike those commonly found in humans in that it uses the highly activated PEP molecule's phosphate group to phosphorylate the transport protein and then transfer the phosphate to the sugar moiety, completing the transport cycle. Molecules of ATP would seem to be the most likely source of high-energy phosphate groups since the phosphoryl group of PEP has a high free energy of hydrolysis (Postma *et al.* 1993); the product, however, of the de-phosphorylated PEP is pyruvate, which can then shuttle back into metabolism decreasing the energetic cost of phosphoryl hydrolysis. The transport mechanism also reduces that free energy cost by using PEP to phosphorylate the transport protein and to accomplish transport, finally transferring the same phosphate to the translocated sugar (Postma *et al.* 1993). Such ATP-driven transporters typically hydrolyze at least a single ATP molecule to facilitate transport and another ATP molecule to phosphorylate the sugar substrate so the sugar can enter the metabolism (Postma *et al.* 1993).

PTS transporters consist of Enzyme I (EI), which is often a highly soluble component that accomplishes the phosphorylation of carbohydrates transported by PTS in a given organism (Postma *et al.* 1993). Enzymes II (EIIs) are specific to the carbohydrate transported by the individual PTS system and consist of three components denoted A, B, and C. The EIIA and EIIB can be either free or membrane bound and serve as phosphate transfer centers, and EIIC forms the translocation channel and seems to account for substrate specificity (Pas and Robillard 1988).

The maltose PTS systems exhibit inducer exclusion of the transport and metabolism of other sugars such as lactose, melibiose, and the carbohydrate alcohol glycerol, which are PTS substrates and non-PTS substrates (Saier and Roseman 1972). This action is believed to occur *via* the binding of maltose PTS EIIA to target transporters which has been demonstrated *in vitro* and indicated *in vivo* (Saier and Roseman 1976). In general, PTS systems are also capable of regulating adenylate cyclase in *E. coli* cells that have been toluenized (Peterkofsky and Gazdar 1975). The role that maltose transport may play in limiting the ability of host microbial flora to utilize carbohydrates such as lactose, glucose, and glycerol has not been well investigated as it pertains to potential human disease of the gastrointestinal tract, but may be a promising avenue for research.

7.4 Mutations that Confer Maltose Transport in other Sugar Transporters

It is well established that solute transporters of the major facilitator superfamily share conserved amino acid sequences, similarities in protein structures, and common evolutionary origins in bacteria, plants, fungi, and humans

(Henderson *et al.* 1993). Thus, it has been thought for many years that study of these transport systems in lower organisms, such as bacteria, pertain to the molecular biology and biochemistry of their homologous transport systems in humans, as well. Hence, bacterial transport systems are useful models for the study of human solute transport across the membrane. The bacterial sugar transporters to be discussed below do not utilize maltose as a 'natural' transportable substrate. However, various laboratories have studied mutations within each of these transporters in which maltose binding and transport has been conferred (Brooker and Wilson 1985; Varela and Wilson 1996). These studies are briefly summarized below.

7.4.1   The Lactose Permease of *E. coli* and Maltose

The milk sugar lactose is a galactopyranoside molecule consisting of a glucose monomer linked to a galactose monomer via a β-1,4 glycosidic bond to form 4-O-β-D-galactopyranosyl-D-glucose. Lactose enters bacterial cells by being transported across the IM of *E. coli via* a secondary active transporter called the lactose permease, LacY (Varela and Wilson 1996), the gene of which is contained with the so-called *lac* operon, a regulated and inducible locus that confers lactose transport and utilization in *E. coli* (Rickenberg *et al.* 1956). The sugar substrate specificity profile for LacY is diverse in that both α-galactoside and β-galactoside sugars, plus synthetically-modified derivatives, such as *O*-nitrophenylgalactoside (ONPG), thiomethylgalactoside (TMG), and thiodigalactoside (TDG) (Varela and Wilson 1996), make good substrates for transport across the membrane *via* LacY.

Shuman and Beckwith (1979) were first the first investigators to isolate bacterial mutants harboring LacY in which enhanced transport of maltose was conferred. Subsequent sequencing of the *lacY* gene from these mutants, and DNA from newer maltose-positive mutants isolated by Brooker and Wilson (1985), revealed alterations in the transmembrane domains of LacY (Varela and Wilson, 1996). In particular, Alanine-177 of LacY was changed to either Threonine or Valine, and Tyrosine-236 was changed to Asparagine, Histidine, Phenylalanine, or Serine (Brooker and Wilson 1985). Alanine-177 resides within helix 6, and Tyrosine-236 resides in helix 7 of the lactose permease, suggesting that the amino acids that bind maltose lie within the integral membrane region of the permease. Furthermore, Alanine has a non-polar aliphatic side chain whereas Tyrosine has a bulky aromatic side chain. Since Alanine and Tyrosine bind and transport the α-galactosides, β-galactosides and synthetic derivatives mentioned above, but not maltose, it indicates that minor alterations in amino acid residues confer significant changes in sugar selection. Likewise, because maltose transport is enhanced with mutations to Valine (has a non-polar aliphatic side chain), Histidine (positively charged side chain), Phenylalanine (aromatic side chain), and Asparagine, Threonine, and Serine (each has polar uncharged side chains), these side chain biochemical properties are necessary for the binding of maltose and translocation across the membrane. These latter amino acids are, therefore, good candidates for use in the designing of novel or distinct proteins in order to interact with maltose for other industrial and biomedical purposes.

7.4.2 The Melibiose Permease of *Enterobacter cloacae* and Maltose

The sugar melibiose, 6-O-α-D-galactopyranosyl-D-glucose, consists of a galactose monomer and a glucose monomer linked together in a 1,6 α-linkage. The gene encoding the melibiose permease from *E. cloacae* was cloned in the laboratory of Tsuchiya (Okazaki *et al.* 1997a) and found to be homologous to LacY (Okazaki *et al.* 1997b). In work from our laboratory, we found mutants harboring MelY that fermented maltose better than cells with wild-type MelY (Shinnick *et al.* 2003). After sequencing the *melY* gene on DNA isolated from the mutants, we found that Leucine-88, Leucine-91 and Alanine-182 of MelY each had a Proline amino acid in their places, instead. These mutations in MelY each conferred enhanced transport of maltose in host cells harboring the altered melibiose permeases (Shinnick *et al.* 2003). Prolines that lie within intramembraneous α-helices of membrane-bound proteins (*e.g.*, in transporters) will introduce kinks, or bends, in these α-helices (Varela *et al.* 1995). Such molecular configurations are thought to provide conformational changes that occur during the transport of solutes through transporters (Green *et al.* 2000). Further, it is interesting to note that Alanine-177 of LacY is conserved as Alanine-182 of MelY (Shinnick *et al.* 2003; Varela and Wilson 1996). This conservation of the amino acid Alanine in transporters with distinctive sugar selection profiles suggests that evolution has preserved this locus within distinctive transporters for the purposes of binding and facilitating transport of structurally different sugars, depending on the type of amino acid residue that occupies this particular site.Thus, this residue location within homologous and closely-related transporters will be an important location for future modulation to confer transport of desirable sugar substrates, such as maltose.

7.4.3 The Raffinose Permease of *E. coli* and Maltose

The sugar raffinose, O-α-D-galactopyranosyl-(1 → 6)-α-D-glucopyranosyl β-D-fructofuranoside, is a trisaccharide sugar; it consists of a monomer of galactose, glucose, and fructose. Alternatively, one may choose to view this raffinose sugar as a combination of a sucrose molecule with a glucose, or a melibiose with a fructose. Raffinose transport activity in *E. coli* was first studied by Schmid and Schmitt in 1976. The *rafB* gene that codes for the raffinose transporter (RafB) was found to reside in a naturally-occurring plasmid in an *E. coli* isolate and was cloned by Aslanidis *et al.* in 1989. In subsequent work conducted in our laboratory, a search for bacterial mutants harboring RafB that either grew better than cells with wild-type RafB on maltose minimal media, or that fermented maltose faster than the wild-type on indicator media, resulted in the isolation of several candidate mutants for further study

(Van Camp *et al.* 2007). Upon nucleotide sequencing of the *rafB* genes from these candidate mutants, three classes of mutations were found. The first type of mutation, called MT-1, showed Valine-35 of RafB changed to Alanine. The second type of mutation, MT-2, had Isoleucine-391 changed to Serine. The third type of mutation, MT-3, had three mutations: Serine-138 changed to Aspartate; Serine-139 changed to Leucine; and Glycine-389 changed to Alanine (Van Camp *et al.* 2007). All three types of mutations exhibited enhancement of maltose transport across the membrane (Van Camp *et al.* 2007).The replacement residues have side chains that are non-polar aliphatic (Alanine, Leucine and Valine), polar uncharged (Serine), or negatively-charged (Aspartate). Thus, such residues confer structural configurations and biochemical features that permit enhanced molecular interactions with maltose such that its transport across the membrane proceeds in an energy-driven mode.

Summary Points

- This chapter focuses on the metabolism and transport of the sugar maltose.
- The molecular mechanisms for the metabolism of maltose and transport across the membrane are thought to be evolutionarily conserved across many taxa.
- Maltose as a dietary source for man can be found largely in vegetables, fruits, and grains.
- The transport of maltose across the cellular membrane involves both passive and active transport systems that are homologous between lower and higher organisms.
- The selection of maltose as a substrate for transport can be induced by minor sequence alteration in homologous sugar transport systems.

Key Facts of Maltose Chemistry and Biochemistry

- Maltose is a disaccharide sugar consisting of 2 glucose molecules attached by a glycosidic bond and represents an important metabolite for living organisms.
- Starch is a common storage form for maltose.
- Many aspects of the metabolic and transport machinery for maltose share a common evolutionary origin, ranging from bacteria to man.
- Maltose is liberated from starch using a variety of enzymes, including α-amylase, β-amylase, glucoamylase, and debranching enzymes.
- Passive transport systems for maltose consist of LamB as a key model protein for study.
- Active transport systems of maltose include both primary and secondary transport systems.
- Primary active transport systems for maltose include MalFGK$_2$E.
- A secondary active maltose transport system is CscB.

- Maltose group translocation systems include the maltose PTS.
- Mutations in a variety of other sugar transporters have resulted in the enhancement of maltose transport.

Definition of Words and Terms

Active transport: movement of solutes, *e.g.*, maltose, across the membrane against the solute gradient such that the solutes accumulate on one side of the membrane. This process will utilize either ATP (primary active transport) or an ion gradient (secondary active transport) to mediate this solute accumulation against the solute gradient.

Amylase: a general term given to several distinct classes of enzymes, such as α-amylase and β-amylase, which partake in the degradation of starch to produce smaller sugar subunits.

Amylose: linear polymer of glucose molecules linked together by α-1,4-glycosidic bonds.

Amylopectin: a form of starch in which glucose molecules are linked in chains (α-1,4 glycosidic bonds) and branches (α-1,6 glycosidic bonds).

Exoenzymes: enzymes that are synthesized inside a cell and exported to the extra-cellular matrix, where they are active.

Glycosidic bond: a chemical bond between an anomeric carbon atom of one sugar and an oxygen atom from a hydroxyl group (–OH), to form an *O*-glycosidic linkage to another sugar molecule. Glycosidic bonds may be in the α- or β-configuration and are numbered according to the carbon numbers of the attached sugars.

Limit dextrins: short dextrins with many 1,6-glycosidic bonds produced by the incomplete digestion of amylopectins by α-amylase.

Maltodextrins: short polymers of 7 or 8 glucose molecules linked by α1,4-glycosidic bonds.

Passive transport: movement across the membrane of solutes down the solute concentration gradient until equilibrium, *i.e.*, solute concentrations are the same on both sides of the membrane, is reached. No biological energy is expended in this passive transport process.

Starch: a macromolecular carbohydrate consisting of chains of monosaccharide polymers, often with branches.

Transmembrane domain: a portion of an integral membrane protein that traverses the biological membrane.

Transport machinery: integral membrane proteins that serve to translocate water-soluble solutes across the biological membrane.

List of Abbreviations

ABC	ATP-binding cassette
ATP	adenosine-5′-triphosphate
CscB	chromosomally-encoded sucrose permease B

IM	inner membrane
LacY	lactose permease transport system from *Escherichia coli*
LamB	maltoporin
OM	outer membrane
OmpR	osmoregulator
NBD	nucleotide binding domain
MAL	maltose utilization locus from the chromosome of *Saccharomyces cerevisiae*
MalFGK$_2$E	maltose/maltodextrin transport system
MalE	periplasmic maltose-binding protein
MalF	maltose transport subunit F
MalG	maltose transport subunit G
MalK	ATPase subunit
MalT	maltose regulon
MaltP	maltose secondary active transporter from *Bacillus subtilis*
MALX1	maltose transporter from *Saccharomyces cerevisiae*
MALX2	maltase enzyme from *Saccharomyces cerevisiae*
MALX3	transcriptional activator from *Saccharomyces cerevisiae*
MelY	melibose permease transport system from *Enterobacter cloacae*
ONPG	*O*-nitrophenylgalactoside
PTS	phosphoenolpyruvate-dependent phosphotransferase system
RafB	raffinose permease transport system from *Escherichia coli*
TDG	thiodigalactoside
TMG	thiomethylgalactoside
TMD	transmembrane domain

References

Aslanidis, C., Schmid, K., and Schmitt, R., 1989. Nucleotide sequences and operon structure of plasmid-borne genes mediating uptake and utilization of raffinose in *Escherichia coli. Journal of Bacteriology.* 17: 6753–6763.

Boos, W., and Shuman, H., 1998. Maltose/maltodextrin system of *Escherichia coli*: transport, metabolism, and regulation. *Microbiology and Molecular Biology Reviews.* 62: 204–229.

Brooker, R.J., and Wilson, T.H., 1985. Isolation and nucleotide sequencing of lactose carrier mutants that transport maltose. *Proceedings of the National Academy of Sciences USA.* 82: 3959–3963.

Charron, M.J., Dubin, R.A., and Michels, C.A., 1986. Structural and functional analysis of the MALI locus of *Saccharomyces cerevisiae. Molecular and Cellular Biology.* 6: 3891–3899.

Green, A.L., Anderson, E.J., and Brooker, R.J., 2000. A revised model for the structure and function of the lactose permease. Evidence that a face on transmembrane segment 2 is important for conformational changes. *Journal of Biological Chemistry.* 275: 23240–23246.

Henderson P.J., Roberts, P.E., Martin, G.E., Seamon, K.B., Walmsley, A.R., Rutherford, N.G., Varela, M.F., and Griffith, J.K., 1993. Homologous sugar-transport proteins in microbes and man. *Biochemical Society Transactions*. 21: 1002–1006.

Jones, P.M., O'Mara, M.L., and George, A.M., 2009. ABC transporters: a riddle wrapped in a mystery inside an enigma. *Trends in Biochemical Sciences*. 34: 520–531.

Klebba, P.E., 2002. Mechanism of maltodextrin transport through LamB. *Research in Microbiology*. 153: 417–424.

Lin, X.M., Yang, J.N., Peng, X.X., and Li, H., 2010. A novel negative regulation mechanism of bacterial outer membrane proteins in response to antibiotic resistance. *Journal of Proteome Research*. 9: 5952–5959.

Lu, Y., and Sharkey, T.D., 2006. The importance of maltose in transitory starch breakdown. *Plant, Cell & Environment*. 29: 353–366.

Naumov, G.I., Naumova, E.S., and Michels, C.A., 1994. Genetic variation of the repeated MAL loci in natural populations of *Saccharomyces cerevisiae* and *Saccharomyces paradoxus*. *Genetics*. 136: 803–812.

Nigam, P., and Singh, D., 1995. Enzyme and microbial systems involved in starch processing. *Enzyme and Microbial Technology*. 17: 770–778.

Okazaki, N., Jue, X.X., Miyake, H., Kuroda, M., Shimamoto, T., and Tsuchiya, T., 1997a. A melibiose transporter and an operon containing its gene in *Enterobacter cloacae*. *Journal of Bacteriology*. 179: 4443–4445.

Okazaki, N., Jue, X.X., Miyake, H., Kuroda, M., Shimamoto, T., and Tsuchiya, T., 1997b. Sequence of a melibiose transporter gene of *Enterobacter cloacae*. *Biochimica et Biophysica Acta*. 1354: 7–12.

Pas, H.H., and Robillard, G.T., 1988. Enzyme II^{Mtl} of the *Escherichia coli* phosphoenolpyruvate-dependent phosphotransferase system: identification of the activity linked cysteine on the mannitol carrier. *Biochemistry*. 27: 5515–5519.

Peng, Y., Kumar, S., Hernandez, R.L., Jones, S.E., Cadle, K.M., Smith, K.P., and Varela, M.F., 2009. Evidence for the transport of maltose by the sucrose permease, CscB, of *Escherichia coli*. *Journal of Membrane Biology*. 228: 79–88.

Peterkofsky, A., and Gazdar, C., 1975. Interaction of enzyme I of the phosphoenolpyruvate:sugar phosphotransferase system with adenylate cyclase of *Escherichia coli*. *Proceedings of the National Academy of Science*. 72: 2920–2924.

Postma, P.W., Lengeler, J.W., and Jacobson, G.R., 1993. Phosphoenolpyruvate: Carbohydrate Phosphotransferase Systems in Bacteria. *Microbiological Reviews*. 57: 543–594.

Raben, N., Plotz, P., and Byrne, B.J., 2002. Acid alpha-glucosidase deficiency (glycogenosis type II, Pompe disease). *Current Molecular Medicine*. 2: 145–166.

Ranquin, A., and Van Gelder, P., 2004. Maltoporin: sugar for physics and biology. *Research in Microbiology*. 155: 611–616.

Ray, R.R., and Nanda, G., 1996. Microbial beta-amylases: biosynthesis, characteristics, and industrial applications. *Critical Reviews in Microbiology*. 22: 181–199.

Rickenberg, H.V., Cohen, G.N., Buttin, G., and Monod, J., 1956. La Galacto-side perméase of d' *Escherichia coli. Annales de l'Institut Pasteur.* 91: 829–857.

Saier, M.H., Jr., and Roseman, S. 1972. Inducer exclusion and repression of enzyme synthesis in mutants of *Salmonella typhimurium* defective in enzyme I of the phosphoenolpyruvate:sugar phosphotransferase system. *Journal of Biological Chemistry.* 247: 972–975.

Saier, M.H., Jr., and Roseman, S., 1976. Sugar transport. The *crr* mutation: its effect on repression of enzyme synthesis. *Journal of Biological Chemistry.* 251: 6598–6605.

Schirmer, T., and Cowan, S.W., 1993. Prediction of membrane spanning β-strands and its application to maltoporin. *Protein Science.* 2: 1361–1363.

Schirmer, T., Keller, T.A., Wang, Y.F., and Rosenbusch, J.P., 1995. Structural basis for sugar translocation through maltoporin channels at 3.1Å resolution. *Science.* 267: 512–514.

Schmid, K., and Schmitt, R., 1976. Raffinose metabolism in *Escherichia coli* K12. Purification and properties of a new alpha-galactosidase specified by a transmissible plasmid. *European Journal of Biochemistry.* 67: 95–104.

Shinnick, S.G., Perez, S.A., and Varela, M.F., 2003. Altered substrate selection of the melibiose transporter (MelY) of *Enterobacter cloacae* involving point mutations in Leu-88, Leu-91, and Ala-182 that confer enhanced maltose transport. *Journal of Bacteriology.* 185: 3672–3677.

Shuman, H., and Beckwith, J., 1979. *Escherichia coli* K-12 mutants that allow transport of maltose via the β-galactoside transport system. *Journal Bacteriology.* 137: 365–373.

Tangney, M., Buchanan, C J., Priest, F.G., and Mitchell, W.J., 1992. Maltose uptake and its regulation in *Bacillus subtilis. FEMS Microbiology Letters.* 97: 191–196.

Van Camp, B.M., Crow, R.R., Peng, Y., and Varela, M.F., 2007. Amino acids that confer transport of raffinose and maltose sugars in the raffinose permease (RafB) of *Escherichia coli* as implicated by spontaneous mutations at Val-35, Ser-138, Ser-139, Gly-389 and Ile-391. *Journal of Membrane Biology.* 220: 87–95.

Varela, M.F., and Wilson, T.H., 1996. Molecular biology of the lactose carrier of *Escherichia coli. Biochimica et Biophysica Acta.* 1276: 21–34.

Varela, M.F., Sansom, C.E., and Griffith, J.K., 1995. Mutational analysis and molecular modeling of an amino acid sequence motif conserved in anti-porters but not symporters in a transporter superfamily. *Molecular Membrane Biology.* 12: 313–319.

Wang, X., Conway, P.L., Brown, I.L., and Evans, A.J., 1999. *In vitro* utilization of amylopectin and high-amylose maize (amylomaize) starch granules by human colonic bacteria. *Applied and Environmental Microbiology.* 65: 4848–4854.

CHAPTER 8

Fructose Chemistry

DAVID J TIMSON

School of Biological Sciences, Queen's University Belfast, Medical Biology
Centre, 97, Lisburn Road, Belfast, BT9 7BL, UK
Email: d.timson@qub.ac.uk

8.1 Introduction: Structure and Stereochemistry

The word "fructose" is derived from "*fructus*" (Latin), meaning fruit.
This reflects the compound's natural occurrence in many different types of
fruit. Fructose is of importance in the diet, especially in Western nations.
This is due to the extensive use of sucrose (a disaccharide of glucose and fructose)
as a sweetener and the increasing use of high fructose corn syrups in processed
foods. Fructose is a hexose monosaccharide with the molecular formula
$C_6H_{12}O_6$. Like all hexose monosaccharides, fructose can exist in a non-cyclic,
straight chain form. In this form, the compound has a ketone group at carbon-2
in the chain. There are two enantiomers of fructose; biochemists have traditionally named these the so-called L- and D-forms (see Figure 8.1 and Key Facts
section). These two stereoisomers are distinguished by the configuration at
carbon 5 and are named based on their similarity to either the L- or D-form of
glyceraldehyde. In nature, D-fructose is found almost exclusively.

In solution and in the crystalline form, the straight chain form makes up a tiny
minority of the overall population of molecules (typically less than 1% [Sinnott
2007]). The majority take up various cyclic forms. These cyclic forms arise due to
intra-molecular reactions between the carbonyl group and hydroxyl groups
within fructose. Two main outcomes are possible. If the carbonyl group reacts
with the hydroxyl attached to carbon 5, a five-membered, furanose ring is formed

Food and Nutritional Components in Focus No. 3
Dietary Sugars: Chemistry, Analysis, Function and Effects
Edited by Victor R Preedy
© The Royal Society of Chemistry 2012
Published by the Royal Society of Chemistry, www.rsc.org

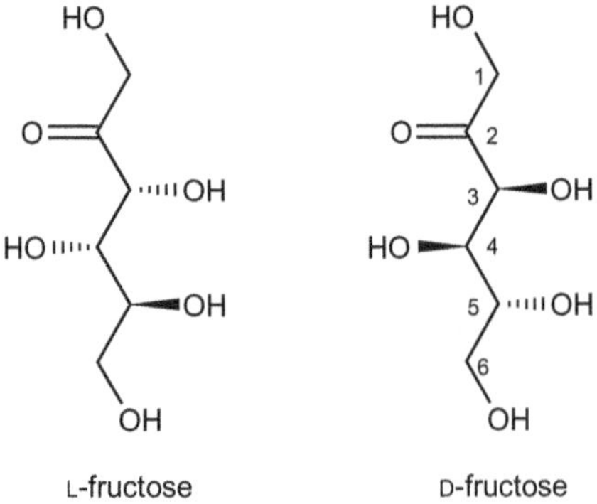

Figure 8.1 The D- and L-isomers of the straight chain form of fructose. The carbon atoms are numbered according to convention in the biologically more common form, D-fructose.

containing four carbon atoms and one oxygen (see Figure 8.2 and Key Facts). Alternatively, the carbonyl group can react with the hydroxyl group attached to carbon six, giving rise to a six-membered, pyranose ring containing five carbons and one oxygen (Figure 8.2 and Key Facts). Many textbooks give the misleading impression that the furanose form predominates in solution at room temperature and near-neutral pH. In fact, the pyranose form accounts for three-quarters of fructose molecules at 27 °C in aqueous solution (Sinnott 2007).

When the carbonyl group reacts internally, it creates another chiral centre. There are, thus, two possible stereoisomers resulting from these reactions. These are known as the α- and β-anomers. Consequently, there are four main stereoisomers of D-fructose: α-D-fructopyranose, β-D-fructopyranose, α- D-fructofuranose and β-D-fructofuranose (Figure 8.2 and Key Facts). Each of these stereoisomers has different optical rotatory properties; therefore, interconversions between them can be monitored by measurement of the optical rotation in a polarimeter. These interconversions between the α- and β-forms, and between the pyranose and furanose forms generally occur at relatively high rates in aqueous solution. Consequently, dissolving a pure form of any of these stereoisomers in water will result in an equilibrium mixture of the various forms. Interconversions between the various forms (mutarotation) can be catalysed by either acid or base and always occur via a straight-chain intermediate. In many cells, the rate of mutarotation is insufficient to support supply of the required stereoisomer for metabolic reactions. To overcome this problem, mutarotase enzymes catalyse interconversion – typically with very high turnover numbers ($> 10\,000$ s^{-1}). Many mutarotase enzymes have quite broad specificity and to date there have been no reports of specific fructose mutarotases. However, some broad specificity mutarotases can catalyse the isomerisation of fructose (*e.g.* the *Aspergillus niger* enzyme [Kinoshita *et al.* 1991]). Given that fructose is relatively rarely tested as a substrate, it is likely that more mutarotases are able to catalyse reaction with this compound.

The acid-promoted mutarotation of monosaccharides requires the initial protonation of the ring oxygen of the ring form. This introduces a positive charge

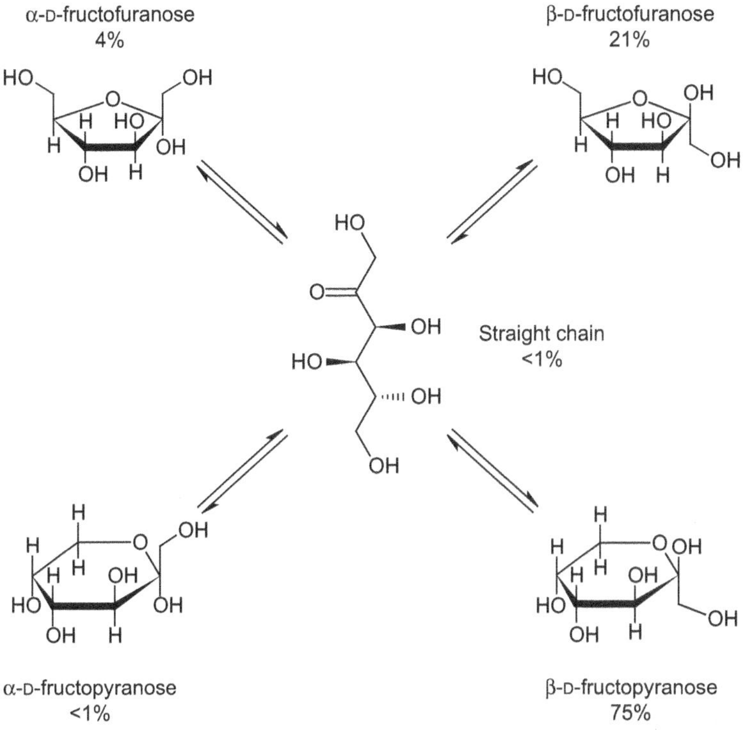

α-D-fructofuranose
4%

β-D-fructofuranose
21%

Straight chain
<1%

α-D-fructopyranose
<1%

β-D-fructopyranose
75%

Figure 8.2 The cyclic forms of D-fructose and their pathways of interconversion. Conversion between the α- and β-anomers requires passage through a straight-chain intermediate. The percentage abundance of the various stereoisomers at 27 °C is given below their structures (abundance values are from Sinnott, 2007).

into the ring, destabilising it. Consequently, the ring breaks between the oxygen and carbon 1, resulting in a straight chain form of the molecule. In this form, there is free rotation about carbon 1, so the hydroxyl group can reorientate. Therefore, reversal of the ring-opening event results in a random mixture of α- and β-forms, the ratio of which will be determined by the relative thermo-dynamic stabilities of the two forms (Figure 8.3). Enzyme catalysed mutarota-tion occurs *via* a similar mechanism in which a histidine group in the enzyme first donates a proton to the oxygen in the monosaccharide, and a glutamate or aspartate (acting as a general base) removes a proton from the C_1-OH. This opens the ring, permitting free rotation about carbon-1; reversal of the proton donation and acceptance steps reseals the ring (Thoden *et al.* 2003).

8.2 Chemical Reactions of Fructose

Although there have been extensive investigations into the chemical reactivity of glucose, much less work has been carried out on fructose. Nevertheless, many of the reactions are broadly similar in mechanism and, therefore, they will only be reviewed briefly here.

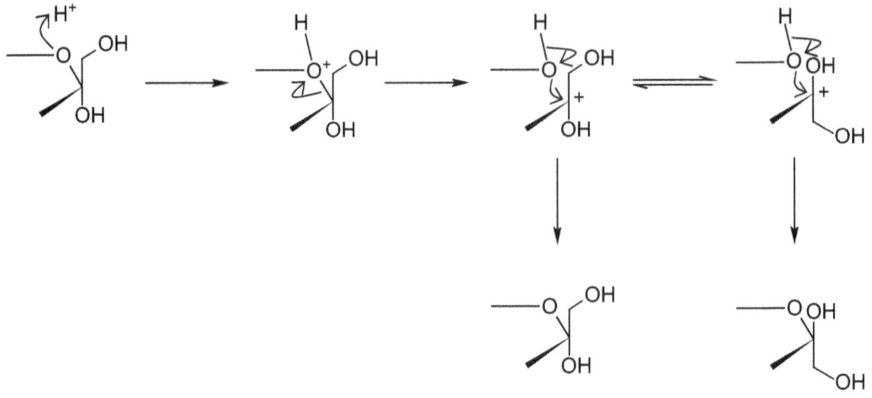

Figure 8.3   The mutarotation of hexose monosaccharides. In a general acid catalysed reaction (a), the process is initiated by the donation of a proton to the ring oxygen of the monosaccharide. Once the ring is opened there is free rotation of the groups around carbon-1 and, thus, the ring can be resealed in either configuration. When the reaction is catalysed by a mutarotase enzyme, the general acid is a protonated histidine in the active site and the reaction is further assisted by the removed of a proton from C_1-OH.

8.2.1   Oxidation and Reduction

The complete oxidation of fructose in oxygen yields carbon dioxide and water. This is the ultimate consequence of the biochemical oxidation of the compound in living organisms, albeit one which occurs in many, discrete metabolic steps (see below for details of the reactions which involve fructose). Reaction with other strong oxidising agents, such as permanganate, dichromate or and vanadate (V) ions yields a mixture of products including shorter chain monosaccharides, sugar acids, formic (methanoic) acid, formaldehyde and, ultimately, carbon dioxide (Figure 8.4a). The reaction mechanism is complex and involves free radicals. Furthermore, the kinetics and products can be controlled by the pH of the reaction mixture with acidic conditions generally promoting the rate of reaction (Fandis 1986). Fructose can be reduced by sodium borohydride to give an equimolar mixture of sorbitol and mannitol (Abdek-Akhar *et al.* 1951) (Figure 8.4b).

8.2.2   Modifications of the Hydroxyl Groups: Acetylation and Methylation

Reaction of fructose with acetic anhydride results in the acetylation of all free hydroxyl groups (Figure 8.5a). The reaction is often carried out in pyridine which acts as both a catalyst and solvent. Alternatively, a variety of metal ion catalysts can be used. The fully acetylated derivative of fructose is an important starting point for further chemical transformations of the molecule.

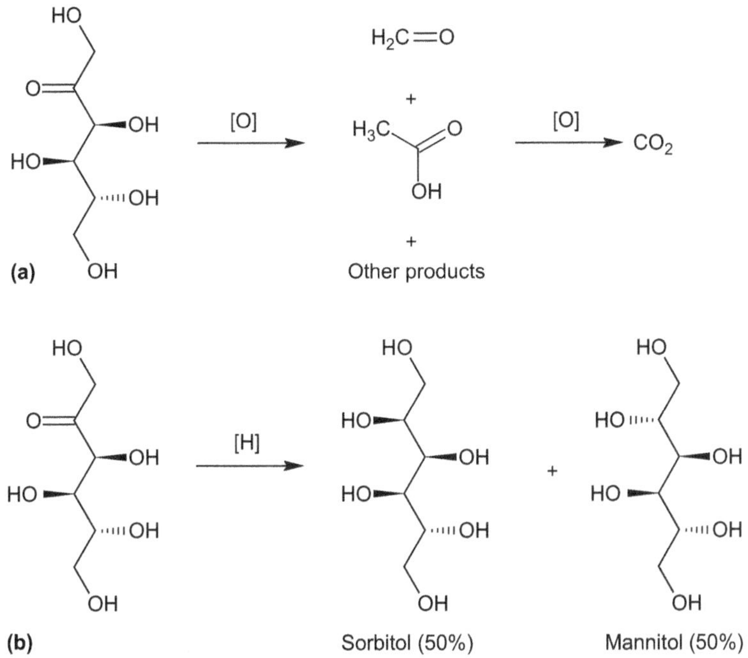

Figure 8.4 The oxidation and reduction of fructose. (a) Oxidation with strong oxidising agents (represented here as [O]) yields a complex mixture of products. The ultimate product is carbon dioxide. (b) Reduction with, for
example, sodium borohydride (represented here as [H]) results in an
equimolar mixture of the sugar alcohols sorbitol and mannitol.

Fructose can be methylated at the anomeric hydroxyl group by reaction with
methanol in dilute acid (Figure 8.5b). A mixture of α- and β-pyranose and α-
and β-furanose products are formed, known as methyl fructopyranosides and
methyl fructofuranosides respectively. The furanose products are formed more
quickly, but the pyranose ones are more stable thermodynamically (Bethell and
Ferrier 1973).

8.3 Key Metabolic Reactions Involving Fructose

The glycolytic pathway is the main, central, initial pathway for the oxidation of
glucose in energy metabolism. The first enzyme of this pathway, hexokinase,
has a strong preference for glucose as a substrate. Therefore, in cells which have
high concentrations of glucose, alternative pathways are required for the initial
metabolism of other monosaccharides, such as fructose. The aim of these
pathways is to convert these other monosaccharides to intermediates in the
glycolytic pathway. Thereafter, the carbon atoms can continue down this
pathway and, under aerobic conditions, enter the tricarboxylic acid cycle.
In higher organisms, the metabolism of glucose and other sugars occurs

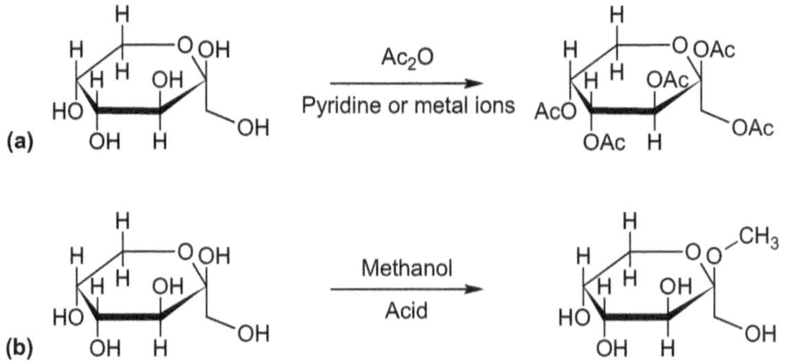

Figure 8.5 Modifications of the hydroxyl groups in fructose. (a) Acetylation can be achieved at all free hydroxyls by the action of acetic anhydride (Ac_2O). (b) Methylation at the anomeric hydroxyl can be achieved by reaction with methanol in acid. The most stable product, β-methyl fructopyranoside, (Bethell and Ferrier 1973) is shown here.

simultaneously and often in the same cell. In many bacteria and fungi, the metabolism of monosaccharides is stringently controlled at the genetic level. Often, glucose is the preferred carbon and energy source and when this compound is present, the genes encoding the enzymes for the degradation of alternative sugars are strongly repressed. Only under conditions of low, or zero, glucose are these genes expressed and, therefore, elaborate control mechanisms are required to ensure that these changes in gene expression occur rapidly and under the appropriate conditions.

8.3.1 Phosphorylation of Fructose Catalysed by Fructokinase

Fructokinase (ketohexokinase; EC 2.7.1.3) catalyses the phosphorylation of fructose at position 1 at the expense of the phosphate donor, ATP. This reaction is the main pathway for the utilisation of fructose in mammalian tissues which have high glucose concentrations and also in micro-organisms, which induce genes encoding the enzymes of fructose metabolism in response to metabolic conditions (see above). Defects in the gene encoding fructokinase can lead to the hereditary disease essential fructosuria (OMIM #229800) (Bonthron *et al.* 1994). The main symptom of the disease is the appearance of high concentrations of fructose in the blood and urine following ingestion of fructose, or di- and polysaccharides containing fructose residues. The condition is considered to be essentially benign.

Human fructokinase is an elongated dimer (Figure 8.6a) (Gibbs *et al.* 2010). The enzyme has one active site per subunit; however, these are not in the same conformation in the crystal structure – one is closed and catalytically competent, whereas the other is open. A potassium ion is required at the active site for catalytic activity. The mechanism of phosphorylation is believed to involve the abstraction of a proton from the C_1-OH of the sugar substrate by an aspartate

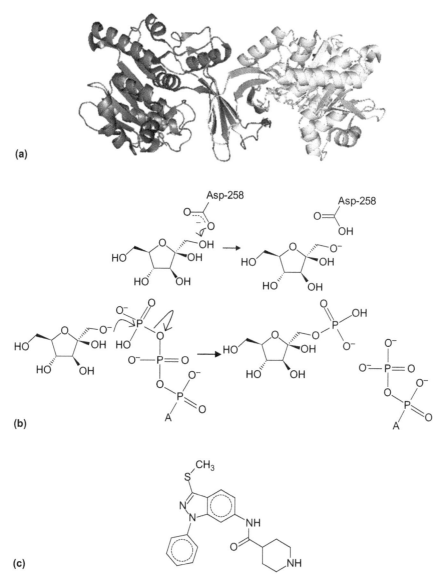

Figure 8.6 Fructokinase. (a) The overall fold of the dimeric enzyme with the two subunits shown in light and dark grey. The figure was produced from PDB file 3NBV (Gibbs *et al.* 2010) using PyMol (www.pymol.org). (b) The mechanism of this enzyme involves an initial attack by an aspartate residue and conversion of the C_1 hydroxyl group to an alkoxide ion which attacks the γ-phosphorus of ATP. A represents the adenosine moiety of ATP. (c) A promising lead compound for the treatment of non-insulin dependent diabetes by the inhibition of fructokinase. Its IC_{50} is 330 nM (Gibbs *et al.* 2010).

residue (Asp-258 in the human enzyme). This introduces a negative charge into the sugar, facilitating nucleophilic attack on the γ-phosphorus of ATP (Figure 8.6b). The enzyme only binds to fructose in the furanose form. The enzyme is considered as a potential target for the treatment of non-insulin dependent diabetes and a number of compounds have been discovered which inhibit the enzyme, including one promising lead (Figure 8.6c) (Gibbs *et al.* 2010).

Fructose 3-phosphate has also been detected in human cells. This appears to be something of a metabolic "dead end" with less that 30% of the compound disappearing in cultured cells incubated overnight in the absence of fructose (Petersen *et al.* 1992). The enzyme responsible catalysing this phosphorylation is likely to be fructosamine-3-kinase, the main physiological role of which is proposed to be the removal of fructose moieties from the surfaces of proteins (Delpierre *et al.* 2002).

8.3.2 Further Metabolism of Fructose by the Fructokinase Pathway

Fructose 1-phosphate is not an intermediate in the glycolytic pathway. Therefore, further metabolism is required before the carbon atoms can be processed by this route. The cleavage of fructose 1-phosphate to dihydroxyacetone phosphate and glyceraldehyde is catalysed by fructose 1-phosphate aldolase (aldolase B; EC 4.1.2.3). This isoform of the enzyme also catalyses the cleavage of fructose 1,6-*bis*phosphate in the main glycolytic pathway and is largely found in the liver. It is defects in this enzyme which are the primary cause of the genetic disease hereditary fructose intolerance (HFI) (Reviewed in Bouteldja and Timson 2010). Human aldolase B is a tetrameric enzyme (Figure 8.7a). Its complex mechanism requires the opening of the ring of fructose 1-phosphate to a straight chain form. A lysine residue in the active site then attacks the fructose 1-phosphate molecule at carbon 2, converting the ketone group into a Schiff's base, which is covalently linked to the enzyme through the lysine side chain. Cleavage of the bond between carbons 3 and 4 releases glyceraldehyde and protonation of the resulting carbanionic species followed by reversal of the Schiff base formation releases dihydroxyacetone phosphate (Choi *et al.* 2001).

Dihydroxyacetone phosphate enters glycolysis; however, glyceraldyhyde must be phosphorylated to glyceraldehyde 3-phosphate in a reaction catalysed by triose kinase (triokinase; EC 2.7.1.28). This enzyme, which also catalyses the phosphorylation of dihydroxyacetone, has been surprisingly poorly studied. No structure is available and only limited kinetic data have been published (Frandsen and Gunnet 1971). Putative, mammalian dihydroxyacetone kinases have been identified (Cabezas *et al.* 2005) and it is possible that these enzymes are the same ones which were purified from mammalian tissues and shown to have glyceraldehyde kinase activity in the 1970s. No experimentally determined structure of the putative human dihydroxyacetone kinase has been obtained; a molecular model (based on the *Lactococcus lactis* structure) is shown in Figure 8.7b.

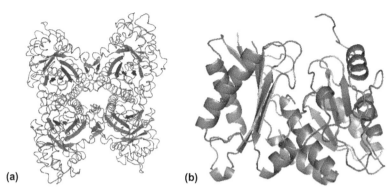

(a) **(b)**

Figure 8.7 The structures of enzymes responsible for the further metabolism of fructose 1-phosphate. (a) The human Aldolase B enzyme forms a tetramer; as visualised here, each subunit occurs at the vertex of a square. The figure was constructed using PDB file 1QO5 (Dalby *et al.* 2001). (b) A model of the human putative dihydroxyacetone kinase (EAW73939). This enzyme may also be responsible for catalysing the triose kinase reaction. The model is based on the *Lactococcus lactis* dihydroxyacetone kinase structure (PDB ID: 2IU6) (Christen *et al.* 2006) and was constructed using Phyre (http://www.sbg.bio.ic.ac.uk/~phyre/) (Kelley and Sternberg 2009).

8.3.3 Phosphorylation of Fructose Catalysed by Hexokinase

Under conditions of low glucose concentration, hexokinase (EC 2.7.1.1) can catalyse the phosphorylation of fructose directly to produce the glycolytic intermediate fructose 6-phosphate. This reaction is important in sperm cells, for which fructose is the main source of metabolic energy. No high resolution structure of hexokinase with fructose bound is available. However, to produce the product, fructose 6-phosphate, the hydroxyl group on carbon 6, must be available and not participating in the ring structure. Therefore, only the furanose forms of fructose are potential substrates for the hexokinase catalysed reaction. This is in contrast to the preferred substrate, glucose. This aldose sugar has a free C_6-OH group in its pyranose forms, and more than 99% of the molecules are in these configurations in aqueous solution at 30 °C (Sinnott 2007). Crystal structures confirm that glucose is bound in the pyranose form (Figure 8.8a) (Aleshin *et al.* 1998). Given that fructose cannot bind and participate in catalysis in this configuration, this suggests that there is considerable flexibility in the recognition sites of hexokinase enzymes.

Hexokinases are often subject to complex, allosteric regulation. Many organisms, including humans, have several isoforms to respond to different metabolic needs. The mechanism of hexokinase is similar to that of fructokinase. An aspartate residue (Asp-209 in human hexokinase I) acts as a general base and abstracts a proton from the C_6-OH of the substrate. This process is assisted by a positively charged lysine residue (Lys-173 in human hexokinase I), which lies on the opposite side of the hydroxyl to the aspartate residue

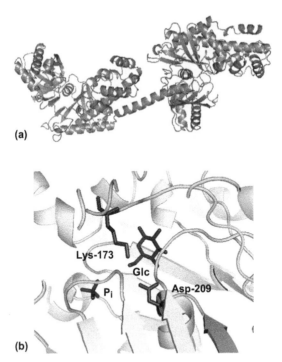

(a)

(b)

Figure 8.8   Hexokinase, an enzyme which can catalyse an alternative route for the
phosphorylation of fructose. (a) The overall fold of human hexokinase I
shows that the enzyme is an elongated monomer. (b) In the active site
critical aspartate and lysine residues lies on opposite sides of the C_6-OH in
order to facilitate removal of the proton. This structure contains glucose
(Glc) and a phosphate ion (P_i) in the active site. These figures were made
from PDB file 1HKC (Aleshin *et al*. 1998) using PyMol.

(Figure 8.8b) (Aleshin *et al*. 1998). The negatively charged intermediate can
then attack the γ-phosphorus of ATP, resulting in transfer of a phosphate
group to the sugar.

8.4   The Synthesis of Fructose

8.4.1   Chemical Synthesis of Fructose

The presence of several chiral centres in monosaccharides means that their
chemical synthesis from simple precursors is a challenging task. Furthermore,
the wide availability of fructose from natural sources means that there is no
compelling commercial need to develop such processes. Nevertheless, some
routes have been published, including one which achieves a 71% yield of
fructose by reaction of (*S*)-1:2,3-*O*-isopropylidene-L-glyceraldehyde (which can
be obtained from ascorbic acid) with dihydroxyacetone in the presence of
cinchonine (Figure 8.9) (Markert and Mahrwald 2008).

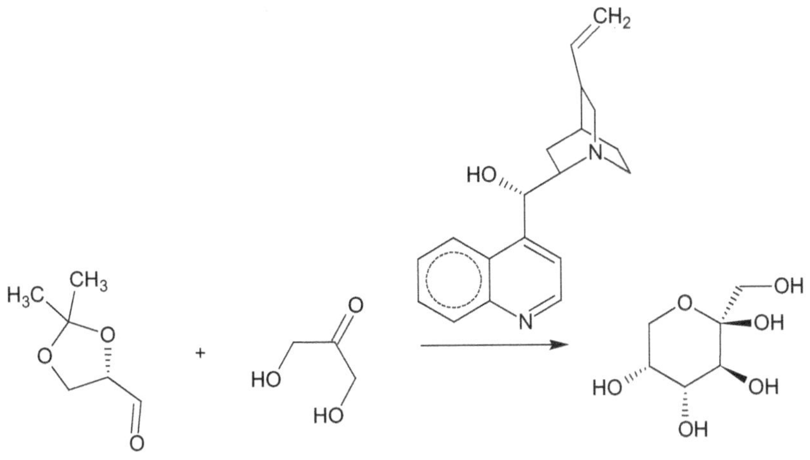

Figure 8.9 The chemical synthesis of fructose. The compound can be synthesied from (*S*)-1:2,3-*O*-isopropylidene-L-glyceraldehyde and dihydroxyacetone in the presence of cinchonine (Markert and Mahrwald 2008).

8.4.2 Biosynthesis of Fructose

Carbohydrate synthesis from simple precursors such as carbon dioxide is only possible in plants and some microbes. In plants, fructose 1,6-*bis*phosphate is synthesised in the Calvin cycle by the reaction of glyceraldehyde 3-phosphate and dihydroxyacetone phosphate. The reaction is a reversal of one of the steps of glycolysis but, unlike glycolysis, it occurs in the stroma of the chloroplast and requires a different isoform of the enzyme aldolase to catalyse it.

Although animals cannot synthesise fructose from carbon dioxide, they can convert glucose to fructose. In this pathway, glucose is reduced to its corresponding sugar alcohol sorbitol using NADPH as a redox cofactor in a reaction catalysed by aldol reductase (EC 1.1.1.21). Polyol dehydrogenase (EC 1.1.1.14) then catalyses the oxidation of sorbitol using NAD^+ as a cofactor (Figure 8.10). This reaction is particularly important in tissues with high concentrations of aldose reductase, such as the lens cells of the eye.

8.5 Non-enzymatic Biochemical Reactions of Fructose

Although fructose rarely occurs as a subunit in the carbohydrate parts of mammalian glycoproteins, the non-enzymatic reaction of fructose, and other sugars, with proteins results in the covalent incorporation of fructose residues. This can occur *in vivo* and also during, for example, cooking.

8.5.1 Glycation and the Maillard Reaction

Although glycation reactions with glucose have been extensively studied, compared to those with fructose, it is now clear that fructose is capable of

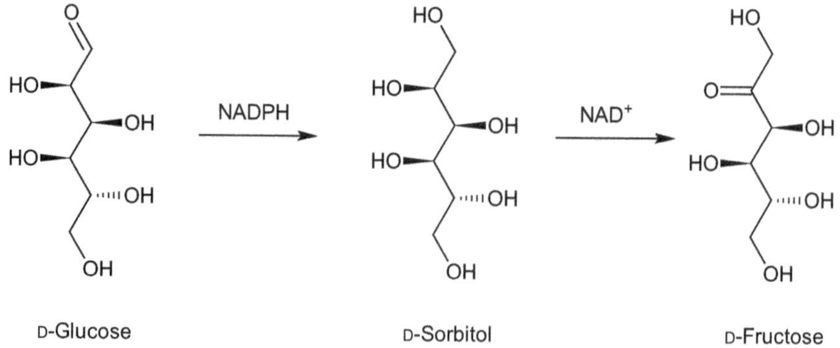

Figure 8.10   The conversion of glucose to fructose in animals. The first reaction is catalysed by aldose reductase and the second by polyol dehydrogenase.

undergoing non-enzymatic reactions with proteins. These reactions have been termed "fructation" (Suárez *et al.* 1989). Non-enzymatic reactions with fructose may be considerably faster than those with glucose (Dillis 1993). In healthy individuals, the plasma concentration of glucose far exceeds that of fructose and thus reaction with glucose dominates. However, in diabetics, the concentrations of the two monosaccharides can be approximately equal and in these individuals, fructose is expected to have a much greater impact (Schalkwijk *et al.* 2004). As with glucose, glycation occurs between the free carbonyl group in the straight chain form of the monosaccharide and amine groups in the protein. These amine groups almost always come from the ε-amino group of lysine side chains; however, the α-amino groups of free N-termini can also react in this way. The initial reaction is the formation of a Schiff's base between the carbonyl and the amino groups. This product is not stable and undergoes further re-arrangements with the result that multiple, different products are formed over time. Unlike the reactions with glucose, the major products following Schiff base formation are not the result of the so-called Amadori rearrangements, but instead are Heyns products (Dillis 1993). However, Amadori rearrangements are also possible (Suárez *et al.* 1989). These Heyns and Amadori products further decompose and also react with additional fructose molecules to produce reactive intermediates which can, ultimately, crosslink the proteins. These reactions are not well characterised chemically, but give rise to protein products which are often yellow-brown in colour and fluorescent. These advanced glycation endproducts (AGE) produced from the Maillard reaction with fructose are immunologically distinct (and, therefore, presumably also chemically different) from those produced by reaction with glucose (Takeuchi *et al.* 2010). These reactions are important in the browning of food-stuffs during cooking and also may have pathological implications, especially for patients with abnormally high concentrations of fructose in their blood or other tissues. Interestingly, it has been noted that the expression of fructokinase (the first enzyme of the fructose metabolic pathway, see above) is particularly high in mammalian tissues, which also have high fructose levels.

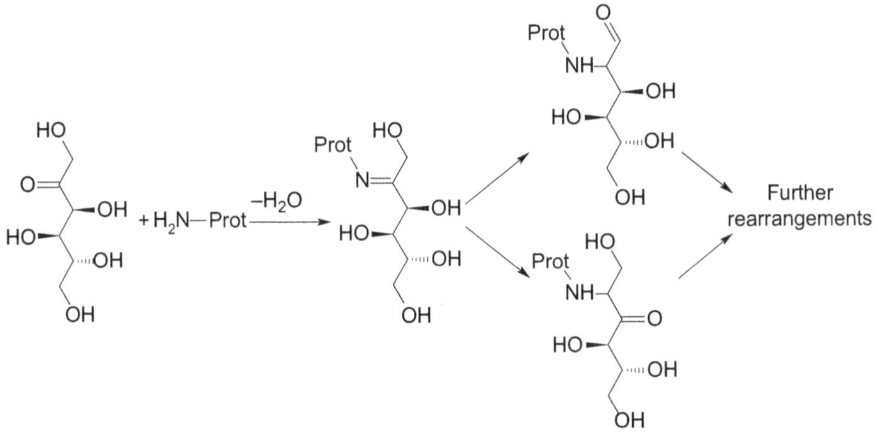

Figure 8.11 A simplified scheme for the Maillard with fructose. Initial reaction between a lysine side chain and the ketone group in fructose results in a Schiffs base. This rearranges to form the two possible Heyns products (Suárez *et al.* 1989). Further rearrangements of these products give rise to the advanced glycation end-products, protein crosslinking and the distinctive yellow-brown colour and flavours of many cooked foods. Prot represents the protein containing the amino group.

This suggests that mammals may have evolved rapid clearance of fructose not only as a means to extract energy from the compound, but also to remove it from cells before it can cause damage by the Maillard reaction (Dillis 1993). A simplified scheme for the Maillard reaction with fructose is shown in Figure 8.11. At elevated temperatures, reaction with asparagine is also possible. This is because asparagine can undergo internal rearrangements into a reactive form (Mottram *et al.* 2002). One potential, final product from this reaction is the neurotoxin acrylamide. However, fructose is reported to be less likely to result in acrylamide formation than glucose (Koutsidis *et al.* 2008).

8.6 Di- and Polysaccharide Synthesis Involving Fructose

8.6.1 Non-enzymatic Synthesis

The abundance of sucrose from natural sources means that, like fructose, the chemical synthesis is not considered commercially viable. Synthesis is possible by the reaction of 1,2-anhydro-α-D-glucopyranose triacetate and 1,3,4,6-tetra-*O*-acetyl-D-fructose (Figure 8.12) (Lemieux and Huber 1953).

8.6.2 Biosynthesis of Sucrose

Sucrose is a disaccharide consisting of one glucose and one fructose residue joined together *via* an oxygen atom linking carbon 1 of the glucose and carbon 2

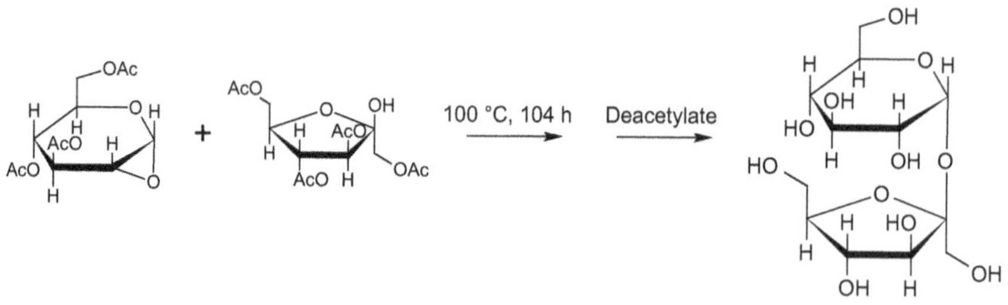

Figure 8.12   The chemical synthesis of sucrose. This can be achieved from glucose and fructose derivatives (Lemieux and Huber 1953).

of the fructose. The fructose moiety is in the β-furanose configuration. Green plants and some micro-organisms can synthesise sucrose as part of the "dark" (*i.e.* non-light-requiring) reactions of photosynthesis. This biosynthesis does not start with the underivatised monosaccharides. Glucose is "activated" by reaction with uridine triphosphate (UTP) to produce UDP-glucose, catalysed by the enzyme UDP-glucose pyrophosphorylase (EC 2.7.7.9). UDP-monosaccharides, sometimes referred to as Leloir sugars after their discoverer Louis Leloir, are commonly used as precursors in the biosynthesis of di-, oligo- and polysaccharides. In the synthesis of sucrose, UDP-glucose reacts with fructose 6-phosphate. This reaction produces sucrose phosphate and is catalysed by the enzyme sucrose phosphate synthase (EC 2.4.1.14). Although, no structures of this enzyme from plants have been determined, the structure of the enzyme from the bacterium *Halothermothrix orenii* has been (Chua *et al.* 2008). This enzyme is a monomer consisting of two domains (Figure 8.13). The fructose 6-phosphate substrate binds to the N-terminal domain and UDP-glucose is believed to bind to the C-terminal domain. This brings the two substrates together, on opposite sides of the cleft such that a hydrogen bond can form between the β-phosphate of UDP-glucose and the C_2-OH of fructose 6-phosphate. This weakens the O-H bond, increasing the negative charge on this oxygen atom and enabling nucleophilic attack on carbon 1 of the glucose moiety of UDP-glucose. This results in cleavage of the glucose from the UDP and the formation of a sucrose molecule (Chua *et al.* 2008).

8.6.3   Biosynthesis of Inulin

Inulin is a polysaccharide composed of fructose units, except for one terminal residue, which is glucose. The fructose residues are entirely in the furanose form and are linked through the hydroxyl groups attached to carbons 1 and 2 (Figure 8.14). The polymer functions as a storage polysaccharide in some plants and microbes and has a highly mobile, flexible structure (Sinnott 2007). It is synthesised from sucrose molecules. An initial reaction, catalysed by sucrose-sucrose 1-fructosyltransferase (1-SST, EC 2.4.1.99), transfers the fructose residue from one sucrose molecule to another resulting in a trisaccharide

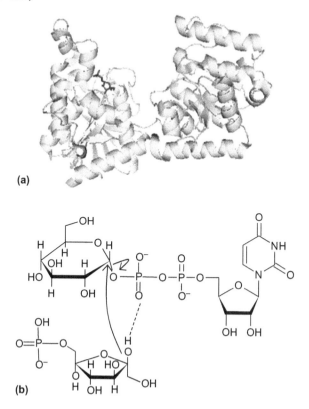

Figure 8.13 Sucrose phosphate synthase. (a) The structure of the enzyme revealed two domains, with a deep cleft between them. Here, one substrate (fructose 6-phosphate) is shown (dark grey) bound to the N-terminal domain. The figure was produced from PDB file 2R66 (Chua *et al.* 2008) using PyMol. (b) The proposed mechanism of action involves the formation of a hydrogen bond between the C_2-OH of fructose 6-phosphate (top) and the β-phosphate of UDP-glucose (bottom) shown here as a dashed line. This enables a nucleophilic attack on C_1 of the glucose moiety cleaving it from UDP and joining to the fructose 6-phosphate.

(glucose-fructose-fructose) and glucose. The transfer of further fructose residues from sucrose molecules is catalysed by fructan 1-fructosyltransferase (1-FFT, EC 2.4.1.100). In some microbial species branching of the polymer also occurs.

8.7 Degradation of Polysaccharides Containing Fructose Residues

Most di-, oligo- and polysaccharides can be degraded to their monosaccharide components by acid hydrolysis. This hydrolysis is often assisted by acid, base or heat. In the digestion of foods, hydrolysis generally occurs under more controlled, gentler conditions and is catalysed by enzymes.

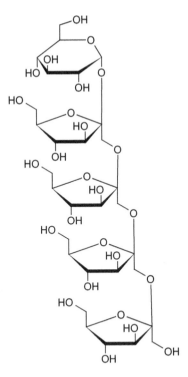

Figure 8.14   The structure of part of an inulin molecule. Here the initial α-D-glucose
residue (top) and four β-D-fructose residues are shown. For simplicity,
the hydrogen atoms on the sugar residues have been omitted. To prevent
overlapping of parts of the figure, the bonds joining the sugar residues
have been made unnaturally long.

8.7.1   Non-enzymatic Hydrolysis

Sucrose is essentially stable in solution at neutral pH. However, at lower pH
values, the rate of acid catalysed hydrolysis is significant. This reaction is
important in digestion of sucrose as some molecules are hydrolysed by this
route in the acidic environment of the stomach (the remainder being split
enzymatically – see below). The reaction mechanism requires the protonation
of the glycosidic oxygen. This results in splitting of the bonds linking the two
monosaccharide residues together (Figure 8.15) (Moiseev *et al.* 1976).

  Inulin can also be hydrolysed under acidic conditions. The reaction
mechanism is similar to that for the acid hydrolysis of sucrose. Since proto-
naton can occur at any of the glycosidic oxygen atoms, the initial products
are shorter oligosaccharides. Base hydrolysis is also possible. However, this
requires a free ketone group and thus can only occur at fructose residues at the
end of the polymer. Consequently, it releases free fructose molecules for each
round of hydrolysis, but the rate of hydrolysis is typically much slower than
that catalysed by acid. Acid catalysed (pH 2.0–4.2) hydrolysis is generally used

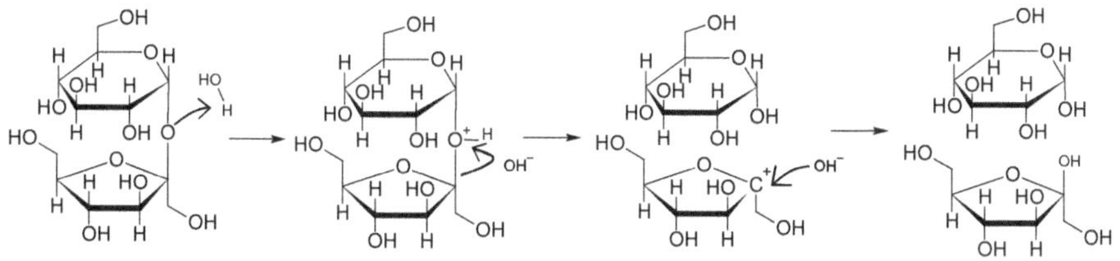

Figure 8.15 Hydrolysis of sucrose. The mechanism of acid hydrolysis of sucrose to give glucose and fructose.

in industry (Barclay *et al.* 2010). This reaction is potentially important for the release of fructose from the polymer for use in biofuel fermentations.

8.7.2 The Enzymatic Hydrolysis of Sucrose: The Invertase Reaction

Sucrose is a major component of human diets, especially in Western countries. Like all disaccharides, it must be broken down to its component mono-saccharides prior to absorption by the cells of the gut. This digestion is cata-lysed by the enzyme, sucrase-isomaltase (EC 3.2.1.10), which is covalently linked to the membrane of the brush border epithelial cells of the intestine. Mutations in the gene encoding this enzyme can cause the genetic disease congentital sucrase-isomaltase deficiency (OMIM #222900). Patients with this disease cannot absorb dietary sucrose and suffer bloating and diarrhoea if this sugar is ingested. The enzyme has an interesting overall structure, with two highly similar domains with an active site in each domain. The N-terminal domain contains the isomaltase activity and the C-terminal domain, which can be cleaved from the remainder of the enzyme, the sucrase activity. The structure of the N-terminal, isomaltase domain of the human enzyme has been solved (Sim *et al.* 2010) (Figure 8.16a). The two domains have approximately 40% sequence similarity and are, therefore, expected to have similar structures. The mechanism of reaction has not been fully elucidated, but is expected to involve an oxocarbonium ion in the transition state (Sinnott 1990).

Plants, fungi and some bacteria can also catalyse the hydrolysis of sucrose; however, they use a different enzyme called invertase (EC 3.2.1.26). Inhibition of these enzymes in crop plants may result in increased yields and, therefore, there are considerable efforts to identify safe and effective inhibitors of this enzyme. The enzyme has a two-domain structure, with the N-terminal domain containing the sucrose binding site (Figure 8.16b) (Lammens *et al.* 2008). The catalytic mechanism requires two carboxylate-containing side chains, one to act as a nucleophile and one to act as a general acid/base. The nucleophile attacks the fructose ring at position 2 and the general acid protonates the oxygen linking the two monosaccharide units. This cleaves the glycosidic linkage and releases the glucose molecule. The fructose moiety is covalently linked to the enzyme, through the nucleophilic carboxylate side chain. The fructose molecule

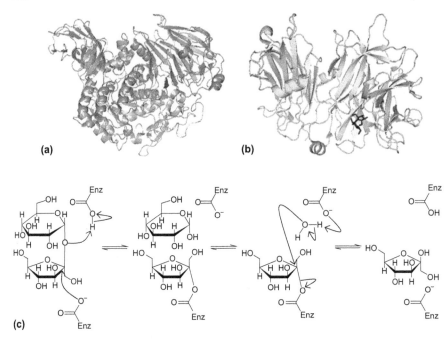

Figure 8.16 The structures and mechanism of sucrose hydrolysing enzymes. (a) The structure of the isomaltase domain of human sucrase-isomaltase. This domain is similar in sequence to the sucrase domain, and so is anticipated to have a similar structure. The figure was made using PDB file 2LPO (Sim *et al.*, 2010) and PyMol. (b) The structure of invertase from *Arabidopsis thaliana*. A bound sucrose molecule is shown in dark grey towards the lower right of the figure. The figure was made using PDB file 2QQW (Lammens *et al.* 2008) and PyMol. (c) The catalytic mechanism of invertase involves a frutosyl-enzyme intermediate. Redrawn after Lammens *et al.* 2008. Enz represents the remainder of the enzyme molecule.

is released following attack by a water molecule (assisted by the general base) on carbon 2 of the ring (Figure 8.16c) (Lammens *et al.* 2008; Sinnott 1990).

8.7.3 Enzymatic Degradation of Inulin

Although the hydrolysis of inulin can be catalysed by some invertases, there are also specialised inulinases. Exo-inulinase (EC 3.2.1.80) catalyses the removal of fructose residues at the ends of the polymer chains. Its overall structure is similar to that of invertase (Nagem *et al.* 2004). Endo-inulinases (EC 3.2.1.7) also exist. These enzymes catalyse the hydrolysis of inulin polymers into shorter oligomers. Although no structure of an enzyme from this group has been determined, homology modelling suggests that they also have overall folds similar to invertase (Basso *et al.* 2010).

8.8 Conclusions

While the chemistry of fructose has many features in common with that of other monosaccharides, such as glucose, it has some special features mainly influenced by the ketone group in the straight chain form and the relative ease with which fructose can form furanose rings. When ingested as part of foods, special metabolic pathways are required to metabolise it to glycolytic intermediates and, similarly, organisms which synthesise this sugar have enzymes specific for this purpose. The importance of fructose in non-enzymatic modifications of proteins *in vivo* and during cooking has been somewhat overlooked; however, given the increasing levels of fructose in (especially) Western diets its contribution to these processes merits further investigation.

Summary Points

- Fructose is a six-carbon monosaccharide containing one ketone group
- In aqueous solution fructose forms a mixture of five-membered (furanose) and six-membered (pyranose) rings
- Chemical reactions of fructose are similar to those of glucose
- In fructose-rich cells, the fructokinase pathway is required to transform the molecule into glycolytic intermediates
- Plants and some microbes can synthesise fructose from carbon dioxide and water
- Animals can convert glucose to fructose *via* sorbitol
- Fructose can undergo the Maillard reaction (fructation) with amine groups in proteins
- Fructation is especially important in the cooking of fructose-rich foods and in diabetic patients
- Inulin is a polysaccharide made up of fructose residues and one, terminal glucose residue
- Inulin degradation occurs at the fructose terminus and in the middle of the chain

Key Facts

Fructose stereochemistry

1. A carbon atom bonding to four different groups gives rise to a stereo-centre or chiral centre
2. These chiral centres cause the molecule to be optically active and affect the polarisation of light passing through solutions of the compound
3. In the straight chain form of the compound, fructose has four chiral centres
4. The stereochemistry at carbon-5 determines whether the compound will be L- or D-fructose

5. This naming is based on the configurations in glyceraldehyde, a three carbon aldose carbohydrate
6. Alteration of the stereochemistry at any of the other centres results in a different monosaccharide, *i.e.* it is no longer fructose
7. The formation of furanose or pyranose rings results in the formation of another chiral centre
8. This new chiral centre is called the anomeric carbon
9. The two possible configurations of the anomeric carbon are called α and β

Furanose and pyranose rings

1. In aqueous solution the majority of fructose molecules are in a ring form
2. Furanose rings are five-membered with four carbon atoms and one oxygen
3. Furanose rings are named after furan (C_4H_4O)
4. Furanose rings are formed by internal hemiketal formation: in fructose the ketone group reacts with the hydroxyl on carbon 5
5. Pyranose rings are six-membered with five carbon atoms and one oxygen
6. Pyranose rings are named after pyran (C_5H_6O)
7. Pyranose rings are formed by internal hemiacetal or hemiketal formation: in fructose the ketone group reacts with the hydroxyl on carbon 6

Definitions

Aldose: A monosaccharide containing an aldehyde group

Anomeric carbon: In either the furanose or pyranose form, the carbon atom which participates in the joining together of the straight chain form

D-fructose: The stereoisomer of fructose which resembles D-glyceraldehyde at carbon-5

Fructation: Glycation reactions with fructose

Furanose ring: A five membered ring containing four carbon atoms and one oxygen

Glycation: The non-enzymatic reaction of monosaccharides with free amine groups, especially in amino acids and proteins

Glycolysis: The metabolic pathway which converts glucose to pyruvate with the concomitant production of two molecules of ATP and two molecules of NADH per glucose molecule

Inulin: A polymer of fructose residues with one, terminal glucose residue

Ketose: A monosaccharide containing a ketone group

Mutarotation: Interconversion between the α- and β-anomers

OMIM: Online Mendelian Inheritance in Man, an online database of genetic diseases maintained here: http://www.ncbi.nlm.nih.gov/omim

Pyranose ring: A six membered ring containing five carbon atoms and one oxygen

Sorbitol: The sugar-alcohol produced by the mild reduction of fructose (and also glucose)

Straight chain form: The non-cyclic form of a monosaccharide carbohydrate. (Note that the constraints on the bonds and the dynamics of a molecule in solution mean that it is unlikely that the atoms are arranged in a perfect straight line.)

List of Abbreviations

Ac_2O	Acetic anhydride
AGE	Advanced glycation end-product
ATP	Adenosine triphosphate
C_1-OH	The hydroxyl group attached to carbon 1
C_2-OH	The hydroxyl group attached to carbon 2
EC	Enzyme Commission (number)
HFI	Hereditary fructose intolerance
NAD^+	Nicotinamide adenosine dinucleotide (oxidised form)
NADPH	Nicotinamide adenosine dinucleotide phosphate (reduced form)
OMIM	Online Mendelian inheritance in man
UDP	Uridine diphosphate
UTP	Uridine triphosphate

References

Abdek-Akher, M., Hamilton, J.K. and Smith, F., 1951. The reduction of sugars with sodium borohydride. *Journal of the American Chemical Society* 73: 4691–4692.

Aleshin, A.E., Zeng, C., Bartunik, H.D., Fromm, H.J. and Honzatko, R.B., 1998. Regulation of hexokinase I: crystal structure of recombinant human brain hexokinase complexed with glucose and phosphate. *Journal of Molecular Biology* 282: 345–357.

Barclay, T., Ginic-Markovica, M., Cooper, P. and Petrovsky, N., 2010. Inulin - a versatile polysaccharide with multiple pharmaceutical and food chemical uses. *Journal of Excipients and Food Chemistry* 1: 27–50.

Basso, A., Spizzo, P., Ferrario, V., Knapic, L., Savko, N., Braiuca, P., Ebert, C., Ricca, E., Calabrò, V. and Gardossi, L., 2010. Endo- and exo-inulinases: enzyme-substrate interaction and rational immobilization. *Biotechnology Progress* 26: 397–405.

Bonthron, D.T., Brady, N., Donaldson, I.A. and Steinmann B., 1994. Molecular basis of essential fructosuria: molecular cloning and mutational analysis of human ketohexokinase (fructokinase). *Human Molecular Genetics* 3: 1627–1631.

Bouteldja, N. and Timson, D.J., 2010. The biochemical basis of hereditary fructose intolerance. *Journal of Inherited Metabolic Disease* 33:105–112.

Cabezas, A., Costas, M.J., Pinto, R.M., Couto, A. and Cameselle, J.C., 2005. Identification of human and rat FAD-AMP lyase (cyclic FMN forming) as ATP-dependent dihydroxyacetone kinases. *Biochemical and Biophysical Research Communications* 338: 1682–1689.

Choi, K.H., Shi, J., Hopkins, C.E., Tolan, D.R. and Allen, K.N., 2001. Snapshots of catalysis: the structure of fructose-1,6-*bis*phosphate aldolase covalently bound to the substrate dihydroxyacetone phosphate. *Biochemistry* 40: 13868–13875.

Christen, S., Srinivas, A., Bähler, P., Zeller, A., Pridmore, D., Bieniossek, C., Baumann, U. and Erni, B., 2006. Regulation of the Dha operon of *Lactococcus lactis*: a deviation from the rule followed by the TetR family of transcription regulators. *Journal of Biological Chemistry* 281: 23129–23137.

Chua, T.K., Bujnicki, J.M., Tan, T.C., Huynh, F., Patel, B.K. and Sivaraman, J., 2008. The structure of sucrose phosphate synthase from *Halothermothrix orenii* reveals its mechanism of action and binding mode. *Plant Cell* 20: 1059–1072.

Dalby, A.R., Tolan, D.R. and Littlechild, J.A., 2001. The structure of human liver fructose-1,6-*bis*phosphate aldolase. *Acta Crystallography D: Biological Crystallography* 57: 1526–1533.

Delpierre, G., Collard, F., Fortpied, J. and van Schaftingen, E., 2002. Fructosamine 3-kinase is involved in an intracellular deglycation pathway in human erythrocytes. *Biochemical Journal* 365: 801–808.

Dillis, W.L., 1993. Protein fructosylation: fructose and the Maillard reaction. *American Journal of Clinical Nutrition* 58: 779S–787S.

Fandis, A.G., 1986. Metal-ion oxidation reactions of monosaccharides: A kinetic study. *Carbohydrate Research* 146: 97–105.

Frandsen, E.K. and Grunnet, N., 1971. Kinetic properties of triokinase from rat liver. *European Journal of Biochemistry* 23: 588–592.

Bethell, G.S. and Ferrier, R.J. 1973. Studies with radioactive sugars: Part IV. the methanolysis of D-fructose and L-sorbose. *Carbohydrate Research.* 31: 69–80.

Gibbs, A.C., Abad, M.C., Zhang, X., Tounge, B.A., Lewandowski, F.A., Struble, G.T., Sun, W., Sui, Z. and Kuo, L.C., 2010. Electron density guided fragment-based lead discovery of ketohexokinase inhibitors. *Journal of Medicinal Chemistry* 53: 7979–7991.

Kelley, L.A. and Sternberg, M.J.E., 2009. Protein structure prediction on the web: a case study using the Phyre server. *Nature Protocols* 4: 363–371.

Kinoshita, S., Kadota, K. and Taguchi, H. 1991. Purification and properties of aldose 1-epimerase from *Aspergillus niger*. *Biochimica et Biophysica Acta* 662: 285–290.

Koutsidis, G., de la Fuente, A., Dimitriou, C., Kakoulli, A., Wedzicha, B.L. and Mottram, D.S., 2008. Acrylamide and pyrazine formation in model systems containing asparagine. *Journal of Agriculture and Food Chemistry* 56: 6105–6112.

Lammens, W., le Roy, K., van Laere, A., Rabijns, A. and van den Ende, W., 2008. Crystal structures of *Arabidopsis thaliana* cell-wall invertase mutants in complex with sucrose. *Journal of Molecular Biology* 377: 378–385.

Lemieux, R.U. and Huber, G., 1953. A chemical synthesis of sucrose. *Journal of the American Chemical Society* 75:4118.

Markert, M. and Mahrwald, R., 2008. Total syntheses of carbohydrates: organocatalyzed aldol additions of dihydroxyacetone. *Chemistry, A European Journal* 14: 40–48.

Moiseev, Y.V., Khalturinskii, N.A. and Zaikov, G.E., 1976. The mechanism of the acid-catalysed hydrolysis of glucosides. *Carbohydrate Research* 51: 23–37.

Mottram, D.S., Wedzicha, B.L. and Dodson, A.T., 2002. Acrylamide is formed in the Maillard reaction. *Nature* 419: 448–499.

Nagem, R.A., Rojas, A.L., Golubev, A.M., Korneeva, O.S., Eneyskaya, E.V., Kulminskaya, A.A., Neustroev, K.N. and Polikarpov, I., 2004. Crystal structure of exo-inulinase from *Aspergillus awamori*: the enzyme fold and structural determinants of substrate recognition. *Journal of Molecular Biology* 344: 471–480.

Petersen, A., Kappler, F., Szwergold, B.S. and Brown, T.R., 1992. Fructose metabolism in the human erythrocyte. Phosphorylation to fructose 3-phosphate. *Biochemical Journal* 284: 363–366.

Schalkwijk, C.G., Stehouwer, C.D. and van Hinsbergh, V.W., 2004. Fructose-mediated non-enzymatic glycation: sweet coupling or bad modification. *Diabetes/Metabolic Research and Reviews* 20: 369–382.

Sim, L., Willemsma, C., Mohan, S., Naim, H.Y., Pinto, B.M. and Rose, D.R., 2010. Structural basis for substrate selectivity in human maltase-glucoamylase and sucrase-isomaltase N-terminal domains. *Journal of Biological Chemistry* 285: 17763–17770.

Sinnott, M., 1990. Catalytic mechanisms of enzymic glycosyl transfer. *Chemical Reviews* 90: 1171–1202.

Sinnott, M.L., 2007. *Carbohydrate Chemistry and Biochemistry*. RSC Publishing. Cambridge. UK, 748pp.

Suárez, G., Rama, R., Oronsky, A.I. and Gawinowicz, M.A., 1989. Nonenzymatic glycation of bovine serum albumin by fructose: comparison with the Maillard reaction by glucose. *Journal of Biological Chemistry* 264: 3674–3679.

Takeuchi, M., Iwaki, M., Takino, J., Shirai, H., Kawakami, M., Bucala, R. and Yamagishi, S., 2010. Immunological detection of fructose-derived advanced glycation end-products. *Laboratory Investigation* 90: 1117–1127.

Thoden, J.B., Kim, J., Raushel, F.M. and Holden, H.M., 2003. The catalytic mechanism of galactose mutarotase. *Protein Science* 12: 1051–1059.

CHAPTER 9

Sucrose Chemistry

LEONARDO M. MOREIRA,*[a] JULIANA P. LYON,[b]
PATRÍCIA LIMA,[c] VANESSA J. S. V. SANTOS[d] AND
FABIO V. SANTOS[e]

[a] Leonardo Marmo Moreira, Universidade Federal de São João Del Rei, Praça Dom Helvécio, 74, Fábricas, São João del Rei, MG, Brazil; [b] Juliana Pereira Lyon, Department of Natural Sciences, Federal University at São João del Rei, Praça Dom Helvécio, 74, Fábricas, São João del Rei, MG, Brazil; [c] Patrícia Lima, Universidade Federal de São João Del Rei, Praça Dom Helvécio, 74, Fábricas, São João del Rei, MG, Brazil; [d] Vanessa Jaqueline da Silva Vieira dos Santos, Health Science Center, Federal University at São João del Rei, Divinópolis, MG, Brazil; [e] Professor Fabio Vieira dos Santos, Center of Health Science, Federal University at São João del Rei, Divinópolis, MG, Brazil
*Email: leonardomarmo@gmail.com

9.1 Introduction

Sucrose is the most abundant organic molecule produced at the industrial scale from renewable sources (Queneau *et al.* 2004). Sucrose is usually known as "table sugar" due to its widespread employment in the world of nutrition. It corresponds to the major transport carbohydrate in higher plants and is synthesised solely by plants and other oxygenic photosynthetic organisms (Wind *et al.* 2010). The properties of this class of organic compounds are of great interest because of their biological and commercial roles. Chemically speaking, sucrose is a carbohydrate; however, it is known by several names, including saccharide, hexose, hydrate of carbon, and sugar.

Food and Nutritional Components in Focus No. 3
Dietary Sugars: Chemistry, Analysis, Function and Effects
Edited by Victor R Preedy
© The Royal Society of Chemistry 2012
Published by the Royal Society of Chemistry, www.rsc.org

9.1.1 General Features of Carbohydrates

The word "saccharide" comes from Greek (*sákkharon*) and means "sugar". Indeed, the peculiar taste characteristics of carbohydrates is an important property for their employment as a food. In this context, it is important to mention that there is a correlation between the number of hydroxyls in the molecular structure of the carbohydrate and the organoleptic property associated with its taste.

Carbohydrates are frequently associated with compounds or a combination of compounds that possess the formula $C_n(H_2O)_x$, where n can be a different value in relation to x. For this reason, this class of compounds is still known as hydrates of carbon. However, it is important to remember that there are carbohydrates that contain heteroatoms, such as nitrogen (N), sulphur (S) and phosphorus (P). Carbohydrates may be monosaccharides, disaccharides, oligosaccharides or polysaccharides. Monosaccharides, such as glucose, fructose and galactose, are polyhydroxy aldehydes or polyhydroxy ketones, and they precisely fit the formula $C_n(H_2O)$. Monosaccharides are carbohydrates that contain between 3 and 8 carbon atoms in their molecular structures, but the more widespread carbohydrates contain 5 or 6 carbon atoms.

Carbohydrates are organic compounds that perform two basic organic functions: the alcohol organic function and the aldehyde, or ketone, organic function.Thus, polyhydroxy aldehydes and polyhydroxy ketones are similar compounds in terms of their great number of hydroxyl groups. They are mutually differentiated as a function of the structural position of their carbonyl groups. Thus, when the carbonyl (C=O) group is found in the extremity of the carbonic chain, it will constitute the termination H-C=O and characterise the compound as an aldehyde. Therefore, a mixed function involving an alcohol with several hydroxyl groups and an aldehyde constitutes a carbohydrate that is a polyhydroxy aldehyde. However, when the carbonyl (C=O) is not encountered at the extremity of the carbonic chain, the organic compound is a ketone. The respective carbohydrate would then constitute a mixed function involving an alcohol organic function with various hydroxyl groups and a ketone group, which is a carbohydrate that is a polyhydroxy ketone.

Polyhydroxy aldeydes and polyhydroxy ketones are interesting organic compounds with significant polarities. Indeed, despite the carbonic structures of these compounds, the presence of a great number of hydroxyls makes these carbohydrates significantly hydrophilic.

9.1.2 Carbohydrates as an Energy Source

Carbohydrates constitute the main reserve of energy for living beings, whereas the lipids are the principal reservoir of energy for living beings.It is important to note that the accumulation of carbohydrates in living organisms is very limited. For instance, in humans, glycogen is a carbohydrate resource that is concentrated predominantly in the liver. Thus, a large part of the carbohydrate excess ingested in the nutrition process is not accumulated in the respective organism in this chemical form, *i.e.*, carbohydrates (polyhydroxy aldehydes or polyhydroxy ketones). In vegetables, the carbohydrate that corresponds to glycogen in animal tissues is amide, which also functions as an energetic resource.

A significant content of the total quantity of carbohydrates that is in excess to the necessities of nutrition – which is dependent on the organic metabolism of each human – is transformed into lipids to accumulate as an energetic reservoir for the physiological activities of the organism. Thus, excess corporal mass is typically characterised by the substantial presence of different types of fats and not carbohydrates. In fact, fat compounds such as triacylglycerol mainly function as the most efficient resource for energy generation. One gram of triacylglycerol generates approximately 38 kJ, which is significantly higher than the 18 kJ g^{-1} obtained from carbohydrates or 17 kJ g^{-1} generated by proteins (Mahan and Myers 1987). This difference is due to the higher thermal energy generated by the triacylglycerol because this compound is almost completely composed of hydrocarbons. However, in proteins and particularly in carbohydrates, the hydrocarbon framework is already in a partially oxidised form (Mahan and Myers 1987).

Nevertheless, the ingestion of carbohydrates, as well as the ingestion of lipids, such as triacylglycerols and fatty acids, constitutes one of the main causes of medical problems related to obesity worldwide. In spite of the fact that the caloric content of carbohydrates is lower than that inherent to fats, the priority of the saccharides in terms of metabolism, in comparison to fats, has indeed made the exaggerated ingestion of carbohydrates a relevant question for public health. Consequently, different diet strategies with the aim of loosing corporal mass have frequently included a decrease in fat ingestion and lower carbohydrate ingestion.

9.2   Sucrose

Sucrose is a disaccharide formed by ligation between two monosaccharides, glucose and fructose. The ligation between two monosaccharides to form a disaccharide is termed glycosidic ligation or a glycosidic bond. This ligation is formed through the loss of a water molecule, which is similar to the process of forming a peptidic ligation to form dipeptides. In fact, the glycosidic ligation is fundamental to the formation of the large macromolecules of carbohydrates as the peptidic ligation is a basic prerequisite to the formation of polypeptide chains to form proteins.

Indeed, the reducing end of fructose (C2) is linked to the reducing end of glucose (C1) to make a non-reducing sucrose. The limited chemical reactivity that results from this combination is advantageous for the transport and storage of molecules (Wind *et al.* 2010).

Monosaccharides can be identified in cyclic (Figure 9.1) or linear forms (Figure 9.2). However, the cyclic form is predominant, and the main form found when the carbohydrate is in solution.

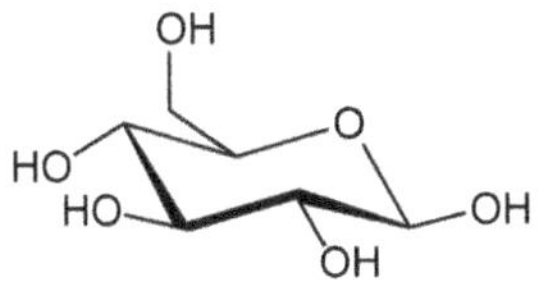

Figure 9.1 Structural arrangement of a monosaccharide in a cyclic molecular configuration.

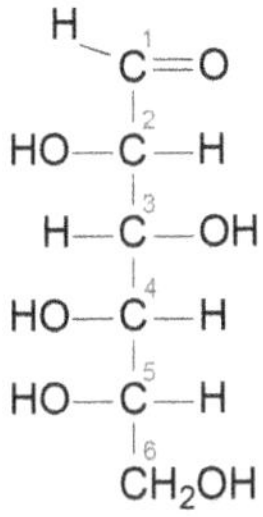

Figure 9.2 Fischer projection of the structural formula of glucose.

Hermann Emil Fischer (1852–1919) was a German chemist and the recipient of the Nobel Prize for Chemistry in 1902. He is known for discovering Fischer esterification and proposed the well-known Fischer projection to didactically represent, in a bidimensional form, the tridimensional molecular structure of carbohydrates (Figure 9.2).

9.2.1 Sucrose Monosaccharides

9.2.1.1 Glucose

Glucose is a six carbon monosaccharide that consists of a polyhydroxy aldehyde; that is, it contains a COOH group (Figure 9.2). Its six carbons are numbered from 1 to 6 with C1 being the aldose carbon.

The bonding patterns of the hydrogens and hydroxyl groups around each carbon atom are very important to the structures of carbohydrates. Glucose has four asymmetric carbon atoms.Therefore, these respective carbon atoms present four different ligand groups, which cause an alteration of a polarised light plane. The spatial arrangement of the hydroxyl groups (OH) and hydrogen atoms (H) of these carbon atoms is important because the reciprocal orientations of these OH and H can vary for all carbon atoms. This variation can generate a great number of isomer compounds with significant, distinctive physico-chemical properties, which are caused by several factors, such as the number of hydrogen bonds, van der Waals interactions, and crystal structures between groups.

In agreement with Fischer's proposal, structural formulas for sugar molecules are often written in a vertical arrangement with the aldehyde or the ketone group at or near the top. When written in this particular manner, the position of the OH on the last asymmetric carbon atom indicates whether it is a "D" sugar or an "L" sugar. "D" stands for dextro and "L" stands for levo in regards to the deviation of a polarised light plane. If the OH on the last asymmetric carbon atom is on the right, it is a "D" sugar (*e.g.*, D-glucose). Each of these sugars is identified as a "D" saccharide because the OH on the last asymmetric carbon atom is found on the right side of the respective carbon. Therefore, similar to D-glucose, these sugars are termed D-galactose, D-glucose, and D-fructose because these carbohydrates present the OH group attached to the last asymmetric carbon atom on the right side, according to their Fisher projections. On the other hand, if the hydroxyl group is ligated to the last asymmetric carbon atom positioned on the left side, it is an "L" sugar. Figure 9.1 presents the cyclic molecular structure of glucose, which permits the identification of the spatial position of the important constituent atoms of this carbohydrate.

9.2.1.2   Fructose

Fructose is a six carbon monosaccharide that is also known as fruit sugar because it is present in fruits and other foods. This molecule was first described by Augustin-Pierre Dubrunfaut in 1847. Fructose is a polyhydroxy ketone, *i.e.*, its carbonyl group is not found in the terminal position of the carbon chain (Lee *et al.* 2009). Figure 9.3 shows the cyclic structure of fructose.

9.2.1.3   Sucrose Structure

Figure 9.4 displays the molecular structure of sucrose, which has a molecular formula of $C_{12}H_{22}O_{11}$. In particular, this spatial arrangement involves the use of the alpha form of D-glucose and the beta form of D-fructose, which are ligated through the loss of a water molecule (glycosidic ligation). The α-D-glucose is in the conventional orientation (the 6C is up). However, the β-D-fructose is shown in two orientations. In the standard orientation, the 6C is up on the left side, and the -2-OH is up on the right. Thus, it will be the respective -2-OH group that bonds to the -1-OH of the α-D-glucose, which implies that the β-D-fructose molecule must be inverted.

The glycosidic bond of the sucrose is termed an α-β-1-2 bond because it involves an alpha-OH from the glucose and a beta-OH from the sucrose.

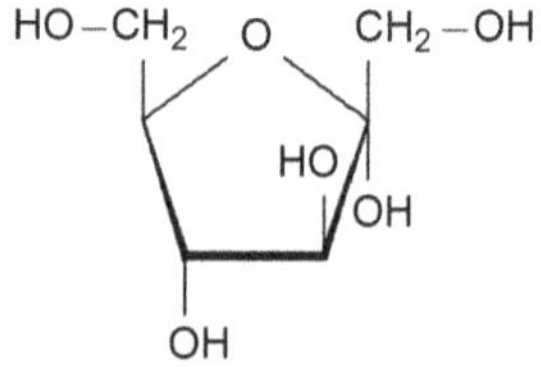

Figure 9.3   Chemical structure of fructose.

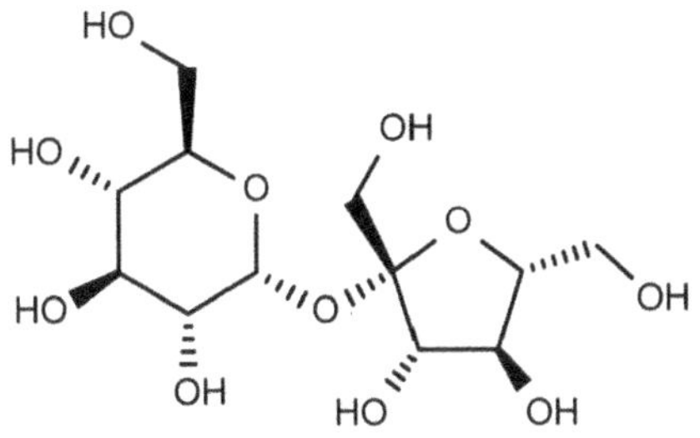

Figure 9.4 Chemical structure of sucrose.

Therefore, the tridimensional arrangement of the sucrose molecule involves the number 1-carbon on the glucose to the number 2-carbon on the fructose.

Several instrumental tools focused on sucrose analysis, especially spectroscopic methods, have been applied extensively because its molecular shapes help to determine its properties. In fact, the structure-function relationship of carbohydrates has been the focus of several approaches because slight structural changes, such as those associated with different isomers (same molecular formula), can generate biological activities quite distinct from each other.

The reducing end of fructose is linked to the reducing end of glucose to make non-reducing sucrose, which leads to a limited chemical activity. This limitation is possibly associated with its chemical reactivity, which is correlated with the relatively high redox potential of sucrose as well as the stereochemical impairment that the spatial configuration provokes in relation to eventual attacks of reactive molecules, such as free radicals and/or reactive oxygen species (ROS) and reactive nitrogen species (RNS). This property is positive in relation to the transport and storage of sucrose (Wind *et al.* 2010). Indeed, carbohydrates are known as important reducing compounds in biological media, which has been the target of interest in studies focused on oxidative stress.

9.3 Sucrose as a Sweetener

Another interesting aspect of the structure-activity relationship of sucrose is associated with the relationship between its sweetness and its chemical structure, which is not completely understood. In fact, the relationship between the number of hydroxyls and the sweetness of a sugar has been thought to correlate with the sweeter sugars containing a higher number of hydroxyl groups. Thus, an increase in the number of hydroxyls would be associated with an increase in sweetness, which is a relevant organoleptic property of carbohydrates. However, this relationship is still the target of discussion and is a relevant subject of research being developed by the food industry.

In this context, it is important to note that "artificial sweeteners", which are also known as "sugar substitutes", are substances that are used instead of sucrose to sweeten foods and beverages. Because artificial sweeteners are many times sweeter than table sugar, smaller amounts are needed to create the same level of sweetness. Saccharin, aspartame, cyclamate, acesulphame potassium, sucralose and neotame are some of the main artificial sweeteners used. The complex and worldwide obesity epidemic has led to the development and popularisation of the use of "sugar substitutes" due to the caloric impact of carbohydrates on the daily diet.However, in agreement with the National Cancer Institute, questions about artificial sweeteners and cancer have arisen from early studies that indicate cyclamate, in combination with saccharin, causes bladder cancer in laboratory animals. However, results from subsequent carcinogenicity studies (studies that examine whether a substance can cause cancer) of these sweeteners have not provided clear evidence for an association with cancer in humans. Similarly, studies of other FDA-approved sweeteners have not demonstrated clear evidence of an association with cancer in humans. Regardless, these worries have precluded a more pronounced increase in the commercial use of "sugar substitutes" and maintained the current high level of sugar use even in wealthier countries.

9.4 Sucrose Synthesis and Metabolism

Sucrose is synthesised in the cytosol of photosynthetic organisms utilising fixed carbon, starch reserves or lipids (Wind *et al.* 2010). The enzyme sucrose-phosphate synthase catalyses the conversion of UDP-glucose and fructose-6-phosphate into sucrose-6-phosphate and sucrose phosphate. Sucrose phosphatase then converts sucrose-6-phosphate into sucrose. An alternative pathway consists of the conversion of UDP-glucose and fructose to sucrose by the enzyme sucrose synthase (Sauer 2007). The catabolism of sucrose may occur in plant cytosol, mitochondria, chloroplasts or vacuoles (Barratt *et al.* 2009; Gerrits *et al.* 2001; Grennan and Gragg 2009; Szarka *et al.* 2008).

Sucrose cleavage is very important to multicellular plants as a source of carbon and for the hexose-based sugar signals in important structures (Koch 2004). Sucrose cleavage in plants is catalysed by invertase (sucrose + H_2O → glucose + fructose) or sucrose synthase (sucrose + UDP → fructose + UDP-glucose) (Koch 2004).

In humans, the hydrolytic cleavage of sucrose is performed by the enzyme β-fructosidase and takes place at the surface of intestinal epithelial cells.

9.5 Inverted Sugar

Inverted sugar is the denomination of a mixture of glucose and fructose that is obtained by splitting the glycosidic bond of sucrose through the loss of one molecule of water. When compared with its precursor, inverted sugar possesses a sweeter taste, has less tendency to crystallise and its products have a higher tendency to retain humidity. Therefore, inverted sugar is valuable for culinary purposes, especially in bakeries. It is important to note that a higher tendency to retain humidity is associated with a higher possibility of hydrogen bond formation due to a higher accessibility of water molecules to all hydroxyl sites of two monosaccharide units. Furthermore, the "inverted sugar" presents a higher number of hydroxyl sites as a function of the reconstitution of the hydroxyl group associated with the glycosidic bond.

The hydrolysis of sucrose into glucose and fructose can be induced by heating an aqueous solution of sucrose. However, in industry, this reaction is accelerated by the addition of catalysts such as sucrases and invertases, to obtain better economic viability. The mechanism involved in the original reaction is different from that obtained through catalysis. Acids can also accelerate the conversion of sucrose into inverted sugar, and this mechanism is also employed. It is interesting to note that acid catalysed solutions are neutralised when the desired level of inversion is reached.

The name "inverted sugar" is employed due to the behaviour of this compound when visualised by a polarimeter. The asymmetric carbons present in carbohydrates cause a deviation in a plane of polarised light. When the light passes through a solution of pure sucrose, the light is rotated to the right. On the other hand, when a solution is converted to a mixture containing sucrose, fructose and glucose, the rotation is reduced until there is a complete inversion to the left. This complete inversion is achieved when the solution is completely converted to glucose and fructose. Thus, a solution of sucrose has an specific rotation of $+66.5°$. When it is converted into glucose (rotation of $+52.7°$) and fructose (rotation of $-92°$), it undergoes a change from $+66.5°$ to $-39°$.

Because the components of inverted sugar can be used as an energy source by several microorganisms, the solutions that constitute inverted sugar are also degraded. However, inverted sugar possesses a lower water activity compared to sucrose, which is another desirable property for use in the food industry. Products made with inverted sugar have a longer shelf life because their lower water activities make the nutrients in these foods less available to microorganisms.

9.6 Biological Functions of Sucrose

Together with the basic and general biological functions of carbohydrates, which are the main resource of energy for living organisms, there are several other biological functions performed by these organic compounds. Carbohydrates are important constituents of nucleic acids. In fact, deoxyribonucleic acid (DNA) and ribonucleic acid (RNA) are composed of a sugar (ribose or deoxyribose), phosphate group and nitrogen base. Carbohydrates are associated with cellular recognition through a compound known as glycocalyx. Carbohydrates are related strictly to the stabilisation of the tridimensional structures of proteins. Indeed, the spatial arrangement of the tertiary structures of proteins depends on interactions with different neighbouring compounds, especially carbohydrates.

In this context, it is important to note the different roles for sucrose in living beings. For example, sucrose is fundamental to the life cycle of the microorganism *Streptococcus mutans*, which is uniquely responsible for tooth decay.

9.7 Conclusions

The ingestion of sucrose and the public health problems associated with diabetes, obesity and other chronic health difficulties have motivated studies focused on this highly relevant carbohydrate.Nevertheless, sucrose is still an extraordinarily nutritious constituent of a great number of foods. Thus, research focused on sucrose chemistry is interesting and necessary for the improvement of the general health condition of the worldwide population. In addition, the future of sucrose applications economically affects a great number of families worldwide and requires a multidisciplinary and interdisciplinary study with regards to the substitution of this important compound. The questions about the possible risks of the systemic use of "sugar substitutes" as a main sweetening product remain a concern of specialists and professionals in this area of study. Therefore, studies focused on sucrose chemistry will continue to be auspicious research for several more years.

Summary Points

- Sucrose is the most abundant organic molecule produced at the industrial scale from renewable sources.
- Sucrose is a disaccharide formed by glycosidic ligation between the two monosaccharides glucose and fructose.
- Glucose is a polyhydroxy aldehyde, that is, a monosaccharide composed of six carbons, and fructose is a polyhydroxy ketose, a monosaccharide composed of six carbons.
- Inverted sugar has an important role in the food industry. This compound consists of a mixture of glucose and fructose and is obtained by splitting the glycosidic bond of sucrose.
- Sucrose is an extraordinarily nutritious constituent of a great number of foods; however, its ingestion is associated with public health problems, including diabetes, obesity and other chronic health diseases.

Key Facts

- Sucrose is most commonly extracted from sugar cane or the sugar beet.
- First, molasses, which consists of a brown syrup containing sucrose and other byproducts, is produced. This product is rich in calcium and iron.
- Raw sugar is the first part of the crystals collected during sucrose production. It contains impurities and has only passed through half of the stages of refinement.
- Refined sugar or white sugar appears after the dissolution, concentration and recrystallisation of raw sugar, and the refined sugar is only composed of sucrose.

- Brown sugar is obtained through the addition of molasses to the refined sugar, or it may also simply be sugar with residual molasses. The flavour and colour of molasses are natural, and it contains 91 to 96% sucrose.

Definitions of Words and Terms

Carbohydrates: Relevant biochemical compounds that are the main natural source of energy for living beings. Carbohydrates are basically polyhydroxy aldehydes or polyhydroxy ketones.

Isomers: Compounds with the same molecular formula but with different physico-chemical properties due to the function of the different spatial distributions of their atoms.

Triacylglycerol: A relevant lipid that constitutes the principal reservoir of energy for organisms because its hydrolysis generates more calories compared to carbohydrates or proteins.

Hermann Emil Fischer: Hermann Emil Fischer (1852–1919) was a German chemist and recipient of the Nobel Prize for Chemistry in 1902. He is known for discovering Fischer esterification and proposing the Fischer projection to didactically represent the tridimensional molecular structure of carbohydrates in a bidimensional form. Furthermore, Fischer developed several different lines of research and is considered by various authors to be one of the most important chemists and/or biochemists of the early twentieth century.

Reactive Oxygen Species (ROS): A group of compounds that contain oxygen atoms with significant chemical reactivities in their molecular structures. These types of compounds commonly produce free radicals and provoke significant oxidative stress.

Reactive Nitrogen Species (RNS): A group of compounds that contain nitrogen atoms with significant chemical reactivities in their molecular structures. These types of compounds commonly produce free radicals and provoke significant oxidative stress.

Artificial sweeteners: These compounds are also known as "sugar substitutes" and are substances often used instead of sucrose to sweeten foods and beverages. Because artificial sweeteners are many times sweeter than table sugar, smaller amounts are needed to create the same level of sweetness. The major artificial sweeteners include saccharin, aspartame, cyclamate, acesulphame potassium, sucralose and neotame. As a function of the caloric impact of carbohydrates in daily diets, the complex and general problems regarding obesity worldwide have provoked the development and popularisation of "sugar substitutes". Furthermore, controlling the symptoms of diabetes is one of the major factors that has motivated the employment of these "sugar substitutes".

Inverted sugar: An inverted sugar is the denomination for a mixture of glucose and fructose, which is obtained by splitting sucrose. When compared to sucrose, an inverted sugar possesses a sweeter taste, has less of a tendency to crystallise and its products have a higher tendency to retain humidity. Therefore, inverted sugar is valuable for applications for the preparation of several types of foods, such as for culinary purposes and especially in bakeries.

Glycocalyx: An external coat on many cells that is composed of glycoproteins and polysaccharides. It is used for cellular communication.

Diabetes: A metabolic disease characterised by an elevation in blood sugar. It is caused by a failure in the production of insulin by the pancreas. Insulin is the hormone that regulates the glucose level in blood.

List of Abbreviations

DNA   deoxyribonucleic acid
FDA   Food and Drug Administration
N   nitrogen atoms
H   hydrogen atoms
OH   hydroxyl groups
P   phosphorus
RNA   ribonucleic acid
RNS   reactive nitrogen species
ROS   reactive oxygen species
S   sulphur atoms
UDP   uridine diphosphate glucose

References

Barratt, D.H., Derbyshire, P., Findlay, K., Pike, M., Wellner, N., Lunn, J., Feil, R., Simpson, C., Maule, A.J., Smith, A.M., 2009. Normal growth of *Arabidopsis* requires cytosolic invertase but not sucrose syntase. *Proceedings of the National Academy of Science USA* 106: 13124–13129.

Gerrits, N., Turk, S.C., van Dun, K.P., Hulleman,S.H., Visser, R.G., Weisbeek, P.J., Smeekens, S.C., 2001. Sucrose metabolism in plastids. *Plant Physiology* 125: 926–934.

Grennan, A.K., Gragg, J., 2009. How sweet it is: identification of vacuolar sucrose transporters. *Plant Physiology* 150: 1109–1110.

Koch, K., 2004. Sucrose metabolism: regulatory mechanisms and pivotal roles in sugar sensing and plant development. *Current Opinion in Plant Biology* 7: 235–246.

Lee, O., Bruce, W.R., Dong, Q., Mehta, R., O'Brien, P.J., 2009. Fructose and carbonyl metabolites as endogenous toxins. *Chemico-Biological Interactions* 178: 332–339.

Mahan, B.M. and Myers, R.J. 1987. *University Chemistry*, 4 ed. The Benjamin/ Cummings Publishing Company, United States of America, 1076p.

Queneau, Y., Fitremann, J., Trombotto, S., 2004. The chemistry of unprotected sucrose: The selectivity issue. *Comptes Rendus Chimie* 7: 177–188.

Sauer, N., 2007. Molecular physiology of higher plant sucrose transporters. *FEBS Letters* 581: 2309–2323.

Szarka, A., Horemans, N., Passarella, S., Tarcsay, A., Orsi, F., Salgó, A., Bánhegyi, G., 2008. Demonstration of an intramitochondrial invertase activity and the corresponding sugar transporters of the inner mitochondrial membrane in Jerusalem artichoke (*Helianthus tuberosus* L.) tubers. *Planta* 228: 765–775.

Wind, J., Smeeekens, S., Hanson, J., 2010. Sucrose: Metabolite and signaling molecule. *Phytochemistry* 71: 1610–1614.

CHAPTER 10
Lactose Chemistry

LEE D. HANSEN AND JENNIFER B. NIELSON*

Department of Chemistry and Biochemistry, Brigham Young University,
Provo, Utah, 84602 USA,
*Email: jnielson@chem.byu.edu

10.1   Introduction

(+)-Lactose, also known as lactobiose, milk sugar, β-D-galactopyranosyl-(1→
4)-D-glucopyranose, β-D-galactopyranosyl-(1→4)-D-glucose and β-D-galactosyl-
(1→4)-D-glucose, is a sugar found in relatively high concentrations (2–8%) in
mammalian milk. Lactose, obtained in large quantities as a byproduct from
production of cheese and other dairy products, is used in many manufactured
foods and as an excipient in pharmaceuticals. The bland flavor and low
sweetness of lactose makes it useful as a binder and extender without detracting
from the flavor of the product.

  Lactose is a disaccharide of galactose and glucose with the following α- and
β-anomeric molecular structures (Figure 10.1). The α- and β-anomers differ in
the stereochemistry of the –OH group at carbon 1, also known as the anomeric
carbon. The β-anomer is sweeter than the α-anomer and an equilibrium mixture of both anomers is approximately 1/3 as sweet as a solution of sucrose. In
aqueous solutions at equilibrium at room temperature, lactose exists as a 40:60
mixture of the α- and β-anomers. The β-anomer is more stable, because the
OH group on the anomeric carbon is in the equatorial position and is better
solvated. The α-anomer is destabilized by steric interactions between the axial
anomeric –OH and axial hydrogens on the ring. The ratio of α- to β- is higher
though than would be predicted for cyclohexanol (16:84 mixture of

Food and Nutritional Components in Focus No. 3
Dietary Sugars: Chemistry, Analysis, Function and Effects
Edited by Victor R Preedy
© The Royal Society of Chemistry 2012
Published by the Royal Society of Chemistry, www.rsc.org

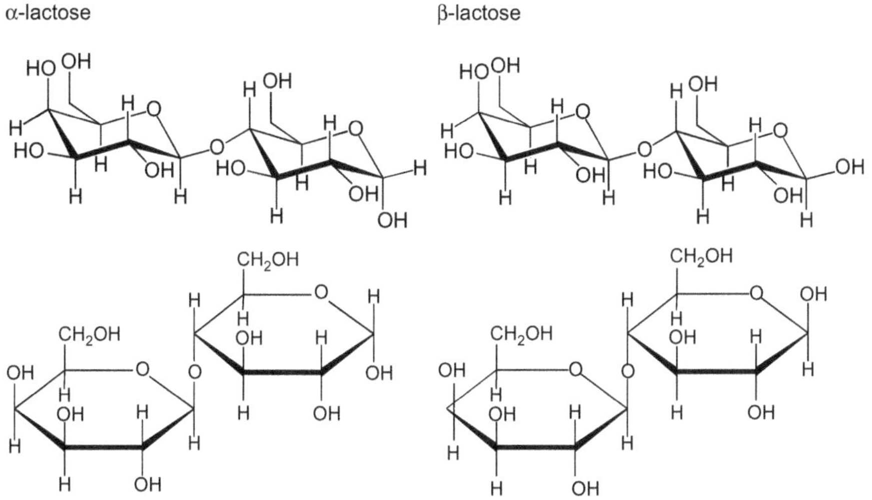

α-lactose   β-lactose

Figure 10.1 Structures of lactose. Lactose is a disaccharide composed of two smaller sugar molecules, galactose (on the left) and glucose (on the right). Lactose has two structural isomers, labeled α and β, that differ in the orientation of the hydrogen atom and the hydroxyl group on the glucose portion of the molecule. Two ways of representing the structures are shown.
(Unpublished.)

axial:equatorial), because there is some stabilization of the α-anomer from overlap of the lone pair electrons of the ring oxygen with an empty anti-bonding orbital of the axial –OH, a phenomenon known as the anomeric effect. In solution, the α and β-anomers readily interconvert by mutarotation, and as temperature increases, the ratio of α-anomer to β-anomer concentration increases (Listiohadi *et al.* 2005).

10.2   Crystal Structures and Properties

Crystallization of lactose is complex because of the presence of anomers and polymorphs. Polymorphs are different crystal structures in the solid state that have identical properties when dissolved in an equilibrium solution. In the solid state, in addition to an amorphous form, there are four polymorphs of lactose; α-lactose monohydrate (Lα · H$_2$O), β-lactose (Lβ), and two forms of anhydrous α-lactose (Lα$_S$ and Lα$_H$) (Gänzle *et al.* 2008; Kirk *et al.* 2007, Listiohadi *et al.* 2005 and 2008).

Crystallization from water below 93.5 °C produces Lα · H$_2$O, the commonly available polymorph. The pure Lβ polymorph can be prepared by purification of the equilibrium 40:60 mixture obtained by rapid drying of an aqueous solution. Lα$_S$ and Lα$_H$ are obtained by dehydration of Lα · H$_2$O. Lα$_S$ is stable, but Lα$_H$ is hygroscopic. Some of the properties of the pure polymorphs are given in Table 10.1 (Gänzle *et al.* 2008).

Table 10.1 Properties of the polymorphs of lactose. Crystalline lactose exists
in three different crystal structures or polymorphs designated as
$L\alpha \cdot H_2O$, $L\alpha$, and $L\beta$. These polymorphs have different physical
properties. Data are selected from several secondary sources such
as handbooks and data tables.

Formula	
$L\alpha \cdot H_2O$	$C_{12}H_{24}O_{12}$
$L\alpha$ and $L\beta$	$C_{12}H_{22}O_{11}$
Molecular mass	
$L\alpha \cdot H_2O$	360.31
$L\alpha$ and $L\beta$	342.30
Appearance	white
Density	
$L\alpha \cdot H_2O$	1.525 g/cm^3
$L\alpha$	
$L\beta$	1.59 g/cm^3
Melting point	
$L\alpha \cdot H_2O$	loses water at 130 °C, melts at 201–2 °C
$L\alpha$	222.8 °C
$L\beta$	253 °C
Enthalpy of formation	
$L\alpha \cdot H_2O$	−2519 kJ/mole
$L\alpha$	
$L\beta$	−2218 kJ/mole
Enthalpy of combustion	
$L\alpha \cdot H_2O$	−5632 kJ/mole
$L\alpha$	
$L\beta$	−5648.4 kJ/mole

Methods for distinguishing different lactose polymorphs include X-ray
diffraction patterns, NMR, Raman and IR spectra, and thermal properties
(Kirk *et al.* 2007; Listiohadi *et al.* 2005 and 2008; Shah *et al.* 2006,). Determination of the relative amounts of polymorphs in mixtures is more challenging; methods for quantification in mixtures include powder X-ray diffraction
(Kirk *et al.* 2007; Snyder and Bish 1989), solid state NMR, Raman and IR
spectrophotometry (Kirk *et al.* 2007), differential scanning calorimetry, and
isothermal calorimetry (Khalef *et al.* 2010; Shah *et al.* 2006). A summary and
comparison of available methods is given in Listiohadi *et al.* (2005 and 2008).

The form of lactose produced during food processing and storage depends on
temperature, water activity, and the nature of the food (Gänzle *et al.* 2008). The
crystalline form affects the texture and flavor of the product. The texture
imparted by lactose depends on the size of the crystals and the flavor depends
on the rate of solubility and the anomeric form. Amorphous lactose has a soft,
smooth texture, but can crystallize during processing or storage and impart an
undesirable, gritty texture to foods. The solubility of lactose is of interest in
predicting the effects of food processing and storage. But such prediction is
difficult because of mutarotation and polymorphism and because lactose
readily forms supersaturated aqueous solutions. Large crystals can form in a

supersaturated solution to give the food the grainy texture mentioned above (Gänzle *et al.* 2008). One method to produce a smooth texture is to control the crystallization with temperature, the rate of agitation, and by adding small seed crystals of lactose. Crystallization to an anhydrous form can release water trapped in amorphous lactose and increase the water activity of dry milk powders (Thomsen 2005). In dairy products frozen at temperatures between 0 °C and −15 °C, anhydrous lactose can crystallize in its hydrate form, Lα · H₂O. The milk proteins are concentrated by this removal of water and can aggregate, imparting off-flavors and changing the texture of the product. Several methods, such as freezing the food quickly to temperatures lower than −20 °C, have been noted in the literature (Muir 1990).

10.3   Reactions of Lactose

Lactose, like most disaccharides, is subject to hydrolysis of the glycosidic linkage. Other important reactions at the anomeric carbon include oxidation, reduction, and acid-base reactions of aldehydes, since the –CHO functional group is available through mutarotation. Many of these reactions lead to changes in texture, taste, or stability of food products that contain lactose. Some of the modified-lactose products of these reactions are considered value-added products and are commercially important. Other reactions have been developed to degrade lactose in whey waste, since the disposal of whey is a major economic and environmental concern in the production of cheese.

10.3.1   Hydrolysis and Transglycosylation

Lactose is hydrolyzed by the enzyme lactase (a β-galactosidase) to galactose and glucose. A person that is lactose-intolerant frequently does not have the lactase enzyme and cannot digest dairy products. Hydrolysis of lactose commercially has therefore become an important reaction in producing low-lactose foods that can be eaten by consumers who are lactose-intolerant. The sweeter monosaccharides are also used in sweetening syrups for ice cream and other foods.

In the hydrolysis reaction, water "splits" (lysis) the disaccharide into galactose and glucose (Figure 10.2). The reaction using an acid catalyst was reported in the literature in 1925 (Whittier 1925), but has been of little commercial use because of the harsh acid and temperature conditions. However, Coté *et al.* (2004) report a process using mineral acids that produces a product suitable for fermentation to ethanol. Most industrial processes of hydrolysis of lactose use β-galactosidases from various organisms as the catalyst (Gänzle *et al.* 2008, Siso 1996). The structure and mechanism of the enzyme have been well documented (Smart 1991).

β-galactosidase enzymes also catalyze transglycosylation reactions that are similar in chemistry to hydrolysis. In a transglycosylation reaction, a sugar monomer is transferred from one saccharide to another saccharide. For

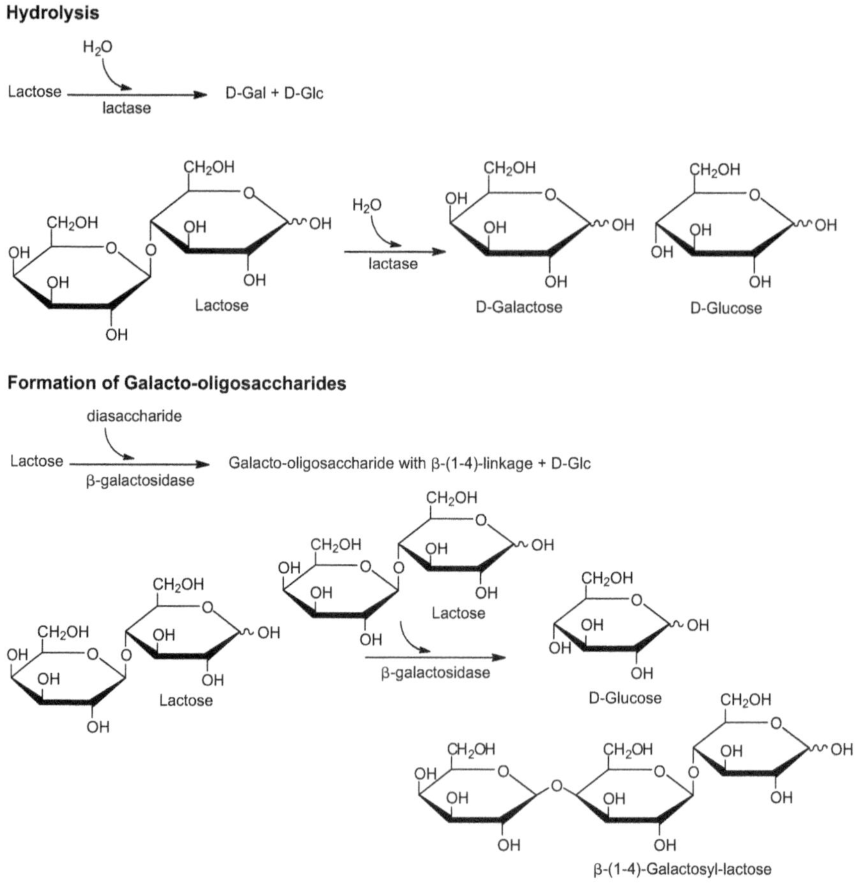

Hydrolysis

Formation of Galacto-oligosaccharides

Figure 10.2 Hydrolysis of lactose by water to produce galactose (D-Gal) and glucose (D-Glc) and the general reaction for the formation of galacto-oligo-saccharides with a specific example of lactose as the disaccharide nucleophile. Hydrolysis of lactose, which is a disaccharide, to two monosaccharides, galactose and glucose, is an important reaction of lactose in both biology and industry. Reaction of lactose with itself to form galactosyl-lactose trisaccharides, is an important reaction that occurs during processing of milk products at elevated temperatures.

example, in transgalactosylation, a β-galactosidase enzyme can transfer a galactose from lactose to a disaccharide or oligosaccharide to form galacto-oligosaccharides, GOS. A monomer of glucose is still created. This reaction is analogous to hydrolysis, but instead of water as the nucleophile, a disaccharide or other molecule with a hydroxyl group acts as the nucleophile to form another glycosidic linkage, such as β-(1→4) or β-(1→6). β-Galactosidase is labeled as a hydrolase enzyme, but it is more correctly classified as a transferase enzyme. Hydrolysis of lactose is then a special galactosyl reaction where water is the galactose acceptor (Splechtna *et al.* 2006). One example of transgalactosylation is the formation of the trisaccharide β-(1→4)-galactosyl-lactose. The diasaccharide nucleophile is another molecule of lactose.

The formation of GOS has historically been a problem in the production of lactose and its monosaccharide hydrolysis products (Mahoney 1998). GOS crystallize in large crystals and can cause an undesirable grainy texture in foods. Since hydrolysis of GOS is slower than lactose, it has been suggested that GOS causes symptoms associated with lactose intolerance. Also because the hydrolysis of GOS takes longer, the production costs are higher. The traditional method for estimating processing time for hydrolysis has been based on monitoring the concentration of the monosaccharides produced, but this has been proven to be inaccurate because of the formation of GOS.

GOS and other oligosaccharides found in milk (Messer and Urashima 2002; Urashima *et al.* 2001) have received much attention over the last 20 years because of their possible health benefits (Mussatto and Mancilha 2007; Schaafsma 2008). GOS can be classified as prebiotic compounds. Prebiotic compounds stimulate the growth of good bifidobacteria in the colon, as opposed to probiotics such as *L. acidophilus* in yogurt that are eaten to introduce more bacteria into the colon. GOS also act as dietary fiber and there is good evidence of their beneficial effects in absorbing minerals and preventing diarrhea. Breast milk has a high concentration of GOS and they are credited with these effects in babies. Some countries, *e.g.* Japan, have incorporated GOS into food products for many years based on the research evidence of GOS as prebiotic compounds (Tomomatsu 1994). Other useful properties of GOS include their ability to hold moisture and not dry out completely, but their water content is low enough that bacteria and fungi growth can be controlled (Crittenden and Playne 1996). Modification of lactose into galacto-oligo-saccharides is potentially one of the most useful reactions because of the possibility of value-added products and because it would help alleviate the whey waste problem in cheese production.

Another important transglycosylation reaction is the transfer of fructose from a sucrose molecule to lactose to form the product lactosucrose. This fructosyl transfer reaction is catalyzed by the enzyme β-fructofuranosidase and creates a new β-(2→1) glycosidic bond between the fructose and the glucose end of α-lactose (Figure 10.3). The enzyme also hydrolyzes sucrose and lactosucrose, so the concentration of lactosucrose that can be produced from lactose is very dependent on starting sucrose concentrations. Lactosucrose is 0.3–0.6 times the sweetness of sucrose, is often used as a bulking agent in foods, and also has beneficial prebiotic properties.

10.3.2   Isomerization

An isomerization reaction is one in which the atoms of a molecule are rearranged; the product has the same molecular formula as the reactant. One of the most common isomers of lactose is allolactose, which has a β-(1→6) glycosidic linkage instead of β-(1→4). The isomerization reaction is catalyzed by

Formation of Lactosucrose

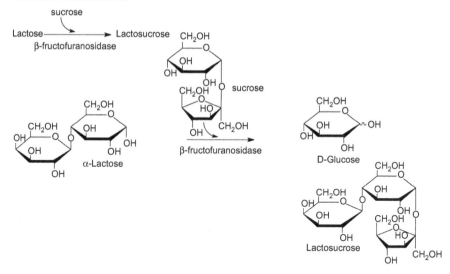

Figure 10.3 Formation of lactosucrose from lactose and sucrose. Reaction of lactose with sucrose, another disaccharide, produces lactosucrose, a synthetic trisaccharide. Lactosucrose is less sweet than sucrose and is used as a bulking agent in foods.

β-galactosidase. Allolactose is an inducer for the lac operon, which controls for the production and transport of lactose.

One of the most important isomerization products is the synthetic sugar lactulose, 4-O-β-D-galactopyranosyl-D-fructofuranose (or β-D-galactopyranosyl-(1 → 4)-D-fructofuranose). Lactulose has widespread use in pharmaceuticals, most commonly as a medicine for chronic constipation and hepatic encephalopathy. A small amount of lactulose is produced in the sterilization of milk and the concentration of lactulose has been used to monitor the heat processing of milk. Commercial production, however, requires alkaline reaction conditions (Aider and Halleux 2007). The isomerization of the glucose pyranose ring in lactose to the fructose furanose ring occurs through the open-chain glucose aldehyde end in mutarotation. The reaction is also known as a tautomerization reaction where the aldose is converted to the ketose *via* an enediol intermediate (Figure 10.4).

10.3.3 Maillard Reaction

Another important reaction in lactose chemistry is the reaction with amino acids to form Maillard products (Muir 1996). End products of the Maillard reaction can negatively alter the properties of dairy products. For example, Maillard products may be the cause of bitter flavors sometimes formed in cheeses during aging, although the bitter taste was previously blamed on the

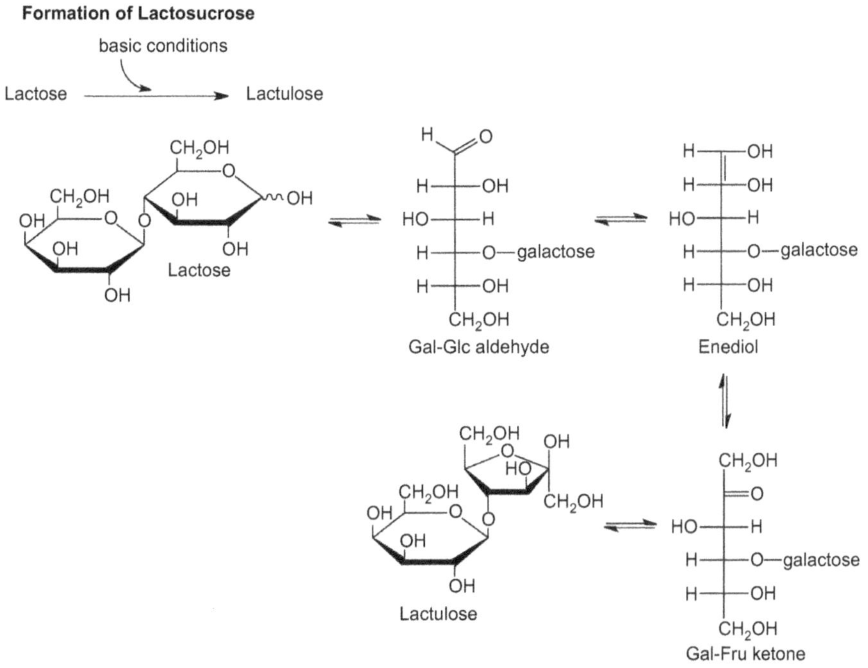

Formation of Lactosucrose

basic conditions

Lactose ⟶ Lactulose

Figure 10.4 Formation of lactulose from the isomerization of lactose in base. Lactose → Gal-Glc aldehyde (open chain aldehyde form of Lactose) → enediol intermediate → Gal-Fru ketone (open chain ketone form of Lactulose) → Lactulose. Lactulose is produced when a basic solution of lactose is heated. The mechanism of the reaction is shown in this figure. Lactulose is a synthetic sugar used in several pharmaceutical applications.

presence of unwanted products of bacterial metabolism (Lawrence *et al.* 1976). The end products are also intensely colored and even small amounts can impart a brownish color to products. In addition, since lysine in milk proteins is reactive, the milk then has less nutritional value since there is less of this essential amino acid (Muir 1990).

Lactose initially forms an imine or Schiff base with amino acids, which rearranges to early Maillard products known as Amadori products (Figure 10.5). The Amadori products are often stable and do not affect the properties of the milk. But upon prolonged heating, many of these compounds further react to form molecules that alter the milk properties as suggested above or form harmful products. For example, Maillard products formed from lactose and asparagine can release acrylamide when heated (Stadler 2005). Acrylamide, a known carcinogen, is produced from the decarboxylation of the imine and then beta-elimination (Figure 10.6). Other carcinogens formed during thermal degradation are furan (Vranova and Ciesarova 2009) and the flavoring agent maltol (Anwar-Mohamed and El-Kadi 2007). Maillard products formed from

Formation of Maillard Products

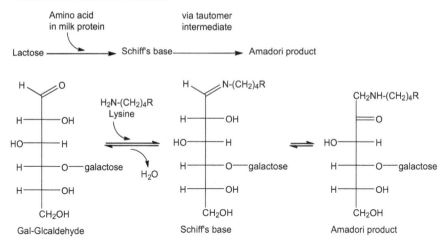

Amino acid
in milk protein

via tautomer
intermediate

Lactose ⟶ Schiff's base ⟶ Amadori product

Gal-Glcaldehyde   Schiff's base   Amadori product

Figure 10.5   General reaction for the formation of early Maillard products. A specific
example of lactose reacting with lysine in a milk protein is illustrated.
Lactose reacts to form a Schiff's base intermediate which isomerizes to an
Amadori product *via* an intermediate similar to the enediol shown in
Figure 10.4. R is the rest of the milk protein.

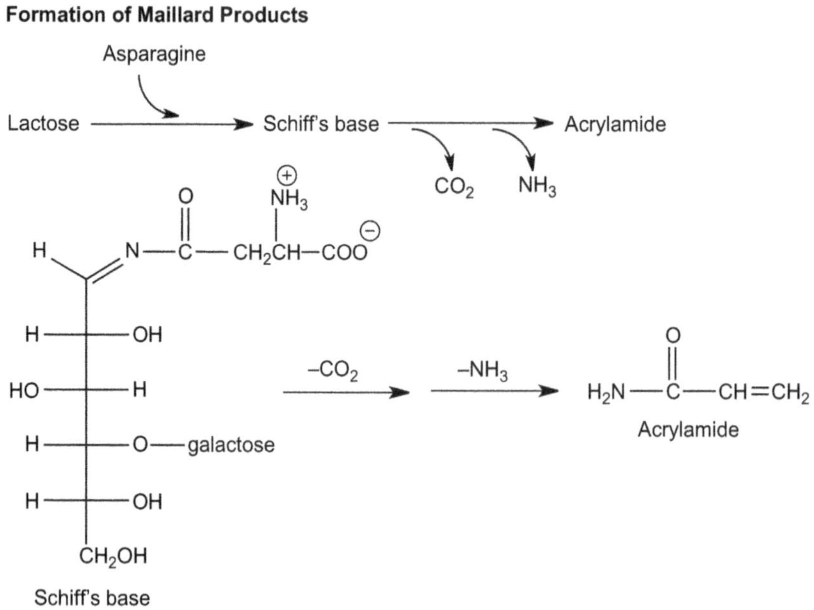

Formation of Maillard Products

Asparagine

Lactose ⟶ Schiff's base ⟶ Acrylamide

CO_2 NH_3

$-CO_2$ $-NH_3$ H_2N—C—CH=CH$_2$

Acrylamide

Schiff's base

Figure 10.6   Reaction of lactose with asparagine to form acrylamide. The amino acid
asparagine can be thermally degraded to acrylamide, a neurotoxin and
probable carcinogen, *via* a Schiff base intermediate with lactose. The first
step in the reaction to form the Schiff base is not shown.

lactose do not show high mutagenicity properties as have been found for the Maillard products of the monosaccharides glucose and fructose, and, to a lesser extent, the disaccharide lactulose (Brands *et al.* 2000).

Other late Maillard products are formed from the degradation of Amadori compounds and eventually become melanoidins, which discolor the milk product. The structures of melanoidins are known to have high molecular weights and contain nitrogen, but are not well characterized. Addition of GOS can be used to control the intensity of browning due to Maillard reactions in heat processed foods (Mussatto and Mancilha 2007).

Maillard compounds are often difficult to analyze in milk products, but methods such as HPLC, and electrospray mass spectrometry have been successfully used to identify both early and late amine-free Maillard products in milk (French *et al.*, 2002; Jones *et al.* 1998; Peschetsrieder *et al.* 1999).

10.3.4   Fermentation

Fermentation is an ancient process used to increase the shelf-life, add to the taste, and improve the digestibility of milk. Fermentation of lactose to lactic acid, most commonly done with *Lactobacillus*, *Lactococcus*, and *Leuconostoc*, is the reaction used to produce yogurt and other fermented milk products (Lawrence *et al.* 1976). These products are of particular interest for consumers who are lactose intolerant. Depending on the micro-organism and conditions, fermentation of lactose can also produce pyruvate, lactate, acetate, ethanol, and a variety of other compounds (Audic *et al.* 2003; Yang and Silva 1995). Lactose can be fermented directly to ethanol by *Kluyveromyces fragilis* (Siso 1996) and indirectly by hydrolysis followed by fermentation with *Saccharomyces cerevisiae*.

10.4   Disposal

Disposal of excess whey and lactose is an ongoing environmental problem (Durham *et al.* 2004; Giacomo *et al.* 1996; Nickerson 1979), and development of uses for whey and lactose has been of interest for hundreds of years (Yang and Silva 1995). Europe and North America produce about 1.5 million tons annually (Gänzle *et al.* 2008). Worldwide, several million tons of lactose are produced annually (Siso 1996). About half is used for food production, mainly in animal feed. The other half is discarded, much of it without treatment. Discarded onto soil or into water bodies, oxygen is rapidly depleted, the soil or water is acidified, and the system becomes eutrophic.

Some of the beneficial uses for excess whey that have been suggested (Audic *et al.* 2003; Collet *et al.* 2004; Coté *et al.* 2004; Lifran *et al.* 2000; Mawson 1994; Siso 1996; Yang and Silva 1995) are production of methane fuel and hydrogen fuel by anaerobic digestion, production of bioethanol for fuel, and production of single cell protein by growth of bacteria on whey. Lactose can also be used to make adhesives, polymers, and a variety of organic chemicals (Lifran *et al.*

2000; Yang and Silva 1995). Modification of lactose into galacto-oligo-saccharides and similar products are important reactions because of the possibility of value-added products and alleviation of the problem of whey disposal (Gänzle *et al.* 2008; Mussatto and Mancilha 2007; Schaafsma 2008).

Summary Points

- Lactose is a disaccharide of galactose and glucose found in high concentrations in mammalian milk and whey from cheese-making.
- Lactose exists as α- and β-anomers and four crystalline polymorphs; α-lactose monohydrate, β-lactose, and two forms of anhydrous α-lactose.
- The important reactions of lactose in food chemistry are hydrolysis, transglycosylation, isomerization, and Maillard reactions.
- Fermentation of lactose to lactic acid is the reaction used to produce yogurt and other fermented milk products, but fermentation of lactose can also produce pyruvate, lactate, acetate, ethanol, and a variety of other compounds.
- Despite the possibilities of value-added products, disposal of excess whey and lactose is an ongoing environmental problem.

Key Facts

- Lactose is a sugar that makes up 2–8% of the mass of mammalian milk.
- Large quantities of lactose are obtained as a byproduct of production of cheese and other dairy products.
- Lactose is used in many manufactured foods for both people and animals and as an excipient in pharmaceuticals.
- Lactose is about one-third as sweet as sucrose.
- Lactose crystallizes in four different forms or polymorphs. The crystalline form affects the flavor and texture of foods with added lactose.
- Lactose is a disaccharide that can be hydrolyzed to produce two monosaccharides, galactose and glucose.
- Lactose can react with itself or with other sugars to produce oligosaccharides with prebiotic properties.
- At high temperatures or over long periods of time, lactose reacts with milk proteins to produce compounds that are highly colored, bitter and sometimes harmful.
- During fermentation of milk to make yogurt, lactose is converted into lactic acid.
- Disposal of lactose and whey into water and onto land acidifies the water or soil and rapidly uses up the oxygen, causing the system to become eutrophic.
- Lactose can be used to make methane and other fuels, protein, adhesives, polymers, and a variety of organic chemicals.

Definition of Words and Terms

Aldehyde. An organic compound with a –CHO group on a terminal carbon.

Anomers. Stereoisomers of a cyclic saccharide that differ only in orientation at the anomeric carbon, which is the carbon at which the sugar forms the ring from its open-chain structure.

Disaccharide. A sugar composed of two monosaccharides.

Enzyme. Protein that catalyzes a reaction or reactions.

Eutrophic. A low oxygen environment.

Excipient. An inactive material added to drugs as a binder and extender.

Fermentation. Biological process that converts a sugar into lactic acid.

Galacto-oligosaccharide. A short polymer of sugar molecules with galactose on the end of the polymer.

Glycosidic linkage. The chemical bond between two sugar molecules.

Hydrolysis. Splitting of a chemical bond by addition of a water molecule, one of the atoms that was making the original bond gains a hydrogen and the other atom gains an OH group from the water.

Isomerization. Rearrangement of the atoms in a molecule with no loss or gain of any atoms.

Maillard reaction. A series of chemical reactions that begins with reaction of a reducing sugar with an amine group, usually an amino acid in a protein. The ultimate product depends on the amine, the sugar, and the conditions.

Monosaccharide. A simple sugar molecule, typically with 3 to 7 carbon atoms.

Mutarotation. The conversion of one anomer of a sugar molecule into the other anomer.

Oligosaccharide. A short polymer of sugar molecules, usually containing 3–10 monosaccharides.

Oxidation. Removal of electrons or hydrogen atoms or addition of oxgen to a molecule, ion or atom.

Polymorphs. Different crystal structures of the same molecular formula.

Reduction. Addition of electrons or hydrogen atoms or removal of oxygen from a molecule, ion or atom.

Saccharide. A sugar. The general formula is $C_n(H_2O)_n$ with $n \geq 3$.

Transglycosylation. Transfer of a sugar molecule from one oligosaccharide to another sugar molecule or oligosaccharide.

Trisaccharide. A sugar polymer composed of three monosaccharides.

Whey. The liquid byproduct of cheese-making, the major constituents are water, lactose and salts.

β-Galactosidase. An enzyme that catalyzes hydrolysis of galactose from an oligosaccharide.

List of Abbreviations

| Fru | fructose |
| Gal | galactose |

Glc   glucose
GOS   galacto-oligosaccharides
HPLC   high performance liquid chromatography
IR   infra-red
$L\alpha \cdot H_2O$ α-lactose monohydrate
$L\alpha_S$ and $L\alpha_H$ anhydrous α-lactose
$L\beta$ β-lactose
NMR   nuclear magnetic resonance

References

Aider, M., and Halleux, D., 2007. Isomerization of lactose and lactulose production: review, 2007. *Trends Food Sci. & Tech.* 18: 356–364.

Anwar-Mohamed, A., and El-Kadi, A.O., 2007. Induction of cytochrome P450 1a1 by the food flavoring agent, maltol. *Toxicol. Vitro.* 21: 685–690.

Audic, J-L., Chaufer, B., and Daufin, G., 2003. Non-food applications of milk components and dairy co-products: A review. *Lait.* 83: 417–438.

Brands, C.M.J., Alink, G.M., van Boekel, M.A.J.S., and Jongen, W.M.F., 2000. Mutagenicity of heated sugar - casein systems: Effect of the Maillard reaction. *J. Agric. Food Chem.* 48: 2271–2275.

Collet, C., Adler, N., Schwitzguebel, J-P., and Peringer P., 2004. Hydrogen production by *Clostridium thermolacticum* during continuous fermentation of lactose. *Int. J. Hydrogen Energy.* 29: 1479–1485.

Coté, A., Brown, W.A., Cameron, D., and van Walsom, G.P., 2004. Hydrolysis of lactose in whey permeate for susequesnt fermentation to ethanol. *J. Dairy Sci.* 87: 1608–1620.

Crittenden, R.G., and Playne, M.J., 1996. Production, properties, and applications of food-grade oligosaccharides. *Trends Food Sci. & Tech.* 71: 353–361.

Durham, R.J., Sleigh, R.W., and Hourigan, J.A., 2004. Pharmaceutical lactose: a new whey with no waste. *Australian J. Dairy Tech.* 59: 138–141.

French, S.J., Harper, W.J., Kleinholz, N.M., Jones, R.B., and Green-Church, K.B., 2002. Maillard reaction induced lactose attachment to bovine beta-lactoglobulin: Electrospray ionization and matrix-assisted laser desorption/ionization examination. *J. Agric. Food Chem.* 50: 820–823.

Gänzle, M.G., Haase, G., and Jelen, P., 2008. Lactose: Crystallization, hydrolysis and value-added derivatives. *Int. Dairy J.* 18: 685–694.

Giacomo, G.D., Re, and G.D., Spera, D., 1996. Milk whey treatment with recovery of valuable products. *Desalination.* 108: 273–276.

Jones, A.D., Tier, C.M., and Wilkins, J.P.G., 1998. Analysis of the Maillard reaction products of beta-lactoglobulin and lactose in skimmed milk powder by capillary electrophoresis and electrospray mass spectrometry. *J. Chromatography A.* 822: 147–154.

Khalef, N., Pinal, R., and Bakri, A., 2010. Limitations of amorphous content quantification by isothermal calorimetry using saturated salt solutions to control relative humidity: Alternative methods. *J. Pharma. Sci.* 99: 2080–2089.

Kirk, J.H., Dann, S.E., and Blatchford, C.G., 2007. Lactose: A definitive guide to polymorph determination. *Int. J. Pharma.* 334: 103–114.

Lawrence, R.C., Thomas, T.D., and Terzaghi, B.E., 1976. Reviews of the progress of dairy science: Cheese starters. *J. Dairy Res.* 43: 141–193.

Lifran, E.V., Hourigan, J.A., Sleigh, R.W., and Johnson, R.L., 2000. New wheys for lactose. *Food Australia.* 52: 120–125.

Listiohadi, Y., Hourigan, J.A., Sleigh, R.W., and Steele, R.J., 2005. Properties of lactose and its caking behaviour. *Australian J. Dairy Tech.* 60: 2005, 33–52.

Listiohadi, Y., Hourigan, J.A., Sleigh, R.W., and Steele, R.J., 2008. Moisture sorption, compressibility and caking of lactose polymorphs. *Int. J. Pharma.* 359: 123–134.

Mahoney, R.R., 1998. Galactosyl-oligosaccharide formation during lactose hydrolysis: a review. *Food Chem.* 63: 147–154.

Mawson, A.J., 1994. Bioconversions for whey utilization and waste abatement. *Bioresource Tech.* 47: 195–203.

Messer, M., and Urashima, T., 2002. Evolution of milk oligosaccharides and lactose. *Trends Glycosci. and Glycotech.* 14: 153–176.

Muir, D.D., 1990. Lactose. *J. Soc. Dairy Tech.* 43: 33–43.

Muir, D.D., 1996. The shelf-life of dairy products: 3. Factors influencing intermediate and long life dairy products. *J. Soc. Dairy Tech.* 49: 67–72.

Mussatto, S.I., and Mancilha, I.M., 2007. Non-digestible oligosaccharides: A review. *Carbohydrate Polymers.* 68: 587–597.

Nickerson, T.A., 1979. Lactose chemistry. *J. Agric. Food Chem.* 27: 672–677.

Pischetsrieder, M., Gross, U., and Schoetter, C., 1999. Detection of Maillard products of lactose in heated or processed milk by HPLC/DAD. Z. Lebensmittel-Untersuchung Forschung A – *Food Res. Tech.* 208: 172–177.

Schaafsma, G., 2008. Lactose and lactose derivatives as bioactive ingredients in human nutrition. *Int. Dairy J.* 18: 458–465.

Shah, B., Kakumanu, V.K., and Bansal, A.K., 2006. Analytical techniques for quantification of amorphous/crystalline phases in pharmaceutical solids. *J. Pharma. Sci.* 95: 1641–1665.

Siso, M.I.G., 1996. The biotechnological utilization of cheese whey: A review. *Bioresource Tech.* 57: 1–11.

Smart, J. B., 1991. Transferase reactions of the β-galactosidase from *Streptococcus thermophilus. Appl. Microbiol. Biotech.* 34: 495–501.

Snyder, R.L., and Bish, D.L., 1989. Quantitative analysis. In: Bish, D.L. and Post, J.E. (eds) *Modern Powder Diffraction, Reviews in Mineralogy.* Washington, D.C.: Mineralogical Society of America. pp. 101–144.

Splechtna, B., Nguyen, T., Steinbo, M., Kulbe, K., Lorenz, W., and Haltrich, D., 2006. Production of prebiotic galacto-oligosaccharides from lactose using β-galactosidases from lactobacillus reuteri. *J. Agric. Food Chem.* 54: 4999–5006.

Stadler, R.H., 2005. Acrylamide formation in different foods and potential strategies for reduction. In: Friedman, M., Mottram, D. (eds) Chemistry and

Safety of Acrylamide in Food. *Advances in Experimental Medicine and Biology.* 561: 157–169.

Thomsen, M.K., Lauridsen, L., Skibsted, L.H., and Risbo, J., 2005. Two types of radicals in whole milk powder. Effect of lactose crystallization, lipid oxidation, and browning reactions. *J. Agric. Food Chem.* 53: 1805–1811.

Tomomatsu, H., 1994. Health effects of oligosaccharides. *Food Tech.* 48: 61–65.

Urashima, T., Saito, T., Nakamura, T., and Messer, M., 2001. Oligosaccharides of milk and colostrum in non-human mammals. *Glycoconjugate J.* 18: 357–371.

Vranova, J., and Ciesarova, Z., 2009. Furan in food-A review. *Czech J. Food Sci.* 27: 1–10.

Whittier, E.O., 1925. Lactose. *Chem. Rev.* 2: 85–125.

Yang, S.T., and Silva, E.M., 1995. Novel products and new technologies for use of a familiar carbohydrate, milk lactose. *J. Dairy Sci.* 78: 2541–2562.

Analysis

CHAPTER 11

Characterization of Sugars, Cyclitols and Galactosyl Cyclitols in Seeds by GC

RALPH L. OBENDORF,*[a] MARCIN HORBOWICZ[b] AND
LESŁAW BERNARD LAHUTA[c]

[a] Professor of Crop Physiology, Seed Biology, Department of
Crop and Soil Sciences, Cornell University, 617 Bradfield Hall, Ithaca,
NY 14853-1901, United States of America; [b] Siedlce University of Natural
Sciences and Humanities, Faculty of Natural Sciences, Department
of Plant Physiology and Genetics, Prusa 12, 08-110 Siedlce, Poland;
[c] University of Warmia and Mazury in Olsztyn, Department of Plant
Physiology and Biotechnology, Faculty of Biology, Oczapowskiego 1A/103A,
10-719 Olsztyn, Poland
*Email: rlo1@cornell.edu

11.1 Introduction

Low molecular weight water-soluble carbohydrates are present in maturing
seeds and have important physiological roles. They are accumulated during
seed maturation (Górecki *et al.* 2001; Peterbauer and Richter 2001), and may
accumulate to about 15% of total seed dry mass and up to 25% of the
embryonic axis dry mass of mature soybean seeds. Water-soluble carbohydrates, such as sucrose, galactosides of sucrose, or galactosides of cyclitols
(*myo*-inositol, D-pinitol, D-*chiro*-inositol, or other cyclitols), are non-reducing
and are generally considered to be compatible solutes or protective factors of

Food and Nutritional Components in Focus No. 3
Dietary Sugars: Chemistry, Analysis, Function and Effects
Edited by Victor R Preedy
© The Royal Society of Chemistry 2012
Published by the Royal Society of Chemistry, www.rsc.org

cytoplasmic membranes, proteins, and other important bio-molecules against damage during loss of water in desiccation of seeds (Hoekstra *et al.* 2001; Leprince and Buitink 2010; Morse *et al.* 2011; Obendorf 1997). Water-soluble carbohydrates also affect the vitality of seeds during storage (Dey 1990). The study of different factors which affect the accumulation and content of soluble carbohydrates may contribute to a deeper understanding of their role. Described in this review are methods of quantitative analysis of soluble carbohydrates using gas chromatography (GC), a technique that seems to be optimal for such studies (Horbowicz and Obendorf 2005). Due to the higher resolution and sensitivity, analysis by GC has advantages over alternatively used high performance liquid chromatography (Kadlec *et al.* 2001).

11.2 Water-soluble Carbohydrates in Seeds

Seed soluble carbohydrates are predominantly monomers and small molecular weight oligomers (dimers, trimers, tetramers, pentamers), including sucrose and galactosides of sucrose or cyclitols (Table 11.1). Sucrose (**14**) and *myo*-inositol (**12**) are present in all seeds. The raffinose family oligosaccharides (RFO) are galactosyl oligomers of sucrose (Figure 11.1) commonly present in many seeds (Horbowicz and Obendorf 1994; Kuo *et al.* 1988). The galactosyl oligomers of *myo*-inositol (**12**) include galactinol (**23**), digalactosyl *myo*-inositol (DGMI, **30**), trigalactosyl *myo*-inositol (TGMI, **35**), and sometimes tetragalactosyl *myo*-inositol (**38**; proposed by Courtois and Percheron 1971; Table 11.1). Other cyclitols present in seeds of specific species include D-pinitol or 1D-3-*O*-methyl-*chiro*-inositol (**5**), D-ononitol or 1D-4-*O*-methyl-*myo*-inositol (**9**), D-*chiro*-inositol (**10**), and D-bornesitol or 1D-1-*O*-methyl-*myo*-inositol (**11**) (Horbowicz and Obendorf 1994; Obendorf *et al.* 2005). *myo*-Inositol is optically inactive. D-Pinitol, D-ononitol, and D-*chiro*-inositol are dextrarotary (+), whereas D-bornesitol is levorotary (−). When a sufficient quantity of a cyclitol or other soluble carbohydrate is available, optical rotation may be determined with a polarimeter to establish chirality (Obendorf *et al.* 2005). When limited quantities are available, chirality may be established by GC analysis using a commercially available chiral capillary column (Chirasil-Val) (Horbowicz *et al.* 1998; Obendorf *et al.* 2000; Steadman *et al.* 2001). Typically, when free cyclitols are present in seeds, galactosides of the cyclitol also may be present, but not always. Galactosyl oligomers of D-pinitol (**5**) present in many seeds may include galactopinitol A (**17**, GPA), galactopinitol B (**18**, GPB), ciceritol (**26**, digalactosyl pinitol A, DGPA), trigalactosyl pinitol A (**32**, TGPA) and rarely tetragalactosyl pinitol A (**37**, tetraGPA) (Table 11.1; Figure 11.2B; Horbowicz and Obendorf 1994; Szczeciński *et al.* 2000). Higher galactosyl oligomers of galactopinitol B (**18**) are rarely present in seeds. There is one report of a trigalactosyl pinitol B (mimositol, α-D-galactopyranosyl-(1 → 6)-α-D-galactopyranosyl-(1 → 6)-α-D-galactopyranosyl-(1 → 2)-1D-3-*O*-methyl-*chiro*-inositol) (Ganter *et al.* 1991), but the occurrence of a digalactosyl pinitol B has not been reported. In a few species, galactosyl oligomers of D-ononitol (**9**) may include galactosyl

Table 11.1   Identification of compounds of chromatogram peaks by number. This table identifies compounds illustrated in Figures 11.2, 11.3, and 11.4 by peak number, trivial name, and chemical name, and provides relevant comments about each compound. Source: Obendorf, R. L., Horbowicz, M. and Lahuta, L. B. 2011. Unpublished.

No.	Trivial name	Chemical name	Comment
1	xylitol	D-xylitol	internal standard 1
2,3[a]	fructose	D-fructofuranoside	reducing sugar
4,6,8[a]	galactose	D-galactopyranoside	reducing sugar
5	D-pinitol	1D-3-*O*-methyl-*chiro*-inositol	*O*-methyl-cyclitol
7,11[a]	glucose	D-glucopyranoside	reducing sugar
9	D-ononitol	1D-4-*O*-methyl-*myo*-inositol	*O*-methyl-cyclitol
10	D-*chiro*-inositol	D-*chiro*-inositol	cyclitol
11	D-bornesitol	1D-1-*O*-methyl-*myo*-inositol (overlap with glucose)	*O*-methyl-cyclitol
12	*myo*-inositol	*myo*-inositol	cyclitol
13	phenyl β-D-glucose	phenyl β-D-glucopyranoside	internal standard 2
14	sucrose	α-D-glucopyranosyl-(1 → 2)-β-D-fructofuranoside	non-reducing sugar
15,16[a]	maltose	α-D-glucopyranosyl-(1 → 4)-α-D-glucopyranoside	reducing sugar
17	galactopinitol A (GPA)	α-D-galactopyranosyl-(1 → 2)-1D-4-*O*-methyl-*chiro*-inositol	galactosyl cyclitol
18	galactopinitol B (GPB)	α-D-galactopyranosyl-(1 → 2)-1D-3-*O*-methyl-*chiro*-inositol	galactosyl cyclitol
19	lathyritol (galactosyl D-bornesitol)	α-D-galactopyranosyl-(1 → 3)-1D-1-*O*-methyl-*myo*-inositol	galactosyl cyclitol
20	fagopyritol A1	α-D-galactopyranosyl-(1 → 3)-1D-*chiro*-inositol	galactosyl cyclitol
21	galactosyl ononitol	α-D-galactopyranosyl-(1 → 3)-1D-4-*O*-methyl-*myo*-inositol	galactosyl cyclitol
22	fagopyritol B1	α-D-galactopyranosyl-(1 → 2)-1D-*chiro*-inositol	galactosyl cyclitol
23[b]	galactinol	α-D-galactopyranosyl-(1 → 1)-1L-*myo*-inositol	galactosyl cyclitol
24	digalactosyl glycerol	α-D-galactopyranosyl-(1 → 6)-α-D-galactopyranosyl-(1 → 1)-D-glycerol	
25	raffinose	α-D-galactopyranosyl-(1 → 6)-α-D-glucopyranosyl-(1 → 2)-β-D-fructofuranoside	non-reducing sugar
25	1-kestose	α-D-glucopyranosyl-(1 → 2)-β-D-fructofuranosyl-(1 → 2)-β-D-fructofuranoside	overlap with raffinose
26	ciceritol (digalactosyl pinitol A; DGPA)	α-D-galactopyranosyl-(1 → 6)-α-D-galactopyranosyl-(1 → 2)-1D-4-*O*-methyl-*chiro*-inositol	galactosyl cyclitol
27	fagopyritol A2	α-D-galactopyranosyl-(1 → 6)-α-D-galactopyranosyl-(1 → 3)-1D-*chiro*-inositol	galactosyl cyclitol

Table 11.1 (*Continued*)

No.	Trivial name	Chemical name	Comment
28	digalactosyl ononitol	α-D-galactopyranosyl-(1→6)-α-D-galactopyranosyl-(1→3)-1D-4-*O*-methyl-*myo*-inositol	galactosyl cyclitol
29	fagopyritol B2	α-D-galactopyranosyl-(1→6)-α-D-galactopyranosyl-(1→2)-1D-*chiro*-inositol	galactosyl cyclitol
30	digalactosyl *myo*-inositol (DGMI)	α-D-galactopyranosyl-(1→6)-α-D-galactopyranosyl-(1→1)-1L-*myo*-inositol	galactosyl cyclitol
31	stachyose	α-D-galactopyranosyl-(1→6)-α-D-galactopyranosyl-(1→6)-α-D-glucopyranosyl-(1→2)-β-D-fructofuranoside	non-reducing sugar
32	trigalactosyl pinitol A (TGPA)	α-D-galactopyranosyl-(1→6)-α-D-galactopyranosyl-(1→6)- α-D-galactopyranosyl-(1→2)-1D-4-*O*-methyl-*chiro*-inositol	galactosyl cyclitol
33	fagopyritol A3	α-D-galactopyranosyl-(1→6)-α-D-galactopyranosyl-(1→6)-α-D-galactopyranosyl-(1→3)-1D-*chiro*-inositol	galactosyl cyclitol
34	fagopyritol B3	α-D-galactopyranosyl-(1→6)-α-D-galactopyranosyl-(1→6)-α-D-galactopyranosyl-(1→2)-1D-*chiro*-inositol	galactosyl cyclitol
35	trigalactosyl *myo*-inositol (TGMI)	α-D-galactopyranosyl-(1→6)-α-D-galactopyranosyl-(1→6)-α-D-galactopyranosyl-(1→1)-1L-*myo*-inositol	galactosyl cyclitol
36	verbascose	α-D-galactopyranosyl-(1→6)-α-D-galactopyranosyl-(1→6)-α-D-galactopyranosyl-(1→6)-α-D-glucopyranosyl-(1→2)-β-D-fructofuranoside	non-reducing sugar
37	tetragalactosyl pinitol A (tetra-GPA)	α-D-galactopyranosyl-(1→6)-α-D-galactopyranosyl-(1→6)- α-D-galactopyranosyl-(1→6)-α-D-galactopyranosyl-(1→2)-1D-4-*O*-methyl-*chiro*-inositol	galactosyl cyclitol
38	tetragalactosyl *myo*-inositol (tetra-GMI)	α-D-galactopyranosyl-(1→6)-α-D-galactopyranosyl-(1→6)-α-D-galactopyranosyl-(1→6)-α-D-galactopyranosyl-(1→1)-1L-*myo*-inositol	galactosyl cyclitol (not shown)
?	unknown		
S	solvents		

[a]The anomeric forms are captured during TMS-derivatization.
[b]Galactinol (**23**) is also known as α-D-galactopyranosyl-(1→3)-1D-*myo*-inositol, a name preferred by many biochemists.

ononitol (**21**; Richter *et al.* 1997) and digalactosyl ononitol (**28**; Peterbauer *et al.* 2003) (Table 11.1). A galactosyl oligomer of D-bornesitol (**11**), lathyritol or α-D-galactopyranosyl-(1→3)-1D-1-*O*-methyl-*myo*-inositol (**19**), was identified in sweet pea (*Lathyrus odoratus* L.) seeds (Table 11.1, Figure 11.3D;

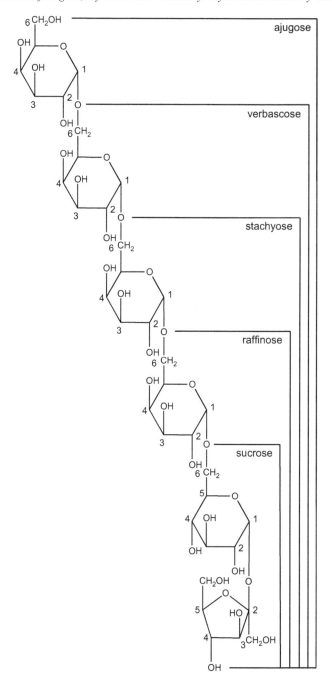

Figure 11.1 Structures of raffinose family oligosaccharides (RFO) present in seeds, including raffinose, stachyose, verbascose, and ajugose. Raffinose, stachyose, verbascose, and ajugose, respectively, are mono-, di-, tri-, and tetra-galactosides of sucrose.

Source: Obendorf, R.L., Horbowicz, M. and Lahuta, L.B. 2011. Unpublished.

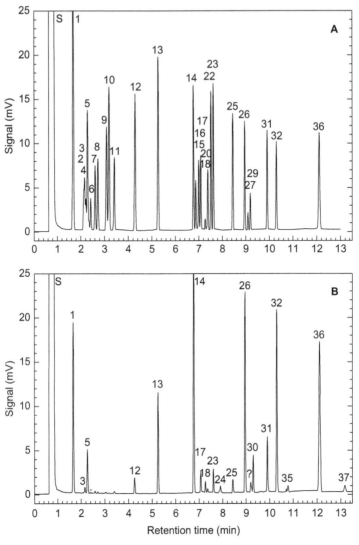

Figure 11.2 Gas chromatograms of (**A**) selected standard compounds, and (**B**) soluble carbohydrates in winter (hairy) vetch (*Vicia villosa* Roth) seeds. This figure demonstrates the rapid, high-resolution separation of (**A**) selected standard compounds, and (**B**) vetch seed sugars, cyclitols (cyclic polyols), and galactosyl cyclitols (galactose derivatives of cyclitols) by gas chromatography. Numerical identification of compounds: **S**, solvents; **1**, xylitol (internal standard 1); **2**, **3**, fructose; **4**, **6**, **8**, galactose; **5**, D-pinitol; **7**, **11**, glucose; **9**, D-ononitol; **10**, D-*chiro*-inositol; **11**, D-bornesitol (overlaps with glucose), **12**, *myo*-inositol; **13**, phenyl β-D-glucopyranoside (internal standard 2); **14**, sucrose; **15**, **16**, maltose; **17**, galactopinitol A (GPA); **18** glactopinitol B (GPB); **19**, galactosyl D-bornesitol (not shown); **20**, fagopyritol A1; **21**, galactosyl ononitol (not shown); **22**, fagopyritol B1; **23**, galactinol (galactosyl *myo*-inositol); **24**, digalactosyl glycerol; **25**, raffinose, 1-kestose (overlaps with raffinose); **26**,

Obendorf *et al.* 2005). Higher galactosyl oligomers of D-bornesitol have not been reported. The galactosyl oligomers of D-*chiro*-inositol (**10**) may include fagopyritol B1 (**22**), fagopyritol B2 (**29**), and fagopyritol B3 (**34**). In addition, common buckwheat (*Fagopyrum esculentum* Moench) seeds also contain fagopyritol A1 (**20**), fagopyritol A2 (**27**) and fagopyritol A3 (**33**) (Table 11.1, Figure 11.4B; Horbowicz *et al.* 1998; Obendorf *et al.* 2000; Steadman *et al.* 2001; Szczeciński *et al.* 1998).

11.2.1   Extraction of Water-soluble Carbohydrates from Seeds

Seeds or seed parts (3 to 300 mg of axis, cotyledons, embryo, endosperm, seed coat, pericarp or fruit coat) may be frozen in liquid nitrogen and finely pulverized with a pestle and mortar pre-cooled with liquid nitrogen. Water-soluble carbohydrates may be extracted with boiling water (Peterbauer *et al.* 2003) or hot water (Ruiz-Matute *et al.* 2007). Contaminating proteins, hydrolytic enzymes, membranes, cell walls or cell wall components may also be present in water extracts. Extraction with aqueous ethanol (ethanol:water, 1:1, v/v) or methanol can decrease the contamination and activity of hydrolytic enzymes. An internal standard added to the extraction medium allows a more accurate quantitation of the extracted compounds (Horbowicz and Obendorf 1994). Compounds used as internal standards should not normally be present in seeds and should be presented on the chromatogram with retention times distinctly different than the naturally occurring compounds. Examples of suitable internal standards include D-xylitol (**1**), phenyl β-D-glucopyranoside (**13**), or phenyl α-D-glucopyranoside. Extracts containing an internal standard are centrifuged (15 000 × *g* for 20–30 min). Passing the cleared aqueous ethanol supernatants through a 10 000 molecular weight cut-off filter can remove many of the contaminating proteins (Obendorf and Kosina 2011). Heating the aqueous ethanol extract to 80 °C may inactivate hydrolytic enzymes and reduce potential degradation of galactosyl cyclitols and oligosaccharides. Certain artifacts may occur upon heating of acidic plant extracts, including isomerization of *myo*-inositol (Sasaki *et al.* 1988; Taguchi *et al.* 1997). To minimize the potential for artifacts, aliquots of the filtered extracts are placed in silanized glass inserts and dried under a stream of nitrogen gas at room temperature. Freeze-drying may be used with larger samples of extracts. If a rotary evaporator is used to dry plant extracts, lowering the evaporative temperature to 40 °C (Peterbauer *et al.* 2003; Ruiz-Matute *et al.* 2007) may reduce the potential for artifacts. Alternatively, extracts may be dried by the freeze-drying method. The dry residues

---

ciceritol (digalactosyl pinitol A); **27**, fagopyritol A2; **28**, digalactosyl ononitol (not shown); **29**, fagopyritol B2; **30**, digalactosyl *myo*-inositol (DGMI); **31**, stachyose; **32**, trigalactosyl pinitol A (TGPA); **33**, fagopyritol A3 (not shown); **34**, fagopyritol B3; **35**, trigalactosyl *myo*-inositol (TGMI); **36**, verbascose; **37**, tetragalactosyl pinitol A (tetra-GPA); **38**, tetragalactosyl *myo*-inositol (not shown).
Source: Lahuta, L.B. 2011. Unpublished.

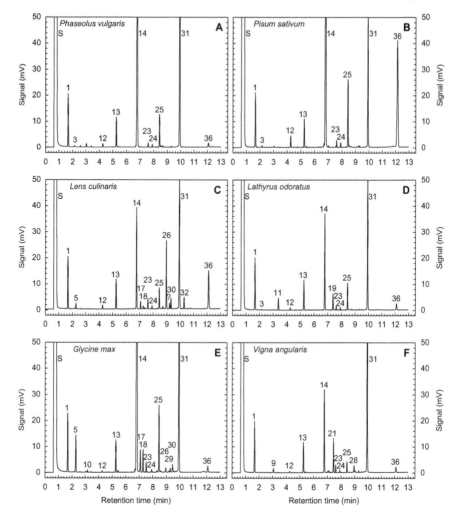

Figure 11.3   Gas chromatograms of soluble carbohydrates extracted from seeds of six legume species (**A**) garden bean (*Phaseolus vulgaris* L.), (**B**) smooth seeded garden pea (*Pisum sativum* L.), (**C**) lentil (*Lens culinaris* Medik), (**D**) sweet pea (*Lathyrus odoratus* L.), (**E**) soybean (*Glycine max* (L.) Merrill), and (**F**) Adzuki bean (*Vigna angularis* (Willd.) Ohwi and Ohashi). This figure demonstrates the separation of soluble carbohydrates from seeds of six species showing differences in RFO, cyclitol and galactosyl cyclitol compositions by gas chromatography. Numerical identification of compounds: **S**, solvents; **1**, xylitol (internal standard 1); **3**, fructose; **5**, D-pinitol; **9**, D-ononitol; **10**, D-*chiro*-inositol; **11**, D-bornesitol, **12**, *myo*-inositol; **13**, phenyl β-D-glucopyranoside (internal standard 2); **14**, sucrose; **17**, galactopinitol A (GPA); **18** glactopinitol B (GPB); **19**, galactosyl D-bornesitol; **21**, galactosyl ononitol; **23**, galactinol; **24**, digalactosyl glycerol; **25**, raffinose; **26**, ciceritol; **28**, digalactosyl ononitol; **29**, fagopyritol B2; **30**, digalactosyl *myo*-inositol (DGMI); **31**, stachyose; **32**, trigalactosyl pinitol A (TGPA); **36**, verbascose; **?**, unknown.
Source: Lahuta, L.B. 2011. Unpublished.

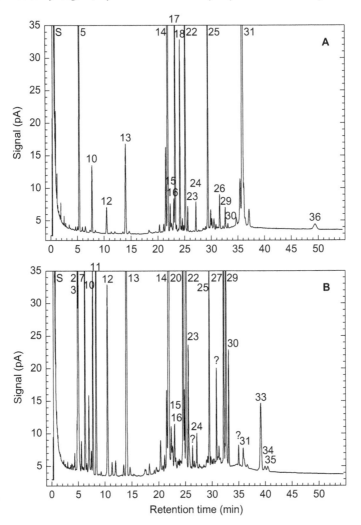

Figure 11.4 Gas chromatograms (0.25 film thickness) of soluble carbohydrates extracted from (**A**) soybean (*Glycine max* (L.) Merrill) seed axis, and (**B**) buckwheat (*Fagopyrum esculenum* Moench) seed bran milling fraction. This figure demonstrates the longer retention times required for separation of soluble carbohydrates from seeds of soybean and buckwheat (using a column with 0.25 μm film thickness; contrast with Figures 11.2 and 11.3 using a column with 0.10 μm film thickness) showing differences in RFO, cyclitol and galactosyl cyclitol compositions by gas chromatography. Numerical identification of compounds: **S**, solvents; **2,3**, fructose; **5**, D-pinitol; **7, 11**, D-glucose; **10**, D-*chiro*-inositol; **12**, *myo*-inositol; **13**, phenyl β-D-glucopyranoside (internal standard 2); **14**, sucrose; **15, 16**, maltose; **17**, galactopinitol A (GPA); **18** glactopinitol B (GPB); **20**, fagopyritol A1; **22**, fagopyritol B1; **23**, galactinol; **24**, digalactosyl glycerol; **25**, raffinose; **26**, ciceritol; **27**, fagopyritol A2; **29**, fagopyritol B2; **30**, digalactosyl *myo*-inositol (DGMI); **31**, stachyose; **33**, fagopyritol A3; **34**, fagopyritol B3; **35**, trigalactosyl *myo*-inositol (TGMI); **36**, verbascose; **?**, unknown.
Source: Obendorf, R.L. 2011. Unpublished.

are stored overnight above P_2O_5 to remove traces of water. The dry residues containing soluble carbohydrates are then derivatized with trimethylsilyl(TMS)-imidazole:pyridine (1:1, v/v) at 80 °C for 45 min in sealed vials in preparation for gas chromatography. Carbohydrates must be transformed into volatile derivatives for separation and analysis by gas-liquid chromatography. Many popular and emerging methods for derivatization of carbohydrates for GC and GC-MS analysis have recently been compared in detail (Ruiz-Matute *et al.* 2011; Ye *et al.* 2006). Mature dry seeds may contain small to undetectable amounts of reducing sugars, including fructose (**2,3**), glucose (**7,11**), and maltose (**15,16**) (Table 11.1). Free galactose (**4,6,8**) is seldom detected in extracts of mature, dry seeds unless there is hydrolysis of the galactosyl oligomers during extraction. The anomeric forms (α- or β-linkage at the anomeric carbon in ring forms, and sometimes the straight chain form) are captured during TMS-derivatization resulting in multiple peaks for the derivatization products of reducing sugars (Horbowicz and Obendorf 1994). The water-soluble carbohydrates present in low concentrations in plant tissues cannot be analyzed by other common techniques (such as HPLC) due to the lower sensitivity limits for detection and determination.

11.2.1.1 Alternative Method for Extraction and Purification of Soluble Carbohydrates

Extraction with a mixture of methanol:chloroform:water (12:5:3, v/v/v) at 60 °C for 30 min (Hoch *et al.* 2003) and then centrifugation at 20 000 × *g* at 4 °C for 20 minutes allows separation of seed extracts into two fractions: an upper fraction of methanol-water, containing soluble carbohydrates, and a lower fraction of chloroform, containing non-polar compounds. Additionally, some inorganic and organic ions can be removed by adding 200 μL of a 1:1 (v/v) mixture of ion-exchange resins (DOWEX 50W × 8, H^+ and DOWEX W2 × 8, formate) to 400 μL of the methanol-water fraction and vigorously shaking (1200 rpm, 30–60 min). After centrifugation (20 000 *g* at room temperature for 10 min), the clear supernatant (without additional filtration) can be dried in a rotary speed vacuum evaporator (Lahuta 2006). To reduce the potential for artifacts, the temperature during evaporation should be 40 °C or lower.

11.2.2 Natural Sources of Standards not Commercially Available

GC analysis of soluble carbohydrates requires comparison with standard compounds. Many water-soluble carbohydrates present in seeds, especially monomers, may be purchased commercially for use as standards (Kadlec *et al.* 2001). Commercially available water-soluble carbohydrates include glucose (**7,11**), fructose (**2,3**), galactose (**4,6,8**), maltose (**15,16**), sucrose (**14**), galactinol (**23**), raffinose (**25**), stachyose (**31**), verbascose (**36**), *myo*-inositol (**12**), D-pinitol (**5**), D-ononitol (**9**), D-*chiro*-inositol (**10**), galactinol (**23**), and others. Many water-soluble carbohydrates present in seeds are not commercially available and must be isolated and purified from natural sources (Kadlec *et al.* 2001).

Galactopinitols including galactopinitol A (**17**), galactopinitol B (**18**), ciceritol (**26**), TGPA (**32**) and tetra-GPA (**37**) may be isolated from *Vicia villosa* seeds (Szczeciński *et al.* 2000). Triglalactosyl pinitol B may be isolated from seeds of *Mimosa scabrella* (Ganter *et al.* 1991). Galactosyl ononitol (**21**) and diga-lactosyl ononitol (**28**) may be isolated from seeds of *Vigna angularis* (Peterbauer *et al.* 2003; Richter *et al.* 1997). Fagopyritols (**20, 22, 27, 29, 33, 34**) may be isolated from seeds of *Fagopyrum esculentum* (Obendorf *et al.* 2000; Steadman *et al.* 2001; Szczeciński *et al.* 1998). DGMI (**30**) and TGMI (**35**) may be isolated from seeds of *Fagopyrum esculentum* or *Vicia* species. D-Bornesitol (**11**) and galactosyl D-bornesitol (**19**) may be isolated from seeds of *Lathyrus odoratus* (Obendorf *et al.* 2005). Some soluble carbohydrates in seeds remain to be identified.

11.2.3 Identification of Compounds Forming Unknown Peaks

Identification of unknown carbohydrates, which are present on gas chroma-tograms as small peaks, can be approached systematically. Retention times can provide a clue to the identity of the compound. If small amounts of the unknown soluble carbohydrate can be purified, hydrolysis of the unknown carbohydrate by 3 N trifluoroacetic acid followed by derivatization and GC analysis of the hydrolysis products can provide identification of the monomeric components and the relative ratio of specific monomers in the compound. In the case of an oligomer, enzymatic hydrolysis of the compound using a specific hydrolyase such as α-D-galactosidase or β-D-galactosidase can provide evidence about the anomeric linkage(s). If a sufficient amount of unknown is available, analysis by HPLC or HPLC-MS or HPLC-MS-MS may be possible. Smaller amounts of the derivatized unknown may be analyzed by GC-MS, however, results from such analyses are not always reliable. Water-soluble carbohydrates from seeds produce many common MS-fragments, although the magnitudes of abundance peaks may vary. Therefore, MS-fragments are not always reliable indicators for identification of unknown soluble carbohydrates.

A more reliable method of structural identification of unknown carbohy-drates is by NMR analysis. A larger amount (10 to 30 mg) of purified (95% pure) compound is needed for NMR analysis, especially for ^{13}C-NMR. Typi-cally, a combination of ^{1}H-NMR and ^{13}C-NMR experiments, including interactive experiments, can provide useful information for the resolution of the molecular structures of water-soluble carbohydrates from seeds (Obendorf *et al.* 2005; Peterbauer *et al.* 2003; Richter *et al.* 1997; Steadman *et al.* 2001; Szczeciński *et al.* 2000).

11.3 GC Analysis of Water-soluble Carbohydrates

High resolution gas chromatography is a preferred method of analysis of the water-soluble carbohydrates in seeds (Horbowicz and Obendorf 1994). Fifteen to thirty different soluble carbohydrates may be separated with good resolution on a single chromatogram in less than 15 min (Figures 11.2 and 11.3).

Chromatograms represented in Figures 11.2 and 11.3 were run on a Shimadzu (Tokyo, Japan) gas chromatograph (GC 2010) equipped with a flame ionization detector (FID), split-mode injection port, and a Phenomenex (Torrance, CA, USA) Zebron ZB-1 capillary column (15 m length, 0.25 mm inside diameter, 0.1 μm film thickness, 100% dimethylpolysiloxane). Two hundred μL of TMSI (trimethylsilyl imidazole):pyridine (1:1, v/v) was added to the dry residue obtained from extract equivalent to 10 mg of dry seed tissues and heated at 80 °C for 50 min. The injection port was operated in split mode (1:10) at 335 °C using a cup splitter injection port sleeve. Sample injection included 1 μL dichloromethane, 1 μL air and 1 μL sample of derivatized soluble carbohydrates. The detector was operated at 350 °C. Detector gases were air at 500 mL min^{-1} and hydrogen at 50 mL min^{-1}. The makeup gas was helium at 20 mL min^{-1}. The carrier gas was helium with initial column flow of 1.18 mL min^{-1} controlled at a linear velocity of 40 cm s^{-1}. The column heating program was initial temperature of 150 °C for 1.5 min, increased from 150 °C to 175 °C at 6 °C min^{-1}, increased from 175 °C to 335 °C at 30 °C min^{-1}, and held at 335 °C for 3.5 min. Total resolution time was 13.5 min. Equilibration time prior to injection of the next sample was 1.5 min.

Chromatograms represented in Figure 11.4 were run on a Hewlett Packard 6890 Plus gas chromatograph (Agilent Technologies, Palo Alto, CA, USA) equipped with a flame ionization detector (FID), split-mode injection port, and an Agilent HP-1MS capillary column (15 m length, 0.25 mm inside diameter, 0.25 μm film thickness). The injection port was operated in split mode (1:20) at 335 °C using a long-cup laminar cup splitter liner (Catalog #20802, Restek International, intltechsupp@restek.com). Sample injection volume was 5 μL (Figure 11.4A) or 1 μL (Figure 11.4B) followed by eight rinses of the syringe with dichloromethane and three rinses with isopropyl alcohol. Direct on-column injection is preferred by some researchers (Traitler *et al.* 1984). The detector was operated at 350 °C. Detector gases were air at 400 mL min^{-1} and hydrogen at 30 mL min^{-1}. The makeup gas was nitrogen at 30 mL min^{-1}. The carrier gas was ultra high purity nitrogen with initial column flow of 2.5 mL min^{-1} constant flow (linear velocity 70 cm s^{-1}). The column heating program was initial temperature of 150 °C for 0 min, increased from 150 °C to 200 °C at 3 °C min^{-1}, increased from 200 °C to 325 °C at 7 °C min^{-1}, and held at 325 °C for 20 min. Total resolution time was 54.5 min. Equilibration time prior to injection of the next sample was 2.0 min.

11.4 Species Differences in Water-soluble Carbohydrates

Use of standard compounds is essential for the correct identification and quantification of compounds during analysis by gas chromatography. Figure 11.2A illustrates the rapid and high-resolution separation of 25 standard compounds including two compounds used as internal standards. The rapid, high-resolution separation of soluble carbohydrates extracted from winter (hairy) vetch seeds shows sucrose, RFO, and galactopinitols as the dominant compounds

(Figure 11.2B). Verbascose is the dominant RFO. This vetch seed contains both galactopinitol A (**17**, GPA) and galactopinitol B (**18**, GPB). Ciceritol (digalactosyl pinitol A, **26**, DGPA) and trigalactosyl pinitol A (**32**, TGPA) are the dominant galactopinitols present. Higher galactosyl oligomers of galactopinitol B are not detected. Of special interest is the resolution of a rarely present tetragalactosyl pinitol A (**37**, tentative identification; Szczeciński *et al.* 2000).

Contrasting compositions of soluble carbohydrates present in seeds of six different species are illustrated in Figure 11.3. All seeds have sucrose (**14**) and *myo*-inositol (**12**), and usually small amounts of galactinol (**23**). Both garden bean (Figure 11.3A) and garden pea (Figure 11.3B) have sucrose and RFO, but garden pea has more verbascose (**36**), noted for its potentially high flatulence production, than garden bean. Extracts of lentil seeds have sucrose and galactopinitols (Figure 11.3C). In contrast to lentil which has more ciceritol (**26**) than GPA (**17**) and GPB (**18**) (Figure 11.2B), soybean has more GPA (**17**) and GPB (**18**) than ciceritol (**26**) (Figure 11.3E). Sweet pea (Figure 11.3D) shows D-bornesitol (**11**, 1D-1-*O*-methyl-*myo*-inositol) and α-D-glactopyranosyl-1D-1-*O*-methyl-*myo*-inositol (**19**, a galactosyl bornesitol, lathyritol; Obendorf *et al.* 2005) while Adzuki bean (Figure 11.3F) shows D-ononitol (**9**, 1D-4-*O*-methyl-*myo*-inositol), galactosyl ononitol (**21**, α-D-galactopyranosyl-(1 → 3)-1D-4-*O*-methyl-*myo*-inositol; Richter *et al.* 1997) and digalactosyl ononitol (**28**, α-D-galactopyranosyl-(1 → 6)-α-D-galactopyranosyl-(1 → 3)-1D-4-*O*-methyl-*myo*-inositol; Peterbauer *et al.* 2003), compounds unique to seeds of these species.

The effect of column film thickness on retention times is illustrated in Figure 11.4. The TMS-derivatives of a soybean seed extract chromatographed on a column with 0.10 µm film thickness is illustrated in Figure 11.3E with retention times of compounds less than 13 min. By contrast, the TMS-derivatives of a soybean seed extract chromatographed on a column with 0.25 µm film thickness is illustrated in Figure 11.4A with retention times of compounds greater than 50 min. Part of the longer retention times demonstrated in Figure 11.4A may be a slower increase in column temperature, but the major effect on retention time was the film thickness of the column. Uniquely, seeds of buckwheat accumulate small amounts of RFO (raffinose and stachyose) but large amounts of fagopyritols including fagopyritol A1 (**20**, α-D-galactopyranosyl-(1 → 3)-1D-*chiro*-inositol), fagopyritol A2 (**27**, α-D-galactopyranosyl-(1 → 6)-α-D-galactopyranosyl-(1 → 3)-1D-*chiro*-inositol), fagopyritol A3 (**33**, α-D-galactopyranosyl-(1 → 6)-α-D-galactopyranosyl-(1 → 6)-α-D-galactopyranosyl-(1 → 3)-1D-*chiro*-inositol), fagopyritol B1 (**22**, α-D-galactopyranosyl-(1 → 2)-1D-*chiro*-inositol), fagopyritol B2 (**29**, α-D-galactopyranosyl-(1 → 6)-α-D-galactopyranosyl-(1 → 2)-1D-*chiro*-inositol), and fagopyritol B3 (**34**, α-D-galactopyranosyl-(1 → 6)-α-D-galactopyranosyl-(1 → 6)-α-D-galactopyranosyl-(1 → 2)-1D-*chiro*-inositol), a composition unique to this species (Figure 11.4B; Horbowicz *et al.* 1998; Obendorf *et al.* 2000; Steadman *et al.* 2001). In addition, galactinol (**23**, α-D-galactopyranosyl-(1 → 1)-1L-*myo*-inositol), digalactosyl *myo*-inositol (**30**, DGMI, α-D-galactopyranosyl-(1 → 6)-α-D-galactopyranosyl-(1 → 1)-1L-*myo*-inositol), and trigalactosyl *myo*-inositol (**35**, TGMI, α-D-galactopyranosyl-(1 → 6)-α-D-galactopyranosyl-(1 → 6)-α-D-galactopyranosyl-(1 → 1)-1L-*myo*-inositol) are also detected.

Many seeds contain a large array of compounds that make up the background of chromatograms. If sufficiently concentrated, some of these compounds may form small visible peaks on chromatograms (see peaks **34** and **35** in Figure 11.4B). Correct identification of these small peaks is sometimes problematic and may require purification and structural characterization of the specific compound by NMR. For example, in the paper by Horbowicz *et al.* (1998), peak "O" in Figure 11.1 was incorrectly identified as fagopyritol B3. When the compound corresponding to this peak was subsequently purified and analyzed by NMR (Steadman *et al.* 2001), it was determined to be fagopyritol A3 (α-D-galactopyranosyl-(1→6)-α-D-galactopyranosyl-(1→6)-α-D-galactopyranosyl-(1→3)-1D-*chiro*-inositol) (peak **33**, Figure 11.4B, Table 11.1). Fagopyritol B3 (α-D-galactopyranosyl-(1→6)-α-D-galactopyranosyl-(1→6)-α-D-galactopyranosyl-(1→2)-1D-*chiro*-inositol) (peak **34**) is a subsequent but much smaller peak on chromatograms, followed closely by TGMI (peak **35**) (Figure 11.4B). All three compounds chromatograph closely together. We recently verified the structures of fagopyritol B3 (**34**), digalactosyl *myo*-inositol (DGMI, **30**), and trigalactosyl *myo*-inositol (TGMI, **35**) by nuclear magnetic resonance (NMR) spectroscopy (Gui *et al.* unpublished).

The retention time sometimes can suggest the identity of a soluble carbohydrate represented by a peak, but it is not always a reliable identification. The uncertainty is due to hundreds of chemicals present in extracts from plant tissues. Like soluble carbohydrates, other chemicals can be silylated such as phenolic acids or flavonoids. These compounds may then form peaks on chromatograms. Some extraction and derivatization procedures may permit a precise identification of soluble carbohydrates. If the procedures are more complicated, the final results may produce significant errors. Therefore, the correct identification of a compound on a GC chromatogram may be difficult to resolve, especially in the case of small peaks.

When unsure about the identity of a compound corresponding to a small peak, additional experiments may be conducted: (a) use a longer column or a column with a thicker stationary phase; (b) decrease the flow rate of the carrier gas; (c) alter the column temperature program; (d) use HPLC-MS if a high concentration is available, but in the case of small peaks, a more sophisticated analysis such as HPLC-MS-MS may be needed; (e) in the case of TMS-derivatized phenolics, GC-MS can be useful.

11.5 Conclusions

High resolution gas chromatography is a powerful method for the separation, identification and quantification of soluble carbohydrates present in dry, germinated and processed seeds used for human and animal nutrition. This method can be used for rapid identification of changes in soluble carbohydrates in seeds as a result of plant breeding or genetic transformation.

Summary Points

- Soluble carbohydrates accumulate as part of the natural process of seed development and maturation.
- Soluble carbohydrates are important contributors to desiccation tolerance, permitting seeds to dry to very low water contents while preserving viability and structural integrity of intracellular biological components in the dry state.
- In edible seeds, seed soluble carbohydrates are dietary components.
- Some seed soluble carbohydrates may have health benefits for treatment of non-insulin dependent diabetes mellitus and polycystic ovary syndrome.
- High resolution gas chromatography can facilitate the identification and quantitation of 15 to 30 soluble carbohydrates on a single chromatogram in less than 15 minutes.
- Soluble carbohydrates require chemical derivatization before they can be separated by gas chromatography.
- Examples of seeds with unique compositions of soluble carbohydrates are illustrated.
- Suggested approaches to the analysis and identification of minor soluble carbohydrates in seeds are discussed.

Key Facts

1. High resolution gas chromatography is a technique to separate, identify, and quantify naturally occurring compounds, including minor soluble carbohydrates in seeds.
2. Components of a gas chromatograph include an injector port, a separation capillary column, a flame ionization detector, and a cylinder with mobile gas.
3. Soluble carbohydrates are extracted from seed tissues using a mixture of ethanol and water.
4. Carbohydrates are converted into their trimethylsilyl (TMS) derivatives by applying a mixture of trimethylsilylimidazole and pyridine following removal of the extraction solvents, especially traces of water.
5. A solution containing a mixture of TMS derivatives of carbohydrates to be analyzed is injected into the injection port.
6. The TMS-carbohydrates pass through the capillary column at different retention times due to variability in their partitioning rates between the mobile gas phase and the stationary phase covering the inner wall of the capillary column.
7. High resolution gas chromatography is a sensitive analytical technique for identifying and quantifying specific soluble carbohydrates, such as cyclitols and their galactosides, naturally present in many seeds.
8. High resolution gas chromatography can facilitate the identification and quantification of up to thirty soluble carbohydrates on a single run in less than 15 minutes.

Definitions of Words and Terms

Anomers. An anomeric carbon is the carbon in a monosaccharide, such as glucose or galactose, about which rotation occurs. The anomeric carbon is joined to two oxygen atoms, each with a single bond. The anomeric carbon is carbon-1 in glucose or galactose or carbon-2 in fructose (the reducing site). This rotation forms two distinct configurations called α- and β-anomers. Carbohydrates can spontaneously change between the α- and β-configurations (stereoisomers with different structures and biological properties). During chemical reactions, such as derivatization, both the α- and β-configurations of a reducing sugar are captured in approximately equal amounts resulting in multiple peaks on a GC chromatogram for the derivatized products of a reducing sugar. In plant seeds and other biological systems, enzymes are usually stereospecific forming either the α-configuration or the β-configuration when linking two monosaccharide units together.

Cyclitol. A cyclic polyol that is non-reducing; typically, a six-carbon ring carbohydrate with one hydrogen and one hydroxyl on each carbon. Some cylitols, such as D-ononitol or D-pinitol, may have a methyl ether (OMe) substituted for a hydroxyl on a carbon at a specific position in the molecular structure.

Chromatogram. A recording of the separation of individual components from a mixture of compounds by gas chromatography.

Derivatization. The chemical transformation of a non-volatile carbohydrate or other compound into a form that is volatile, enabling the separation of the transformed product by gas chromatography.

Desiccation tolerance. The ability to tolerate the loss of water to very low concentrations without irreversible damage and to resume normal cellular activity upon rehydration.

Gas chromatography. An analytical technique used to separate a mixture of compounds into individual components based on differential volatilities, size, and shape; soluble carbohydrates must be derivatized, permitting compounds to be volatilized at elevated temperatures. A mobile gas phase separates the derivatized soluble carbohydrates based on their differential adsorption to the stationary phase (liquid or solid).

Galactosyl cyclitol. A galactoside of a cyclitol.

Non-reducing sugar. A sugar unable to be oxidized to form carboxyl groups. Sucrose, cyclitols, galactosyl cyclitols and raffinose family oligosaccharides are non-reducing soluble carbohydrates, relatively non-reactive, and believed to be useful for protecting plant seeds and other biological systems during stressful environments.

Nuclear magnetic resonance. A powerful analytical technique, most commonly ^{1}H-NMR and ^{13}C-NMR, used to deduce the molecular structure of a specific soluble carbohydrate in solution based on detection of absorption and re-emission of electromagnetic radiation by magnetic nuclei at characteristically specific resonance frequencies when the solution is subjected to a magnetic field.

Oligosaccharides. Low molecular weight oligomers of sugar monomers such as the raffinose family oligosaccharides.

Optical rotation. Solutions of chiral molecules such as sugars or other specific soluble carbohydrates can turn the plane of linearly polarized light when passed through the solution. A turn in a clockwise direction is dextrarotary, and a turn in a counter clockwise direction is levarotary, information useful for identifying the isomeric form or chirality of the molecule.

Orthodox seeds. Seeds which become desiccation tolerant during maturation.

Raffinose family oligosaccharides. The mono-, di-, tri-, tetra-, and sometimes higher galactosyl oligomers of sucrose including raffinose, stachyose, verbascose, and ajugose present in many orthodox seeds.

Reducing sugar. Reducing sugars are monosaccharides, such as glucose or galactose with aldehyde groups or fructose with a keto group, that can be converted to an aldehyde or some disaccharides such as maltose with an unreacted (exposed) reducing end, and can be oxidized to form carboxyl groups.

List of Abbreviations

DGMI	digalactosyl *myo*-inositol
DGPA	digalactosyl pinitol A (ciceritol)
FID	flame ionization detector
GC	gas chromatography
GC-MS	gas chromatography – mass spectrometry
GPA	galactopinitol A
GPB	galactopinitol B
^{13}C-NMR	carbon-13 nuclear magnetic resonance
^{1}H-NMR	hydrogen nuclear magnetic resonance
HPLC	high pressure liquid chromatography
HPLC-MS	high pressure liquid chromatography – mass spectrometry
HPLC-MS-MS	high pressure liquid chromatography – mass spectrometry – mass spectrometry
NMR	nuclear magnetic resonance
min	minute
RFO	raffinose family oligosaccharides
tetra-GMI	tetragalactosyl *myo*-inositol
tetra-GPA	tetragalactosyl pinitol A
TGMI	trigalactosyl *myo*-inositol
TGPA	trigalactosyl pinitol A
TMS	trimethylsilyl
TMSI	trimethylsilylimidazole

References

Courtois, J.E., and Percheron, F., 1971. Distribution of monosaccharides, oligosaccharides and polyols. In: Harborne, J. B., Boulter, D., and Turner, B. L., (eds.) *Chemotaxonomy of the Leguminosae*. Academic Press, New York, pp. 207–229.

Dey, P.M., 1990. Oligosaccharides. In: Dey, P.M. (ed.) *Methods in Plant Biochemistry*, Volume 2, Carbohydrates. Academic Press, New York, pp. 189–218.

Ganter, J.L.M.S., Correa, J., Reicher, F., Heyraud, A., and Rinaudo, M., 1991. Low molecular weight carbohydrates from *Mimosa scabrella* seeds. *Plant Physiology and Biochemistry*. 29: 139–146.

Górecki, R.J., Fordonski, G., Halmajan, H., Horbowicz, M., Jones, R.G., and Lahuta, L.B., 2001. Seed Physiology and Biochemistry. In: Hedley, C.L. (ed.) *Carbohydrates in Grain Legume Seeds: Improving Nutritional Quality and Agronomic Characteristics*. CAB International, Wallingford, UK, pp. 117–143.

Hoch, G., Richter, A., and Körner, C., 2003. Non-structural carbon compounds in temperate forest trees. *Plant, Cell and Environment*. 26: 1067–1081.

Hoekstra, F.A., Golovina, E.A., and Buitink, J., 2001. Mechanisms of plant desiccation tolerance. *TRENDS in Plant Science*. 6: 431–438.

Horbowicz, M., Brenac, P., and Obendorf, R.L., 1998. Fagopyritol B1, O-α-D-galactopyranosyl-$(1 \rightarrow 2)$-D-*chiro*-inositol, a galactosyl cyclitol in maturing buckwheat seeds associated with desiccation tolerance. *Planta*. 205: 1–11.

Horbowicz, M., and Obendorf, R.L., 1994. Seed desiccation tolerance and storability: Dependence on flatulence-producing oligosaccharides and cyclitols – review and survey. *Seed Science Research*. 4: 385–405.

Horbowicz, M., and Obendorf, R.L., 2005. Fagopyritol accumulation and germination of buckwheat seeds matured at 15, 22, and 30 °C. *Crop Science*. 45: 1264–1270.

Kadlec, P., Bjergegaard, C., Gulewicz, K., Horbowicz, M., Jones, A., Kintia, P., Kratchanov, C., Kratchanova, M., Lewandowicz, G., Soral-Smietana, M., Sorensen, H., and Urban, J., 2001. Carbohydrate chemistry. In: Hedley, C.L. (ed.) *Carbohydrates in Legume Seeds: Improving Nutritional Quality and Agronomic Characteristics*. CAB International, Wallingford, UK. pp. 15–59.

Kuo, T.M., VanMiddlesworth, J.F., and Wolf, W.J., 1988. Content of raffinose oligosaccharides and sucrose in various plant seeds. *Journal of Agricultural and Food Chemistry*. 36: 32–36.

Lahuta, L.B., 2006. Biosynthesis of raffinose family oligosaccharides and galactosyl pinitols in developing and maturing seeds of winter vetch (*Vicia villosa* Roth). *Acta Societatis Botanicorum Poloniae*. 75: 219–227.

Leprince, O., and Buitink, J., 2010. Desiccation tolerance: From genomics to the field. *Plant Science*. 179: 554–564.

Morse, M., Rafudeen, M.S., and Farrant, J.M., 2011. An overview of the current understanding of desiccation tolerance in the vegetative tissues of higher plants. *Advances in Botanical Research*. 57: 319–347.

Obendorf, R.L., 1997. Oligosaccharides and galactosyl cyclitols in seed desiccation tolerance (Review Update). *Seed Science Research*. 7: 63–74.

Obendorf, R.L., Steadman, K.J., Fuller, D.J., Horbowicz, M., and Lewis, B.A., 2000. Molecular structure of fagopyritol A1 (O-α-D-galactopyranosyl-$(1 \rightarrow 3)$-D-*chiro*-inositol) by NMR. *Carbohydrate Research*. 328: 623–627.

Obendorf, R.L., McInnis, C.E., Horbowicz, M., Keresztes, I., and Lahuta, L.B., 2005. Molecular structure of lathyritol, a galactosyl bornesitol from *Lathyrus odoratus* seeds, by NMR. *Carbohydrate Research*. 340: 1441–1446.

Obendorf, R.L., and Kosina, S.M., 2011. Soluble carbohydrates in soybean. In: Ng, T.B. (ed.) *Soybean - Biochemistry, Chemistry and Physiology*. In Tech Open Access Publisher, Rijeka, Croatia. pp. 201–228, ISBN: 978-953-307-219-7. Available at: http://www.intechopen.com/articles/show/title/soluble-carbohydrates-in-soybean, Accessed 29 December 2011.

Peterbauer, T., and Richter, A., 2001. Biochemistry and physiology of raffinose family oligosaccharides and galactosyl cyclitols in seeds. *Seed Science Research*. 11: 185–198.

Peterbauer, T., Brereton, I., and Richter, A., 2003. Identification of a diga-lactosyl ononitol from seeds of adzuki bean (*Vigna angularis*). *Carbohydrate Research*. 338: 2017–2019.

Richter, A., Peterbauer, T., and Brereton, I., 1997. The structure of galactosyl ononitol. *Journal of Natural Products*. 60: 749–751.

Ruiz-Matute, A.I., Hernández-Hernández, O., Rodríguez-Sánchez, S., Sanz, M.L., and Martínez-Castro, I., 2011. Derivatization of carbohydrates for GC and GC–MS analyses. *Journal of Chromatography B*. 879: 1226–1240.

Ruiz-Matute, A.I., Montilla, A., del Castillo, M.D., Martínez-Castro, I., and Sanz, M.L., 2007. A GC method for simultaneous analysis of bornesitol, other polyalcohols and sugars in coffee and its substitutes. *Journal of Separation Science*. 30: 557–562.

Sasaki, K., Hicks, K.B., and Nagahashi, G., 1988. Separation of eight inositol isomers by liquid chromatography under pressure using a calcium-form, cation-exchange column. *Carbohydrate Research*. 183: 1–9.

Steadman, K.J., Fuller, D.J., and Obendorf, R.L., 2001. Purification and molecular structure of two digalactosyl D-*chiro*-inositols and two trigalactosyl D-*chiro*-inositols from buckwheat seeds. *Carbohydrate Research*. 331: 19–25.

Szczeciński, P., Gryff-Keller, A., Horbowicz, M., and Obendorf, R.L., 1998. NMR investigation of the structure of fagopyritol B1 from buckwheat seeds. *Bulletin of the Polish Academy of Sciences, Chemistry*. 46: 9–13.

Szczeciński, P., Gryff-Keller, A., Horbowicz, M., and Lahuta, L.B., 2000. Galactosylpinitols isolated from vetch (*Vicia villosa* Roth) seeds. *Journal of Agricultural and Food Chemistry*. 48: 2717–2720.

Taguchi, R., Yamazaki, J., Tsutsui, Y., and Ikezawa, H., 1997. Identification of *chiro*-inositol and its formation by isomerization of *myo*-inositol during hydrolysis of glycosylphosphatidylinositol-anchored proteins. *Archives of Biochemistry and Biophysics*. 342: 161–168.

Traitler, H., Del Vedovo, S., and Schweizer, T.F., 1984. Gas chromatographic separation of sugars by on-column injection on glass capillary column. *Journal of High Resolution Chromatography and Chromatography Communications*. 7: 558–562.

Ye, F.T., Yan, X.J., Xu, J., and Chen, H., 2006. Determination of aldoses and ketoses by GC-MS using differential derivatisation. *Phytochemical Analysis*. 17: 379–383.

CHAPTER 12

Dietary Sugars: TLC Screening of Sugars in Urine and Blood Samples

JOSÉ RAMÓN ALONSO-FERNÁNDEZ*[a] AND
VINOOD B. PATEL[b]

[a] Laboratorio de Tría Neonatal en Galicia – Detección Precoz de Enfermidades Metabólicas, Laboratorio de Metabolopatías, Hospital Clínico (CHUS), Departamento de Pediatría, Universidade de Santiago de Compostela, Choupana, s/n, E-15706 Santiago de Compostela, Spain
[b] Senior Lecturer in Clinical Biochemistry, Department of Biomedical Sciences, School of Life Sciences, University of Westminster, 115 New Cavendish Street, London W1W 6UW, United Kingdom
*Email: joseramon.alonso@usc.es

12.1 Introduction

The urine and blood samples referred to in this chapter may be of two kinds: bulk samples of these body fluids, or samples impregnated in paper. The former are suitable for screening only a quite limited population. Sugars in such samples may be detected and identified by TLC prior to their quantitative or semiquantitative determination by other techniques, such as densitometry. The presence of unphysiological sugars, or of physiological sugars in unphysiological amounts, is indicative of inappropriate ingestion or of altered biochemistry due to a genetic defect or other underlying pathology. Samples

Food and Nutritional Components in Focus No. 3
Dietary Sugars: Chemistry, Analysis, Function and Effects
Edited by Victor R Preedy
© The Royal Society of Chemistry 2012
Published by the Royal Society of Chemistry, www.rsc.org

of blood and urine impregnated in paper can be used for mass screening of large populations – generally newborn infants. In this case, identification of individual sugars by thin layer chromatography (TLC) is usually performed only if a preliminary nonspecific test for sugars or reducing agents has proved positive. Throughout this chapter, it is borne in mind that mass screening for a particular disorder is only justified if it allows the detection of patients in time for the application of an existing effective treatment.

12.2 Forerunners of TLC Methods

A nonspecific test for sugars was probably first used in mass screening in 1949, when the anthrone test was applied to bulk blood and urine to screen for diabetes mellitus in adults (Fetz and Lester 1950). In the same year, Kahn (1949) proposed a blood sugar screening method based on the technique of Hagedorn, Halström and Jensen. Designed for ease of use in the field by relatively unqualified personnel, this method employed reagents in solid tablet form. One tablet was for precipitating protein; a second consisted of potassium ferricyanide, the substrate to be reduced by any reducing sugars; a third was potassium iodide for reduction of excess ferricyanide; and a fourth was for acidification. However, this method seems never to have been employed in practice.

While solid reagent tablets facilitate testing in the field, absorption of a blood or urine sample in paper increases its stability and facilitates its being sent for testing in a central laboratory. This latter approach, developed by Drey (1950), was applied by Olmsted *et al.* (1953) to mass screening of urine samples for diabetes using Benedict's reagent. Packer *et al.* (1958) reported that for detection of diabetes, the use of Benedict's reagent in either solution or solid tablet form was inferior to testing a urine sample specifically for glucose using paper strips impregnated with glucose oxidase. However, when it is necessary to screen large populations for a wider variety of sugar-related pathologies it is preferable to use a two-tier approach in which tests that are specific for individual sugars are employed only after a nonspecific test for altered sugar metabolism. This was the view of Helen Berry, who for the purposes of screening newborn infants developed a procedure for taking urine samples into paper that was simpler than Drey's (Berry *et al.* 1958) and then used aniline phthalate for the nonspecific detection of sugars in the paper-borne samples (Berry 1959).

Berry's paper impregnation method is still the procedure employed by newborn screening programmes that continue to screen for sugar anomalies. However, the nonspecific screening stage is now generally a test for reducing substances such as those published by Alonso-Fernández *et al.* (1979, 1987) or Bradley (1975), although the anthrone test has also been employed (Harper 1969). The anthrone test and those of Alonso-Fernández all involve elution of the sugars from the paper bearing the sample. Bradley's adaptation of

Benedict's test does not, but like the anthrone test is less clear than the Alonso-Fernández test.

For the second stage of the two-tier approach, TLC may be employed. Chromatography, especially when multiple detectors or sequential visualization stains are employed, is a so-called "multiplex" technique. Though in the newborn screening community this term is generally used to contrast tandem mass spectrometry with techniques that in a given run determine just a single biochemical marker of a single disorder, it is equally applicable to chromatography, which for this reason among others has served newborn screening so well (Alonso-Fernández and Colón 2009, 2010b). TLC continues to be employed for the identification of multiple substances in a wide variety of samples of diverse origin; the identification of sugars in blood and urine is of interest in both human and veterinary medicine.

The separation of sugars in blood and urine by ascending paper chromatography was described by Woolf (1951). The mobile phase was a mixture of ethyl acetate and pyridine saturated with water, and chromatograms were left to run overnight. Ammoniacal silver nitrate was discarded as a visualization agent because it is reduced to a brown product not only by all common sugars except sucrose, but also by other reducing substances. This problem could be avoided by running duplicate chromatograms, one of which was sprayed with aniline phthalate (which stains the aldohexoses glucose, galactose, lactose and maltose yellow-brown, pentoses red-brown, and hexosamines yellow, but does not stain fructose or sucrose) and the other with naphthoresorcinol (which stains the ketoses fructose and sucrose red-brown and most other aniline-phthalate-stainable sugars a light grey-blue); together, these two agents visualize all the common sugars. Aniline phthalate can be replaced by *p*-anisidine, benzidine, α- or β-naphthylamine, *m*-phenylenediamine or dimethylaniline; and naphthoresorcinol by α-naphthol, orcinol, resorcinol, anthraquinone or α-naphthylethylenediamine. A number of these reagents, such as benzidine, can in fact react with both ketoses and aldoses, making it in principle unnecessary to obtain duplicate chromatograms, but some sugars are difficult to identify by position alone. A better alternative is sequential staining: spraying with a solution of *p*-aminodiphenylamine phthalate in glacial acetic acid, followed by 4 minutes' heating at 80 °C, stains aldohexoses brown and pentoses red-brown while failing to stain ketoses; after which, spraying with a solution of *p*-aminodiphenylamine hydrochloride in trichloroacetic acid, followed by 90 seconds at 100 °C, stains ketoses bright red while leaving the aldoses with their previous colours (but note that aldoses also stain red if the first round of spraying and heating is omitted). The products of reactions between sugars and amines exhibit UV fluorescence, the wavelength and intensity of which depends more on the amine than the sugar. Aniline affords particularly intense fluorescence, allowing the detection of aniline-phthalate-stainable sugars at concentrations several times lower than can be detected by observation of visible colour alone. This fluorescence moreover lends useful qualitative support to conclusions reached on the basis of visible spots.

12.3   Early TLC Methods

The year 1964 saw the publication of two TLC methods for detection of clinically relevant sugars. Kaeser and Masera (1964) used a stationary phase prepared from a suspension of Merck silica gel G in 0.1N boric acid, 5:3:2 isopropanol/butanol/water as mobile phase, and 50:1 0.2% naphthoresorcinol/ 95% sulphuric acid as visualization agent, with urine samples demineralized as per Woolf (1953). The Rfs were 0.105 for palatinose, 0.153 for fructose, 0.165 for ribose, 0.223 for lactose, 0.283 for galactose, 0.365 for glucose and maltose, and 0.394 for sucrose. Cotte *et al.* (1964) used two simultaneous TLC procedures to detect eight sugars in blood and urine. Urine samples were stored in the cold after addition of chloroform to prevent glycolytic fermentation, while blood was collected on sodium fluoride and immediately precipitated with zinc hydroxide; both urine and blood were desalted with Permutite C50 resin, and after determination of reducing substances the samples were brought to a concentration of 0.5–1.0 g/L for TLC. The stationary phases for TLC were prepared from suspensions of Merck silica gel G in either 0.02 M sodium acetate (procedure 1) or 0.1N boric acid (procedure 2); the mobile phase was in both cases 6:2:2 methyl ethyl ketone/boric acid/methanol; and the visualization agent was either 1:2:100 anisaldehyde/sulphuric acid/glacial acetic acid (procedure 1) or a 1:1 mixture of 20% trichloroacetic acid and a 0.2% solution of naphthoresorcinol in 95° ethanol (procedure 2). Plates for both procedures were loaded with 10 µL of sample and run simultaneously in the same tank, taking about 1 h to run 14 cm. The Rf order for procedure 1 was lactose < maltose < sucrose < fructose, while for procedure 2 it was fructose < galactose < glucose < xylose. Procedure 1 afforded good separation of disaccharides, and was particularly useful for separation of sucrose from fructose, which were poorly resolved by procedure 2; while procedure 2 afforded better separation of monosaccharides. Identification was assisted by spot colours.

Another TLC method for detection of some clinically important carbohydrates in plasma and urine was described by Szustkiewicz and Demetriou (1971), who used commercial 20 × 20 cm plates bearing a 250 µm layer of silica gel. The plates were scored 14 cm from the loading point (1 cm from the base of the plate) and activated by spraying lightly with the mobile phase (13:33:4 pyridine/ethyl acetate/water) and drying for 2 h at 105 °C. Glucose, galactose, ribose, lactose, arabinose, fructose and xylulose standards were prepared in 1:1 isopropanol/water. Urine samples that had given positive Benedict tests were treated with NaF and applied directly to the plates (any known to contain lactose were also pretreated with BioRad AG501 × 8D resin); blood was collected on NaF and the plasma was loaded on the plates following ultrafiltration. After the solvent front had reached the score line under conventional ascending chromatography, the plates were dried in an oven for 10 minutes at 50 °C and re-run in the same tank with the remaining mobile phase. After this second run they were air-dried for 5 minutes at 80 °C, sprayed with aniline/ phthalic acid, dried for 10 minutes at 105 °C, and examined under near ultraviolet light. Though this method did not include standards for such clinically relevant sugars as maltose, sucrose, xylose or raffinose, it was expected to be useful given suspicion of diabetes mellitus, essential pentosuria, food-related pentosuria, ribosuria, intolerance to lactose, galactosaemia, and other conditions involving abnormal carbohydrate metabolism.

Noting that the method of Kaeser and Masera (1964) afforded spots that were well separated but diffuse, Kraffczyk and Helger (1972) developed a method for urine (also applicable to faeces) that was two-dimensional (for better separation of the many mono- and disaccharides that can appear in these matrices) and that, unlike most previous methods for urine samples, required neither prior desalting nor extraction. The stationary phase consisted of 10×10 cm silica gel plates obtained by cutting commercial 20×10 cm plates in half. The urine sample, diluted 1:1 with isopropanol, was loaded at one corner of the plate, the internal standard was loaded on top of the sample, and the plate was run twice with the same mobile phase as in the method developed for plasma (see previous paragraph), each run taking 150–180 minutes. The plate was then run twice in the second dimension with 17:3 acetonitrile/water as mobile phase, each run taking 25–30 minutes. After each of the four runs, the plate was dried for 2–3 minutes in an oven at 80 °C, and after the last run visualization was performed in the same way as in the method developed for plasma. The first sugars to appear were sucrose, fructose and palatinose (as red spots), followed by fucose (also red); all other spots were blue, and all spots became grey if heated for longer than 7 minutes. This method allowed the detection and probable identification of the hexoses glucose, galactose, fructose, and fucose, the heptoses seduheptulose and mannoheptulose, the disaccharides lactose and sucrose, the tetrose D-threitol, and a number of pentoses, including D-xylose. The limit of detection was about 10 mg/dL.

12.4   Galactosaemia Studies

In the context of research into galactosaemia, Klethi (1973a) analysed urine by a two-tier TLC method. At both levels, untreated urine samples were loaded onto 5×10 cm Kodak 511V polycarbonate plates that had been pretreated with 0.2 M phosphate buffer of pH 6.8, and were subjected to two 2-hour runs, with a drying stage in between, using 5:1:1 isopropanol/ethyl acetate/water as mobile phase. On the first tier, spots were visualized by treatment with periodic acid and benzidine, which is highly sensitive but not specific for sugars. Samples of urine testing positive on the first tier proceeded to the second, on which the visualization agent was anisidine. The urine of galactosaemics always contained galactose, galactitol and galactonic acid, none of which are ever found in the urine of healthy individuals. The method was employed to monitor the efficacy of a galactose-free diet for treatment of galactosaemia.

Since carriers of galactosaemia due to galactokinase deficiency are liable to suffer cataract, Klethi (1973b) devised a method for rapid investigation of galactose chemistry in patients who developed cataracts at an early age. Heparinized blood was haemolysed by successive freeze-thaw cycles, a sample of haemolysate was incubated with [^{14}C]-galactose for 4 min at 37 °C in an appropriate medium, the reaction was quenched by heating in a water bath for 1 minute, the sample was centrifuged, and the supernatant was subjected to ascending paper chromatography with a 15:6 mixture of 95° ethanol and 1 M sodium acetate (pH 3.8) as mobile phase. Good separation of radioactive galactose, galactose-1-phosphate and uridine diphosphate galactose (UDPgal) was achieved in index samples by running for 5–6 hours (20–22 cm), while only radioactive galactose appeared in samples from heterozygotes for galactokinase deficiency.

12.5   Oligosaccharides

In many lysosomal storage diseases, patients excrete large amounts of pathological oligosaccharides. This is generally the result of metabolites with sugar moieties undergoing defective catabolism due to lysosomal enzyme deficiencies, though in some cases it is accessory proteins such as membrane transport proteins that are defective or missing. Humbel and Collart (1975) studied oligosaccharides in the urine of patients with a group of hereditary glycoprotein storage diseases including mannosidosis, fucosidosis, type 1 GM1 gangliosidosis, and aspartylglucosaminuria. Samples were run overnight on 20 × 20 cm silica gel plates (Merck 5715) using freshly prepared 2:1:1 *n*-butanol/acetic acid/water as mobile phase. The dried plates were sprayed with a 2 g/L solution of orcinol in 200 g/L sulphuric acid, heated for 10 minutes at 100 °C, and examined against backlight. All samples from the targeted patients exhibited spots at lower Rf than those of normal urine sugars: mannosidosis afforded an intense red-brown spot, fucosidosis a pronounced brown-yellow spot together with others at lower Rf, and aspartylglucosaminuria a similar low-Rf spot (identified by earlier authors as due to glycoasparagines), while keratan sulphate remained at the origin. None of these spots appeared in chromatograms of urines from patients with mucolipidosis, sulphatidosis, or a variety of mucopolysaccharidoses.

To improve the TLC of neutral oligosaccharides in the diagnosis of mannosidosis, Friedman *et al.* (1978) treated urine samples with ion exchange resins to remove charged species, which eliminated the characteristic oligosaccharides of all lysosomal storage diseases except mannosidosis and GM1 gangliosidosis. The eluate was loaded on Whatman LK5DF silica gel plates with a concentration zone, and run with *n*-propanol/water made up in proportions ranging from 8:3 for large oligosaccharides to 7:1 for mono- and disaccharides. Primary amines were visualized with ninhydrin, sialic acid with resorcinol/HCl, and sugars with a modified form of the reagent used by Humbel and Collart (1975), a 2 g/L solution of orcinol in a 200 g/L solution of sulphuric acid in methanol. Urines from mannosidosis patients showed 11 spots, including the monosaccharides glucose, galactose and mannose. When run with 7:1 *n*-propanol/water, Rgs relative to glucose (1.0) were as follows: mannose, 1.19; galactose, 0.79; lactose, 0.40; and raffinose 0.19 (traces of xylose - but not

N-acetylglucosamine – were also observed in the urine of mannosidosis patients).

A different way of improving the Humbel-Collart method for oligo-saccharides in urine was proposed by Sewell (1979). Untreated urine, together with lactose, fructose and raffinose standards, was loaded 1.5 cm from the bottom edge of a Merck 5715 20 × 20 cm silica gel plate, run overnight with freshly prepared 2:1:1 *n*-butanol/glacial acetic acid/water, dried thoroughly, run again for 4 hours in the same direction with 4:5:3 nitromethane/*n*-propanol/water, dried, sprayed with the Humbel-Collart visualization agent (Humbel and Collart 1975), heated for 10 minutes at 100 °C, and examined against backlight. Stachyose was separated in two components, and sialyllactose in six, but raffinose appeared as a single spot (which was why it was chosen as a reference compound).

Noting that of the 23 physiological carbohydrates that could be found in normal urine, ten required sure identification, Vaysse *et al.* (1980) used TLC for rapid (3-hour) qualitative analysis of carbohydrates in serum, urine and faeces. Serum samples were deproteinized by dripping onto isopropanol, urine was desalted with ion exchange resin, and faeces were homogenized in 0.9% sodium chloride solution, centrifuged, and desalted in the same way as urine. Samples were loaded onto acid-resistant Schleicher & Schüll F1500 silica gel plates, run for 2 hours (20 cm) with 6:3:1 ethyl acetate/pyridine/water, dried for 10 minutes at 100 °C in a ventilated oven, sprayed with naphthoresorcinol in sulphuric acid solution, and heated for another 10 minutes at 100 °C. Rfs increased in the order stachyose, raffinose, melibiose, trehalose, lactose, maltose, sucrose, galactose, glucose, mannose, fructose, arabinose, fucose, xylose, ribose, rhamnose. Spot colours – also reported – facilitated discrimination between close pairs such as fructose (Rf 0.50, violet) and mannose (Rf 0.49, blue). Sensitivity could be increased by repeated sample loading to ensure diagnosis of the main pathological melliturias (essential fructosuria, intolerance to sucrose, congenital galactosaemia, *etc.*).

Having observed that screening for mannosidoses by one-dimensional TLC of urine is complicated by interferences from physiological, dietary and med-icinal oligosaccharides, Abeling *et al.* (1996) proposed a dual method in which desalted urine samples were run on different plates with 2:1:1 *n*-butanol/acetic acid/water (mobile phase I) and 85:1:15 isopropanol/acetic acid/water (mobile phase II). The plates were Merck 5553 silica gel 60 plates with an aluminium base, and both were run twice and sprayed with orcinol in sulphuric acid for visualization of spots. On plates run with mobile phase II, urines from patients with α-mannosidosis showed a spot just above lactose that was absent from urines from patients with post-prandial syndrome, a condition that proved indistinguishable from α-mannosidosis if the isopropanol was replaced with *n*-propanol. β-Mannosidosis and excretion of oligosaccharides related to blood group A were differentiated by the appearance, in the latter case, of a narrow band at the same Rf as lactose on the plates run with mobile phase I.

For rapid diagnosis of Pompe's disease (acid maltase deficiency), Blom *et al.* (1983) slightly modified the Humbel-Collart method (Humbel and Collart

1975) by allowing the solvent front to reach the upper edge of the chromato-graphy plate (which meant an even longer run than in the original Humbel-Collart method). Urines were desalted with BioRad 200–400 mesh AG50W-X8 and 100–200 mesh AG3 × 4A resins, the eluate was concentrated to dryness and redissolved in an equal volume of double-distilled water, and the quantity of sample loaded was calculated in mL as 2F/C, where the parameter F depended on age and C was creatinine content. Like Dhondt *et al.* (1982) before them, Blom *et al.* (1983) noted the possibility of false positives due to lactobionate in parenteral feed.

Not all urinary oligosaccharides are pathological. In particular, that 80% of the population who secrete blood group antigens A, B or O(H) in saliva and other exocrine fluids, also excrete a related group of neutral oligosaccharides in urine. These were targeted by Kuriyama *et al.* (1990) in a TLC procedure that introduced a number of improvements over previous analyses of neutral oli-gosaccharides. Finally, though addressed to phytochemists, a short monograph on TLC of carbohydrates by Sassaki *et al.* (2008) serves to summarize the developments sketched above. Silica gel G60, the most widespread stationary phase, affords faster runs than cellulose, and resolution and sensitivity are improved by the use of HPTLC plates and, for monosaccharides, pretreating the plates with 0.02 M boric acid/sodium borate buffer. Since carbohydrates are strongly hydrophilic, with great affinity for both silica gel and cellulose, their TLC requires highly polar mobile phases. Of the many mobile phases employed with silica gel, Sassaki *et al.* recommended 4:2:2:1 ethyl acetate/*n*-propanol/acetic acid/water for TLC of monosaccharides, and 7:3 *n*-propanol/water for oligosaccharides. Polysaccharides cannot be separated by TLC.

12.6   A TLC Procedure for Screening for Sugars in Blood and Urine

The following procedure, developed over 30 years by the first author and coworkers (Alonso-Fernández *et al.* 1981; Alonso-Fernández *et al.* 2010a), is followed in their laboratory for screening newborn infants for disorders that alter sugar levels in blood or urine. It may also be applied to older children and adults.

12.6.1   Sample Collection

Samples are subjected to analysis in dry solid form, impregnated in Whatman 903 paper. In the case of newborn infants, blood samples are obtained by pricking the side of the heel and letting the emerging blood fall onto the paper, pricking a finger or earlobe, or drawing from a vein. (N.B.: anticoagulants can interfere with certain analyses). The paper should be impregnated until the affected areas on the two sides of the paper are equal; if the paper slip only has a target circle printed on one side, blood should be dripped onto the other side until the target area is filled. Dripping onto both sides of the paper is incorrect, since it may leave an insufficiently impregnated zone between the two sides.

Once impregnated, the paper is laid horizontal to dry. The area from which blood is obtained, be it heel, finger, earlobe or elsewhere, should previously have been warmed under a stream of warm water or by vigorous massage, and should then have been disinfected with an antiseptic that evaporates without trace. Pricking or vein puncture should be performed with all due care. The first drop of fluid emerging from a prick site should be discarded, since it contains fluid other than blood.

Urine samples are collected from newborns by placing a slip of Whatman 903 paper on the genitals and holding it in place with a napkin. This should be done immediately before the blood sample is taken – the stress induced by the latter causes the infant to urinate, after which the paper can be withdrawn and laid horizontal to dry. To avoid interferences, the infant should be clean and free of creams, oils, talcs or any other cosmetic. If it defecates during the operation as well as urinating, the sample should be discarded, the infant cleaned, and the operation repeated with another paper slip (each newborn screening kit generally contains two); even though the first paper may not have been soiled with faeces, it may have been contaminated with faecal material taken up by the urine. In the case of older children and adults, the paper slip should be dipped in urine collected in some clean vessel - this will invariably ensure the uptake of more urine than is achieved by urinating directly onto the paper slip or using a dropper or pipette to transfer urine from the collecting vessel to the paper.

The urine- and blood-impregnated paper slips require about 4 hours to dry completely at room temperature. They should not be dried in the sun, with hot air, or by placing on a hot surface, since the samples are altered by UV light and temperatures higher than room temperature. Elevated temperatures precipitate and denaturalize proteins such as biotinidase or gal-1-P-uridyltransferase, as do vestiges of alcohol used for prick site disinfection, and precipitated protein can hinder the subsequent elution of analytes such as thyrotropin.

12.6.2   Concentration of Samples

It is not uncommon for samples to require concentration or dilution prior to analysis. In the case of urine samples that have arrived at the laboratory as bulk liquid found to have a low creatinine concentration, the sample paper can be dipped repeatedly in the bulk (and left to dry between dips); this procedure is analogous to freeze-drying (used by some authors) with the inclusion of an excipient, the paper. The procedure followed to concentrate urine received in sample paper is as follows for an illustrative concentration factor of about 9.6. A disc 5 mm in diameter is cut from the sample paper and, with the help of tweezers, is placed at the bottom of a 1.5 mL Eppendorf tube against the sector of the wall that will form the floor of the tube when centrifuged. Six more discs, each 6 mm in diameter, are cut from the sample paper and placed in a round-bottomed test tube with an inside diameter of at least 8 mm. 500 μL of water is added to the test tube with an automatic pipette, and the test tube is stoppered and shaken in a vortex shaker for 5 minutes. The six discs are then manoeuvred to the bottom of the tube with a stainless steel rod, the tube is centrifuged for

3 minutes at 3000 rpm, and the supernatant is transferred to the Eppendorf tube with an automatic pipette. Another 500 μL of water is added to the tube with the six 6 mm discs and the whole cycle (shaking, centrifuging, transfer) is repeated. The Eppendorf tube, duly balanced with another tube containing a 5 mm disc and approximately 1 mL of water, is then vacuum-centrifuged to dryness, at which point the 5 mm disc it contains has a urine content approximately 9.6 times its original content. The concentration factor can be increased by repeating the whole procedure with the same 5 mm disc, or varied more generally by changing the number of 6 mm discs placed in the test tube or the size of the discs that in this example were 6 mm in diameter. If sample concentration is thought likely to impair chromatogram quality due to the concentration of salts, an appropriate quantity of ion exchange resin can be added to the test tube between the insertion of the 6 mm discs and the addition of water; at the end of the operation this resin can be regenerated, after removal of the discs, by washing *in situ* in the test tube, which can then be used for another sample.

To dilute a urine sample borne in a 5 mm paper disc, this disc is placed in an Eppendorf tube and shaken with 100 μL of water, 50 μL of the resulting solution is transferred to another Eppendorf tube containing a 5 mm disc of clean (unimpregnated) Whatman 903 paper, and this second tube is vacuum-centrifuged to dryness. If the tube with the original disc and the remaining 50 μL of solution is used as counterpoise, both discs will now have a urine content approximately half that of the original disc.

The procedure for concentrating paper-borne blood samples is similar to the procedure for urine, except that the 5 mm disc is clean (unimpregnated), and the eluent added to the round-bottomed test tube is 1:1 isopropanol/water; for the volumes, *etc.*, of the above example, this affords a concentration factor of approximately 8.6. Furthermore, the isopropanol fixes pigments and proteins to the discs while allowing unimpeded elution of sugars. 1:1 Isopropanol/water is likewise used for dilution of paper-borne blood samples by a procedure analogous to that described above for urine.

12.6.3   Loading Samples onto the TLC Plate

Blood or urine samples, with or without concentration or dilution as above, are transferred from their 5 mm discs to the 2.5 cm concentration zone of a 20×10 cm silica gel plate (Merck 1.11844.0001), as follows. Discs of samples and standards are placed with their centres 1 cm from the lower edge of the plate (opposite their identification numbers), and are covered with a 2×10 cm glass slip 2 mm thick that is held in place with bulldog clips. This "sandwich" is placed for 1 minute in a tank containing freshly prepared 23:30:12 isopropanol/ methyl ethyl ketone/30% ammonia to a depth of less than 1 cm (eluent more than 1 day old gives rise to horizontal artifact bands in the chromatogram). Salts are not eluted by this solvent, in which they have very low solubility and mobility. When 1 minute has elapsed, the plate is withdrawn from the tank and placed on a stand where the portion of the concentration zone that is not covered by the glass slip is exposed for 3 minutes to a stream of room temperature air from a hair drier at maximum output. Hot air should not be used, because it leads to the appearance in the chromatogram of horizontal bands similar to those caused by non-fresh eluent. The glass slip is then removed, and the air stream is used to remove the sample discs (if air alone does not suffice, the tip of a hypodermic needle can be used to help the discs fall). All the above operations are performed in a fume hood.

12.6.4   The TLC Cell

The front and back walls of each cell are glass plates 1 cm thick. They are separated by a glass sheet 2 mm thick that has been cut to form a right-angled U-plate with a distance of 105 mm between its vertical limbs, so that the internal horizontal cross-section of the cell is for most of its height a 105×2 mm rectangle. This leaves approximately 0.4 mm between the front wall and the silica gel layer of a TLC plate with its back side wholly in contact with the back wall. The reservoir for the mobile phase is formed by $115 \times 10 \times 7$ mm horizontal recesses cut in the internal faces of the front and back walls with their lower edges 10 mm above the top of the horizontal portion of the U-plate (which constitutes the bottom of the cell), together with half discs 10 mm in diameter cut from the vertical limbs of the U-plate so as to coincide with the ends of the recesses. The height of the reservoir above the bottom of the cell ensures that the TLC plate stands in at least 10 mm of mobile phase until the mobile phase is almost exhausted, while the half discs cut in the U-plate prevent the shape of the solvent front from being affected by the meniscus at the edge of the surface of the mobile phase in the reservoir.

The height of the back wall above the bottom of the cell is the same as the length of the TLC plate. The front wall and the U-plate are some 2 cm shorter in the larger of the cells and 1 cm shorter in the smaller cells. Their top edges bear a silicone rubber "bed" 1 mm thick, and silicone rubber guide rails installed on the back wall where it overtops the front wall ensure that when the TLC plate is inserted in the cell its stationary phase face makes no contact with the front wall. Once the TLC plate has been inserted, the cell is closed by a glass lid 2 mm thick that lies with its long edge flush against the stationary phase (to which end the guide rails terminate 2 mm short of the bed). On either side of the TLC plate, lateral to the guide rails, the back wall bears a 2 mm thick silicone rubber pad and a rectangle cut from a TLC plate of the same kind as is used in the cell; pressing the lid against these rectangles ensures that it neither cuts into the stationary phase nor leaves a gap by lying askew. The chamber of the closed cell only communicates with the outside *via* the stationary phase.

12.6.5   Procedure

Before the insertion of the TLC plate, a precise volume of mobile phase (1.5, 2.5, 3.0 or 5.0 mL in the first author's experiments) is injected with a digital pipette. The recommended mobile phase is 15:3:4 acetone/ethyl acetate/water, which in a well-stoppered flask is stable for several weeks at 2–8 °C. The tank is then closed with a silicone rubber slab 100 mm wide and 2 mm thick that is slid down between the guide rails as though it were a TLC plate, and to equilibrate the mobile phase with the atmosphere inside the cell, the cell is placed for 3 minutes in a forced-air chamber thermostatted at 25.5 °C. The TLC plate is then introduced into the cell, the cell is closed with the glass lid, and the plate is run until the mobile phase is exhausted.

On a 20 × 10 cm plate (Merck 1.11844.0001), less mobile sugars are separated in about 4 hours using 5 mL of the recommended mobile phase. The sugars with known limits of detection and Rt values under these conditions are galactosamine (Rt 0.02, limit of detection 25 mg/dL), glucosamine (0.04, 100 mg/dL), maltoheptaose (0.05, 90 mg/dL), stachyose (0.08, <50 mg/dL), maltohexaose (0.09, 100 mg/dL), maltopentaose (0.11, 92 mg/dL), mal-totetraose (0.17, 90 mg/dL), glucuronic acid (0.26, 80 mg/dL), raffinose (0.27, <50 mg/dL), maltotriose (0.34, 80 mg/dL), sialic acid (0.35, 100 mg/dL), lac-tose (0.42, <50 mg/dL), lactulose (0.44, <50 mg/dL), galactose (0.65, <50 mg/dL), sucrose (0.71, <50 mg/dL), glucose (0.76, <50 mg/dL), fructose (0.81, <50 mg/dL), arabinose (0.90, <50 mg/dL), ribose-xylose (1.02, <50 mg/dL), and 3-O-methylglucose (1.03, <50 mg/dL); see Figure 12.1 for chromatograms run with some of these.

More mobile sugars are separated in 2 hours using 3 mL of the recommended mobile phase; in this case, the sugars with known limits of detection and Rt values are stachyose (0.02, <50 mg/dL), raffinose (0.11, <50 mg/dL), isomaltose (0.14, 25 mg/dL), lactulose (0.14, <50 mg/dL), lactose (0.21, <50 mg/dL), trehalose (0.25, 40 mg/dL), palatinose (0.29, <25 mg/dL), cellobiose (0.30, 75 mg/dL), galactose (0.33, <50 mg/dL), mannitol (0.34, 25 mg/dL), sucrose (0.37,

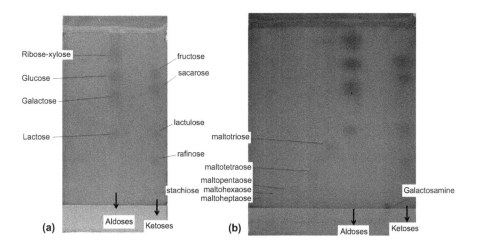

Figure 12.1 TLC of less mobile sugars. Samples are run 18 cm over 4 hours using 5 mL of mobile phase. (a) separation of mixtures of aldose and ketose standards. (b) separation of aldose and ketose standards, together with samples of galactosamine and various malto-oligoses.

<50 mg/dL), glucose (0.40, <50 mg/dL), mannose (0.43, 40 mg/dL), fructose (0.43, <50 mg/dL), arabinose (0.48, <50 mg/dL), fucose (0.55, 25 mg/dL), ribose-xylose (0.59, <50 mg/dL), xylulose (0.68, <50 mg/dL), 3-O-methylglucose (0.71, <50 mg/dL), and rhamnose (0.74, <50 mg/dL) (Figure 12.2 shows some of these). Note that many sugars with intermediate mobility can be run under either set of conditions. Glucose and galactose can be identified in about 40 minutes using 1.5 mL of the recommended mobile phase in the smaller tank.

Spots are visualized by dipping the TLC plate in a 0.1 M solution of ammonium monovanadate in 1 M sulphuric acid, and then heating for 5 minutes in an oven at 120 °C (microwaving has been tried but was not a success). Figures 12.1–12.5 show illustrative chromatograms. Blue spots result from the reduction of vanadium(V) to vanadium(IV), further reduction (caused by greater quantities of reducing substance or greater reducing power) affords first white spots and then, upon reduction to vanadium(III), black spots. Non-reducing disaccharides and oligosaccharides are visualized because the sulphuric acid medium hydrolyses them to reducing species. Spot colour and intensity remain essentially unaltered for about 24 hours. Observation of an intense glucose spot in the chromatogram of a urine sample leads to the running of the corresponding blood sample; an intense glucose spot in this latter chromatogram almost invariably indicates (transient) neonatal diabetes mellitus.

12.6.6   Differential Diagnosis of Galactosaemias

TLC of blood samples is insufficiently sensitive to detect galactosaemias in 3-day-old newborns, but can be used for differential diagnosis of galactosaemias originally detected using urine samples. For this purpose, a 10 × 10 cm HPTLC plate (Merck 1.13748.0001) is employed, and a slightly different sample loading method. The sample and standard discs are placed on the concentration zone with their centres 1.25 cm from the bottom edge, the eluent is 1:1 isopropanol/water, and the whole loading procedure comprises four cycles, in each of which the sample disc "sandwich" is dipped for a few seconds in the eluent, dried thoroughly with hot air, and allowed to cool to room temperature before the next dip in eluent. The transferred samples and standards are run for 1 hour in the smaller cell using 2.5 mL of the recommended mobile phase. With the patient's sample are run a sample from a healthy control newborn (all newborns, including galactosaemia patients, have detectable glucose levels) and standards for galactose-1-phosphate, lactose, galactose, glucose, and uridine diphosphate (UDP) galactose, which run to Rt 0.05, 0.28, 0.59, 0.69 and 0.87, respectively (galactose standards of 5, 10 and 50 mg/dL may be run for semi-quantitative analysis; the cut-off is 10 mg/dL). Paper-borne standards are prepared by immersing Whatman 903 paper in the corresponding solution, allowing it to dry in a horizontal position, and cutting the required 5 mm disc. Spots are visualized by dipping the TLC plate in a solution of 1 g of aniline-2-sulphonic acid in 50 mL of 2 M orthophosphoric acid, heating for 10 minutes in

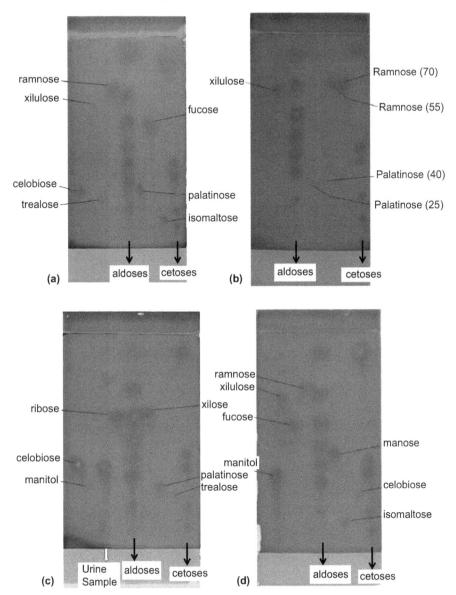

Figure 12.2   TLC of more mobile sugars. Samples are run 18 cm over 2 hours using 3 mL of mobile phase. All plates show mixtures of aldose and ketose standards (mg/dL), together with samples of: (a) rhamnose, xylulose, cellobiose, trehalose, fucose, palatinose and isomaltose; (b) xylulose, rhamnose (two concentrations) and palatinose (two concentrations); (c) ribose, cellobiose, mannitol, xylose, palatinose, trehalose and urine; (d) rhamnose, xylulose, fucose, mannitol, mannose, cellobiose and isomaltose.

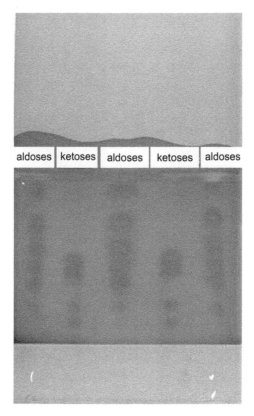

| aldoses | ketoses | aldoses | ketoses | aldoses |

Figure 12.3   Identification of glucose and galactose. Samples are run 9 cm over 40 minutes using 1.5 mL of mobile phase.

an oven at 120–130 °C, allowing to cool, and examining under ultraviolet light. Classical galactosaemia (galactose-1-phosphate uridyl transferase deficiency) gives rise to spots for both galactose and galactose-1-phosphate, galactokinase deficiency only to a galactose spot, and UDP-epimerase deficiency to spots for UDP-galactose, galactose-1-phosphate, and galactose (which become visible in that order in chromatograms of serial samples taken at intervals of a few days). This procedure is based on the work of Fujimoto *et al.* (1983) and Bóveda *et al.* (1989).

By way of illustration of some of the pitfalls of TLC, Figure 12.6 shows three chromatograms with contaminated areas that might mistakenly be interpreted as indicative of pathology. The advantages of examining urine samples are illustrated by the fact that glucose was not detected by TLC of blood in either the case corresponding to the chromatogram of Figure 12.7, which shows glucosuria caused by congenital nephrotic syndrome, or in that of Figure 12.5, in which the glucosuria reflected in the chromatogram disappeared spontaneously.

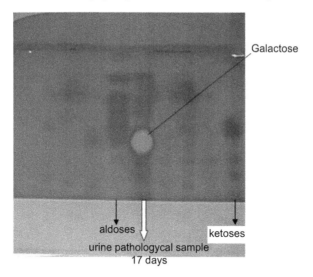

Figure 12.4 A urine chromatogram for a patient with galactosaemia (lane 6 from the left). Lane 5 contains a mixture of aldose standards, the rightmost lane a mixture of ketose standards.

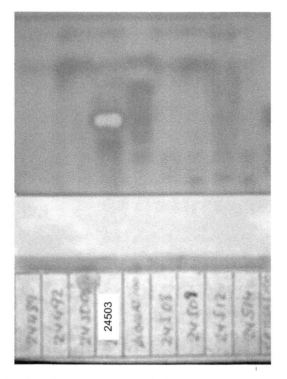

Figure 12.5 A urine chromatogram profile for a patient with glucosuria (patient 24503).

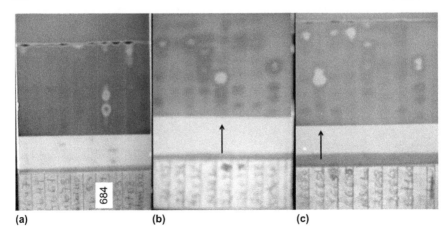

(a)   (b)   (c)

Figure 12.6   Contaminated chromatograms. (a) The intense lactose and galactose
spots of sample 684 (lane 7 from the left) might lead it to be mistaken for a
case of galactosaemia, but failed to appear in a second subsample from the
same patient. (b) The lane to the right of the aldose standard mixture
(arrowed) presents a similar pattern. (c) The second lane from the left
(arrowed) shows contamination with galactose and glucose.

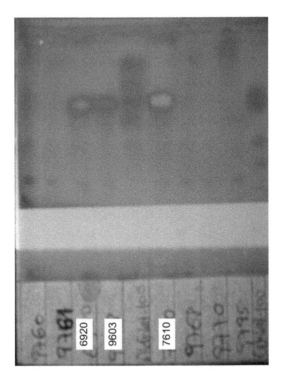

Figure 12.7   A chromatogram with lanes for a newborn infant with elevated urinary
glucose due to congenital nephrotic syndrome. In all three cases (6920,
7610 and 9603), proteinuria was detected by a modified form of the
method of Louis I. Woolf (Woolf and Giles 1956).

Summary Points

- This chapter reviews techniques of thin layer chromatography (TLC) that employed to detect and identify sugars in blood and urine.
- The emphasis is on methods used for screening of newborn infants or other populations for disorders characterized by altered blood and/or urinary sugar profiles.
- The treatment of both bulk liquid samples and dry, mass screening samples impregnated in paper is discussed, paying attention to stationary phases, mobile phases, visualization agents, the sugars separated, and causative disorders.
- A procedure is recommended for TLC of blood and urine received as dry paper-borne samples, with details of sample concentration and dilution, elution onto the silica gel chromatography plate, chromatography with 15:3:4 acetone/ethyl acetate/water, and visualization with vanadium(V) in sulphuric acid or, for differential diagnosis of galactosaemias, with aniline-2-sulphonic acid in orthophosphoric acid.
- TLC chromatograms can be used to identify and characterize sugar disorders, such as galactosaemia.

Key Facts

1. TLC (Thin Layer Chromatography) is a chromatographic technique in which the stationary phase (silica gel for sugars) forms a layer that adheres to and is supported by a plate of inert material. A liquid mobile phase in a reservoir that communicates with the stationary phase is drawn through the latter by capillary forces ("high pressure TLC", in which the stationary phase is sandwiched between two plates and the movement of the mobile phase is assisted by a pressure gradient, has not to our knowledge been used to separate sugars).

2. In radial TLC, samples are loaded at points on the circumference of a circle at the centre of a horizontal TLC plate. The mobile phase enters the stationary phase *via* a wick or capillary touching the centre of the circle, and spreads radially towards the edges of the plate.

3. In two-dimensional TLC, sample is loaded only at one corner of the TLC plate, and after performance of a linear TLC run the plate is turned 90° and subjected to a second linear TLC run perpendicular to the direction of the first.

4. Visualization, or staining, is the process by which the components of the sample that have been separated by TLC are made visible to the human eye on the TLC plate. In visualization by fluorescence quenching, the stationary phase includes a fluorescent reagent.

5. In linear TLC, the position of a component spot on a TLC plate is generally recorded as a "retardation factor" (Rf), which is the distance run by the component divided by the distance run by the

solvent front, both of which distances are measured on a line perpendicular to the base of the plate that passes through the point at which the corresponding sample was deposited. In other TLC modalities, appropriate modifications of this definition are applicable. Rfs are sometimes expressed as percentages. Alternative position measures that can be used when Rfs cannot be calculated (as in continuous TLC) are obtained by replacing the denominator of Rf (the distance run by the solvent front) with the distance run by a reference component such as glucose (which affords the measure Rg) or lactose (Rl), or by the distance between the loading point and the lid of a continuous TLC tank (Rt or cRf). Rg is particularly robust against variability in run conditions.

6. TLC can be hampered by interference from non-target components of samples during both the chromatography run proper (development) and visualization. For example, salts in urine samples can give rise to poorly defined sugar spots, and urea can mask sugar spots if it is stained by the agent used for their visualization.

Definitions of Words and Terms

Biochemical markers are endogenously produced compounds, the presence or altered concentration of which in biological samples indicates a given pathology.

Deproteinization is the removal of proteins.

Desalting is the removal of salts from a sample such as urine.

Eluents are liquids that upon passage through a solid phase dissolve and transport certain components of that phase.

Fructosuria is the presence of an abnormally high concentration of fructose in urine.

Galactosaemia is the presence of an abnormally high concentration of galactose in the blood.

Galactosuria is the presence of an abnormally high concentration of galactose in urine.

HPTLC (high performance TLC) is TLC performed with a high-quality plate.

Inborn errors of metabolism consist in the inherited or congenital mutation of a gene encoding an enzyme. The enzyme is normally lacking or defective.

Maltodextrins are products of the hydrolysis of starch that consist of fewer than 20 dextrose units.

Mellituria is the presence of any kind of sugar in the urine.

List of Abbreviations

Rf   retardation factor
TLC   thin layer chromatography
UDP   uridine diphosphate

Acknowledgements

A more extensive review based on all the accessible literature in English, French, Italian, Portuguese and Spanish up to the year 2010, and from which this chapter was extracted, is available from the first author on request. The technical assistance of Pilar Villar is gratefully acknowledged.

References

Abeling, N.G.G.M., Rusch, H., and van Gennip, A.H., 1996. Improved selectivity of urinary oligosaccharide screening using two one-dimensional TLC systems. *J Inher Metab Dis* 19: 260–262.

Alonso-Fernández, J.R., Fraga, J.M., and Barreiro Mosquera, J.L., December 1979. Un método sencillo para la detección de azucares y reductores en orinas utilizadas en el cribado de errores metabólicos [A simple method for detection of sugars and reducing substances in urine samples used for screening for errors of metabolism]. Paper presented at the Third National Meeting of the Spanish Society of Clinical Chemistry, La Manga del Mar Menor, Murcia.

Alonso-Fernández, J.R., Bóveda, M.D., Parrado, C., Peña J., and Fraga, J.M., 1981. Continuous thin layer chromatography of sugars of clinical interest in samples of urine impregnated on paper. *J Chromatogr* 217: 357–366.

Alonso-Fernández, J.R., Castiñeiras, D.E., Parrado, C., Fraga, J.M., and Peña, J., 1987. Galactose newborn screening: Test for reducing sugars in urine samples impregnated on paper. In: BL Terrel Jr. (Ed.), *Advances in Neonatal Screening,* Elsevier Science Publishers B.V. (Biomedical Division), Amsterdam, pp. 233–238.

Alonso-Fernández, J.R., and Colón, C., 2009. The contributions of Louis I Woolf to the treatment, early diagnosis and understanding of phenyketonuria. *J Med Screening* 16: 205–211.

Alonso-Fernández, J.R., Carpinteiro, M.I., Baleato, J., and Fidalgo, J., 2010a. Vertical sandwich-type continuous/evaporative TLC with fixed mobile phase volume for separating sugars of clinical relevance in paper-borne urine and blood samples in newbornn screening. *J Clin Lab Anal* 24: 106–112.

Alonso-Fernández, J.R., and Colón, C., 2010b. Newborn screening in Spain, with particular reference to Galicia: Echoes of Louis Woolf. *Mol Genet Metab* 101: 95–98.

Berry, H.K., Sutherland, B., Guest, G.M., and Warkainy, J., 1958. Simple method for detection of phenylketonuria. *JAMA* 167: 2189–2190.

Berry, H.K., 1959. Procedures for testing urine specimens dried on filter paper. *Clin Chem* 5: 603–608.

Blom, W., Lutyn, J.C., Kelholt-Dijkam, H.H., Uijmans, and Loonen, M.C.B., 1983. Thin-layer chromatography of oligosaccharides in urine as a rapid indication for the diagnosis of lysosomal acid maltase deficiency (Pompe's disease). *Clin Chim Acta* 134: 221–227.

Bóveda, M.D., Alonso-Fernández, J.R., Fraga, J.M., and Peña, J., 1989. Simultaneous elution from sample-paper and loading in thin layer chromatography for differential diagnosis of galactosemias. In: B.J. Schmidt, A.J. Diament, NS Login-Grosso (Eds.), *Current Trends in Infant Screening. Proceedings of the 7th international Screening Symposium, Sao Paulo, Brazil, 6-9 November 1988*, Excerpta Medica, Amsterdam-New York-Oxford, pp. 181–185.

Bradley, D.M., 1975. Screening for inherited metabolic disease in Wales using urine-impregnated filter paper. *Arch Dis Child* 50: 264–268.

Cotte, J., Mathieu, M., and Collombel., 1964. Méthode d'identification des sucres sanguins et urinaires par chromatographie en couche mince. *Path Biol* 12(11–12/13–14): 747–749.

Dhondt, J.L., Fariaux, J.P., Cartigny, B., and Michalski, J.C., 1982. Lacto-bionic acid: A pitfall in screening for oligosaccharidurias. *J Inher Metab Dis* 5(Suppl 1): 7.

Drey, N.W., 1950. A simple container for mailing of urine samples. *Am J Clin Path* 20(3): 297–298.

Fetz, R.H., and Lester, M.P., 1950. The anthrone blood sugar method adapted to diabetes case finding in a multiple screening program. *Public Health Report* 65(51): 1709–1718.

Friedman, R.B., Williams, M.A., Moser, H.W., and Kolodny, E.H., 1978. Improved thin-layer chromatographic method in the diagnosis of mannosi-dosis. *Clin Chem* 24(9): 1576–1577.

Fujimoto, A., Aono, S., and Oura, T., 1983. A simple, new method for dif-ferential diagnosis of galactosemia. In: H. Naruse, M. Irie (Eds.), *Neonatal Screening*, Excerpta Medica, Amsterdam-Oxford-Princeton, pp. 254–255.

Humbel, R., and Collart, M., 1975. Oligosaccharides in urine of patients with glycoprotein storage diseases. I. Rapid detection by thin-layer chromato-graphy. *Clin Chim Acta* 60: 143–145.

Kaeser, H., and Masera, G., 1964. Zur dünnschichtchromatographischen Trennung einiger einfacher, klinisch wichtiger Zucker in biologischen Material [Separation of certain simple clinically important sugars from biological materials by thin-layer chromatography]. *Schweizerische Medizi-nische Wochenschrift* 94(5): 158–160.

Kahn, R.B., 1949. Blood Sugar "Screening Method" (Wilkerson-Heftmann). *Cinc J Med* 30(10): 557–558.

Klethi, J., 1973a. Galactosémie: détection du déficit en galactose-phosphate-uridyltransferase. *Comp Rendus des Seances de la Societe de Biologie et de Ses Filiales* 167(10): 1461–6.

Klethi, J., 1973b. Galactosémie: detection du déficit en galactokinase. *Comp Rendus des Seances de la Societe de Biologie et de Ses Filiales* 167(6–7): 1048–1051.

Kraffczyk, F., Helger, R., 1972. Thin-layer chromatographic screening test for carbohydrate anomalies in plasma, urine and faeces. *Clin Chim Acta* 42: 303–308.

Kuriyama, M., Hiwatari, R.-I., and Igata, A., 1990. Thin-layer chromato-graphy of urinary neutral oligosaccharides: The detection of blood group-related oligosaccharides and screening for lysosomal storage disease. *Ohoku J Exp Med* 161: 335–341.

Olmsted, W.H., Drey, N.W., Agress, H., and Roberts, H.K., 1953. Mass screening for diabetes. The use of a device for the collection of dried urine specimens and testing for sugars [St. Louis Dreypak]. *Diabetes* 2(1): 37–42.

Packer, H., and Ackerman, R.F., 1958. A comparison of Benedict's and the glucose oxidase tests for the mass detection of diabetes. *Diabetes* 7(4): 312–315.

Sassaki, G.L., de Souza, L.M., Cipriani, T.R., and Iacomini, M., 2008. TLC of Carbohydrates. In: M Waksmundzka-Hajnos, J Sherma, T Kowalska (Eds.), *Thin Layer Chromatography in Phytochemistry*, Chromatographic Science Series Vol. 99, CRC Press, pp. 255–276.

Sewell, A.C., 1979. An improved method for urinary oligosaccharide screening. *Clin Chim Acta* 92: 411–414.

Szustkiewicz, C., and Demetriou, J., 1971. Detection of some clinically important carbohydrates in plasma and urine by means of thin-layer chromatography. *Clin Chim Acta* 32: 355–359.

Vaysse, J., Pilardeau, P., and Garnier, M., 1980. Analyse qualitative chromatographique des oses sériques, urinaires et fécaux. *Pathologie Biologie* 28(3): 206–208.

Woolf, L.I., 1951. Paper Chromatography. *Great Ormund St. J.* 1: 61–91.

Woolf, L.I., 1953. Use of ion-exchange resins in paper chromatography of sugars. *Nature* 171(4358): 841.

Woolf, L.I., and Giles, H.Mc.C., 1956. Urinary excretion of amino acids and sugar in the nephrotic syndrome. A chromatographic study. *Acta Paediatr.* 45(5): 489–500.

CHAPTER 13

Analysis of Dietary Sugars in Beverages by Gas Chromatography

O. HERNÁNDEZ-HERNÁNDEZ,[a] F.J. MORENO[b]
AND M.L. SANZ*[a]

[a] Instituto de Química Orgánica General (IQOG), CSIC, Juan de la Cierva 3, 28006 Madrid, Spain; [b] Instituto de Investigación en Ciencias de la Alimentación (CIAL), CSIC-UAM, Nicolás Cabrera, 9, Campus de la Universidad Autónoma de Madrid, 28049 Madrid, Spain
*Email: mlsanz@iqog.csic.es

13.1 Introduction

In this chapter, the term "beverage" will cover all liquids for human consumption considered in the Official Methods of Analysis (AOAC). Hot drinks such as coffee or tea-based drinks will be also mentioned.

Dietary carbohydrates include sugars (usually named low molecular weight carbohydrates, LMWC), oligosaccharides and polysaccharides, all of them being found in beverages at different levels. However, GC analyses of intact high molecular weight carbohydrates (HMWC) can not be directly carried out because their low volatility. Furthermore, previous hydrolysis of HMWC with a subsequent derivatization of the released monosaccharide is required, providing a structural characterization of the molecule. Therefore, in this chapter a special focus on the analysis of mono-, di- and trisaccharides will be

Food and Nutritional Components in Focus No. 3
Dietary Sugars: Chemistry, Analysis, Function and Effects
Edited by Victor R Preedy
© The Royal Society of Chemistry 2012
Published by the Royal Society of Chemistry, www.rsc.org

done, although structural analysis of high molecular weight carbohydrates present in beverages will be also considered.

The analysis of sugars by GC in all beverages has some steps in common (Figure 13.1):

(i) *Sample preparation of beverages:*
   Sample preparation normally includes degasification processes (if necessary), filtration, dilution and/or evaporation. Fractionation of some carbohydrates is occasionally required for their subsequent analysis.
(ii) *Derivatization of carbohydrates:*
   It is a mandatory step for GC analysis of carbohydrates. The active hydrogen of hydroxyl groups, responsible for their low volatility is replaced by non-polar groups usually by alkylation, acylation or silylation, increasing their volatility. It has to be carefully controlled considering that there are a high number of functional groups in the molecule, the lability of some molecules and, in certain cases, the steric

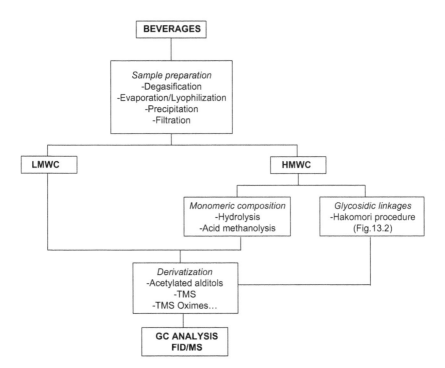

Figure 13.1   Diagram of required steps for the gas chromatographic analysis of carbohydrate in beverages. Required steps for the GC analysis of low molecular weight carbohydrates (LMWC) and high molecular weight carbohydrates (HMWC) in beverages based on sample preparation, derivatization and analysis.
Unpublished.

hindrance. The presence of different tautomeric forms in solution gives rise to complex chromatograms.

(iii) *GC analysis:*

Although separations of carbohydrates of beverages have been carried out on packed columns some years ago, the resolution highly improves with the use of capillary columns. The most common stationary phases are based on polysiloxanes (silicones) considering their good thermal stability and high permeability towards solutes. Oven temperatures depend on the volatility of the derivatives formed although they usually range from 100–200 °C for monosaccharides, 200–270 °C for disaccharides and 250–320 °C for trisaccharides. Higher oligosaccharides can be occasionally analysed by GC using high-temperature columns (up to 400–480 °C) where silicone is substituted by carborane skeleton and even the polyimide clad of fused silica capillaries is replaced by aluminium clad.

Both qualitative and quantitative analysis can be done by GC. Regarding detection, flame ionization detectors (FID) are usually used or mass spectrometry (MS) is coupled to the gas chromatograph. It is worth noting that MS is a useful tool to help for characterization of chemical structures, considering that commercial standards are not always available. However, the fragmentation pattern of close isomers is also very similar and often retention time data (I or I^T) becomes decisive for the carbohydrate identification.

13.2 Beer

Beer is made by fermentation, produced by yeast from *Saccharomyces* genus, of starch mainly from malted barley, although wheat, rice, maize, oats, sorghum can also be used. Hops are used to give flavour and act as preservatives and, depending on the manufacturer, other ingredients such as sugar syrups can be added (Ferreira 2009).

Carbohydrates are the main non-volatile constituents (3–4%) present in beer generated from the enzymatic hydrolysis of polysaccharides (starch, arabinoxylan and β-glucan) and sucrose during malting process. The presence of some low molecular weight carbohydrates such as glucose, fructose, glucobioses, lactose, sucrose, trehalose, maltulose, raffinose, kestose, stachyose, etc. has been described. Isomaltooligosaccharides up to degree of polymerization (DP) 10, maltooligosaccharides and fragments of arabinoxylans and glucans can be also found (Ferreira 2009; Vinogradov and Bock 1998).

Scarce information about carbohydrate analyses by GC is found and it is mainly orientated to the structural analysis of HMWC such as arabinoxylans. Nevertheless, the characterization of LMWC could be carried out following the procedures described below for other kind of beverages.

Sample preparation for the analysis of carbohydrates in beer is based on the degasification of the sample by ultrasound and the elimination of alcohols, mainly ethanol, by evaporation under vacuum and a water bath with temperature not higher than 80 °C. In some cases, the samples are lyophilized or even directly used with an appropriate dilution.

Structural analysis of HMWC can be followed by GC determining the monomeric composition and the glycosidic linkages.

13.2.1 Determination of Monomeric Composition

In order to determine the glycosidic composition and arabinoxylan content in beer, hydrolysis followed by derivatization is required.

Hydrolysis can be carried out using 250 μL of 2.0 M trifluoroacetic acid (TFA) at 121 °C for 1 hour adding *myo*-inositol as internal standard.

Among the different derivatization procedures the formation of acetylated alditol derivatives is common (Schwarz and Han 1995; Han 2000). Hydrolyzed samples are reduced with 0.5 mL of sodium borodeuteride (NaBD$_4$) in dimethylsulfoxide (DMSO) (20 mg mL^{-1}) at 40 °C for 90 min followed by the formation of the *O*-acetylated alditols with 0.5 mL of acetic anhydride in the presence of 0.1 mL of 1-methylimidazole. The per-*O*-acetylated alditols are washed with water and methylene chloride and the organic phase is evaporated and re-dissolved in acetone prior to their injection in GC port.

GC analysis is carried out using fused silica capillary columns (*i.e.* 30 m × 0.25 mm *i.d.* and 0.20 μm film thickness) and a elution program starting at 80 °C hold for 3 min followed by an increase of temperature up to 170 °C at 30 °C min^{-1} and to 240 °C at a rate of 4 °C per min^{-1} and finally hold for 8.4 min. Arabinoxylan is calculated using the formula: total arabinoxylan = (% arabinose + % xylose) × 0.88 and the content is around 0.2–0.3% (w/v) (Han 2000).

13.2.2 Determination of Glycosidic-Linkages of Arabinoxylans

Partially-*O*-methylated partially-*O*-acetylated alditols (PMAA) derivatives are commonly used for the determination of glycosidic linkages of HMWC by GC-MS. That is the case of the analysis of glycosidic linkages of arabinoxylans present in beer. Figure 13.2 shows a scheme of the methylation procedure proposed by Han (2000) and based on a modification of the Hakomori (1964) method. Dried samples are dissolved in 250 μL of DMSO and treated with *n*-butyllithium (250 μL) under nitrogen for 40 min. Methyl iodide (300 μL) is added and stir for 1 h for obtaining the per-methylated sugars which are extracted with chloroform. Methylated samples are dried, hydrolyzed, reduced and acetylated as indicated before for the determination of glycosidic composition. Similar GC conditions as those showed above are used. The identification of glycosidic linkages should be done either using the corresponding standards or comparing retention times and mass spectrum data in the bibliography. Figure 13.3 shows the GC and GC-MS chromatograms of the PMAA derivatives of carbohydrates present in beer. The different resolution in both profiles is due to the use of two different column phases: poly(80%

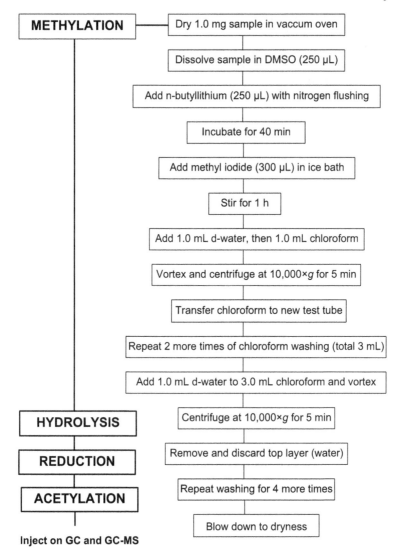

Figure 13.2   Steps for glycosidic linkage determination of high molecular weight carbohydrates (HMWC). Flow chart of methylation procedure for partially-*O*-methylated, partially-*O*-acetylated alditols (PMAA) derivatives of arabinoxylans in beer.
Data are from Han (2000) with permission from Elsevier.

biscyanopropyl/20% cyanopropylphenyl siloxane) in Figure 13.3A and a poly(90% biscyanopropyl/10% cyanopropylphenyl siloxane) in Figure 13.3B. Moreover, sensitivities of FID and MS are different and it could affect to the detection of carbohydrates. Major peaks corresponded to terminal gluco-pyranose (20–26 mol%), 4-xylopyranose (10–12 mol%), 4-glucopyranose (33–39 mol%) and 2,3,4-xylopyranose (8–9 mol%) although other monosaccharide

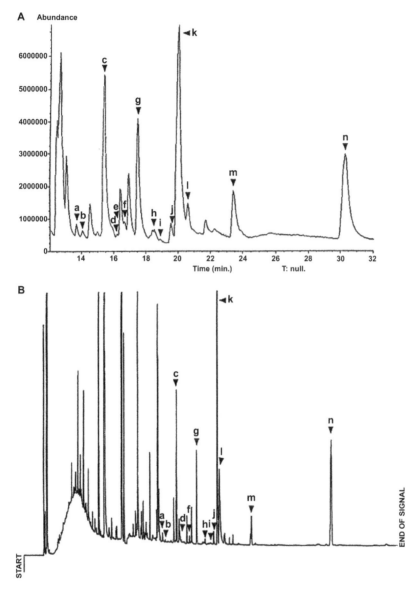

Figure 13.3 Gas chromatogram of PMAA derivatives of carbohydrates present in beer. (A): GC-MS; (B): GC. Peak a: terminal-arabinofuranose (t-Ara-f), b: terminal-xylopyranose (t-Xyl-p), c: terminal-glucopyranose (t-Glu-p), d: 3-arabinofuranose (3-Ara-f), e: terminal-galactopyranose (t-Gal-p), f: 4-arabinopyranose (4-Ara-p), g: 2-xylopyranose (2-Xyl-p) and 4-xylopyranose (4-Xyl-p) (1:4), h: 2-galactopyranose (2-Gal-p), i: 4-galactopyranose (4-Gal-p), j: 6-glucopyranose (6-Glu-p), k: 4-gluco-pyranose (4-Glu-p), l: 2,4-xylopyranose (2,4-Xyl-p) and 3,4-xylopyranose (3,4-Xyl-p) (1:1), m: 2,3,4-xylopyranose (2,3,4-Xyl-p) and n: inositol (internal standard).

Data are from Han (2000) with permission from Elsevier.

units were detected at lower levels (<5 mol%). The high amount of terminal glucopyranose was attributed to the hydrolysis of dextrins (maltose or iso-maltose) and the presence of 2,3,4-xylopyranose corresponded to the branch regions of arabinoxylans (Hans 2000).

13.3 Wines

Carbohydrates in wines contribute to their taste conferring sweetness and to the aroma through the formation of volatile compounds. They are the fermentation source during winemaking, being converted into ethanol, carbon dioxide and different by-products. Carbohydrates in wines mainly come from the action of enzymes during winemaking (either from hydrolysis or transglycosylation) although some of them could come from grape juice. Others can also be provided from the contact with oak wood in those wines aged in barrels.

Wines are composed by different kinds of carbohydrates which in terms of their analysis can be divided into LMWC and polysaccharides. Differences in the composition and quantitative levels are mainly depending on the kind of wine (white, rosé and red wines).

13.3.1 Low Molecular Weight Carbohydrates (LMWC)

Among LMWC, the most abundant are the monosaccharides: hexoses such as glucose, galactose, fructose and mannose and pentoses such as arabinose, ribose and xylose or methyl-pentoses such as rhamnose. Main disaccharides are α,α-trehalose, cellobiose, sophorose, laminaribiose and gentiobiose, whereas the presence of lactose has also been reported, although its identity could not be confirmed. Other disaccharides containing fructose units have been detected. Free glycosides (β-glyceryl-glycosides and β-ethyl-glycoside), sugar alcohols (erythritol, threitol, ribitol, arabitol, *myo*-inositol, *chiro*-inositol, *scyllo*-inositol) and sugar acids (gluconic and galacturonic acids) are also present.

In most cases, sample preparation just consists in the evaporation under vacuum of wines (0.5 mL) previously mixed with an internal standard (phenyl-β-D-glucoside). Only in specific cases such as the analyses of sugar alcohols where reducing carbohydrates could interfere in their determination, the use of ion-exchange columns has been proposed. For this procedure 1 mL of wine containing an internal standard (*i.e.* perseitol) is passed through a 10 mL Amberlite IRA-400 column and eluted with 125 mL deionized water at a flow of 1 mL min^{-1}. Samples are then evaporated under vacuum (Santa-Maria *et al.* 1985). Nevertheless, recent works determine sugar alcohols without a previous fractionation from reducing sugars (Carlavilla *et al.* 2006).

Different derivatization procedures have been used before GC analysis of wines:

- Acetylation by dissolving the dried samples in 1 mL of pyridine and treating with 1 mL of acetic anhydride for 1 h at 100 °C. Acetylation reagent was then evaporated and the residue was dissolved in 0.1 mL dichloromethane. This procedure has been used for the analysis of poly-alcohols in wine (Santa-Maria *et al.* 1985).

– Trimethylsilylation by dissolving the dried sample in 100 µL of pyridine and adding 100 µL of trimethylsilyl imidazol (TMSI) and 100 µL of tri-methylchlorosilane (TMCS) shaking after each addition (Carlavilla *et al.* 2006) or by addition of *N,O*-bis(trimethylsilyl)trifluoroacetamide (BSTFA) containing 1% of TMCS (Liddle 1982) or 1:1:1:1 TMCS:TMSI:bis(trimethylsilyl)acetamide(BSA):methyl-trimethylsilyl-trifluoroacetamide (MSTFA) at 80 °C for 60 min (Sponholz and Ditrich 1985) or 200 µL anhydrous DMSO and 150 µL TMSI at 60 °C for 30 min (Olano and Gomez-Cordovés 1983). Derivatized carbohydrates were recovered with 100 µL of hexane and the excess of reagent was separated with 200 µL of water. In this derivatization reaction different compounds are obtained for each tautomeric form of reducing sugars and GC analysis gives different peaks, obtaining complicated chromatograms for complex sugar mixtures.

– Formation of oximes previous to trimethylsilylation by a two step deri-vatization procedure has also been carried out to avoid the formation of multiple peaks for reducing carbohydrates. Only two peaks corresponding to the syn (*E*) and anti (*Z*) forms are formed for reducing sugars, whereas a single peak is obtained for non-reducing carbohydrates which do not form oximes. The dried sample is dissolved in 350 µL of 2.5% hydroxylamine chloride and heated at 70 °C for 30 min. After cooling, samples are silylated with 350 µL of HMDS and 35 µL of TFA (Ruiz-Matute *et al.* 2009). In order to eliminate the pellet formed, a centrifugation (8,000 rpm for 10 min) or filtration step can be carried out. Trimethylsilyl oximes (TMSO) are also appropriate derivatives for the identification of carbohydrates by GC-MS, considering the easiness of interpretation of MS fragmentation pattern of carbohydrates with different glycosidic linkages or mono-saccharide composition (aldoses/ketoses) compared with other derivatives. As an example, Figure 13.4 shows characteristic fragmentation patterns of TMSO disaccharide standards with different glycosidic linkages. A high abundance of fragment ion at m/z 319 is characteristic of aldose oximes substituted in C2 such as kojibiose (α-(1-2)-glucopyranosyl-glucose; Figure 13.4A), whereas a high relative abundance of the fragment ion at m/z 307 is indicative of disaccharides with reducing ketose substituted in C1 or C3 such as turanose (α-(1-3)-glucopyranosyl-fructose; Figure 13.4B). The ion at m/z 538 which represents the whole oxime chain; however, no other ions of known structural origin can be observed for maltose (α-(1-4)-gluco-pyranosyl-glucose; Figure 13.4C) (Sanz *et al.* 2002).

GC analyses are usually carried out using methylsilicone capillary columns, although packed columns have also been used. Oven temperature gradient depends on the carbohydrates to be analysed. A common gradient was pro-posed by Carlavilla *et al.* (2006): 100 °C for 1 min, then programmed at 200 °C at a heating rate of 30 °C min^{-1} and 270 °C at 15 °C min^{-1} and keep for 20 min.

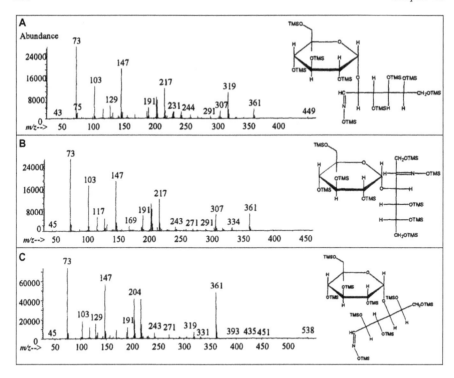

Figure 13.4 Mass spectra of trimethylsilyl oximes (TMSO) of disaccharides with different glycosidic linkages and monosaccharide units. Mass spectra of trimethylsilyl oximes (TMSO) of kojibiose (A), turanose (B), maltose (C). Data are from Sanz *et al.* (2002) with permission from Springer.

Nitrogen or helium are used as carrier gases depending on the detector used (FID or MS, respectively). A GC profile of LMWC previously converted into their TMSO is shown in Figure 13.5. Free disaccharides and other glycosides were identified in white, rosé and red wines. In general, white wines had less disaccharides ($< 50 \, \text{mg L}^{-1}$) than rosé and red wines ($80\text{–}139 \, \text{mg L}^{-1}$) (Ruiz-Matute *et al.* 2009).

13.3.2 Polysaccharides

Polysaccharides contribute to the increased viscosity and stability of wines. Rhamnogalacturonans (RG) I and II, arabinogalactan-proteins, mannoproteins and mannans can be found.

Their analysis requires a previous isolation which can be done by centrifugation at 10,000 *g* for 15 min and adding 5 volumes of 95% ethanol in acid medium. Then, polysaccharides were recovered by centrifugation at 10,000 *g* for 15 min and washing with 75% ethanol. These carbohydrates cannot be analysed in its intact form by GC considering their high molecular weight, as previously commented. Therefore, hydrolysis is necessary and this step can be

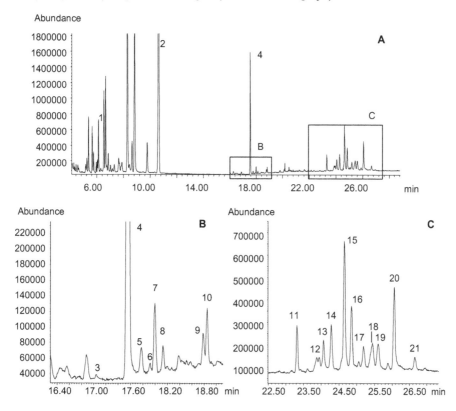

Figure 13.5 Gas chromatographic profile of trimethylsilyl oxime carbohydrates in a red wine. Gas chromatographic profile of carbohydrates previously converted into their trimethylsilyl oximes from a red wine. (1) β-ethyl-glucoside, (2) *myo*-inositol, (3) β-glyceryl-galactoside 1, (4) β-phenyl-glucoside (internal standard), (5) α-glyceryl-glucoside 1, (6) α-glyceryl-glucoside 2, (7) unknown glyceryl-glycoside, (8) β-glyceryl-galactoside 2, (9) β-glyceryl-glucoside 1, (10) β-glyceryl-glucoside 2, (11) α,α-trehalose, (12) fructose disaccharides + epicatechin, (13) lactose *E*, (14) lactose *Z* + cellobiose *E* + catechin, (15) cellobiose *Z* + laminaribiose *E*, (16) laminaribiose *Z*, (18) unknown disaccharides, (19) sophorose + unknown disaccharides, (20) gentiobiose *E*, (21) gentiobiose *Z*.
Data are from Ruiz-Matute *et al.* (2009) with permission from Elsevier.

done with 30% formic acid at 100 °C for 48 h (Pueyo *et al.* 1995). Acidic methanolysis using 0.5 mL of MeOH 0.5 M HCl (which is prepared using 140 μL of acetyl chloride and 4 mL of dried methanol) at 80 °C for 16 h is also used to obtain methyl glycosides (Ayestarán *et al.* 2004). The samples are then evaporated to dryness. The free monosaccharide units can be derivatized previous to their GC analysis as indicated above.

In some works, polysaccharides are previously fractionated to separated one from the others using different chromatographic techniques such as anion-exchange, size-exclusion or affinity chromatography (Vidal *et al.* 2003).

Structural analysis which involves the characterization of monomer composition and glycosidic linkages can be also determined by GC following the Hakomori procedure (Hakomori 1964) as shown for arabinoxylans in beer (Figure 13.2). First of all, polysaccharides such as RGI and RGII should be submitted to cation-exchange chromatography on Amberlite IR 10 H^+ to be converted in their H^+ form to be soluble in DMSO required for permethylation (Vidal *et al.* 2003).

Neutral monosaccharides such as arabinose (6–42%), xylose (0–9%), mannose (38–89%) and galactose (4–26%) have been quantified in wines after hydrolysis of polysaccharides (Pueyo *et al.* 1995).

13.4   Spirit Drinks

According to the Spirit Drinks Regulation (EC) No 110/2008 of the European Union "spirit drinks" means an alcoholic beverage: (i) intended for human consumption; (ii) possessing particular organoleptic qualities; (iii) having a minimum alcoholic strength of 15% vol.; (iv) having been produced either by the mixture of a spirit drink with one or more other spirit drinks, and/or ethyl alcohol or distillates both of agricultural origin, and/or other alcoholic beverages, and/or drinks, or directly. In the last case, spirits can be produced by the distillation, with or without added flavourings, of naturally fermented products; by the maceration or similar processing of distillates or plant materials in ethyl alcohol both of agricultural origin; and/or by the addition of flavourings, sugars or other sweetening products listed in the regulation.

The most common spirit drinks are liqueurs or distilled beverages such as rum, whisky, brandy, vodka and gin. Liqueurs contain a minimum alcoholic strength of 15% vol. and a sugar content of $100\,g\,L^{-1}$. In contrast, the above-mentioned distilled spirits have a minimum alcoholic strength of 36% vol. and are produced by distillation from fermented sugar-containing agricultural products. In addition, distilled spirits shall not be sweetened or flavoured, nor contain any additives other than caramel as a means to adapt colour. In consequence, carbohydrates in distilled spirits are only present at trace level. Nevertheless, oak-aged spirits such as whisky can also contain some carbohydrates, mainly monosaccharides, derived from toasting process of casks. Thus, Martin and Eib (1968) determined the levels of α- and β-glucose and α,β-fructose in whisky by GC-FID analysis using sorbitol as an internal standard. Before analysis, sample preparation was based on the evaporation of the distilled spirits (5–25 mL) to dryness at 60 °C in a forced air oven for 4 h, and the dried residue was dissolved in 3 mL pyridine and 1 mL of 3.82 mg sorbitol mL^{-1} pyridine and stand for 1 h. For the derivatization step, 600 µL each of TMCS and HMDS were finally added. Analyses were carried out in a glass column (1.83 m $\times$ 0.6 cm), containing 3% of SE-30 as stationary phase on 80–100 mesh Gas Chrom Q. Injector temperature was 180 °C, and the elution was isothermal at 150 °C. Total sugar amounts ranged from 11 to 219 mg $100\,mL^{-1}$.

Later on, Black and Andreasen (1974) quantitated the levels of a wide range of monosaccharides such as arabinose, xylose, fructose, glucose, galactose, and rhamnose, as well as fucose and mannose at trace levels, in neutral spirits and whiskies aged for 8 and 12 years, respectively, in new and charred white oak barrels. Erythritol was used as internal standard and after the evaporation of 25 mL of spirits, TMS ether derivatives were formed by adding 1 mL of a mixture consisting of HMDS : TMCS : pyridine (3 : 1 : 9). A stainless steel column (5.48 m × 0.2 cm) packed with 3 JXR on 100–120 mesh Gas-Chrom Q was used and injector temperature was 270 °C. The oven temperature was programmed from 100 °C to 200 °C at 2 °C min^{-1}. In addition to the TMS ether derivatives, these authors also successfully prepared TMS-aldono-1,4-lactones, alditol acetates and trifluoro-acetates derivatives of monosaccharides. Likewise, as the evolution of sugars in distilled spirits may be important for process control, Black and Andersen (1974) also studied the development of individual sugars upon the aging process. Thus, they found that all monosaccharides showed a linear increase over the 8-year maturation period of neutral spirits, whilst for the whisky, fructose exhibited a linear increase throughout the 12-year aging period, but arabinose, xylose, rhamnose, glucose, and galactose increased linearly only for the first 6–7 years maduration. In consequence, total sugar contents in neutral spirits aged for 8 years were 292.2 mg L^{-1} with arabinose 35%, glucose 38%, xylose 12%, fructose 9%, rhamnose 4% and galactose 3%, whereas the total sugar contents of whiskies matured for 12 years resulted to be 360.7 mg L^{-1} with a similar sugar composition consisting of arabinose 39%, glucose 34%, xylose 13%, fructose 7%, rhamnose 3% and galactose 3%.

Regarding liqueurs, Coll Hellín *et al.* (1993) optimized a GC-FID method for resolving and quantifying sugars and polyalcohols in apple liqueurs. For the sample preparation, sugars and polyalcohols were isolated by adding to 5 mL of liqueur, 0.5 mL zinc acetate solution and 0.5 mL potassium ferrocyanide, forming zinc ferrocyanide, which is a clearing agent. After shaking, the mixture was made up to 100 mL with distilled water and then filtered. The filtrate was evaporated to dryness at 40 °C and the TMS derivatives were prepared by using HMDS as explained above. All assayed liqueurs had a high sugar content (above 100 g L^{-1}), being the major sugars β-fructose, α-glucose, β-glucose and sucrose, although minor amounts of sorbitol and a keto-monosaccharide such as sorbose could also be determined. The presence of sorbitol was attributed to the fact that it is a characteristic sugar of fruits from the *Rosaceae* family which includes apples, plums, peaches, pears, *etc.* These authors also hypothesized that the presence of sorbose could be due to the oxidation of sorbitol during production of the liqueur or of the sugar solution. Likewise, one of the studied samples contained minor amounts of isomaltose, α-maltose and β-maltose which was indicative of the production of this liqueur from a starch syrup.

13.5   Juices

Glucose, fructose and sucrose are the most abundant carbohydrates present in fruit juices, but other carbohydrates such as inositols and polysaccharides are also found.

The analysis of LMWC of juices by GC and GC-MS requires a sample preparation mainly based on the removal of suspended particles by centrifugation (10,000 rpm, 20 min, 5 °C) and filtration by cellulose-acetate membranes (Sanz *et al.* 2004; Villamiel *et al.* 1998). Previous to the derivatization procedure, the samples are diluted in order to obtain a soluble solid content between 5.5 and 10° Brix. In this sense, some authors have reported a dilution of 0.5 mL of juice to 25 mL with 70% methanol (Sanz *et al.* 2004; Villamiel *et al.* 1998). Addition of an internal standard (*i.e.* 2 mL 1% phenyl-β-glucoside) is recommended for quantitative analyses. Samples should be lyophilized or evaporated under vacuum at 40–60 °C to dryness.

Finally, sample derivatization is carried out prior to their injection into the GC port. Formation of TMS derivatives is a useful procedure for the analysis of carbohydrates in juices (Villamiel *et al.* 1998), taking into account they are not complex mixtures and the appearance of multiple peaks for each sugar does not mean a notable problem for coelutions (Figure 13.6). The disadvantage of TMS derivatives is the stability (they are not stable for more than 24 h after the addition of hexane and water for their recovery). On the contrary, TMSO are more stable and can be stored for 1 year at 4 °C. Although reagents and conditions for carrying out these derivatization procedures have been indicated before for the analysis of LMWC in wines, some specific cases are indicated here. Low *et al.* (1999) compared different derivatization procedures for the detection of adulterations in apple juices by GC. Silylation was carried out using 0.5 mL of TMSI in pyridine (1:4, v/v) at 75 °C for 1 h or at 22 °C for 12 h or at 50 °C for 1 h or 0.5 mL of methyl-trimethylsilyl-heptafluorobutyramide (MSHFBA) in 1-methylimidazole (95:5, v/v). Moreover, the equilibration of the lyophilised sample in 0.4 mL of dry pyridine at 50 °C for 20 min prior to silylation was also assayed. The MSHFBA reagent gave in general, higher coefficients of variations (CVs) for the samples. However, the equilibrium in dry pyridine followed by the silylation with TMSI resulted in a higher reproducibility (lower CVs) and the best way to detect adulterations with invert sugar in these juices.

Methyl silicone and 5% phenylmethyl silicone capillary column are normally used for GC and GC-MS as previously indicated. Elution programs can be similar for both stationary phases. Villamiel *et al.* (1998) used a methyl silicone capillary column (30 m × 0.25 mm) for the analysis of TMS derivatives of orange juice by GC (Figure 13.6) and GC-MS. The oven was programmed from 200 °C to 270 °C at a heating rate of 20 °C min^{-1} with an initial holding of 10 min. Concentrations varied from 1.19–1.62 g L^{-1} for *myo*-inositol to 43.1–50.0 g L^{-1} for sucrose in fresh juices.

13.6 Soft Drinks

Methods based on density or refractive index measurements have been normally used for the estimation of the total sugar content of soft drinks (Ramasami *et al.* 2004). Likewise, enzymatic, liquid chromatography, or

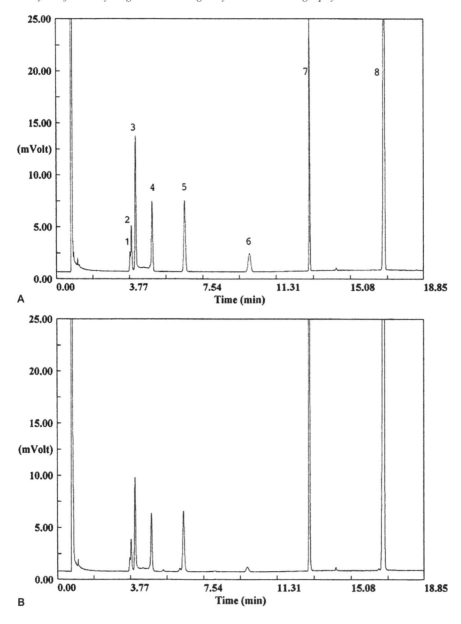

Figure 13.6 Gas chromatographic profiles of trimethylsilyl carbohydrates of a standard mixture and a commercial orange juice. Chromatograms of carbohydrates as their trimethyl silyl ethers of: (A) standard mixture and (B) commercial orange juice. Peaks: 1, 2 and 3: fructose; 4 and 5: glucose; 6: *myo*-inositol, 7: β-phenyl-glucoside (internal standard) and 8: sucrose. Data are from Villamiel *et al.* (1998) with permission from Springer-Verlag.

infrared spectroscopy methods provide information about the total content of sugars and the specific concentration of each carbohydrate (Vonach *et al.* 1997). In contrast, analysis of sugars in soft drinks has rarely been determined by gas chromatography. Nonetheless, Elena *et al.* (1998) measured the concentration of sugars in thirty commercial soft drinks readily available in Spain by GC-FID and using a TRB-1 (100% Dimethyl polysiloxane, 15 m × 0.53 mm) column. Prior to GC analysis, samples were evaporated to dryness under a nitrogen stream and, then, they were dissolved with 10 µL of anhydrous pyridine and derivatized with 15 µL of BSA : TMCS : TMSI, 3 : 2 : 3 for 30 min at 60 °C. After centrifugation at 14,000 rpm for 5 min, 0.2 µL of the supernatant were injected and the oven temperature was programmed from 175 °C (6.5 min), then at 10 °C min^{-1} to 240 °C (5 min). *Myo*-inositol and turanose were used as internal standards for the quantification of mono- and disaccharides, respectively. The main determined sugars were glucose, fructose and sucrose and the total sugar contents ranged from ~5 to 18 g 100 mL^{-1}.

13.7   Hot Drinks

Hot drinks are mainly referred to both coffee and tea extracts. Sucrose, glucose, fructose, mannose and galactose are the main free mono- and disaccharides in coffee extracts, whereas small amounts of rhamnose, fucose, arabinose, xylose and polyalcohols (mannitol, *myo*-inositol and bornesitol) have also been reported. Some polysaccharides (galactomannans and arabinogalactans) can be also present.

Glucose, fructose, sucrose, arabinose, ribose and some soluble polysaccharides (hemicelluloses or pectic substances) are present in tea brew.

First of all, sample preparation requires a filtration process using paper filters followed by the removal of proteins and fats in the case of coffee and their surrogates by treating with methanol. Filtrated solution (3 mL) is stirred with 25 mL of methanol and stand for 1 h; the methanolic solution is evaporated to dryness and derivatised for the determination of LMWC (Ruiz-Matute *et al.* 2007). Silylation has been used for derivatization of the samples and in some cases a previous oximation step is required to avoid overlaps. Derivatization conditions and GC analyses are similar to those indicated before.

Polysaccharides can be isolated using dialysis, size exclusion chromatography or precipitations with alcohols. Determination of glycosidic linkages is carried out as indicated above following the methylation procedure (Hakomori 1964).

13.8   Conclusions

Previous to GC analyses of carbohydrates in beverages, sample preparation and derivatization are required. However, sample preparation is usually simple and not time consuming, whereas derivatization is a mandatory and crucial step for these analyses, being silylation the most extended procedure.

GC methods for the analysis of carbohydrates in beverages are accurate and sensitive; however, their use is not generally extended. Refractive index measurements, enzymatic and colorimetric methods, infrared spectroscopy or even liquid chromatography are commonly used for these samples. However, separation provided by GC and information from its coupling with MS offer both qualitative and quantitative advantages for the characterization of individual carbohydrates of low molecular weight. Structural analysis to determine oligo- and polysaccharide glycosidic linkages and monosaccharide composition is also afforded.

Summary Points

o This chapter is focused on the analysis of dietary carbohydrates of beverages by gas chromatography (GC).
o Qualitative and quantitative analyses require three different steps: sample preparation, derivatization and GC determination.
o Sample preparation is mainly based on degasification, filtration, dilution, evaporation or fractionation.
o Derivatization is a crucial step for GC analysis to make carbohydrates volatiles.
o The most common derivatives are the trimethylsilyl and trimethylsilyl oximes for low molecular weight carbohydrates and the partially-*O*-methylated partially-*O*-acetylated alditols (PMAA) derivatives for high molecular weight carbohydrates.
o Methyl silicone columns are usually used for the separation of these compounds.
o Retention data and mass spectrometric fragmentation are necessary for the identification of carbohydrates.

Key Facts

Key Facts of Beverages Sample Preparation

1. The first step prior to analysis of carbohydrates present in beverages is the sample preparation.
2. Sample preparation for carbohydrates analysis by GC includes clean-up and fractionation.
3. The type of technique used for sample preparation largely depends on the sample matrix.
4. The most common sample preparation technique in beverages is the filtration and/or centrifugation, allowing the removal of suspended particle.
5. Another common technique is the evaporation or concentration, if heating is used, the temperature should be lower than 80 °C.
6. Sample preparation should not change the structure or quantity of carbohydrates present in beverages.

Key Facts of Carbohydrate Derivatization

1. Carbohydrates are not volatile compounds, and that is why it is mandatory to derivatize prior to GC analysis.
2. Derivatization is a substitution of the active hydrogens of hydroxyl groups by other non-polar groups increasing the volatility of carbohydrates at atmospheric pressure.
3. Samples injected in GC should be volatile and stable at the injector temperature, which is normally between 200 and 300 °C.
4. The most common derivatizing agents used to analyze carbohydrates are: trimethylsilyl ethers, trimethylsilyl oxime, per-acetylated and methylated.
5. Derivatization reactions have to be complete and stable during the analysis.

Key Facts of Gas Chromatographic Analysis

1. Gas chromatography is an analytical technique based on the separation of volatile compounds into a column, using as mobile phase gasses such as helium, nitrogen or hydrogen.
2. Separations on GC can be carried out either in isotherm mode or programmed temperature. This parameter has to be optimized for the analysis of any compound, in order to obtain a high resolution in the shortest possible time.
3. GC can be coupled to different detectors, being the most commonly used the FID and the mass spectrometers.
4. Capillary columns are normally used in analytical GC. Currently, a great variability of phases is commercially available, but the most common ones are of polysiloxane.
5. Only LMWC, previously derivatized, can be analysed by GC. However, oligosaccharides and polysaccharides (HMWC) are also analyzed by GC but with a previous hydrolysis step in order to obtain LMWC.
6. Only retention data and abundance can be obtained from a GC chromatogram. Moreover, if the GC is coupled to mass spectrometry (MS) structural information is obtained from the corresponding mass spectra.

Definitions of Words and Terms

Beverages. All liquids suitable for human consumption with the exception of water which does not contain carbohydrates. It includes alcoholic beverages such as beer, wine and spirits and non-alcoholic beverages such as juices, soft drinks and hot drinks.

Low Molecular Weight Carbohydrates (LMWC). LMWC refers to those carbohydrates with molecular weight equal or lower than $504 \, \text{g} \, \text{mol}^{-1}$. It is mainly associated to mono-, di- and trisaccharides constituted by one, two or three monomeric units, respectively. Free glycosides, sugar alcohols and sugar acids are also included.

High Molecular Weight Carbohydrates (HMWC). HMWC are constituted by more than four monosaccharide units and can be divided into oligo-saccharides (degree of polymerization from 4 to 10) and polysaccharides (degree of polymerization equal or higher than 11).

Fractionation. It is a sample preparation technique to separate the target compounds from other interferences previous to their analysis. Ion exchange chromatography, size exclusion chromatography, *etc.* are some examples.

Filtration. It is a clean-up technique used to eliminate solids dispersed in a liquid solution using membranes or filters of different materials (*i.e.* cellulose acetate, nylon, glass, *etc.*) where the liquid is permeable and the solid is retained.

Derivatization. It is a chemical modification of the active hydrogen of hydroxyl groups which are replaced by non polar groups increasing their volatility.

Silylation. It is a derivatization procedure based on the substitution of the hydrogen of functional groups by alkylsilyl groups. The most common procedure is the trimethylsilylation.

Oximation. It is a derivatization procedure based on the conversion of the aldehyde or the keto group of a carbohydrate into an oxime. Usually, 2.5% hydroxylamine hydrochloride is used as oximation reagent.

Methylation. It is an alkylation process based on the substitution of the hydrogen of functional groups by a methyl group.

Acetylation. It is a derivatization procedure based on the substitution of the hydrogen of functional groups by an acetyl group.

Structural analysis. It is the characterization of complex carbohydrates based on the determination of the chemical structure and sequence of mono-saccharide residues and their glycosidic linkages. It is constituted by four steps: methylation, hydrolysis, reduction and acetylation.

Hydrolysis. It is a chemical process based on the division of a molecule (oligo-or polysaccharide) in the presence of water.

Gas chromatography (GC). GC is an analytical technique for the separation of mixtures of volatile compounds. It is constituted by a liquid stationary phase placed in a column where the analytes are retained depending on its affinity and by an inert gas (nitrogen, hydrogen or helium) or mobile phase which transports the analytes through the stationary phase. High temperatures are also used to help to the transport of analytes to a detector where the signal is registered.

Flame ionization detector (FID). FID is a common detector used in GC based on the use of a flame and two electrodes. The current between the electrodes increases when a compound eluted from the column reaches the flame.

Mass Spectrometry (MS). MS is a powerful analytical technique composed by an ion-source which produces ions from each compound, a mass analyzer to separate the ions depending on their mass to charge ratio (m/z) and a detector which collect these ions, resulting in a specific mass spectrum for each analyzed compound.

Abbreviations

AOAC	Official Methods of Analysis
BSA	bis(trimethylsilyl)acetamide
BSTFA	Bis(trimethylsilyl)trifluoroacetamide
CVs	Coeficients of variation
DMSO	Dimethylsulfoxide
DP	Degree of Polymerization
FID	Flame Ionization Detector
GC	Gas Chromatography
HMDS	Hexamethyldisilazane
HMWC	High Molecular Weight Carbohydrates
HPLC	High Performance Liquid Chromatography
I	Kovats retention indices
I^T	Linear retention indices
LMWC	Low Molecular Weight Carbohydrates
MeOH	methanol
MS	Mass Spectrometry
MSHFBA	methyl-trimethylsilyl-heptafluorobutyramide (MSHFBA)
MSTFA	methyl-trimethylsilyl-trifluoroacetamide
PMAA	Partially-O-methylated partially-O-acetylated alditols
RG	Rhamnogalacturonans
TFA	Trifluoroacetic Acid
TMCS	Trimethylchlorosilane
TMS	Trimethylsilyl
TMSI	Trimethylsilyl imidazole
TMSO	Trimethylsilyl oximes

References

AOAC, *Official methods of Analysis of the Association of Official Analytical Chemists,* 1984. Sidney Williams, fourteen edition. Arlington, Virginia, US.

Ayestarán, B., Guadalupe, Z. and León, D. 2004. Quantification of major grape polysaccharides (*Tempranillo* v.) released by maceration enzymes during the fermentation process. *Analytica Chimica Acta.* 513: 29–39.

Black, R.A. and Andreasen, A.A., 1974. Gas-liquid chromatographic determination of monosaccharides and glycerol in aged distilled spirits. *Journal of the AOAC.* 57: 111–117.

Carlavilla, D., Villamiel, M., Martínez-Castro, I. and Moreno-Arribas, M.V., 2006. Occurrence and significance of quercitol and other inositols in wines during oak Word aging. *American Journal of Enology and Viticulture.* 57: 468–473.

Coll Hellín, L., Gutiérrez Ruiz, L. and Zapata Revilla, A., 1993. Gas chromatographic determination of sugars and polyalcohols in apple liqueurs. *Zeitschrift für Lebensmittel-Untersuchung und-Forschung.* 196: 49–52.

Elena, M., Pérez, M., Jansà, M., Deulofeu, R., Esmatjes, E., Schinca, N., Mas, E., Molina, R. and Ballesta, A.M., 1998. Content of carbohydrates and trace elements composition in a group of non alcoholic refreshments. *Medicina Clinica*. 110: 365–369.

European Commission (EC), 2008. Regulation (EC) No 110/2008 of the European Parliament and of the Council of 15 January 2008. On the definition, description, presentation, labelling and the protection of geographical indications of spirit drinks and repealing Council Regulation (EEC) No 1576/89, L/39, 16–54. *Official Journal of the European Union*. Available at: http://www.mee.government.bg/ind/doc/LexUriServ.pdf.

Ferreira, I.M.P.L.V.O., 2009. Beer Carbohydrates. In: Preedy, V. (ed.) *Beer in Health and Disease Prevention*, San Diego, USA, pp. 291–298.

Hakomori, S., 1964. A rapid permethylation of glycolipid and polysaccharide catalyzed by methyl sulfinyl carbonion in dimethyl sulfoxide. *Journal of Biochemistry*. (Tokyo). 55: 205–208.

Han, J.Y., 2000. Structural characteristics of arabinoxylan in barley, malt and beer. *Food Chemistry*. 70: 131–138.

Liddle, P.A.P. 1982. Glass capillary chromatography in the wine and spirit industry. *Analytical Processes*. November: 515–516.

Low, N.H., McLaughlin, M., Hofsommer, H.J. and Hammond, D.A., 1999. Capillary gas chromatographic detection of invert sugar in heated, adulterated, and adulterated and heated apple juice concentrates employing the equilibrium method. *Journal of Agricultural and Food Chemistry*. 47: 4261–4266.

Martin, G.E. and Eib, N.K., 1968. Determination of sugars in distilled spirits by GLC. *Journal of the AOAC*. 51: 925–927.

Olano, A. and Gomez-Cordovés, M.C., 1983. Componentes no volátiles em vinagres: inositol, trehalosa, polifenoles totales y catequinas. *Anales de Bromatología*. XXXV: 171–176.

Pueyo, E., Olano, A. and Polo, M.C., 1995. Neutral monosaccharides composition of the polysaccharides from musts, wines and cava wines. *Revista Española de Ciencia y Tecnología de Alimentos*. 35: 191–201.

Ramasami, P., Jhaumeer-Laulloo, S., Rondeau, P., Cadet, F., Seepujak, H. and Seeruttun, A., 2004. Quantification of sugars in soft drinks and fruit juices by density, refractometry, infrared spectroscopy and statistical methods. *South African Journal of Chemistry*. 57: 24–27.

Ruiz-Matute, A.I., Montilla, A., del Castillo, M.D., Martínez-Castro, I. and Sanz, M.L., 2007. A GC method for simultaneous analysis of bornesitol, other polyalcohols and sugars in coffee and its substitutes. *Journal of Separation Science*. 30: 557–562.

Ruiz-Matute, A.I., Sanz, M.L., Moreno-Arribas, M.V. and Martínez-Castro, I., 2009. Identification of free disaccharides and other glycosides in wine. *Journal of Chromatography A*. 1216: 7296–7300.

Santa-Maria, G., Olano, A. and Tejedor, M., 1985. Quantitative determination of polyalcohols in wine and vinegar by gas chromatography. *Chromatographia*. 20: 197–200.

Sanz, M.L., Sanz, J. and Martínez-Castro, I., 2002. Characterizatin of O-trimethylsilyl oximes of disaccharides by gas chromatography-mass spectrometry. *Chromatographia.* 56: 617–622.

Sanz, M.L., Villamiel, M. and Martínez-Castro, I., 2004. Inositols and carbohydrates in different fresh fruit juices. *Food Chemistry.* 87: 325–328.

Schwarz, P.B. and Han, J.Y., 1995. Arabinoxylan content of commercial beers. *Journal of the American Society of Brewing Chemists.* 53: 157–159.

Sponholz, W.R. and Ditrich, H.H., 1985. Zuckeralkohole und myo-inosit in weinwn und sherries. *Vitis.* 24: 97–105.

Vidal, S., Williams, P., Doco, T., Moutounet, M. and Pellerin, P., 2003. The polysaccharides of red wine: total fractionation and characterization. *Carbohydrate Polymers.* 54: 439–447.

Villamiel, M., Martínez-Castro, I., Olano, A. and Corzo, N., 1998. Quantitative determination of carbohydrates in orange juice by gas chromatography. *Zeitschrift für Lebensmittel-Untersuchung und–Forschung.* 206: 48–51.

Vinogradov, E. and Bock, K., 1998. Structural determination of some new oligosaccharides and analysis of the branching pattern of iso-maltooligosaccharides from beer. *Carbohydrate Research.* 309: 57–64.

Vonach, R., Lendl, B. and Kellner, R., 1997. Hyphenation of ion exchange high-performance liquid chromatography with fourier transform infrared detection for the determination of sugars in nonalcoholic beverages. *Analytical Chemistry.* 69: 4286–4290.

CHAPTER 14

UV Spectrophotometry Method for Dietary Sugars

ANA C. A. VELOSO,[a] LÍGIA R. RODRIGUES,[b]
LUÍS G. DIAS[c] AND ANTÓNIO M. PERES*[d]

[a] IBB – Institute for Biotechnology and Bioengineering, Center of Biological Engineering, DEQB - Departamento de Engenharia Química e Biológica, Instituto Superior de Engenharia de Coimbra, Instituto Politécnico de Coimbra, Rua Pedro Nunes, Quinta da Nora, 3030-199, Coimbra Portugal; [b] IBB – Institute for Biotechnology and Bioengineering, Centre of Biological Engineering, Department of Biological Engineering, University of Minho, Campus de Gualtar, 4710-057 Braga, Portugal; [c] CIMO – Mountain Research Centre, Departamento de Ambiente e Recursos Florestais, Escola Superior Agrária, Instituto Politécnico de Bragança, Campus de Santa Apolónia, Apartado 1172, 5301-855 Bragança, Portugal; [d] CIMO – Mountain Research Centre, LSRE – Laboratory of Separation and Reaction Engineering - Associate Laboratory LSRE/LCM, Departamento de Produção e Tecnologia Vegetal, Escola Superior Agrária, Instituto Politécnico de Bragança, Campus de Santa Apolónia, Apartado 1172, 5301-855 Bragança, Portugal
*Email: peres@ipb.pt

14.1 Introduction

Dietary carbohydrates are a diverse group of compounds with a range of chemical, physical and physiological properties. They include sugars, complex carbohydrates or starch, and fiber.

Food and Nutritional Components in Focus No. 3
Dietary Sugars: Chemistry, Analysis, Function and Effects
Edited by Victor R Preedy
© The Royal Society of Chemistry 2012
Published by the Royal Society of Chemistry, www.rsc.org

The analysis of dietary sugars is crucial since they are essential in the human diet, being consumed in large amounts (Williams *et al.* 2009). Additionally, they are key players in healthy diets that promote, for example, weight loss. However, some sugars are incompletely absorbed and so they may act as dietary triggers for several clinical symptoms (Fernandez and Vergara-Jimenez 2006). On the other hand, some oligosaccharides, lactulose, lactosucrose, among others, are of major importance as they stimulate the specific growth/activity of beneficial colonic bacteria (MacFarlane *et al.* 2008). Therefore, sugars have a huge economic importance in the food industry and a great health impact.

Measuring dietary carbohydrates requires the use of a systematic approach that explores characteristic differences among them, enabling the determination of each carbohydrate (Englyst *et al.* 2007).

Since sugars have no obvious UV-Vis absorption or fluorescence properties, the use of spectrophotometric methods for quantitative analysis usually requires the production of an absorbing by-product by acid hydrolysis followed by chromogenic reaction or enzymatic hydrolysis (Zhang *et al.* 2010). Alternatively, absorbing photodegradated by-products can be obtained by UV irradiation (Li and He 2007). The main drawback is the difficulty or impossibility of quantifying each sugar, since most oxidized UV-Vis absorbing compounds have a maximum absorbance at the same wavelength, allowing only the measurement of total sugar content. The use of UV-Vis spectrum together with multicomponent exploitation methods may overcome this issue (Dias *et al.* 2009).

Many other analytical methods have been proposed, such as the redox titration reference method and chemical oxidation of reducing carbohydrates; chromatographic techniques; FTIR methods; enzymatic electrochemical (bio)sensors; immunologic methods; electrochemical detection and front-face fluorescence method (Amine *et al.* 2000; Khanmohammadi *et al.* 2009; Li and He 2007; Ni *et al.* 2003; Pham *et al.* 2011; Wang *et al.* 2011; Zhang *et al.* 2010).

Although GC and HPLC are the most widespread tools for sugar analysis, they are demanding technical and economical methodologies.

So, spectrophotometry is becoming progressively more accepted for analysis of (bio)chemical substances (Moeawski 2006), namely sugars. Direct and indirect spectrophotometry has been proposed based on raw or derivative spectroscopy data, which has gained increased attention, due to the emergence of large number of software packages that enable automatic derivative tools. First- and higher-order derivatives combined with chemometrics techniques allow improving qualitative and quantitative analysis (Vogt 2005). The most common multivariate tools within this field are PLS and ANN. The goal of these techniques is to establish empirical models derived from experimental data, which allow estimating one or more properties of a given system (Rao and Biju 2005).

14.2   Dietary Sugars

Dietary carbohydrates are classified according to their DP, the type of linkage and character of the individual monomers. Table 14.1 summarizes the major

Table 14.1 Major dietary carbohydrates. Carbohydrates classification according to the degree of polymerization (DP) detailing some of the subgroups components.

Class (DP)	Subgroup: components
Sugars (DP = 1–2)	*Monosaccharides*: glucose, galactose, fructose *Disaccharides*: sucrose, lactose, trehalose *Polyols*: sorbitol, mannitol, lactitol, xylitol
Oligosaccharides (DP = 3–9) (short-chain carbohydrates)	*Malto-oligosaccharides (α-glucans)*: maltodextrines *Other oligosaccharides (non-α-glucans)*: raffinose, stachyose, fructo and galacto-oligosaccharides, polydextrose, inulin
Polysaccharides (DP ≥ 10)	*Starch (α-glucans)*: amylose, amylopectin, modified starches *Non-starch polysaccharides (non-α-glucans)*: cellulose, hemicellulose, pectins, hydrocolloids

classes of dietary sugars (Cummings and Stephen 2007; Roberfroid 2005). A chemical approach involving separation, identification and quantification of the different classes of carbohydrate present in food, forms the basis for an understanding of their physiological and health effects.

14.2.1 Physicochemical Properties

The physicochemical properties of dietary carbohydrates depend on their primary structures and, less frequently, on their higher-order structures. Monosaccharides are comprised of single saccharide units, which constitutes the building blocks of naturally occurring di-, oligo- and polysaccharides. Due to their high level of conformational flexibility, mono- and oligosaccharides typically do not form stable secondary or higher order structures when dissolved in a solvent. Dietary carbohydrate physicochemical differences give rise to distinct properties, including solubilities and susceptibility to digestive enzymes. Unmodified mono-, oligo- and polysaccharides, except for cellulose and starch, are generally soluble in aqueous solvents. Polysaccharides are among the most viscous natural products, many being used in the food industry as thickeners. While polysaccharides do not lower the surface tension of aqueous solutions, their affinity for the oil-water interface gives them great value as emulsifiers (Englyst *et al.* 2007; Linhardt and Bazin 2001). Most carbohydrates are not easily crystallized, with the exception of sucrose and cellulose, mainly due to their conformational flexibility and the presence of multiple structural forms of reducing sugars. Carbohydrates are generally fairly stable molecules. However, reducing sugars contain an aldehyde or hemiacetal function, thus being sensitive to oxidation. Furthermore, reducing sugars are frequently base-sensitive, and oligo- and polysaccharides also contain acid-sensitive glycosidic linkage. Many polysaccharides can form stable secondary structures in solution, which can often be disrupted by elevating the temperature, changing the pH, or through the addition of denaturants (Izydorczyk 2005; Linhardt and Bazin 2001). Each dietary carbohydrate presents specific properties that influence their use in the food processing industry as reviewed by many authors (Cummings and Stephen 2007; Izydorczyk 2005; Linhardt and Bazin 2001; Torres *et al.* 2010).

14.2.2 Physiological Effects

A wide range of physiological effects that may be important to health have been reported for the dietary carbohydrates. While carbohydrates are mainly substrates for energy metabolism; they can affect satiety; control blood glucose and insulin; lipid metabolism and; through fermentation, exert a major control on colonic function, including the metabolism and balance of the gut microbiota (prebiotic effect). Also, they may be immunomodulatory and influence calcium absorption. Thus, carbohydrates have been implied in the control of body weight, diabetes and ageing, cardiovascular disease, bone mineral density, cancer, constipation and resistance to gut infection (Cummings and Stephen 2007).

Carbohydrates are often classified as available or unavailable. Available carbohydrates are those that are hydrolyzed by enzymes of the GI system to monosaccharides that are absorbed in the small intestine and enter the pathways of carbohydrate metabolism. Unavailable carbohydrates are not hydrolyzed by endogenous enzymes, although may be fermented by the gut microbiota in the large intestine to varying extents. These include the non-digestible oligosaccharides, resistant starch and non-starch polysaccharides.The process of fermentation is metabolically less efficient than absorption in the small intestine and these carbohydrates provide the body with less energy. One of the most significant developments with regard to the gut microbiota has been the demonstration that specific dietary carbohydrates (prebiotics) selectively stimulate the growth of beneficial bacteria (Roberfroid 2005). Prebiotics include fructans (inulin and fructo-oligosaccharides (FOS)), galacto-oligosaccharides (GOS) and other oligosaccharides (Macfarlane *et al.* 2008; Torres *et al.* 2010). Several benefits have been attributed to prebiotics, ranging from reducing the risk of GI infections, improving laxation and stimulating the GI immune system (Roberfroid 2005).

14.2.3 Methodology for Dietary Carbohydrates Analysis

In most of the applications of dietary carbohydrates in food processing and labeling, it is of major relevance to accurately measure their amounts. From an analytical perspective, the simplest situation is where there is only one type of carbohydrate present in a sample with minimal interfering compounds. Nevertheless, in most of the raw materials and food products, various types of carbohydrates are present in a sample with other compounds including soluble lipids, proteins and minerals. The heterogeneity of this group of compounds can make difficult the analysis of the total carbohydrate content of a complex sample.

Before analyzing for any class of carbohydrate the sample must be prepared to remove interferents. Fats, proteins, pigments, vitamins, minerals, and many other compounds must be removed prior to analysis (Brummer and Cui 2005).

Generally, samples are dried (vacuum, atmospheric pressure or in a freeze dryer) and ground first, followed by a fat removal step (hexane or chloroform). Low molecular weight carbohydrates, such as mono- and disaccharides, can then be extracted using hot 80% ethanol (Muir *et al.* 2007). The extract will be rich in mineral salts, pigments, and organic acids, as well as low molecular weight sugars and proteins, while the residue will mainly contain proteins and high molecular weight carbohydrates (cellulose, pectin and starch). Protein is removed from samples using a protease; water soluble polysaccharides can be extracted using water and separated from insoluble material by centrifugation or filtration. An enzymatic treatment with amylase and/or amidoglucosidase can be used, if starch is not the compound of interest. In this case, starch is hydrolyzed to glucose, which can be separated from high molecular weight polysaccharides by dialysis or by precipitation with ethanol.

14.2.3.1   Mono- and Disaccharide Analysis

The mono- or disaccharides concentration in a sample can be determined using GC, HPLC or enzymatic methods. Unlike enzymatic methods, which tend to be specific only for one type of monosaccharide, chromatographic techniques provide information on one or several monosaccharides in a sample. GC analysis is an established technique; it requires small sample sizes and is very sensitive. Enzymatic methods are highly specific, usually rapid and sensitive to low sugar concentrations.

14.2.3.2   Oligosaccharide Analysis

Oligosaccharides can also be determined by GC or HPLC methods. These methods work well for purified preparations, but in complex foods or diets, partial acid or enzymatic hydrolysis and determination of liberated mono-saccharides is an alternative for specific determination. In addition, size exclusion techniques (size exclusion or gel permeation HPLC) can be used to separate oligosaccharide mixtures by size.

Quantifying fructan levels in foods is very challenging, as food contains a complex mix of these compounds of varying DP, from 2 to 60 units (Muir *et al.* 2007), which can be performed using HPLC or GC. Sometimes, an alternative approach involving the enzymatic hydrolysis of fructans prior to the chroma-tographic analysis is performed.

14.2.3.3   Dietary Fibre Analysis

Three methods for dietary fibre analysis have undergone extensive testing in recent years. The enzymatic-gravimetric AOAC methods are derived from methods aiming at simulating the digestion in the human small intestine to isolate an undigested residue as a measure of dietary fibre. The Englyst method measures the non-starch polysaccharide (NSP) specifically, either as individual monomeric components by GC (or HPLC) or colourimetrically as reducing substances (total NSP). Accordingly, dimethyl sulfoxide is used initially to ensure a complete removal of starch, and lignin is not determined. The Uppsala method requires measuring the neutral sugars, uronic acids (pectic material), and Klason lignin, and summing these components to obtain a dietary fiber value. The sample is first subjected to an enzymatic digestion to remove starch. Starch hydrolysates and low molecular weight sugars are separated from soluble fiber using an ethanol precipitation leaving a residue containing both soluble and insoluble fiber. Neutral sugars are determined after derivitisation as their alditol acetates by GC, uronic acids are assayed colourimetrically, and Klason lignin is determined gravimetrically.

14.3   UV-Vis Spectophotometric Methods for Dietary Sugars

Several analytical techniques have been described for individual or total sugar quantifications. However, it is still lacking a rapid, inexpensive and sensitive method for their detection (Pham *et al.* 2011). Spectrophotometric methods present reasonable investment, operation and maintenance costs, and the possibility of being used by non-skilled operators (Zhang *et al.* 2010). As a result, UV–Vis spectrophotometry is used for the quantitative determination of organic and inorganic constituents in a wide range of sample matrices.

UV-Vis spectrophotometry is a quantitative analytical technique related with the absorption of near-UV (190–390 nm) or visible (390–1100 nm) radiation by chemical species in solution. Under controlled experimental conditions (*e.g.* using diluted solutions) the amount of radiation absorbed can be directly related to the concentration of the analyte in solution (Beer's law), allowing quantifying organic (primarily in the near-UV) and inorganic (primarily in the visible) species (Worsfold 2005).

UV-Vis spectrophotometry has been used to measure sugars in solution based on direct or indirect absorbance readings. Single- or multivariate calibration models, based on absorbance readings at one or more wavelengths, and using raw or derivative spectroscopy data have been established for individual or total sugars quantification. Also, derivative pre-treatment of the spectroscopy data combined with multivariate modeling based on the full spectra readings, is often applied. This procedure is more crucial when trying to apply direct UV spectrophotometric methods (Dias *et al.* 2009) or indirect kinetic differential UV-Vis spectrophotometric techniques (Ni *et al.* 2003), although they have also been used with more classical colourimetric-spectrophotometric methods (Khanmohammadi *et al.* 2009; Wang *et al.* 2011).

14.3.1   Derivative Spectroscopy and Chemometric Analysis

A natural trend of UV-Vis spectrophotometry is the use of multicomponent analysis, since it is increasingly important to simultaneously determine the concentrations of one or more species, with unequal concentrations, in the presence of one or more interferents, with highly overlapping spectra (Rao and Biju 2005; Smith 2002). When the direct determination of an analyte is difficult due to the presence of interfering species instead of eliminating them, the use of multivariate calibration could enable the quantification of both interferents and analyte (Li and He 2007).

The chosen method should consider the complexity of the spectra, the intended accuracy, the speed of analysis and the analyst skills. Multicomponent methods can deal with overlapping spectrum issues between analytes and interferents using the full spectra (instead of that recorded at a specific wavelength) or using derivative spectroscopy approaches (Smith 2002). This last technique allows background correction and resolution enhancement, even in the presence of overlapping spectra, namely by eliminating matrix interferences (Vogt 2005; Wang *et al.* 2011).

Nowadays, the calibration techniques used mainly depend on the software available (*e.g.* SPSS, JMP, Minitab, Systat, Unscrambler, Statistica). However, despite the remarkable mathematical power of these algorithms, they will not compensate poorly made standards, sloppy technique or low quality spectra data (Smith 2002). PLS and ANN are the most used multivariate techniques.

14.3.1.1 Partial Least Squares Regression Models

PLS is a factor analysis technique frequently used due to the quality of the models obtained, the ease of implementation and the availability of software packages. It allows fast determination of different components in a given mixture without a prior separation that is otherwise necessary due to signals overlapping, by using conventional, derivative or even kinetic signals (Khanmohammadi *et al.* 2009).

Moreover, PLS is used as a data reduction technique. The spectral data and the property or assay data are used together in an iterative way to establish a model, ensuring that the estimated regression factors are significant allowing predicting the property of interest (chemical, physical or sensory attributes) of other samples (Rao and Biju 2005). PLS calibration models can be based on entire spectra data and be established without knowing exactly where the components of the solution absorb, as long as they absorb somewhere in the spectral regions that are used in the model. Also, this technique is very useful when the number of observations is scarce and the problem of multicollinearity among the predictor variables exists, since the derived principal component scores are uncorrelated. The main disadvantage of this technique is its mathematical complexity. Many trial calibrations using different conditions (wavelength region(s), algorithm, pre-processing and number of factors) may be required (Andrade-Garda 2009).

PLS as other multivariate techniques needs a calibration step where the models relating spectra data and analyte concentrations are deduced from a set of standards or samples, followed by a prediction step, in which the concentrations of unknowns are estimated from the raw or derivative sample spectrum (Ni *et al.* 2003). Two common PLS approaches can be used. Cali-brations can be generated for one component at a time or for multiple com-ponents simultaneously. The number of PLS factors that significantly contribute to the model variability can be identified through a full cross-vali-dation "leave one-out" method (Dias *et al.* 2009). Cross-validation is a model evaluation method that gives an indication of how well the model will perform when used to predict unknown samples. Data are partitioned into subsets such that the analysis is initially performed on a single subset, while the other sub-set(s) are retained for subsequent use in confirming and validating the initial analysis. The initial subset of data is called the training set; the other subsets are called validation or testing sets (Khanmohammadi *et al.* 2009).

14.3.1.2   Artificial Neural Networks

Recently, a non-linear regression methodology that works with rules rather than with well-defined and fixed algorithms has been introduced to handle complex data sets (*e.g.* spectra with strong nonlinearities). One of these com-putation techniques is ANNs, which are particularly applied in multivariate regressions.

To deal with interferences and increase accuracy and precision leads to a point where many of the researcher's difficulties cannot be resolved using simple, univariate, linear regression methods or linear multivariate regression methods.

Although PLS can handle nonlinearities and yields surprisingly good results, it usually requires the inclusion of a large number of factors (or latent vari-ables), which may hamper robustness and model stability.

In contrast, ANNs are flexible since they do not require an underlying mathematical model. They accept any linear or nonlinear relation between the predictors (spectra) and the predictand(s) (*e.g.* concentration(s) of the ana-lyte(s)). Further, noisy data or random variance can be easily dealt with by ANNs. They are robust, allowing small spectral deviations from the calibration data set, which are usual in spectrometric applications. Also, ANNs have relevant generalization capabilities, being able to extract learning patterns from known data sets and giving accurate predictions for new unknown samples (Andrade-Garda 2009). However, the procedure usually implies a great amount of computer work to establish a reliable model.

Usually, the structure of the ANN algorithm comprises three types of layers: an input, a hidden and an output layer. The experimental data are centered and normalized, and used as the input data. The input nodes transfer the weighted input signals to the nodes in the hidden layer that are the same nodes of the output layer. To determine the optimal number of hidden nodes, an iterative trial-and-error strategy is usually applied. The learning process for developing a neural network can be either supervised or unsupervised if it needs or not target outputs as teacher, respectively. Two different models can be used according to whether it is intended to obtain calibration models for one or more components at a time: single or multiple ANN (Figure 14.1), respectively.

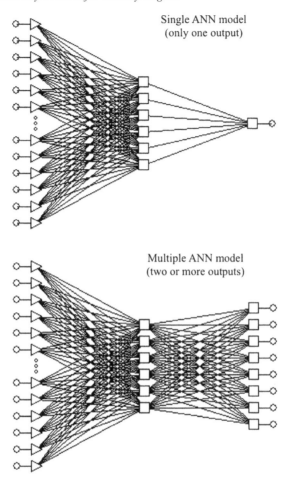

Single ANN model
(only one output)

Multiple ANN model
(two or more outputs)

Figure 14.1 Artificial neural networks three-layer networks. Diagram illustrating input, hidden and output layers of single and multiple ANN models.

In general, the network is trained using supervised back propagation learning algorithms. MLP and RBF are the most common type of networks. For each network tested, an automatic search for an effective sub-set of the specified variables is usually performed, being the network complexity automatically fixed by the software (Dias *et al.* 2009; Ni *et al.* 2003).

The implementation of a truly validated ANN model requires splitting the experimental database into two independent subsets, one for calibration (training and verification) and the other for validation purposes. Thus, ANNs are initially trained on the first subset and their performance monitored using the second subset to avoid obtaining a model that suffers from over-fitting, which occurs when the network learned to model the noise, but not the underlying non-linear function that relates input to output variables. Finally, the latter subset is used to evaluate the predictive performance of the network, since the data included in this subset are not used in the ANN development (Dias *et al.* 2009). Data splitting should consider the type of goodness-of-fit errors used for calibration and validation procedures. If the mean-squared sum error is used, usually, one-fourth to one-third of the entire sample database should be set aside for validation purposes (Good and Hardin 2003).

14.3.2 UV-Vis Spectrophotometry for Indirect Sugar Analysis

UV-Vis spectrophotometric techniques are commonly used for indirect sugars analysis, mainly due to their low cost. However, these methods are highly prone to interferences due to the presence of other carbohydrates in the sample (Zhang *et al.* 2010). Nevertheless, several indirect UV-Vis spectrophotometric methods have been proposed for sugar analysis, namely colourimetric-spectro-photometric methods (Pham *et al.* 2011; Zhang *et al.* 2010; Wang *et al.* 2011), UV irradiation-spectrophotometric method (Roig and Thomas 2003), enzymatic-coloured-spectrophotometric techniques (Amine *et al.* 2000) and kinetic-coloured-differential spectrophotometric approaches (Ni *et al.* 2003).

Among the colourimetric methods, several reagents have been used to obtain coloured compounds, which absorbances are measured at a single- or multi-wavelengths, allowing the indirect quantification of individual or total sugar contents in a sample. Visualization reagents for obtaining coloured products after acidic hydrolysis (H_2SO_4, H_3PO_3 or HCl) include anthrone, iodine, resorcinol, phenol, 3,5-dinitrosalicylic acid (DNS), orcinol and cysteine/carbazole (Pham *et al.* 2011; Zhang *et al.* 2010; Wang *et al.* 2011).

Amine *et al.* (2000) proposed an enzymatic-spectrophotometric assay to quantify lactulose in pasteurized milk samples, based on the use of b-galacto-sidase fructose dehydrogenase. However, some interference from glucose and galactose (from lactose hydrolysis) can occur. Also, enzymes costs and lack of stability limit the generalized use of enzymatic methods (Li and He 2007).

Anthon and Barrett (2002) developed a high sensitivity UV-Vis spectro-photometric method for quantifying reducing sugars, without proteins inter-ference. The method is based on the colourimetric determination of aldehydes, obtained from sugars, after oxidization.

Roig and Thomas (2003) used a UV/UV method to measure total sugars in fruit juices, wine and soft drinks. The procedure is based on the UV detection of oxidized by-products obtained from UV photodegradation of carbohydrates. However, identical calibration curves were obtained for standard solutions of glucose, fructose, lactose or sucrose. Therefore, if all these sugars are present in a sample, the quantification of each one is not feasible.

Ni *et al.* (2003) developed a kinetic differential spectrophotometric method for the simultaneous quantification of glucose, fructose and lactose in food samples. The method is based on the different kinetic rates of oxidized analytes. PLS and ANNs models were established to analyze synthetic mixtures of the three reducing sugars based on raw and first-derivative kinetic data, allowing quantifying the three reducing sugars in several food samples. It was found that the method was not affected by the presence of starch, sugar and most amino acids, except for cysteine. On the other hand, since xylose and maltose have similar kinetic properties as glucose, the method would require the removal of these sugars from the samples before analysis.

Khanmohammadi *et al.* (2009) developed an UV-Vis-PLS method based on second-derivative spectra data recorded between 490–800 nm to predict trehalose and sucrose contents in synthetic mixtures and olive leaf samples.

Zhang *et al.* (2010) described a novel, sensitive and simple spectrophotometric method to quantify lactulose in syrups, biological fluids or dairy products. The method was based on the hydrolysis of lactulose under acidic conditions, followed by the reaction of the hydrolyzed product with cysteine-hydrochloride-tryptophon reagent. Possible interferences of lactose and galactose were minimized.

Pham *et al.* (2011) proposed an UV-Vis method to quantify glucose and xylose, based on the dehydration reaction of glucose and/or xylose with hydrochloric acid to produce 5-(hydroxymethyl)-2-furfuraldehyde and 2-furaldehyde (furfural), respectively. To avoid interferences from other substances that absorb light in the same region as furfural, a new complex, formed by reacting furfural with orcinol in an acidic medium was used.

Wang *et al.* (2011) developed an alternative spectrophotometry method to determine amylose contents in different starches. The methodology is based on iodine colourimetry, involving the classical reaction between a-1,4-glucans and iodine to form a blue complex. To minimize the recurring problems reported for iodine-based techniques (interference from lipids or amylopectin), an orthogonal-function spectrophotometry was used, based on multi-wavelength spectral data.

14.3.3 UV Spectrophotometry for Direct Sugar Analysis in Controlled Multicomponent Samples

The indirect UV-Vis based methods successfully used for sugar quantification in real samples usually require a pre-treatment step to obtain sugar by-products with different UV-Vis absorbances. This step can be time-consuming or require the use of specific reagents or instruments, which may increase the final cost of the analysis. Therefore, even if sugars are normally low adsorbing molecules, the possibility of developing a direct UV spectrophotometric method represents a merit task.

Recently, a direct UV methodology based on derivative spectra data and chemometric approaches (multiple PLS and ANN) has been proposed by the authors for the simultaneous quantification of lactose and GOS in fermentation samples (Dias *et al.* 2009). This technique directly relates the spectra profile measured with the sugars contents, without the formation of any coloured sugar-chromophore or photodegradated sugar-compound. The proposed methodology requires a minimal sample preparation step, namely sample filtration to remove biomass and other solid in suspension from the fermentation samples. However, the range of applicability of the models is limited to fermentation samples obtained in identical experimental conditions of those used to develop the chemometric models. Therefore, the end users of this direct UV based technique must guarantee that the sample matrix is reproducible, obtained under similar and controlled experimental conditions and the sugar(s) content(s) that is (are) intended to be measured is (are) much higher than those of other existing sugars in solution, which could act as interferents.

Although sugars are low UV-Vis absorbing molecules, their raw spectra profiles show slight differences (Figures 14.2, 14.3 and 14.5), which can be used to establish direct measuring methods. Those differences can be enhanced after data treatment, namely using first- or higher order derivatives. As shown in Figure 14.2, the raw spectra obtained for aqueous standard solutions of fructose, glucose and galactose are similar (especially the last two). Thus, the development of a UV spectrophotometric method for quantifying these monosaccharides would be difficult or even not feasible. However, by taking a derivative spectroscopy approach (Figure 14.2), differences between these sugars spectra are enhanced, especially in some spectral regions. This may enable the establishment of a direct UV-chemometric-technique, based on multi-wavelength derivative spectroscopy data, for the quantification of individual or total monosaccharide content, if the standard solutions could model the real samples features.

A similar situation can be observed for the raw UV spectra of some disaccharides, such as lactose, maltose and sucrose. As can be seen from Figure 14.3, based on the UV spectra of aqueous standard solutions, recorded in the range of 190–210 nm, it would not be an easy task to distinguish each disaccharide. Hence, the use of UV spectrophotometry to quantify any of these sugars in the presence of the others would be impossible. Nevertheless, the first-derivative spectra already shows enhanced differences between lactose and the other two disaccharides and, from the second-derivative spectra more evident differences can be found between the three disaccharides. This finding could enable their individual quantification based on a combined UV-spectrophotometric-chemometric method using the second-derivative data.

Indeed, the authors successfully established a direct methodology for simultaneously quantifying lactose and GOS in fermentation samples. The spectra data obtained by Dias *et al.* (2009) only showed a significant absorption in the range of 190–210 nm. The absorbances were corrected by subtracting the spectra of deionised water recorded each day, which allowed a background correction by reducing base-line effects. The results obtained enabled the use of two UV-chemometric methods (Figure 14.4), one combining raw spectra data and a multiple PLS model, and the other combining the second-derivative spectroscopy data and a multiple ANN model. The last approach showed an acceptable predictive performance and so it could be used as a practical tool for monitoring those sugars content during a fermentation process. However, it should be emphasized that to obtain such satisfactory results, this technique must be applied to a well-established, routinely controlled fermentation process.

Monitoring the consumption and/or production of sugars during fermentation is a key step for process optimization during microbial cell growth, namely when the production of complex saccharides such as GOS is envisaged. Among the fermentative production of other prebiotics, it would also be of

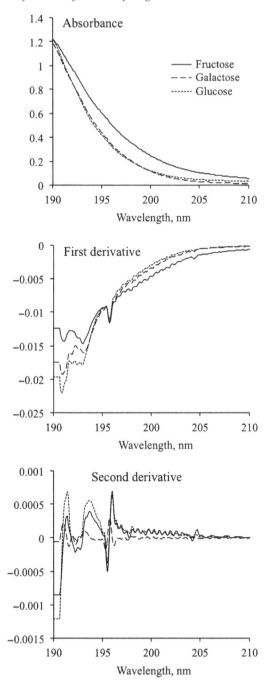

Figure 14.2   UV spectra for standard solutions of monosaccharides. UV spectra (190–210 nm) for solutions of fructose, galactose and glucose: raw, first- and second-derivative spectra (unpublished data).

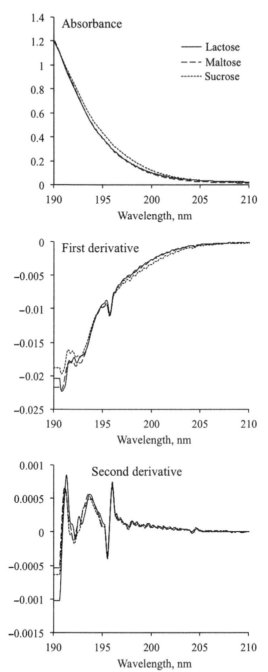

Figure 14.3 UV spectra for standard solutions of disaccharides. UV spectra (190–210 nm) for solutions of lactose, maltose and sucrose: raw, first- and second-derivative spectra (unpublished data).

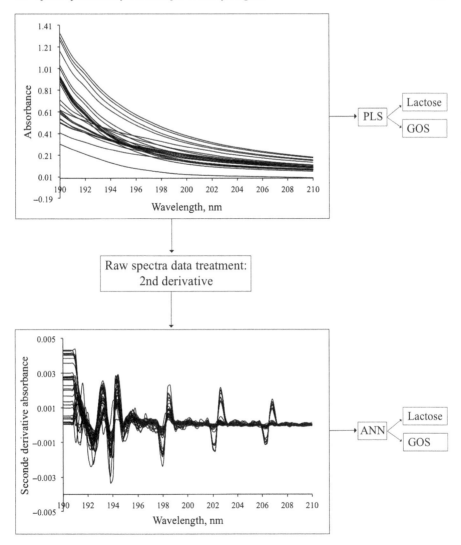

Figure 14.4   Direct UV-chemometric models for quantifying lactose and galacto-oligosacharides. Procedure for the simultaneous quantification of sugars in fermentation samples (Dias *et al.* 2009).

major interest to develop a similar simple, rapid and direct UV spectrophotometric approach for monitoring the production of individual or total FOS (*e.g.* kestose and nystose) using sucrose as the carbon source (Rocha *et al.* 2009). In this case, besides sucrose and FOS, glucose and fructose must also be considered, since they are formed during the fermentative process and can act as interferents, affecting the UV readings. Thus, the spectral features of aqueous standard solutions containing glucose, fructose, sucrose, kestose and nystose were studied. The results (Figure 14.5) showed that although the raw UV spectra are very similar, the application of derivative techniques could

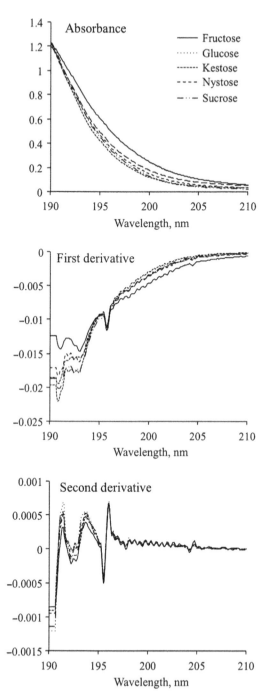

Figure 14.5 UV spectra for standard solutions of monosaccharides, disaccharides and oligosaccharides. UV spectra (190–210 nm) for solutions of fructose, glucose, sucrose, kestose and nystose produced during the fermentative production of FOS: raw, first- and second-derivative spectra (unpublished data).

enhance spectral differences in some regions of the spectra. So, it is possible to foresee the development of a direct UV-chemometric method based on derivative spectroscopy data for monitoring the production of FOS in fermentative processes. However, it is important to notice that matrix issues were not taken into account in these experiments, so real samples must be used in the future to validate this hypothesis.

In conclusion, the direct UV approach can be used for measuring sugar contents within practical applications, namely if the samples are obtained from routine industrial processes where the matrix, experimental conditions and entire system are well controlled and follow a pre-established operative follow-sheet, such as the industrial production of prebiotic sugars using a fermentative technology.

14.4 Conclusions and Future Prospects

This chapter provides detailed information on the most relevant aspects of recent application of UV-Vis spectrophotometry for measuring individual or total sugar contents, with a special focus on food samples. UV-Vis methodologies can be used for sugar measurements independently of the matrix and therefore with a broad spectrum of possible applications. The challenge of developing direct UV methods is establishing a simple analytical tool that can reduce the analysis time and the amount of solvents used in pre-treatment steps required by the majority of the available and more sophisticated analytical techniques used in this field.

Summary Points

- This chapter focuses on UV-Vis spectrophotometry and its applications for direct quantification of dietary sugars.
- Chemometric tools, namely multivariate models, based on full raw or derivative spectra data, are a feasible way to enhance the power of spectrophotometric techniques.
- Derivative spectroscopy can be used to minimize or solve overlapping or interference between signals of different compounds present in mixtures, suppressing background overlaying signals.
- UV-Vis-based techniques are advantageous; especially the new UV direct approach based on standard solutions or real samples containing sugars.
- The potential of this direct, low-cost, fast, sensitive and user-friendly integrated green analytical approach to measure sugars can only be settled if a further effort is carried out by the scientific community.

Key Facts

- Dietary sugars have no obvious natural UV-Vis absorption.
- Direct UV-Vis spectrophotometry can be used to quantify sugars in standard and real samples.

- Sample pre-treatment is not required for the methodology based on multiple-wavelength data.
- Multivariate models enhance the power of spectrophotometric techniques.
- Derivative spectroscopy minimizes interference between signals of different compounds present in mixtures.

Definitions of Key Terms

Chemometrics. Science that applies statistical and mathematical methods to chemical data, relating measurements with the state of the system.
Multivariate Statistical Analysis. Statistical approach that encompasses the simultaneous observation and analysis of more than one variable.
Partial Least Squares Method. Statistical method used to find a linear regression model by projecting the predicted and observable variables into a new orthogonal space.
Artificial Neural Networks. Non-linear statistical models or computational tools inspired by the structure and/or functional aspects of biological neural networks.
Derivative Spectroscopy. Technique based on the differentiation of spectra, mainly for spectral discrimination, spectral resolution enhancement and quantitative analysis.

List of Abbreviations

ANN   Artificial Neural Network
AOAC   Association of Official Analytical Chemists
DP   Degree of Polymerization
FOS   Fructo-oligosaccharides
FTIR   Fourier Transform Infrared Spectroscopy
GC   Gas Chromatography
GI   Gastrointestinal
GOS   Galacto-oligosaccharides
HPLC   High Performance Liquid Chromatography
MLP   Multilayer Perceptron
NSP   Non Starch Polysaccharides
PLS   Partial Least Squares
RBF   Radial Basis Function
UV   Ultraviolet
UV-Vis   Ultraviolet-Visible

References

Amine, A., Moscone, D., Bernardo, R.A., Marconi, E., and Palleschi, G. 2000. A new enzymatic spectrophotometric assay for the determination of lactulose in milk. *Analytica Chimica Acta*. 406: 217–224.

Andrade-Garda, J.M. 2009. *Basic Chemometric Techniques in Atomic Spectroscopy*. RSC Analytical Spectroscopy Monographs No. 10. RSC.

Anthon, G.E., and Barrett, D.M. 2002. Determination of Reducing Sugars with 3-Methyl-2-benzothiazolinonehydrazon. *Analytical Biochemistry*. 305: 287–289.

Brummer, Y., and Cui, S.W. 2005. Understanding carbohydrate analysis. In: Cui, S.W. (ed.) *Food Carbohydrates: Chemistry, Physical Properties and Applications*. Taylor & Francis, Boca Raton, pp. 67–104.

Cummings, J.H., and Stephen, A.M. 2007. Carbohydrate terminology and classification. *European Journal of Clinical Nutrition*. 61 (Suppl 1): S5–S18.

Dias, L.G., Veloso, A.C.A., Correia, D.M., Rocha, O., Torres, D., Rocha, I., Rodrigues, L.R., and Peres, A.M. 2009. UV spectrophotometry method for the monitoring of galacto-oligosaccharides production. *Food Chemistry*. 113: 246–252.

Englyst, K.N., Liu, S., and Englyst, H.N. 2007. Nutritional characterisation and measurement of dietary carbohydrates. *European Journal of Clinical Nutrition*. 61 (Suppl 1): S19–S39.

Fernandez, M.L., and Vergara-Jimenez M. 2006. Associations between carbohydrate intake and risk for coronary heart disease, insulin resistance and the metabolic syndrome. In: Landow, M.V. (ed.) *Trends in Dietary Carbohydrates Research*. Nova Science Publishers, Inc., pp. 55–70.

Good, P.I., and Hardin, J.W. (ed.), 2003. *Common Errors in Statistics (and how to avoid them)*. John Wiley & Sons, Inc, Hoboken, NJ.

Izydorczyk, M. 2005. Understanding the chemistry of food carbohydrates. In: Cui, S.W. (ed.) *Food Carbohydrates: Chemistry, Physical Properties and Applications*. Taylor & Francis, Boca Raton, pp.1–65.

Khanmohammadi, M., Moeini, M., Garamarudi, A.B., Sotudehnia, A., and Zarrabi, M. 2009. Simultaneous determination of sucrose and trehalose in oliveleaves by spectrophotometry utilizing partial least squares method. *Acta Physiologiae Plantarum*. 31: 865–869.

Li, B., and He, Y. 2007. Simultaneous determination of glucose, fructose and lactose in food samples using a continuous-flow chemiluminescence method with the aid of artificial neural networks. *Luminescence*. 22: 317–325.

Linhardt, R.J., and Bazin, H.G. 2001. Properties of carbohydrates. *Glycoscience*. 1: 53–61.

Macfarlane, G.T., Steed, H., and Macfarlane, S. 2008. Bacterial metabolism and health-related effects of galacto-oligosaccharides and other prebiotics. *Journal of Applied Microbiology*. 104: 305–344.

Moeawski, R.Z. 2006. Spectrophotometric applications of digital signal processing. *Measurement Science and Technology*. 17: R117–R144.

Muir, J.G., Shepherd, S.J., Rosella, O., Rose, R., Barrett, J.S., and Gibson, P.R. 2007. Fructan and free fructose content of common Australian vegetables and fruit. *Journal of Agriculture and Food Chemistry*. 55: 6619–6627.

Ni, Y., Huang, C., and Kokot, S. 2003. A kinetic spectrophotometric method for the determination of ternary mixtures of reducing sugars with the aid of artificial neural networks and multivariate calibration. *Analytica Chimica Acta.* 480: 53–65.

Pham, P.J., Hernandez, R., French, W.T., Estill, B.G., and Mondala, A.H. 2011. A spectrophotometric method for quantitative determination of xylose in fermentation medium. *Biomass and Bioenergy.* 35: 2814–2821.

Rao T.P., and Biju V.M. 2005. *Spectrophotometry: Organic Compounds. In Encyclopedia of Analytical Science* (Second Edition). Elsevier, pp. 358–366.

Roberfroid, M.B. 2005. Introducing inulin-type fructans. *British Journal of Nutrition.* 93(Suppl. 1): S13–S25.

Rocha, O., Rocha, I., Nobre, C., Dominguez, A., Torres, D., Rodrigues, L.R., Teixeira, J.A., and Ferreira, E.C. 2009. A dynamical model for the fermentative production of fructooligosaccharides. *Computer Aided Chemical Engineering.* 27: 1827–1832.

Roig, R., and Thomas, O. 2003. Rapid estimation of global sugars by UV photodegradation and UV spectrophotometry. *Analytica Chimica Acta.* 477: 325–329.

Smith, B.C. 2002. *Quantitative Spectroscopy: Theory and Practice.* Academic Press. Elsevier.

Torres, D., Gonçalves, M.P.F., Teixeira, J.A., and Rodrigues, L.R. 2010. Galacto-Oligosaccharides: Production, Properties, Applications, and Significance as Prebiotics. *Comprehensive Reviews in Food Science and Food Safety.* 9: 438–454.

Vogt, F. 2005. Spectrophotometry: Derivative Techniques. In: *Encyclopedia of Analytical Science* (Second Edition). Elsevier, pp 335–343.

Wang, J.P., Yu, B., Xu, X.M., Yang, N., Jin, Z.Y., and Kim, J.M. 2011. Orthogonal-function spectrophotometry for the measurement of amylose and amylopectin contents. *Food Chemistry.* 127: 102–108.

Williams, H.D., Ward, R., Hardy, I.J., and Melia, C.D. 2009. The extended release properties of HPMC matrices in the presence of dietary sugars. *Journal of Controlled Release.* 138: 251–259.

Worsfold, P.J. 2005. Spectrophotometry: Overview. In: *Encyclopedia of Analytical Science* (Second Edition). Elsevier, pp. 318–321.

Zhang, Z., Wang, H., Yang, R., and Jiang, X. 2010. A novel spectrophotometric method for quantitative determination of lactulose in food industries. *International Journal of Food Science and Technology.* 45: 258–264.

CHAPTER 15

Extraction and Quantification of Sugars and Fructans from Vegetable Matter

KATHERINE COOLS AND LEON A. TERRY*

Plant Science Laboratory, Cranfield Health, Vincent Building, Cranfield University, Bedfordshire, MK43 0AL, United Kingdom, *Email: l.a.terry@cranfield.ac.uk

15.1 Introduction

15.1.1 Structure of Sugars and Fructans

Non-structural carbohydrates (NSC) can be categorized into three main groups; monosaccharides, oligosaccharides and polysaccharides (Lee *et al.* 1970). Monosaccharides are one-unit sugars consisting of either five or six carbons named pentoses or hexoses, respectively. The oligosaccharides are made up of two (disaccharides) or more sugar units linked by glycosidic bonds. The last group is the polysaccharides which consist of long chain sugar units (Lee *et al*. 1970). The aldehyde or ketone group of an open-chain monosaccharide can react with one of its hydroxyl groups to form a ring structure which is the predominant form when in an aqueous form (Hermanson 2008). Sugars can be classified as reducing or non-reducing by whether the ring structure can open to yield a free aldehyde or ketone group. Fructose and glucose are both reducing sugars since they possess an aldehyde group when in the linear structure. However, the

Food and Nutritional Components in Focus No. 3
Dietary Sugars: Chemistry, Analysis, Function and Effects
Edited by Victor R Preedy
© The Royal Society of Chemistry 2012
Published by the Royal Society of Chemistry, www.rsc.org

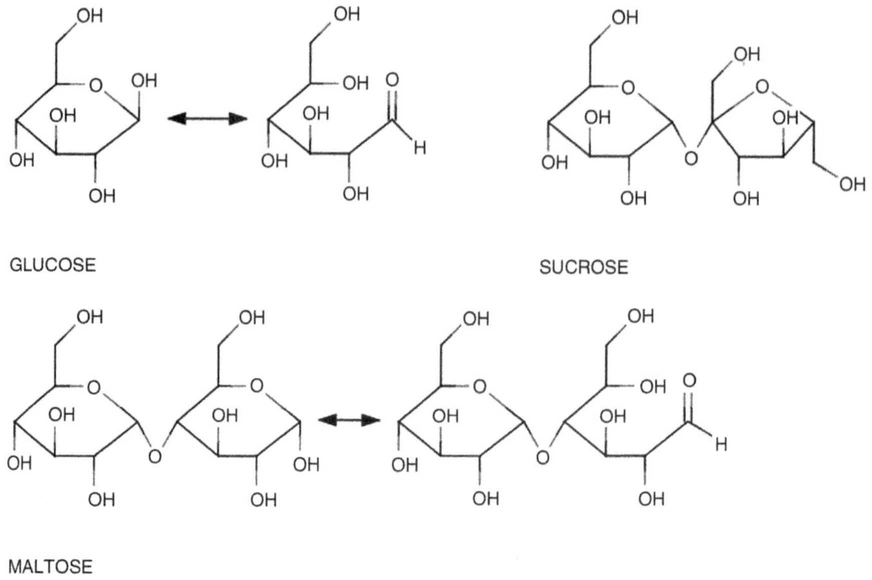

GLUCOSE

SUCROSE

MALTOSE

Figure 15.1 Glucose and maltose in the open chain and cyclic form (reducing sugars) plus sucrose in the cyclic form (non-reducing sugar). Unpublished diagram demonstrating reducing (glucose and maltose) and non-reducing (sucrose) sugars in the open chain and cyclic form.

disaccharide sucrose, for example, is not a reducing sugar since neither ring structure is able to open to free the aldehyde group as the groups are complexed with the glycosidic bond (Figure 15.1). Not all oligosaccharides are non-reducing since the aldehyde group can be positioned on the linked side or at the end of the molecule as demonstrated by maltose in Figure 15.1.

Fructans are oligo- and polysaccharides which can vary widely in their degree of polymerization from three to a few hundred. Fructans can take many forms; the most simple is the linear inulin form which is found in chicory and onions (Ritsema and Smeekens 2003). The units of fructans consist of a series of fructosyl polymers with sucrose as its base (Figure 15.2; Suzuki and Cutcliffe 1989). In onions, the neo-series of fructans consists of a sucrose unit at the centre with two β(1-2)-linked fructose chains attached to either side of the sucrose unit (Ritsema and Smeekens 2003). The term fructooligosaccharides (FOS) is often used when referring to shorter chain fructans mainly kestose, nystose and fructofuranosylnystose which consist of 3–4 sugar units (Downes and Terry 2010).

15.1.2 Biosynthesis

The disaccharide sucrose is converted into glucose and fructose by the enzyme sucrose invertase, a reaction which can be reversed by the enzyme sucrose synthase (Winter and Huber 2000). A group of transferase enzymes are responsible for the production of fructan molecules in vegetables. The enzyme

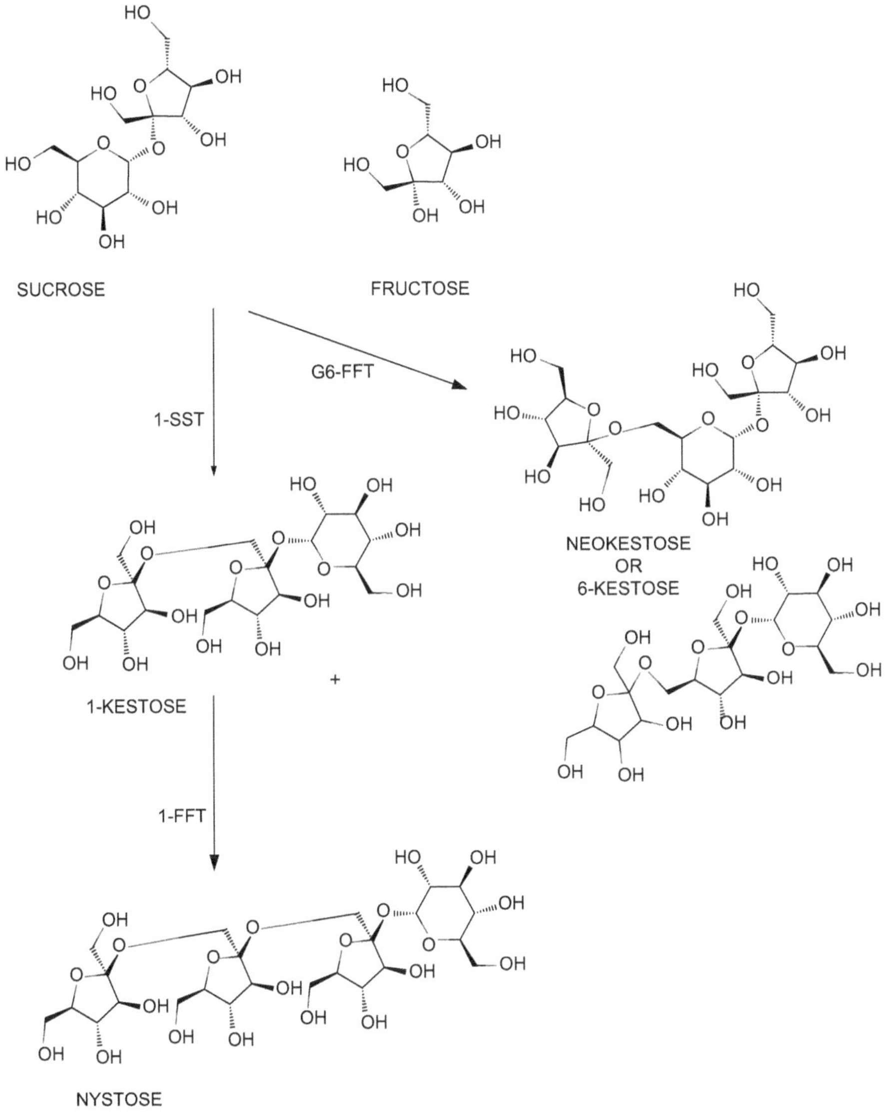

Figure 15.2 Simple linear and neo-series inulin fructans and the enzymes which catalyse production. Unpublished diagram showing the enzymes responsible for the synthesis of fructooligosaccharides.

sucrose:sucrose 1-fructosyltransferase (1-SST) catalyses the production of 1-kestose from sucrose and fructose. For the production of larger fructan molecules, the enzyme fructan:fructan 1-fructosyltransferase (1-FFT) is required. For more complex inulin compounds such as the neo-series in onions, where two chains of fructoses are attached to the sucrose base, the enzyme fructan:fructan G6-fructosyltransferase (G6-FFT) is required (Figure 15.2). Fructan breakdown is catalysed by fructan exohydrolases (FEH) which release fructose monomers from the end of each fructosyl chain (Ritsema and Smeekens 2003).

15.1.3   Distribution of Sugars and Fructans

The simple sugars; fructose, glucose and sucrose are found in a wide variety of vegetables including onions, green beans and potatoes (Downes *et al.* 2010; Foukaraki *et al.* 2010; Sánchez-Mata *et al.* 2002). Fructans are found in very high concentrations in Jerusalem artichokes (Pontis 1990); however, they are also found in onion (Cools *et al.* 2011), asparagus (Martin and Hartmann 1989) and globe artichokes (Ritsema and Smeekens 2003). In a study of 60 vegetables, onion, garlic, shallot and leek were among the top six in terms of fructan concentration (a range of 1.8–17.4 g 100 g^{-1} FW edible portion) (Muir *et al.* 2007). Non-structural carbohydrates can account for a large amount of the vegetable matter dry weight. For example, fructose, glucose, sucrose and fructans account for more than 65% of the dry weight of some onion varieties (Darbyshire and Henry 1978) (Table 15.1).

Table 15.1   Common vegetables containing high concentrations of sugars and fructooligosaccharides (mg g^{-1} FW) (after Muir *et al.* 2009). Reprinted with permission from Muir *et al.* (2009). Copyright 2009 American Chemical Society.

Food	% Moisture	Fructose	Glucose	Kestose	Nystose
Asparagus	84	31.6	27.5	3.4	0.9
Artichoke, globe	83	1.8	21.9	tr	tr
Artichoke, Jerusalem	75	nd	20.8	nd	tr
Beetroot	84	1.5	3.2	2.2	1.1
Broccoli	85	3.8	9.4	7.1	0.8
Brussel sprouts	81	0.6	4.1	5.5	nd
Cabbage, common	86	13.3	29.6	4.6	nd
Cabbage, savoy	86	9.7	18.0	3.9	nd
Capsicum, green	87	23.7	40.0	nd	nd
Capsicum, red	86	32.3	46.6	nd	nd
Chicory	91	1.5	3.2	1.2	0.5
Chilli, red	82	7.2	12.6	3.4	0.9
Cucumber	89	11.6	20.5	nd	nd
Fennel, bulb	88	6.7	22.5	1.6	1.5
Garlic	61	3.5	11.6	7.1	2.1
Lettuce, radicchio	88	6.4	7.7	2.3	4.4
Mushroom, button	89	0.1	2.0	1.9	0.8
Onion, white	84	13.8	33.6	2.6	1.3

nd, analysed but not detected; tr, trace amounts detected.

15.1.4 Sugar and Fructan Measurement

Sugars are fed into the citric acid cycle *via* glycolysis to provide energy. The measurement of sugar concentrations is therefore important for identifying changes in metabolism as a result of different developmental stages or pre- or postharvest treatments and conditions. For example, Chope *et al.* (2012) found that monosaccharide to disaccharide ratio was a marker for sprouting in onion. Sugar content is also important from a consumer point of view since high levels of monosaccharides give a sweet taste. It is therefore important to measure sugar content in response to preharvest and postharvest factors and treatments to ascertain whether there is an effect on carbohydrate profile. Changes in the carbohydrate profile of onion bulbs are important for taste preference as concentrations of fructose and glucose are positively correlated with likeability and sweetness (Terry *et al.* 2005).

Fructans and FOS are reserve carbohydrates which are hydrolysed into simple sugars, then fed into the citric acid cycle. As described above, these polysaccharides are measured in vegetable tissue to investigate the effect of preharvest and postharvest conditions on the metabolism or developmental stage. In addition, fructans and FOS have reported health promoting properties, therefore, vegetables with high concentrations are of nutritional interest (Terry 2011). Polysaccharides are not digested in the upper intestine, and are a source of energy for bacteria producing β-fructosidases in the caeco-colon. Fructans have a reported prebiotic effect, whereby it is believed to promote proliferation of beneficial bacteria like *Bifidobaceteria* (Bielecka *et al.* 2002) and *Lactobacilli* (probiotic bacteria), which in turn result in a decrease in the population of potentially harmful bacteria (Roberford 2007). The beneficial colonic microbiota produce short-chain fatty acids (*e.g.* lactate and butyrate) which lower pH, thus favouring increased absorption of mineral cations (such as Ca and Mg) from the gut into the bloodstream. Changes in colonic bacteria may reduce carcinogen activation in the colon and stimulate the immune system. Animal studies have also shown benefits for glucose metabolism (increased insulin secretion and changes to hormone metabolism) (Brewster 2008). It has been suggested that fructans with a higher degree of polymerisation (such as those present in garlic and leek) are less likely to induce undesirable gastrointestinal side effects (Muir *et al.* 2007).

15.2 Sugar and Fructan Extraction

There have been several extraction procedures reported for the quantification of sugars and fructans in vegetable matter (Davis *et al.* 2007) and these have been summarised in Table 15.2. Alkaline extractions are common when anion exchange chromatography is adopted as the method of quantification (see section 15.3.2). A method described by Shiomi *et al.* (1997) extracted sugars using 70% (v/v) ethanol and a small amount of calcium carbonate, creating an alkaline extraction solvent, although the exact amount of calcium carbonate (CaCO₃) was not verified. This method was again used by Benkeblia *et al.*

Table 15.2 Summary of the extraction and quantification methods of sugars and fructans measured in postharvest onion bulbs (after Davis et al. 2007 with modification). Reprinted with permission from Davis et al. (2007). Copyright 2007 American Chemical Society.

NSC^1 Analytes	Extraction Method	Bulb Section	Cultivar	Storage conditions	Quantification Method	Range Measured	Reference
Fructose Glucose Sucrose Fructan	FD tissue in 80% (v/v) EtOH at 75 °C	Bulb base, inner scale, outer scale	Hystar Hysam Centurian	16 °C 17 weeks	Total CHOs – anthrone-H_2SO_4 method F^7, G^8, S^9 – HPLC-PED[10]; CarboPak PA1 column; isocratic elution in 150mM NaOH for 30 min; low rate $1\,ml\,min^{-1}$ Fructan = total CHOs–(F+G+S)	$20\text{–}160\,mg\,g^{-1}$ DW $40\text{–}270\,mg\,g^{-1}$ DW $90\text{–}150\,mg\,g^{-1}$ DW $10\text{–}260\,mg\,g^{-1}$ DW	Pak et al. (1995)
Fructose Glucose Sucrose Fructan	50 g fresh tissue in 75 ml 96% (v/v) EtOH at 80 °C	Outer, middle and inner scales	Hyton Hyduro	1 °C; 75–80% RH 40 weeks	HPLC-RID[11]; ion exchange column; 60 °C; mobile phase 10–4N NaOH in water; flow rate $0.5\,ml\,min^{-1}$ Fructan – Hydrolysed sugars (2N TFA[12] at 40 °C for 60 mins) minus non-hydrolysed sugars	$26\text{–}192\,mg\,g^{-1}$ DW $85\text{–}144\,mg\,g^{-1}$ DW $85\text{–}188\,mg\,g^{-1}$ DW $89\text{–}419\,mg\,g^{-1}$ DW	Hansen (1999)
F + G + S Glucose	5 g FD tissue in 50 ml water for 30 min at 100 °C	Inner bud tissue	Rouge Amposta[H]	18 °C; 70% RH 24 weeks	HPLC-DRD[13]; Polyspher CH-CA column, 80 °C, mobile phase water, flow rate $0.5\,ml\,min^{-1}$	$400\text{–}650\,mg\,g^{-1}$ DW $110\text{–}220\,mg\,g^{-1}$ DW	Benkeblia and Salselet-Attou (1999)
Fructose Glucose Sucrose Fructan	1 g FD tissue in 70% (v/v) EtOH reflux for 10 min Fructan – aliquot treated with inulinase to release F and G	Dry skin, outer leaves, top, bottom, inner part	Hysam[H]	Not stored	HPLC-RID; Aminex cation exchange column, 85 °C, mobile phase water, flow rate $0.5\,ml\,min^{-1}$	$2.5\text{–}80\,mg\,g^{-1}$ DW $30.3\text{–}163.3\,mg\,g^{-1}$ DW $5.4\text{–}103.1\,mg\,g^{-1}$ DW $0.8\text{–}316.3\,mg\,g^{-1}$ DW	Jaime et al. (2000)

Component	Extraction	Tissue	Cultivar	Storage conditions	Analysis	Values	Reference
Fructose Glucose Sucrose Fructan	As above	Inner fleshy leaves	Cultivars with range of DM (n = 5)	0°C; 60–65% RH 24 weeks	As above	43.1–241.8 mg g⁻¹ DW 34.9–263.8 mg g⁻¹ DW 31.7–130.6 mg g⁻¹ DW 40.2–458.1 mg g⁻¹ DW	Jaime *et al.* (2001)
Fructose Glucose Sucrose Fructan	10–100 mg FD tissue refluxed in 80% EtOH, 1 h, concentrated and redissolved in 1 ml water	Whole bulb	Cultivars with range of DM (n = 11)	Not stored	HPLC-ELSD	No absolute values stated	Kahane *et al.* (2001)
F + G + S Fructan	F, G, S - 5 g FD tissue in 50 ml water for 30 min at 100 °C Fructan – 1 g FD tissue in 80 ml 70% (v/v) EtOH reflux for 10 min	Bulb tissue	Rouge Amposta^H	4 °C; 85% RH 10 °C; 80% RH 20 °C; 65% RH 24 weeks	F, G, S – HPLC-DRD Polyspher CH-CA column, 80 °C, mobile phase water, flow rate 0.5 ml min⁻¹ Fructan - HPLC-PAD; carbohydrate column PA1 (CarboPak); gradient elution with NaOH and Na-acetate, flow rate 1 ml min⁻¹	30–70 mg g⁻¹ FW 25.59–68.13 mg g⁻¹ DW	Benkeblia *et al.* (2002)
Fructose Glucose Sucrose	1 g fresh tissue in 10 ml 80% (v/v) CH₃CN in H₂O for 3 min	Outer fleshy part, inner part	Tropea^H	5 °C; 30% RH 25 °C; 66% RH 30 °C; 50% RH 6 weeks	HPLC-RID; column Hypersil 5 APS 2, mobile phase 80% (v/v) CH₃CN in H₂O, flow rate 0.5 ml min⁻¹	175.2–177.5 mg g⁻¹ DW 252.4–276.2 mg g⁻¹ DW 53.7–148.7 mg g⁻¹ DW	Gennaro *et al.* (2002)
F + G + S	5 g FD tissue homogenised in 50 ml water for 30 min at 100 °C	Bulb tissue	Rouge Amposta^H	18 ± 1 °C; 65 ± 1% RH 2 weeks	As above	50–55 mg g⁻¹ FW	Benkeblia and Varoquaux (2003)
Fructose Glucose Sucrose Fructan	60 mg FD tissue in 7 ml water saturated with CaOH at 100 °C for 15 mins.	Bulb tissue	Sherpa^H	0.5, 1.0 or 21% O₂ <0.3% CO₂ 36 weeks	HPAEC-PAD[14]; Flow rate 1 ml min⁻¹. Total G and S measured after hydrolysis of extract with HCl	40–200 mg g⁻¹ DW 100–180 mg g⁻¹ DW 80–140 mg g⁻¹ DW 20–320 mg g⁻¹ DW	Ernst *et al.* (2003)
Total soluble sugars	5 g FD tissue homogenised in 50 ml water, 30 min boiling water bath	Inner bud tissue	Rouge Amposta^H	20 °C; 65% RH 8 weeks	HPLC-PAD carbohydrate column PA1 (CarboPak); gradient elution with NaOH and Na-acetate, flow rate 1 ml min⁻¹	5–23 mg g⁻¹ FW	Benkeblia and Shiomi (2004)

Table 15.2　(Continued)

NSC[1] Analytes	Extraction Method	Bulb Section	Cultivar	Storage conditions	Quantification Method	Range Measured	Reference
Fructose Glucose Sucrose Fructan	F, G, S – 5 g FD tissue in 50 ml water for 30 min in boiling water bath Fructan – 10 g FD tissue in 80% (v/v) EtOH reflux for 10 min	Bulb tissue	Jaune d'Espagne[H]	4°C; 85% RH 10°C; 80% RH 20°C; 65% RH 24 weeks	F,G,S – HPLC-DRD Polysher CH-CA column, 80°C, mobile phase water, flow rate 0.5 ml min^{-1} Fructan - HPLC-PAD carbohydrate column PA1 (CarboPak); gradient elution with NaOH and Na-acetate, flow rate 1 ml min^{-1}	156–190 mg g^{-1} DW 176–215 mg g^{-1} DW 66–93 mg g^{-1} DW 255–295 mg g^{-1} DW	Benkeblia et al. (2004a)
Fructose Glucose Sucrose	10 g fresh tissue in 80 ml 70% (v/v) EtOH reflux for 10 min	Bulb tissue	Tenshin[H]	10 ± 1 °C; 70 ± 1% RH 20 °C; 55% RH 25 weeks	HPLC-PAD; carbo-hydrate column PA1 (CarboPak); gradient elution with NaOH and Na-acetate, flow rate 1 ml min^{-1}	2–22 mg g^{-1} FW 3–17 mg g^{-1} FW 3–18 mg g^{-1} FW	Benkeblia et al. (2004b)
Fructose Glucose Sucrose Fructan	G, F, S – 10 mg FD tissue in 1 ml 62.5% (v/v) MeOH/55°C Fructan –10 mg FD tissue in 1 ml water/ 80 °C for 15 min	Quadrant	PLK[H]Grano[L]	Not stored	F, G, S – HPLC-ELSD[15]; Rezex monosaccharide column, 85 °C, mobile phase water, flow rate 0.6 ml min^{-1} Fructan – Enzyme assay kit	3.0–14.9 mg g^{-1} FW 18.1–20.3 mg g^{-1} FW 4.7–12.0 mg g^{-1} FW 4.6–50.8 mg g^{-1} FW	O'Donoghue et al. (2004)
Sucrose Fructan	10 g fresh tissue in 80 ml 70% (v/v) EtOH reflux for 10 min	Bulb tissue	Tenshin[H]	10 ± 1 °C; 70 ± 1% RH 20 °C; 55% RH 24 weeks	HPLC-PAD; carbo-hydrate column PA1 (CarboPak); gradient elution with NaOH and Na-acetate, flow rate 1 ml min^{-1}	3–18 mg g^{-1} FW 2–25 mg g^{-1} FW	Benkeblia et al. (2005a)

Sugar	Extraction	Tissue	Cultivar	Storage	Analysis	Concentration	Reference
Fructose Fructan	As above	Fleshy bulb tissue	Tenshin[H]	15±1°C; 45±1% RH 24 weeks	HPLC-PAD carbohydrate column PA1 (CarboPak); gradient elution with NaOH and Na-acetate, flow rate 1 ml min[-1]	11.5–13.62 mg g[-1] FW 8–26 g g[-1] FW	Benkeblia et al. (2005b)
Fructose Glucose Sucrose	50 mg FD tissue in 50 ml 80% (v/v) EtOH reflux for 1h	Equatorial slice	SS1[L] Buffalo[L] Shakespeare	Not stored	HPLC-ELSD, Novapak-NH2 reverse phase column, mobile phase acetonitrile-water (80:20, v/v), flow rate 2 ml min[-1]	46–260 mg g[-1] DW 65–241 mg g[-1] DW 57–126 mg g[-1] DW	Terry et al. (2005)
Fructan	1 g FD tissue in 100 ml water 80°C for 15 min	Bulb tissue	Renate Ailsa Craig SS1[L]	2±1°C 5% O2, 3% CO2 12–32 weeks	Enzyme assay kit	25–290 mg g[-1] DW	Chope et al. (2006)
Fructose Glucose Sucrose Fructan	G, F, S – 150 mg FD tissue in 3 ml 62.5% (v/v) MeOH at 50°C for 15 min. Fructan –10 mg FD tissue in 1 ml water at 80°C for 15 min	Bulb quadrant	Renate Carlos SS1[L]	2±1°C 5% O2, 3% CO2 or air 6 weeks	F, G, S – HPLC-ELSD. Fructan – Enzyme assay kit	20–350 mg g[-1] DW 100–360mg g[-1] DW 30–130 mg g[-1] DW 25–360 mg g[-1] DW	Chope et al. (2007a)
Fructose Glucose Sucrose Fructan	As above	Bulb quadrant	SS1[L]	4°C, 12°C or 20°C 6–13 weeks	As above	125–310 mg g[-1] DW 130–320 mg g[-1] DW 40–100 mg g[-1] DW 25–65 mg g[-1] DW	Chope et al. (2007b)
Fructose Glucose	150 mg FD skin tissue in 3 ml 62.5% MeOH at 55°C for 15 mins	Bulb skin	Sherpa[H] Wellington[H] Red Baron[H]	1°C 32 weeks	HPLC-ELSD	0.2–4.07 mg g[-1] FW 6.29–31.06 mg g[-1] FW	Downes et al. (2009)
Fructose Glucose Sucrose	150 mg FD flesh tissue in 3 ml 62.5% MeOH at 55°C for 15 mins	Bulb quadrant	Sherpa[H] Wellington[H]	1°C 38 weeks	As above	20–210 mg g[-1] DW 65–180 mg g[-1] DW 80–225 mg g[-1] DW	Downes et al. (2010)

Table 15.2 (Continued)

NSC[1] Analytes	Extraction Method	Bulb Section	Cultivar	Storage conditions	Quantification Method	Range Measured	Reference
Fructose	150 mg FD flesh tissue in 2.25 ml water at 75 °C for 10 mins then the addition of 3.75 ml MeOH at 55 °C for 15 mins	Bulb quadrant	Red Baron[H]	Not stored	As above	75 mg g^{-1} DW	Downes and Terry (2010)
Glucose						165 mg g^{-1} DW	
Sucrose						130 mg g^{-1} DW	
Kestose						125 mg g^{-1} DW	
Nystose						80 mg g^{-1} DW	
DP5-9[16]						140 mg g^{-1} DW	
Fructose	As above	Bulb quadrant	Sherpa[H]	1 °C 35 weeks	As above	30–220 mg g^{-1} DW	Cools et al. (2011)
Glucose						150–240 mg g^{-1} DW	
Sucrose						95–250 mg g^{-1} DW	
Kestose						70–125 mg g^{-1} DW	
Nystose						25–90 mg g^{-1} DW	
DP5-8						15–150 mg g^{-1} DW	

[H]High dry matter; [L]Low dry matter; [1]NSC, non-structural carbohydrates; [2]CHOs, carbohydrates; [3]PLK, Pukehole Longkeeper; [4]DM, dry matter; [5]FD, freeze-dried; [6]FID, flame ionisation detector; [7]F, fructose; [8]G, glucose; [9]S, sucrose; [10]PED, pulsed electrochemical detector, [11]RID, refractive index detector, [12]TFA, trifluoroacetic acid, [13]DRD differential refractometer detector, [14]PAD, pulsed amperometric detector, [15]ELSD, evaporative light scattering detector, [16]DP, degrees of polymerisation.

(2004a,b) and Benkeblia *et al.* (2005a,b) although again the exact amount of $CaCO_3$ used was not disclosed. In other literature, the alkaline extraction solution was achieved using a saturated calcium hydroxide ($Ca(OH)_2$) water solution (Ernst *et al.* 1998). The problem with using strong alkaline solutions is that certain oligosaccharides can suffer from epimerisation and degradation, which becomes apparent when the carbohydrates are eluted. This problem can be reduced by using a lower concentration of hydroxide ion and compensating with higher acetate and/or alternative competing ion (Lee 1996). The above methods also adopted different extraction solvents with Ernst *et al.* (1998) twice boiling in water for 15 mins and Shiomi *et al.* (1997) boiling in 70% (v/v) ethanol for 10 mins. It has been previously found that boiling in a high (70–80%) ethanol solution for 1 h does not extract sugars as effectively as incubating at a lower temperature (70 °C) for 2 h (Davis *et al.* 2007). Davis *et al.* (2007) compared methods of soluble sugar extractions in onion and concluded methanol/water based mixtures (O'Donoghue *et al.* 2004) were more efficient at extracting monosaccharides from onions than ethanol/water based solutions. It was also noted that differences in extraction efficiencies can be influenced by the ratio of glucose, fructose, sucrose concentrations which differ between cultivars (Davis *et al.* 2007).

To confirm that a methanol solution was more efficacious at extracting sugars from freeze-dried onion powder than ethanol solution, Downes and Terry (2010) compared extraction using 80% (v/v) ethanol for 30 mins at 70 °C (Vågen and Slimestad 2008) with water for 10 mins at 70 °C then 62.5% methanol for a further 15 mins at 55 °C. Results showed that the methanol extraction contained higher fructose, glucose, sucrose, kestose, nystose and individual fructans of degrees of polymerisation from 5–8. Although both Davis *et al.* (2007) and Downes and Terry (2010) have found methanol to be a superior extraction solution, many studies have been published since 2007 which still use ethanol to extract sugars from onion tissue (Rodríguez-Galdón 2009; Vågen and Slimestad 2009).

15.3   Quantification

15.3.1   Quantification of Total Sugars and Fructans

In addition to the extraction of fructans for individual quantification, methods for the quantification of total fructans have also been reported. Total fructan extracts can be mixed with an acid (1M HCl) to hydrolyse fructans into their individual units (Yasin and Bufler 2007). Comparison between the sugar content before and after hydrolysis gives an indication of the total fructan content. The hydrolysis of the glycosyl bonds in fructans can also be separated enzymatically. The enzyme inulinase which hydrolyses the β(1-2)-linked fructose chains in inulin, was added to fructans suspended in water then incubated at 57.5 °C for 30 mins (Rodriguez Galdon *et al.* 2009).

Total soluble solid (TSS) quantification, using a refractometer, can give an indication of the total amount of sugars in a vegetable juice sample. This method quantifies the amount of dissolved solids, which include sugars, proteins and organic acids and is therefore not an accurate method for total sugar content (Chope *et al.* 2006). Crowther *et al.* (2005) found that TSS measurement did not relate to the perceived sweetness of some onion cultivars.

15.3.2 Quantification using High Pressure Liquid Chromatography

High Pressure Liquid Chromatography (HPLC) is used extensively to quantify sugars in vegetable matter. O'Donoghue *et al.* (2004) used a Rezex RCM monosaccharide column (Phenominex, Torrance, CA) with an isocratic water mobile phase to elute monosaccharides and disaccharides. The advantages of this column is that separation can be achieved using just a water mobile phase, therefore avoiding costly disposal of solvents. Also, the run time of each sample can be as short as 15 minutes, resulting in a high turnover of sample numbers. Disadvantages of this column is that the peak shapes are particularly wide and therefore peak overlap can be a problem depending on the number of monosaccharides in the vegetable matter. To reduce the pressure in the system when using a Rezex monosaccharide column (maximum pressure 40 bar) and to rapidly elute the compounds, a high operating column temperature is used (70–80 °C). Slimestad and Vågen (2006) found that column temperature had a profound effect on the detection of glucose using a Prevail Carbohydrate ES column (Alltech, UK). Increasing the column temperature from 25 °C to 70 °C suppressed the glucose signal of glucose by $> 98\%$. Even an increase from 25 °C to 45 °C reduced the glucose signal by 41%. The same experiment was conducted on an amino-bonded silica column yet the same trend was not observed, so this phenomenon appears to be column specific. The Prevail Carbohydrate ES column is a very good column for the quantification of sugars and fructans as fructans of different degrees of polymerisation can be separated up to approximately 9 units.

Acetonitrile is a common mobile phase for the elution of sugars from vegetable matter and can be used in conjuncture with the Prevail Carbohydrate ES (Vågen and Slimestad 2008). In 2009, in the light of the global economic problems, plastic production declined resulting in a world-wide shortage of acetonitrile, a co-product of acrylonitrile production. Downes and Terry (2010) developed an alternative method replacing the mobile phase acetonitrile with ethanol. Although it was possible to separate all onion fructans, it was not possible to separate fructose and sucrose. The new method was not an improvement on the original method, however, it does represent an alternative if, for whatever reason, acetonitrile is not readily available in the future.

15.3.3 Quantification using Anion Exchange Chromatography

High performance/pH anion exchange chromatography (HPAEC) involves extracting in an alkaline solution using calcium hydroxide ($Ca(OH)_2$;

Ernst *et al.* 1998) or calcium carbonate (CaCO$_3$; Shiomi *et al.* 1997) which converts the hydroxyl groups on the oligosaccharides into oxyanions (Ernst *et al.* 1998). Making the carbohydrate molecules negatively charged allows them to interact with the positively charged stationary phase. Using an increasing gradient of a competing ion such as acetate or nitrate, separation of smaller, more weakly charged carbohydrates to highly charged polysaccharides can be achieved. This method is very effective for the separation of different anomeric and positional isomers of simple linear inulins and neo-series inulins (Lee 1996). Schütz *et al.* (2006) quantified fructans in artichoke using HPAEC-pulsed amperometric detector (PAD) and successfully separated monomers, oligomers and polymers up to 79 degrees of polymerisation in one run. The column used was a CarboPac PA-100 operated at 25 °C and a gradient mobile phase of increasing completing ion 500mM NaOAc in 225mM NaOH.

15.3.4 Gas Chromatography

The quantification of sugars using gas chromatography (GC) requires the preparation of volatile derivatives. A method described by Sweeley *et al.* (1963) details the conversion of sugars into trimethylsilyl (TMS) derivatives which can then be analysed by GC. Salama *et al.* (1990) extracted sugars from onion powder using 80% (v/v) ethanol at 50°C and then obtained TMS derivatives using the above described method. The sugar concentrations obtained by Salama *et al.* (1990) are in the range of those reported using HPLC (Cools *et al.* 2011).

15.3.5 Detectors

Sugars and fructans do not absorb UV or visible wavelengths, therefore they cannot be detected using a photodiode array. Sugars and fructans can be measured using an evaporative light scattering detector (ELSD) which nebulises the sample and evaporates the mobile phase. The amount of light scattered as a result of particles in the nebulised sample is then detected and presented as a peak. The ELSD does not produce a linear calibration curve but rather a sigmoidal or exponential curve (Mathews *et al.* 2004). Therefore, either a non-linear curve is applied or the log10 of the both the peak area and calibration concentrations can be calculated to achieve a linear curve (Downes and Terry 2010). Another popular detection method for sugars and fructans is the PAD. PAD requires alkaline conditions therefore is often coupled with anion exchange chromatography which also requires high alkalinity (Zook and LaCourse 1995).

Refractive index detection (RID) is another detector that can be employed to detect sugars and fructans. RID is not as sensitive as ELSD, however, the advantage is that the sample is not destroyed when using an RID therefore further analysis can be performed on the eluted sample (*e.g.* fraction collection). Other disadvantages with the RID includes drift in baseline when using a gradient; Foukaraki *et al.* (2010; 2011) used an RID to detect sugars in potato, however, baseline drift was not a problem as an isocratic HPLC water mobile phase was adopted with a Rezex monosaccharide column.

Davis *et al.* (2007) measured fructan content with matrix-assisted laser desorption ionisation – time of flight (MALDI-TOF) originally described by Wang *et al.* (1999). Comparison between extraction methods (Davis *et al.* 2007) was investigated with onion powder extracted in 62.5% methanol at 55 °C for 15 mins (O'Donoghue *et al.* 2004) having better resolution of fructans compared with samples extracted in 80% (v/v) ethanol at 70 °C for 2 h (Viola and Davies 1992).

15.4 Conclusions

In conclusion, many methods for the extraction of sugars and fructans have been reported in vegetable matter, although more research is required on the most efficacious method, so that sugar and fructan extraction can be standardised. Comparisons between extraction methods have been reported in onion and methanol has been found to be superior to ethanol as an extraction solvent; however, this has not been applied to quantification using HPAEC. Quantification of sugars and fructans can be achieved using GC, normal phase HPLC and HPAEC. This said, one of the most effective methods for the separation of individual FOS anomeric and positional isomers and fructans of high degrees of polymerisation remains HPAEC coupled with a PAD.

Summary Points

- Sugars can be classed as mono- (one unit), oligo- (2 or more units; short chain), or polysaccharides (long chain).
- Sugars can be loosely categorised into reducing and non-reducing sugars. Reducing sugars yield an aldehyde group when in the open chain form.
- Fructooligosaccharides (FOS) are short chain fructans for example kestose, nystose and fructofuranosylnystose.
- Fructans usually collectively refers to FOS and long chain polysaccharides.
- Sucrose:sucrose 1-fructosyltransferase (1-SST) catalyses the production of 1-kestose from sucrose and fructose. Fructan:fructan 1-fructosyltransferase (1-FFT) is required for longer chain fructans whereas fructan:fructan G6-fructosyltransferase (G6-FFT) is required for more complex fructans such as 6-kestose and neokestose.
- Methanol solution has been shown to be a more efficacious solvent for the extraction of sugars and fructans compared with ethanol.
- Sugars and fructans can be separated using high pH/pressure anion exchange chromatography (HPAEC), normal phase chromatography and gas chromatography.
- Sugars and fructans can be quantified using pulsed amperometric detection (PAD), refractive index detection (RID), evaporative light scattering detection (ELSD) or mass spectrometry (MS).

Definitions of Words and Terms

DP. The degree of polymerization is the number of monomeric units in an oligomer or polymer.

ELSD. An evaporative light scattering detector (ELSD) is used in conjunction with an HPLC and detects compounds with little or no UV absorption. It measures photons scattered due to particles which remain following evaporation of the mobile phase.

FID. A flame ionisation detector is used in conjunction with GC. Particles are ionised by a hydrogen air flame and these ions are then detected.

FOS. Fructooligosaccharides are fructans with low degrees of polymerization.

MALDI-TOF. Matrix assisted laser desorption/ionisation – time of flight is an ionisation technique in conjunction with mass spectrometry. Desorption of the sample is carried out using a UV laser which is highly absorbed on a matrix surface. The sample is then ionised and passed to the TOF where ions are identified based on their velocity which is dependent on mass-to-charge ratio.

PAD. A pulsed amperometric detector is used in conjunction with HPLC. Compounds are detected on the surface of an electrode. Different potentials are applied as pulses which clean the electrode.

RID. A refractive index detector is used to measure compounds which have little or no UV absorption and is used with HPLC. The increase in refractive index between the sample and mobile phase results in an imbalance in the flow cell creating a signal.

Key Facts of High Pressure Liquid Chromatography

- HPLC is a technique used to separate compounds in a mixture which can then be detected, identified, quantified and/or purified.
- Compounds are passed through a column which can contain different **stationary phases** depending on the chemistry of the target compounds.
- The compounds are pumped through the column by a **mobile phase** which usually consists of a solvent/water mix.
- Compounds are eluted from the column at different **retention times** which is dependent on the stationary phase, mobile phase and flow rate.
- A compound is detected as a chromatographic peak, the area of which is an indicator of concentration when compared with a standard curve of known concentration.
- Unknown compounds can be identified through mass spectroscopy.

Abbreviations

1-FFT	fructan:fructan 1-fructosyltransferase
1-SST	sucrose:sucrose 1-fructosyltransferase
$CaCO_3$	calcium carbohydrate
$Ca(OH)_2$	calcium hydroxide

CHOs	carbohydrates
DM	dry matter
DP	degrees of polymerization
DRD	differential refractometer detector
ELSD	evaporative light scattering detector
F	fructose
FD	freeze-dried
FEH	fructan exohydrolase
FID	flame ionisation detector
FOS	fructooligosaccharide
G	glucose
G6-FFT	fructan:fructan G6-fructosyltransferase
GC	gas chromatography
H	High dry matter
HCl	hydrochloric acid
HPAEC-PAD	High performance anion exchange chromatography – pulsed amperometric detector
HPLC	high performance liquid chromatography
L	Low dry matter;
MALDI-TOF	Matrix assisted laser desorption ionisation – time of flight
NaOAc	sodium acetate
NaOH	sodium hydroxide
nd	not detected
NSC	non-structural carbohydrates
RID	refractive index detector
S	sucrose
PAD	pulsed amperometric detector
PED	pulsed electrochemical detector
PLK	Pukehole Longkeeper
RID	refractive index detector
TFA	trifluoroacetic acid
TMS	trimethylsilyl
tr	trace amount
UV	ultraviolet

References

Benkeblia, N., and Selselet-Attou, G., 1999. Effects of low temperatures on changes in oligosaccharides, phenolics and peroxidase in inner bud of onion *Allium cepa* L. during break of dormancy. *Acta Agriculturæ Scandinavica - Section B* 49: 98–102.

Benkeblia, N., Varoquaux, P., Shiomi, N., and Sakai, H., 2002. Storage technology of onion bulbs cv. Rouge Amposta: effects of irradiation, maleic hydrazide and carbamate isopropyl, *N*-phenyl (CIP) on respiration rate and carbohydrates. *International Journal of Food Science and Technology* 37: 169–175.

Benkeblia, N., and Varoquaux, P., 2003. Effect of nitrous oxide (N_2O) on respiration rate, soluble sugars and quality attributes of onion bulbs *Allium cepa* cv. Rouge Amposta during storage. *Postharvest Biology and Technology* 30: 161–168.

Benkeblia, N., and Shiomi, N., 2004. Chilling effect on soluble sugars, respiration rate, total phenolics, peroxidase activity and dormancy of onion bulbs. *Scientia Agricola* 61: 281–285.

Benkeblia, N., Onodera, S., and Shiomi, N., 2004a. Effect of gamma irradiation and temperature on fructans (fructo-oligosaccharides) of stored onion bulbs *Allium cepa* L. *Food Chemistry* 87: 377–382.

Benkeblia, N., Onodera, S., Yoshihira, T., Kosaka, S., and Shiomi, N., 2004b. Effect of temperature on soluble invertase activity, and glucose, fructose and sucrose status of onion bulbs (*Allium cepa*) in store. *International Journal of Food Sciences & Nutrition* 55: 325–331.

Benkeblia, N., Onodera, S., and Shiomi, N., 2005a. Variation in 1-fructo-exohydrolase (1-FEH) and 1-kestose-hydrolysing (1-KH) activities and fructo-oligosaccharide (FOS) status in onion bulbs. Influence of temperature and storage time. *Journal of the Science of Food and Agriculture* 85: 227–234.

Benkeblia, N., Ueno, K., Onodera, S., and Shiomi, N., 2005b. Variation of fructooligosaccharides and their metabolizing enzymes in onion bulb (*Allium cepa* L. cv. Tenshin) during long-term storage. *Journal of Food Science* 70: S208–S214.

Bielecka, M., Biedrzycka, E., and Majkowska, A., 2002. Selection of probiotics and prebiotics for synbiotics and confirmation of their *in vivo* effectiveness. *Food Research International* 35: 125–131.

Brewster, J.L., 2008. *Onions and Other Vegetable Alliums*, 2nd edn. Cab International, Wallingford, UK.

Chope, G.A., Terry, L.A., and White, P.J., 2006. Effect of controlled atmosphere storage on abscisic acid concentration and other biochemical attributes of onion bulbs. *Postharvest Biology and Technology* 39: 233–242.

Chope, G.A., Terry, L.A., and White, P.J., 2007a. The effect of the transition between controlled atmosphere and regular atmosphere storage on bulbs of onion cultivars SS1, Carlos and Renate. *Postharvest Biology and Technology* 44: 228–239.

Chope, G.A., Terry, L.A., and White, P.J., 2007b. The effect of 1-methylcy-clopropene on the physical and biochemical characteristics of onion cv. SS1 during storage. *Postharvest Biology and Technology* 44: 131–140.

Chope, G.A., Cools, K., Hammond, J.P., Thompson, A.J., and Terry, L.A., 2012. Physiological, biochemical and transcriptional analysis of onion bulbs during storage. *Annals of Botany* 109: 819–831.

Cools, K., Chope, G.A., Hammond, J.P., Thompson, A.J., and Terry, L.A., 2011. Ethylene and 1-MCP differentiallt regulate gene expression during onion (*Allium cepa* L.) sprout suppression. *Plant Physiology*, 156: 1639–1652.

Darbyshire, B., and Henry, R.J., 1978. The distribution of fructans in onions. *New Phytologist* 81: 29–34.

Davis, F., Terry, L.A., Chope, G.A., and Faul, C.F.J., 2007. Effect of extraction procedure on measured sugar concentrations in onion (*Allium cepa* L.) bulbs. *Journal of Agricultural and Food Chemistry* 55: 4299–4306.

Downes, K., Chope, G.A., and Terry, L.A., 2009. Effect of curing at different temperatures on biochemical composition of onion (*Allium cepa* L.) skin from three freshly cured and cold stored UK-grown onion cultivars. *Postharvest Biology and Technology* 54: 80–86.

Downes, K., and Terry, L.A., 2010. A new acetonitirle-free mobile phase for LC-ELSD quantification of fructooligosaccharides in onion (*Allium cepa* L.). *Talanta* 82: 118–124.

Downes, K., Chope, G.A., and Terry, L.A., 2010. Postharvest application of ethylene and 1-methylcyclopropene either before or after curing affects onion (*Allium cepa* L.) bulb quality during long term cold storage. *Postharvest Biology and Technology* 55: 36–44.

Ernst, M.K., Chatterton, N.J., Harrison, P.A., and Matitschka, G., 1998. Characterization of fructan oligomers from species of the genus *Allium* L. *Journal of Plant Physiology* 153: 53–60.

Ernst, M.K., Praeger, U., and Weichmann, J., 2003. Effect of low oxygen storage on carbohydrate changes in onion (*Allium cepa* var. *cepa*) bulbs. *European Journal of Horticultural Science* 68: 59–62.

Foukaraki, S.G., Chope, G.A., and Terry, L.A., 2010. Ethylene exposure after dormancy break is as effective as continuous ethylene to control sprout growth in some UK grown potato cultivars. *28th International Horticultural Congress* 22–27th August 2010, Lisbon, Portugal.

Foukaraki, S.G., Chope, G.A., and Terry, L.A., 2011. 1-MCP application before continuous ethylene storage suppresses sugar accumulation in the UK grown potato cv. Marfona. *4th Postharvest Unlimited* 23–26th May 2011, Leavenworth, WA, USA.

Gennaro, L., Leonardi, C., Esposito, F., Salucci, M., Maiani, G., Quaglia, G., and Fogliano, V., 2002. Flavonoid and carbohydrate contents in Tropea red onions: Effects of homelike peeling and storage. *Journal of Agriultural and Food Chemistry* 50: 1904–1910.

Hansen, S.L., 1999. Content and composition of dry matter in onion (*Allium cepa* L.) as influenced by developmental stage at time of harvest and long-term storage. *Acta Agriculturæ Scandinavica - Section B* 49: 103–109.

Hermanson, G.T., 2008. *Bioconjugate Techniques*. Acedemic Press, London, UK, 36–49.

Jaime, L., Martinez, F., Martin-Cabrejas, M.A., Molla, E., Lopez-Andreu, F.J., Waldren, K.W., and Esteban, R.M., 2000. Study of total fructan and fructooligosaccharide content in different onion tissues. *Journal of the Science of Food and Agriculture* 81: 177–182.

Jaime, L., Martin-Cabrejas, M.A., Molla, E., Lopez-Andreu, F.J., and Esteban, R.M., 2001. Effect of storage on fructan and fructooligosaccharide of onion (*Allium cepa* L.). *Journal of Agriultural and Food Chemistry* 49: 982–988.

Kahane, R., Vialle-Guérin, E., Boukema, I., Tzanoudakis, D., Bellamy, C., Chamaux, C., and Kik, C., 2001. Changes in non-structural carbohydrate composition during bulbing in sweet and high-solid onions in field experiments. *Environmental and Experimental Botany* 45: 73–83.

Lee, C.Y., Shalleberger, R.S., and Vittum, M.T., 1970. Free sugars in fruits and vegetables. *Food Sciences: Food Science and Technology* In: New York's Food and Life Sciences Bulletin. 1: 1–12.

Lee, Y.C., 1996. Carbohydrate analyses with high-performance anion-exchange chromatography. *Journal of Chromatography A* 720: 137–149.

Mathews, B.T., Higginson, P.D., Lyons, R., Mitchell, J.C., Sach, N.W., Snowden, M.J., Taylor, M.R., and Wright, A.G., 2004. Improving quantitative measuremeents for the evaporative light scattering detector. *Chromatographia* 60: 625–633.

Martin, S., and Hartmann, H., 1989. The content and distribution of the carbohydrates in asparagus. *Acta Horticulturae* 271: 443–449.

Muir, J.G., Shepherd, S.J., Rosella, O., Rose, R., Barett, J.S., and Gibson, P.R., 2007. Fructan and free fructose content of some common Australian vegetables and fruit. *Journal of Agricultural and Food Chemistry* 55: 6619–6627.

Muir, J.G., Rose, R., Rosella, O., Liels, K., Barrett, J.S., Shepherd, S.J., and Gibson, P.R., 2009. Measurement of short-chain carbohydrates in common Australian vegetables and fruits by high-performance liquid chromatography (HPLC). *Journal of Agricultual and Food Chemistry* 57: 554–565.

Pak, C., Vanderplas, L.H.W., and Deboer, A.D., 1995. Importance of dormancy and sink strength in sprouting of onions (*Allium cepa* L.) during storage. *Physiologia Plantarum*, 277–283.

Ritsema, T., and Smeekens, S., 2003. Fructans: beneficial for plants and humans. *Current Opinion in Plant Biology* 6: 223–230.

Roberford, M.B., 2007. Inulin-type fructans: Functional food ingredients. *The Journal of Nutrition* 137: 2493S–2502S.

Rodríguez Galdón, B., Tascón Rodríguez, C., Rodríguez Rodríguez, E.M., and Díaz Romero, C., 2009. Fructans and major compounds in pnion compounds (*Allium cepa* L.). *Journal of Food Composition and Analysis* 22: 25–32.

O'Donoghue, E.M., Somerfield, S.D., Shaw, M., Bendall, M., Hedderly, D., Eason, J., and Sims, I., 2004. Evaluation of carbohydrates in Pukehohl Longkeeper and Grano cultivars of *Allium cepa*. *Journal of Agricultural and Food Chemistry* 52: 5383–5390.

Pontis, H.G., 1990. Fructans In: Dey, P.M., Dixon, R.A. (eds) *Biochemstry of storage carbohydrates in green plants*. Academic Press, New York, USA. 205–227.

Salama, A.M., Hicks, J.R., and Nock, J.F., 1990. Sugar and organic acid changes in stored onion bulbs treated with maleic hydrazide. *HortScience* 25: 1625–1628.

Sánchez-Mata, M.C., Cámara-Hurtado, M., and Díez-Marqués, C., 2002. Identification and quantification of soluble sugars in green beans by HPLC. *European Food Research and Technology* 214: 254–258.

Schütz, K., Muks, E., Carle, R., and Schieber, A., 2006. Separation and quantification of inulin in selected artichoke (*Cynara scolymus* L.) cultivars and dandelion (*Taraxacum officinale* WEB. ex WIGG.) roots by high-performance anion exchange chromatography with pulsed amperometric detection. *Biomedical Chromatography* 20: 1295–1303.

Shiomi, N., Onodera, S., and Sakai, H., 1997. Fructo-oligosaccharide content and fructosyltransferase activity during growth of onion bulbs. *New Phytologist* 136: 105–113.

Slimestad, R., and Vågen, I.M., 2006. Thermal stability of glucose and other sugar aldoses in normal phase high performance liquid chromatography. *Journal of Chromatography A* 1118: 281–284.

Sweeley, C.C., Bentley, R., Marita, M., and Wells, W.W., 1963. Gas-liquid chromatography of trimethylsilyl derivatives of sugars and related substances. *Journal of the American Chemistry Society* 85: 2497–2507.

Terry, L.A., Law, K.A., Hipwood, K.J., and Bellamy, P.H., 2005. Non-structural carbohydrate profiles in onion bulbs influence taste preference. *Fructic '05* 12–16th September 2005, Montpellier, France.

Terry, L.A., 2011. *Health Promoting Properties of Fruits and Vegetables*. CABI, Wallingford, Oxfordshire, UK.

Vågen, I.M., and Slimestad, R., 2008. Amount of characteristic compounds in 15 cultivars of onion (*Allium cepa* L.) in controlled field trials. *Journal of the Science of Food and Agriculture* 88: 404–411.

Viola, R., and Davies, H.V., 1992. A microplate reader assay for rapid enzymatic quanitification of sugars in potato tubers. *Potato Research* 35: 55–58.

Wang, J., Sporns, P., and Low, N.H., 1999. Analysis of food oligosaccharides using MALDI-MS. *Journal of Agricultural and Food Chemistry* 47: 1549–1557.

Winter, H., and Huber, S.C., 2000. Regulation of sucrose metabolism in higher plants: localization and regulation of activity of key enzymes. *Critical Reviews in Biochemistry and Molecular Biology* 35: 253–289.

Yasin, H.J., and Bufler, G., 2007. Dormancy and sprouting in onion (*Allium cepa* L.) bulbs. I. Changes in carbohydrate metabolism. *Journal of Horticultural Science and Biotechnology* 82: 89–96.

Zook, C.M., and LaCourse, W.R., 1995. Pulsed amperometric detection of carbohydrates in fruit juices following high performance anion exchange chromatography. *Current Separations* 14: 48–52.

CHAPTER 16

Determination of Dietary Sugars by Ion Chromatography and Electrochemical Detection: a Focus on Galactose, Glucose, Fructose and Sucrose

DONATELLA NARDIELLO, CARMEN PALERMO, MAURIZIO QUINTO AND DIEGO CENTONZE*

Department of Agro-Environmental Sciences, Chemistry and Plant Defense, Inter-departmental Research Center BIOAGROMED, University of Foggia, Via Napoli 25, 71100 Foggia, Italy
*Email: centonze@unifg.it

16.1 Analysis of Carbohydrates: Historical Background

Carbohydrates play a key role in physiological and pathological processes, as energetic and information-carrying molecules and consequently, their determination is relevant in the clinical field. Furthermore, the analytical determination of dietary sugars is essential in biotechnology and in the analysis of food, beverages and pharmaceuticals for patients with heart or metabolic diseases, as well as for consumers on a low-calorie diet. Unfortunately, for years the analysis of carbohydrates has been challenging due to the lack of

Food and Nutritional Components in Focus No. 3
Dietary Sugars: Chemistry, Analysis, Function and Effects
Edited by Victor R Preedy
© The Royal Society of Chemistry 2012
Published by the Royal Society of Chemistry, www.rsc.org

chromophore and fluorophore groups, and difficulties in both their separation and detection have taken over. The determination of saccharides has been accomplished by high-performance liquid chromatography, which includes derivatization followed by UV-visible absorption or fluorescence emission detection of the corresponding derivatives. Nevertheless, either pre- or post-column derivatization suffer from several disadvantages, such as relatively long analysis times, low stability of sugar-derivatives, formation of multiple adducts, and reagent interference. The chromatographic approaches for the under-ivatized glucidic compounds are based on indirect UV detection, low-wave-length UV detection or refractive index detection, using silica-based amino bonded or polymer-based metal-loaded cation exchange columns. Refractive index and low-wavelength UV detection methods are sensitive to eluent and matrix composition, thus precluding the use of gradients and requiring exten-sive sample clean-up before injection. Capillary electrophoresis has also been used for the determination of underivatized carbohydrates, coupled to UV detection at 195 nm or indirect absorption. Anyway, problems associated with incomplete separation or peak tailing and detection drawbacks at low con-centration levels are not easily overcome.

In 1974, there was a key turning-point with the successful marriage between ion chromatography and electrochemical detection, represented by the intro-duction of the first electrochemical detector for HPLC. Ever since, the com-bination of such techniques has promoted the development of new analytical methods for the determination of carbohydrates in a variety of complex matrices, from biological fluids to plant and food samples (Andersen and Sørensen 2000; Bernal *et al.* 1996; Cai *et al.* 2005; Cataldi *et al.* 1998b; Cataldi *et al.* 2000a; Cataldi *et al.* 2000b; Cataldi *et al.* 2003; Corradini *et al.* 2004; Kaine and Wolnik 1998). Within a few years, the high-performance anion-exchange chromatography (HPAEC) coupled with pulsed amperometric detection (PAD) has become the method of choice to detect the glucidic compounds. Nowadays, the most widely used detection system for carbohy-drates is still based on electrochemical detection, although the recent scientific activity is generally focused on mass spectrometry and also for the glucidic compounds some papers have been reported. In addition to the direct deter-mination of sugars without the necessity of derivatization, electrochemical techniques are characterized by lower detection limits at less equipment cost, all the while ensuring good detection selectivity.

16.2 Analysis of Carbohydrates by Ion Chromatography and Electrochemical Detection

16.2.1 Chromatographic Separation of Galactose, Glucose, Fructose and Sucrose by HPAEC

The chromatographic separation of sugars take plays by anion exchange mechanisms which exploit the weakly acidic nature of carbohydrates

(Cataldi *et al.* 2000a) that, at high pH, are partially ionized (pk$_a$ of fructose, glucose, galactose and sucrose in water at 25 °C are 12.03, 12.28, 12.39, and 12.62, respectively). A series of anion-exchange columns, with polymer-based stationary phases, have been specifically designed by Dionex for the separation and analysis of saccharides (series CarboPac). The CarboPac PA1, packed with a nonporous and pellicular resin based on polystyrene/divinylbenzene and functionalized with alkyl quaternary ammonium group, is dedicated to the rapid analysis of monosaccharides, disaccharides and some oligosaccharides, thus resulting suitable for the separation of glucose, fructose, galactose and sucrose. The CarboPac MA1 column, packed with a macroporous polymeric resin with higher ion exchange capacity and specifically recommended for the determination of alditols, also generates excellent neutral monosaccharide separations, although retention times are longer than on the PA1. Examples of sugar separations are available in the Applications section of Dionex Technical Notes (Dionex Corporation c). Polymeric anion exchange columns (Hamilton RCX-30, Metrosep Carb 1-250 or 1-150) supplied by Methrom are also commercially available for isocratic and gradient separation of simple and complex carbohydrates. Retention of carbohydrate can be manipulated by changing the sodium hydroxide concentration of the eluent (pH 12–13), which influences the elution order. Low ionic strenght eluents (less than 20 mM, usually 10–16 mM) in isocratic conditions are required for the simultaneous determination of closely-related monosaccahrides, exhibiting very similar retention behavior, to improve the separation selectivity. The quality of an anion-exchange separation strongly depends on the alkaline running eluent, which typically contains an unknown amount of carbonate, coming from atmospheric carbon dioxide. Carbonate ion occupies the anion-exchanging sites of the column, thus interfering with the retention of sugar anions and causing a progressive decrease of retention on successive injections. To minimize carbonate contamination and guarantee good retention time repeatability, which is essential for a correct peak assignment, several strategies have been adopted. First of all, it is fundamental to prepare daily fresh eluents with carbonate-free 50% (w/w) sodium hydroxide solutions and, during the chromatographic runs, it is also recommended to keep an inert gas atmosphere (N$_2$ or He) on the eluent solution. Nevertheless, these precautions are insufficient to totally avoid carbonate contamination, above all when dilute sodium hydroxide concentrations are employed. In order to put the column resin in the hydroxide form and restore the ion-exchange capacity, a possible solution is a post-column addition of 200 mM NaOH, after each run, followed by a column equilibration step before injection, with a consequent increase of the total analysis time. Otherwise, it is possible to use an on-line electrochemically alkaline eluent generator for producing KOH based eluents, through the water electrolysis, using a dedicated and commercially available system supplied by Dionex (Dionex Corporation b). Another very practical solution is represented by the introduction of barium or strontium salts (usually 1–2 mM barium or strontium acetate) in the mobile phase for the chemical suppression of carbonate ions (Cataldi *et al.* 1997b). These divalent cations are able to form carbonate salts (BaCO$_3$ and SrCO$_3$), which are weakly soluble and precipitate in the bottom of the eluent container, thus providing a chemical removal of carbonate ions, without the need of filtering the solution. The presence of Ba(II) or Sr (II) in the mobile phase also influences the selectivity of separation, in addition to the reproducibility of retention, as confirmed by repeated injections of several monosaccharides and alditols. No extensive re-equilibration step after each run is required and the column is flushed for 30–60 min with the carbonate-free 200 mM NaOH solution, just at the beginning of each working day, before equilibration. It is strongly recommended to make the addition of barium or strontium acetate a few hours or the day before using the eluent solution. Furthermore, at the end of each working session the pump system should be washed with 0.5 HCl and then with water to avoid problems with piston wear and seals (Cataldi *et al.* 1998a).

16.2.2   Pulsed Amperometric Detection: Triple Step PAD Waveform for Galactose, Glucose, Fructose and Sucrose

The most relevant aspects of pulsed amperometry are fully described by LaCourse (LaCourse 1997) in a specific book, and in several articles, where the background of the electrochemical detection of carbohydrates was reported (LaCourse and Johnson 1991; LaCourse and Johnson 1993). Amperometric detection of carbohydrates is performed at a thin-layer detection cell, consisting of a gold working electrode and a Ag|AgCl reference electrode with the cell body, serving as the counter electrode. At high pH values, carbohydrates are electrocatalytically oxidized at the surface of the gold electrode by the application of a positive potential. The current generated is proportional to the analyte concentration, but this oxidation process at constant potential is characterized by the fouling of the electrode surface due to the by-products' adsorption. Consequently, for a stable and reproducible signal response, the fouled electrode surface need to be periodically cleaned, prior to each measurement. Removal of adsorbed by-products and electrode conditioning is performed by the pulsed amperometric detection, developed by Johnson and co-workers (LaCourse and Johnson 1991; LaCourse and Johnson 1993; LaCourse 1997), which uses a multistep potential waveform (a series of potentials applied for defined periods of time) to detect analytes and to clean and reactivate the electrode surface, with a repeated sequence performed at a frequency of 0.5–2.0 Hz. The well-established potential waveform (See Figure 16.1, panel A) for the electro-oxidation of carbohydrates on gold working electrodes consists of 3 steps: the direct oxidation of saccharides at the oxide-free electrode surface by applying a detection potential (E_{DET}) in the range 0–200 mV vs Ag|AgCl, used as reference electrode; then the oxidative cleaning by raising the potential to 650–800 mV (oxidation potential, $E_{OX} \gg E_{DET}$) to fully oxidize the gold surface and to cause desorption of sugar oxidation by-products; finally, the reductive restoration of the oxide-free surface by lowering the potential at about −300 mV (reduction potential, E_{RED}) where the

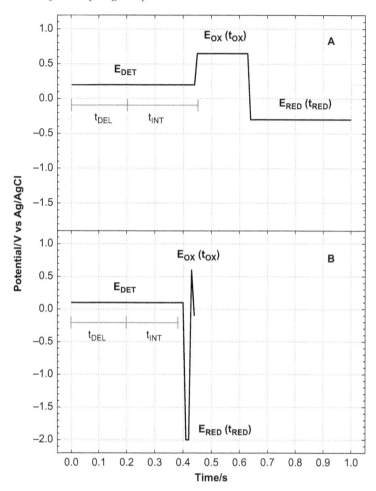

Figure 16.1 Triple step (A) and quadruple step (B) potential waveform for the amperometric detection of galactose, glucose, fructose and sucrose at a gold working electrode. A positive potential (E_{DET}) is applied to promote the oxidation of carbohydrates along the surface of a gold working electrode. The current is measured only during a short sampling interval (t_{INT}), after a delay period (t_{DEL}), that allows the charging current originating from the potential impulse to decay. Subsequently, the electrode surface is reactivated by the application of a repeating potential sequence. (A) Triple step waveform. The electrical potential is stepped to a more positive value (E_{OX}) to fully oxidize the gold surface and desorb the contaminants. The potential is then stepped to a negative value (E_{RED}) to reduce the gold oxide, thereby regenerating the original clean working electrode surface.
(B) Quadruple step waveform. Following the detection period, the potential is dropped to a highly negative value (E_{RED}) ensuring electrode cleansing, while shortening the duration of the subsequent high electrical potential that is necessary to oxidize the surface (E_{OX}) and regenerate the electrode surface.
Unpublished.

gold-oxide film is striped away. The three potentials are applied for fixed intervals of time called t_{DET}, t_{OX}, and t_{RED}, respectively. The carbohydrate oxidation current is measured by integrating over time (t_{INT}) after a delay (t_{DEL}) that allows the charging current due to the application of potential impulse to decay.

16.2.2.1 Optimisation of Potential-Time Settings for the Pulsed Amperometric Detection of Carbohydrates

The traditional method used for studying redox systems and establishing potential-time settings is cyclic voltammetry (CV), in which a triangular waveform is cycled between two potential limit values. It is well recognized that electrochemical detection of carbohydrates on a gold working electrode is most favourable under alkaline conditions, which, moreover, are essential for the chromatographic separation of sugars. The voltammetric behaviour of sugar compounds has been widely described by Johnson and LaCourse (LaCourse and Johnson 1993). In alkaline solutions based on NaOH, in the absence of analyte, the positive scan is characterized by an anodic wave in the region of ca. 0.2 to 0.7 V due to the gold oxide formation, followed by the oxygen evolution at potentials greater than 0.9 V. A cathodic peak is observed on the negative scan at about 0.1 V corresponding to the reduction of the oxide formed during the positive scan. When glucose is present, typical current–potential (i–E) curves show an oxidation peak at about 0.25 V, which is also the potential corresponding to the onset of gold oxide. The glucose oxidation current decreases as gold oxidation increases, demonstrating the inhibition of glucose oxidation by gold oxide. This inhibiting effect is also confirmed by the absence of signal in the range 200–800 mV on the reverse scan. It is more important to choose a determination potential that gives maximum response for carbohydrates with little or no gold oxide formation. Following cathodic dissolution of the gold oxide at 0.1 V, the reactivity of the electrode surface is renewed (LaCourse and Johnson 1993).

Within a pulsed waveform, duration and potential values of each step may significantly affect the sensitivity and reproducibility of response. Therefore, the waveform optimisation should be performed by hydrodynamic voltammetry, injecting sugar standard solutions and plotting the current signal as a function of the parameter to be optimised (duration and potential values), since the optimal values cannot be accurately chosen just on the basis of the CV data. For the electrochemical detection of glucose, fructose and galactose, the maximum response in terms of signal-to-noise usually occurs in the range 0.0–0.2 V. The oxidation potential (E_{OX}) must be high and long enough to achieve a full surface oxidation with consequent desorption of adsorbed impurities, reactants, and detection products; this potential cannot be too high in order to avoid O_2 evolution. Then, the third potential, E_{RED}, must be quite low to give a rapid and complete reduction of gold oxide on the electrode surface. Optimal potential-time settings for the amperometric detection of glucose, fructose, galactose and sucrose are detailed in Table 16.1.

Table 16.1 Potential-time settings of the triple-step PAD waveform for the determination of galactose, glucose, fructose and sucrose.[a]

Potential (V vs. Ag\AgCl)		Time (ms)	
Parameter	Optimised value	Parameter	Optimised value
E_{DET}	0 to 0.25	t_{DET}	400–500
		t_{DEL}	150–200
		t_{INT}	200–250
E_{OX}	0.65 to 0.8	t_{OX}	80–190
E_{RED}	−0.15 to −0.9	t_{RED}	50–390

[a]The amperometric detection is performed in a 3-electrode flow-through detection cell, consisting in a gold working electrode, a Ag\AgCl reference electrode and a titanium counter electrode. At the surface of the gold electrode, carbohydrates are electrocatalytically oxidized by a positive potential (E_{DET}) applied for a total time period of t_{DET}, with digital sampling of the amperometric current during the period t_{INT}, following a holding period of t_{DEL}, that allows the charging current due to the application of potential impulse to decay. The electrode surface is renewed within a pulsing potential waveform (applied potential vs. time) that continually cleans and reactivates the working electrode by an oxidative cleaning (E_{OX}) to cause desorption of sugar oxidation products and a reductive restoration (E_{RED}) to strip away the formed gold-oxide film. Unpublished.

Very recently, the use of a 4-step potential waveform (Rocklin *et al.* 1998; Jensen and Johnson 1997) for the electrochemical detection of carbohydrates on gold working electrodes (Figure 16.1B) has been suggested (Dionex Corporation a). Recommended pulse sequence is characterized by negative rather than positive potentials for the oxidative electrode cleaning, ensuring a higher data collection rate (2 Hz) and improving the long-term reproducibility of the response. Nevertheless, compared to the classical 3-step waveform, the 4 potential sequence presents some small disadvantages, such as a reduced sensitivity and higher noise levels, due to a greater interference of the dissolved oxygen.

In Table 16.2, an overview of analytical applications related to carbohydrates analyses by anion-exchange chromatography and pulsed amperometric detection at a gold working electrode is given, and typical chromatographic experimental conditions for the determination of galactose, glucose, fructose and sucrose are described.

16.2.3 Constant Potential Amperometric Detection of Carbohydrates at Modified Electrodes

In the last twenty years, researchers have developed several strategies to control the characteristics of the electrode interface in order to wide the applications of the electrochemical analysis. In this contest, chemically modified electrodes (CMEs) represent a powerful tool and a valid alternative to metal electrodes (Baldwin and Thomsen 1991; Opallo and Lesniewski, 2011). CMEs are electrodes prepared by attaching a chemical modifier to the electrode surface, which changes substantially the electrochemical response toward a specific analyte. Redox modifier species (M_{ox}) convert electrocatalytically the analyte and give on the electrode surface the reduced form (M_{red}); the current

Table 16.2 Experimental conditions for the determination of galactose, glucose, fructose and sucrose by HPAEC-PAD.[a]

Sugar	Separation column; mobile phase composition	Potential waveform	Application	Reference
galactose glucose fructose sucrose	Dionex CarboPac PA1 (250 × 4 mm i.d.) + guard column (50 × 4 mm i.d.); 12 mM NaOH + 1 mM Ba(CH$_3$COO)$_2$	E_{DET}: 0.25 V (440 ms) E_{OX}: 0.80 V (180 ms) E_{RED}: −0.25 V (360 ms)	Plant tissues (leaves and roots from olive)	Cataldi et al. 2000b
glucose fructose sucrose	Dionex CarboPac PA1 (250 × 4 mm i.d.) + guard column (50 × 4 mm i.d.); 150 mM NaOH	E_{DET}: 0.05 V (400 ms) E_{OX}: 0.75 V (190 ms) E_{RED}: −0.15 V (390 ms)	Milk-based and soy-based infant formula	Kaine and Wolnik 1998
galactose glucose fructose	Dionex CarboPac PA1 (250 × 4 mm i.d.) + guard column (50 × 4 mm i.d.); 10 mM NaOH + 1 mM Ba(CH$_3$COO)$_2$	E_{DET}: 0.05 V (440 ms) E_{OX}: 0.80 V (180 ms) E_{RED}: −0.22 V (360 ms)	Milk and cheese	Cataldi et al. 2003
glucose fructose sucrose	Dionex CarboPac MA1 (250 × 4 mm i.d.) + guard column (50 × 4 mm i.d.); 500 mM NaOH + 1 mM Sr(CH$_3$COO)$_2$	E_{DET}: 0.05 V (450 ms) E_{OX}: 0.65 V (190 ms) E_{RED}: −0.15 V (340 ms)	Fruits and vegetables	Cataldi et al. 1998b
glucose fructose sucrose	Dionex CarboPac MA1 (250 × 4 mm i.d.) + guard column (50 × 4 mm i.d.); A: 1M NaOH; B: water, gradient elution	E_{DET}: 0.05 V (400 ms) E_{OX}: 0.75 V (190 ms) E_{RED}: −0.15 V (390 ms)	Candy, dessert, chewing gum, chocolate, etc.	Andersen and Sørensen 2000

Sugars	Column and eluent	Detection potentials	Application	Reference
glucose fructose sucrose	Dionex CarboPac PA100 (250 × 4 mm i.d.) + guard column (50 × 4 mm i.d.); A: water, B: 0.6 M NaOH, C: 0.5M sodium acetate solution, gradient elution	E_{DET}: 0.10 V (500 ms) E_{OX}: 0.60 V (80 ms) E_{RED}: −0.60 V (50 ms)	Fructo-oligosaccharides in bacterial cultures	Corradini *et al.* 2004
galactose glucose fructose sucrose	Dionex CarboPac PA10 (250 × 4 mm i.d.) + guard column (50 × 4 mm i.d.); A: water, B: 0.2 NaOH, gradient elution	E_{DET}: 0.25 V (500 ms) E_{OX}: 0.70 V (120 ms) E_{RED}: −0.90 V (160 ms)	Wine and coffee	Bernal *et al.* 1996
glucose	Dionex CarboPac PA1 (250 × 4 mm i.d.) + guard column (50 × 4 mm i.d.); A: water, B: 0.5 M NaOH, gradient elution	E_1: 0.10 V (400 ms) E_2: −2.0 V (10 ms) E_3: 0.6 (10 ms) E_4: −0.1 (60 ms)	Serum	Cai *et al.* 2005

[a]For the chromatographic separation of sugars, anion exchange mechanisms take advantage of the weakly acidic nature of carbohydrates. Anion-exchange columns, packed with polystyrene/divinylbenzene resin, functionalized with alkyl quaternary ammonium group, have been specifically designed by Dionex for the separation and analysis of carbohydrates. Alkaline running eluent based on NaOH are used for the simultaneous determination of monosaccahrides. For the amperometric detection at a gold working electrode, a repeating triple or quadrupole step potential waveform is applied at a frequency of 0.5–2.0 Hz. Unpublished.

generated by the modifier reoxidation is proportional to the analyte concentration. Compared to simple electrodes, CMEs provide higher rate for the electrode reaction and exhibits low detection limit, high stability, selectivity and sensitivity for different compounds. These requirements are fundamental for the analytical applications as detectors in liquid chromatography, flow injection analysis or batch determinations. The recommended terminology, definitions and the preparation approaches of CMEs have been described in a IUPAC report (IUPAC Recommendations 1997).

Several analytical applications are based on chemically modified electrodes including the determination of carbohydrates, and quite recent updates have been reviewed (Zen et al. 2003).

Recently, Han et al. (2010) have employed mesoporous platinum electrodes for the amperometric determination of glucose, fructose, and sucrose in anion exchange chromatography. The voltammetric behavior of mesoporous Pt electrode and the effect of mobile phase concentration have been studied. The mesoporous Pt electrode exhibits high catalytic activity and quite good long-term stability and reproducibility, at a constant potential used in place of a complex pulsed potential waveform.

Notably studies on Ni, Co, Cu and Au have received more attention. Vidotti et al. (2009) reported a nickel hydroxide modified electrodes for the amperometric determination of glucose, fructose, lactose and sucrose in ion exchange chromatography. The electrocatalytic performances (optimum amount of electrodeposited nickel hydroxide, oxidation peak potential, sensitivity) were studied by voltammetric analysis in the absence and presence of different sugar concentrations.

The use of activated barrel plating nickel electrodes (Ni-BPE) have been described (Sue et al. 2008) for the amperometric determination of carbohydrates (glucose, fructose, sucrose, and maltose) by FIA and HPAEC. In this study, the interaction between NiOOH and the terminal -OH functional groups of sugars was believed to be essential for the detection, and the characterization of the activated Ni-BPE was performed by spectroscopic (XPS) measurements. The authors point out that the detection limit for glucose is better than those obtained by PAD, cyclic chronopotentiometry with gold electrodes, or at Cu electrodes. These results suggest that nickel hydroxide modified electrodes have a great potential in the carbohydrate detection; nevertheless, their use has not really been subjected to further investigations.

Cu has emerged as a possible alternative modifier for the development of CMEs for carbohydrates (Dong et al. 2007). An interesting application for the simultaneous determination of alditols and sugars based on Cu_2O-carbon composite electrode (Cu_2O-CCE) has been proposed (Cataldi et al. 1997a). This promising approach use Cu_2O as an electrocatalyst, incorporated into a graphite powder-polyethylene composite matrix. The detection of this sensing electrode is based on the measurement of anodic current generated by electrocatalytic oxidation of the substrate at the electrode surface, probably through the formation of Cu^{III} sites and the involvement of the adsorbed hydroxyl racical as the oxygen transfer species.

Table 16.3 Experimental conditions for the determination of galactose, glucose, fructose and sucrose by HPAEC and electro-chemical detection at chemically modified electrodes.[a]

Sugar	Separation column; mobile phase composition	CME/Potential	Application	Reference
glucose fructose sucrose	Hamilton RCX-10 (250 × 4 mm i.d.) 80 mM NaOH	Mesoporous platinum electrode E_{DET}: 0.80 V	Apple juice	Han *et al.* 2010
glucose fructose sucrose	Dionex CarboPac MA1 (250 × 4 mm i.d.) + guard column (50 × 4 mm i.d.); 0.48 or 0.50 M NaOH + 0.5 mM $Ca(OH)_2$ or $Ba(OH)_2$	Cu_2O-carbon composite electrode E_{DET}: 0.45 V	Fruit juice	Cataldi *et al.* 1997a
glucose fructose sucrose	Dionex CarboPac PA1 (250 × 4 mm i.d.) + guard column (50 × 4 mm i.d.); 30 mM NaOH	Cobalt-modified electrode E_{DET}: 0.50 V	Wine	Casella and Contursi 2003
glucose fructose sucrose	Hamilton PRP-X100 (150 × 4.1 mm i.d.) 100 mM NaOH	Activated barrel plating nickel electrode E_{DET}: 0.55 V	Honey	Sue *et al.* 2008

[a]The electrochemical detection of carbohydrates is performed at constant positive potential, applied at chemically modified electrodes, following anion-exchange chromatographic separations. Unpublished.

The modification of a glassy carbon electrode by a cobalt oxyhydroxide film was suggested by Casella and Contursi (2003) as an amperometric sensor for alditols and carbohydrates in red and white wines. In alkaline medium the high-valence states of the cobalt oxyhydroxides show interesting electrochemical activity towards the electrooxidation of carbohydrates, such as glucose, fructose, galactose, mannitol and xylitol.

The growing interest in the field of nanotechnologies has allowed the development of new promising modifiers based on metal nanoparticles. Kurniawan *et al.* proposed a gold nanoparticles (layer-by-layer) modified gold electrode as an electrochemical sensor for saccharides (Kurniawan *et al.* 2006). The deposition of gold nanoparticles on the surface of gold electrode resulted in new electrochemical properties, such as a much higher sensitivity of response. Gold electrodes coated by gold nanoparticles can be used for the development of robust CMEs that are insensitive to heating and can be used even in aggressive chemicals.

In Table 16.3 are summarized the most significant analytical applications of CMEs in the carbohydrates determination following anion-exchange chromatographic separations.

Summary Points

- In this chapter, an overview of analytical methods for the determination of galactose, glucose, fructose and sucrose by ion chromatography and electrochemical detection is reported.
- For years, the analytical determination of saccharides has been accomplished by High Performance Liquid Chromatography and refractive index, low-wavelength UV detection, or coupled with pre- and post-colum derivatization. Nowadays, the methods of choice for the determination of carbohydrates are based on anion-exchange chromatography and electrochemical detection.
- The chromatographic separation of sugars based on anion exchange mechanisms take advantage of the weakly acidic nature of carbohydrates that, at high pH, are partially ionized.
- The electrochemical detection of sugars in alkaline media is performed at gold working electrodes by applying a positive potential and measuring the current, generated from the oxidation process.
- In order to clean and reactivate the electrode surface, contaminated by sugar oxidation products, a repeating sequence of potentials over time is applied.
- The well-established potential waveform for the electro-oxidation of carbohydrates on gold working electrodes consists of 3 steps: the sugars oxidation-detection, the oxidation of the gold electrode surface and, finally the reduction of the gold oxides.
- The electrochemical behavior of sugars in alkaline media can be studied by cyclic voltammetry, then duration and potential values of each step are optimised.

- Recently, four-step potential waveforms have been proposed for the amperometric detection of sugars at gold working electrodes.
- An alternative electrochemical detection of carbohydrates involves chemically modified electrodes, which allow a sensitive detection at a constant potential without any fouling of the electrode surface.

Key Facts of the Analytical Methods for the Determination of Carbohydrates

- Carbohydrates, which have a relevance in clinical and food fields, lack of chromophore and fluorophore groups, and then for years their analytical determination has been challenging.
- The introduction of the first electrochemical detector for HPLC (High Performance Liquid Chromatography), in 1974, has led to the development of new analytical methods for the determination of carbohydrates after separation by ion chromatography. This is an example of a successful cooperative research effort by scientists (Iowa State University) and industry (Dionex Corp.).
- Electrochemical detection has developed into a mature detection technique for the determination of glucidic compounds, following high performance anion-exchange chromatography.

Key Facts of Anion Exchange Chromatography Separations of Carbohydrates

- In the separation and analysis of saccharides, a series of anion-exchange columns, with polymer-based stationary phases, have been specifically designed by Dionex (CarboPac series).
- The CarboPac PA1, packed with a non-porous and pellicular resin based on polystyrene/divinylbenzene and functionalized with alkyl quaternary ammonium group, is suitable for the separation of glucose, fructose, galactose and sucrose.
- Retention of carbohydrate can be manipulated by altering the sodium hydroxide concentration of the eluent (pH 12–13), usually in the range 10 mM–1 M.

Key Facts of the Electrochemical Detection of Carbohydrates

- At high pH, carbohydrates are electrocatalytically oxidized at the surface of the gold electrode by application of a positive potential. Carbohydrates can be detected and quantified by measuring the electrical current or the charge generated by their oxidation.

- The oxidation process at a single constant potential is characterized by fouling of electrode surface by accumulated final oxidation products. Therefore, the poisoned electrode surfaces need to be periodically cleaned, prior to each current measurement.
- For a stable and reproducible signal response, pulsed amperometric detection is based on multistep potential waveforms to detect analytes and to clean and reactivate the electrode surface with a repeated sequence.
- Carbohydrates can be electrochemically detected at constant positive potential, applied at chemically modified electrodes.

Definitions of Words and Terms

Ion Chromatography: Also called ion-exchange chromatography, based on ionic interactions between analytes to be separated (ions and polar molecules) dissolved in a mobile phase and a stationary phase, constituted by resins containing ionic functional groups.

High-Performance Anion Exchange Chromatography: Powerful ion-exchange chromatography technique developed for the separation of anions and weakly acidic compounds at high pH, where they are partially ionised.

Alkaline eluent: Mobile phase, usually based on NaOH, KOH or carbonate buffers, for the chromatographic separation of anions and polar acidic molecules by High-Performance Anion Exchange Chromatography. Based on analyte charge and column retention, the concentration of the alkaline eluent can range, usually, from 10 mM to1 M. Typically, at low concentrations, the alkaline eluent contains an unknown amount of carbonate, coming from the atmospheric carbon dioxide, which could negatively affect the quality of separations.

Electrochemical detection: Detection technique based on electrochemical reactions that take place at the electrode surface. The current generated from the redox process is proportional to the analyte concentration, and represents the signal reported as a function of time in the chromatogram.

Working electrode: In an electrochemical detection cell, the electrode where the oxidation or reduction of the analyte takes place. In order to enable the redox process of the analyte the potential is kept constant at a selected useful value.

Cyclic Voltammetry: Electrochemical technique, generally used to study the electrochemical properties of an analyte in solution. At the surface of a working electrode in an electrochemical cell, the potential is ramped linearly versus time, following a triangular profile, with potentials cycling between two fixed limit values. The current, generated at the working electrode from the redox processes, is plotted as a function of the applied voltage.

Thin-layer detection cell: Flow-through electrochemical detection cell, consisting of three electrodes (the working electrode, the auxiliary counter electrode and the reference electrode) and the electronic circuitry required for monitoring, recording, and processing the detector signal. There are several types of flow-through detection cells, characterized by geometrical properties, such as the length, diameter, and shape of its detection channel.

Potential waveform: Sequence of applied potentials at solid electrodes over time, at frequency ranging from 0.5–2.0 Hz, which is generally appropriate for detection in High-Performance Liquid Chromatography.

Pulsed Amperometric Detection: Based on multistep potential waveform to alternate the operation of analyte anodic detection with the procedure for regenerating the working electrode by an oxidative cleaning and a reductive reactivation of the electrode surface.

Chemically Modified Electrode: Conventional conducting or semiconducting electrode bound to or coated with selected monomolecular, multi-molecular, ionic or polymeric thin film, imparting desirable electrochemical properties.

List of Abbreviations

HPLC	High Performance Liquid Chromatography
IC	Ion Chromatography
HPAEC	High Performance Anion-Exchange Chromatography
RI	Refractive index
UV	Ultra-Violet
ED	Electrochemical Detection
CV	Cyclic Voltammetry
PAD	Pulsed Amperometric Detection
E_{DET}	Determination Potential
E_{OX}	Oxidation Potential
E_{INT}	Integration Potential
t_{DET}	Determination Time
t_{INT}	Integration Time
t_{DEL}	Delay Time
t_{OX}	Oxidation Time
t_{RED}	Reduction Time
CME	Chemically Modified Electrode

References

Andersen, R., and Sørensen, A., 2000. Separation and determination of alditols and sugars by high-pH anion-exchange chromatography with pulsed amperometric detection. *Journal of Chromatography A*. 897: 195–204.

Baldwin, R.P., and Thomsen, K.N., 1991. Chemically modified electrodes in liquid chromatography detection: A review. *Talanta*. 38: 1–16.

Bernal, J.L., Del Nozal, M.J., Toribio, L., and Del Alamo, M., 1996. HPLC analysis of carbohydrates in wines and instant coffees using anion exchange chromatography coupled to pulsed amperometric detection. *Journal of Agricultural and Food Chemistry*. 44: 507–511.

Cai, Y., Liu, J., Shi, Y., Liang, L., and Mou, S., 2005. Determination of several sugars in serum by high-performance anion-exchange chromatography with pulsed amperometric detection. *Journal of Chromatography A.* 1085: 98–103.

Cataldi, T.R.I., Centonze, D., Casella, I.G., and Desimoni, E., 1997a. Anion-exchange chromatography with electrochemical detection of alditols and sugars at a Cu_2O-carbon composite electrode. *Journal of Chromatography A.* 773: 115–121.

Cataldi, T.R.I., Centonze, D., and Margiotta, G., 1997b. Separation and Pulsed Amperometric Detection of Alditols and Carbohydrates by Anion-Exchange Chromatography Using Alkaline Mobile Phases Modified with Ba(II), Sr(II), and Ca(II) Ions. *Analytical Chemistry.* 69: 4842–4848.

Cataldi, T.R.I., Campa, C., Margiotta, G., and S.A. Bufo., 1998a. Role of Barium Ions in the Anion-Exchange Chromatographic Separation of Carbohydrates with Pulsed Amperometric Detection. *Analytical Chemistry.* 70: 3940–3945.

Cataldi, T.R.I., Margiotta, G., and Zambonin, C.G., 1998b. Determination of sugars and alditols in food samples by HPAEC with integrated pulsed amperometric detection using alkaline eluents containing barium or strontium ions. *Food Chemistry.* 62: 109–115.

Cataldi, T.R.I., Campa, C., and De Benedetto, G.E., 2000a. Carbohydrate analysis by high-performance anion-exchange chromatography with pulsed amperometric detection: The potential is still growing. *Fresenius Journal of Analytical Chemistry.* 368: 739–758.

Cataldi, T.R.I., Margiotta, G., Iasi, L., Di Chio, B., Xiloyannis, C., and Bufo, S.A., 2000b. Determination of sugar compounds in olive plant extracts by anion-exchange chromatography with pulsed amperometric detection. *Analytical Chemistry.* 72: 3902–3907.

Cataldi, T.R.I., Angelotti, M., and Bianco, G., 2003. Determination of mono- and disaccharides in milk and milk products by high-performance anion-exchange chromatography with pulsed amperometric detection. *Analytica Chimica Acta.* 485: 43–49.

Casella, I.G., and Contursi, M., 2003. Carbohydrate and alditol analysis by high-performance anion-exchange chromatography coupled with electrochemical detection at a cobalt-modified electrode. *Analytical and Bioanalytical Chemistry.* 376: 673–679.

Corradini, C., Bianchi, F., Matteuzzi, D., Amoretti, A., Rossi, M., and Zanoni, S., 2004. High-performance anion-exchange chromatography coupled with pulsed amperometric detection and capillary zone electrophoresis with indirect ultra violet detection as powerful tools to evaluate prebiotic properties of fructooligosaccharides and inulin. *Journal of Chromatography A.* 1054: 165–173.

Dionex Corporation a. Application Note N. 21. Available at: http:// www. dionex.com/en-us/webdocs/5023-TN21.pdf. Accessed 5 May 2011.

Dionex Corporation b. Application Note N. 40. Available at: http://www. dionex.com/en-us/webdocs/5023-TN40.pdf. Accessed 5 May 2011.

Dionex Corporation c. Application Note N. 70. Available at: http://www.dionex.com/en-us/webdocs/5023-TN20.pdf. Accessed 05 May 2011.

Dong, S., Zhang, S., Cheng, X., He, P., Wang, Q., and Fang, Y., 2007. Simultaneous determination of sugars and ascorbic acid by capillary zone electrophoresis with amperometric detection at a carbon paste electrode modified with polyethylene glycol and Cu_2O. *Journal of Chromatography A*. 1161: 327–333.

Han, J.H., Choi, H.N., Park, S., Chung, T.D. and Lee, W.Y., 2010. Mesoporous Platinum Electrodes for Amperometric Determination of Sugars with Anion Exchange Chromatography. *Analytical Sciences*. 26: 995–1000.

International Union of Pure and Applied Chemistry (IUPAC), 1997. Chemically modified electrodes: recommended terminology and definitions. *Pure & Applied Chemistry*. 69: 1317–1323.

Jensen, M.B., and Johnson, D.C., 1997. Fast Wave Forms for Pulsed Electrochemical Detection of Glucose by Incorporation of Reductive Desorption of Oxidation Products. *Analytical Chemistry*. 69: 1776–1781.

Kaine, L.A., and Wolnik, K.A., 1998. Detection of counterfeit and relabeled infant formulas by high-pH anion- exchange chromatography-pulsed amperometric detection for the determination of sugar profiles. *Journal of Chromatography A*. 804: 279–287.

Kurniawan, F., Tsakova, V., and Mirsky, V.M., 2006. Gold nanoparticles in nonenzymatic electrochemical detection of sugars. *Electroanalysis* 18: 1937–1942.

LaCourse, W.R., and Johnson, D.C., 1991. Optimisation of waveforms for pulsed amperometric detection (PAD) of carbohydrates following separation by liquid chromatography. *Carbohydrate Research*. 215: 159–178.

LaCourse, W.R., and Johnson D.C., 1993. Optimisation of waveforms for pulsed amperometric detection of carbohydrates based on pulsed voltammetry. *Analytical Chemistry*. 65: 50–55.

LaCourse, W.R., 1997. *Pulsed Electrochemical Detection in High-performance Liquid Chromatography*. John Wiley & Sons, Inc. New York, USA.

Opallo, M., and Lesniewski, A., 2011. A review on electrodes modified with ionic liquids. *Journal of Electroanalytical Chemistry*. 656: 2–16.

Rocklin, R.D., Clarke A.P., and Weitzhandler, M., 1998. Improved Long-Term Reproducibility for Pulsed Amperometric Detection of Carbohydrates via a New Quadruple-Potential Waveform. *Analytical Chemistry*. 70: 1496–1501.

Sue, J.W., Hung, C., Chen, W.C., and Zen, J.M., 2008. Amperometric Determination of Sugars at Activated Barrel Plating Nickel Electrodes. *Electroanalysis*. 20: 1647–1654.

Vidotti, M., Cerri, C.D., Carvalhal, R.F., Dias, J.C. Mendes, R.K., Córdoba de Torresi, S.I., and Kubota, L.T., 2009. Nickel hydroxide electrodes as amperometric detectors for carbohydrates in flow injection analysis and liquid chromatography. *Journal of Electroanalytical Chemistry*. 636: 18–23.

Zen, J.M., Kumar A.S., and Tsai, D.M., 2003. Recent Updates of Chemically Modified Electrodes in Analytical Chemistry. *Electroanalysis*. 15: 1073–1087.

CHAPTER 17

Assay of Glucose Using Near Infrared (NIR) Spectroscopy

MOHAMMED BENAISSA,* AMNEH MBAIDEEN AND
BILAL AHMAD MALIK

Electronic and Electrical Engineering Department, The University of
Sheffield, Mappin Street, Sheffield, S1 3JD, United Kingdom
*Email: m.benaissa@sheffield.ac.uk

17.1 Introduction

The analysis of food items is vital for ensuring their quality and for providing accurate information of nutrients such as sugars for dietary and health purposes. Glucose is one of the most important sugars and is found in many food items; it is often defined as nature's fuel due to its role as a source of energy in diverse living organisms.

Traditionally, wet chemistry methods were used in food analysis, but these could be replaced by near infrared (NIR) spectroscopy due to the advancement in this technology over the last few decades. NIR is a non-invasive and non-destructive method of analysis with little or no sample preparation which results in rapid, concurrent, safer analysis. Numerous researchers have attempted to utilize these advantages to food science by applying NIR along with multivariate calibration and pre-processing techniques (Osborne 2006).

The NIR spectra of components can be collected by using a Fourier Transform Infrared FTIR spectrometer that measures the absorbed or transmitted

Food and Nutritional Components in Focus No. 3
Dietary Sugars: Chemistry, Analysis, Function and Effects
Edited by Victor R Preedy
© The Royal Society of Chemistry 2012
Published by the Royal Society of Chemistry, www.rsc.org

NIR radiation as a function of the wavenumber or wavelength. FTIR spectrometers tend to provide high SNR, high resolution and precision, and high opticals throughout with a short scanning time and that is why they have found their way into several fields such as chemistry, biochemistry, medicine, industry and environmental applications (Griffiths *et al.* 2007). NIR spectroscopy refers to the technology that uses electromagnetic radiation in the range of 4000–12800 cm^{-1} (2.5–0.78 μm) to investigate the properties of chemical systems.

Most biological components have absorption bands in the NIR range, with a unique response (fingerprint) to the NIR signals. The variation of intensity of the reflected or absorbed NIR radiation is correlated with the variation of the entire components concentration. The penetration depth of the NIR radiation varies and specifically depends on the wavelength range used, the power of the radiation, and the type of sample.

The NIR spectrum of glucose possesses three absorbance bands in both the combination (2–2.5 μm) and the overtone (1.54–1.82 μm) regions of the NIR spectrum centered at 2.115, 2.273, 2.326 μm and at 1.73, 1.69 and 1.61 μm respectively. The absorbability of glucose in the shortwave region (0.7–1.33 μm) is low. Most recent glucose analysis studies used the combination region (Chen *et al.* 2004) where absorbability of water is low and the glucose absorbance band is narrower (particularly at 2.273 μm). However, the NIR spectra of glucose in many samples in practice are weak in intensity, broad and overlapped. Furthermore, the quality of the collected spectra is influenced by several factors such as:

- The high background spectra of water.
- Baseline variations resulting from the instruments and ambient variation.
- High frequency noise, such as the detector noise.
- High overlapping of the spectra of other components because the structure of the glucose molecules and bonds are similar to many other constituents of samples.
- The light scattering which is anisotropic, inhomogeneous, and non linear resulting from the surfaces and other constituents in a sample. The scattering reduces the SNR due to the degradation of the optical signal.
- The low concentration of glucose in many samples compared to other constituents.

Therefore, sophisticated interpretation (*i.e.* chemometrics) methods and data processing methods are required for the qualitative and quantitative analysis of glucose. The results could be further improved by combining chemometrics methods with digital signal processing methods (DSP). The signal representation as absorbance spectra, interferogram or in frequency domain is selected such that it provides interpretatable information about the analyte of interest.

17.2   Beer Lambert's Law

Based on Beer Lambert's law, the absorbance spectrum A of the sample is related to the transmittance spectrum and the concentration of the chemical component as follows:

$$A = -\log T = -\log\left(\frac{I}{I_o}\right) = \varepsilon c L \tag{1}$$

where I is the intensity of the transmitted light through the sample. I_o is the intensity of the incident light, c is the concentration, L *is* the optical path length, and ε is the molar absorpativity. In the quantitative analysis, it is preferred to convert the transmittance spectra to the absorbance spectra, since the absorbance is related linearly to the concentration, as shown in equation (1).

Additionally, Beer Lambert's law states that for a homogenous sample composed of m components, the measured spectrum at each wavelength is the linear summation of the absorbance spectra of the m components:

$$A = \sum_{i=1}^{m} \varepsilon_i c_i b_i \tag{2}$$

Where ε_i, c_i and b_i are the absorpativity, concentration and optical pathlength for each component respectively. For a homogeneous sample, the optical pathlength and absorpativity are fixed for each component, thus the absorbance spectrum can be defined as the summation of the pure component spectra $S_i, i = 1, 2, ..., m$ weighted by their concentration:

$$A = c_1 S_1 + c_2 S_2 + ... + c_m S_m \tag{3}$$

In the quantitative and qualitative analysis methods of NIR spectroscopy, several samples should be measured in order to extract the required information about the analyte of interest. Therefore, Beer Lambert's law for multi-components and multi-samples (n samples) measured at L points (wavelengths) can be expressed as:

$$A = CS \tag{4}$$

where $A \in \Re^{n \times L}$, $C \in \Re^{n \times m}$, and $S \in \Re^{m \times L}$.

17.3   Multivariate Calibration Methods

Multivariate regression algorithms are widely used in chemometrics to extract the concentration of the desired analyte from the collected spectra of the samples. The main function of multivariate regression is to find the strength of the relation between the absorbance spectra and the concentration of the analyte of interest, and to extract the information that is related to the analyte of interest. Multivariate methods satisfy this goal by generating a calibration model that is related to the response and predictor variables, which can be used later on to estimate the response of new variables. This involves two phases: the calibration phase and the testing phase. In the calibration phase, training data are used to generate the calibration model that regresses the training data spectra against their corresponding known concentrations. Then, the tested data are used to test the capability of the calibration model to predict the concentration. Finally, the calibration model can be used to predict the concentration of unknown samples. The performance, robustness and fitting of this model can be investigated by using training data or by using external data.

Generally, the performance of the calibration model can be evaluated by computing the standard error of prediction (SEP), the coefficient of determination R^2 (correlation coefficients), and the mean percent error (MMPE) of the testing data:

$$SEP = \left(\sum_{i=1}^{n_p} (c_i - \hat{c}_i)^2 / n_p \right)^{1/2} \tag{5}$$

$$MMPE\% = \frac{1}{n_p} \sum_{i=1}^{n_p} \frac{|c_i - \hat{c}_i|}{c_i} \times 100 \tag{6}$$

$$R^2 = 1 - \frac{\sum_{i=1}^{n_p} (c_i - \hat{c}_i)^2}{\sum_{i=1}^{n_p} (c_i - \bar{c})^2} \tag{7}$$

where c_i is the actual concentration of the test data, $\hat{c}_i$ is the predicted concentration, and n_p is the total number of test samples. The SEP measures the ability of the calibration model to predict the concentration of the analyte of interest for new data.

Three multivariate methods are used widely used to generate the calibration model: multiple linear regression (MLR), principle component regression (PCR), and partial least square regression (PLS).

17.3.1 Multiple Linear Regression (MLR)

MLR is used to model the relation between the concentration and the absorbance spectra based on Beer Lambert's law in an inverse way:

$$C = A\beta \tag{8}$$

where A is the measured raw spectra matrix, C is the concentration of the analyte of interest $\in \Re^{n \times 1}$, and $\beta \in \Re^{m \times 1}$ is known as the regression coefficients vector. The regression coefficients β are computed based on the Least Squares method (Burns and Ciurczak 2008).

$$\beta = (A^T A)^{-1} A^T C \tag{9}$$

The regression vector β should be correlated with the analyte of interest. Therefore, the utility of the MLR calibration model can be evaluated by investigating the similarity between the regression vector and the pure component spectrum of the analyte of interest.

The MLR method has the advantage that it treats the baseline variation as an impurity and therefore the model is not affected seriously by the baseline variation. The main problem with MLR is that it is a 'wavenumber' dependent method, in which the absorbance spectra should be measured at the wavenumbers that have more information about the analyte of interest, and the number of samples should be greater than the wavenumbers. However, the NIR spectra in practice have high overlapping and it is common to measure the NIR spectra with a number of variables (a number of wavenumbers) greater than the number of samples. Furthermore, increasing the number of samples may produce collinearity and nonlinearity, which cause the MLR calibration model to be unstable since it does not work well with collinearity and nonlinearity. Many regression models, such as PCR and ICR, have been developed based on MLR.

17.3.2 Principle Component Analysis (PCA)

PCA (Joliffe 2002 and Abdi *et al.* 2010) is used widely in chemistry for data clustering, data classification, filtering, feature extraction, outlier detection, data compression, and to reduce the correlation between variables. PCA finds a new axis that is composed of the linear combination of the original data in which the new axes are the principal of the data covariance matrix. The new axes are known as principal components, loading factors, loading vectors or Eigenvectors. The loading factors are not the pure component spectra of the mixture constituents, but they model all the variations that are due to the chemical components and their interaction, in addition to the instrumental and environmental variations and noise.

The first important property of the loading factors is that they are uncorrelated and orthogonal; therefore PCA can be used to analyse the NIR spectra even if they have collinearity. Another important property is that the loading factors are sorted according to their contribution in the variance of the original data, which means that the first loading factor will capture the maximum amount of the total variance of the original data among the other factors. The second factor then captures the second highest amount of the total variance, and so on. Contribution of the last loading factors will be low and they usually model the noise. Therefore, these factors may be neglected to suppress most of the noise and to remove the redundancy from the variables. Once the loading factors are determined they are used with the original data to compute the scores of the absorbance spectra. The projection amounts of the *m'th* spectrum in the original data on the entire loading vectors are called score vectors. The scores of each data set are orthogonal, uncorrelated and span the whole range of the data; therefore the pseudoinverse of the score matrix can be computed.

Thus, based on the definition of the loading vectors and scores, the original spectra can be represented as:

$$A = T.P' \qquad (10)$$

where the rows of T are the score vectors, and the columns of P are the loading vectors of the spectra matrix A.

17.3.3   Principal Component Regression (PCR)

PCR was proposed to overcome the MLR limitations on one hand and to combine the advantages of the PCA and MLR on the other. It is implemented in two steps; in the first step the scores of the data matrix are computed using PCA, then the scores are regressed against the concentration of the analyte of interest to generate the MLR calibration model instead of the original absorbance spectra.

Thus, the PCR calibration model can be given as:

$$c = T\beta_{mlr} \qquad (11)$$

where c is the concentration of the analyte of interest, T is the matrix of scores. and β_{mlr} is the regression coefficients vector that relates the concentration of the analyte to the scores of their corresponding spectra. Based on equation 6, β_{mlr} is given as:

$$\beta_{mlr} = (T'T)^{-1}T'c \qquad (12)$$

The scores of the data matrix can be defined as $T = AP$. Thus, the regression coefficients β_{mlr} can be converted to predict the concentration of the analyte of interest from the absorbance spectra as follows:

$$c = AP\beta_{mlr} = A(P\beta_{mlr})$$
$$c = A\beta \qquad (13)$$

where the elements of β are the regression coefficients that can be used to detect the glucose concentration from the new samples without computing the PCA model of the testing spectra. P are the loading factors of the training data. The model can be generalized to predict the concentration of the analyte of interest for new samples with unknown concentration

$$c_{new} = (A_{new} - \bar{A})\beta + \bar{c} \qquad (14)$$

where $\bar{A}$ and $\bar{c}$ are the average values of the columns of the spectra and their corresponding concentrations of the training data, respectively.

Compared to the MLR model, the PCR calibration model can be realized regardless of the selected wavelengths, and it is reliable against collinearity. The PCR model is robust and stable against noise if the number of observations is high, since it can eliminate loading factors that have low variance and low influence on the change of the absorbance spectra. On the other hand, PCR has some limitations that may degrade the performance of the calibration model. For instance, sometimes the analyte of interest only provides a small contribution to the variation of the spectra A and the information related to it may be eliminated during the decomposition process, especially if the signal to noise ratio of the measured spectra is low. Hence, the quality of the PCR model is sensitive to the number of loading factors that are used to generate the calibration model. Secondly, the PCR algorithm decomposes the absorbance spectra matrix A only, without using the prior information about the analyte concentrations. Thirdly, in the presence of outlier data the principle components may not represent the maximum variation of the information correctly because the covariance matrix is sensitive to outlier data.

17.3.4 Partial Least Square Regression (PLS)

Partial Least Square Regression has become the most widely used technique for quantitative analysis of NIR spectra as it overcomes most of the serious drawbacks of other multivariate techniques. It decomposes the concentration and the absorbance spectra simultaneously, and uses the prior information of concentration in the decomposition of the spectra, and *vice versa.*

Compared to PCR regression, PLS is more robust, fast, and requires a lower number of factors to establish the calibration model (Geladi and Kowalski 1986; Haaland and Thomas 1988). In contrast to PCR, the PLS model extracts and sorts their loading factors according to their covariance with the scores of the absorbance matrix and the concentration vector. Thus, information related to the analyte of interest's concentration will be enhanced even if it has a low contribution in the absorbance spectra, because the chosen factors weigh the spectra according to their correlation with the concentration.

PLS decomposes both the absorbance and concentration matrices to their scores and loading factors as follows:

$$A = T.P' + E = \sum t_h p'_h + E \tag{15}$$

$$C = U.Q' + F^* = \sum u_h.q'_h + F^* \tag{16}$$

where U and Q are the scores and loadings of the concentration matrix respectively, h is the number of factors, and E and F^* are the residuals. The next step is to construct the relationship between the absorbance data and concentration by computing the weighting factors w using the NIPALS algorithm (Andersson 2009). The NIPALS algorithm first sets an initial vector u as any vector from A:

$$
\begin{aligned}
w'_{old} &= u'A \,/\, u'u \\
w'_{new} &= w'_{old} \,/\, \left\|w'_{old}\right\|
\end{aligned}
\tag{17}
$$

The first score of the absorbance matrix t and the concentration matrix u, in addition to the loading factor of the concentration matrix q, are computed from the weighting factor as follows:

$$t = A w_{new} / w'_{new} w_{new} \tag{18}$$

$$q' = t'c / t'.t \tag{19}$$

$$q'_{new} = q'_{old} / \|q'_{old}\| \tag{20}$$

$$u = cq / q'q \tag{21}$$

where q_{old} is the value of the loading vector from the previous iteration and q_{new} is its present value. The present value of t is compared with its value in the previous iteration. If the difference is small and within the allowed error, then one goes to the next step, otherwise one goes to equation (18).

When the value of the absorbance spectra is converged then the loading factors of the absorbance spectra can be computed from the following formula:

$$p' = t'A / t't$$
$$p'_{new} = p'_{old} / \|p'_{old}\| \tag{22}$$

The final values of the scores and the weighting factors are given as:

$$t_{new} = t_{old} \|p'_{old}\| \tag{23}$$

$$w'_{new} = w'_{old} \|p'_{old}\| \tag{24}$$

Finally, the regression coefficients of the model are given as:

$$\beta = \frac{u't}{t't} \tag{25}$$

If more factors are required, the computations are repeated by replacing the absorbance matrix A and the concentration matrix c with their residuals E and F^* respectively in equations Equations (17–22):

$$E_h = E_{h-1} - t_h p'_h, \quad E_0 = A \tag{26}$$

$$F_h^* = F_{h-1}^* - u_h q'_h, \quad F_0^* = c \tag{27}$$

where h is the loading factor index. Once the calibration is generated it can be used to predict the concentration of analyte for new data, as already explained.

The PLS model can extract an infinite number of factors; however, the first few factors have information about the analyte of interest while the last factors may be the result of noise. Furthermore, increasing the number of factors may produce overfitting, which may lead to degradation of the performance of the calibration model (Faber and Rajk 2007). Therefore, the factors that have information about the analyte of interest should be determined to prevent overfitting and allow a high capability of prediction. Several methods are used to compute the number of loading factors, such as those based on computing the standard error of calibration (SEC), the cross-validation method, and those based on minimizing the residual F^*.

17.3.5 Performance Comparison

The performance of the 3 models reviewed above is illustrated using practical experimental data from samples that were prepared by dissolving glucose, urea and triacetin in a buffer solution. Thirty samples were prepared with different concentrations, selected with reference to their concentrations in blood serum, ranging from 20 mg/dL to 500 mg/dL for glucose, 0 to 50 mg/dL for urea, and 10 to 190 mg/dL for triacetin. A Fourier transform spectrometer (FTIR cary 5000 version 1.09) was used to obtain the spectrum from these prepared samples. From each sample, three spectra were collected, and a total of 90 NIR spectra were collected from the spectrometer in this manner. The wavelength region chosen for collecting the spectra was 2100 nm to 2400 nm, with a spectral resolution of 1 nm. The absorbance spectra of the buffer solution were used as reference. These experiments were carried in a non-controlled environment. The spectra were divided randomly into two sets for the calibration and validation of the model. The calibration model was built using the first set containing the three replicate spectra of 20 samples. The calibrated model was tested using the second set containing the triplicate spectra of 10 samples.

The MLR model results are illustrated in Figure 17.1 with a correlation coefficient R^2 of 0.870 and RMSEP of 103 mg/dL. An improvement to 0.885 and 49.4 mg/dL for R^2 and RMSEP respectively is achieved with a PCR model with 6 principal components as shown in Figures 17.2 and 17.3. The best results however, as shown in Figures 17.4 and 17.5 for the same data, are obtained when the PLS model is used; the PLS with 6 latent variables R^2 to 0.97 and decreased the RMSEP to 27.5 mg/dL. The results are summarized in Table 17.1 and show that MLR is not suitable for calibration and prediction of the dataset used and PLS performs better than PCR. The data was pre-processed using first derivative in all these models.

17.4 Pre-processing Techniques

The capability of the multivariate model to predict the concentration of the analyte of interest in the measurements is sensitive to the quality of the collected spectra. The practical data are influenced by several negative factors such as noise, baseline variations, and scattering. Therefore, the quality of the raw spectra should be improved first by using pre-processing methods or using spectrophotometers and sensors that have the ability to measure the raw spectra with a high signal to noise ratio. The goal of preprocessing is to prevent the parasitic variations from impacting on the prediction capability of the

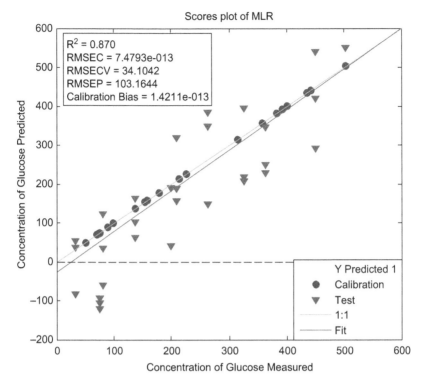

Figure 17.1 Glucose prediction performance using the MLR model. This figure illustrates the poor prediction performance of the MLR model for the experimental data set considered as indicated by the performance parameters R^2, SEC and SEP.

multivariate model by removing the high frequency noise and baseline variations, enhancing the appearance resolution of the collected spectra, removing the high background, selecting the appropriate spectral range, and compensating the negative influence of temperature variation. Smoothing, digital bandpass filtering and derivatives are examples of techniques employed in preprocessing NIR spectra.

17.4.1 Mean Centering of a Data Matrix

It is considered the first step in the construction of the PLS, and PCR models. The main objective of it is to improve the accuracy of the model by preventing the mean from dominating the first extracted latent factors. Mean centering is achieved by subtracting the average of each column from its corresponding column:

$$\bar{A}_j = A_j - \frac{1}{n}\sum_{i=1}^{n} a_{ij} \qquad j = 1, ..., m \tag{28}$$

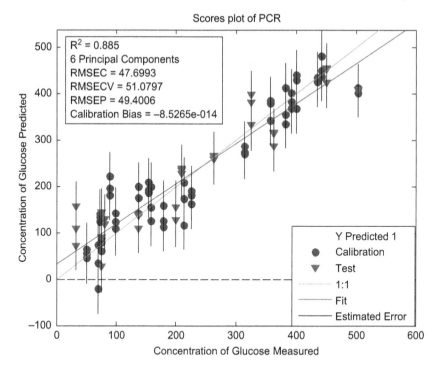

Figure 17.2 Glucose prediction performance using the PCR model. This figure illustrates the improved prediction performance of the PCR model, compared to MLR, for the same experimental data set considered as indicated by the performance parameters R^2, SEC and SEP.

where $\bar{A}_j$ is the *j'th* centered column of A, and a_{ij} is the element of the *i'th* row and *j'th* column. Usually, mean centering is followed by scaling, particularly when the variables are measured with different units. It gives great weight to the bands that have high variation. All the points of such a variable are divided by its corresponding standard deviation, so that its standard deviation is equal to one.

17.4.2 Smoothing

The purpose of smoothing is to enhance the signal to noise ratio of the raw spectra by suppressing the high frequency noise components of the spatial spectrum given that most of the important information of the biological components lies in the low and mid band range of the spectrum. Various methods have been developed and introduced to smooth the spectra, including the moving window average smoothing filter (MWA), the Savitzky-Golay filter, and the frequency domain based filtering (Chau *et al.* 2004; Gorry 1990; Savitzky and Golay 1964).

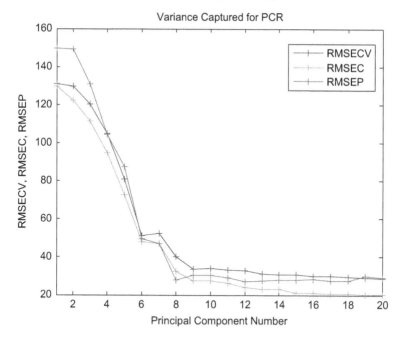

Figure 17.3 Glucose prediction performance parameters *versus* number of components for the PCR model. This figure illustrates the relationship between the key prediction performance parameters (SEC, SECV and SEP) and the number of the principal components required when using the PCR model. This is used to identify the optimal model.

In MWA, a window of size $2t+1$ is defined. The smoothed points of the $1 \times m$ raw spectrum are defined as:

$$a_i^* = \frac{1}{2t+1} \sum_{-t}^{t} a_{i+j} \qquad (29)$$

where a_i^* is the $i'th$ smoothed point, a_{i+j} is the raw spectrum points, $i=t+1$, $t+2,\ldots,m\text{-}t\text{-}1$ and $j=\text{-}t,\ldots,0,\ldots,t$. Then, the index i is increased by 1, which means the window is moved by one to calculate the smoothed point $i+1$. Each time the window is moved by one point until the original spectrum is scanned and smoothed. However, MWA is unable to smooth the first and last t points of the raw spectrum and its performance is strongly related to the window size.

The Savitzky-Golay Filter is widely used in chemometrics, and links the moving average window and the polynomial regression methods. It weighs the moving average filter output and regresses it against a local polynomial with order k. Consequently, the original spectrum is better fitted. The Savitzky-Golay filter suppresses the high-frequency components, including noise, without distorting the spectrum features such as the peaks, their relative amplitudes and width.

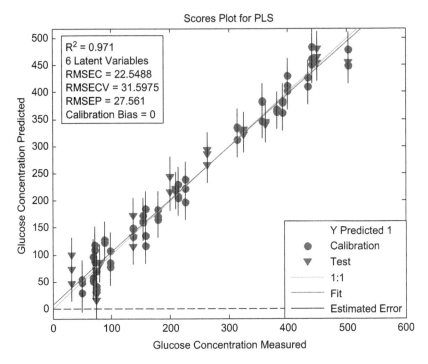

Figure 17.4 Glucose prediction performance using the PLS model. This figure illustrates the prediction performance of the PLS model, which is shown to be better than both MLR, and PCR for the same experimental data set considered as indicated by the performance parameters R^2, SEC and SEP.

However, the Savitzky-Golay filter has to optimize the polynomial order and the window size so that the original spectra are filtered with low distortion. Increasing the window size and using a low order polynomial, which is known as strong filtering, will improve the signal to noise ratio of the filtered spectra, but at the same time it will introduce a distortion in the smoothed signal. On the other hand, in weak filtering resulting from increasing the polynomial order, the filtering operation will preserve part of the noise and produce low distortion. The two end sides of the spectrum will be lost in the smoothing operation in a manner that is similar to the MAW.

17.4.3 Derivative

The derivative (Chau *et al.* 2004, Kauppinen *et al.* 1981) can be used to detect subtle information or features in the raw spectra. It enhances the appearance resolution by locating the peaks of the spectra and resolves them to their principal components. The first derivative is used to locate the peaks. The second derivative, which is the most commonly used, eliminates the baseline

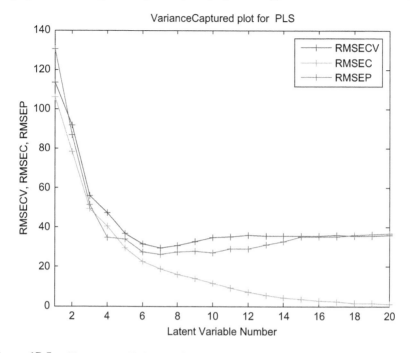

Figure 17.5 Glucose prediction performance parameters *versus* number of components for the PLS model. This figure illustrates the relationship between the key prediction performance parameters (SEC, SECV and SEP) and the number of latent variables required when using the PLS model to select the optimal case.

Table 17.1 Comparative performance summary of the MLR, PCR (6 principal components) and PLS (6 latent variables) models using the same experimental data set.

	R^2	RMSEC (mg/dL)	RMSECV (mg/dL)	RMSEP (mg/dL)
MLR	0.870	$7.4793e^{-013}$	34.1042	103.1644
PCR	0.885	47.6993	51.0797	49.4006
PLS	0.971	22.5488	31.5975	27.561

variation and constant background, sharpens the peaks, and reduces the overlapping. It produces inverse peaks at the original peaks or shoulders and will forward high frequency noise, as well as high frequency components that will be amplified. However, the noise dominates the high frequency components of the NIR spectra degrading the SNR at the output of the differentiator. The approach that is used to compute the derivative will determine the amount and features of produced noise in the differentiated spectra; the raw spectra should be smoothed prior to computing the derivative to suppress most of this noise.

The simple difference method, the Moving Window Polynomial Least Squares Fitting or the frequency domain techniques may be used.

The simple difference definition of the first derivative of a discrete spectrum a_i is expressed as:

$$a_i' = \frac{a_{i+1} - a_i}{\lambda_{i+1} - \lambda_i} \tag{30}$$

where λ_i are the sampling wavelength, $i = 1, 2, \ldots, m\text{-}1$, and m is the number of variables (wavelengths). For equal differences between the sampling wavelengths, the first derivative can be simplified as follows:

$$a_i' = a_{i+1} - a_i$$

The second derivative can be obtained by using the same definition and replacing a_i with its first derivative a_i', and so on. Despite its simplicity, the simple difference method has the main drawback that the peaks, maximum or minimum, are shifted relative to the raw spectra. In addition, the number of points of the differentiated spectrum is less than the raw spectrum. This method is suitable for spectra with high resolution.

The Moving Window Polynomial Least Squares Fitting method (MWPLSF) is based on the Savitzky-Golay smoothing filter. In this method, the *n'th* derivative is obtained by performing the smoothing operation against the *n'th* ordinary derivative of the polynomial, instead of using the polynomial directly as in the smoothing. Once the weighting coefficients are obtained they are used to get the differentiated spectra. The MWPLSF method will not introduce shifting in the peaks of the differentiated spectra but will set the two sides of the differentiated spectrum to zero.

The most practical method used to compute the derivative of the spectra is that using the Fourier transform. Compared to the MWPLSF method, the Fourier differentiator is simple, direct, fast and does not require complex tables. Furthermore, it does not produce a shifting in the peaks or suppress part of the spectrum ends. For example, the first derivative can be computed by multiplying the imaginary part of the Fourier transform of the spectrum by a linear ramp, then the inverse Fourier transform of the multiplier output is computed to get the differentiated spectrum in the wavelength domain.

17.4.4 Digital Bandpass Filtering (DBPF)

Digital bandpass filtering is simply a set of mathematical algorithms that modify the characteristics of the spectra profile in the frequency domain in a desirable and controlled way either as a time domain filtering process or as a frequency domain process.

NIR spectra are an example of frequency encoded signals. Information about a spectrum is present in the frequency components of its Fourier transform which reside in the mid-band range while the noise and baseline variations reside in the high frequency range and the low frequency range of the

Fourier domain respectively. Therefore, the quality of the raw spectra can be improved by using a suitable digital bandpass filter that can smooth the spectra by suppressing most of the noise components, and at the same time remove the baseline variations resulting in an improved performance of the calibration model (Fredric *et al.* 1997; Mattu *et al.* 1997). A number of techniques can be adopted to optimize the parameters of the filter and ensure convergence to global rather than local optima.

17.5 The Potential of Advanced Digital Signal Processing Techniques

Some recent work has shown the potential of combining advanced signal processing techniques with multivariate methods in the quantitative and qualitative analysis of Glucose spectra using NIR spectroscopy. For example, in Al-Mbaideen and Benaissa (2011) frequency self deconvolution (FSD) has been combined with partial least square regression (PLS) to obtain more information about glucose, to improve its appearance resolution, and to remove the baseline variations. The results showed significant improvement (70%) in the standard error of prediction SEP compared to the standard PLS method. In Al-Mbaideen and Benaissa (2011) a new method based on sub-band decomposition is adopted for coupling digital band pass filtering with Regression to improve the quality of the raw glucose absorbance spectra.

17.6 Examples of Application of NIR Spectroscopy in the Analysis of Glucose in Food Items

Many researchers investigated NIR spectroscopy as a method for rapid analysis of the sugar compositions in food items (Giangiacomo and Dull 1986; Rambla *et al.* 1997; Rodriguez-Saona *et al.* 2001, Xie *et al.* 2009).

Li *et al.* (1996) used NIR Spectroscopy and multivariate calibration for the quantitative analysis of glucose, fructose, sucrose, and citric and malic acids in orange juices using enzymatic assays as references. Stepwise multiple linear regression (SMLR) and PLS regression were used after preprocessing to create the calibration models for spectra of dry extract 128 orange juice samples captured in transmission mode in the region 1.1–2.5 μm. The authors concluded that the prediction performance of the calibration models is adequate when pre-processing is used with PLS slightly outperforming SMLR.

Liu *et al.* (2006) concluded that FT-NIR analysis of constituent sugar concentrations in Intact Apples offered more flexibility and was far quicker than high-performance liquid chromatography (HPLC). Savitsky-Golay second derivative, multiplicative scatter correction (MSC), and mean centering were employed as pre-processing options with PLS regression. Their best models for Glucose analysis were obtained from both the combination and the short-wave regions and used the second derivative with 9 factors (error of prediction 0.187%).

Chen Jie Yue and Ryuji (2007) used PLS regression to develop the calibration model for analyzing the sugar content of raw Ume juice. Pre-processing techniques in terms of normalization, first derivative, second derivative and multiplicative scatter correction (MSC) were used to try to remove the variation in the spectra. Their best model for Glucose used the second derivative with 9 PLS factors in the 1.1–1.85 μm range (error of prediction 0.068%).

Queji *et al.* (2010) reported that glucose in apple Pomace was determined by models developed in the NIR region using PLS regression with six factors yielding an average error lower than 7.4%. For improving their performance, further analysis was carried out for the regression coefficients of the best developed models.

17.7   Conclusion

This chapter gave an overview of the key processing techniques adopted in the qualitative and quantitative analysis of glucose using NIR spectroscopy. A brief overview of reported work applied to the glucose analysis in food was also presented. Advances in instrumentation and combining advanced digital signal processing techniques with chemometric techniques have a strong potential for overcoming many practical problems and help exploit the proven advantages of NIR spectroscopy to provide the food industry with rapid and specific tools for analysis of sugars in food for the reliable assessment of its constituents' quality and safety. Reliable and simple online quantification of glucose in food items is very desirable for nutrition and dietary purposes.

Summary Points

- This chapter focuses on the techniques adopted for the quantitative and qualitative analysis of glucose from NIR spectra; a brief review of reported work on application to food/nutrition is given.
- Glucose is best analysed in the combination region of the NIR spectrum (2–2.5 μm)
- Multivariate calibration techniques are crucial for the analysis of NIR data.
- PLS regression is by far the most efficient and most widely employed technique.
- Pre-processing is necessary in practice.
- Band pass filtering is the most efficient technique for pre-processing of raw spectra prior to regression.
- Combining advanced digital signal processing techniques with multivariate calibration (ideally in an adaptive way) would yield significant improvements in performance.

Key Facts about NIR Spectroscopy

- NIR Spectroscopy is a vibrational spectroscopy that uses electromagnetic radiation in the IR range from 4000–12800 cm^{-1} (2.5–0.78) μm. The NIR spectrum can be divided into three regions; first overtone region 6500–5500 cm^{-1} (1.54–1.82 μm), combination region 5000–4000 cm^{-1} (2–2.5 μm), and short wave length region 14286–7500 cm^{-1} (0.7–1.33 μm)

- The NIR response results from overtone and combination vibration of the fundamental vibrations of the C-H, O-H, and N-H bonds. The variation of intensity of the reflected or absorbed NIR radiation is correlated with the variation of the entire components concentration and the penetration depth of the NIR radiation in a sample specifically depends on the used wavelength range, power of radiation and the type of sample.

- Most biological components have absorption bands in the NIR range and an NIR response of a component can be considered as its fingerprint that is identified by its spectrum profile, wavelength range, and the number of peaks and their positions.

- NIR spectroscopy can be used to study the physical and biological properties, in addition to the molecular structure of components. However, the NIR spectra of biological components are weak in intensity, broad and overlapped with nearly similar response to the NIR radiation since they are composed from different forms of C-H-N-O bonds. Practical factors such as instrumental noise, baseline variations and light scattering can have detrimental effect on the quality of the collected spectra. Therefore, chemometrics along with advanced digital signal pre-processing techniques are crucial in the quantitative and qualitative analysis of NIR spectra.

Definitions

Chemometrics: refers to mathematical and statistical procedures and techniques that are employed in the qualitative and quantitative analysis to extract information from chemical data.

Pre-processing: This is the general term used to describe all those techniques which are used for improving the quality of the data before building the calibration model for example by removing the high frequency noise and baseline variations, by enhancing the appearance resolution of the collected data, by selecting the appropriate spectral range, and by compensating the negative influence of temperature variation.

Digital Signal Processing (DSP): refers to the digital algorithms and techniques that are used to manipulate and study signals to enhance their quality or accuracy, to compress them, or to extract specific information from them; the signals can be in the time domain, frequency domain, wavelet domain, spatial domain, autocorrelation domain, or spectral domain where each domain provides different information.

Linear Regression: It is a statistical measure which attempts to determine the strength of relationship between one and one or more dependent variables. A straight line best fits the relationship in this regression.

Correlation Coefficient (R^2): The Correlation Coefficient R measures how well the prediction agrees with the reference. R has the value from -1.0 to 1.0 whereas R^2 has the value from 0 to 1.0. Where 0 indicates there is no correlation and 1.0 means the perfect linear relationship.

Cross-validation: It is a validation technique based on the calibration data only. The estimate based on this technique is called root mean square error of cross-validation (RMSECV).

Prediction testing: This type of validation is based on splitting the data set into two sets, one for building the calibration model and the other for testing/validation.

List of Abbreviations

BPF	Band Pass Filter
DSP	Digital Signal Processing
FFT	Fast Fourier Transform
FSD	Frequency Self Deconvolution
FTIR	Fourier Transform Infrared Spectroscopy
HPLC	High-Performance Liquid Chromatography
MLR	Multiple Linear Regression
MMPE	Monitoring Mean Percent Error
MSC	Multiplicative Scatter Correction Error
MWA	Moving Window Average
MWPLSF	Moving Window Polynomial Least Squares Fitting
NIPALS	Nonlinear Iterative Partial Least Squares
NIR	Near Infrared
PCA	Principal Component analysis
PCR	Principal Component Regression
PLS	Paratial Least Square Regression
R^2	The Square of Correlation Coeffiecient
SEC	Standard Error of Calibration
SEP	Standard Error of Prediction
SMLR	Stepwise Multiple Linear Regression
SNR	Signal to Noise Ratio

References

Abdi, H., and Williams, L.J., 2010. "Principal Component Analysis," Wiley Interdisciplinary Reviews: Computational statistics, Vol. 2, 4, pp. 433–459.

Andersson, M., 2009. "A comparison of nine PLS1 algorithms," *J. Chemometrics.*, 23, pp. 518–529.

Al-Mbaideen, A., and Benaissa, M., 2011. "Frequency Self Deconvolution in the Quantitative Analysis of NIR Spectra," *Analytica Chimica Acta.*, Vol. 705(1–2), pp. 135–47.

Al-Mbaideen, A., and Benaissa, M., 2011. "Coupling Subband Decomposition and Independent Component Regression for Quantitative NIR Spectroscopy," *Chemometrics and Intelligent Laboratory Systems*, Volume 108, Issue 2, 112–122.

Burns, D.A., and Ciurczak, E.W., 2008. "Handbook of Near Infrared Analysis," Taylor and Frances Group, Third edition.

Chau, F., Liang, Y., Gao, J., and Shao, X., Mar 2004. "Chemometrics: From Basics to Wavelet Transform," John Wiley & Sons, Inc, Volume 164, 23.

Chen, J., Arnold, M.A., and Small, G.W., 2004. "Comparison of Combination and First Overtone Spectral Regions for Near-Infrared Calibration Models for Glucose and Other Biomolecules in Aqueous Solutions, " *Anal. Chem.*, 76, pp. 5405–5413.

Chen Jie Yu, Z.H., and Ryuji, M., 2007. "Visible and Near Infrared Spectroscopy for Rapid Analysis of the Sugar Composition of Raw Ume Juice," *Food Science and Technology Research,* vol. 13.

Faber, N.M., and Rajk, R., 2007. "How to avoid over-fitting in multivariate calibration—The conventional validation approach and an alternative," *Analytica Chimica Acta* 595, pp. 98–106.

Fredric, R., Ham, M., Kostanic, I., Cohn, G.M., and Gooch, B.R., June 1997. "Determination of glucose concentrations in an aqueous matrix from NIR spectra using optimal time domain filtering and partial least squares regression," *IEEE Transactions on Biomedical Engineering*, Volume 44, Issue 6, pp.75–485.

Gorry, P.A., 1990. "General least-squares smoothing and differentiation by the convolution (Savitzky-Golay) method," *Anal. Chem.* 62 (6), pp. 570–573.

Geladi, P., and Kowalski, B.R., 1986. "Partial Least Squares Regression: A Tutorial," *Analytical Chimica Acta*, 185, pp. 1–17.

Giangiacomo, R., and Dull, G., May 1986. "Near Infrared Spectrophotometric Determination of Individual Sugars in Aqueous Mixtures", *Journal of Food Science*, Volume 51, Issue 3, pages 679–683.

Griffiths, P.R., and De Haseth J.A., 2007. "Fourier Transform Infrared Spectrometry," Second Edition, Wiley-Interscience A John Wiley & Sons Inc.

Haaland, D.M., Thomas, E.V., 1988. "Partial Least-Squares Methods for Spectral Analyses. 1. Relation to Other Quantitative Calibration Methods and the Extraction of Qualitative Information," *Anal. Chem.*, 60, pp. 1193–1202.

Joliffe, T., 2002. "Principal Component analysis," Springer, second edition.

Kauppinen, J.K., Moffatt, D.J., Manisch, H.H., and Cameron, D.G., 1981. "Fourier Transform in the Computation of Self Deconvoluted and First Order Derivative Spectra of Overlapped Band Contours," *Anal. Chem*, 53, pp. 1454–1457.

Li, W., Goovaerts, P., and Meurens, M., 1996. "Quantitative Analysis of Individual Sugars and Acids in Orange Juices by Near-Infrared Spectroscopy of Dry Extract," *Journal of Agricultural and Food Chemistry*, Vol. 44, pp. 2252–2259.

Liu, Y., Ying, Y., Yu, H., and Fu, X., 2006. "Comparison of the HPLC method and FT-NIR analysis for quantification of glucose, fructose, and sucrose in intact apple fruits," *Journal of Agricultural and Food Chemistry*, Volume: 54, Issue: 8, pp. 2810–2815.

Mattu, M.J., Small, G.W., and Arnold, M.A., 1997. "Determination of Glucose in a Biological Matrix by Multivariate Analysis of Multiple Bandpass Filtered Fourier Transform Near-Infrared Interfeograms," *Analy Chem*, Vol. 69, issue 22, pp. 4695–4702.

Osborne, B.G., September 2006. "Near-infrared spectroscopy in food analysis", *Encyclopedia of Analytical Chemistry*, DOI: 10.1002/9780470027318.a1018, Wiley online library.

Queji, M.D., Wosiacki, G., Cordeiro, G.A., Peralta-Zamora, P.G., and Nagata, N., 2010. "Determination of simple sugars, malic acid and total phenolic compounds in apple pomace by infrared spectroscopy and PLSR," *International Journal of Food Science & Technology,* vol. 45, pp. 602–609.

Rambla, F.J., Garrigues, S., and de la Guardia, M., 1997. "PLS-NIR determination of of total sugar, glucose, fructose, and sucrose in aqueous solutions of fruit juices", *Analytica Chimica Acta*, 344, 41–53.

Rodriguez-Saona, L.E., Fry, F.S., McLaughlin, M.A., and Calvey, E.M., November 2001. "Rapid analysis of sugars in fruit juices by FT-NIR spectroscopy", *Carbohydrate Research*, Vol. 336, Issue 1, pp. 63–74.

Savitzky, A., and Golay, M.J.E., 1964. "Smoothing and Differentiation of Data by Simplified Least Squares Procedures," *Anal. Chem.*, 36 (8), pp. 1627–1639.

Xie, L., Ye, X., Liu, D., and Ying, Y., 2009. "Quantification of glucose, fructose and sucrose in bayberry juice by NIR and PLS", *Food Chemistry*, Volume: 114, Issue: 3, pp. 1135–1140.

CHAPTER 18

A New Liquid Chromatographic-Mass Spectrometric Method to Assess Glucose Kinetics In Vivo Using $^{13}C_6$ D-glucose as a Tracer

DAVID S. MILLINGTON,*[a] HAOYUE ZHANG,[a] CRIS A. SLENTZ[b] AND WILLIAM E. KRAUS[b]

[a] Department of Pediatrics, Biochemical Genetics Laboratory, Duke University Medical Center, 801-6 Capitola Drive, Durham, NC 27713, USA; [b] Department of Cardiology, Duke Center for Living, 1300 Morreene Road, Pmb 3022, Durham, NC 27705-4509, USA
*Email: dmilli@duke.edu

18.1   Background

Metabolic studies *in vivo* are required to understand the mechanisms of impairment or dysfunction of key molecules, such as glucose, and to assess the impact of diet, exercise and other interventions designed to correct such dysfunction. Stable isotope labeling is one of the most valuable tools available to the researcher for this type of study. There is a large body of literature on the use of stable isotopes in glucose research. In particular, the labeled intravenous glucose tolerance test (IVGTT) in conjunction with computer modeling has

Food and Nutritional Components in Focus No. 3
Dietary Sugars: Chemistry, Analysis, Function and Effects
Edited by Victor R Preedy
© The Royal Society of Chemistry 2012
Published by the Royal Society of Chemistry, www.rsc.org

been developed and utilized as a noninvasive method to quantify indices of glucose metabolism. Using this approach, investigators have shown that circulating glucose is cleared under the action of two distinct forces, glucose mediated glucose control (glucose effectiveness) and insulin mediated glucose disposal (insulin sensitivity) (Bergman 2007). The model considered to best represent the true physiological state is the two-pool glucose (TPG) model, also called two-compartment minimal model (Caumo and Cobelli 1993, Vicini *et al.* 1997). In this scenario, the glucose distribution volume is divided into an accessible, rapidly equilibrating pool and a slowly equilibrating exchangeable pool that is assumed to be insulin sensitive. The use of isotope labeled glucose allows separation of the insulin effect on glucose disposal and changes in hepatic glucose production. Typically, a large glucose bolus containing the tracer is infused over a short time period, then blood samples are collected at multiple time points over the ensuing 4-hr period.

Another important parameter, the fasting state hepatic glucose output (HGO), is of interest. Current methods of assessing fasting HGO increasingly involve the use of stable isotopes and generally utilize a priming dose followed by a constant infusion for 90–120 minutes, by which time the isotope enrichment has reached steady state and the rate of glucose production can be determined (Hovorka *et al.* 1997). A common isotope used for HGO is $[6,6-^2H^2]$-D-glucose and typically requires a 6.0 mg/kg priming dose with a 0.1 mg/kg/min constant infusion (Tigas *et al.* 2002).

The traditional analytical method for determination of isotope enrichment is GC/MS with selected ion monitoring after conversion of glucose to a stable derivative such as the pentaacetate or α-D-glucofuranose cyclic 1,2:3,5-bis(butylboronate)-6-acetate (Avogaro *et al.* 1989). Typically, if $^2H_2$-labeled glucose is used, at least 8% of it is required in the bolus for IVGTT to enable accurate determination of isotope ratios. It is possible to reduce the amount of stable isotope required by 15-fold, essentially to tracer levels, by using $^{13}C_1$-D-glucose as a tracer and GC-IRMS to determine isotopic enrichment (Bluck *et al.* 2005; Clapperton *et al.* 2002). The specialized equipment required is, however, not widely available. An economically viable alternative is to employ $^{13}C_6$-D-glucose as a tracer, which both minimizes the impact of possible carbon recycling and enables a reduction of the amount of tracer to less than 1% of the IVGTT bolus without requiring specialized analytical equipment.

Instead of utilizing GC-MS, we adopted a novel approach using UPLC-ESIMS, which is fast becoming standard equipment in clinical diagnostic and research laboratories. This requires the formation of a suitable derivative for ESI. The reaction of glucose and other reducing sugars to form Schiff's base derivatives is well documented (Delpathado *et al.* 2005), and the butyl-PABA derivative formed by the reaction with butyl 4-aminobenzoate and sodium cyanoborohydride is both chemically stable and has excellent properties for ESI (Harvey 2000; Poulter and Burlingame 1990). We have successfully used this method to analyze glucose tetrasaccharides in blood and urine using both HPLC and HPLC-ESI-MS (An *et al.* 2000; Young *et al.* 2003). The latter assay is now used clinically to monitor patients with Pompe disease, a severe disorder of glycogen catabolism, that are receiving enzyme replacement therapy (Young *et al.* 2009). We reasoned that a similar method should simplify the analysis of glucose isotopes used in IVGTT and HGO studies. This chapter focuses on the new method and its application in pilot investigations of both types of isotope labeling study. The labeled IVGTT study using $^{13}C_6$-D-glucose has been previously published (Zhang *et al.* 2009).

18.2   Summary of Procedures and Methods

18.2.1   The Labeled IVGTT Protocol

The IVGTT is performed essentially as described previously (Bergman 2007; Caumo and Cobelli 1993). Briefly, a bolus of 300 mg/kg D-glucose plus 2 mg/kg $^{13}C_6$-D-glucose is infused intravenously over a period of approximately 1 min and 3 mL blood specimens are collected in purple-top tubes (containing ethylenediaminetetraacetic acid as the anticoagulant) from a separate line at times t = −1, 2, 3, 4, 5, 6, 10, 12, 15, 20, 30, 40, 60, 80, 100, 120, 140, 150, 180, 210, 240 min relative to the bolus (t = 0 min). The specimens are centrifuged and the plasma collected. Specimens are assayed for glucose and insulin by standard clinical laboratory methods, and a separate aliquot of each (1 mL) is retained for isotope analysis by mass spectrometry. Plasma should be kept frozen at −20°C or below prior to analysis.

18.2.2   The HGO Infusion Protocol

The procedure is performed essentially as described previously (Hovorka *et al.* 1997; Tigas *et al.* 2002). A priming bolus of 2 mg/kg $^{13}C_6$-D-glucose is delivered intravenously over a period of approximately 0.5 min followed immediately by a constant infusion of $^{13}C_6$-D-glucose at a rate of 0.02 mg/Kg/min over a period of 2 hr. Three mL blood specimens are collected in blood collection tubes (containing ethylenediaminetetraacetic acid as the anticoagulant) from a separate line just before the priming bolus and at intervals of 60, 75, 90, 105 and 120 min after the start of the constant infusion. Specimens are assayed for glucose and insulin by standard clinical laboratory methods, and a separate aliquot of each (1 mL) is retained for isotope analysis by mass spectrometry. Plasma is kept frozen at −20°C or below prior to analysis.

18.2.3   Sample Preparation and Analysis by Mass Spectrometry

Details of the methodology for sample preparation and analysis have been published (Zhang *et al.* 2009). Briefly, a 100 μL aliquot of each plasma sample is deproteinized by adding 500 μL methanol and vortex mixing for 20 s, followed by centrifugation at 2000 g for 5 min. Supernatants are evaporated under nitrogen and the dried extracts incubated with the derivatization reagent (butyl 4-aminobenzoate and sodium cyanoborohydride) for 1 hr at 80 °C. The plasma

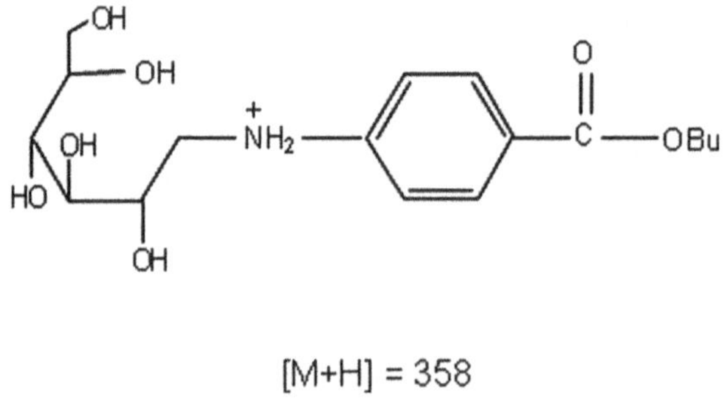

[M+H] = 358

Figure 18.1   Structure of the protonated molecule of the butyl-4-aminobenzoate derivative of glucose. This glucose derivative is easily formed and very stable. The basic nitrogen atom is readily protonated in solution and sensitive to ESI-MS.

glucose is thereby converted to the derivative shown in Figure 18.1. Derivatized samples are purified using solid phase extraction (SPE) cartridges and the eluates stored at 4 °C prior to analysis. Typically, 20 μL of eluate is diluted with 150 μL of mobile phase (10 mM ammonium acetate in methanol:water 80:20 (v/v)), and 5 μL of this mixture is injected into the UPLC-MS system. The UPLC is operated in isocratic mode, and the quadrupole mass analyzer is operated in selected ion recording (SIR) mode. The total analysis time for each sample is approximately 2.2 min.

The derivatized glucose isotope signals corresponding to $A+2$ (m/z 360), $A+3$ (m/z 361) and $A+6$ (m/z 364) are collected by selected ion recording during the elution profile, where A is the most abundant isotope signal of the protonated derivative (m/z 358 – see Figure 18.1). The selection of $A+2$ as the glucose signal is based on its comparable order-of-magnitude intensity with the $A+3$ signal and $A+6$ signal at peak value. The $A+3$ signal is monitored to check for possible recycling of the $^{13}C_6$-D-glucose.

18.2.4   IVGTT: Estimation of $^{13}C_6$ Glucose Concentration

The $^{13}C_6$ glucose concentration in each specimen is calculated according to equation (1) below:

$$g^* = G_t \times ([A+6]/[A+2]) \times 100)/(([A+6]/[A+2]) \times 100 + 3759.4) \qquad (1)$$

Here, g^* and G_t are the $^{13}C_6$-D-glucose concentration (calculated) and total glucose concentration (measured at time t) in mg/L respectively, $[A+2]$ and $[A+6]$ are the absolute intensities of the monitored isotopes of the glucose derivative, and 3759.4 is the theoretical sum of the peak intensities of A, $A+1$, $A+2$, $A+3$, $A+4$, $A+5$ in natural abundance normalized to $A+2$ peak intensity (arbitrarily 100). We make the reasonable assumption that there is zero natural abundance of $A+6$.

18.2.5　HGO: Estimation of Fasting HGO

The calculation of HGO by primed infusion is based on the formula u/r, where u is the rate of infusion of $^{13}C_6$-D-Glucose and r is the isotopic enrichment of the natural glucose at steady state. The formula used to determine r is given in equation (2) below:

$$r = 1.16[(A+6/A+2)_M]/3759.4 \tag{2}$$

where $(A+6/A+2)_M$ is the mean isotopic abundance ratio for the A + 6 and A + 2 isotopes of the glucose derivative at steady state, after subtracting the small residual (baseline) value of A + 6, if any is observed. The factor 1.16 is required to account for the skewed natural isotope abundance distribution (Rosenblatt *et al.* 1992), and the factor 3759.4 is the theoretical sum of natural isotope intensities normalized to the A + 2 peak intensity (arbitrarily 100).

18.3　Results

18.3.1　IVGTT Results

Figure 18.2 is an example showing the UPLC-ESIMS selected ion current chromatograms for the A + 2 and A + 6 isotopes of the protonated molecular

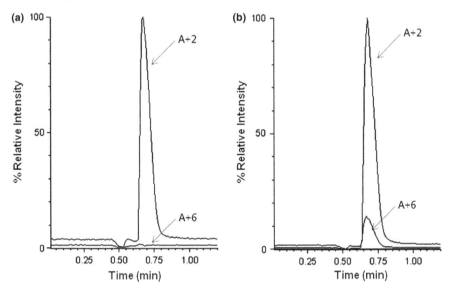

Figure 18.2　Selected ion monitoring traces corresponding to A + 2 and A + 6 isotope signals of the glucose derivative in plasma specimens (a) drawn at baseline and (b) drawn 5 min after bolus infusion. These chromatograms of the glucose butyl-4-aminobenzoate derivative ion A + 2 (m/z 360) and $^{13}C_6$-D-glucose butyl 4-aminobenzoate derivative ion A + 6 (m/z 364) are generated by UPLC-ESI-MS in selected ion recording (SIR) mode. The signal of A + 6 isotope ion is essentially zero at baseline (a) and is markedly elevated, close to its peak value, at 5 min after bolus infusion (b).

ion ($[M + H]^+$) cluster from a normal subject at times 0 and 5 min respectively. The $^{13}C_6$-D-glucose value (A + 6 isotope) at baseline is essentially zero, and markedly increases, close to its peak value, at 5 min. The A + 6 isotope signal is still clearly discernible even 240 min after infusion. The retention time on UPLC-MS is 0.7 min and the cycle time for the entire analysis is only 2.2 min.

Figure 18.3 shows a typical $^{13}C_6$-D-glucose clearance curve during IVGTT measured by the isotope abundance method. For most subjects, the $^{13}C_6$-D-glucose starts to decrease around 10 min after infusion and continues until 240 min. The labeled glucose concentrations do not reach baseline concentration even at the end of the IVGTT time course. The imprecision in these measurements, determined by replicate analysis, is less than 0.6% (CV) until at least the 40-minute mark and increases to about 1.7% by time 240 min. These errors are much lower than previously reported using other methods, probably because of the use of the $^{13}C_6$ isotope that has negligible natural abundance in unlabeled glucose. It is noteworthy that a consequence of the higher precision in the isotope abundance calculations, the minimum number of data points required to compute the modeling parameters is much less than the number actually collected, which implies that essentially the same results are achievable with about one-third of the number of blood draws used in this and previous studies.

Figure 18.3 also shows a plot of the endogenous glucose concentration at each time point in the same subject, calculated from equation (3) (below):

$$G_e(t) = G_t(t) - 151^*g(t) \tag{3}$$

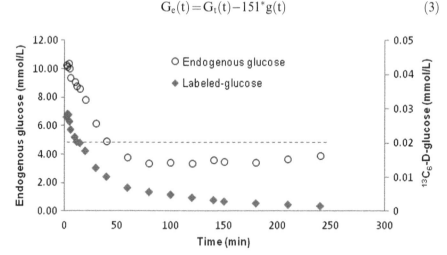

Figure 18.3 Time course of $^{13}C_6$-D-glucose disposal and calculated endogenous glucose from one subject undergoing the labeled IVGTT procedure. The figure shows that the concentration profiles of $^{13}C_6$-D-glucose and calculated endogenous glucose after the labeled IVGTT bolus infusion of both D-glucose (300 mg/Kg) and $^{13}C_6$-D-glucose (2 mg/Kg). The peak values occur at about 6 min, then the $^{13}C_6$-D-glucose decreases smoothly while the calculated endogenous glucose falls below its basal value (represented by the dotted line) and slowly recovers. Neither parameter recovers to its baseline value within the time limit of the experiment (240 min).

Here, $G_e(t)$, $G_t(t)$ and $g(t)$ are the endogenous glucose, total blood glucose and labeled glucose concentrations at time t, respectively. $151*g(t)$ is the sum of the infused cold glucose and labeled glucose in the blood stream at time t estimated from the bolus ratio. The endogenous glucose decreases as expected after glucose infusion but does not recover completely to the baseline concentration, even after 240 min. These data reflect changes in hepatic glucose production during the IVGTT. Recycling of the $^{13}C_6$ glucose isotope *via* the Cori cycle would be expected to produce $^{13}C_3$-glucose, which would result in an increased signal ratio of $[A+3]/[A+2]$. For most of the time course of the study, up to at least 180 min, the signal ratio remains constant (within the experimental error), then shows a slight tendency to increase. It is possible that suppression of the HGO by the large glucose bolus might account for the apparent lack of recycled labeled glucose, and therefore it is deemed unnecessary to account for it in the calculations.

Several parameters from 10 Caucasian male subjects, estimated using a Two-Pool Glucose (TPG) modeling program (Boston *et al.* 2007), are summarized in Table 18.1, where S_i^{2*} and S_G^{2*} represent insulin sensitivity and glucose effectiveness respectively and PCR is the glucose plasma clearance rate (PCR) (Zhang *et al.* 2009). These parameters are compared with those previously reported in the literature from Caucasian males using GC-MS in conjunction with other stable isotope labeled forms of glucose (Table 18.1). Differences in the values are relatively minor and may be due to differences in diet and other genetic and environmental factors. More details of this study and its application have been recently published (Surwit *et al.* 2009).

Table 18.1   Comparison of metabolic parameters estimated by two pool model in different studies. The data shown are comparisons of the study using the new LC-MS method (Zhang et al. 2009: 10 Caucasian male adults) with previously published reports on Caucasian males using methods based on GC-MS. In Cobelli *et al.* (1997): 14 young adults; [a]Nishida *et al.* (2002): 12 middle-aged, sedentary men; [b]Nishida *et al.* (2002): middle-aged trained men; Boston *et al.* (2007): 15 healthy young Nordic men. Notes: S_i^{2*} represents insulin sensitivity; S_G^{2*} glucose effectiveness; EPG_b basal endogenous glucose production. Mean ± SE: population mean and standard error.

Reference	S_I^{2*}	S_G^{2*}	EGP_b
	$10^2 ml.kg^{-1}.min^{-1}\mu U^{-1}.ml$	$ml.kg^{-1}.min^{-1}$	$mg.kg^{-1}.min^{-1}$
Cobelli *et al.* (1997)	13.83 ± 2.54	0.85 ± 0.14	–
[a]Nishida *et al.* (2002)	11.9 ± 2.4	0.60 ± 0.05	1.59 ± 0.05
[b]Nishida *et al.* (2002)	24.6 ± 3.0	0.81 ± 0.08	1.82 ± 0.08
Boston *et al.* (2007)	18 ± 1	1.19 ± 0.04	1.9 ± 0.09
Zhang, *et al.* (2009)	22.4 ± 2.8	1.6 ± 0.2	2.6 ± 0.2

18.3.2 HGO Results

Analogous to the results from the IVGTT study shown in Figure 18.2, the signal for the $A + 6$ isotope of the glucose derivative at steady state corresponds almost exclusively to the infused $^{13}C_6$-D-glucose and far exceeds the baseline signal at $A + 6$ prior to the infusion. Any small baseline signal is nevertheless accounted for in the simple calculation of isotopic enrichment equation (2). Recycling of the infused $^{13}C_6$-D-glucose is observed as $^{13}C_3$-glucose by monitoring the signal at $A + 3$ during the LC-MS analysis. There is an increasing elevation of the ratio of $[A + 3]/[A + 2]$ during the experiments that reaches about 14% by the 90-minute time point compared with the zero time point in all subjects. This corresponds to approximately 3.5% of the amount of the infused isotope-labeled glucose. On theoretical grounds, the contribution of recycling $^{13}C_3$-glucose back to $^{13}C_6$ glucose would be infinitesimal.

The clinical value of the measurement of fasting HGO by the new method can be assessed by comparing the assay imprecision with measurements of HGO performed on the same subjects at different time points. In Table 18.2 are shown the HGO values determined from each of 4 healthy subjects (two male and two female) on two occasions that were from two to three months apart. The range of HGO values for these subjects is 1.82–2.63 and compares with previously reported value for healthy adults (Hovorka *et al.* 1997). The individual assays were initially performed (occasion 1) at time points 60, 75, 90, 105 and 120 min post infusion. Because the mean variation in isotopic enrichment all time points is within the assay imprecision of less than 3% (CV), subsequent collections (occasion 2) were made at 45, 60, 75 and 90 min from the start of infusion. The first data point in each set was not used in the HGO calculations, and one data point for subject HGO-001 (occasion 2) is missing owing to lack of a blood draw. The isotope enrichment profiles of two subjects using these different collection protocols are shown in Figure 18.4. These results indicate that steady state isotopic enrichment is achieved within 60 min, probably even within 45 min. The imprecision of the mean value for each HGO measurement is in the same order of magnitude as the assay imprecision, from which we may

Table 18.2 HGO values in mg/kg/min (mean $\pm$ SD (n)) determined on two separate occasions in four subjects. RSD is the relative standard difference (%) between the two determinations performed at least one month apart on each subject. The number of specimens used to compute the HGO for each patient is in parenthesis. The differences observed in some of these subjects exceed the methodological error and are probably due to physiological changes.

Subject:	HGO-001	HGO-002	HGO-003	HGO-004
Occasion 1	2.09 ± 0.07 (4)	2.32 ± 0.07 (4)	2.19 ± 0.04 (4)	1.97 ± 0.05 (4)
Occasion 2	2.41 ± 0.03 (2)	2.63 ± 0.01 (3)	1.82 ± 0.06 (3)	1.98 ± 0.06 (3)
RSD%	15.31	13.36	-16.89	0.51

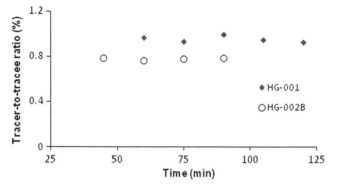

Figure 18.4   Isotope enrichment profiles, represented as tracer-to-tracee ratio, during a constant infusion of $^{13}C_6$-D-glucose for two subjects undergoing the HGO procedure. In this example are shown the isotope enrichment profiles from two healthy subjects that underwent a primed, constant infusion of $^{13}C_6$-D-glucose. In one example (HG-001, Table 18.2) the blood specimens were drawn at time points 60, 75, 90, 105 and 120 min after the priming dose and in the other case (HG-002B, Table 18.2), blood was drawn at 45, 60, 75 and 90 min. These examples show that tracer-to-tracee ratio was stable after 60 min and that only 3 blood draws, at times 0, 60 and 90 min should be required to determine HGO.

conclude that differences that exceed these values, between subjects and for different occasions in a given subject, are the result of physiological variations or changes. Furthermore, the data indicate that the HGO procedure could be reduced to 1.5 hr and a minimum of three blood samples, say at 0, 60 and 90 min post-infusion, without compromising the results.

The differences in HGO observed between the two time points (Table 18.2) reveal a significant increase in HGO for subjects 1 and 2 and a significant reduction for subject 3. The significance of these changes is unclear. Further studies are required to accurately determine the within-patient HGO variability.

18.4   Discussion and Future Direction

The new methods described here to calculate glucose parameters from IVGTT and to determine fasting hepatic glucose output (HGO) are significantly more practical, efficient and less costly than those previously reported. These efficiencies are realized by the low cost of the isotope (approximately \$80/g) and the relative simplicity of the specimen work up and analysis. For example, by using $^{13}C_6$-D-glucose, we are able to perform labeled IVGTT on a person of average body weight (80 kg) using only 0.16 g of the isotope tracer diluted with 24 g unlabeled glucose. Because the cost of this tracer is less than \$14 per patient, it is no longer a factor of significance in this type of study. The ability to detect this tracer and calculate its relative abundance using mass spectrometry at less than 1% in the infusate, which is further diluted in the body, is greatly facilitated by the lack of a perceptible endogenous signal corresponding to $^{13}C_6$-D-glucose at baseline. Although the minimum number of data points required for the TPG model has not yet been determined accurately, in future IVGTT studies with this isotope we anticipate collecting only about one-third of the number of specimens used in the current research, which will save both cost and inconvenience to the subjects. With regard to HGO determination, the time factor of 1.5 hrs to complete the infusion experiment and the minimum requirement of only 3 blood draws implies that a second procedure, such as the IVGTT, could be conducted sequentially without serious inconvenience to the subject. The ability to determine fasting HGO in a straightforward and minimally invasive manner will facilitate metabolic research studies in pre-diabetic and other individuals enrolled in clinical trials.

Although a tandem mass spectrometer was utilized in these studies, it is not required for the assay. A LC-MS system with a single quadrupole is adequate, and is more comparable in cost to a GC-MS system. The use of a UPLC system in place of standard HPLC is a time-saving convenience rather than a necessity for glucose studies. The high chemical stability of the glucose butyl-PABA derivative permits batching of prepared samples for subsequent analysis in 96-well microtiter plates, such that the sample flow on the mass spectrometer can be optimized. Derivatized samples stored for more than 12 months at either $-80\,^{\circ}C$ or $-20\,^{\circ}C$ show no significant differences on re-analysis.

We have also been able to show, using the $^{13}C_6$-D-glucose isotope that on the relatively short time scale of these studies, glucose recycling does not significantly impact the results.

In this chapter, we have focused on the method rather than on the interpretation of patient results. We have established proof of principle by showing that results from control subjects are similar to those previously published. The reader is referred to recent publications from this group for further information regarding application of the IVGTT method (Surwit *et al.* 2009; Zhang *et al.* 2009). We anticipate using the HGO method in patients enrolled in the ongoing STRRIDE-PD (Studies Targetting Risk Reduction Interventions through Defined Exercise in Pre-Diabetics) clinical trial (Slentz *et al.* 2005), where the assessment of HGO and its role in the pathophysiology of impaired fasting glucose is of major importance. The subjects in this study will have impaired fasting glucose or high normal fasting glucose (95–125 mg/dL) and will undergo a three-month control or run-in period followed by randomization into one of four different exercise or exercise plus diet interventions. It is intended to publish the outcomes of this trial when the results have been completed and analyzed.

Finally, there is no reason why the principles established for studies of glucose metabolism should not be extended to other sugars of biological interest. Any reducing sugar, many of which are now available as uniformly labeled ^{13}C analogues, should form a stable derivative with the reagent used in these glucose studies.

Summary Points

- This chapter focuses on the use of $>99\%$ $^{13}C_6$-enriched glucose as a tracer to study glucose metabolism *in vivo*.
- Two commonly used glucose labeling experiments are described: the isotope labeled intravenous glucose tolerance test (IVGTT) and the constant glucose isotope infusion test to determine hepatic glucose output (HGO).
- The method for measurement of the glucose isotope abundances is based on the conversion of glucose to its butyl-*p*-aminobenzoate and analysis by UPLC-MS.
- The analytical process is considerably less complex and technologically challenging that previously published methods based on either $^{13}C_1$ or $^2H_2$ enriched glucose tracers in combination with GC-MS.
- The isotope enrichment determinations and subsequent calculations are greatly simplified by the fact the tracer is virtually undetectable in endogenous glucose.
- The imprecision of the isotope calculations at each time point (approximately 2 to 3% coefficient of variation) enables accurate calculations with fewer sample collections than those required by previously reported methods.
- The reduced cost and invasiveness of this new method, coupled with its straightforward analytical procedure using standard equipment, will facilitate studies of this type with large patient cohorts.

Key Facts of Glucose Metabolism

1. Glucose levels in the blood are generally maintained within the normal range by the action of the hormones glucagon, which releases stored glucose from hepatic glycogen stores as required, and insulin, which mediates the disposal or transport of glucose into cells for metabolic processing.
2. Diabetes Mellitus (DM) is a serious disorder caused by a failure to maintain blood glucose levels by insulin mediated disposal.
3. An increasingly common form of DM, referred to as DM type II, is posing a serious health risk to millions of individuals, especially in Western societies.
4. Risk factors for DM type II include overweight, lack of exercise and dietary excess of fat and/or carbohydrate. It is thought that the condition can be alleviated and possibly reversed with medical intervention, especially if in the early onset stage (pre-diabetic).
5. Clinical trials that include weight loss, exercise programs, dietary changes and other interventions require determination of measurable parameters of glucose metabolism to assess their effectiveness.
6. Certain parameters of glucose metabolism, such as its rate of production by the liver (hepatic glucose output – HGO), are accessible *in vivo* only by stable isotope labeling studies.

Key Facts of Liquid Chromatography and Mass Spectrometry

1. Liquid chromatography coupled to mass spectrometry is arguably the most powerful analytical tool for the analysis of specific organic components of body fluids. It is widely used in clinical research and diagnostic laboratories.

2. Compounds in blood that are targeted for analysis are generally purified or enriched prior to analysis by relatively simple procedures such as protein precipitation and solid phase extraction. Sometimes they are also converted to derivatives that enhance their detection by the mass spectrometer.

3. Depending on the concentration of the compound(s) of interest, prepared samples may be directly infused into the mass spectrometer or, more often, subjected to prior separation by on-line liquid chromatography (LC).

4. The LC-MS system consists of a pump that delivers solvent to a column packed with particles that interact with compounds introduced into in the flowing solvent. The competing forces of the solvent flow and particle interactions result in the separation of components on the column. The eluate from the column flows continually into the mass spectrometer (MS), which is set up to detect the targeted component(s) based on their molecular weight and chemical structure.

5. The ionization mechanism used in LC-MS systems is called electrospray ionization (ESI), which permits solutions to flow directly into the MS *via* an interface that produces a fine mist of electrically charged droplets, which eventually become de-solvated by the action of the vacuum within the MS leaving charged solute molecules with their structures intact.

6. Columns used for LC can be of "high performance" (HPLC) or "ultra performance" (UPLC), depending on the diameter and other properties of the particles in the columns, that affect their separating power or resolution.

7. In general, UPLC separations take much less time and provide significantly higher sensitivity for the same mixture components than HPLC.

Definitions and Explanations of Key Terms

Intravenous glucose tolerance test (IVGTT): a procedure whereby a large bolus of glucose is infused intravenously over a short time period (1 min) and blood samples are drawn at intervals to determine glucose and insulin levels. The test is generally used to identify patients with insulin resistance.

Isotope-labeled IVGTT: a variation on the IVGTT procedure whereby patients in the fasted, resting state are infused with a large bolus of glucose enriched with a stable isotope labeled form of glucose. The physiological response to this challenge is assessed by using computer modeling and measurements of glucose, insulin and isotope enrichment at numerous time points after the infusion. More detailed assessment of glucose kinetics is possible than with the standard IVGTT.

HGO (Hepatic Glucose Output): a physiological parameter that corresponds to the rate of glucose production by the liver. It is usually measured in the fasted state.

Two Compartment Model: a mathematical model that represents the two major physiological compartments into which glucose is distributed in the body. The model is not necessarily strictly correct, but is considered to be a reasonable approximation to the true physiological state.

Primed isotope infusion: A procedure whereby a stable isotope enriched form of a biologically important molecule is infused intravenously first as a bolus, then at a constant steady rate for a period of up to 2 hrs. Blood samples are drawn at baseline and at several intervals thereafter and the isotope enrichment is calculated in each sample. The objective is to reach a steady state of isotope enrichment in the blood as quickly as possible, and from that measurement, the rate of appearance in the blood for that molecule and other important parameters can be calculated.

Isotope enrichment: In the context of *in vivo* isotope labeling studies, this term refers to the excess of a stable isotope over its natural abundance in a given molecule as a result of admixture with an isotopically labeled form of that molecule. The calculation requires mass spectrometric analysis of isotope abundances in the molecular ion complex at baseline and after administration of the labeled compound.

$^{13}C_6$-D-glucose: This is a form of glucose that is uniformly labeled with ^{13}C at 99% isotopic purity. It is produced cost-effectively along with many other uniformly ^{13}C-enriched molecules in bioreactors, wherein organisms grow in an atmosphere of almost 100% $^{13}CO_2$. Each carbon atom in glucose is naturally labeled with only about 1.1 % ^{13}C. Thus, there is virtually no overlap of the major isotope of $^{13}C_6$-glucose with the isotope cluster of natural glucose, making it easy to measure with high precision at low enrichment by mass spectrometry.

Gas chromatography-mass spectrometry (GC-MS): A method of analysis of mixtures in the gas phase. The method is limited to relatively low molecular weight biomolecules, that are usually extracted by solvent extraction, purified and rendered sufficiently volatile by making suitable derivatives.

Liquid chromatography-mass spectrometry (LC-MS): A method of analysis of mixtures in the solution (liquid) phase, that is amenable to the analysis biomolecules with a large range of molecular weight including complex proteins.

Selected ion recording (SIR): A method of collecting mass spectrometric signals specific to one or more target molecules during their analysis by GC-MS or LC-MS. By this means, the MS is utilized in its most sensitive and specific mode, as a compound-specific detector.

List of Abbreviations

A	The base peak of the protonated molecular ion cluster
A + N	The Nth isotope peak of the protonated molecular ion cluster
butyl-PABA	n-butyl-*para*-amino benzoate
CV	Coefficient of variation
DM	Diabetes Mellitus
GC-IRMS	Gas chromatography-isotope ratio mass spectrometry
GC-MS	Gas chromatography – mass spectrometry
ESI	Electrospray ionization
HGO	Hepatic glucose output
HPLC	High performance liquid chromatography
IRB	Internal review board
IVGTT	Intravenous glucose tolerance test
LC-MS	Liquid chromatography-mass spectrometry
LC-ESIMS	Liquid chromatography-electrospray ionization mass spectrometry
M + H	The protonated molecular ion
MeCN	Acetonitrile
MeOH	Methyl alcohol
m/z	mass-to-charge ratio
PCR	Plasma clearance rate
SIR	Selected ion recording
SPE	Solid phase extraction
TPG	Two-pool glucose
UPLC	Ultra-performance liquid chromatography
UPLC-MS	Ultra-performance liquid chromatography-mass spectrometry

Acknowledgements

The labeled IVGTT research was supported by grant no R01-HL-076020 from the National Heart, Lung and Blood Institute and by the GCRC grant no 913. This study was also supported by the Program Project Grant no 5P01-HL-036587-19 and the Behavioral Medicine Research Center at Duke University School of Medicine.

Support for the HGO research was from NIH Grant R01DK081559.

References

An, Y., Young, S.P., Hillman, S.L., Van Hove, J.L., Chen, Y.T., and Millington, D.S., 2000. Liquid chromatographic assay for a glucose tetrasaccharide, a putative biomarker for the diagnosis of Pompe disease. *Analytical Biochemistry*. 287: 136–143.

Avogaro, A., Bristow, J.D., Bier, D.M., Cobelli, C., and Toffolo, G. 1989. Stable-label intravenous glucose tolerance test minimal model. *Diabetes*. 38: 1048–1055.

Bergman, R.N., 2007. Orchestration of glucose homeostasis: from a small acorn to the California oak. *Diabetes*. 56: 1489–1501.

Boston, R.C., Stefanovski, D., Henriksen, J.E., Ward, G.M., and Moate, P.J. 2007. AKA-TPG: a program for kinetic and epidemiological analysis of data from labeled glucose investigations using the two-pool model and database technology. *Diabetes Technology and Therapeutics*. 9: 99–108.

Bluck, L.J., Clapperton, A.T., and Coward, W.A., 2005. 13C- and 2H-labelled glucose compared for minimal model estimates of glucose metabolism in man. *Clinical Science (London)*. 109: 513–521.

Caumo, A., and Cobelli, C., 1993. Hepatic glucose production during the labeled IVGTT: estimation by deconvolution with a new minimal model. *American Journal of Physiology*. 264: E829–841.

Clapperton, A.T., Coward, W.A., and Bluck, L.J., 2002. Measurement of insulin sensitivity indices using 13C-glucose and gas chromatography/ combustion/isotope ratio mass spectrometry. *Rapid Communications in Mass Spectrometry*. 16: 2009–2014.

Dalpathado, D.S., Jiang, H., Kater, M.A., and Desaire, H., 2005. Reductive amination of carbohydrates using NaBH(OAc)3. *Analytical and Bioanalytical Chemistry*. 381: 1130–1137.

Harvey, D.J., 2000. Electrospray mass spectrometry and fragmentation of N-linked carbohydrates derivatized at the reducing terminus. *Journal of the American Society for Mass Spectrometry*. 11: 900–915.

Hovorka, R., Eckland, D.J., Halliday, D., Lettis, S., Robinson, C.E., Bannister, P., Young, M.A., and Bye, A., 1997. Constant infusion and bolus injection of stable-label tracer give reproducible and comparable fasting HGO. *American Journal of Physiology*. 273: E192–E201.

Nishida, Y., Tokuyama, K., Nagasaka, S., Higaki, Y., Fujimi, K., Kiyonaga, A, *et al.*, 2002. S(G), S(I), and EGP of exercise-trained middle-aged men estimated by a two-compartment labeled minimal model. *American Journal of Physiology*. 283: E809–816.

Poulter, L., and Burlingame, A.L., 1990. Desorption mass spectrometry of oligosaccharides coupled with hydrophobic chromophores. In: McCloskey, J.A. (ed.) *Methods in Enzymology: Mass Spectrometry*, Vol 193. San Diego: Academic Press, Inc. pp. 666–688.

Rosenblatt, J., Chinkes, D., Wolfe, M., and Wolfe, R.R., 1992. Stable isotope tracer analysis by GC-MS, including quantification of isotopomer effects. *American Journal of Physiology*. 263: E584–E596.

Slentz, C.A., Aiken, L.B., Houmard, J.A., Bales, C.W., Johnson, J.L., Tanner, C.J., Duscha, B.D., and Kraus, W.E., 2005. Inactivity, exercise, and visceral fat. STRRIDE: a randomized, controlled study of exercise intensity and amount. *Journal of Applied Physiology*. 99: 1613–1618.

Surwit, R.S., Lane, J.D., Millington, D.S., Zhang, H., Feinglos, M., Minda, S., Kuhn, C.M., Boston, R.C., and Georgiades, A., 2009. Hostility and minimal model of glucose kinetics in african american women. *Psychosomatic Medicine* 71: 646–651.

Tigas, S.K., Sunehag, A.L., and Haymond, M.W., 2002. Impact of duration of infusion and choice of isotope label on isotope recycling in glucose homeostasis. *Diabetes.* 51: 3170–3175.

Vicini, P., Caumo A., and Cobelli, C., 1997. The hot IVGTT two-compartment minimal model: indexes of glucose effectiveness and insulin sensitivity. *American Journal of Physiology.* 273: E1024–1032.

Young, S.P., Stevens, R.D., An, Y., Chen, Y.T., and Millington, D.S., 2003. Analysis of a glucose tetrasaccharide elevated in Pompe disease by stable isotope dilution-electrospray ionization tandem mass spectrometry. *Analytical Biochemistry.* 316: 175–180.

Young, S.P., Zhang, H., Corzo, D., Thurberg, B.L., Bali, D., Kishnani P.S., and Millington, D.S., 2009. Long-term monitoring of patients with infantile onset Pompe disease on enzyme replacement therapy using a glucose tetrasaccharide biomarker. *Genetics in Medicine* 11: 536–541.

Zhang, H., Stevens, R.D., Young, S.P., Surwit, R., Georgiades, A., Boston, R., and Millington, D.S., 2009. A convenient liquid chromatographic-mass spectrometric method for assessment of glucose kinetics in vivo using 13C6 glucose as a tracer. *Clinical Chemistry.* 55: 527–532.

CHAPTER 19

Self Monitoring of Blood Glucose (SMBG)

TOMOMI FUJISAWA

Department of Nephrology and Metabolism, Sakai City Hospital,
1-1-1, Minamiyasui-cho, Sakai-Ward, Sakai-City, Osaka, 590-0064, Japan
Email: fujisawa-t@sakai-hospital.jp

19.1 Background

There has been a continuous marked increase in the number of patients with diabetes mellitus – a disease that is known to be associated with significant morbidity and mortality due to its complications. Strict glycemic control has been shown to be effective for preventing or delaying the development and progression of these complications, and there has been progress in improving metabolic control of diabetes to this end. This improvement depends not only on the development of antidiabetic medications, but also on determining and implementing practices that enhance the ability of patients to achieve optimal control. Among the factors contributing to those practices, self-monitoring of blood glucose (SMBG) has been shown to play a significant role in achieving optimal glycemic control. SMBG is an effective technique that allows patients and the healthcare team to assess the effectiveness of actions towards control.

In the present review, clinical aspects of SMBG are discussed, including blood specimens, devices, sampling procedure, choice of SMBG system, and clinical benefits, as well as its limitations.

Food and Nutritional Components in Focus No. 3
Dietary Sugars: Chemistry, Analysis, Function and Effects
Edited by Victor R Preedy
© The Royal Society of Chemistry 2012
Published by the Royal Society of Chemistry, www.rsc.org

19.2   Blood Samples/Skin Puncture Site

The glucose concentration can differ among different blood specimens, usually showing a higher value in arterial blood than in capillary blood, and venous blood showing the lowest value (Table 19.1). In any blood specimen, plasma glucose concentration is higher than whole blood glucose concentration, as the glucose concentration in plasma is higher than that in erythrocytes. Thus, the difference is affected by the hematocrit of the specimen. In addition, the difference in glucose concentration among blood specimens can be affected by several other factors. During a glucose challenge test, glucose concentration in capillary blood was reported to be higher than that in venous blood, while they are comparable in a fasting state.

As a source of blood specimens, whole blood samples, collected by skin puncture from the finger are recommended, except in infants, in whom the heel is recommended as a puncture site. Generally, the flexor surface of the end of the finger is used, and the middle and ring fingers are preferred (Figure 19.1). Practically, it is necessary to avoid previously punctured sites and infected or edematous sites.

More importantly, capillary samples collected from other sites, for example, the forearm, should not be used, especially when hypoglycemia is suspected. This is because, during rapid changes in blood glucose, the change in the monitored glucose level in the arm was delayed compared to that in the finger, which could lead to delayed detection of hypoglycemia (Jungheim and Koschinski 2002). Thus, when testing with the express purpose of detecting hypoglycemia, the fingertip should be the test site, although collection of blood from the forearm was reported to be less painful in diabetic individuals.

19.3   Monitoring Device and Strips

SMBG is feasible with a blood glucose monitoring system, which consists of a portable instrument and reagents used for *in vitro* monitoring of glucose concentration in blood, together with a puncture device and needles/lancets to obtain blood samples (Figure 19.1).

Table 19.1   Glucose concentration according to different blood specimens. Glucose concentration values are different, when different blood specimens are used. Possible clinical situations and methods for each blood specimen are also listed.

Specimens	*Arterial Blood*	*Capillary Blood*	*Venous Blood*
Clinical settings	At intensive care unit	Daily practice as SMBG	At clinic/hospital
Method to obtain samples	From an indwelled arterial catheter	Skin puncture with a lancing device/needle(s)	Using a syringe
Concentration value	Higher	Middle	Lower

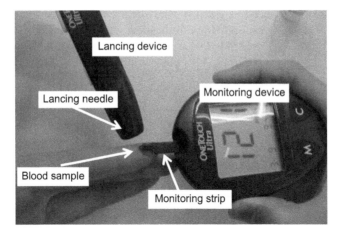

Figure 19.1 Actual view for performing SMBG. This photo describes the actual performance of SMBG. Blood sample was obtained by skin puncture from the finger, using a lancing device with a lancing needle. Then, the blood sample was applied to a monitoring strip attached to the monitoring device. SMBG: self monitoring of blood glucose.

19.3.1 Glucose Readings Monitored by SMBG

Regarding the results of SMBG, while the glucose level in capillary blood samples is measured, monitoring systems may express the results as either the glucose concentration in blood or the equivalent glucose concentration in plasma. As discussed above, glucose concentration in plasma is higher than that in whole blood, and it is of note that currently available SMBG devices/systems report a calibrated "plasma-equivalent" result of glucose concentration, even though whole blood samples are subjected to monitoring. When these SMBG systems are used, the "plasma-equivalent" results are 10 to 12% higher than those monitored with other SMBG systems reporting whole blood glucose readings.

19.3.2 Accuracy of SMBG Results

Due to technological improvement of the electrochemical sensor, the amount of blood sample required has been reduced to 0.3 µL, resulting in less pain. In addition, improvement of the sensor has made the results more accurate. Regarding the accuracy of glucose meters required for diabetes management, there appear to be no evidence-based criteria. However, several recommendations and analytical performance goals have been proposed (Table 19.2). The ADA stipulates the most stringent goals, which were considered unattainable at that time they were proposed (ADA 2011).

How accurate are the *actual* SMBG data with the SMBG systems available to date? Compared with laboratory plasma glucose results, finger-prick blood glucose levels measured using six different meters had an absolute

Table 19.2 Recommended goals of analytical performance for accuracy of SMBG results. Major published recommendations required for accuracy of SMBG results. SMBG: self monitoring of blood glucose, NCCLS: National Committee for Clinical Laboratory Standards, and ADA: American Diabetes Association.

Recommendations	Goals
The Australia Standards (ISO15197.2) and the NCCLS	±20% at glucose concentrations >4.2 mmol/L (75.6 mg/dL)
	Within 0.83 mmol/L (15 mg/dL) bias at glucose concentrations <4.2 mmol/L
ADA	an analytical error of <5%, or, a total (user plus analytical) error of <10%

plasma-equivalent difference of 0.232 ± 0.69 to 0.725 ± 0.62 mmol/L (mean ± SD), and error ranged from 6.1 to 15.8%. Moreover, two out of the six meters were affected by anaemia. Thus, currently used SMBG devices are not optimally accurate when compared with plasma glucose measurement. This point should be recognized as a limitation of SMBG systems.

From the clinical viewpoint, the accuracy of SMBG systems needs to be checked regularly. According to the International Diabetes Federation (IDF) guidelines, SMBG use requires an easy procedure for patients to regularly monitor the performance and accuracy of their glucose meter (IDF 2008).

19.3.3 Enzyme/co-enzyme and Clinical Attention to SMBG

Regarding the methods for glucose assay adopted in SMBG systems, the details are reviewed in other chapters of this book. Here, I would like to share several issues requiring attention in the use of SMBG, especially in clinical settings. There are two major enzymes used in the chemical reaction for assaying glucose in SMBG systems; glucose oxidase (GOD) and glucose dehydrogenase (GDH) (Table 19.3). The former uses the oxidase reaction in which oxygen is used as a substrate to receive an electron from the oxidase, whereas the latter has less specificity for glucose (Table 19.3). Therefore, if SMBG is based on a GOD method, glucose readings in individuals receiving oxygen would be negatively affected by the oxygen dissolved in blood. In contrast, when an SMBG system with the GDH method is used, the "glucose" results could be falsely elevated by cross reactions with other sugars than glucose, such as maltose, galactose and xylose. Hence, attention should be paid to glucose "readings", as they would be falsely high, in individuals administered intravenous maltose-containing agents, such as immunoglobulins, or in patients under intraperitoneal dialysis, if they use icodextrin-containing dialysate. It is of note that cross-reaction to a non-glucose substrate is dependent on the co-enzyme for the GDH reaction used in the glucose meter. While the results are affected by xylose, but not maltose and galactose, when FAD-dependent GDH or NAD-dependent GDH is used as a co-enzyme, with a sensor adopting PQQ-dependent GDH as a

Table 19.3 Effect of substrates/condition on glucose concentration according to properties adopted in systems for SMBG. Glucose readings would be affected by substrates/condition, according to two major enzymes used in the chemical reaction for assaying glucose in SMBG systems. # Maltose-included samples would show falsely high values only when PQQ (pyrroloquinolonine-quinone)-dependent glucose dehydrogenase (EC 1.1.5.2) is used as a co-enzyme. SMBG: self monitoring of blood glucose, GOD: Glucose oxidase, GDH: Glucose dehydrogenase.

	Enzymes	
	GOD	*GDH*
Maltose	no	affected[#] (falsely high)
Xylose	no	affected (falsely high)
Oxygen	affected (falsely low)	no
	Measurement Methods	
	Electrometry	*Colorimetry*
Ascorbate	affected (falsely high)	no
Strong light	no	affected (falsely low)

co-enzyme, maltose, galactose and xylose could affect the results. Many cases of adverse events, including death, have been reported, especially with measurement with PQQ-GDH-based test strips (Frias *et al.* 2010). Therefore, healthcare professionals and patients must pay attention to possible false glucose results, if an SMBG system with an enzyme and its co-enzyme is used to monitor an individual with one of the conditions described above.

19.4   Puncture Device and Lancets/Needles for SMBG

To perform SMBG, a puncture device is also needed to obtain capillary blood samples, while the puncture/lancing device is usually separate from the monitoring device in most SMBG systems (Figure 19.1). Over the last decade, both puncture devices and needles/lancets have been improved, aiming at less pain accompanying the procedure. The lancing device has been ameliorated to reduce the "bang" oscillation which can enhance subjective pain. In addition, the lancet/needle has also been improved, in terms of its sharpness and the shape of the tip. Thus, considerable improvements in monitoring and lancing devices as well as needles have diminished the pain accompanying SMBG, to promote patients' adherence to SMBG.

19.5   Choice of SMBG Systems

Is there an SMBG system that is best for all? Since the beneficial effect of SMBG is limited to those with good SMBG compliance, appropriate choice of

Table 19.4   Independent predictors for choosing SMBG system in inpatients with diabetes mellitus, according to age-group. Independent variables identified as significant factor associated with the final preference score (= preference for use hereafter) were listed. The independent variables included in the analysis, but not identified as significant factors, were size of monitoring strip, pain associated with skin puncture, size of SMBG device for operation, and operation for correcting the monitoring device for a set of strips (Fujisawa *et al.* 2008).

Total
   Display of monitoring device (p < 0.0001)
   Operation to attach monitoring strip(s) to device (p = 0.0004)
   Operation of lancing device (p < 0.05)

Older (≥ 60 years)
   Display of monitoring device (p = 0.001)
   Operation for detaching used needle (p = 0.01)

Younger (< 60 years)
   Operation of lancing device (p < 0.0001)
   Operation for applying blood to strip (p = 0.001)
   Blood volume needed (p = 0.04)

SMBG system in each individual is important. In our previous study in diabetic inpatients (Fujisawa *et al.* 2008), we comparatively monitored different kinds of SMBG systems with questionnaires on multiple items; its operation, size of the device, pain, and preference for future use. Regression analysis revealed that factors affecting the preference for a SMBG system were diverse according to the age group (Table 19.4). In non-elderly patients, important factors for choosing an SMBG system were the lancing device operation, blood application operation, and blood volume needed, all of which are related to skin lancing. These findings highlight the importance of performance of the lancing device and its needle for younger patients in choosing an SMBG system. In contrast, the display and the needle-detaching operation were more important in elderly patients. Hence, when an elderly patient initiates SMBG, a monitoring device with a large display and a puncture device with a simple and/or safe procedure for detaching the used needle may well be recommended. Thus, these findings are directly applicable to the clinical management of diabetes, in helping patients choose a suitable SMBG system, as the proper choice of an SMBG system in each individual should facilitate the patient's compliance with SMBG.

For individuals whose visual acuity is so limited/lost due to diabetic or non-diabetic eye disease as to have difficulties in reading SMBG data, several SMBG instruments provide a voice-reading system with its specific apparatus.

19.6   Procedures for SMBG

Before puncturing skin, standard precautions are necessary; *i.e.*, disinfection of the skin site using an alcohol swab. To avoid mixing the blood with alcohol, the alcohol should be allowed to dry, otherwise the monitor may display a falsely lower glucose level.

Even when the amount of blood sampled from the puncture site is limited, one should not apply too strong, repeated pressure to obtain the required volume, as it may cause false results.

It is noteworthy that technical skill in blood glucose monitoring could affect the accuracy of SMBG, and especially in the hands of non-specialized users there may be significant day-to-day variation in technique and meter performance. Hence, it is important to evaluate each patient's technique for monitoring, initially and at regular intervals thereafter (ADA 2011).

It is possible that a family member or healthcare provider may have to perform blood glucose testing instead of the diabetic individual; for example, in the case of severe hypoglycemia or coma. In such a case, SMBG systems may well be used for non-self monitoring, and special cautions should be taken to avoid transmitting blood-borne pathogens of the diabetic individual. Care homes in England experienced an outbreak of hepatitis B, which was revealed to be caused by inappropriate use of a lancing device for multiple patients that was intended for self-use by a single patient. It is important to confirm that the lancing device and needle are for the patient. In addition, given several reports of penetration of a lancing needle through the patient's earlobe into the finger of a healthcare provider that supported the earlobe, the patient's finger, but not the earlobe, should be used as a puncture site in the case of non-self glucose monitoring. Moreover, on handling a lancet that has already been used, care should be taken to avoid transmission of pathogens by the lancet, especially to healthcare providers. To this end, some SMBG puncture devices have protection systems so that the lancet/needle cannot be touched again after its single use.

19.7 Clinical Significance of SMBG

For optimal blood glucose control in the management of diabetes, the results of SMBG are useful for adjusting nutrition and physical activity as well as anti-diabetic medications, especially insulin. Appropriate use of SMBG results can reduce glucose fluctuation and therefore lessen the risk of hypoglycemia. Specifically, identification of "hypoglycemia" by SMBG can prompt patients to take sugar to rescue them from this state, and will be valuable for the prevention of future hypoglycemia through adequate action.

19.7.1 Clinical Benefit of SMBG in Glycemic Control in Insulin-treated Subjects

Major clinical studies of insulin-treated patients, including DCCT (Table 19.5), showing the benefits of intensive glucose control on complications have included SMBG as a part of multifactorial interventions. Insulin adjustment is vital for successful use of insulin, and SMBG allows patients to assess whether glycemic targets are achieved. In addition, patients can evaluate their individual response to therapy, which is feasible only with SMBG data. Thus, SMBG

Table 19.5   Key facts of DCCT. Clinical aspects of DCCT.

A multicenter, randomized trial, to compare intensive with conventional diabetes
   therapy regarding their effects on the diabetic complications in type 1 diabetes mellitus,
   which established a clinical benefit of glycemic control to reduces the development and
   progression of diabetic complications.

provides medical value in insulin-treated patients, especially in those with
multiple injections or pump therapy, and has been regarded as a component of
effective therapy.

   In contrast to the strong recommendation of performing SMBG for patients
using multiple insulin injections or insulin pump therapy, for patients using
once-daily insulin, it has not been well established whether SMBG has clinical
merits. In the ADA recommendations, SMBG may be useful as a guide to the
success of therapy (ADA 2011).

19.7.2   Clinical Benefit of SMBG on Glycemic Control in Non-insulin-treated Subjects

A number of studies addressing the clinical benefits of SMBG in subjects not
receiving insulin treatment have been reported, which have shown inconsistent
results. In this chapter, some of the studies addressing possible optimal use of
SMBG in non-insulin-treated diabetic individuals are discussed.

   In the Auto-Surveillance Intervention Active (ASIA) study (Guerci *et al.*
2003), a randomized controlled trial for 24 weeks, the effect of SMBG on
glycemic control was investigated in type 2 diabetic patients treated with oral
hypoglycemic agents (OHAs). While the HbA1c level in the non-SMBG group
(n = 344) changed from $8.9 \pm 1.3\%$ to $8.4 \pm 1.4\%$, that in the SMBG group
(n = 345) decreased from $9.0 \pm 1.3\%$ to $8.1 \pm 1.6\%$. Thus, performing SMBG
six times a week was associated with a reduction in HbA1c level of 0.28%. This
study appears valid, with sufficient statistical power. However, the relatively
high drop-out rates, 48% in the SMBG group and 40% in the control group,
limit the general applicability of the findings.

   Soumerai reported a study of patients treated with sulfonylurea, an insulin
secretagogue, in which an increase in SMBG frequency from 0.5 to 2.0 tests per
week was associated with a significant 0.6% reduction in HbA1c, in those with
a baseline HbA1c over 10%. This study suggested an obvious beneficial effect of
SMBG to improve glycemic control in individuals with poor glycemic control.

   In a cohort study of diabetic individuals from a diabetes registry, Karter
et al. investigated longitudinal glycemic control of new-users and ongoing-users
of SMBG. Among 16 091 patients initiating SMBG (new users), the reduction
in HbA1c level was greater in those with more frequent SMBG practice compared to that in non-users, regardless of diabetes therapy. Particularly, in those
receiving OHAs, initiating SMBG was associated with improved adherence to
medication. Thus, new use of SMBG could accord better glycemic control
through a better attitude of patients. In contrast, among 15 347 ongoing users

of SMBG (prevalent users), a change in SMBG frequency was inversely associated with a change in HbA1c, only in those receiving pharmacological medication(s). A change in SMBG frequency of one strip per day resulted in subsequent inverse changes in HbA1c of 0.16% in patients taking oral agents and 0.12% in insulin-treated patients. Such a dose-response relationship diminished over a change of two or three strips per day. Thus, in diabetic subjects who already perform SMBG, an increase in its frequency may be associated with a slight decrease in HbA1c level.

The Fremantle Diabetes Study was an observational community-based study in Australia (Davis *et al.* 2006). Patients with type 2 diabetes who reported their SMBG status at study entry (n = 1286) were cross-sectionally analyzed. There was no significant difference (p = 0.12) in HbA1c between SMBG users (7.3%; interquartile range: 6.4 to 8.8) and nonusers (7.6%; 6.4 to 8.9). The majority of patients (70%) reported performing SMBG, with a median of four tests per week (interquartile range 2 to 7). Of the study population, 531 patients who had undergone annual reviews over 5 years were subjected to longitudinal analysis (Davis 2006). In this longitudinal cohort, the proportion of SMBG users increased over time (trend p < 0.001), from 75.2% at entry to 85.5% at the third review, and diet-treated patients decreased and those on OHA or insulin increased during the study period. Throughout the period, the HbA1c level did not differ between patients using SMBG and those who did not. However, the study participants were recruited, irrespective of duration of diabetes and, at baseline, the majority had already used SMBG, which increased during the follow-up period. Thus, the study design might not have been adequate to extract the effect of SMBG on glycemic control.

In the Fremantle Diabetes Study (Davis 2006), the minority (30%) were not using SMBG at recruitment. The reasons for not using SMBG were no education on how to perform SMBG (45%), no motivation to start or continue SMBG (31%), fear of finger pricks (9%), and physical or psychological problems preventing use (5%). Thus, education on SMBG as well as adequate healthcare action in motivating patients to initiate and continue SMBG are important. SMBG was shown to be more prevalent in those with shorter disease duration, especially in those who had received diabetes education, suggesting that recent diagnosis and diabetes education are motivational. According to the IDF guidelines on SMBG in non-insulin-treated type 2 diabetes (IDF 2008), SMBG should be considered at the time of diagnosis to enhance the understanding of the disease, as a part of patient education, and to facilitate timely treatment initiation and titration optimization.

In contrast, the ROSSO study (Martin *et al.* 2006) retrospectively investigated the relationship between initiating SMBG at the diagnosis of diabetes and disease-related morbidity and mortality, in non-insulin-treated patients with type 2 diabetes mellitus in Germany. In this epidemiological cohort study with a mean follow-up period of 6.5 years, SMBG users were at decreased risk of non-fatal endpoints and death. Cox regression analysis identified SMBG as an independent predictor of reduced morbidity and mortality, with adjusted hazard ratios of 0.68 (95% CI: 0.51–0.91, p = 0.009) and 0.49 (95%CI: 0.31–0.78, p = 0.003), respectively. Notably, when analysis was limited to those who were not receiving insulin, SMBG use was still associated with a better outcome for both endpoints. Since this study was conducted at primary care practices randomly selected throughout the country, the results may be clinically applicable to general practice in diabetes mellitus.

In the DiGEM study (Farmer *et al.* 2007), an open parallel group randomized trial, the effect of self-monitoring alone and self-monitoring with instruction in incorporating the results into self care on glycemic control was studied. A total of 455 non-insulin-treated patients with type 2 diabetes mellitus whose mean HbA1c was 7.5% were divided into three groups; control group (n = 152) receiving usual care with measurement of HbA1c every three months, less-intensive SMBG group (n = 150) being advised to contact their doctor for interpretation of SMBG data, and more-intensive SMBG group (n = 151) with additional training. The patients allocated to the more-intensive SMBG group were given training in interpretation and application of their SMBG results to motivate and maintain adherence to a healthy lifestyle. At 12 months, the differences in HbA1c level between the three groups were not statistically significant (p = 0.12). From these results, there was no evidence supporting the glycemic benefit of SMBG even with self-management training on glycemic control in non-insulin-treated individuals with reasonably well-controlled diabetes. In addition, the persistent use of SMBG (at least twice a week for 12 months) was less frequent in the more-intensive SMBG group (52%) than in the less-intensive SMBG group (67%). Thus, the intensified group had a higher drop-off rate in terms of performing a preset SMBG frequency, which may be explained by the speculation that patients with good glycemic control do not need active encouragement to use or adhere to SMBG. This study, however, reported interesting results on lipid control; there was a significant reduction in the cholesterol level in the less-intensive SMBG group and more-intensive SMBG group, compared to the control group. This finding is consistent with an increased intensity of self management by SMBG, possibly through improved adherence to diet or regular medications. Thus, even in patients with well-controlled glycemia, SMBG could be valuable for improving risk factors.

The ESMON study was a prospective randomized controlled trial in patients with newly diagnosed type 2 diabetes mellitus (O' Kane *et al.* 2008). The effect of SMBG on glycemic control and psychological indices was studied. The control group (n = 88), with an initial HbA1c of 8.6%, received a structured core education program. In addition, individuals randomised to the SMBG group (n = 96), with mean HbA1c of 8.8%, were provided with a single glucose monitor and instructions for use and, at each visit, received ongoing support and advice in the appropriate interpretation of and response to the SMBG data. Both groups attained a satisfactory HbA1c level of 6.9%, without any significant difference in HbA1c level between the SMBG and control groups. The intensive advancement of medications soon after the diagnosis of diabetes appeared to have dimmed the possible additive effect of SMBG. In particular, the depression subscale score in the wellbeing questionnaire was higher in the SMBG group compared to that in the control group. Thus, routine use of

SMBG by unselected patients could worsen patients' moods and thus have a negative effect on quality of life.

From these two studies published in *BMJ*, the DiGEM study (Farmer *et al.* 2007) and the ESMON study (O' Kane *et al.* 2008), no significant difference in HbA1c level was observed between those with and without random SMBG, questioning whether or not SMBG is beneficial for achieving better glucose control in non-insulin-treated patients with type 2 diabetes.

To address whether SMBG is of medical value in non-insulin-treated subjects, several meta-analyses have been reported, most of which pointed out the diversity of the study subjects, frequency and timing of monitoring, patient education, counseling by healthcare providers, and concomitant medication intensification. Among them, one updated meta-analysis (Poolsup *et al.* 2009) addressed the medical value of SMBG in non-insulin-treated patients with type 2 diabetes. The summary result indicated a significant reduction in HbA1c with SMBG compared with non-SMBG, with a pooled mean difference of -0.24% (95% confidence interval: -0.34 to -0.14%, $p < 0.00001$). However, in many of the studies in the meta-analysis, the effect of SMBG alone could not be isolated, as effects of education, counseling and medication appeared to be involved. In the meta-analysis, SMBG was shown to be effective in the subgroup of patients whose baseline HbA1c was above 8%.

After the meta-analysis, two interesting reports on the clinical use of SMBG were published, which indicated how SMBG could be utilized for more clinical benefits.

The St Carlos Study (Duran *et al.* 2011), a randomized clinic-based interventional study, investigated the effects of SMBG in patients with newly diagnosed type 2 diabetes. While in the control group, treatment was initiated and adjusted according to the simple algorithm of the HbA1c level (HbA1c-based group), the intervention group used SMBG data (SMBG-based group) for initiation and adjustment of treatment according to a simple algorithm based on the SMBG values (pharmacological program), together with teaching to adapt the lifestyle to obtain better glycemic control (educational program). After a one-year follow-up period, the SMBG-based group had earlier and more frequent pharmacological changes, a greater reduction in HbA1c and BMI, as well as better choice of food types, compared to those in the HbA1c-based control group, suggesting empowerment of self-management by SMBG. Notably, life satisfaction was more improved in the SMBG-based group, which was in contrast to the previous findings of higher anxiety and a more depressive state in the SMBG group (O' Kane *et al.* 2008). The reason why the results were diverse regarding the effects of SMBG on mental/psychological conditions is unclear, but it is possible that a proper framework for earlier intensification/adjustment treatment by adopting a simple algorithm would have motivated patients, with increased, rather than decreased, satisfaction.

More recently, the Structured Testing Program (STeP) study (Polonski *et al.* 2011) was reported in which the effect of a comprehensive, structured SMBG intervention package on blood glucose control was investigated in poorly-controlled, non-insulin-treated type 2 diabetes. In this cluster-randomized study, both the active control group and the structured testing group were provided with free glucose meters and strips, but only in the structured testing group, patients had a validated tool to record 7-point SMBG profile on three consecutive days and were trained how to address problematic glycemic patterns through changes in lifestyle. In addition, health providers of the structured testing group received training to interpret the recorded data, viewed the SMBG results every 3 months, and adjusted treatment as needed. Over 12 months, both groups showed a reduction in HbA1c, as a primary outcome, but this was more marked in the structured testing group (− 0.3% in intent-to-treat analysis), compared to the active control group. The structured SMBG group had more frequent recommendations for treatment change. Interestingly, general wellbeing was improved in both groups, without a difference between the groups. Thus, as long as appropriately used, structured SMBG would improve glycemic control and facilitate timely/aggressive treatment changes in insulin-naïve type 2 diabetic subjects, without decreasing general wellbeing. These results emphasize the importance of adequate SMBG timing and its recording system, as well as structured educational and pharmacological programs for both patients and healthcare professionals for interpretation and decision making. At the same time, these results also pointed to the importance of an appropriate feedback system to transfer the obtained SMBG data to utilization for changing lifestyle as well as for medication modification, to achieve clinical benefits of SMBG in non-insulin-treated subjects. According to the IDF clinical guidelines, SMBG should also be considered as a part of ongoing diabetes self-management education to assist people with diabetes to better understand their disease and provide a means to actively and effectively participate in its control and treatment, modifying behavioral and pharmacological interventions as needed, in consultation with their healthcare provider (IDF 2008).

Taking these results together, SMBG is beneficial for diabetic patients with insulin treatment, and probably valuable for those with poor glycemic control. However, testing *per se* is not enough to get better glycemic control. To make SMBG an effective self-management tool, the SMBG results should be reviewed and acted on by healthcare providers and/or patients with diabetes to actively modify behavior and/or adjust treatment, as suggested in the IDF Clinical Guidelines (IDF 2008). However, it is recommended only when individuals with diabetes and/or their healthcare providers have the knowledge, skills, and willingness to incorporate SMBG monitoring and therapy into their diabetes care plan.

19.7.3 Cost-effectiveness of SMBG

Since SMBG is a kind of medical procedure and the present monitoring strips are costly, it is necessary to consider the cost-effectiveness balance of SMBG, from both economic and medical points of view. Provided that SMBG is beneficial for glycemic control and therefore reduces diabetic complications, the total cost of treatment and management of complications prevented by the adequate utilization of SMBG is ideally greater than the cost spent performing SMBG. There are a number of reports addressing this issue, which has been fairly and briefly summarized (IDF 2008). It is difficult, however, to measure the social loss resulting from complications, and it is hardly possible to assess the total balance between the cost and benefits related to SMBG. While the medical value of SMBG has been established for patients with type 1 diabetes, as discussed above, it is still controversial whether SMBG is beneficial for patients not treated with multiple insulin injections or not receiving insulin at all. Thus, it is yet to be addressed whether the use of SMBG is ultimately beneficial for society. It is likely that the answer will remain unclear until more effective SMBG strategies are devised.

19.7.4 Effects of SMBG on Psychological Distress

Besides the economic issue, the psychological issue is of another concern, given that SMBG causes patients pain and inconvenience. Several studies aimed to address the psychological or mental distress associated with SMBG. As reviewed, however, the effects of SMBG on wellbeing or a depressive state among studies have been diverse (O' Kane *et al.* 2008; Polonski *et al.* 2011), which may reflect the complexity of the integrated psychological impact of several factors; duration of diabetes, glycemic control, educational and economic conditions, the means to incorporate SMBG data into treatment intensification/adjustment, performance and inconvenience of SMBG, pain, etc. The medical evidence thus far has not fully addressed whether the use of SMBG is beneficial or harmful psychologically, which is very likely to depend on the patient and the diabetic condition. At present, the indication for SMBG, especially in non-insulin-treated subjects, should be individually decided with careful consideration of the balance between possible distress and benefits, which could negatively affect the mood and self-management in some individuals. According to the present ADA recommendations, SMBG frequency and its timing should be directed according to the needs and goals of each patient (ADA 2011).

19.8 Future Perspectives of SMBG

As reviewed, there are definitive clinical benefits of SMBG in subjects with diabetes who are insulin-treated, and probable benefits in non-insulin-treated patients. Although clinical benefit was evident in individuals with poor glycemic control, it is necessary to clarify the effect of SMBG in those with moderate or good control, and to devise more effective SMBG strategies for those not taking insulin. A larger prospective study is awaited to clarify the efficacy of SMBG on clinical outcomes other than HbA1c, such as diabetic complications, although benefit of glycemic control by SMBG, if any, in preventing macrovascular complications will take a longer time.

A continuous glucose monitoring system (CGMS), which is currently used in limited countries/institutions, could be widely utilized for more fine assessment of the daily profile of blood glucose levels. It is expected that SMBG systems with an easier and more comfortable procedure together with less pain would facilitate patients' compliance with SMBG. Several novel strategies for non-invasive monitoring systems for blood glucose are currently underway.

Establishment of a training system for physicians/care-teams for more effective interpretation of SMBG data and for engaging patients in diabetes self-management is urgently needed.

Finally, SMBG is painful, time-consuming and costly; therefore, its application should be balanced with its benefit. Otherwise it will not be continued with clinical value. In the present situation with the available SMBG systems, the clinical benefits of SMBG appear to largely depend on the patient's willingness to achieve optimal control and healthcare professionals' enthusiasm for supporting diabetic individuals, both of which interact mutually.

Summary Points

- This chapter focuses on the clinical issues related with self monitoring of blood glucose (SMBG).
- SMBG has contributed to optimal blood glucose control for preventing or delaying the development and progression of diabetic complications.
- Clinical issues related with SMBG, which diabetic individuals and healthcare providers are expected to know in the management of diabetes, are discussed in this chapter.
- SMBG results are affected by the condition of the blood samples.
- SMBG results could have false values, depending on the substrates/ condition of the blood samples and enzymes used in the monitoring systems.
- Clinical factors affecting the preference for a SMBG system were diverse between elderly and non-elderly.
- To perform SMBG safely and accurately, there are several technical points that need attention.
- The clinical benefits of SMBG are apparent in individuals with a multiple insulin regimen, but are controversial in those with less frequent insulin injections and non-insulin users.
- SMBG is probably valuable for diabetic individuals with poor glycemic control.
- To make SMBG an effective self-management tool, the SMBG results should be reviewed and acted on by healthcare providers and/or diabetic patients to actively modify behavior and/or adjust treatment.
- The economic and psychological issues of SMBG are of concern, given that monitoring strips are costly and SMBG causes patients pain and inconvenience.
- The clinical benefits of SMBG appear to largely depend on the patient's willingness to achieve optimal control and healthcare professionals' enthusiasm for supporting diabetic individuals.

Key Facts of Self Monitoring of Blood Glucose (SMBG)

Key features of SMBG including the procedure, devices, recommendations for patients with diabetes according to diabetes therapy, and issues to be considered. SMBG: self-monitoring of blood glucose.

1. Strict glycemic control has been shown to be effective for preventing or delaying the development and progression of diabetic complications.
2. Self monitoring of blood glucose (SMBG) has contributed to the optimal blood glucose control in the management of diabetes.To perform SMBG, skin puncture is necessary to obtain blood samples, usually with a puncture device and needles/lancets.
3. The collected blood samples were subjected to the electrochemical sensor adopted in portable SMBG systems.
4. Monitoring and lancing devices have been improved to diminish the pain accompanying SMBG, which promoted patients' adherence to SMBG.
5. SMBG results are useful for adjusting nutrition and physical activity as well as anti-diabetic medications, especially insulin.
6. According to the ADA recommendations, SMBG should be performed three or more times a day in patients receiving multiple insulin injections or insulin pump therapy.
7. SMBG may be useful for patients using once-daily insulin.
8. Clinical benefits of SMBG in subjects not receiving insulin treatment have been controversial.
9. Cost-effectiveness and psychological distress accompanying SMBG should be considered.
10. SMBG frequency and its timing should be directed according to the needs and goals of each patient.

Definitions of Words and Terms

Self monitoring of blood glucose. Useful strategy to achieve optimal blood glucose control in the management of diabetes mellitus. To perform SMBG, lancing device and needle/lancet(s) as well as monitoring device and strip(s) are needed (Figure 19.1).

Blood glucose concentration. Blood glucose concentration is measured as a glucose level in a volume of blood and expressed in mmol/L or mg/dL. Clinically, glucose concentration could be different according to blood specimens (Table 19.1). Also, glucose concentration in plasma is higher than that in whole blood.

HbA1c. HbA1c is clinically used as a marker for glycemic control, and shown as a percentage (%) of glycated Hb (known as HbA1c) in a total Hb. Clinically, HbA1c is considered to reflect mean glucose levels for approximately 2 months before the time of measurement.

Lancing device. A puncture device used to obtain capillary blood samples from skin for SMBG, which requires a lancing needle/lancet.

Lancing needle/lancet(s). Lancing needle, or lancet, is attached to the lancing device, and used to puncture skin. From safety points of view, the needle/lancet is limited for single use, intended for self-use by a single patient.

Monitoring device. A portable instrument used for *in vitro* monitoring of glucose concentration in capillary blood for SMBG.

Monitoring strip. A monitoring strip, which contains reagents for *in vitro* monitoring of glucose concentration, is attached to the monitoring device and used for SMBG.

Glucose oxidase (GOD) method. GOD method is one of the two major methods adopted in SMBG systems to assay glucose levels. This method uses the oxidase reaction in which oxygen is used as a substrate to receive an electron from the oxidase. Therefore, glucose readings in individuals receiving oxygen would be negatively affected, when monitored by this method (Table 19.3).

Glucose dehydrogenase (GDH). GDH method is the other method adopted in SMBG systems to assay glucose levels. Since this chemical reaction has less specificity for glucose, the "glucose" readings could be falsely elevated by cross reactions with other sugars than glucose, such as maltose, galactose and xylose (Table 19.3).

Diabetes control complication trial (DCCT). DCCT is performed in USA with individuals with type 1 diabetes, and established clinical benefits of glucose control for preventing the development and delaying the progression of diabetic eye and kidney diseases. This study also demonstrated clinical benefits of SMBG to control glycemia.

List of Abbreviations

SMBG	self monitoring of blood glucose
NCCLS	National Committee for Clinical Laboratory Standards
ADA	American Diabetes Association
IDF	International Diabetes Federation
GOD	glucose oxidase
GDH	glucose dehydrogenase
DCCT	diabetes control complication trial
OHA	oral hypoglycemic agents
CGMS	continuous glucose monitoring system

References

American Diabetes Association (ADA)., 2011. Standards of medical care in diabetes-2011. *Diabetes Care* 34: S11–S61.

Davis, W.D., Bruce, D.G., and Davis, T.M.E., 2006. Is self-monitoring of blood glucose appropriate for all type 2 diabetic patients?; The Fremantle Diabetes Study. *Diabetes Care* 29: 1764–1770.

Durán, A., Martín, P., Runkle, I., Pérez, N., Abad, R., Fernández, M., Del Valle, L., Sanz, M.F., and Calle-Pascual, A.L., 2011. Benefits of self-monitoring blood glucose in the management of new-onset type 2 diabetes mellitus: The St Carlos Study, a prospective randomized clinic-based interventional study with parallel groups. *Journal of Diabetes* 2: 203–211.

Farmer, A., Wade, A., Goyder, E., Yudkin, P., French, D., Craven, A., Holman, R., Kinmonth, A.L., and Neil, A., 2007. Impact of self-monitoring of blood glucose in the management of patients with non-insulin treated diabetes: open parallel group randomised trial. *British Medical Journal* 335: 132–139.

Frias, J.P., Lim, C.G., Ellison, J.M., and Montandon, C.M., 2010. Review of adverse events associated with false glucose readings measured by GDH-PQQ-based glucose test strips in the presence of interfering sugars. *Diabetes Care* 33: 728–729.

Fujisawa, T., Ikegami, H., Kasayama, S., Matsuhisa, M., Yamasaki, Y., Miyagawa, J., Funahashi, T., and Shimomura, I., 2008. Age-dependent difference in factors affecting choice of system for self monitoring of blood glucose. *Diabetes Research and Clinical Practice* 79: 103–107.

Guerci, B., Drouin, P., Grangé, V., Bougnères, P., Fontaine, P., Kerlan, V., Passa, P., Thivolet, Ch., Vialettes, B., and Charbonnel, B. (ASIA Group), 2003. Self-monitoring of blood glucose significantly improves metabolic control in patients with type 2 diabetes mellitus; the Auto-Surveillance Intervention Active (ASIA) study. *Diabetes Metabolism* 29: 587–594.

International Diabetes Federation (IDF), 2008. Guideline on self-monitoring of blood glucose in non-insulin treated type 2 diabetes. Available at: http://www.idf.org/webdata/docs/SMBG_EN2.pdf. Accessed 04 Dec 2011.

Jungheim, K., and Koschinski, T., 2002. Glucose monitoring at the arm; risk of delays of hypoglycemia and hypoglycemia detection. *Diabetes Care* 25: 956–960.

Martin, S., Schneider, B., Heinemann, L., Lodwig, V., Kurth, H.J., Kolb, H., and Scherbaum, W.A., 2006. Self-monitoring of blood glucose in type 2 diabetes and long-term outcome: an epidemiological cohort study. *Diabetologia* 49: 271–278.

O'Kane, M.J., Bunting, B., Copeland, M., and Coates, V.E., (ESMON study group), 2008. Efficacy of self monitoring of blood glucose in patients with newly diagnosis type 2 diabetes (EMSON study): randomised controlled trial. *British Medical Journal* 336: 1174–1177.

Polonsky, W.H., Fisher, L., Schikman C.H., Hinnen, D.A., Parkin, C.G., Jelsovsky, Z., Petersen, B., Schweitzer, M., and Robin S., 2011. Structured self-monitoring of blood glucose significantly reduces A1c levels in poorly controlled, noninsulin-treated type 2 diabetes mellitus. *Diabetes Care* 34: 262–267.

Poolsup, N., Suksomboon, N., and Rattanasookchit, S., 2009. Meta-analysis of the benefits of self-monitoring of blood glucose on glycemic control in type 2 diabetes patients: an update. *Diabetes Technology and Therapeutics* 11: 775–784.

CHAPTER 20

The Glucose Oxidase-Peroxidase Assay for Glucose

MARY BETH HALL

USDA-Agricultural Research Service, United States Dairy Forage Research Center, 1925 Linden Drive, Madison, Wisconsin 53706, USA
Email: marybeth.hall@ars.usda.gov

20.1 Introduction

Measurement of glucose by enzymatic-colorimetric methods based on the reactions of glucose oxidase and peroxidase (GOP) was begun in the 1950s and has been in common use since that time (Keston 1956). Before the advent of the GOP assays, glucose was measured with chemical methods such as reducing sugar assays, or condensation reactions involving strong acid and phenol or anthrone. The extensive conversion from chemical to GOP assays speaks to the superiority of the enzymatic approach in its greater specificity for and sensitivity to glucose, ease of use, and decreases in the types of compounds such as proteins and non-glucose carbohydrates that interfere with the assay. Chromatographic methods can offer similar specificity for glucose, but GOP methods require less time for sample preparation and allow more rapid analysis and greater throughput of samples. Applications for GOP methods have been diverse, from determining free glucose in blood and urine, to quantifying free or enzymatically released glucose from starch, glycogen, or maltooligosaccharides in plant, microbial, or animal tissues and in human foods or animal feeds. Knowledge of key control points and interfering substances in the GOP assays allow the analyst to productively utilize these assays on varied matrices.

Food and Nutritional Components in Focus No. 3
Dietary Sugars: Chemistry, Analysis, Function and Effects
Edited by Victor R Preedy
© The Royal Society of Chemistry 2012
Published by the Royal Society of Chemistry, www.rsc.org

20.2   Chemistry

The two chemical reactions involved in the GOP assay are a glucose oxidase (EC 1.1.3.4) catalyzed oxidation of D-glucose to produce gluconic acid and H_2O_2, followed by a peroxidase (EC 1.11.1.7)-mediated decomposition of H_2O_2 and oxidation of the chromogen to produce a light-absorbing complex that is measured spectrophotometrically (Reaction Schemes [20.1] and [20.2]). Glucose oxidase acts specifically on the β-anomer of D-glucose. Mutarotation of α-D-glucose to β-D-glucose ultimately allows all of the D-glucose present to participate in the reaction. Increasing temperature and acidity or use of mutarotase (EC 5.1.3.3) can increase the rate of mutarotation; their use has been suggested to enhance the rate or extent of reaction (Hudson and Dale 1917; Okuda and Miwa 1971). However, given adequate time and temperature of incubation, colorimetric responses of α- and β-D-glucose anomers converge, with both achieving the same maximal absorbance (Karkalas 1985). What constitutes "adequate time" depends upon the conditions of the specific GOP assay used.

$$\text{β-D-Glucose} + H_2O + O_2 \xrightarrow{\text{Glucose Oxidase}} \text{D-Gluconic Acid} + H_2O_2 \quad (20.1)$$

$$H_2O_2 + \text{Chromogen} \xrightarrow{\text{Peroxidase}} 2\,H_2O + \text{Coloured Product} \quad (20.2)$$

The GOP assays have varied primarily in the oxygen-accepting chromogen used, with *o*-dianisidine, *o*-tolidine, 4-aminoantipyrine (4-aminophenzaone), *p*-hydroxybenzoic acid, adrenaline, and others used in this role. Issues with sensitivity and development of the coloured complex (adrenaline) and potential carcinogenicity (*o*-dianisidine and *o*-tolidine) led to searches for more effective and less toxic chromogens (Trinder 1969a, 1969b). The successful introduction of 4-aminophenazone as a chromogen (CAS number: 83-07-8, EC number 201-452-3; synonym: 4-aminoantipyrine) by Trinder (1969b) decreased carcinogenicity concerns, whilst maintaining the sensitivity and accuracy of the GOP assay (Pennock *et al.* 1973).

20.3   Factors Affecting Method Performance

Over the decades, the GOP assays changed as choices of chromogens and run conditions were modified to enhance accuracy, repeatability, and throughput, and to decrease errors caused by interfering substances. Common factors crucial to the success of the GOP assays are: (1) accuracy of volumetric additions in an assay that relies on concentrations and the ratio of sample to GOP reagent (GOPr); (2) execution of the entire, single protocol chosen for the assay; (3) purity and activity of enzymes; (4) reading of samples within the recommended timeframes after incubation; (5) an appropriate equation to describe the standard curve; and (6) knowledge of potentially interfering materials in the samples. The array of GOP assays are affected similarly by these factors. In this discussion, data generated through evaluation of the method developed by Trinder (1969b) as modified by Karkalas (1985) and by

Hall and Keuler (2009) will be used to describe the performance of a typical GOP assay.

20.3.1   Volumetric Additions and Precision

A point on careful attention to volumetric additions might be considered banal, but cannot be overemphasized for this assay. With limits of determination of approximately 0.5% of the range of the glucose standards and a coefficient of variation of 0.3%, small differences in glucose concentrations can be detected, and errors or inconsistencies in volume additions become glaringly obvious. Use of repeating positive displacement pipettes or dispensers for additions of GOPr can enhance pipetting accuracy.

Additionally, maintaining the desired volume ratios of sample solution to GOPr (SS:GOPr) is important because it influences how the assay behaves. Doubling the concentration of GOPr was shown to give more rapid colour evolution, but colour intensity also diminished more rapidly, leading to inaccurate results (Karkalas 1985). Different ratios of sample solution to GOPr incubated under the same conditions responded differently to post-incubation handling: when cooled in the dark for 10 min after incubation, samples containing 0.1:3.0 SS:GOPr had reduced absorbances compared to samples that were read on the spectrophotometer immediately after incubation; the 0.5:2.5 SS:GOPr preparations were unaffected by post-incubation treatment (Hall and Keuler 2009). By making accurate volumetric additions, the analyst can take advantage of the sensitivity of the assay, whilst avoiding known and unknown pitfalls that are introduced by changing reagent concentrations.

20.3.2   Adherence to Protocol

Adherence to the entirety of a selected GOP protocol only bears mention as the array of GOP assays do not necessarily have interchangeable segments even when the procedures are nearly identical. The aforementioned effect of post-incubation cooling on samples that differed only in SS:GOPr is a case in point – the ratio that showed no ill effects of post-incubation cooling was in the original protocol that included cooling (Karkalas 1985); the ratio that showed reduced absorbance was a modification of the original procedure (Hall and Keuler 2009). The impact of incubation temperature in the GOP assay speaks to this same point. By increasing incubation temperatures to 50 °C from 35 °C, incubation time was reduced from 45 to 20 min and gave comparable results. However, incubation at 60 °C for 20 min reduced absorbance by 13% (Hall and Keuler 2009). Careful control of incubation temperatures assures that the reaction will go to maximum colour development within the allotted time.

20.3.3   Enzyme Activity and Purity

Activity and purity of enzymes are understandable determinants of success or failure for an enzymatic assay. With each new lot of enzyme, recalculation of the amount of enzyme product required for the GOPr is necessary to account for what may be considerable variation in units of activity per unit of weight or volume among enzyme preparations or lots of product. Proper storage and handling of the enzymes is essential to providing the desired amount of enzymic activity. For example, the dry glucose oxidase and peroxidase enzymes that may be stored with desiccant at $-20\,^{\circ}\text{C}$ should be allowed to come to ambient temperature before opening the vials. This minimizes the amount of moisture that enzymes take up when vials are opened, and decreases the potential for enzymes to degrade during storage.

Regarding enzyme purity, the certificate of analysis and the manufacturer's specifications indicate suitability of the product for the GOP assay. Presence of catalase in the glucose oxidase enzyme preparation must be limited or it will degrade some portion of the evolved H_2O_2 and depress absorbance values. Catalase has a much lower Km for H_2O_2 than does peroxidase (93 and <5 millimolar, respectively; BRENDA 2008), so a small amount can be tolerated. Catalase presence did not appear to affect results when the ratio of peroxidase to catalase was 460:1 and the maximal concentration of glucose with the GOPr was 32.3 millimolar (Hall and Keuler 2009).

20.3.4 Stability of Absorbance

As in many colorimetric assays, the colour generated in GOP assays has a limited lifespan, and samples must be read before the decline in absorbance is appreciable ($>2\%$). Even with similar chemistries, rate of decline can depend upon run conditions (Table 20.1). Absorbance with a higher ratio of SS:GOPr incubated at a cooler temperature (0.5:2.5, incubated at $35\,^{\circ}\text{C}$ for 45 min) declined at 0.3 to 0.7% per 10 min, whereas the lower ratio of SS:GOPr incubated at a higher temperature (0.1:3.0, incubated at $50\,^{\circ}\text{C}$ for 20 min) declined at 0.7 to 0.8% per 10 min (Hall and Keuler 2009). Note that absorbance declined to a greater degree with greater concentrations of glucose, whereas the 0 µg glucose mL^{-1} solution actually increased slightly in absorbance. Reading all samples within 25 to 30 min should be adequate for these specific GOP assays; timeframe for reading samples in GOP assays using other chemistries or run conditions should be tested by the analyst. The number of samples plus glucose standards that can be read in the allowable time span dictates the number of samples that can be analysed in a run.

20.3.5 Nonlinear Equation Describes Standard Curve

A nonlinear equation most accurately describes the standard curve in the GOP assay. This defies the generally accepted criterion that colorimetric assays meet the requirement of the Beer-Lambert-Bouguer law: absorbance is directly proportional to concentration of the analyte being measured. In acceptable colorimetric assays, concentrations of the analyte are measured in the linear portion of the curve at lower concentrations where the law is met. Measurements at higher analyte concentrations where the curves become overtly nonlinear and

Table 20.1 Delayed reading changes absorbance of glucose solutions.[a]

Sample solution, mL:GOP reagent, mL	Incubation conditions	Glucose μg mL^{-1}	Post-incubation Time Delay to Reading, min				
			0	15	30	45	60
0.5:2.5	35 °C for 45 min	0	0.000	0.001	0.001	0.002	0.003
		40	0.239	0.238	0.237	0.234	0.236
		60	0.359	0.357	0.354	0.352	0.351
		100	0.594	0.586	0.582	0.579	0.579

			Post-incubation Time Delay to Reading, min				
			0	10	20	30	40
0.1:3.0	50 °C for 20 min	0	0.000	0.002	0.004	0.000	0.001
		400	0.455	0.453	0.452	0.445	0.444
		600	0.674	0.671	0.671	0.662	0.656
		1000	1.114	1.107	1.098	1.087	1.079

[a]Absorbance measured at 505 nm.
Absorbance values of glucose solutions decrease with time; reading samples before 30 minutes passes is recommended. Reprinted with modification from *Journal of AOAC INTERNATIONAL*, Hall, M.B. and Keuler, N.S., Factors affecting accuracy and time requirements of a glucose oxidase-peroxidase assay for determination of glucose. 92 (1): 50–60, 2009. Copyright 2009 by AOAC INTERNATIONAL.

where absorbance and concentration are clearly not proportional are avoided. How then, could nonlinearity be acceptable in the GOP assay?

The GOP assays have long been considered to adhere to proportionality of absorbance and glucose concentration. The relationship between absorbance and glucose concentration only becomes grossly nonproportional at greater than 150 μg of glucose in the reaction mixture containing 0.5:2.5 or 0.1:3.0 SS:GOPr (Figure 20.1) (Hall and Keuler 2009; Karkalas 1985). The GOP assays have shown extremely high R^2 ranging from 0.9996 to 1.0000 for the linear forms of the standard curves. However, intercepts of these curves were not zero, and predictions of the glucose concentrations of the standard solutions used to produce them were not precisely correct (Hall and Keuler 2009). Investigations of factors that could affect mutarotation and invoke catalase interference did not apparently explain the deviations that influenced the results of Trinder (1969a), Karkalas (1985) and Hall and Keuler (2009). However, a comparison of equations showed that a quadratic rather than linear equation consistently gave a better fit to the standard curve as evidenced by an increase in R^2, decreases in the root mean squared error and in the sum of squared residuals, and significance of the quadratic term (Table 20.2). The quadratic equation is not necessarily the "true" form of the relationship, but gave a clear improvement over linear functions. The pattern of values generated with the GOP assay appear to be inherently, slightly nonlinear. The previously presumed linearity of the GOP standard curve probably had its basis in the very high R^2 of the linear equations and that the absorbance per unit of glucose differed only in the fourth or fifth decimal places among glucose concentrations (Table 20.3).

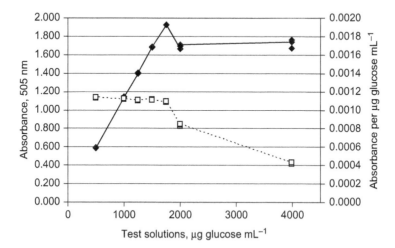

Figure 20.1 The curve of absorbance by glucose concentration (◆) becomes grossly nonlinear at concentrations of more than 1500 μg glucose mL^{-1} when absorbance per μg glucose mL^{-1} (□) declines. Results of triplicate samples from one GOP assay run are shown (0.1:3.0 ratio of standard solution, mL : GOP reagent, mL).
(From data of Hall and Keuler 2008).

Table 20.2 Fit of linear and quadratic equations to standard curves and effect of time elapsed between standard solution preparation to absorbance reading; prepared with 0.1:3.0 standard solution, mL:GOP reagent, mL.

	Curve Form[a]	Intercept	Coefficient for Abs[b]	Coefficient for Abs^2	R^2	RMSE[c]	Quadratic Term p-value	Sum of Squared Residuals
Trinder (1969a) Time[d]	L	−5.782	505.536		0.9996	7.03		197.5
	Q	0.559	472.30	16.36	1.0000	1.48	<0.01	6.56
Fresh glucose solutions in water, 7 point standard curves								
45 min	L	−4.267	899.68		0.9997	4.743		890.0
45 min	Q	0.342	871.47	25.9859	0.9999	3.402	<0.01	451.3
140 min	L	−1.794	899.60		0.9997	5.067		1027.0
140 min	Q	3.338	867.66	29.6173	0.9999	3.342	<0.01	435.6
380 min	L	−0.318	899.60		0.9998	4.502		810.8
380 min	Q	3.028	878.92	19.1439	0.9998	3.802	<0.01	563.8
Benzoic acid solutions, 5 point standard curves								
1 d	L	−4.054	903.22		0.9998	4.851		305.9
1 d	Q	0.007	873.80	26.6310	0.9999	3.307	<0.01	131.2
2 d	L	−5.232	900.70		0.9998	5.183		349.2
2 d	Q	0.137	861.88	35.0511	1.0000	1.972	<0.01	46.7
3 d	L	−4.669	900.55		0.9999	3.964		204.2
3 d	Q	−0.405	869.83	27.6968	1.0000	1.061	<0.01	13.5

[a]L = Linear, Q = quadratic.
[b]Abs = absorbance measured at 505 nm.
[c]RMSE = root mean squared error.
[d]Time from standard solution preparation to GOP analysis and reading absorbance of samples.
Quadratic equations show a better fit to the GOP assay standard curve data than do linear equations. Reprinted from *Journal of AOAC INTERNATIONAL*, Hall, M.B. and Keuler, N.S., Factors affecting accuracy and time requirements of a glucose oxidase-peroxidase assay for determination of glucose. 92 (1): 50–60, 2009. Copyright 2009 by AOAC INTERNATIONAL.

Table 20.3 Absorbance per µg glucose mL^{-1} measured in glucose standard solutions; prepared with 0.1:3.0 standard solution, mL:GOP reagent, mL.

Glucose µg mL^{-1}	Abs per glucose µg mL$^{-1\,a}$	SD^b × 10^{-6}
250	0.00116	4.0
500	0.00113	7.2
750	0.00113	4.7
1000	0.00112	0.6
p-value of quadratic term^c	0.05	

^{a}Absorbance at 505 nm divided by glucose µg mL^{-1} of the standard solution.
^{b}standard deviation.
^{c}From a regression line describing the absorbance per unit of glucose data.
Absorbance per unit of glucose changes with glucose concentration; the relationship between glucose concentration and absorbance is not linear. Reprinted with modification from *Journal of AOAC INTERNATIONAL*, Hall, M.B. and Keuler, N.S., Factors affecting accuracy and time requirements of a glucose oxidase-peroxidase assay for determination of glucose. 92 (1): 50–60, 2009. Copyright 2009 by AOAC INTERNATIONAL.

Linear or not, the value of this assay is in its ability to predict glucose content with the desired accuracy. The impact of the form of the standard curve on accuracy becomes evident with evaluation of predicted *vs.* actual values. To demonstrate, 5 point linear and quadratic standard curves were generated using 5 standard solutions, and a 2 point linear standard curve was generated using the lowest and highest glucose standard solutions (according to a commercially recommended modification of AOAC Method 996.11; AOAC, 2005). The standard curves were used to predict the glucose concentrations of the standards using the measured absorbance values of the standards that had been used to generate the curves. Actual glucose concentrations of the standard solutions were subtracted from predicted values to give the deviations for each standard in each standard curve. Glucose concentrations predicted from the quadratic standard curves gave the least deviation from actual values (Figure 20.2). Both the 5 point and 2 point standard curves gave curvilinear deviations from correct values. The values from the 5 point curve over or underpredicted at different concentrations of glucose, whereas the 2 point curve overpredicted in the middle of the range and accurately predicted the lowest and greatest glucose concentrations. The standard curve deviations were converted to the errors they would introduce into the results of a starch assay ([deviation glucose mg ml^{-1} × 50 mL final volume × 0.9 factor to convert free glucose to a starch basis]/[1000000 µg g^{-1} × 0.1 g sample × 900 g kg^{-1} dry matter]). Deviations in predicted starch content follow the same pattern as the glucose standards data (Figure 20.3). Sample weight and dilution factors may also have great effect on accuracy; they are divisors or multipliers of the glucose µg mL^{-1} deviations, decreasing or increasing their impact on predicted glucose content of the sample.

On a practical basis, interpretation of single measures may be little affected whether linear or quadratic equations are used. However, the greater deviations created with linear equations can skew interpretation of results or mask

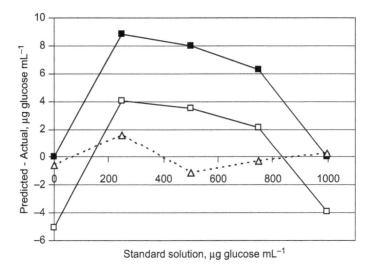

Figure 20.2 Predicted minus actual concentrations of glucose in standard solutions show the deviations of standard curves from correct values. The quadratic curve shows less deviation/is more accurate than the linear standard curves. Values were predicted with a linear equation calculated from 5 standard solutions (□), with a linear equation calculated from 2 standard solutions (■), and with a quadratic equation calculated from 5 standard solutions (△).
(From data of Hall and Keuler 2008).

differences when values are used for comparison, such as for starch contents among grain varieties or efficiency of yield of ethanol from starch.

Use of linear standard curve equations may also affect accuracy by compensating for or adding to other errors in assays which incorporate a GOP method. For example, maltulose formation during analysis of starch at neutral pH should decrease the amount of starch recovered (Dias and Panchal 1987). However, overestimation of glucose concentrations in the central portion of linear standard curves may provide a compensating error, allowing values for purified starches to measure closer to 100% starch, when the actual amount of released glucose would dictate a lower starch value. Although prediction errors associated with linear equations are relatively small, they can be decreased further using quadratic standard curves, which is preferred.

20.3.6 Interference

A variety of compounds have been shown to interfere with the GOP assay as applied to blood samples, and the interference can vary by method and chromogen used (Pennock *et al.* 1973). Of interferences likely to be found in foods, animal feeds, and plant materials, hydrophilic antioxidants that are naturally occurring, generated during processing, or added are a concern. Ascorbic acid (used as a model hydrophilic antioxidant) giving a concentration of 0.10 μmol

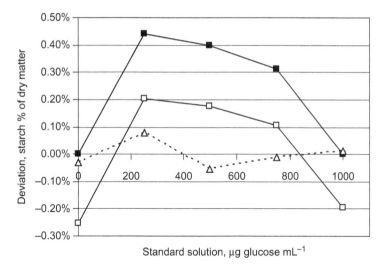

Figure 20.3 Effect of the glucose standard curves in Figure 20.2 on prediction of starch as a percentage of sample dry matter. Linear standard curves over predict starch content in the middle of the range. Values were predicted with a linear equation calculated from 5 standard solutions (□), with a linear equation calculated from 2 standard solutions (■), and with a quadratic equation calculated from 5 standard solutions (△). (From data of Hall and Keuler 2008).

mL^{-1} in the final diluted sample depressed absorbance values in the GOP assay using 0.1:3.0 SS:GOPr (Table 20.4) (Hall and Keuler 2009). Accordingly, analysis of samples that introduce greater concentrations of hydrophilic anti-oxidants should be avoided because free glucose levels will be underestimated. Most high starch or leafy vegetable foods have been shown to have hydrophilic antioxidant levels equivalent to <10 μmol per 0.1 g of dry matter. Exceptions include foods high in phenolic compounds (*e.g.* beets and red sorghum grain with antioxidant contents of equivalent to approximately 23 and 14 μmol of ascorbic acid) (Wu *et al.* 2004). Other antioxidants/reducing substances such as reductones produced during nonenzymatic browning may also interfere (Hall and Mertens 2008). It is the responsibility of the analyst to determine whether matrices and amounts of sample used in an assay contain unacceptably high concentrations of hydrophilic antioxidants. If through dilution and reduction in sample size the antioxidant contribution cannot be decreased below $0.10\ \mu mol\ mL^{-1}$ of sample solution, an alternative to GOP methods of glucose analysis should be used.

As with any colorimetric assay, solution clarity is essential, as finely particulate suspended matter increases absorbance values. Clarification of solutions through filtration may not adequately address this issue as fine material may still pass through even fine filters. Centrifugation can be a useful approach for clarification. Greater relative centrifugal force and longer times of centrifugation may be required for more recalcitrant samples. For example, in a starch

Table 20.4   Effect of ascorbic acid additions on absorbance of glucose samples
carried through a starch analysis and analyzed with a GOP assay;
prepared with 0.1:3.0 sample solution, mL:GOP reagent, mL.[a]

Ascorbic Acid, μmoles[b]	Absorbance, 505 nm	% of 0 μmole Ascorbic Acid Addition
0	1.080	100.0
1	1.084	100.4
2.5	1.079	99.9
5	1.080	100.0
10	1.072	99.3
20	1.047	97.0
30	1.036	96.0
50	0.982	91.0
SED[c]	0.0049	

[a]Values are least squares means.
[b]Concentration of 10 μmol ascorbic acid addition in final diluted sample solution = 0.098 μmol mL^{-1}.
[c]standard error of the difference.
At concentrations greater than 0.098 μmol mL^{-1} ascorbic acid begins to depress absorbance values
due to its antioxidant effect. Reprinted with modification from *Journal of AOAC INTERNA-
TIONAL*, Hall, M.B. and Keuler, N.S., Factors affecting accuracy and time requirements of a
glucose oxidase-peroxidase assay for determination of glucose. 92 (1): 50–60, 2009. Copyright 2009
by AOAC INTERNATIONAL.

assay, $1000 \times g$ for 10 min may be adequate to clarify most animal feed samples,
however, centrifugation at up to $12,000 \times g$ for 10 min is needed with some very
finely ground materials or those containing finely particulate microbial matter.
Fecal samples typically require stronger centrifugation.

20.4   A Glucose Oxidase-Peroxidase Method for Glucose Determination

20.4.1   Preparation of and Considerations on Use of Glucose Standard Solutions

Among the issues to be considered in preparing standard solutions are that
methods for weighing are often more verifiably accurate than are those for
volumetric additions, and that errors made with a stock solution will be carried
unnoticed to all solutions prepared from it unless the error is grossly obvious.
Individual preparation of separate glucose standard solutions in 0.2% benzoic
acid allows weighing of dry glucose to be used instead of volumetric transfers
of a stock solution. This allows ready detection of improperly made standard
solutions, as errors in solution concentrations become apparent as deviations
from nominal linearity with the other standards. The use of 0.2% benzoic acid
as the solvent rather than water also resolves other issues, by allowing solutions
to be prepared in advance and used over the course of months, by preventing
microbial predation on glucose, and by slightly acid conditions and advanced
preparation obviating concerns about equilibration of α- and β-anomers by the

time of analysis. The 0.2% benzoic acid solutions work well in GOP assays that use 4-aminoantipyrine as a chromogen with phenol or *p*-hydroxybenzoic acid, but should be evaluated for compatibility with other GOP chemistries.

Use of standard solutions with greater concentrations of glucose is recommended (maximum of 1000 µg glucose mL^{-1} rather than 100 µg glucose mL^{-1}). When used with 0.1:3.0 SS:GOPr the greater glucose concentrations typically reduce the number of dilutions required for sample solutions to fall within the standard curve. Detection limits are approximately 0.5% of the range of glucose standards (5.12 µg glucose mL^{-1} for 0.1:3.0 SS:GOPr) with excellent repeatability of absorbance values on triplicate samples (coefficient of variation = 0.31%; Hall and Keuler 2009).

20.4.1.1 Reagents

0.2% benzoic acid solution – Add 0.2 g of benzoic acid (ACS reagent, solid, >99.5% purity) 100 mL^{-1} of desired solution volume to a flask. Bring to desired volume with good quality distilled or deionized water. Add stir bar, stopper flask, and stir overnight at ambient temperature to dissolve.

Purified glucose – powdered crystalline glucose, purity >99.5%. Determine dry matter (measured by drying at 105 °C in a forced-air oven for 2 hours or other appropriate method) and percentage purity as provided by the manufacturer in the certificate of analysis.

20.4.1.2 Standard Solution Preparation

For standard solutions to be used as 0.1:3.0 SS:GOPr: Separately weigh approximately 62.5, 125, 187.5, and 250 mg of glucose and record weight to 0.0001 g. Rinse each portion of glucose from weigh paper into a separate 250 mL volumetric flask with 0.2% benzoic acid solution and swirl to dissolve. Note that glucose does not dissolve instantaneously; do not bring to volume until all glucose is dissolved. Bring each standard to 250 mL volume with 0.2% benzoic acid solution to give 4 independent standard solutions. The 0.2% benzoic acid solution with no added glucose serves as the 0 µg glucose mL^{-1} standard. Multiply weight of glucose by dry matter percentage and percentage purity as provided by the manufacturer's certificate of analysis and divide by 250 mL to calculate actual glucose concentrations. The standards are nominally 250, 500, 750, and 1000 µg glucose mL^{-1}, but actual glucose concentrations should be used in all calculations. Prepare solutions three days before use. Standard solutions may be stored in sealed, amber bottles at room temperature for 6 months.

20.4.1.3 Time from Standard Preparation to Use: Effect of Mutarotation

As previously mentioned, glucose oxidase reacts specifically with the β-anomer of D-glucose. A solution of D-glucose contains 36% α-anomer and 64%

β-anomer at equilibrium. Although the β-anomer reacts more rapidly with GOPr, given adequate reaction time all of the α-anomer is also oxidized (Karkalas 1985; Pomeranz and Meloan 1987). However, α-D-glucose is the usual form in which glucose crystallizes (Hudson and Dale 1917), so the effects of glucose source and time from preparation of standard solutions to solution equilibrium should be considered for their impact on standard solution values in GOP assays.

The effect of time elapsed from standard preparation to absorbance measurement can depend on assay and standard solution preparation procedures. Average absorbance readings of glucose standards prepared with powdered crystalline D-glucose in 0.2% benzoic acid solution increased over the course of three days for 0.5:2.5 SS:GOPr, but increased only numerically on the first day for 0.1:3.0 SS:GOPr (Table 20.5). When glucose solutions were freshly prepared in water, average absorbance values declined slightly between 45 min and 380 min after preparation (Table 20.5). Given that the standard solutions made with water were maintained at a similar temperature through the course of analyses so that density did not change appreciably, the decline in absorbance suggests that glucose was disappearing from the solutions. Microbial predation

Table 20.5 Effect of time elapsed between preparation of glucose standards and time of absorbance reading (Hall and Keuler 2009).

		Sample Solution, mL:GOP Reagent, mL	
Standard Solutions	Time[a]	0.1:3.0	0.5:2.5
Prepared fresh in H_2O[b]	Min	Abs[c]	Abs[c]
	45	0.639	0.298
	140	0.636	0.296
	380	0.634	0.295
	SED[d]	0.0031	0.0007
	p-value of time	0.15	<0.01
Prepared in 0.2% benzoic acid[b]	Day	Abs[c]	Abs[c]
	1	0.557	0.293
	2	0.560	0.294
	3	0.560	0.296
	SED[d]	0.0019	0.0009
	p-value of time	0.27	0.02

[a]Time from preparation of standard solutions until reading absorbance in GOP assay. Both ratios of SS:GOPr were incubated at 50 °C for 20 min.
[b]Standard solutions in water were prepared fresh on the day of analysis. Standard solutions had concentrations appropriate to the SS:GOPr.
[c]Absorbance read at 505 nm with standard solutions run in triplicate.
[d]Standard error of the difference of the absorbances for the effect of time.
Decline in absorbance with standards prepared in water may be due to microbial predation on glucose. Increase in absorbance of standards prepared in benzoic acid solution may relate to equilibration of anomers.

is a potential culprit. Based on these results, the recommendation is to prepare glucose standard solutions in 0.2% benzoic acid three days in advance of use in a GOP assay.

20.4.1.4 Effect of Number of Standard Solutions Used

The sense that the greater the number of standard solutions the more precise the measurement of the analyte should be balanced with the actual reward for the work of making additional standard solutions and the penalty of replacing test samples with standards in a run. Using standard curves generated over three days, actual minus predicted values for glucose concentrations of standard solutions showed no significant effect of using 3, 4, or 5 standard solutions to construct the quadratic standard curve ($P > 0.89$; 0 µg glucose mL^{-1} standard included in all curves) (Table 20.6). Standard curves based on 5 standard solutions did have the numerically smallest residuals, but residuals for all curves were small. Even though the accuracy was acceptable, using three

Table 20.6 Effect of number of glucose standards used for calculation of quadratic standard curves on accuracy of prediction of glucose concentrations in the standards.[a]

		Actual Minus Predicted Glucose $\mu g\,mL^{-1}$ Sample Solution, mL:GOPOD Reagent, mL	
Day of Analysis[b]	# of Glucose Standards	0.1:3.0	0.5:2.5
2	3	− 0.243	0.064
2	4	0.161	− 0.044
2	5	<0.001	<0.001
3	3	0.170	0.047
3	4	− 0.113	− 0.032
3	5	<0.001	<0.001
9	3	− 0.615	− 0.046
9	4	0.542	0.032
9	5	0.017	<0.001
SED[c]		0.645	0.087
Contrasts[d]		p-values	
3 v 4 & 5		0.56	0.26
4 v 5		0.80	0.61

[a]Values are the average least squares mean for each standard curve.
[b]Analysis of standard solutions was replicated on 3 days.
[c]standard error of the difference.
[d]Statistical contrasts of results from use of 3 vs 4 and 5 standard solutions, and between use of 4 or 5 standard solutions to calculate quadratic standard curves.
Three, 4 or 5 standard solutions may be used to generate standard curves, however, use of 5 is recommended based on numerically smallest deviations and not over-fitting the quadratic standard curve.
Reprinted with modification from *Journal of AOAC INTERNATIONAL*, Hall, M.B. and Keuler, N.S., Factors affecting accuracy and time requirements of a glucose oxidase-peroxidase assay for determination of glucose. 92 (1): 50–60, 2009. Copyright 2009 by AOAC INTERNATIONAL.

glucose concentrations to fit a quadratic equation is overfitting the data and is not recommended. Four or five concentrations of glucose may be used to prepare the standard curve, with five preferred.

20.4.2   Preparation of a Glucose Oxidase-Peroxidase Reagent

The preparation of a GOPr that does not contain known carcinogens and gives very good specificity and sensitivity to glucose is as follows (Karkalas 1985).

Dissolve 9.1 g of Na_2HPO_4 and 5.0 g of KH_2PO_4 in ca 300 mL good quality distilled or deionized water in a 1 L volumetric flask. Use water to rinse chemicals from weighing vessel and neck of flask into the bulb of the flask. Swirl to dissolve completely. Add 1.0 g phenol (ACS grade, solid) and 0.15 g 4-aminoantipyrine. Use water to rinse chemicals into bulb of flask. Swirl to dissolve completely. Add glucose oxidase (7000 U) and peroxidase (7000 U), rinse enzymes into flask with water, swirl gently to dissolve without causing excessive foaming. Bring to 1 L volume with water. Seal and invert repeatedly to mix. Filter solution through a glass fiber filter with 1.6 μm retention. Stored in a sealed amber bottle at 4 °C. Reagent life is 1 month. Determine a standard curve for the reagent to verify efficacy of the reagent before use in sample determinations.

20.4.3   Determination of Standard Curve and Glucose Content of Test Samples

A standard curve is prepared with each run of a GOP assay. Test samples should contain no greater than 1000 μg glucose mL^{-1} in order to fall within the standard curve. Test sample solutions generated from another assay to be analysed by the GOP assay should include reagent blanks and test samples, and possibly control samples of known materials.

Pipette 0.1 mL of glucose standard solutions (nominally 0, 250, 500, 750, and 1000 μg glucose mL^{-1}) and test sample solutions in duplicate into the bottom of 16 × 100 mm glass culture tubes. Add 3.0 mL of GOPr to each tube. Use of a positive displacement repeating pipette aimed against wall of tube to add GOPr will mix the solutions well. Alternatively, add GOPr by another means and vortex tubes. Cover/seal the tops of tubes with a sheet of plastic film. Incubate in a 50 °C water bath for 20 min. Set the spectrophotometer to read absorbance at 505 nm. Zero the spectrophotometer with the 0 μg glucose mL^{-1} and read absorbances of all standards and samples. All readings should be completed within 30 min of the end of incubation.

Calculate the quadratic equation describing the relationship of glucose μg mL^{-1} (response variable) and absorbance (abs) at 505 nm (independent variable). The equation will have the form quadratic coefficient × abs × abs + linear coefficient × abs + intercept. Use this standard curve to calculate glucose μg mL^{-1} in sample solutions from sample solution absorbance minus reagent blank absorbance.

20.5 Conclusion

The GOP assays have been demonstrated to have great sensitivity and specificity for glucose, as well as exhibiting very good ease of use and excellent repeatability. These characteristics readily account for the widespread use of these assays for detection of free glucose in many applications with foods and biological samples. With care taken to strictly follow protocols, assure accurate volumetric additions, and take into account potential issues with interferences, the GOP assays can provide reliable analyses for free glucose.

Summary Points on Glucose Oxidase-Peroxidase Assays for Glucose

- GOP assays are very specific for glucose and have fewer interferences compared to other chemical assays for glucose because of the action of glucose oxidase, which reacts only with the β-anomer of D-glucose.
- The variety of GOP assays primarily differ in choice of chromogen, and those may differ in carcinogenicity.
- With a limit of determination of approximately 0.5% of the range of the glucose standards and coefficient of variation (standard deviation $\times$ mean^{-1}) of 0.3% for absorbance readings of a standard glucose solution, the methods are very sensitive and repeatable.
- Accurate volumetric additions and strict adherence to procedure in any given GOP protocol are essential to the success of the assays.
- Purity and proper handling of the glucose oxidase and peroxidase are essential to providing the desired enzymatic activity in the assay; minimization of contamination with catalase reduces interference.
- Sample absorbance declines with time and excessive addition of the GOP reagent, so samples must be read within the timeframe recommended in the GOP assay and recommended ratios of sample solution to GOP reagent must be used.
- The relationship of absorbance to glucose concentration is slightly nonlinear, with the 4 to 5 point standard curves better described by quadratic than by linear equations.
- Hydrophilic antioxidants found in foods or feeds can depress the absorbances if the antioxidants exceed 0.02 and 0.10 µmol hydrophilic antioxidant mL^{-1} of sample solution in the final diluted sample (for 5:2.5 and 0.1:3.0 SS:GOPr, respectively).
- Preparation of glucose standards in 0.2% benzoic acid rather than water allows the solutions to be prepared in advance, used over the course of months, stored at ambient temperature, prevents microbial predation on glucose, and ensures equilibration of α- and β-anomers by the time of analysis. Compatibility of 0.2% benzoic solutions with the chosen GOP chemistry must be evaluated.

Key Facts about Glucose Oxidase-Peroxidase Assays for Glucose

- First developed in 1956, their specificity and sensitivity for glucose and ease of use accounts for their common, widespread use.
- The assays are very specific for glucose compared to other chemical analyses for carbohydrates because of the action of glucose oxidase, which only reacts with the β-anomer of D-glucose.
- Applications for these assays include measuring free glucose in blood, urine, and plant materials or enzymatically released glucose for starch, glycogen, or maltooligosaccharide analysis of plant, microbial, or animal tissues and in human foods or animal feeds.
- The volumetric assay is so repeatable and sensitive, that it provides an excellent tool for checking on pipetting skills.

Definitions

α- or β-Anomer: in the case of glucose, 2 stereoisomers that differ in arrangement of atoms at the reducing carbon.

Chromogen: a chemical, often light in color, that may be converted to a more strongly colored pigment through chemical reactions.

Coefficient of variation: the standard deviation divided by the mean.

Mutarotation: for glucose, the interconversion of α- and β-anomers when in solution.

Residual: Observed minus predicted values.

Abbreviations

abs	absorbance, measured on a spectrophotometer
GOP	glucose oxidase-peroxidase
Km	concentration of substrate at which a reaction rate is half its maximum speed
L	liter
min	minute
mL	milliliter
nm	nanometer
RMSE	root mean square error
SED	standard error of the difference
GOPr	glucose oxidase – peroxidase reagent
SS:GOPr	ratio of sample solution, mL to GOP reagent, mL

References

BRENDA, 2008. The comprehensive enzyme information system Release 2008.2. Available at: http://brenda-enzymes.org.

Dias, F.F., and Panchal, D.C., 1987. Maltulose formation during saccharification of starch. *Starch/Stärke*. 39: 64–66.

Hall, M.B., and Keuler, N.S., 2009. Factors affecting accuracy and time requirements of a glucose oxidase–peroxidase assay for determination of glucose. *Journal of the Association of Official Analytical Chemists International*. 92: 50–59.

Hall, M.B., and Mertens, D.R., 2008. Technical note: Effect of sample processing procedures on measurement of starch in corn silage and corn grain. *Journal of Dairy Science*. 91: 4830–4833.

Hudson, C.S., and Dale, J.K., 1917. Studies on the forms of d-glucose and their mutarotation. *Journal of the American Chemical Society*. 39: 320–328.

Karkalas, J., 1985. An improved enzymic method for the determination of native and modified starch. *Journal of the Science of Food and Agriculture*. 36: 1019–1027.

Keston, A.S., 1956. Specific colorimetric enzymatic analytical reagents for glucose. Abstracts of Papers, 129th Meeting of the American Chemical Society. 129: 31C.

Official Methods of Analysis, 18th Ed., AOAC International, Gaithersburg, MD, 2005.

Okuda, J., and Miwa, I., 1971. Mutarotase effect on micro determinations of D-glucose and its anomers with β-D -glucose oxidase. *Analytical Biochemistry*. 43: 312–315.

Pennock, J.K., Murphy, D., Sellers, J., and Longdon, J., 1973. A comparison of autoanalyzer methods for the estimation of glucose in blood. *Clinica Chimica Acta*. 48: 193–201.

Pomeranz, Y., and Meloan, C.E., 1987. Enzymatic methods. In: *Food Analysis: Theory and Practice*, 2nd Ed., Van Nostrand Reinhold, New York. p. 527.

Trinder, P., 1969a. Determination of blood glucose using an oxidaseperoxidase system with a non-carcinogenic chromogen. *Journal of Clinical Pathololology*. 22: 158–161.

Trinder, P., 1969b. Determination of blood glucose using 4-amino phenazone as oxygen acceptor. *Journal of Clinical Pathololology*. 22: 246.

Wu, X., Beecher, G.R., Holden, J.M., Haytowitz, D.B., Gebhardt, S.E., and Prior, R.L., 2004. Lipophilic and hydrophilic antioxidant capacities of common foods in the United States. *Journal of Agricultural and Food Chemistry*. 52: 4026–4037.

CHAPTER 21
Glucose Biosensors

AZILA ABDUL-AZIZ

Institute of Bioproduct Development (KL), Universiti Teknologi Malaysia International Campus, 54100 Kuala Lumpur, Malaysia, Email: azila@ibd.utm.my; Bioprocess Engineering Department, Universiti Teknologi Malaysia, 81310 UTM Skudai, Johor, Malaysia, Email: r-azila@utm.my

21.1 Introduction

A biosensor is a self-contained detection device composed of a specific biological recognition element coupled with a chemical or physical transducer (Figure 21.1). The biological recognition element can be in the form of an enzyme, antibody, receptor, microbes or others. The transducers can be in the form of electrochemical, optical, mass or thermal. The transducer will convert the interaction between a specific analyte and the biological element into a measurable signal. Ideally, a biosensor must provide fast, accurate and reliable information on the analyte of interest.

Biosensors have the potential to be used as the analytical tools in medicine, bioprocessing, food industry, agriculture, industrial monitoring and environmental monitoring (Andreescu and Sadik 2004). Due to the huge potential of the biosensor market, the investment for biosensor R&D has been quite large. Global investment for biosensor R&D has been estimated to be worth USD 300 million annually (Alocilja and Radke 2003, Weetall 1999). Hence, significant improvements in selectivity and sensitivity of biosensors have been achieved.

Food and Nutritional Components in Focus No. 3
Dietary Sugars: Chemistry, Analysis, Function and Effects
Edited by Victor R Preedy
© The Royal Society of Chemistry 2012
Published by the Royal Society of Chemistry, www.rsc.org

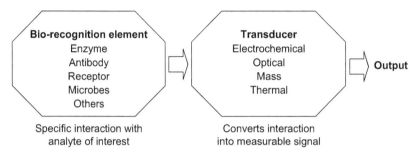

Figure 21.1 Key components of a generic biosensor. The three major elements of a biosensor are a bio-recognition element that interacts selectively with the analyte, a transducer that converts the interaction into a measurable signal and a signal processing system that records the electrical signals. Author's original artwork.

In the food industry, analytical and regulatory problems are becoming more complex. Sensing and quantification of food constituents and contaminants are no doubt essential for consumer protection.Hence, the food industry needs analysis equipment that is portable, fast, needs minimal sample preparation, is sensitive, specific, accurate, capable of providing on-line monitoring and capable of measuring several analytes in parallel. Moreover, they also need to be inexpensive, reliable and robust.

Conventional laboratory techniques such as chromatography, electrophoresis and colorimetry are accurate but they can be time-consuming, complicated and expensive. An alternative to these conventional methods of routine food analysis is a biosensor which can be very specific and thus requires minimal sample pre-processing. A biosensor also has the ability to provide immediate information regarding the samples being tested, allowing quick assessment of a situation. Biosensors are very promising in food analysis as they have the ability to fulfill some analytical demands that conventional methods could not fulfill.

21.2 Requirements for Glucose Analysis in Food

The most researched biosensor for the food industry is the glucose biosensor. This probably stems from the vast amount of research done on clinical glucose measurement to meet market demand as it has been estimated that 90% of biosensor sales are for glucose biosensors for medical applications (Alocilja and Radke 2003). Requirements for glucose analysis in food are slightly different from those of clinical glucose measurement. Issues that plague glucose analysis in food center on analyte concentration, pH, electrochemical interferents, sterilization and multi-analyte biosensing systems.

Enzymes follow the Michaelis Menten equation when responding to variations in substrate concentration. As such, the working range for substrate detection of a biosensor will be below the Michaelis constant, K_m, of the enzyme employed for the sensor. The K_m of glucose oxidase, the enzyme commonly used for glucose detection, is 4.2 mM (Schonberg and Stephen 1994). This restricts the upper limit of glucose detection; however, operating concentration ranges for glucose measurement can be extended by immobilizing glucose oxidase and using a diffusion restricting membrane to avoid enzyme saturation.

Clinical glucose measurements require working ranges of glucose biosensors to be below 40 mM (Churchouse et al. 1986). Extension of the linearity of glucose biosensors to enable clinical glucose measurement has been demonstrated (Abd. Aziz et al. 2006). In food, however, glucose concentration may be as high as 500 mM (Belitz and Grosch 1987), so a larger extension of glucose oxidase's K_m is needed. Glucose biosensors that are able to measure the high concentrations of glucose in food have reported by several authors (Amine et al. 1991, Maines et al. 1996).

pH values of food samples might vary markedly from glucose oxidase's optimum pH, which is 5.6 (Schonberg and Stephen 1994). Citrus fruits are acidic, whereas fruits such as banana, fig and avocado can be alkaline. Loss in glucose oxidase activity has been demonstrated at low pH (Amine et al. 1991, Maines et al. 1996), however a protective outer membrane can be used to protect the enzyme.

One problem with glucose bio-sensing of food samples, which is also faced by clinical glucose measurement, is electrochemical interference from electo-oxidizable substances such ascorbate and urate. In food, ascorbate interference is more significant as up to 3 mM ascorbate concentration has been reported in citrus fruits (Belitz and Grosch 1987). Various techniques have been employed to overcome this problem, such as the use of ascorbate oxidase (Gilmartin et al. 1992) or the use of selective membranes that work through ionic repulsion or size exclusion (Abdul-Aziz and Wong 2011).

Glucose biosensors for in-line process analysis, where the sensor is brought to the sample, need to be sterilized. However, for probes containing biological components, sterilization is not a straightforward process as the process should not only assure sterility but also preserve the functionality of the biological component. Options for the sterilization of enzymatic glucose biosensors include dry heat, steam, gamma irradiation, electron beam irradiation, UV irradiation, ethylene oxide gas sterilization and others (von Woedtke et al. 2002).

Food products are complex in nature. Thus, it is advantageous if a biosensor can detect more than just one analyte. This can be done by using a multi-purpose biosensor or housing different biosensors in one measurement system. These biosensors will be very useful during process control of fermentation or quality analysis of food products.

Other general requirements for commercially viable glucose biosensors for the food industry include long-term stability, accuracy in detecting glucose in real samples, repeatable output signal, no signal drift, continuous sensing, automation capability and possibility of remote detection.

21.3 Glucose Transduction Technologies

Although a variety of glucose transduction technologies have been studied, the most common methods are electrochemical and optical. Thus, these two methods will be discussed in this section.

21.3.1 Electrochemical Glucose Biosensors

Electrochemical glucose biosensors can be based on the change in measured current at a fixed applied voltage (amperometric), the change in measured voltage between electrodes (potentiometric) or the change in the ability of the sensing material to transport charge (conductometric). Of the three, amperometric transduction is the most widely used method in glucose sensing. The underlying principle is electrochemical measurement that is based on the catalytic action of glucose oxidase, explained according to the following equation:

$$\beta\text{-D-glucose} + O_2 + H_2O \xrightarrow{\text{glucose oxidase}} H_2O_2 + \text{D-gluconic acid} \qquad (1)$$

21.3.1.1 First Generation Glucose Biosensors

Glucose can be measured by measuring the reduction current of oxygen consumption eq (2) or oxidation current of the hydrogen peroxide product eq (3).

$$O_2 + 4H^+ + 4e^- \rightarrow 2H_2O \qquad (2)$$

$$H_2O_2 \rightarrow O_2 + 2H^+ + 2e^- \qquad (3)$$

For oxygen-based glucose sensor, a differential set-up is needed to correct for oxygen background variations. Two oxygen working electrodes are needed with one of them covered with glucose oxidase. The advantage of this type of sensor is that oxygen reduction occurs at high negative potential range (-0.5 to -0.7 vs Ag/AgCl) where low electrochemical interference from endogeneous electrochemical interferents can be expected. However, the differential set-up makes it a complicated device.

Hydrogen peroxide-based electrode on the other hand, gives a current which is directly proportional to glucose concentration and is free from background oxygen interference. Thus, no differential set-up is needed. This offers a simpler approach to glucose bio-sensing. This is the basis of the first generation glucose biosensor which focuses on the use of natural oxygen as a co-substrate in the enzymatic glucose oxidation process.

Glucose oxidase requires a redox co-factor, flavin adenine dinucleotide (FAD) that works as the initial electron acceptor. The reaction with glucose will reduce FAD to $FADH_2$ eq (4).

$$GOx(FAD) + \text{glucose} \rightarrow GOx(FADH_2) + \text{gluconic acid} \qquad (4)$$

The co-factor is then regenerated by the reaction of the reduced form of the enzyme with molecular oxygen eq (5).

$$\text{GOx}(\text{FADH}_2) + \text{O}_2 \rightarrow \text{GOx}(\text{FAD}) + \text{H}_2\text{O}_2 \qquad (5)$$

Figure 21.2 shows the schematic diagram of the first generation glucose biosensor.

Measurements for hydrogen peroxide are usually performed at a platinum electrode at a potential of around 0.6 V (vs Ag/AgCl). This results in electrochemical interference from electro-oxidizable endogeneous analytes such as ascorbate and urate which are also oxidizable at this potential.

One method that can be utilized to minimize the interfering effects of most of the electroactive interferents is by employing a perm-selective internal membrane that restricts the access of these analytes toward the electrode surface. Different membranes based on charge, size or polarity have been studied. Among others, the use of poly (2-hydroxyethyl methacrylate) (Abdul-Aziz and Wong 2011), electropolymerized films (Palmisano *et al.* 1993), lipid layers (Wang and Wu 1993) and composite membranes of cellulose and nafion (Zhang *et al.* 1994) as internal perm-selective membranes have been reported.

Electrochemical interference can also be minimized by reducing the oxidation over-potential for hydrogen peroxide. This can be done through the modification of the electrode material. Reduction of oxidation over-potential

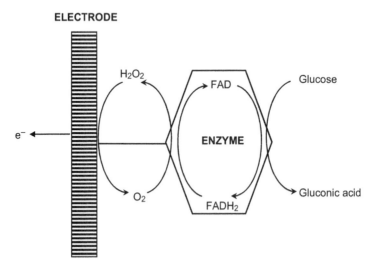

Figure 21.2   A schematic representation of a first generation amperometric glucose biosensor. First generation amperometric glucose biosensors rely on the production and detection of hydrogen peroxide. Oxygen is used as the natural mediator of electron between enzyme and electrode.
Reprinted with modifications and permission from Wang (2008). Copyright 2008, American Chemical Society.

has been demonstrated by employing metal-hexacyanoferrate based transducers (Zhu *et al.* 2006) and metalized carbon electrodes (Wang *et al.* 1995).

Another problem associated with first generation glucose biosensors is dependency on oxygen as a catalytic mediator. This problem is more pronounced in clinical samples. Oxygen deficit can affect the accuracy of glucose concentration detection by changing sensor response and reducing the upper limit of sensor linearity. The sensor will become insensitive to glucose and only respond to changes in oxygen concentration. Approaches to address this issue include the use of mass transport-limiting films that prefer oxygen diffusion compared to glucose (Armour *et al.* 1990) and oxygen-rich electrode materials (Wang *et al.* 2001).

21.3.1.2 *Second Generation Glucose Biosensors*

The problem of oxygen dependence of the first generation glucose biosensors necessitates the search of a different type of electron shuttler from the redox center of the enzyme to the surface of the electrode. Non-physiological electron-accepting mediators have been used to replace oxygen in the second generation glucose biosensors. Instead of hydrogen peroxide, a reduced mediator is formed. This reduced mediator will then be re-oxidized at the electrode, generating amperometric signal. The scheme of reactions is presented by equations 4, 6 and 7.

$$GOx(FAD) + glucose \rightarrow GOx(FADH_2) + gluconic\ acid \qquad (4)$$

$$GOx(FADH_2) + 2Med_{(ox)} \rightarrow GOx(FAD) + 2Med_{(red)} + 2H^+ \qquad (6)$$

$$2Med_{(red)} \rightarrow 2Med_{(ox)} + 2e^- \qquad (7)$$

A variety of synthetic electron mediators such ferrocene derivatives, ferricyanide, quinones and transition-metal complexes can be used for the development of a second generation glucose biosensor. Figure 21.3 illustrates the sequence of events relating to the enzymatic glucose oxidation of the second generation glucose biosensors.

The oxidation of the mediator can be carried out at a significantly lower over-potential than that of hydrogen peroxide, resulting in diminished but not total elimination of electrochemical interference from endogenous species. Problems with this type of glucose biosensor are competition with oxygen, toxicity issues of mediators and diffusion of the mediator from the system.

Despite a faster rate of reaction of the mediator with the enzyme compared to oxygen, competition with oxygen is still not totally eliminated. This can reduce the accuracy of the system. A build-up of hydrogen peroxide in the system is also possible.

ELECTRODE

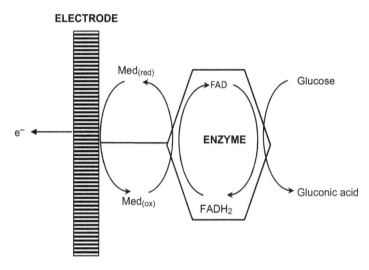

Figure 21.3 A schematic representation of a second generation amperometric glucose biosensor. Second generation amperometric glucose biosensors rely on synthetic mediators to shuttle electrons to the electrode.
Reprinted with modifications and permission from Wang (2008). Copyright 2008, American Chemical Society.

Furthermore, as the molecules are small and diffusive, retaining the mediator near the electrode and enzyme surface is also rather tricky. Different strategies have been employed to improve the system, such as wiring of the enzyme to electrode using electron-conducting redox hydrogel (Gregg and Heller 1990), chemical modification of the enzyme with electron relay groups (Riklin *et al.* 1995) and using nanomaterials as electrical connectors (Lin *et al.* 2005).

21.3.1.3   *Third Generation Glucose Biosensors*

Third generation glucose biosensors are based on direct electron transfer between the enzyme and the electrode in the absence of natural or synthetic mediators. The electron is transferred directly from glucose to the electrode *via* the active site of the enzyme. As the redox active center is embedded in thick protein, successful electron transfer between enzyme and electrode can only be achieved by making sure that the electron-transfer distance is as short as possible. A low operating potential, close to that of the redox potential of the enzyme, is very favorable as it can eliminate electrochemical interferences. This will lead to increased selectivity and sensitivity. Figure 21.4 shows the schematic diagram of enzymatic glucose oxidation mechanism of the third generation glucose biosensors.

Methods of direct electron transfer from glucose to electrode include using conducting organic salts such as tetrathiafulvalene-tetracyanoquinodimethane (TTF-TCNQ) that has tree-like crystal growth (Palmisano 2002), over-oxidized boron doped diamond electrode (Wu and Qu 2006) and the use of mesoporous electrode material that entraps the enzyme (Bao *et al.* 2008). The exact

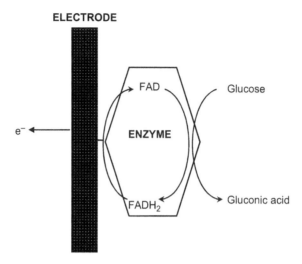

Figure 21.4 A schematic representation of a third generation amperometric glucose biosensor. Third generation glucose biosensors are reagentless and rely on direct electron transfer between enzyme and electrode.
Reprinted with modifications and permission from Wang (2008). Copyright 2008, American Chemical Society.

mechanism of direct electron transfer catalyzed by glucose oxidase remains controversial (Wang 2008).

21.3.2 Optical Glucose Biosensors

Apart from electrochemical biosensors, optical glucose biosensors are also rapidly developing. These types of biosensors are desirable as they are free from electrical and magnetic interferences and are suitable for on-line monitoring. Optical glucose biosensors can be based on different transduction technologies such as absorbance measurement, chemiluminescence or electrogenerated chemiluminescence measurement and fluorescence measurement to detect oxygen, hydrogen peroxide, gluconic acid, glucose oxidase (Wu and Choi 2006) or fluorophore-labeled carbohydrate derivatives (Lakowicz and Maliwal 1993). Table 21.1 lists transduction strategies for the detection of these molecules for glucose sensing.

21.3.2.1 Absorbance Measurement

Absorbance measurement is quite simple and has been the basis of many glucose test kits and test papers. High sensitivity can be achieved with a visible colorimeter or photometer. Colorimetric assay involves the reaction of a chromogen with hydrogen peroxide or proton ion resulting in colored products. The absorbance change of the oxidized chromogen can be followed at a specified wavelength, for example at 460 nm for o-dianisidine.

Table 21.1 Optical glucose bio-sensing strategies based on different target molecules. For optical glucose bio-sensing, optical transducers based on absorbance, chemiluminescence, electrogenerated chemiluminescence or fluorescence can be used to detect different target molecules.

Target Molecule	Transducer	References
Oxygen	Fluorescence	Moreno-Bondi *et al.* 1990
Hydrogen Peroxide	Absorbance	Narang *et al.* 1994
	Chemiluminiscence	Xu and Fang 2004
	Electrogenerated chemiluminiscence	Marquette *et al.* 2003
	Fluorescence	Wolfbeiss *et al.* 2003
Glucose Oxidase	Absorbance	Chudobova *et al.* 1996
	Fluorescence	Trettnak and Wolfbeiss 1989
Gluconic Acid	Absorbance	Piletsky *et al.* 2000
	Fluorescence	Trettnak *et al.* 1989
Flurophore-labeled Carbohydrate Derivatives	Fluorescence	Lakowicz and Maliwal 1993

Apart from colorimetric assay, other methods are also feasible. One method is by following pH change of enzymatic glucose reaction when using pH dependent membrane such as polyaniline. The color of polyaniline is strongly pH dependent. Piletsky *et al.* (2000) designs an absorbance based glucose biosensor by immobilizing glucose oxidase in a matrix containing polyaniline.

Another method is measuring the absorbance change of the conformational changes of glucose oxidase during the enzymatic reaction. During interaction between GOx and glucose, measurable changes in the intrinsic absorbance of the oxidized and/or reduced form of GOx can be detected. Thus, FAD can be used as an indicator where the decrease of the absorbance of FAD in relation to its reduction by glucose can be related to glucose concentration (Chudobova *et al.* 1996).

21.3.2.2   Chemiluminescence or Electrogenerated Chemiluminescence Measurement

Chemiluminiscence (CL) is the emission of light with limited change of temperature as a result of a reaction. Electrogenerated chemiluminescence (ECL) is a controllable form of CL where an electron-transfer reaction at the electrode surface initiates light emission. Usually glucose biosensing using CL or ECL is based on the detection of hydrogen peroxide generated during the enzymatic reaction. Excitation light source is generally not required for this type of sensor.

The sensitivity of the detection depends on the amount of hydrogen peroxide, photon yield efficiency and the accuracy of the optical instrument.

The CL or ECL method employs a luminescent reagent such as luminol for the detection process. The luminescent reaction has to be performed at a neutral or slightly alkaline condition for it to be efficient. In most systems, the flow mode detection method is employed. Examples include the use of a microfluidic sequential injection analysis system utilizing porous silicon flow-through microchips containing glucose oxidase and horseradish peroxidase for glucose CL sensing (Davidsson *et al.* 2004), a poly(dimethylsiloxane)/glass microfluidic system with glucose oxidase packed-bed reactor for CL glucose sensing (Xu and Fang 2004) and a microfluidic system based on glassy carbon electrode for ECL glucose sensing (Marquette *et al.* 2003).

21.3.2.3   *Fluorescence Measurement*

Fluorescence glucose measurement is fairly simple and usually involves the use of a sensitive molecule that conveys glucose concentration information through fluorescence detection. This type of glucose biosensing system is quite promising because it is sensitive to the point that even single molecule detection is possible. Besides that, it also causes little damage to the host system (Pickup *et al.* 2005). Fluorescence based glucose biosensors can be performed through measuring fluorescence intensity or fluorescence decay time. Time resolved fluorescence spectroscopy can be relatively independent of light scattering issues and fluorophore concentration.

Concanavalin A (con A) is a plant lectin with four binding sites for glucose per molecule of the protein (Reeke *et al.* 1975). Concanavalin A glucose biosensors are affinity biosensor based on competitive binding of glucose or labeled carbohydrate derivative such as fluorescein-dextran to lectin and glucose measurement can be done through fluorescence intensity measurement (Mansouri and Schultz 1984) or fluorescence decay time (Lakowicz and Maliwal 1993).

Other types of fluorescence-based glucose biosensor are also possible. Direct measurement of fluorescence intensity has been used to construct a glucose biosensor based on the incorporation of a pH-sensitive dye to monitor pH change due to enzymatic conversion of glucose to gluconic acid (Trettnak *et al.* 1989). Another glucose biosensor employed Eu(III)-tetracycline complex as the fluorescent probe for hydrogen peroxide (Wolfbeiss *et al.* 2003). The change in intrinsic fluorescence of glucose oxidase has been manipulated by Trettnak and Wofbeiss (1989) for glucose determination. A fluorescence based enzymatic glucose biosensor that employs a generic optical oxygen transducer to monitor the depletion of oxygen concentration through dynamic quenching of fluorescence intensity has been described by Moreno-Bondi *et al.* (1990).

21.3.3   Instrumentation

Amperometric biosensors can be designed as two-electrode or three-electrode systems (Abd. Aziz 2001). A two-electrode system, comprising a working electrode and a counter electrode, is simpler but the counter electrode might limit the current. A three-electrode system comprises a working electrode, a counter electrode and a reference electrode. The reference electrode can be used to control the potential of the working electrode with respect to itself while measuring the current between the working and the counter electrodes. Figure 21.5

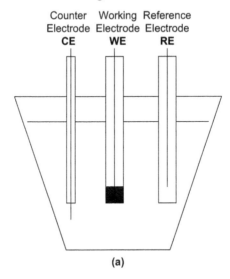

(a)

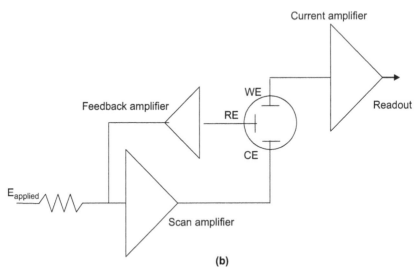

(b)

Figure 21.5 A schematic representation of a three-electrode electrochemical cell (a) and its three-electrode circuit configuration (b). Amperometric glucose biosensors are best designed as three-electrode systems. A three-electrode system comprises a working electrode, a counter electrode and a reference electrode. The reference electrode can be used to control the potential of the working electrode with respect to itself while measuring the current between the working and the counter electrodes.
Reprinted from Abd. Aziz (2001).

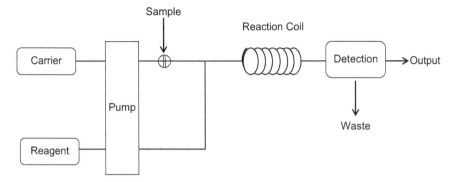

Figure 21.6 A schematic representation of the general principle of flow injection analysis. Flow injection analysis measures analytes in flowing liquids. A defined sample volume is injected into a flowing carrier system consisting of necessary reagents for reaction and the analyte of interest will be detected at the detector.
Reprinted with modifications and permission from Ruzicka and Hansen (2000). Copyright 2000, American Chemical Society.

illustrates a schematic diagram of a three-electrode electrochemical cell and its circuit configuration.

Flow injection analysis (FIA) measures analytes in flowing liquids (Figure 21.6). A defined sample volume is injected into a flowing carrier system, maintained at a suitable pH. The carrier system will also be mixed with any co-factors, co-substrates or other necessary substances. For bio-sensing, the sample passes through a phase containing a bio-recognition element and will be subsequently monitored. Newer versions of FIA utilize sequential injection or bead injection and micro-flow analysis. Advanced technology such as auto-mated valve-based injector, high precision syringe pumps and scanning flow-through detectors have also been incorporated into FIA (Ruzicka and Hansen 2000).

FIA is suitable for measuring glucose as part of on-line or off-line analysis of raw material, product quality or the manufacturing process of food pro-ducts. FIA can be combined with any type of detection device. FIA has the advantages of not returning the sample to its source, minimizing the contact period between a biosensor and its substrate, the possibility of obtaining almost real time data, the possibility of on-line calibration of the sensor and the manipulation of the sample before contact with the sensor (Chen and Karube 1992).

21.4 Commercialization Activities

Around 200 companies were involved in the field of biosensors and bioelec-tronics in 1999 (Weetall 1999) and most of them based their products on existing technologies. As stated before, this market is comprised of a few segments such as medicine, bioprocessing, the food industry, agriculture, industrial monitoring and environmental monitoring, with the bulk of the sales coming from medical applications. The commercialization activities of bio-sensor technology seem to be rather sluggish compared the huge amount of output from research and development activities.

New legislation and regulation may open up new markets for biosensors for the food industry; however, biosensors need to compete in terms of perfor-mance and reliability with established analytical equipment. The major advantages of biosensors compared to established analytical equipment are their specificity, which resulted in minimal sample pre-processing, their ability to provide immediate information regarding the samples being tested and their potential for on-line or in-line analysis.

A glucose biosensor needs to satisfy certain criteria for it to be successfully commercialized for the food industry, as the profit margin of the food industry is relatively low and competition is intense. Requirements for glucose analysis in food have been summarized in section 21.2. Commercialized glucose biosensors can appear in several forms such as auto-analyzers, manual laboratory instruments or portable devices. To date, only a limited number of glucose biosensors are used in the food industry such as Apec glucose analyzer (Danvers), ESAT Glucose Analyzer (Eppendorf), Glucoprocesseur (Solea-Tacussel), Amperometric Biosensor Detector (Universal Detectors), YSI Analyzers (Yellow Springs Instruments) and Toyo Jozo Biosensors (Mello *et al.* 2010). However, it is important to note that this list is not comprehensive. These glucose biosensors are mostly part of a multi-analyte sensing equipment.

21.5   Concluding Remarks

The use of a glucose biosensor results in a glucose monitoring process that is specific, sensitive and rapid. Even though this type of biosensor has been extensively researched for clinical blood glucose monitoring, requirements for the food industry are slightly different. The most important differences would be the need for a much larger extension of the linearity of the sensor and the need of multi-analyte sensing as food matrix is usually more complex and simultaneous detection of several analytes makes more econom-ical sense.

Research and development work for advanced technologies for glucose biosensor elements such as electrodes, membranes and immobilization techni-ques will continue to progress to support the demand for more sophisticated and stable glucose biosensors. In addition, future trends for commercialized glucose biosensors might include the use of non-enzymatic glucose electrode which can offer high sensitivity, high stability, low cost and heat resistance; the use of thermophilic bio-recognition element; the construction of a biosensor in a modular mode that it can be integrated with other analytical systems; the use of wireless technology and miniaturization and integration of the biosensor for simultaneous detection of a wide variety of target molecules.

Summary Points

- This review focuses on glucose biosensors and their applications in the food industry.
- The need for a glucose sensing system that is relatively cheap, fast, requires minimal sample preparation, sensitive, specific and accurate leads to the vast amount of research and development in this area; however, the market for this type of biosensor for the food industry is less developed compared to its clinical counterpart.
- Requirements for glucose analysis in food are slightly different from those of clinical glucose measurement.
- Even though there are a number of transduction technologies available for glucose bio-sensing, the most important ones are electrochemical and optical.
- Amperometry is the most widely used technology for electrochemical glucose biosensors and glucose biosensors based on this technique can be further classified into three generations based on the method of electron shuttling.
- Optical glucose bio-sensing techniques can be based on absorbance, chemiluminescence, electrogenerated chemiluminescence and fluorescence measurements.
- To date, only a limited number of glucose biosensors are used in the food industry and they are mostly part of a multi-analyte sensing equipment.

Key Facts

Key Facts of Amperometric Glucose Biosensors

1. First generation amperometric glucose biosensors rely on the use of natural oxygen as the co-substrate for the regeneration of flavin adenine dinucleotide (FAD) and the production and detection of hydrogen peroxide.
2. Second generation amperometric glucose biosensors depend on synthetic mediators to shuttle electrons to the electrode.
3. Third generation glucose biosensors are reagentless and rely on direct electron transfer between enzyme and electrode.

Key Facts of Optical Glucose Biosensors

1. Optical glucose biosensing techniques can be based on the detection of oxygen, hydrogen peroxide, gluconic acid, glucose oxidase or fluorophore-labeled carbohydrate derivatives.

2. Transducer technologies for optical glucose bio-sensing include absorbance, chemiluminescence, electrogenerated chemiluminescence and fluorescence measurements.
3. Using oxygen as the target molecule, fluorescence measurement can be used to detect glucose.
4. Using gluconic acid as the target molecule, fluorescence or absorbance measurement can be used to detect glucose.
5. Absorbance, chemiluminescence, electrogenerated chemiluminescence or fluorescence measurement can be used to detect glucose using hydrogen peroxide as the target molecule.
6. Fluorescence or absorbance measurement can be used to detect glucose using glucose oxidase as the target molecule.
7. Fluorescence measurement can be used to detect glucose using fluorophore-labeled carbohydrate derivatives as the target molecule.

Definitions of Words and Terms

Absorbance. Absorbance refers to the light absorbing ability of a substance at a specific wavelength.

Amperometry. Amperometry is the measurement of current when a constant potential is applied to the system.

Chemiluminescence. Chemiluminescence is the emission of light with limited change of temperature as a result of a reaction.

Conductometry. Conductometry is the determination of the ability of a material to transport charge.

Electrochemiluminescence. Electrochemiluminescence is a controllable form of chemiluminescence where an electron-transfer reaction at the electrode surface initiates light emission.

Flow injection analysis. Flow injection analysis measures analytes in flowing liquids. A defined sample volume is injected into a flowing carrier system that will also be mixed with any necessary reagents for reaction and the analyte of interest will detected at the detector.

Michaelis constant. Michaelis constant is a substrate's concentration at one-half of the limiting rate of an enzymatic reaction and is a measure of the substrate's affinity for the enzyme. For enzymatic biosensors linear region of biosensor response is below K_m.

Michaelis Menten equation. Michaelis Menten equation explains the relationship between substrate concentration and reaction rate of an enzymatic reaction.

Potentiometry. Potentiometry is the determination of voltage between electrodes.

Sequential injection analysis. Sequential injection analysis is the second generation of flow injection based methods. In SIA, a multi-position valve is used to sequentially inject the reagent to be mixed with the previously injected samples before detection at the detector.

List of Abbreviations

CL	Chemiluminiscence
Con A	Concanavalin A
ECL	Electrogenerated chemiluminescence
FAD	Flavin adenine dinucleotide
FIA	Flow injection analysis
GOx	Glucose oxidase
Km	Michaelis constant
TTF-TCNQ	Tetrathiafulvalene-tetracyanoquinodimethane

References

Abd. Aziz, A., 2001. Amperometric glucose biosensors: systematic material selection and quantitative analysis of performance. PhD Dissertation. Johns Hopkins University.

Abd. Aziz, A., Wong, F.L., and Jusoh, N., 2006. Challenges in the development of amperometric glucose biosensors. In: Lee, C.T., and Muhammad, I.I. (eds). *Special Topics in Bioprocess Engineering*. UTM Press, Skudai, Johor, Malaysia. pp. 43–60.

Abdul-Aziz, A., and Wong, F.L., 2011. Interference elimination of an amperometric glucose biosensor using poly(hydroxyethyl methacrylate) membrane. *Eng. Life Sci.* 11: 20–25.

Alocilja, E.C., and Radke, S.M., 2003. Market analysis of biosensors for food safety. *Biosensor and Bioelectronics*. 18: 841–846.

Amine, A., Patriarche, G.J, Marraza, G., and Mascini, M., 1991. Amperometric detemonation of glucose in indiluted food samples. *Anal. Chim. Acta*. 212: 91–98.

Andreescu, S. and Sadik, O.A., 2004. Trends and challenges in biochemical sensors for clinical and environmental monitoring. *Pure Applied Chem.* 76: 861–878.

Armour, J., Lucisano, J., McKean, B., and Gough, D., 1990. Application of chronic intravascular blood glucose sensor in dogs. *Diabetes*. 39: 1519–1526.

Bao, S., Li, C.M., Zang, J., Cui, X., Qiao, Y., and Guo, J., 2008. New Nanostructured TiO_2 for Direct Electrochemistry and Glucose Sensor Applications. *Advanced Functional Materials*. 18: 591–599.

Belitz, H.D., and Grosch, W. 1987. *Food Chemistry*. Springer-Verlag, Heidelberg.

Chen, C.Y., and Karube, I. 1992. Biosensors and flow injection analysis. *Current Opinion in Biotechnology*. 3: 31–39.

Chudobova, I., Vrbova, E., Kodicek, M., Janovcova, J., and Kas, J., 1996. Fibre optic biosensor for the determination of D-glucose based on absorption changes of immobilized glucose oxidase. *Analytica Chimica Acta*. 319: 103–110.

Churchouse, S.J., Mullen, W.H., Keedy, E.H., Battersby, C.M., and Vadgama, P.M., 1986. Studies on needle glucose electrodes. *Anal. Proc.* 23: 146–148.

Davidsson, R., Genin, F., Bengtsson, M., Laurell, T., and Emneus, J., 2004. Microfluidic biosensong system. Part 1. Development and optimization of enzymatic chemiluminescence micro-biosensors based on silicon microchips. *Lab. Chip.* 4: 481–487.

Gilmartin, M.A. T., Hart, J.P., and Birch, Brian., 1992. Voltammetric and amperometric behaviour of uric acid at bare and surface-modified screen-printed electrodes: studies towards a disposable uric acid sensor. *Analyst.* 117: 1299–1303.

Lakowicz, J.R., and Maliwal, B., 1993. Optical sensing of glucose using phase-modulation fluorimetry. *Analytica Chimica Acta.* 271: 155–164.

Maines, A., Cambiaso, A., Delfino, L., Verreschi, G., Christie, I., and Vadgama, P., 1996. Use of surfactant-modified cellulose acetate for a high-linearity and pH-resistant glucose electrode. *Analytical Communications.* 33: 27–30.

Mansouri, S., and Schultz, J.S., 1984. A miniature optical glucose sensor based on affinity binding. *Biotechnology.* 2: 885–890.

Marquette, C.A., Deguili, A., and Blum, L.J., 2003. Electrochemiluminescent biosensors array for the concomitant detection of choline, glucose, glutamate, lactate, lysine and urate. *Biosensors and Bioelectronics.* 19: 433–439.

Mello, L.D., Ferreira, D.C.M., and Kubota, L.T., 2010. Enzymes as analytical tool in food processing. Panesar, S.P., Marwaha, S. S. and Chopra, H. K. (eds) *Enzymes in food processing: fundamentals and potential applications.* I K International Publishing House., New Delhi, India.

Moreno-Bondi, M.C., Wolfbeis, O.S., Leiner, M.J.P., and Schaffar, B.P.H., 1990. Oxygen optrode for use in a fiber-optic glucose biosensor. *Analytical Chemistry.* 62: 2377–2380.

Narang, U., Prasad, P.N., and Bright, F.V., 1994. Glucose biosensor based on a sol-gel-derived platform. *Analytical Chemistry.* 66: 3139–3144.

Palmisano, F., Centonze, D, Guerrieri, A., and Zambonin, P. 1993. An interference-free biosensor based on glucose oxidase electrochemically immobilized in a non-conducting poly(pyrrole) film for continuous subcutaneous monitoring of glucose through microdialysis sampling. *Biosensors and Bioelectronics.* 8: 393–399.

Palmisano, F., Zambonin, P.G., Centonze, D., and Quinto, M.A., 2002. A disposable, reagentless, third generation glucose biosensor based on over-oxidized poly(pyrrole)/ tetrathiafulvalene-tetracyanoquinodimethane composite. *Analytical Chemistry.* 74: 5913–5918.

Pickup, J.C., Hussain, F., Evans, N.D., Rolinski, O.J., and Birch, D.J.S., 2005. Fluoresence-based glucose sensors. *Biosensors and Biolectronics.*, 20: 2555–2565.

Piletsky, S.A., Panasyuk, T.L., Piletskaya, E.V., Sergeeva, T.A., Elkaya, A.V., Pringsheim, E., and Wolfbeis, O.S., 2000. Polyanlikine-coated microtiter plates for use in longwave optical bioassays. *Fresenius J. Anal. Chem.* 366: 807–810.

Reeke, G.N., Becker, J.W., and Edelman, G.M., 1975. The covalent and three dimensional structure of Concanavalin A. *J. Biol. Chem.* 250: 1525–1547.

Ruzicka, J., and Hansen. E.H., 2000. Flow injection: from beaker to micro-fluidics. *Analytical Chemistry.* 72: 212A–217A.

Schonberg, D., and Stephen, D., 1994. *Enzyme Handbook*. Vol 8. Class 1.13–1.97 – Oxidoreductases, Springer Verlag, Berlin.

Trettnak, W., Leiner, M.J.P., and Wolfbeis, O.S., 1989. Fibre-optic glucose sensor with a pH optrode as the transducer. *Biosensors*. 4: 15–26.

Trettnak, W., and Wolfbeis, O.S., 1989. Fully reversible fiberoptic glucose biosensor based on the intrinsic fluorescence of glucose oxidase. *Analytica Chimica Acta*. 221: 195–203.

von Woedtke, T., Jülich, W.D., Hartmann, V., Stieber, M., and Abel, P.U., 2002. Sterilization of enzyme glucose sensors: problems and concepts. *Biosens Bioelectron*. 17:373–382.

Wang, J., and Wu, H., 1993. Permselective lipid---poly(*o*-phenylenediamine) coatings for amperometric biosensing of glucose. *Analytica Chimica Acta*. 283: 683–688.

Wang, J., Lu, F., Angnes, L, Liu, J., Sakslund, H., Chen, Q., Pedrero, M., Chen, L., and Hammerich, O., 1995. Remarkably selective metallized-carbon amperometric biosensors. *Analytica Chimica Acta*. 305: 3–7.

Wang, J., Mo, J.W., Li, S.F., and Porter, J., 2001. Comparison of oxygen-rich and mediator-based glucose-oxidase carbon-paste electrodes. *Analytica Chimica Acta*. 441: 183–189.

Wang, J., 2008. Electrochemical glucose biosensors. *Chem. Rev*. 108: 814–825.

Weetall, H.H., 1999. Chemical sensors and biosensors, update, what, where, when and how. *Biosensor and Bioelectronics*. 14: 237–242.

Wolfbeiss, O.S., Schaeferling, M., and Duerkop, A., 2003. Reversible optical sensor membrane for hydrogen peroxide using an immobilized fluorescent probe and its application to a glucose biosensor. *Microchim. Acta*. 143: 221–227.

Wu, X.J., and Choi, M.F., 2006. Optical enzyme-based glucose biosensors. In: Geddes, C.D. and Lakowicz, J.R. (eds) *Glucose sensing: topics in fluorescence spectroscopy*. Vol. 11. pp. 201–236. DOI: 10.1007/0-387-33015.1-8.

Wu, J., and Qu, Y., 2006. Mediator-free amperometric determination of glucose based on direct electron transfer between glucose oxidase and an oxidized boron-doped diamond electrode. *Anal. Bioanal. Chem*. 385: 1330–1335.

Xu, Z., and Fanf, Z., 2004. Composite poly(dimethylsiloxane)/glass microfluidic system with an immobilized enzymatic particle-bed reactor and sequential sample injection for chemiluminescence determination. *Analytica Chimica Acta*. 507: 129–135.

Zhang, Y., Hu, Y., Wilson, G.S., Moatti-Sirat, D., Poitout, V., and Reach, G., 1994. Elimination of the acetaminophen interference in an implantable glucose sensor. *Anal. Chem*. 66: 1183–1188.

Zhu, L., Zhai, J., Guo, Y., Tian, C., and Yang, R., 2006. Amperometric Glucose Biosensors Based on Integration of Glucose Oxidase onto Prussian Blue/Carbon Nanotubes Nanocomposite Electrodes. *Electroanalysis*, 18: 1842–1846.

CHAPTER 22
Assay Galactose by Biosensors

MARTIN MING-FAT CHOI*[a] AND HAN-CHIH
HENCHER LEE[b]

[a] Department of Chemistry, Hong Kong Baptist University, 224 Waterloo
Road, Hong Kong SAR, P.R. China; [b] Department of Pathology, Chemical
Pathology Laboratory, Princess Margaret Hospital, 2-10 Princess Margaret
Hospital Road, Hong Kong SAR, P.R. China, Email: leehch@ha.org.hk
*Email: mfchoi@hkbu.edu.hk; mmfchoi@gmail.com

22.1   General Review of Biosensors

The concept of biosensors represents a rapidly expanding field of instruments
to determine the concentration of given substances with rapid, accurate, simple,
compact and possibly inexpensive devices. The availability of these devices is
translated into portability of the analytical tests beyond the laboratory and also
convenience for industries and the clinical setting. The International Union of
Pure and Applied Chemistry (IUPAC) in 2001 defined the term "biosensor" as
"an integrated receptor-transducer device which is capable of providing
selective quantitative or semi-quantitative analytical information using a biological recognition element" (Thevenot *et al.* 2001). Figure 22.1 displays a
general configuration of a biosensor. The biological recognition element
translates information obtained from either biological or non-biological
matrices, usually the concentration of an analyte, from the biochemical domain
into a chemical or physical output signal with a defined sensitivity; whilst the
transducer part of the biosensor transfers this output signal to the electrical
domain.

Food and Nutritional Components in Focus No. 3
Dietary Sugars: Chemistry, Analysis, Function and Effects
Edited by Victor R Preedy
© The Royal Society of Chemistry 2012
Published by the Royal Society of Chemistry, www.rsc.org

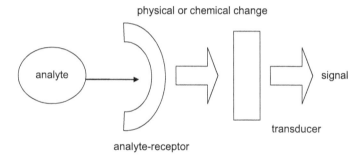

physical or chemical change

analyte

signal

transducer

analyte-receptor

Figure 22.1 General configuration of a biosensor.
Reproduced with permission from Choi 2004, copyright 2004 Springer-Verlag.

Biosensors have been widely applied in the fields of quality assurance in agriculture, food and pharmaceutical industries, monitoring environmental pollutants and biological warfare agents, as well as medical diagnostics. The first biosensor involved immobilization of glucose oxidase on the surface of an amperometric oxygen electrode for measurement of glucose in blood (Clark and Lyons 1962). To date, commercial application in devices for home blood glucose monitoring has been developed (Cass *et al.* 1984), as well as for blood lactate measurement in critical care analyzers (Aduen *et al.* 1994). Other analytes that could be measured by amperometric enzymatic biosensors include cholesterol (Vidal *et al.* 2004), ethanol (Wen *et al.* 2007), pyruvate, alanine, glutamate and glutamine (Moser *et al.* 2002).

22.2 Types of Biosensors

The biosensors can be classified according to the biological specificity-conferring mechanism or the mode of signal transduction. A biosensor can either be based on a reaction catalyzed by macromolecules (biocatalytic recognition element), which are often an enzyme, whole cells or a tissue slice; or it can be based on interaction of the analyte with macromolecules (biocomplexing or bioaffinity recognition element), *e.g.*, an antibody-antigen interaction, or an agonist/antagonist-receptor/channel system (Thevenot *et al.* 2001).

22.2.1 Transducers: Electrochemical and Optical

Biosensors can also be classified according to the transducers which utilize different detection principles. The transducer is an important component in a biosensor through which the measurement of the target analyte(s) is achieved by selective transformation of a biomolecule-analyte interaction into a quantifiable signal. The most common transducers utilized are electrochemical and optical, followed by piezoelectric ones (Luong *et al.* 2008). A wide range of optical and electrochemical instruments have been employed in conjunction with biosensors. Electrochemical biosensors measure the electrochemical changes that occur when chemicals interact with a sensing surface of the detecting electrode. The electrical changes can be based on a change in the measured voltage between the electrodes (potentiometry), a change in the measured current at a given applied voltage (amperometry), a change in the amount of charge passing between two electrodes (coulometry), or a change in the ability of the sensing material to transport charge (conductometry). Since miniaturization is possible, electrochemical biosensors are often used for field monitoring applications and fabrication of implantable biosensor (Luong *et al.* 2008). Many of the biosensors reported in the literature are based on electrochemical transducers for their high sensitivity, simplicity and low cost (Meadows 1996).

Optical biosensors employ optical fibres or planar waveguides to direct light to the sensing film (Luong *et al.* 2008). The measured optical signals include absorbance (Frederix *et al.* 2003), fluorescence (Fierke and Thompson 2001), chemiluminescence (Hinrichs *et al.* 2007), or reflectance (Asanov *et al.* 1998). Optical biosensors are preferred for screening a large number of samples at the same time, yet they lack portability, since they cannot be easily miniaturized for on-site monitoring (Luong *et al.* 2008).

22.2.2 Enzyme Immobilization Techniques

The main reason for enzyme immobilization is to allow the re-use of immobilized-enzyme over an extended period of time, hence leading to cost savings. Furthermore, it allows for easier biosensor manipulation and operation. Numerous methods have been utilized for the immobilization of enzymes in biosensors and generally the techniques comprise either physical or chemical methods. Both methods intend to keep the enzymes on the solid support without being washed out by solutions. Physical immobilization of enzyme includes adsorption on solid supports or polymeric gels, entrapment in inorganic or organic polymeric gels, or confinement within semi-permeable membranes; chemical immobilization of enzymes can be achieved by covalently bonding the enzymes to functionalized solid materials or intermolecular cross-linking of the biomolecules. In general, chemical immobilization can provide longer shelf-life and more stable biosensors; however, the enzymatic activity often is not as satisfactory as in case of immobilized enzymes prepared by physical immobilization.

In the first glucose biosensor, glucose oxidase was entrapped between two membranes (Clark and Lyons 1962). In the design of this electrode, the glucose oxidase solution was physically entrapped between the gas-permeable membrane of the oxygen electrode and a dialysis membrane, which only allowed glucose and oxygen to pass. When glucose was oxidized as catalyzed by glucose oxidase into gluconic acid and hydrogen peroxide, the rate of decrease in oxygen was a function of the glucose concentration. For this method, sufficient amount of oxygen must be present in the sample to support the catalytic reaction to generate a linear relationship between the signal and the analyte. To eliminate the problem of oxygen dependence, an oxygen-rich electrode material can be used to support the reaction (Wang and Lu 1998). Alternatively, mediators other than oxygen which serve as the electron acceptor can be used (Mu 2006; Pandey *et al.* 1997).

Numerous methods of immobilization have been developed since Clark's glucose biosensor. Biological recognition elements with high biological activity can be immobilized in a thin layer at the surface of the transducer by different procedures, *e.g.*, entrapment behind a membrane, entrapment within a polymeric matrix, entrapment within a bilayer lipid membrane, covalent bonding activated by bifunctional groups or spacers, or bulk modification of the entire electrode material (Thevenot *et al.* 2001). Figure 22.2 displays some common enzyme immobilization strategies for biosensors. These immobilized biological elements are superior to their soluble counterparts in application in biosensors for better control of reaction, as well as improved stability and reusability with the aim of reducing the production cost by efficient recycling (Cao 2005). Polymers have previously been popular agents to immobilize enzymes onto the electrode surface, (Davis *et al.* 1995) although natural materials including pig intestine (Guilkbault *et al.* 1983) and eggshell membrane (Wu *et al.* 2005; Xiao and Choi 2002; Zhang *et al.* 2007) were also used. The latest development involved using tomato epidermis, which was found to possess good biocompatibility, gas-permeable and water-impermeable properties, for enzyme immobilization in biosensor construction; this proved to be another successful method in determination of human serum glucose as the results were in good agreement with a commercial blood glucose analyzer commonly used in

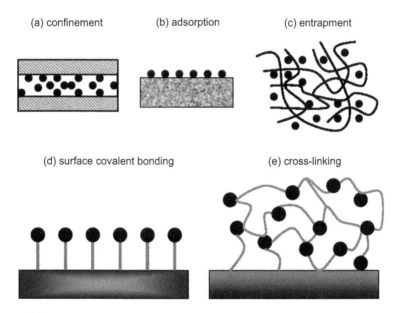

Figure 22.2   Some common enzyme immobilization strategies for biosensors. Reproduced with permission from Choi 2004, copyright 2004 Springer-Verlag.

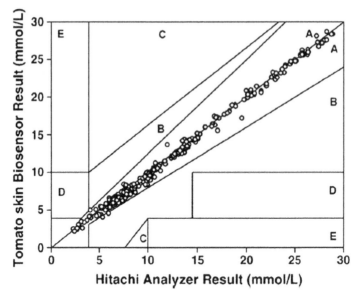

Figure 22.3   The Error Grid analysis of two methods on exposure to various blood
serum samples. Zone definitions: A: clinically accurate; B: deviating from
the reference method by 20% but would lead to benign or no treatment
error; C: deviating from the reference method by 20% and would lead to
unnecessary corrective treatment errors; D: potentially dangerous failure
to detect and treat blood glucose concentrations outside of desired target
range; and E: would result in erroneous treatment.
Reproduced with permission from Han *et al.* 2008, copyright 2008
Elsevier B.V.

hospital as shown in Figure 22.3 (Han *et al.* 2008). The biosensor was shown to
be simple to prepare, fast to respond, highly sensitive, precise and inexpensive.

Recently, it has been found that chitosan is an attractive and useful bio-
material for immobilizing enzyme (Choi 2005). This natural polymer product
possesses the favourable properties of biodegradability, non-toxicity, biocompatibility, excellent film-forming ability, high permeability toward water
and good adhesion, without being noted to inactivate enzymes.

22.3   Development of Galactose Biosensors

Galactose is often measured in the laboratory by polarimetric, fluorometric,
spectrophotometric and chromatographic methods, which are sophisticated,
costly and time-consuming (Gulce *et al.* 2002). In the clinical setting, the assay
is again available only in large hospital laboratories with a usual turn-around
time in weeks. Determination of galactose making use of the oxidation reaction
of the analyte by galactose oxidase was first described in 1962 (de Verdier and
Hjelm 1962). Since then, several attempts have been made to construct biosensors for determination of galactose (Amiss *et al.* 2007; Dahodwala *et al.*

Table 22.1 A summary of galactose biosensors.

Enzyme Immobilization Platform	Transducer	Sample	Detection Limit	Reference
Eggshell membrane	Amperometric	Blood	500 μM	Wen *et al.* 2005
Tomato epidermis	Amperometric	Urine	440 μM	Lee *et al.* 2011
Carbon nanotubes	Amperometric	Blood	25 μM	Tkac *et al.* 2007
Langmuir–Blodgett films	Amperometric	Blood	278 μM	Sharma *et al.* 2004
Pig's small intestine	Amperometric	Fermentation broth	50 μM	Szabó *et al.* 1996
Gelatine	Amperometric	NA	10 mM	Schumacher *et al.*1994
Poly(*N*-glycidylpyrrole-co-pyrrole)	Amperometric	NA	2.0 mM	Şenel *et al.* 2011
Hi-Trap column	Amperometric	NA	1.0 μM	Stoecker *et al.* 1998
Nafion-coated platinum electrode	Amperometric	NA	0.10 mM	Jia *et al.* 2003
Flow injection method	Amperometric	Milk	100 μM	Rajendran and Lrudayaraj 2002

NA: not available.

1976; Gulce *et al.* 2002; Kondakova *et al.* 2007; Rajendran and Lrudayaraj 2002; Taylor *et al.* 1977), with the enzyme galactose oxidase immobilized on different membranes/surfaces, including acetylcellulose, nylon net, silica, polyvinylferrocenium, *etc.* Table 22.1 summarizes the enzyme immobilization platforms and transducers of a wide range of galactose biosensors. Among these, tomato skin-based galactose biosensor has been demonstrated to be a useful, reliable and cost-effective biosensor for clinical assay of galactose.

Here, we describe our experimental work in developing a tomato epidermis-immobilized enzyme-based amperometric biosensor for determination of galactose, an analytically important analyte in the clinical setting as well as in the food industries. The biosensor was tested and validated in the urinary matrix and would also have the potential to apply to other sample types in many fields.

22.3.1 Principles

Galactose oxidase (E.C. 1.1.3.9) is a monomeric 68 kDa copper metalloenzyme with a 639-residue polypeptide chain. It catalyzes the oxidation of primary alcohols to the corresponding aldehydes, coupling substrate oxidation to the reduction of oxygen to hydrogen peroxide (Whittaker 2002), as in the following electrochemical equation:

$$\text{Galactose} + H_2O + O_2 \xrightarrow{\text{galactose oxidase}} \text{galactonic acid} + H_2O_2$$

This enzymatic reaction is strictly stereospecific and regioselective. The strong ability for galactose oxidase to differentiate between glucose and galactose makes the enzyme useful in a wide range of bioanalytical applications (Whittaker 2002). The formation of hydrogen peroxide allows the reaction to be coupled to a number of detection reactions, *e.g.*, peroxidase-catalyzed oxidation of dyes in colorimetric detection.

In this biosensor we describe, the galactose oxidase was immobilized on the epidermis of *Solanum lycopersicum* or common tomato fruits using chitosan together with a layer of dialysis membrane. The oxidation of galactose consumes the oxygen in the sample and this in turn causes depletion in the oxygen tension, which is detected and measured by the oxygen sensor. The decrease in the oxygen tension would correspond to the concentration of the galactose in a fixed amount of sample tested.

22.3.2 Methodology

22.3.2.1 Epidermis of Tomato Fruit

Common tomato (*Solanum lycopersicum*) fruits were purchased from the local market fresh and kept frozen at $-20\,°C$. The epidermis was then removed from the fruit. Excess flesh was carefully removed from the epidermis. The epidermis was then cleansed with copious amount of deionized water. The epidermis was cut into disks of 2 cm diameter. These disks were protected from light and stored at $4\,°C$.

22.3.2.2 Galactose Oxidase

Galactose oxidase (EC 1.1.3.9; specific activity $54\,U\,mg^{-1}$, Sigma-Aldrich, St. Louis, MO, USA) was purchased in lyophilized powder form and stored at $-20\,°C$.

22.3.2.3 Chitosan

Chitosan flakes of medium molecular weight (200–800 cps) were purchased from Sigma-Aldrich and stored at room temperature. Chitosan solution was prepared by dissolving appropriate amounts of chitosan flakes in 10 mL of 0.05 mol/L hydrochloric acid at 80–90 °C, following by cooling to room temperature. The pH was adjusted to 5.4 with 0.10 mol/L sodium hydroxide solution, followed by dilution to 20 mL with deionized water. pH measurements were taken on a combined pH glass electrode (Orion, Chicago, IL, USA). The solutions were stored at room temperature.

22.3.2.4 Dialysis Membrane

Dialysis membrane (MW cutoff 13 000) was purchased from Sigma-Aldrich and trimmed into 2 cm circular disks. The disks were soaked in deionized water overnight before use and stored at room temperature.

22.3.2.5 Phosphate Buffer

Phosphate buffer was prepared by dissolving appropriate amount of disodium hydrogen phosphate ($Na_2HPO_4 \cdot 2H_2O$, MW 177.9 g mol^{-1}; VWR International Ltd., West Chester, PA, USA) and sodium dihydrogen phosphate ($NaH_2PO_4 \cdot 2H_2O$, MW 156.01 g mol^{-1}; Fluka BioChemika, Buchs, Switzerland) with deionized water and stored at room temperature. pH measurements were taken on the Orion combined pH glass electrode.

22.3.2.6 Galactose

D-Galactose in powder form was purchased from British Drug House (Chemicals) Ltd. (Poole, UK) and stored at room temperature. It was dissolved in deionized water to the desired concentration overnight before use to allow the attainment of mutarotational equilibrium.

22.3.2.7 Copper Sulfate

Copper sulfate ($CuSO_4$) in powder form was purchased from Sigma-Aldrich and stored at room temperature.

22.3.2.8 Immobilization of Galactose Oxidase with Chitosan

100 µL of galactose oxidase solution (0.50% w/v) was added to inner surface of each disk of epidermis evenly and left to dry. 0.30% w/v medium molecular weight chitosan solution was prepared. 100 µL of the chitosan solution was added to the disks evenly and left dry for 2 hours, followed by rinsing with 0.10 mol/L pH 6.5 phosphate buffer for 10 minutes. The prepared epidermis disks were then protected from light and stored in 0.10 mol/L pH 6.5 phosphate buffer with 2 mmol/L copper sulfate at 4 °C until use.

22.3.2.9 Assembly of Biosensor and Measurement of Galactose

Figure 22.4 depicts the configuration of a tomato epidermis galactose biosensor. A galactose oxidase-immobilized disk of tomato epidermis was positioned on the surface of a Pasco CI-6542 oxygen sensor (Pasco Scientific, Roseville, CA, USA) and covered with a disk of pre-soaked dialysis membrane. The enzyme-immobilized side of the epidermis was facing towards the dialysis membrane. The two layers were kept in a steady position on the oxygen sensor with an *O*-ring. The electrode was then immersed into a 10 mL aliquot of stirred 0.50 mol/L phosphate buffer solution, pH 7.0 at room temperature (22 ± 1 °C), into which the standard solutions were injected. The dissolved oxygen signal was captured at a sampling rate of 10 s^{-1} and processed by a datalogger system consisting of a ScienceWorkshop 500 interface, serial cables, a power supply, and control software (Pasco Scientific). The data were logged in a personal computer for real-time display and processing.

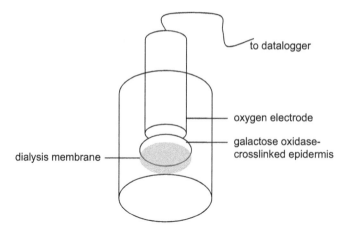

Figure 22.4   Configuration of a tomato epidermis galactose biosensor.

22.3.2.10   *Performance*

Figure 22.5 displays a typical response curve of the galactose biosensor on exposure to various concentrations of galactose. Figure 22.6 depicts the calibration curve of the galactose biosensor, demonstrating a good linear relationship between the change in oxygen signal and concentration of galactose. The linearity range is up to 7.0 mmol/L galactose as tested on three biosensors. The coefficients of variation (obtained from five replicate measurements on each of four biosensors) were 9.4% and 3.2% at 2.0 and 4.0 mmol/L galactose, respectively. The limit of detection based on a signal-to-noise ratio of 3 (determined from 20 replicate blank measurements on a single biosensor) was 0.44 mmol/L. The recoveries obtained from three replicate measurements on a single biosensor at 1.0–5.0 mmol/L ranged from 88 to 110%. The results were reproducible (concentrations > 90% retained) after a period of 30 days, demonstrating that our developed biosensor is relatively stable and suitable for routine clinical testing. The interference tests were carried out with some potential interferents, including L-ascorbic acid, β-D-fructose, β-D-glucose, D-lactic acid, sodium chloride, D-xylose and α-lactose, by exposing the biosensor to interferents in 0.50 mol/L phosphate buffer at pH 7.0. The decrease in the oxygen level was calculated as galactose equivalence, *i.e.*, the concentration of galactose that can produce the same decrease as the interferents. All these interferents tested did not affect the biosensor system, apart from a high urinary concentration of L-ascorbic acid, which caused a positive interference. The threshold used for a significant change was $+/-10\%$ at level tested, *i.e.*, 0.20 mmol/L at 2.0 mmol/L galactose.

Chitosan was used as an agent to cross-link enzyme onto tomato epidermis. It was shown to be an effective cross-linking agent without large disruption of the enzyme function in this system. With the epidermis of tomato fruit being relatively impermeable, the orientation of the enzyme-immobilized epidermis

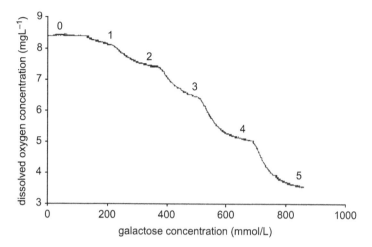

Figure 22.5 A typical response curve of the optimized galactose biosensor on exposure to various concentrations of standard galactose solutions: (0) 0.0, (1) 1.0, (2) 2.0, (3) 3.0, (4) 4.0, and (5) 5.0 mmol/L.
Reproduced with permission from Lee *et al.* 2011, copyright 2010 Elsevier B.V.

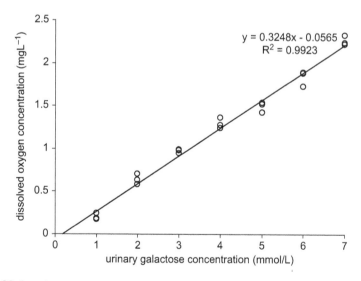

Figure 22.6 Linear calibration curve of galactose biosensor. Galactose standards of 1.0–7.0 mmol/L in urinary matrix were measured in triplicates. (unpublished results from authors).

during assembly of the biosensor became critical in our biosensor. It was observed that when the inner surface (with enzyme immobilized) of the epidermis was facing upwards towards the electrode, the response of the system was unacceptably blunted and slow, and the experiment was only possible when the epidermis was facing downwards. While the former orientation was possible with an enzyme-immobilized eggshell membrane biosensor, it was postulated that when the epidermis of tomato fruit was largely impermeable, glucose diffused slowly across the epidermis to allow oxidation by its oxidase on the other side; when the enzyme was facing towards the phosphate buffer and the substrate, oxygen, with a smaller molecular size, could diffuse from the electrode across the epidermis to oxidize the glucose on the side of the buffer. This orientation, however, might have a larger effect on the shell-life of the biosensor with leaching of the enzyme. Mild decrease of responses of the biosensor was seen over a number of runs.

22.3.3 Advantages and Disadvantages

For galactose oxidase, the medium molecular weight chitosan seemed to offer the best response curve as compared with other cross-linking agents alone or in combinations and thus was selected. A layer of dialysis membrane was used to cover the enzyme-immobilized epidermis for better entrapment of the enzyme. For practical reasons, the biosensor would be operated at room temperature, although the sensitivity would be increased at higher temperatures. The newly constructed biosensor system was making use of galactose oxidase-immobilized epidermis of tomato fruit with 0.30% medium molecular weight chitosan as the cross-linking agent, in 0.50 mol/L phosphate buffer at pH 7.0. The system was evaluated with its performance parameters determined. It had a sensitivity of $0.325 \, \text{mg} \, \text{L}^{-1}$ per mmol/L galactose standard in urine matrix. With a dilution factor of four in our system, the new biosensor has a sensitivity of 44% higher than that of a biosensor similarly fabricated using galactose oxidase immobilized on eggshell membrane with glutaraldehyde as the cross-linking agent (Wen *et al.* 2005). The superior sensitivity may be due to the less disrupted enzyme activity by chitosan as compared with glutaraldehyde.

The lower limit of detection was 0.44 mmol/L galactose and the linearity range was tested up to 7.0 mmol/L. Patients with citrin deficiency or galactossemia could have urinary galactose up to more than 10 mmol/L (Hutchin *et al.* 2009) and thus the new biosensor would be able to differentiate these cases from normal subjects. Since the linearity range is only tested up to 7.0 mmol/L, an appropriate dilution of the clinical urine sample may be required in cases of grossly elevated galactose concentration for accurate quantification. The coefficients of variation were 9.38% and 3.21% at the level of 2.0 mmol/L galactose and 4.0 mmol/L galactose in urine matrix respectively, with adequate precision at clinically important urinary concentrations of galactose.

Recovery testing showed recoveries ranging from 88 to 111% at urine galactose concentration 1.0–5.0 mmol/L. The shelf-life for this biosensor was 30 days, allowing infrequent preparation of the biosensor. The biosensor system was tested to be largely unaffected by a number of commonly encountered potential interferents, apart from L-ascorbic acid. The level is rarely seen in urine samples yet the interference should be eliminated in other sample types as well.

With chitosan as the cross-linking agent, the optimal biosensor displays improved sensitivity, fast response, satisfactory dynamic range, linearity and reproducibility, with minimal clinically significant interferents identified. These favourable characteristics would allow rapid, near-patient and inexpensive determination of galactose in urine samples of patients and aid in diagnosing galactossemia or citrin deficiency.

22.4   Applications of Biosensors in Galactose Assay

The measurement and monitoring of galactose are important in the field of food and fermentation industries. Determination of galactose aids optimization of the manufacturing procedures of the dairy products, as well as the studying of fermentation processes in these products (Gardini *et al.* 2006; Ward *et al.* 2007). A bio-strip was developed for quick analysis of galactose in dairy foods (Sharma *et al.* 2004). Apart from quality assurance and nutritional evaluation, the exclusion of galactose is important in preparing special foods for those who do not tolerate this sugar commonly found in dairy products.

Determination of galactose is equally important in the setting of nutritional monitoring and clinical chemistry. A number of inherited metabolic diseases are signified by an elevation of galactose in blood and urine, mainly those defects of galactose metabolism, including galactokinase, galactose-1-phosphate uridyltransferase and uridine diphosphate galactose 4-epimerase (Brunetti-Pierri *et al.* 2006), which appear to be more common in the Western population. The defects in these three enzymes are known to cause the galactossemia. Prompt diagnosis and treatment usually relieve the symptoms and prevent long-term complications in the patients, which include liver cirrhosis, cataracts, mental retardation and infantile deaths.

On the other hand, hyperexcretion of galactose in urine, usually together with galactitol, is highly suggestive of the highly prevalent citrin deficiency in Asian infants presenting with prolonged neonatal jaundice and conjugated hyperbilirubinemia with intrahepatic cholestasis (Lu *et al.* 2005). Clinical cases have now been reported in Arabic, Pakistani, French Canadian and Northern European origins, suggesting that citrin deficiency is a pan-ethnic disease (Dimmock *et al.* 2009). A prompt diagnosis or exclusion of citrin deficiency definitely aids management of the condition and subsequent genetic counselling, as well as avoiding unnecessary and potentially invasive investigations such as scintigraphic scan or liver biopsy. However, the mainstay of laboratory diagnosis of citrin deficiency relies on plasma amino acid profiling, often performed by high-performance liquid chromatography, which is a special clinical test offered in large hospital laboratories, with a usual turn-around time of weeks.

A rapid, simple and reliable method of determination of galactose will be of great value in aiding with the diagnosis of the above inherited diseases, allowing prompt management and close monitoring of the disease progress and response towards treatment. A well-designed galactose biosensor would be able to provide a convenient solution for the frontline clinicians and their patients.

22.5 Conclusion

Many biosensors studied and tested in recent years could potentially be improved into compact, integrated and/or automated devices (Nakamura and Karube 2003). When a biosensor is usually a reagent-less self-contained integrated device with a compact size, the invention and development of biosensors give an alternative to the traditional time-consuming, expensive and highly sophisticated chromatographic and spectroscopic techniques for quantitative analysis of given analytes in complex biological and environmental samples (Davis *et al.* 1995). The compact size and the often cheaper cost of the instrument enable testing in various locations outside the laboratory, as well as rapid turn-around time, to such an extent that real-time and even *in vivo* monitoring of metabolites in clinical settings by indwelling biosensors has been made possible (Wilson and Gifford 2005). Development of biosensors that respond reversibly and specifically to the concentration or activity of a biologically important analyte continues to be an active area of research for academic and industrial research laboratories. Interest in biosensor research is driven by the continuing need for sensors that are able to make routine measurements in a broad range of areas.

22.6 Future Trends

Scientific knowledge will always advance and new biosensing and transducing techniques will be employed in the enzyme-based biosensors. Monolayer and multilayer arrays of enzymes can be organized on transducers using covalent bonds, affinity interactions, or hydrophobic or hydrophilic interactions. The architecture of enzyme-immobilized layers on transducer elements is envisaged to offer exciting perspectives at the frontiers of biology, chemistry, medicine, and material science. Biosensors are one of the most fruitful, exciting, and interdisciplinary areas of research in analytical science. The concepts of selectivity, reversibility, detection limit, robustness, and shelf-life remain extremely important characteristics of any biosensors. Despite the complexity of sample matrices and the associated problems with real sample measurements, significant progress has already been made in improving reliability and extending capabilities to higher sensitivity and selectivity and faster response time. Looking ahead, the largest market for galactose biosensors is still in clinical settings and food and fermentation industries. The current trend of reducing costs of biological recognition elements and instrumentation will open new and cost-effective analytical horizons and broaden existing areas. Enzyme-based optical biosensors provide novel ways of performing the rapid, remote, in-line determination of a wide variety of analytes in a range of application fields. It is anticipated that enzyme-based biosensors incorporated with nanomaterials will be the next key direction in developing compact, self-contained devices for diversified chemical or biological analysis, which play an increasingly significant role in the modern society.

Summary Points

> ➢ This chapter focuses on the development and applications of a galactose biosensor in the clinical setting.
> ➢ The favourable characteristics of biosensors, including being rapid, accurate, simple and inexpensive, render it a suitable method for determination of galactose in clinical specimens of patients.
> ➢ The level of galactose is significantly elevated in urine and/or serum specimens in patients with inherited metabolic diseases such as galactossemia and citrin deficiency.
> ➢ An amperometric biosensor can be fabricated in relatively low cost with galactose oxidase immobilized onto tomato epidermis.
> ➢ The biosensor is potentially useful in other fields such as food and fermentation industries as well as with suitable validation.

Key Facts

Key Features of Galactossemia

1. Galactossemia literally means the elevation of galactose in blood, which is caused by an inherited disorder of galactose metabolism.
2. The condition is due to an underlying gene defect which causes a deficiency in either one of the following enzymes in the galactose metabolic pathway: galactokinase, uridine diphosphate galactose 4-epimerase and galactose-1-phosphate uridyltransferase.
3. The affected children commonly present with poor feeding, failure to thrive or neonatal jaundice. Sepsis, cataract, liver failure, bleeding problem or mental retardation can also occur as complications.
4. The condition can be life-threatening if untreated.
5. Diagnosis is usually made by measurement of the enzyme activity or the metabolite concentration, and/or studying of the genetic defect.
6. Lactose and galactose should be restricted in diets for these affected patients.

Key Features of Citrin Deficiency

1. Citrin, a protein encoded by the gene *SLC25A13*, is a calcium-dependent mitochondrial aspartate/glutamate carrier which transports the solutes across the mitochondrial membrane.
2. Citrin deficiency, caused by a genetic defect in *SLC25A13*, appears to be more prevalent in the East Asia but is also present in other ethnic groups.

3. The affected children most of the time present with neonatal jaundice with a cholestatic picture, which often resolve spontaneously yet some can have fulminant hepatitis. This phenotype is called intrahepatic cholestasis caused by citrin deficiency.

4. In adulthood some patients can present with psychiatric symptoms, seizures or coma. This phenotype is called adult-onset citrullinemia type II.

5. These patients often show a peculiar interest in protein- and/or lipid-rich food substances and an aversion to carbohydrates. They apparently have susceptibility towards carbohydrate toxicity.

6. Diagnosis is usually made by a combination of the measurements of the amino acids in plasma as well as galactose and/or galactitol in plasma or urine. Genetic confirmation is also feasible by studying the *SLC25A13* gene.

Definitions of Words and Terms

Amperometric biosensor. In an amperometric biosensor, the signal is measured as the change of measured current at a given applied voltage.

Biological recognition element. The biological recognition element senses the interaction with the analyte to be measured. This can be biocatalytic or biocomplexing/bioaffinity in nature.

Biosensor. Short for "bio-selective senor", biosensor is a detection device which combines a selective sensing material of biological origin and a physical or chemical transducer, which gives the signal corresponding to the analyte concentration to be measured.

Chitosan. Originally extracted from the shells of crustaceans, chitosan is a natural polymer with multiple biomedical and industrial applications. Recently it has shown favourable characteristics and promising performance in the field of biosensor development as a cross-linking agent of the enzyme.

Conductometric biosensor. In a conductometric biosensor, the signal is measured as the change in the conductance of the sensing material.

Coulometric biosensor. In a coulometric biosensor, the signal is measured as the amount of charge passing between two electrodes.

Potentiometric biosensor. In a potentiometric biosensor, the signal is measured as the change of measured voltage between the electrodes.

Transducer. The transducer converts the change sensed by the biosensor into a quantifiable signal, usually electrochemical or optical.

References

Aduen, J., Bernstein, W.K., Khastgir, T., Miller, J., Kerzner, R., Bhatiani, A., Lustgarten, J., Bassin, A.S., Davison, L., and Chernow, B., 1994. The use and clinical importance of a substrate-specific electrode for rapid determination of blood lactate concentrations. *Journal of American Medical Association*. 272: 1678–1685.

Amiss, T.J., Sherman, D.B., Nycz, C.M., Andaluz, S.A., and Pitner, J.B., 2007. Engineering and rapid selection of a low-affinity glucose/galactose-binding protein for a glucose biosensor. *Protein Science*. 16: 2350–2359.

Asanov, A.N., Wilson, W.W., and Oldham, P.B., 1998. Regenerable biosensor platform: a total internal reflection fluorescence cell with electrochemical control. *Analytical Chemistry*. 70: 1156–1163.

Brunetti-Pierri, N., Opekun, A.R., and Craigen, W.J., 2006. Two familial cases of high blood galactose of unknown aetiology. *Journal of Inherited Metabolic Disease*. 29: 762.

Cao, L., 2005. Immobilised enzymes: science or art? *Current Opinion in Chemical Biology*. 9: 217–226.

Cass, A.E., Davis, G., Francis, G.D., Hill, H.A., Aston, W.J., Higgins, I.J., Plotkin, E.V., Scott, L.D., and Turner, A.P., 1984. Ferrocene-mediated enzyme electrode for amperometric determination of glucose. *Analytical Chemistry*. 56: 667–671.

Choi, M.M.F., 2005. Application of a long shelf-life biosensor for the analysis of L-lactate in dairy products and serum samples. *Food Chemistry*. 92: 575–581.

Clark, L.C., Jr., and Lyons, C., 1962. Electrode systems for continuous monitoring in cardiovascular surgery. *Annals of the New York Academy of Sciences*. 102: 29–45.

Dahodwala, S.K., Weibel, M.K., and Humphrey, A.E., 1976. Galactose oxidase: applications of the covalently immobilized enzyme in a packed bed configuration. *Biotechnology and Bioengineering*. 18: 1679–1694.

Davis, J., Vaughan, D.H., and Cardosi, M.F., 1995. Elements of biosensor construction. *Enzyme and Microbial Technology*. 17: 1030–1035.

de Verdier, C., and Hjelm, M., 1962. A galactose-oxidase method for the determination of galactose in blood plasma. *Clinica Chimica Acta*. 7: 742–744.

Dimmock, D., Maranda, B., Dionisi-Vici, C., Wang, J., Kleppe, S., Fiermonte, G., Bai, R., Hainline, B., Hamosh, A., O'Brien, W.E., Scaglia, F., and Wong, L.J., 2009. Citrin deficiency, a perplexing global disorder. *Molecular Genetics and Metabolism*. 96: 44–49.

Fierke, C.A., and Thompson, R.B., 2001. Fluorescence-based biosensing of zinc using carbonic anhydrase. *Biometals*. 14: 205–222.

Frederix, F., Friedt, J.M., Choi, K.H., Laureyn, W., Campitelli, A., Mondelaers, D., Maes, G., and Borghs, G., 2003. Biosensing based on light absorption of nanoscaled gold and silver particles. *Analytical Chemistry*. 75: 6894–6900.

Gardini, F., Tofalo, R., Belletti, N., Iucci, L., Suzzi, G., Torriani, S., Guerzoni, M.E., and Lanciotti, R., 2006. Characterization of yeasts involved in the ripening of Pecorino Crotonese cheese. *Food Microbiology*. 23: 641–648.

Guilkbault, G.G., Danielsson, B., Mandenius, C.F., and Mosbach, K., 1983. Enzyme electrode and thermistor probes for determination of alcohols with alcohol oxidase. *Analytical Chemistry*. 55: 1582–1585.

Gulce, H., Ataman, I., Gulce, A., and Yildiz, A., 2002. A new amperometric enzyme electrode for galactose determination. *Enzyme and Microbial Technology*. 30: 41–44.

Han, H., Li, Y., Yue, H., Zhou, Z., Xiao, D., and Choi, M.M.F., 2008. Clinical determination of glucose in human serum by a tomato skin biosensor. *Clinica Chimica Acta*. 395: 155–158.

Hinrichs, K., Gensch, M., Esser, N., Schade, U., Rappich, J., Kroning, S., Portwich, M., and Volkmer, R., 2007. Analysis of biosensors by chemically specific optical techniques. Chemiluminescence-imaging and infrared spectroscopic mapping ellipsometry. *Analytical and Bioanalytical Chemistry*. 387: 1823–1829.

Hutchin, T., Preece, M.A., Hendriksz, C., Chakrapani, A., McClelland, V., Okumura, F., Song, Y.Z., Iijima, M., Kobayashi, K., Saheki, T., McKiernan, P., and Baumann, U., 2009. Neonatal intrahepatic cholestasis caused by citrin deficiency (NICCD) as a cause of liver disease in infants in the UK. *Journal of Inherited Metabolic Diseases*. DOI: 10.1007/s10545-009-1116-x.

Jia, N.Q., Zhang, Z.R., Zhu, J.Z., and Zhang, G.X., 2003. A galactose biosensor based on the microfabricated thin film electrode. *Analytical Letters*. 36: 2095–2106.

Kondakova, L., Yanishpolskii, V., Tertykh, V., and Buglova, T., 2007. Galactose oxidase immobilized on silica in an analytical determination of galactose-containing carbohydrates. *Analytical Sciences*. 23: 97–101.

Lee, H.H.-C., Wong, E.S.-C., Chan, A.Y.-W., and Choi, M.M.F., 2011. Development of a galactose biosensor with galactose oxidase-immobilized epidermis of salanum lycopersicum: potential point-of-care testing for citrin deficiency in high-prevalence areas, *Clinica Chimica Acta*. 412: 391–392.

Lu, Y.B., Kobayashi, K., Ushikai, M., Tabata, A., Iijima, M., Li, M.X., Lei, L., Kawabe, K., Taura, S., Yang, Y., Liu, T.T., Chiang, S.H., Hsiao, K.J., Lau, Y.L., Tsui, L.C., Lee, D.H., and Saheki, T., 2005. Frequency and distribution in East Asia of 12 mutations identified in the SLC25A13 gene of Japanese patients with citrin deficiency. *Journal of Human Genetics*. 50: 338–346.

Luong, J.H., Male, K.B., and Glennon, J.D., 2008. Biosensor technology: technology push versus market pull. *Biotechnological Advances*. 26: 492–500.

Meadows, D., 1996. Recent developments with biosensing technology and applications in the pharmaceutical industry. *Advanced Drug Delivery Reviews*. 21: 179–189.

Moser, I., Jobst, G., and Urban, G.A., 2002. Biosensor arrays for simultaneous measurement of glucose, lactate, glutamate, and glutamine. *Biosensors and Bioelectronics*. 17: 297–302.

Mu, S., 2006. Catechol sensor using poly(aniline-co-o-aminophenol) as an electron transfer mediator. *Biosensors and Bioelectronics*. 21: 1237–1243.

Nakamura, H., and Karube, I., 2003. Current research activity in biosensors. *Analytical and Bioanalytical Chemistry*. 377: 446–468.

Pandey, P.C., Upadhyay, S., and Upadhyay, B., 1997. Peroxide biosensors and mediated electrochemical regeneration of redox enzymes. *Analytical Biochemistry*. 252: 136–142.

Rajendran, V., and Lrudayaraj, J., 2002. Detection of glucose, galactose, and lactose in milk with a microdialysis-coupled flow injection amperometric sensor. *Journal of Dairy Science*. 85: 1357–1361.

Schumacher, D., Vogel, J., and Lerche, U., 1994. Construction and applications of an enzyme electrode for determination of galactose and galactose-containing saccharides. *Biosensors and Bioelectronics*. 9: 85–89.

Şenel, M., Bozgeyik, İ., Çevik, E., and Abasıyanık, M.F., 2011. A novel amperometric galactose biosensor based on galactose oxidase-poly(N-glycidylpyrrole-co-pyrrole). *Synthetic Metals*. 161: 440–444.

Sharmaa, S.K., Singha, S.K., Sehgalb, N., and Kumar, A., 2004. Biostrip technique for detection of galactose in dairy foods. *Food Chemistry*. 88: 299–303.

Stoecker, P.W., Manowitz, P., Harvey, R., and Yacynych, A.M., 1998. Determination of galactose and galactocerebroside using a galactose oxidase column and electrochemical detector. *Analytical Biochemistry*. 258: 103–108.

Szabó, E.E., Adányi, N., and Váradi, M., 1996. Application of biosensor for monitoring galactose content. *Biosensors and Bioelectronics*. 11: 1051–1058.

Taylor, P.J., Kmetec, E., and Johnson, J.M., 1977. Design, construction, and applications of a galactose selective electrode. *Analytical Chemistry*. 49: 789–794.

Thevenot, D.R., Toth, K., Durst, R.A., and Wilson, G.S., 2001. Electrochemical biosensors: recommended definitions and classification. *Biosensors and Bioelectronics*. 16: 121–131.

Tkac, J., Whittaker, J.W., and Ruzgas, T., 2007. The use of single walled carbon nanotubes dispersed in a chitosan matrix for preparation of a galactose biosensor. *Biosensors and Bioelectronics*. 22: 1820–1824.

Vidal, J.C., Espuelas, J., and Castillo, J.R., 2004. Amperometric cholesterol biosensor based on in situ reconstituted cholesterol oxidase on an immobilized monolayer of flavin adenine dinucleotide cofactor. *Analytical Biochemistry*. 333: 88–98.

Wang, J., and Lu, F., 1998. Oxygen-rich oxidase enzyme electrodes for operation in oxygen-free solutions. *Journal of the American Chemical Society*. 120: 1048–1050.

Ward, R.E., Ninonuevo, M., Mills, D.A., Lebrilla, C.B., and German, J.B., 2007. In vitro fermentability of human milk oligosaccharides by several strains of bifidobacteria. *Molecular Nutrition and Food Research*. 51: 1398–1405.

Wen, G., Zhang, Y., Shuang, S., Dong, C., and Choi, M.M.F., 2007. Application of a biosensor for monitoring of ethanol. *Biosensors and Bioelectronics*. 23: 121–129.

Wen, G., Zhang, Y., Zhou, Y., Shuang, S., Dong, C., and Choi, M.M.F., 2005. Biosensors for determination of galactose with galactose oxidase immobilized on eggshell membrane. *Analytical Letters.* 38: 1519–1529.

Whittaker, J.W., 2002. Galactose oxidase. *Advances in Protein Chemistry.* 60: 1–49.

Wilson, G.S., and Gifford, R., 2005. Biosensors for real-time in vivo measurements. *Biosensors and Bioelectronics.* 20: 2388–2403.

Wu, B., Zhang, G., Zhang, Y., Shuang, S., and Choi, M.M.F., 2005. Measurement of glucose concentrations in human plasma using a glucose biosensor. *Analytical Biochemistry.* 340: 181–183.

Xiao, D., and Choi, M.M.F., 2002. Aspartame optical biosensor with bienzyme-immobilized eggshell membrane and oxygen-sensitive optode membrane. *Analytical Chemistry.* 74: 863–870.

Zhang, Y., Wen, G., Zhou, Y., Shuang, S., Dong, C., and Choi, M.M.F., 2007. Development and analytical application of an uric acid biosensor using an uricase-immobilized eggshell membrane. *Biosensors and Bioelectronics.* 22: 1791–1797.

CHAPTER 23

Combined Assays for Lactose and Galactose by Enzymatic Reactions

NISSIM SILANIKOVE* AND FIRA SHAPIRO

Biology of lactation Lab., Institute of Animal Science, Agricultural Research Organization (ARO), PO Box 6, Bet Dagan 50 250, Israel
*Email: nsilanik@agri.huji.ac.il

23.1 Introduction

Lactose (milk sugar) is a disaccharide that is found primary in the milk of various mammals and milk products (Linko 1982). It is formed from the condensation of two monosaccharides, galactose and glucose, which form a β-1→4 glycosidic (covalent) linkage.

The systematic name of lactose is β-D-galactopyranosyl-(1→4)-D-glucose. The glucose can be in either the α-pyranose form or the β-pyranose form, whereas the galactose can only have the β-pyranose form: hence α-lactose and β-lactose refer to anomeric form of the glucopyranose ring alone.

Lactose concentration in the milk of various mammals range 4 to 9%. Human milk has the highest lactose percentage at around 9%. Cow's milk has ~5% lactose and similar values are found in buffalo, yak and sheep milk, whereas those in goat's milk (4.4–4.7%) are somewhat lower, but are not sufficiently low to prevent the problem of lactose intolerance (Silanikove *et al.* 2010).

Food and Nutritional Components in Focus No. 3
Dietary Sugars: Chemistry, Analysis, Function and Effects
Edited by Victor R Preedy
© The Royal Society of Chemistry 2012
Published by the Royal Society of Chemistry, www.rsc.org

Galactose (derived from the Greek word *galaktos*, which means milk) occurs in milk, plants and microbiological products (Sunehag *et al.* 2002). Milk galactose is less sweet than glucose. However, it is considered a nutritive sweetener because it has food energy. Galactose exists in both open-chain and cyclic form. The open-chain form has a carbonyl at the end of the chain. Four isomers are cyclic; two of them with a pyranose (six-membered) ring are predominant in eukaryotes, and two with a furanose (five-membered) ring occurs in bacteria, fungi and protozoa.

23.2 Methodology for the Determination of Lactose and Galactose

Lactose and galactose concentrations, and in fact most organic components in nature, can be determined in biological fluids and food samples by means of HPLC, paper chromatography or gas chromatography, and those are considered in different chapters in this book. These techniques are valuable, particularly in situations where many metabolites can be analyzed simultaneously. However, these techniques typically require either expensive chemicals or instrumentation, are time-consuming and sometimes may lack appropriate sensitivity (Kleyn 1985).

Analysis of lactose and galactose concentration by enzymatic reactions coupled to spectrometric detection of the reaction products is simple, specific and inexpensive (Kleyn 1985). This chapter is associated with the presentation of its principle, recent improvements that make it more effective and of unique advantageous associated with the enzymatic detection in testing dairy foods.

23.3 Principles of the Simultaneous Determination of Lactose and Galactose in Dairy Products by Enzymatic Reactions

Milk lactose is hydrolysed into β-D-glucose and β-D-galactose in infant mammals' small intestines by enzyme called lactase (β-D-galactosidase; β-D-galactoside galactohydrolase, EC 3.2.1.23), which is secreted by the villi of epithelial cells. Splitting of lactose by lactase in a similar manner to the way it occurred in the intestine forms also the first step in the enzymatic determination of lactose (Reaction [1]). Lactase enzymes similar to those produced in the small intestines of humans are produced industrially by fungi of the genus *Aspergillus*, and are available at much lower price (Seyis and Aksoz 2004). Thus, industrially produced enzymes are mostly applied in the enzymatic determination of lactose (Kleyn 1985; Shapiro and Silanikove, 2010, Shapiro *et al.* 2002). Another important factor is the need to use β-lactose as a standard; lactose isolated from cow's milk most likely would fit for that.

Reaction (1)

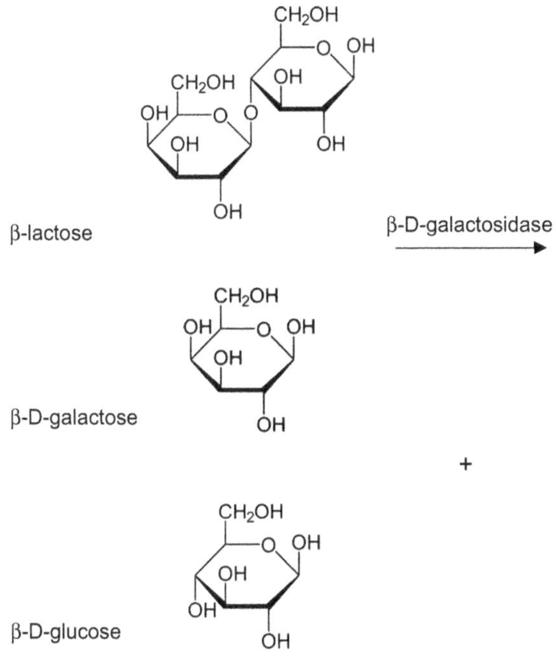

β-lactose

β-D-galactosidase

β-D-galactose

+

β-D-glucose

Most organisms metabolize galactose to glucose 6-P by the Leloir pathway (Frey 1996). However, few species of bacteria, most notably *Pseudomonas saccharophila* and *Pseudomonas fluorescens* have been shown to possess a unique enzyme pathway, called the DeLey-Doudoroff pathway (Wong and Yao 1994). The first reaction in this pathway converts the pyridine nucleotides coenzyme, nicotinamide adenine dinucleotide, commonly abbreviated NAD, into its reduced forms NADH in direct proportion to oxidation of galactose to D-galactono-γ-lactone (D-galactono-1,4-lactone) in the presence of β-galactose dehydrogenase (systemic name, D-galactose:NAD + 1-oxidoreductase; E. C. 1.1.1.48) as described below (Reaction [2]).

Reaction (2)

$$\text{D-galactose} + \text{NAD}^+ \xrightarrow{\text{β-galactose dehydrogenase}} \text{D-galactono-1,4-lactone} + \text{NADH} + \text{H}^+$$

Thus, by adding β-galactose dehydrogenase and NAD^+ to reaction mixture that contain galactose it is possible to quantify it by measuring the increase in NADH absorption at 339, 334 or 365nm (Kleyn 1985). The simultaneous determination of lactose and galactose requires the determination of the samples in two sets. Set 1 starts with Reaction (1) and followed by Reaction (2). Set 2 starts with Reaction (2). Galactose concentration is determined from Reaction (2), using galactose as a standard, whereas lactose concentration from (Reaction [1] + [2]) – Reaction (1), using lactose as a standard (Kleyn 1985; Shapiro *et al.* 2002).

23.4   Simplification of the Colorimetric Assay by Replacing NAD with Thio-NAD

The original above procedure was carried out in single-well spectrophotometers (Kleyn 1985), which is much more time consuming in comparison to determination with microplate readers. Microplate readers have been widely used for the determination of many materials by ELISA, but also have been proved useful in measuring various colorimetric reactions. Theoretically, NAD^+ can be replaced by thio-NAD^+, as described by Reaction (3), with the advantage that the accumulation of thio-NADH can be monitored by the increase in absorbance in the visible range at 405 or 415 nm (Buckwalter and Meyerhoff 1994).

Reaction (3)

$$\text{D-galactose} + \text{thio-NAD}^+ \xrightarrow{\text{β-galactose dehydrogenase}} \text{D-galactono-1, 4-lactone} + \text{thio-NADH} + H^+$$

The results described by Shapiro *et al.* (2002) point out that the procedure for simultaneous detection of lactose and galactose which is based on detection of thio-NADH in the visible range were very similar to those based on detection of NADH in the UV range both in milk and in mammary secretion during involution, in which lactose concentration was only $\sim 10\%$ of its normal value. Plate readers in the UV range are much more expensive than plate readers in the visible range, particularly when using the wavelength (405 nm) most widely available on inexpensive microplate readers used for ELISA determinations in which the indicator enzyme is alkaline phosphatase.

However, whereas the modification described above simplified and improved the rate of analysis of lactose and galactose, it still suffers from the deficiency of applying colorimetric procedure for the determination of metabolites in complex materials such as milk and dairy products. This was the drive for further development.

23.5   The Challenge of Quantifying Metabolites in Milk and Dairy Products

Applying monochromatic absorbance photometry for analysis of foods is frequently difficult and problematic (Nielsen 2003). This is particularly so when relating to products such as milk and yogurt. The following challenges emerge when the aim is to measure metabolites such as lactose and galactose by colorful reactions in these substances: (i) they contain fat droplets of varying size that scatters light in an unpredictable way; (ii) they are opaque and colloidal solution of proteins that scatter and absorb light; (iii) that frequently contain intense colorant (*e.g.*, cacao drink, yogurts with fruits or colorant) that interfere with the monochromatic absorbance.

Overcoming these problems necessitates the use of various pretreatment procedures of samples to minimize the above described problems. However, frequently such pretreatments only partially resolve the problem, in addition to the fact that many of these procedures are cumbersome and time consuming.

23.6 Fluorometry Coupling to Formation of Chromophore and Enzymatic Cycling as a Solution for the Determination of Lactose and Galactose in Milk and Dairy Products

Fluorometry refers to measurement of fluorescence emitted by certain compounds when exposed to intense radiant energy, usually in the ultraviolet range. The atoms of such substances emit fluorescence light when excited. The response of a fluorescent chromophore is specific to the excitation light and emit fluorescent light with a characteristic color and wavelength, allowing identification and quantification of the chromophore in biologic specimens. Fluorometry is a highly sensitive method of analysis, which is free to a large degree from the deficiencies of monochromatic absorbance photometry outlined above, making it suitable for detection of metabolites such as milk and other dairy products (Shapiro and Silanikove 2010).

Fluorescent chromophore (fluorochromophore) may be indigenous (biological) or synthetic molecules. As already mentioned, NAD^+ and NADH absorb strongly in the ultraviolet range due to the adenine base. In addition, when NADH in solution is excited at 340 nm, it emits fluorescence with a peak at 460 nm and a lifetime of 0.4 nanoseconds, whereas the oxidized form of the coenzyme (NAD) does not fluoresce. These properties of NADH have been used to measure dissociation constants, which are useful in the study of enzyme kinetics (Lakowicz *et al.* 1992) and changes in the redox state of living cells, by fluorescence microscopy (Kasimova *et al.* 2006). Fluorescent emission of NADH has been also use to measured metabolites in NAD-depended reaction; however, the accuracy of this method when applied to biological specimens is less satisfactory (Shapiro and Silanikove 2010) most likely because many indigenous molecules emit florescence in this rage, which increases the background noise.

A potential solution to this problem of specifity is to use synthetic fluorochromophore, which emits florescence at specific excitation/emission wavelength. One such candidate is resazurin (7-Hydroxy-3H-phenoxazin-3-one 10-oxide). Resazurin is a blue dye, which itself is nonfluorescent until it is reduced to the pink colored and highly red fluorescent resorufin (7-Hydroxy-3H-phenoxazin-3-one) (Figure 23.1).

The main applications of these substances are their use as an oxidation-reduction indicator in cell viability assays for bacteria and mammalian cells. Resazurin was already in use in 1929 to quantify bacterial content in milk (Pesch and Simmert 1929). It is also widely used as an indicator for cell viability

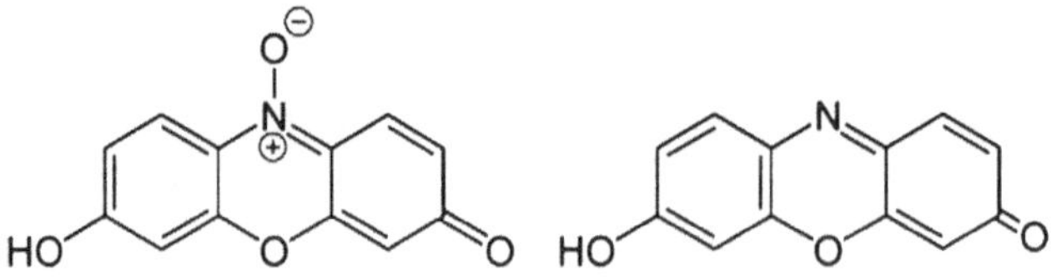

Figure 23.1   Molecular structure of nonfluorescent resazurin (left) and florescent resorufin (right).

in mammalian cell cultures (Anoopkumar *et al.* 2005). The conversion of resazurin to resorufin can be used to measure NADH, metabolites derived from NAD-depended reactions, and the kinetics of the enzyme involved of such reactions (Lakowicz *et al.* 1992). For this, we need an enzyme that will be able to catalyze the following reaction:

$$NADH + H^+ + acceptor(resazurin) \rightleftharpoons NAD^+ + reduced\ acceptor\ (resorufin)$$

One such candidate is the oxidoreductase diaphorase (EC 1.6.99.1) also known as NADPH dehydrogenase; this enzyme has many other commonly used names related to its mode of action. From its common name NADPH dehydrogenase its also appears that it can oxidize NADPH to NADP+. However, this does not form a problem of specifity for our case because the conversion of resazurin to resorufin is specific for the presence of NADH (Shapiro and Silanikove 2010). Diaphorase has the ability to reduce chromogenic acceptors, such as decolorization of 2, 6-dichlorophenolindophenol, tetrazolium dye and resazurin; the last two become chromogenic on reduction. Diaphoreses are a ubiquitous class of flavin-bound enzymes, and members of this group are also produced by bacteria such as *Cl. Kluyveri*. Bacterial diaphorase are commercially available and can serve as relatively cheap source for the enzyme (Shapiro and Silanikove 2010). Consequently, following reaction 1, galactose can be quantified by reaction 4 and lactose as explained for the colorimetric reaction. A schematic presentation of the methodology is presented in Figure 23.2. As can be seen in Figure 23.2, the reaction system occurs according to the principles of enzymatic cycling.

For the simultaneous determination of lactose and galactose by enzymatic reaction coupled to formation of fluorochromophore, stage 1 (described by reaction 2), whereas stage 2 is described by Reaction (4):

Reaction (4)

$$NADH + H^+ \xrightarrow{\text{resazurin diaphorase}} NAD^+ + resorufin$$

In order to start the reaction in the desired direction, the reaction medium needs to contain excess of NAD^+ and appropriate concentration of β-galactose dehydrogenase, whereas the coupling of the conversion of resazurin to resorufin with the conversion of NADH to NAD^+ requires the presence of

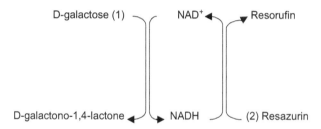

Figure 23.2   The principle of enzymatic cycling reaction in the fluorimetric determination of galactose.

Table 23.1   Concentrations (mean) of L-lactate, D-lactate and lactose in commercial plain (classical) yogurts and modern bio-yogurts.*

Product	L-lactate, mM	D-lactate, mM	D/L ratio, mM	Lactose, mM	D-l/L Ratio
			Plain Yogurts		
A	75.1	122.2	1.64	89.1	0.84
B	116.1	73.0	0.63	92.1	0.79
C	83.1	54.2	0.63	98.1	0.55
D	79.2	29.0	0.37	100.2	0.29
E	87.1	90.1	1.01	85.5	1.05
			Bio-yogurts		
F	96.2	2.0	0.02	128.5	0.02
G	123.2	5.2	0.04	173.0	0.03
H	101.5	6.6	0.04	129.3	0.03
I	123.0	12.0	0.03	130.3	0.02

D/L ratio = D-lactate/L-lactate ratio; D-l/l ratio = D-lactate/lactose ratio.
*Rearranged from Shapiro and Silanikove (2010). The product codes were fully identified in the paper. Briefly, the plain yogurts and the bio-yogurts were produced in Israel; however, the bio-yogurts under license of international brands: F and H by DANONE, G by Muller, and I by Yoplait.

appropriate concentrations of resazurin and diaphorase. An example for appropriate reaction conditions is detailed in Shapiro and Silanikove (2010).

The benefits of enzyme cycle reaction are well known (Lowry 1980) and related to improved sensitivity, which allow to analyze minute amount of substances, or to carry out the reaction in small volume, like the wells of microplate, and thus in a rapid and potentially automatic mode. This principle is exemplified in its successful application in measuring lactose and galactose in milk in yogurts applying minimal preparation of the samples (Table 23.1), attaining accuracy which is not falling or even exceeding those attained by HPLC and in much faster way (Shapiro and Silanikove, 2010).

23.7   Conclusions

Enzymatic methods were found particularly valuable in characterizing the fate of lactose and galactose in yogurts, because their metabolism is stereoisomer specific and thus applying stereoisomer specific enzymes is the only way to distinguish between the isomer products and the final nutritional quality of the product (Shapiro and Silanikove 2010). An example for the efficiency of this method in analyzing classical yogurts and common modern bio-yogurts that contains additions of probiotic bacterial cultures is provided in Table 23.1, rearranged from Shapiro and Silanikove (2010). The results clearly indicate that in the set measured degradation of lactose is more extensive in yogurts then in bio-yogurts and that the level of D-lactate (measured by similar fluorimetric principle) is higher in yogurts then in bio-yogurts. Only enzymatic reactions can distinguish between the presence of D-and L- lactate.

The versatility of present fluorimetric methodology allows applying the same principle for analyzing other test substances in various types of biological fluids and foods (Larsen 2010, 2011; Shapiro and Silanikove 2011).

Summary Points

- Lactose and galactose are important carbohydrate resources and their main dietary source is milk and dairy products.
- It is possible to determine the concentration of lactose (after its splitting to glucose and galactose) and galactose colorimetrically by NAD/NADH depended reaction.
- Use of thio-Nad instead of NAD is a modification allowing the reaction to be carried out in plate reader in the visible range.
- However, applying monochromatic colorimetric analysis in complex media such as milk and dairy products is inherently difficult.
- A solution based on fluorometry, which applies formation of synthetic chromophore and enzymatic cycling has been shown as effective solution for analyzing lactose and galactose in milk and dairy products.
- The main advantages are minimal sample preparation, low reaction volume, high rate of analysis and low cost.
- This approach is versatile and may be applied for a broad spectrum of metabolites, and some recent applications were noted.

Key Facts

- Use of enzymatic reaction for the determination of lactose and galactose is particularly important when the fate of their metabolism is stereoisomer specific.
- An example was provided by comparing the composition of classical (plain) yogurts and modern bio-yogurts (Table 23.1). The results suggest that bio-yogurts are inferior in comparison to classical yogurt in important aspects, such as lower lactose concentration.

References

Anoopkumar, D., Carey, J.B., Conere, T., O'Sullivan, E., Van Pelt, F.N., and Allshire, A., 2005. Resazurin assay of radiation response in cultured cells. *British Journal of Radiology* 78: 945–947.

Buckwalter, K.B., and Meyerhoff, F.F., 1994. Adapting homogenous enzyme-linked competitive binding assays to microtiter plates. *Analyitical Biochemistry* 218: 14–19.

Frey, P.A., 1996. The Leloir pathway: a mechanistic imperative for three enzymes to change the stereochemical configuration of a single carbon in galactose. *FASEB J.* 10:461–470.

Kasimova M.R., Grigiene, J., Krab, K., Hagedorn P.H., Flyvbjerg H., Andersen P.E., and Møller I.M., 2006. The free NADH concentration is kept constant in plant mitochondria under different metabolic conditions. *Plant Cell* 18: 688–698.

Kleyn, D.H., 1985. Determination of lactose by an enzymatic method. *Journal of Dairy Science* 68: 2791–2798.

Lakowicz, J.R. Szmacinski, H., Nowaczyk, K., and Johnson, M.L., 1992. Fluorescence lifetime imaging of free and protein-bound NADH. *Proceedings of the National Academy of Science, U.S.A.* 89: 1271–1275.

Linko, P., 1982. Lactose and Lactitol, In: *Nutritive Sweeteners*, Birch, G.G., and Parker, K.J. (Eds), London & New Jersey: Applied Science Publishers, pp. 109–132.

Lowry, O.H., 1980. Amplification by enzymatic cycling. *Molecular and Cellular Biochemistry* 32: 135–146.

Nielsen, S.S., 2003. *Food Analysis*. Birkhäuser. Kluwer Academic, New York.

Pesch, K.L., and Simmert, U., 1929. Combined assays for lactose and galactose by enzymatic reactions. *Milchw. Forsch.* 8: 551.

Seyis, I., and Aksoz, N., 2004. Production of lactase by *Trichoderma sp. Food Technology and Biotechnology* 42: 121–124.

Shapiro, F., and Silanikove, N., 2010. Rapid and accurate determination of D- and L-lactate, lactose and galactose by enzymatic reactions coupled to formation of a fluorochromophore: Applications in food quality control. *Food Chemistry* 119: 829–833.

Shapiro, F., and Silanikove, N., 2011. Rapid and accurate determination of malate, citrate, pyruvate and oxaloacetate by enzymatic reactions coupled to formation of a fluorochromophore: application in colorful juices and fermentable food (yogurt, wine) analysis. *Food Chemistry*, in press.

Shapiro, F., Shamay, A., and Silanikove, N., 2002. Determination of lactose and d-galactose using thio-NAD + instead of NAD +. *International Dairy Journal* 12: 667–669.

Silanikove, N. Leitner, G., Merin, U., and Prosser, C.G., 2010. Recent advances in exploiting goat's milk: quality, safety and production aspects. *Small Ruminant Research* 89: 110–124.

Sunehag A.L., Louie K., Bier, J.L., Tigas, S., and Haymond, M.W., 2002. Hexoneogenesis in the human breast during lactation. *Journal of Clinical Endocrinology and Metabolisim* 87: 297–301.

Wong, T.Y., and Yao, X.T., 1994. The DeLey-Doudoroff pathway of galactose metabolism in Azotobacter vinelandii. *Apply and Environmental Microbiology* 60: 2065–2068.

CHAPTER 24

Food Sources and Analytical Approaches for Maltose Determination

ELVIRA M. S. M. GASPAR,*[a] JOÃO F. LOPES,[a]
DANIEL GYAMFI[b] AND INÊS S. NUNES[a]

[a] Departamento de Química, CQFB-REQUIMTE, Faculdade de Ciências e Tecnologia, Universidade Nova de Lisboa, Quinta da Torre 2829-516, Caparica, Portugal; [b] Department of Biomedical Sciences, School of Life Sciences, University of Westminster, 115 New Cavendish Street, London W1W 6UW
*Email: elvira.gaspar@fct.unl.pt; elvira.smg@gmail.com

24.1 Introduction

Carbohydrates, hydrates of carbon or saccharides, are considered the most abundant group of natural products. Carbohydrates are produced in plants by the processes of photosynthesis for their own nutritional and physiological needs. They are also constituents of a wide range of biological systems, and the main source of energy supply in most cells. Saccharides are utilized as food by humans and other animals or micro-organisms. They are the main source of metabolic energy for the human body (starch and glycogen), they form the major constituents of the shells of insects, crabs, and lobsters (chitin) and are the supporting tissue of plants (cellulose). They are also present in

Food and Nutritional Components in Focus No. 3
Dietary Sugars: Chemistry, Analysis, Function and Effects
Edited by Victor R Preedy
© The Royal Society of Chemistry 2012
Published by the Royal Society of Chemistry, www.rsc.org

the cell walls of plants and bacteria, and the soft coats of animal cells (Boons *et al.* 1997).

More recently, carbohydrates are described as multifunctional compounds that include chiral polyhydroxy aldehydes, ketones, alcohols and acids that have the empirical formula $(CH_2O)_n$ and their simple derivatives, as well as their polymers having polymeric linkages of the acetal form.

24.2   Food Carbohydrates

Carbohydrates can be classified into two main groups: simple carbohydrates, which means they are molecules that contain only carbohydrates in their structure, and complex carbohydrates, which are covalently bound to lipids, proteins, aglycones, *etc.* (Sznaidman *et al.* 1999).

Based on their molecular size, simple carbohydrates are divided into three groups: monosaccharides, oligosaccharides and polysaccharides. Monosaccharides are the simple sugars which cannot be hydrolysed to smaller sugar molecules. By definition, oligosaccharides are simple polymers of monosaccharides joined by glycosidic linkages; they contain from 2 to 10 monosaccharide units which can be hydrolysed. Polysaccharides are polymers of high molecular weight consisting of more than 10 units (Kennedy and White 1983).

The simplest carbohydrates, monosaccharides and disaccharides, are commonly referred to as sugars. Sugar in Sanskrit means sweet sand and the term is used due to the fact that the lower members of carbohydrates are usually sweet and water soluble (Sznaidman *et al.* 1999). Monosaccharides are usually divided into aldoses – chiral polyhydroxy aldehydes, and ketoses – chiral polyhydroxy ketones. They are classified according to the number of carbon atoms present – trioses ($C_3H_6O_3$), tetroses ($C_4H_5O_4$), pentoses ($C_5H_{10}O_5$) and hexoses ($C_6H_{12}O_6$), *etc.* Due to their polyfunctionality, monosaccharides usually exist in many isomeric forms, both acyclic and cyclic. Most of the cyclic isomers can be isolated in pure form; however, acyclic isomers have been detected only as very minor components in solution, where they coexist in equilibrium with the cyclic forms (Boons *et al.* 1997; Kennedy and White 1983; Sznaidman *et al.* 1999).

A sugar can also be classified as a reducing and non-reducing sugar. A reducing sugar is any sugar that either has an aldehyde group or is capable of forming one in solution through isomerization if it has an open-chain form with an aldehyde group. Sugars with ketone groups in their open-chain form are capable of isomerizing *via* a series of tautomeric shifts to produce an aldehyde group in solution. Therefore, ketone-bearing sugars like fructose are considered reducing sugars but it is the isomer containing an aldehyde group which is reducing, since ketones cannot be oxidized without decomposition of the sugar. Reducing monosaccharides include glucose, fructose, glyceraldehyde, mannose and galactose (Figure 24.1).

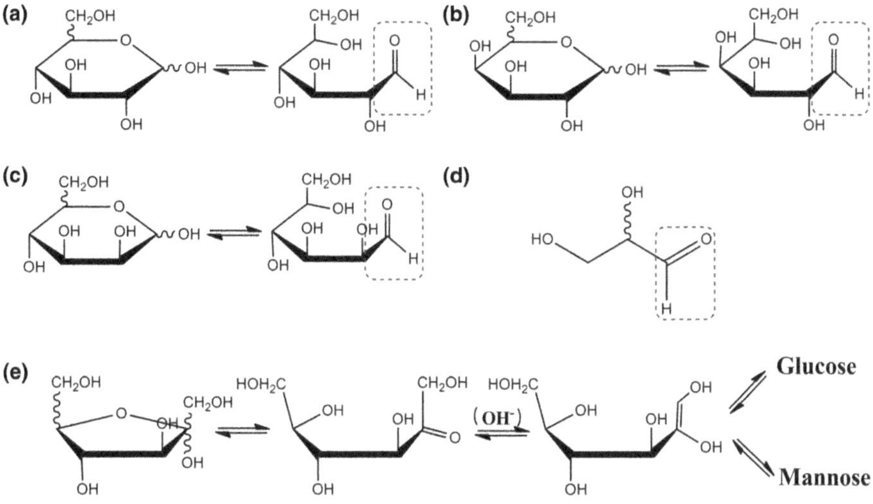

Figure 24.1 Chemical structures of some reducing monosaccharides. Structures of (a) glucose; (b) galactose; (c) mannose; (d) glyceraldehyde and (e) fructose; in the open-chain form, the presence of the aldehyde group is emphasized.

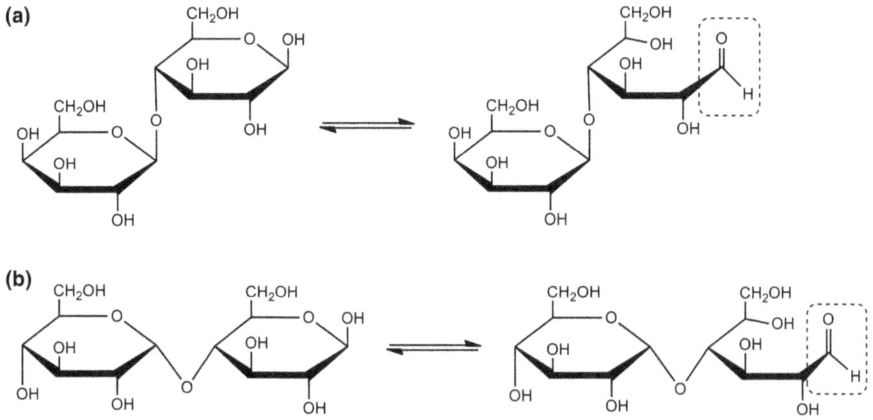

Figure 24.2 Chemical structures of two reducing disaccharides. Chemical structures of (a) lactose and (b) maltose; the presence of aldehyde group is emphasized in the open-chair form.

Many disaccharides, like lactose and maltose (Figure 24.2), also have a reducing form, as one of the two units may have an open-chain form with an aldehyde group.

However, sucrose (Figure 24.3) and trehalose, in which the anomeric carbons of the two units are linked together, are non-reducing disaccharides since neither of the rings is capable of opening. In glucose polymers such as starch and starch-derivatives like glucose syrup, maltodextrin and dextrin, the

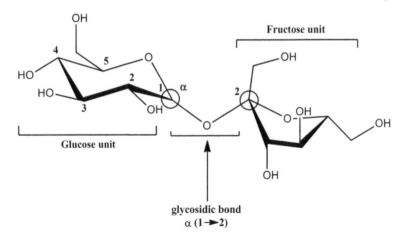

Figure 24.3 Sucrose structure. Chemical structure of sucrose (α-D-glucopyranosyl-(1→2)-β-D-*O*-fructofuranoside), a non-reducing disaccharide.

macromolecule begins with a reducing sugar, a free aldehyde. A more hydrolysed starch contains more reducing sugars. The percentage of reducing sugars present in these starch derivatives is called dextrose equivalent (DE) (Rong *et al.* 2009).

For nutritional purposes, carbohydrates are classified primarily according to the degree of polymerization with subdivisions based on glycosidic linkages and chemistry of individual sugars because these determine physiological properties (Cummings 1995; Cummings *et al.* 1997). Table 24.1 shows a classification of carbohydrates.

The most common dietary sugars are glucose (D-Glucose), fructose (D-fructose), galactose (D-galactose), sucrose, lactose and maltose (Cummings 1995; Cummings *et al.* 1997).

24.2.1 Relevance of Dietary Carbohydrates and Food Sugars

The most important dietary carbohydrates nutritionally are polysaccharides and oligosaccharides, because free monosaccharides are not commonly present in the diet in significant quantities (Sznaidman *et al.* 1999). As can be seen in Table 24.2, most of the dietary sugars are absorbed and provide a ready source of energy.

Free glucose in foods, the glucose moiety of sucrose and the glucose released from maltose and malto-oligosaccharides and from some forms of starch, all appear rapidly in the blood after a meal. Carbohydrates that escape digestion in the small intestine enter the colon where they are fermented with the production of the short chain fatty acids – acetate, propionate and butyrate, which are absorbed and influence systemic glucose and lipid metabolism (Cummings 1995; Wolever 2003). Dietary carbohydrates also have other physiological

Table 24.1 Chemical classification of dietary carbohydrates. Division is based on the degree of polymerization and glycosidic linkages, which affect physiological properties. Relative significance in human diet is also presented.

Carbohydrate Classes	Significance in Diet
Sugars	
Monosaccharides, DP = 1	
Glucose (D-glucose) and Fructose (D-fructose)	Major
Galactose (D-galactose)	Minor, depending on consumption of fermented milks
Dissacharides, DP = 2	
Sucrose (β-D-fructofuranosyl-α-D-glucopyranoside)	Major
Lactose [O-β-D-galactopyranosyl-(1 → 4)-α-D-glucopyranose]	Variable, depending on intake of milk products
Maltose [O-α-D-glucopyranosyl-(1 → 4)-α-D-glucopyranose]	Minor
Oligosaccharides, 3 < DP < 10	
Trisaccharides	Minor, depending on intake of
Tetrasaccharides	glucose syrups and on
Pentasaccharides	amounts of vegetables consumed
Polysaccharides DP > 10	
Starches	Major
Non-starch polysaccharides	
Plant cell wall material (dietary fiber)	Variable, depending on intake
Cellulose [linear β(1 → 4) glucan]	of high extraction cereal
Noncellulosic	products and plant foods
Hemicelluloses and pectic substances	
Isolated polysaccharides	Minor

DP - degree of polymerization.

properties which are relevant to health in diverse ways: as satiety's controller, as regulator of blood glucose and insulin production, in protein glycosylation, in cholesterol lowering, in bile acid dehydroxylation, as a laxative agent and in the selective stimulation of microbial growth (bifidobacteria) (Cummings 1995; Cummings *et al.* 1997).

Sugars have been used industrially to add sweetness to food, to make it pleasant to eat, and they are also used by the food industry to aid in food preservation and to improve upon the flavour and texture of food products. They also have the ability to reduce water activity and consequently act as inhibitors of microbial growth. Fructose, glucose and other reducing sugars are initial substrates in the Maillard reaction (Rizzi 1997), which yields products that add to the aroma and flavour of food. However, similar reactions are possible *in vivo* and may contribute to protein cross-linking in human tissues being associated with lens and premature ageing (Cummings *et al.* 1997).

Table 24.2   Classification and behaviour during digestion of dietary carbohydrates. Division of carbohydrate classes is based on the number
of monosaccharide units (oligosaccharides have between 3 and 10)
and it is also described their behaviour during digestion.

Carbohydrate Classes	Behaviour During Digestion
Sugars	*Absorbed from small intestine* monosaccharides: D-Glucose and D-Fructose disaccharides: Sucrose, maltose, trehalose and Lactose[1] *Poorly absorbed* sugar alcohols[2]: Sorbitol, maltitol and lactitol
Oligosaccharides	*Digested* some are absorbed in small intestine *Resistant* some pass into the large intestine *Fermented* some of them are fermented and stimulate growth of bifidobacteria in colon
Polysaccharides	*Digested* some can be digested in small intestine, but with different rates *Resistant* several are not absorbed in small bowel and continue to the large intestine *Fermented* several can be fermented, and affect essentially the large bowel function, as well as lipid absorption in the large bowel

[1]Lactose is fermented in many populations.
[2]Some sugar alcohols can be partly fermented.

24.2.2   Maltose

Maltose is a crystalline dextrorotatory fermentable sugar; it is a disaccharide
consisting of two D-glucose units joined by a bond known as the glycosidic
bond. Maltose is 4-*O*-α-D-glucopyranosyl-α-D-glucose (Figure 24.4). As
mentioned in Figure 24.2, it is a reducing sugar.

24.2.3   Food Sources

Maltose is found in raw cereals, vegetables and vegetable products, fruits,
almonds and pistachio nuts as well as chicken – as a free sugar – or it can be
derived from hydrolysis of amyloses and amylopectins (starch), which occur in
a variety of food sources. Among cereals, barley is described to contain free
maltose. Brewing industries increase free maltose content by malt production, a
process where the maltose producing amylases convert starch into free maltose,
an essential procedure in beer industry. In fact, maltose is one of the main
disaccharides present in beers, contributing approximately 14% of the total
carbohydrates. Other maltose containing cereals are wheat, kamut and corn,
which are also used in the production of beers. Flours made from these cereals

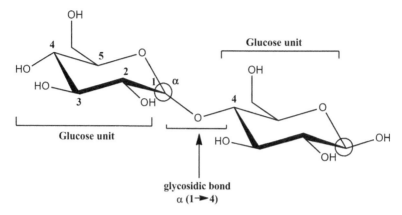

Figure 24.4 Maltose structure. Chemical structure of maltose (α-D-glucopyranosyl-(1 → 4)-α-D-*O*-glucopiranoside) structure.

have also high amounts of maltose; the amount increases during the drying/toasting process, due to decomposition of starch and caramelization of glucose. Bread, being made from flours, also contains appreciable amounts of maltose.

Non-processed vegetables, such as sweet potato, also contain maltose; sweet potato may reach values of 50 g/Kg, depending on geographic origin. Maltose content also increases in sweet potato through cooking, due to caramelization and Maillard reactions. Tomato, broccoli, peas, cucumber, cabbage and potato are among the vegetables that contain maltose. Heating/cooking causes hydrolysis/decomposition of starch, cellulose and other polysaccharides, with the resulting products having higher maltose contents; a good example is ketchup.

With respect to fruits, maltose has been detected in kiwis, cherries, peaches, plums, apricots and watermelons. It can also be found in other fruits, but to a lesser extent. Processed fruit or fruit-based products typically contain more maltose than fruit in its raw form.

Table 24.3 summarizes some maltose-containing foods, in both natural and processed form.

24.2.4 Metabolism and Effects

Maltose is consumed in different food matrices, which show considerable variation in physical properties and chemical composition. In the food and drinks ingested, it may be in free solution, dispersed throughout the food-stuff, or contained within the food matrix either in the cellular structures of plant foods or the emulsions and foamed structures of processed foods. The characteristic of matrices influences maltose's absorption. However, once absorbed, its transport and metabolism converge, and the effects of dietary source become of minor significance (Southgate 1995).

Table 24.3   Food sources of maltose.* Table mentions the main food sources of maltose, both raw and processed or cooked food.

Main Food Sources of Maltose

Food type	Raw	Quantities (g/Kg)	Derived Products	Quantities (g/Kg)
Vegetables	Sweet potatoes	23 to 50	Cooked sweet potatoes	31.2 to 67.1
	Tomatoes	residual	Ketchup	17
			Canned tomato paste	2.8
	Peas	1.7	Cooked peas	1.8
	Cucumbers	0.1	Cucumber pickles	2.9
	Potatoes	residual	Frozen or dehydrated mashed potatoes	0.5 to 5.2
	Cabbages	0.1	Frozen edamame	9 to 12
	Broccoli	2.1	Spinach soufflé	0.3
Fruits	Pears	0.1	Canned pears	9.5 to 19
	Peaches	0.8	Canned peaches	9 to 14.4
	Kiwis	1.9	Canned guanaba nectar	7.6
	Cherries	1.2	Canned Cherries	8.7
	Blackberries	0.7	Canned tamarind nectar	5.7
	Apricots	0.6	Canned mango nectar	5.6
	Plums	0.8	Dried plums	0.6
	Watermelons	0.6	Canned guava nectar	7.4
	Bananas	0.1	Canned applesauce	0.7
Cereals	Kamut	66.3	Noodles	3 to 15.3
	Spelt	51.5	Black beer	Up to 12.0
	Lentils	3.0	Pilsner Beer	0.3 to 3.0
	Wheat	n. a.	White wheat flour	2.2 to 9.0
	Barley	n. a.	Spaghetti	4 to 19.6
	Corn	n. a.	Canned or frozen sweet corn, cooked or not	1 to 2.2
Other	Milk	Up to 0.2	Swiss, parmesan, mozzarella and cheddar cheese	0.8 to 6.2
	Honey	14.4	Cured ham	1.4
	Chicken breasts	1.5	Sausages (different types of meat)	0.6 to 6.3

*Data sources: FoodInfo.us 2011; USDA National Nutrient Database for Standard Reference 2010.

Maltose, as a disaccharide, is too large to cross the mucosal cell membrane and must be hydrolyzed for absorption to take place. Thus the food matrix in which maltose is ingested exerts effects on the rates at which it is supplied to the small intestine.

Maltose is relatively resistant to acid hydrolysis conditions in the stomach (Southgate 1995). The action of α-amylase, which hydrolyzes α-1,4-glycosidic bonds, on dietary starch releases maltose into the small intestine; maltose is then hydrolyzed by the enzyme maltase to form 2 free glucose units, which can be absorbed by the intestinal mucosal cells.

The hydrolysis of maltose by appropriate brush border disaccharidases is rapid, so that the rate of absorption of glucose from maltose is significantly higher than from free glucose. In fact, the glucose released has been detected to appear rapidly in the blood after a meal (Cummings *et al.* 1997).

The concentration of glucose in the blood has a number of implications for health including diabetes, ageing, and possibly cancer (Cummings *et al.* 1997; Hu *et al.* 2010; Jentjens *et al.* 2004). Dietary carbohydrates influence metabolism by at least four mechanisms: nature of the monosaccharide absorbed, amount of carbohydrate consumed, rate of absorption, and colonic fermentation. Decreasing glycaemic responses by reducing carbohydrate intake increases postprandial serum free-fatty acids (FFA) and does not improve overall glycaemic control in diabetic subjects. In contrast, low glycaemic index (GI) diets reduce serum FFA and improve glycaemic control. FAO/WHO recommends maintaining a high-carbohydrate diet and choosing low-GI starchy foods (Wolever 2003).

24.3   Analytical Approaches

A literature search revealed several different analytical approaches to analyze the total sugar content as well as the amount of reducing and individual sugars. In terms of maltose, very different techniques have been used. In this chapter, non-chromatographic and chromatographic methodologies were described separately to achieve a better comprehensive coverage. The idea is to present the most representative methodologies and their principles and to characterize the broader group of available methodologies for maltose analysis.

24.3.1   Non-chromatographic Methodologies

Maltose being a reducing sugar, its determination can be done by colorimetric and other procedures described in the literature for the estimation of reducing sugars (Alves *et al.* 2006; Dubois *et al.* 1956). However, these methodologies are applied to determine the presence of mixtures. For individual maltose analysis, other methodologies have been described.

Enzymic methodologies are the most common methods used in maltose analysis because they do not require large and expensive apparatus, and are appropriate for use in a large number of samples. Using a four enzyme reaction system (which includes maltose 1-epimerase, maltose phosphorylase, β-phosphoglucomutase, and glucose-6-phosphate dehydrogenase), maltose was identified and quantified in samples prepared from potato soluble starch with different concentrations (Shirokane *et al.* 2000). The obtained results were compared with those achieved by High Performance Liquid Chromatography (HPLC). The enzymic methodology demonstrated good sensitivity and reproducibility with a limit of detection (LOD) of 0.02 µmol/mL, relative standard deviation (RSDs) values below 2% and it was reported as a method that did not show any interference in the presence of other disaccharides and monosaccharides. The results showed good correlation with the ones obtained by HPLC ($r = 0.997$), having advantage in terms of simplicity, fastness and the possibility to perform multiple analysis using microlitre plates or automatic analyzers.

In recent years, carbohydrates biosensors have also been extensively studied due to some well appreciated advantages over other methods, such as speed of response (sometimes less than 1 minute), and easier operation and miniaturization. An example of a maltose biosensor based on a screen-printed system using a glucose oxidase electrode and an amyloglucosidase/glucose oxidase electrode allowed the simultaneous detection of maltose and glucose in the same mixture (Ge *et al.* 1998). This enzyme biosensor was reported as having several advantages, such as providing a rapid analysis, using cheap enzymes, being disposable and not requiring the addition of reagents. The sensor was described to show good storage stability (4 months), a wide pH variation (2–8) and good reproducibility, with a linear range of the system for maltose up to 20 mmol/L, and RSDs ranging from 3.5% to 5.29%. Also, the reported working potential of the sensor (0.3 V) minimized the interference of other electroactive species. The dual-electrode system was applied to the determination of maltose and glucose content in some commercial beers from China. The obtained results were cross-checked with those obtained from commercial Fehling titrations solutions (used to determine the total content of sugars), and the described correlation coefficients for both methods were good ($r = 0.997$). The amounts of maltose in the beers studied ranged from 28 to 42 mmol/L.

Another biosensor for maltose analysis has been developed (Odaci *et al.* 2010), using a new bi-enzymatic system by co-immobilizing α-glucosidase and pyranose oxidase. The ideal operational conditions for the biosensor were reported to be the fast linear response for maltose in the concentration range from 0.25 to 2.0 mM, at ideal temperature of 35 °C and pH 6.0, and the optimum working potential -0.70 V. The method performance showed an r^2 of 0.999 and RSDs between 1.75% and 4.4%. The biosensor was tested in the analysis of maltose in beer samples, and it was discovered that the complexity of this matrix had no influence on the biosensing response. The maltose values found in beer varied from 0.97 to 1.490 g/L.

Another biosensor based on a fluorescence/Förster resonance energy transfer (FRET) was reported as an immunosensor for maltose detection (Engström *et al.* 2008). Considerable interest in the development of FRET-based biosensors has grown in the last few years, mainly due to their robustness and applicability for crude samples continuous monitoring. In the specific case of maltose, it is of extreme interest for food manufacturers to control it during food processing. Engström's sensor used the conjugation of maltose-labeled bovine serum albumin and a monoclonal antibody (which had affinity towards free maltose molecules) inside a semipermeable capsule. It was used to detect the presence of maltose through the change of FRET signal which made it possible to monitor this sugar. After optimization of the biosensor construction and its working and calibration parameters, the sensor was applied to the analysis of maltose content in crude samples of oat drinks. The biosensor was described to show good response for maltose, with no detectable cross-reactivity against its structural analog (cellobiose); around 3.74 % w/v of maltose were found in samples.

Based on Fourier transform infrared (FTIR) spectroscopy and chemometrics (partial least squares, principal component regression, among others), a method was developed (Wang *et al.* 2010) and used in the quantification of maltose and other sugars in honeys from different geographic origins. Fourier transform infrared spectroscopy "fingerprints" in the range 1500–800 cm^{-1} were established with the help of standard mixtures of 45 sugars. The method was validated using synthetic blends of sugars and compared with a traditional HPLC methodology. The method performance showed a limit of detection (LOD) near 5 ppm for maltose and a good correlation with HPLC analysis ($r = 0.999$). The authors claim this methodology as faster, less expensive and simpler than other methods, and as appropriate to determine the concentration of maltose and other sugars in multiple regional honey samples, allowing the identification of honeys adulterated with corn syrup.

24.3.2 Chromatographic Methodologies

24.3.2.1 Gas Chromatography

Gas chromatography (GC), due to its characteristic high resolution, high sensitivity and easy coupling to different detectors including mass spectrometers (MS), is theoretically suitable for the analysis of mixtures containing carbohydrates. However, due to the low volatility of carbohydrates, there is an essential need to prepare appropriate derivatives (by derivatization reactions) which enhance their volatility, in order to make it possible to analyze them by GC or GC–MS.

Carbohydrates often occur in complex mixtures, which sometimes require the use of techniques for extraction, clean-up and/or fractionation of carbohydrate-containing matrixes. Due to the volatility of carbohydrates, few studies were found in the literature using GC to analyze maltose.

The main sugars present in several flours, where maltose is an especially important saccharide, were determined by GC using an extensive sample preparation protocol (due to the complexity of the analyzed matrixes), followed by a derivatization step to obtain trimethylsilyl oximes (TMSO) of the corresponding sugars. The analysis (Rufián-Henares *et al.* 2009) was done using a non-polar column, Chrompack CP-SIL 5 CB, and flame ionization detection (FID). The method demonstrated good sensitivity with LOD and limit of quantification (LOQ) in the range of the parts per billion (ppb). The maltose content in the different flours studied was 1.2 g/Kg in wheat flour (refined wheat flour), 2.1 g/Kg in whole wheat flour and 1.4 g/Kg in oat flour, while in corn, soybean and rice flour no maltose was found. After 10 minutes of toasting, maltose content rose to 6.1 g/Kg in wheat flour, 7 g/Kg in whole wheat flour and 4.2 g/Kg in oat flour.

Several carbohydrates in different vegetable families (chicory, spinach, cabbage and others) were identified and quantified using GC-MS (Hernández-Hernández *et al.* 2011). The samples were previously derivatized in order to obtain the volatile trimethylsilyl oxime derivatives, and analysed by GC-MS using a cross-linked methyl silicone column (Teknokroma TRB-1). Maltose, isomaltose, fructose, galactose, glucose and sucrose, are among sugars that were separated with good resolution and quantified in vegetables. The method showed good sensitivity (1.3 and 4 ppm) for maltose's LOD and LOQ, respectively. Maltose content of 1.46 g/Kg was also found in purple yam.

A GC-MS method for the analysis of some sugars and carboxylic acids in atmospheric aerosols has been developed (Pietogrande and Bacco 2011). The authors optimized a derivatization protocol for the silylation of sugars, especially reaction time, temperature, and reagents concentration. The derivatized samples were analysed by GC-MS using a non-polar DB-5 column. The method allowed the identification and quantification of seven sugars in the environmental matrices, including maltose. Maltose analysis showed a good linearity ($r^2 = 0.994$), a good sensitivity (LOD of 3 ng/m^3), RSDs varying from 4 to 10%. Maltose content was present in the range of 3 to 47 ng/m^3.

24.3.2.2   High-performance Liquid Chromatography

Due to their poor volatility, the analysis of carbohydrates has been traditionally carried out by HPLC, a chromatographic method that can provide good qualitative and quantitative results. The main liquid chromatographic systems used for the separation of underivatized carbohydrates can be generalized in: anion-exchange columns and water containing bases or salts as eluent composition; cation-exchange columns and water in the eluent composition; alkyl-bonded silica gel columns and water in the eluent composition; and amine-bonded silica gel columns and water–acetonitrile combinations as eluent.

Carbohydrates lack significant chromophoric and fluorophoric groups. The use of UV/fluorescence detection also requires derivatization with chromophoric/fluorescent compounds. So, refractive index (RI) still remains the most popular detection using HPLC for carbohydrates analysis, despite its many disadvantages, such as its lower sensitivity, temperature and flow-rate dependence, and incompatibility with gradient elution. Another detection that has been progressively adopted is the evaporative light scattering detector (ELSD). It is a semi-universal mass detector which is based on the detection of solute molecules by light scattering after mobile phase and solute nebulization and evaporation.

A simple and direct HPLC-ELSD methodology was validated for the quantification of maltose, maltotriose, maltotetrose, glucose and fructose in beer (Luciana *et al.* 2005). The chromatographic separation was achieved using a 5 µm amine-bonded silica gel chromatographic column and a gradient elution with acetonitrile/water was used as eluent. A good resolution and sensitivity (LODs around 10 ppm) were reported for maltose analysis. Also, a good relative standard deviation ranging between 1.59 and 5.95% ($n = 10$), and recoveries, between 94 and 98.4%, were described. Sample preparation involved only the degassing of beer samples in an ultrasonic bath, followed by dilution in acetonitrile and a subsequent filtration. The quantities found for maltose ranged from 0.35 g/L (standard pilsner, added with high maltose syrup, 4.5% alcohol) to 38.5 ± 0.6 g/L (100% malt, low-alcohol beer).

An HPLC-RI method for the determination of maltose and other sugars in Royal Jelly has been established (Sesta 2006). Unlike honey, other bee products are still lacking official and international quality standards. Besides being among the main components, sugars may be used as quality parameters for the detection of possible adulterations of Royal Jelly with honey or sugars, and therefore, their determination is important for quality control of Royal Jelly. Sesta's method included the use of a 5 μm amine-bonded chromatographic column, acetonitrile/water as eluent, and maintaining the column and the RI detector at 30 °C. The procedure also involved an extensive sample preparation, including dilutions, centrifugations and reactions to eliminate proteins and filtrations. Maltose content in Royal Jelly varied, but in some samples it was found to be 30 g/Kg.

A rapid analysis of underivatized maltose, lactose and sucrose in vegetables matrices by ultra-HPLC with detection by electrospray mass spectrometry (ESI-MS/MS) in the negative mode has been described (Gabbanini *et al.* 2010). The technique also allowed the anomeric discrimination of α- and β-anomers of the analysed sugars. Gabbanini's chromatographic approach used a porous graphitic carbon column, working at low temperature, with an eluent composed of acetonitrile/formic acid buffer in gradient mode, which allowed the separation and detection of maltose and its anomeric forms as formate adducts. The method demonstrated good sensitivity (LOD and LOQ for maltose were 4.8 and 16.0 μg/L, respectively), and RSDs varying from 2.8 to 4.6% and recoveries from 92 to 96%, showing good linearity, precision and accuracy. Maltose was detected in green tea leaf (7.4%, w/w) and also in the composition of anti-inflammatory herbal preparations (12.4%, w/w).

24.3.2.3 Capillary Electrophoresis

In recent years, due to the higher efficiency, faster separation times, easy operation and low sample volumes required, capillary electrophoresis (CE) has evolved into an interesting alternative to HPLC. The attractive features of CE include its small sample requirement and almost zero solvent consumption, its speed, which enables high-throughput chemical analysis of a wide variety of substances, and its versatility, partly because of the range of separation modes available. Sample preparation steps include only dilution and/or filtration, ensuring low costs. The most frequently used modes of CE have been capillary zone electrophoresis (CZE), micellar electrokinetic chromatography (MEKC), isotacophoresis (ITP), capillary electrochromatography (CEC) and capillary gel electrophoresis (CGE).

The use of CE in the separation of carbohydrates has two major difficulties: most carbohydrate molecules lack readily ionizable charged functions, essential for their direct differential migration, and as it was mentioned before, most carbohydrates neither absorb nor fluoresce, as they lack chromophores or fluorophores in their structures, hindering their sensitive detection. To overcome these obstacles, two approaches have been used: carbohydrates are converted to charged species *in situ* by complexation with other ions, such as borate and metal cations, which ensure their differential electromigration and consequent separation in an electric field; and carbohydrates are transformed into UV-absorbing or fluorescent detectable derivatives.

Several works have been published that analyze maltose in different food or natural sources using CE. An interesting review by Cortacero-Ramírez *et al.* (2003) summarizes some of the main CE methods used for the analysis of beer carbohydrates, with a special emphasis on maltose. High Performance CE (HPCE) with indirect absorbance detection (IAD) for the determination of maltose and other carbohydrates in plant and barley extracts has been described. The use of fluorophore-assisted CE (FACE) detection for monitoring the production and quality control of beer, with good analytical results, was also included. CZE was also mentioned to quantify maltose in commercial beer and during the brewing process; it made used of derivatization of carbohydrates with *p*-aminobenzoic acid, and UV detection.

Pre-column CE derivatization combined with *p*-aminobenzoic acid and UV detection was used for the determination of maltose and other carbohydrates in some different types of beer and natural juices (Cortacero-Ramírez *et al.* 2004). The method allowed the complete separation of maltose, glucose, maltotriose, maltotetraose, maltopentaose, maltohexaose and maltoheptaose, showing good analytical performance (LOD, 4.1 mg/L and RSDs values from 1.6% to 9.9% for maltose analysis). Maltose was found in two beers (0.79 g/L in "special beer" and 1.1 g/L in "black beer"), while maltotriose was found in all analysed beers (values ranging from 0.22 to 1.93 g/L).

An optimized electrolyte, comprising of sorbate, cetyltrimethylammonium bromide (CTAB) and NaOH, for CE coupled to UV detection (Jager *et al.* 2007) has been described for the determination of maltose and other carbohydrates in pre-sweetened cereals. Different sample preparation procedures including ultrasonic extraction and solid phase extraction (SPE) were tested and compared. The method considered that carbohydrates can be ionized at a pH above 12 and do not absorbed in the ultraviolet region; so, the optimized electrolyte contained NaOH to provide a highly alkaline medium, sorbate ion as chromophore, and CTAB as flow reversal surfactant. Separation between maltose, fructose, glucose, galactose, lactose, maltotriose and sucrose was achieved. The method showed a good analytical performance, with a maltose LOD around 30 ppm. Maltose was found in four of the six cereal samples analyzed, in a quantity ranging from 0.22 to 0.58% (w/w).

Table 24.4 summarizes the methods for maltose analysis described above.

Table 24.4 Maltose analytical methodologies. Table summarizes the analytical methodologies as well as the sample preparation techniques used in the mentioned food matrices for maltose determination.

Work	Method	Sample Preparation	Matrixes
Ge *et al.* 1998	Biosensor	Dilution and adjustment of pH values	Beers
Odaci *et al.* 2010	Biosensor	Dilution and adjustment of pH values	Beers
Engström *et al.* 2008	Biosensor	Dilution	Oat drinks
Wang *et al.* 2010	FTIR spectroscopy and chemometrics	Dilution, centrifugation and incubation	Honeys
Rufián-Henares *et al.* 2009	GC-FID	Extraction, derivatization, filtration and dilution	Flours
Hernández-Hernández *et al.* 2011	GC-MS	Extraction, derivatization, filtration and dilution	Vegetables
Pietogrande and Cacco 2011	GC-MS	Extraction, derivatization, filtration and dilution	Atmospheric aerosols
Luciana *et al.* 2005	HPLC-ELSD	Degassing, dilution and filtration	Beer
Sesta 2006	HPLC-RI	Dilution, centrifugation, protein elimination reaction and filtration	Royal jelly
Gabbanini *et al.* 2010	UHPLC-ESI-MS/MS	Extraction, dilution and filtration	Herbal plants
Cortacero-Ramírez *et al.* 2003	HPCE-IAD, FACE, CZE-UV	Extraction, dilution, derivatization and filtration, depending on the matrix	Beers, plant and cereals
Cortacero-Ramírez *et al.* 2004	CE-UV	Dilution, derivatization and filtration	Beers and natural juices
Jager *et al.* 2007	CE-UV	Extraction (ultrasonic and SPE), filtration and dilution	Pre-sweetened cereals

24.4 Conclusions

Dietary carbohydrates are a diverse group of substances with broad physiological properties and different levels of importance to health. The simplest carbohydrates (monosaccharides and disaccharides), commonly referred to as sugars, add sweetness to food to make it pleasant to eat and are also used by the food industry to aid in food preservation and to improve upon the flavour and texture of foods.

Maltose is a crystalline fermentable disaccharide sugar, composed of two D-glucose units. It is consumed within different food matrices, especially in cereals and some derived products, which show considerable changes in physical properties and chemical composition. In the foods and drinks ingested it may be in a free solution, dispersed throughout the food-stuff, or contained within the food matrix. The human biological hydrolysis of maltose is rapid, and the glucose released appears rapidly in the human blood after a meal. Knowing that the concentration of glucose in the blood has a number of implications for health including diabetes, ageing, and possibly cancer, maltose analysis is an important subject with clinical relevance.

This chapter describes the most suitable analytical methods with respect to accuracy, low detection limits and simple procedures developed for maltose analyses in food products. Very different techniques, such as non-chromatographic enzymic methodologies and biosensors, and also chromatographic methodologies, such as gas chromatography, high-performance liquid chromatography and capillary electrophoresis were described together with sample preparation procedures, to achieve a broad comprehensive overview.

Taking into account the importance of detecting and analysing maltose in food, this work will contribute immensely to an improved characterization and quantification of its presence in human diet and will make it possible to increase the knowledge of its biological role, relevance and long-term implications of its consumption for human health.

Summary Points

- This chapter focuses on food sources and analysis of maltose.
- Maltose is a disaccharide composed of two D-glucose units.
- Maltose is consumed within different food matrices.
- The human biological hydrolysis of maltose is rapid, and the glucose appears rapidly in the human blood.
- The concentration of glucose in the blood has a number of implications for health including diabetes, ageing, and possibly cancer.
- With regard to the importance of detecting and analysing maltose in food, the description of the most representative analytical methodologies (non-chromatographic and chromatographic methodologies) and their principles are discussed.

Key Facts of Maltose as a Dietary Carbohydrate

1. Maltose exists as a free sugar or can be derived from hydrolysis of amyloses and amylopectins (starch) which occur in different food sources.
2. Maltose is a disaccharide composed by two glucose units.
3. The most important nutritional dietary carbohydrates are polysaccharides and oligosaccharides, which must be hydrolyzed to their constituent monosaccharide units by hydrolytic enzymes.
4. Dietary carbohydrates have very different physiological properties with different importance to health, depending on the site, rate and extent of their digestion or fermentation.

5. The rate and extent of carbohydrates digestion are major factors controlling blood glucose and insulin levels which are risk factors for diabetes, cardiovascular disease, cancer and ageing.

Definitions of Words and Terms

Carbohydrates: multifunctional compounds that include chiral polyhydroxy aldehydes, ketones, alcohols and acids, having the empirical formula $(CH_2O)_n$.

Sugars: the simplest carbohydrates, monosaccharides and disaccharides, usually sweet and water soluble.

Reducing sugar: A reducing sugar is any sugar that either has an aldehyde group or is capable of forming one in solution.

Maltose: is a crystalline fermentable sugar; it is a disaccharide composed by two D-glucose units.

Pectin: is a structural heteropolysaccharide present in citrus fruit composition.

Dextrose equivalent (DE): percentage of reducing sugars present in a food matrix.

Maillard reaction: a nonenzymatic browning reaction similar to caramelization.

RSD: Relative Standard Deviation that is the absolute value of the coefficient of variation.

FTIR: Fourier Transformed Infrared Spectroscopy; is a technique which is used to obtain an infrared spectrum of a compound or chemical mixtures.

Gas Chromatography: is a common type of chromatography used for separating and analyzing volatile and semi-volatile compounds or mixtures.

High Performance Liquid Chromatography: is a type of chromatography used for separating and analyzing non-volatile compounds or mixtures.

Capillary Electrophoresis: an analytical methodology used to separate ionic species by their charge and frictional forces and hydrodynamic radius.

Biosensor: an analytical device for the detection of chemical compounds that combine a biological component with a physicochemical detector component.

List of Abbreviations

CE	capillary electrophoresis
CEC	capillary electrochromatography
CGE	capillary gel electrophoresis
CZE	capillary zone electrophoresis
DE	dextrose equivalent
ELSD	evaporative light scattering detector
ESI-MS	electrospray mass spectrometry
FACE	fluorophore-assisted capillary electrophoresis
FAO	Food and Agriculture Organization
FFA	free-fatty acids
FRET	fluorescence resonance energy transfer

FTIR   Fourier transform infrared spectroscopy
GC   gas chromatography
GI   glycaemic index
HPLC   high performance liquid chromatography
IAD   indirect absorbance detection
ITP   isotacophoresis
LOD   limit of detection
LOQ   limit of quantification
MEKC   micellar electrokinetic chromatography
MS   mass spectrometry
ppb   parts per billion
RI   refractive index
RSD   relative standard deviation
SPE   solid phase extraction
TMSO   trimethylsilyl oximes
UV   ultraviolet
VLDL   very low density lipoprotein
WHO   World Health Organization

Acknowledgements

The authors acknowledge the Portuguese Foundation for Science and Technology for funding through the FCT SFRH/33809/2009 (Inês Nunes) and FCT SFRH/SFRH/BD/40564/2007 (João Lopes).

References

Alves, E.R., Fortes, Paula, R., Borges E.P., and Zagatto, E.A.G., 2006. Spectrophotometric flow-injection determination of total reducing sugars exploiting their alkaline degradation. *Analytica Chimica Acta.* 564: 231–235.

Boons, G.J., Ziegler, T., Heskamp, B., Veeneman, G.H., Polt, R.L., Roy, R., Nepogodiev, S.A., Stoddart, J.F., Nicotra, F., Hounssel, E.F., and Widmaln, G., 1997. Mono- and oligosaccharides: structure, configuration and conformation. In: Boons, G. (ed.) *Carbohydrate Chemistry.* Chapman and Hall, London, UK, pp 1–2.

Cortacero-Ramírez, S., Castro, M., Segura-Carretero, A., Cruces-Blanco, A., and Fernández-Gutiérrez, A., 2003. Analysis of beer components by capillary electrophoretic methods. *Trends in Analytical Chemistry.* 22: 440–455.

Cortacero-Ramírez, S., Segura-Carretero, A., Cruces-Blanco, C., Castro, M., and Fernández-Gutiérrez, A., 2004. Analysis of carbohydrates in beverages by capillary electrophoresis with precolumn derivatization and UV detection. *Food Chemistry.* 87: 471–476.

Cummings, J.H., and Englyst H.N., 1995. Gastrointestinal effects of food carbohydrate. *The American Journal of Clinical Nutrition.* 61: 938–945.

Cummings, J.H., Roberfroid M.B., and members of the Paris Carbohydrate Group, Andersson, H., Barth, C., Ferro-Luzzi, A., Ghoos, Y., Gibney, M., Hermonsen, K., James, W.P.T., Korver, O., Lairon, D., Pascal, G., and Voragen, A.G.S., 1997. A new look at dietary carbohydrate: chemistry, physiology and health. *European Journal of Clinical Nutrition*. 51: 417–423.

Dubois, M., Gilles, K.A., Hamilton, J.K., Rebers, P.A., and Smith, F., 1956. Colorimetric method for determination of sugars and related substances. *Analytical Chemistry*. 28: 350–356.

Engström, H.A., Andersson, P.O., Gregorius, K., and Ohlson, S., 2008. Towards a FRET-based immunosensor for continuous carbohydrate monitoring. *Journal of Immunological Methods*. 333: 107–114.

FoodInfo.us, 2011. Top food sources of maltose, 2011. Available at: http://foodinfo.us/SourcesUnabridged.aspx?Nutr_No=214. Accessed 29 June 2011.

Gabbanini, S., Lucchi, E., Guidugli, F., Matera, R., and Luca Valgimigli. 2010. Anomeric discrimination and rapid analysis of underivatized lactose, maltose and sucrose in vegetable matrices by UHPLC–ESI-MS/MS using porous graphitic carbon. *Journal of Mass Spectrometry*. 45: 1012–1018.

Ge, F., Zhang, X., Zhang, Z., and Zhang, X., 1998. Simultaneous determination of maltose and glucose using a screen-printed electrode system. *Biosensors and Bioelectronics*. 13: 333–339.

Hernández-Hernández, O., Ruiz-Aceituno, L., Sanz, M. L., and Martínez-Castro, I., 2011. Determination of Free Inositols and Other Low Molecular Weight Carbohydrates in Vegetables. *Journal of Agricultural and Food Chemistry*. 59: 2451–2455.

Hu, J., Vecchia, C.L., Gibbons, L., Negri, E., and Mery, L., 2010. Nutrients and risk of prostate cancer. *Nutrition and Cancer*. 62: 710–718.

Jager, A.V., Tonin, F.G., and Tavares, M.F.M., 2007. Comparative evaluation of extraction procedures and method validation for determination of carbohydrates in cereals and dairy products by capillary electrophoresis. *Journal of Separation Science*. 30: 586–594.

Jentjens, R.L.P.G., Venables, M.C., and Jeukendrup, E., 2004. Oxidation of exogenous glucose, sucrose and maltose during prolonged cycling exercise. *Journal of Applied Physiology*. 96: 1285–1291.

Kennedy, J.F., and White, C.A., 1983. *Bioactive Carbohydrates In Chemistry, Biochemistry and Biology*. Ellis Horwood Publishers Ltd., Chichester, UK, 331 pp.

Luciana, C., Nogueira, F.S., Ferreira, I.M., and Trugoa, L.C., 2005. Separation and quantification of beer carbohydrates by high-performance liquid chromatography with evaporative light scattering detection. *Journal of Chromatography A*. 1065: 207–210.

Odaci, D., Telefoncu, A., and Timur, S., 2010. Maltose biosensing based on co-immobilization of α-glucosidase and pyranose oxidase. *Bioelectrochemistry*. 79: 108–113.

Pietogrande, M.C., and Bacco, D., 2011. GC–MS analysis of water-soluble organics in atmospheric aerosol: Response surface methodology for optimizing silyl-derivatization for simultaneous analysis of carboxylic acids and sugars. *Analytica Chimica Acta*. 689: 257–264.

Rizzi, G.P., 1997. Chemical structure of colored Maillard reactions products. *Food Reviews International*. 13: 1–28.

Rong, Y., Sillick, M., and Gregson, C.M., 2009. Determination of dextrose equivalent value and number average molecular weight of maltodextrin by osmometry. *Journal of Food Science*. 74: 33–40.

Rufián-Henares, J.A., Delgado-Andrade, C., and Morales, F.J., 2009. Assessing the Maillard reaction development during the toasting process of common flours employed by the cereal products industry. *Food Chemistry*. 114: 93–99.

Sesta, G., 2006. Determination of sugars in royal jelly by HPLC. *Apidologie*. 37: 84–90.

Shirokane, Y., Ichikawa, K., and Suzuki, M., 2000. A novel enzymic determination of maltose. *Carbohydrate Research*. 329: 699–702.

Southgate, D.A.T., 1995. Digestion and metabolism of sugars. *The American Journal of Clinical Nutrition*. 62: 203–211.

Sznaidman, M., Vasella, A., Fraser-Reid, Madsen, R., Campbell, A.S., Roberts, C.S., Merrit, J.R., Nicolaou, K.C., Bockovich, N.J., Kahne, D., Silva, D., Walker, S., Hendrix, M., Wong, C., Serianni, A.S., Hollenbaugh, D., Bajorath, J., Aruffo, A., Tropper, F., Bednarski, M., Bundle, D.R., Quiocho, F.A., Vyas, N.K., Armspach, D., Gattuso, G., Koniger, R., Stoddart, J.F., and Preiss, J., 1999. Introduction to Carbohydrates. In: Hecht, S.M. (ed.) *Bioorganic Chemistry Carbohydrates*. Oxford University Press, Oxford, UK, pp 1–2.

USDA National Nutrient Database for Standard Reference, Nutrient Data Laboratory, 2010. Available at: http://www.ars.usda.gov/nutrientdata. Accessed 28 June 2011.

Wang, J., Kliks, M.M., Jun, S., Jackson, M., and Li, Q.X., 2010. Rapid Analysis of Glucose, Fructose, Sucrose, and Maltose in Honeys from Different Geographic Regions using Fourier Transform Infrared Spectroscopy and Multivariate Analysis. *Journal of Food Chemistry*. 75: 208–214.

Wolever, T.M.S., 2003. Carbohydrate and the regulation of blood glucose and metabolism. *Nutrition Reviews*. 61: 40–48.

CHAPTER 25

Determination of Maltose in Food Samples by High-temperature Liquid Chromatography Coupled to ICP-AES

AMANDA TEROL, SOLEDAD PRATS, SALVADOR
MAESTRE AND JOSÉ LUIS TODOLÍ*

Faculty of Sciences, Department of Analytical Chemistry, Nutrition and
Food Sciences, University of Alicante, Carretera de San Vicente s/n,
03080 (Alicante), Spain
*Email: jose.todoli@ua.es

25.1 Maltose Chemical Properties and their Importance in the Food Industry

The separation and characterization of carbohydrates (mono and oligo-saccharides) have gained attention in recent years due to their ubiquity in nature and their important roles in biology, industry and consumer products (Verardo *et al.* 2009). Consequently, carbohydrate analysis is important in biochemical, pharmaceutical, agricultural and food sciences.

Maltose has limited application as a sweetener in foods. However, its determination (together with glucose) is of great importance in the fermentation processes. The concentration of malto-oligosaccharides influences the characteristics of beers and therefore their determination is of great practical

Food and Nutritional Components in Focus No. 3
Dietary Sugars: Chemistry, Analysis, Function and Effects
Edited by Victor R Preedy
© The Royal Society of Chemistry 2012
Published by the Royal Society of Chemistry, www.rsc.org

interest. Furthermore, monitoring of sugars in malted cereal grains before, during and after fermentation is highly significant. During malting and mashing operations for the production of wort, polysaccharides (primarily starch) are hydrolyzed by amylases mainly to malto-oligosaccharides, which consist of 3–10 glucose units. Fermentation of wort converts low molecular weight sugars, such as glucose, fructose, sucrose, maltose and maltotriose into alcohol. Due to their low levels in malt, the content of glucose, fructose and sucrose quickly decreases during fermentation. Maltose, the most abundant sugar in wort, requires a longer time to be completely fermented. On the other hand, maltotetraose and higher oligosaccharides are unaffected by most yeast strains. The evaluation of the carbohydrate profile in brewing technology is an important issue, because any in it may influence the sensory characteristics of beer (Gotsick and Benson 1991).

25.2   Analytical Methods for the Determination of Maltose in Food Samples

25.2.1   Sample Preparation

Due to their similar structure, maltose determination is almost always associated with the determination of other disaccharides and monosaccharides present in foods. The most common sugars present in foods are glucose, fructose, galactose, maltose, sucrose and lactose. Aqueous extraction of these sugars is readily achieved following a short period of heating and mixing. It is well known that the solubility of sugars and the insolubility of proteins and polysaccharides in 80% ethanol can be used to further isolate them. The particular sample preparation method may depend on the considered food.

25.2.2   Determination of Maltose in Food Samples

Physical separation of saccharides is achieved by using chromatographic techniques and Capillary Electrophoresis (CE). Biochemical separation is obtained by using the so-called enzymatic methods of analysis (Nollet 2004).

Currently, High Performance Liquid Chromatography (HPLC) is one of the most widely used and important separation techniques for the analysis of soluble carbohydrates. For the detection of low molecular weight carbohydrate, several detectors have been used in combination with HPLC systems: Ultraviolet (UV) absorption, Refractive Index (RI), Mass Spectrometer, Pulsed Amperometric (PAD) (Lee 1996), Evaporative Light Scattering and Charged Aerosol detectors can be found in the literature.

Generally, RI measurement is the most popular detection method for carbohydrates, the HPLC-RI coupling has been used for the determination of maltose in lager beers (Del Pozo-Insfran *et al.* 2004). In the case of the malt extract drinks, Ovaltine and Horlicks, the values of lactose/maltose were expressed as total values because the HPLC method did not fully separate these sugars (Greenway and Kometa 1995). Moreover, RI has several disadvantages, such as the lack of sensitivity and selectivity, signal dependence with temperature and with mobile phase flow rate and the incompatibility with gradient elution.

A pulsed amperometric detection (PAD) can also be used for the determination of sugars in different kinds of food. In conjunction with PAD, the HPLC method is highly sensitive and selective; however, the PAD electrode requires a time-consuming cleaning procedure to remove any interfering agent from the sample matrix (Verardo *et al.* 2009).

To overcome some of these drawbacks the so-called evaporative light scattering detection (ELSD) is widely used as a semi-universal mass detector for HPLC. It is based on the detection of solute molecules by light scattering after nebulization and evaporation of the mobile phase. This detector has been used for the determination of some common carbohydrates in drinks such as fruit juices, grape wines and liquors (Wei and Ding 2000). However, when using HPLC-ELSD with water as mobile phase, a problem often encountered is the co-elution of lactose and maltose, for this reason several injections with different stationary phases can be performed to separate them (Muir *et al.* 2009). Table 25.1 summarizes representative examples in which maltose has been determined in foods.

25.3   What is High-temperature Liquid Chromatography?

Some disadvantages presented by HPLC are overcome when high mobile phase temperatures are used. Despite its advantages, few studies have been carried out concerning the determination of carbohydrates by HTLC (Koizumi 1996, Terol *et al.* 2010). Actually, only one of them has applied this technique to the determination of carbohydrates (maltose included) in foodstuff (Terol *et al.* 2010).

HTLC is a modified version of the HPLC technique where the column is introduced in an oven to take advantage of the effect of high temperatures on the chromatographic separation. Teutenberg (Teutenberg 2010) defines the normal boiling point of water as the lower temperature limit for HTLC. From this point of view, a lower temperature limit would be more appropriate if considering the boiling point of the mobile phase. Teutenberg defines the upper temperature limit as the temperature at which every solvent or solvent mixture will be in the supercritical state. In this case, water limits this region because it has the highest critical temperature among all the employed mobile phases. Thus, HTLC could be located at temperatures between 60 °C (boiling point of methanol) and 374 °C.

25.3.1   General Advantages and Characteristics of HTLC

The kinetics and thermodynamics of the chromatographic process are a function of temperature. Some of the advantages of HTLC are:

- Shortened analysis times. The increase in the temperature causes a decrease in retention times due to the decrease in mobile phase viscosity.
- Improvement in the efficiency and resolution of the chromatographic run. At high temperatures the solute transfer from the mobile phase to the stationary one becomes more efficient. This leads to a flatter van Deemter curve with the possibility of using higher flow rates without hampering the separation efficiency (Vanhoenacker and Sandra 2008).
- Temperature influences the selectivity of the stationary phase (Teutenberg 2010). This is especially useful for polar and ionisable compounds given that ionization equilibria are temperature dependent (Gagliardi *et al.* 2005).
- Lower consumption of organic solvents (green chromatography). By increasing the temperature of the mobile phase, the amount of organic solvent in the mobile phase can be reduced or avoided (Fields *et al.* 2001). At ambient temperature, water is a very weak solvent if used in reversed-phase mode. However, increasing the temperature leads to a partial break-up of the strong hydrogen-bond network among the water molecules (Teutenberg 2010).
- The number of detectors that can be adapted to HTLC is higher than in HPLC. On this subject, HTLC has been adapted to detection techniques such as visible-UV spectrophotometry, fluorimetry, refractometry, ELSD, Flame Ionization Detector, (FID), Mass Spectrometry (MS) and Nuclear Magnetic Resonance (NMR) (Guillarme *et al.* 2005).
- Temperature programming can be used to modify the retention factor and, in some instances, it can replace the solvent gradient (Vanhoenacker and Sandra 2008).

25.3.2   Special Instrumentation

HTLC shows several disadvantages such as the need for special refractory stationary phases and the requirement of special equipment to carry out the experiments, although currently, specific instrumentation for HTLC is commercially available. Figure 25.1 shows a schematic layout of the components required to work in HTLC.

25.3.2.1   Stationary Phases

The traditional stationary phases are not sufficiently stable at the temperatures needed in HTLC. In fact, classical reversed-phase silica columns allow only a relatively small increase (30–40 °C) in temperature over the ambient one

Table 25.1 HPLC applications and experimental conditions in which maltose has been determined together with additional sugars. The data reported are taken from literature. Original table. This table summarizes representative examples in which maltose has been determined in foods, taking into account the mobile and stationary phase, the analysis conditions, detectors, time of retention, analyzed foods and problems found.

Mobile Phase	Stationary Phase	Conditions (Ql, T)	Isocratic or Gradient Separation	Extra Analytes	Detector	Time of Retention of Maltose (min)	Analyzed Food with Maltose	Problems Found	Reference
99% water and 1% methanol	Spherisorb ODS 2 column (5 μm particle size, 250 mm long × 4.6 mm id.)	1 ml/min	Isocratic	Fructose, glucose, lactose, sucrose and raffinose	RI (Waters 410)	± 5 (lactose + maltose)	Malt extract drink, Ovaltine, Horlicks, chocolate (brown)	The peak of lactose and maltose appear overlapped, then the determination of the two sugars was at the same time	(Greenway and Kometa 1995)
Acetonitrile – water (75:25)	Zorbax NH₂ column (250 × 4.6 mm)	1.5 ml/min Ambient T	Isocratic	Glucose and maltotriose	Hewlett Packard 1100 system equipped with a RI detector	11.5	Lager beers		(Del Pozo-Insfran et al. 2004)
150 mM NaOH	Carbopac PA1 (Dionex) pellicula anion-exchange analytical column (250 × 4 mm)	1 ml/min	Isocratic	Glucose, fructose, lactose, sucrose, maltodextrin	Pulsed electrochemical detection system (PED-1) in the integrated amperometry mode	± 23	Milk based, soy based and protein hydrolysate infant formula		(Kaine and Wolnik 1998)
0.080 mol/L NaOH	Anion exchange column: Dionex HPIC-AG6 y Dionex HPIC-AS6	1 ml/min	Isocratic	Mannitol, arabinose, glucose, fructose, lactose, sucrose, raffinose.	PDA	± 9.70	Wine and beer		(Yan et al. 1997)

Table 25.1 (continued)

Mobile Phase	Stationary Phase	Conditions (Ql, T)	Isocratic or Gradient Separation	Extra Analytes	Detector	Time of Retention of Maltose (min)	Analyzed Food with Maltose	Problems Found	Reference
(i) Water with added EDTA (50 mg/L) (ii) 75:25 (v/v) acetonitrile/water	Two different columns: (i) Waters Sugar-Pak column (5 μm, 6.5 × 300 mm) (ii) Waters High-Performance Carbohydrate column, 4 μm, 4.6 × 250 mm	(i) 0.5 mL/min, 90 °C (ii) 1.0 mL/min, 40 °C	Isocratic	Sucrose, lactose, glucose, galactose, fructose	ELSD (Waters 2424)	± 8	Common Australian Vegetables and fruits	(i) With the first column sucrose and maltose overlapped — Maltose was not determined in the samples	(Muir et al. 2009)
Mixture of water – acetonitrile (1:2.6, v/v) containing 0.03% (v/v) ethylenediamine as a modifier and ammonium hydroxide (0.05% v/v) to adjust the pH.	Zorbax Rx-SIL, 250 mm × 4.6 mm I.D., 5 μM	25 °C	Isocratic	Fructose, glucose, sucrose, lactose, raffinose	ELSD	± 14.5	Fruit juices, grape wine and liquor	Maltose was not determined in the studied samples	(Wei and Ding 2000)
Ultra-pure water and is 0.2 M NaOH	Dionex GP 50 (USA) incorporating a Carbopac PA1 (4 _ 250 mm)	0,5 ml/min with a gradient of ultra-pure water and is 0.2 M NaOH	Gradient	Glu, Fru, 12 saccharides and trisaccharides	PAD	25	Honey		(Ouchemoukh et al.)

(Guillarme and Heinisch 2005). Obviously, the thermal stability of stationary phases is a subject of capital importance in HTLC. Nowadays, the available columns stable at high temperatures are made of porous graphite (Terol *et al.* 2010), zirconia (Yan *et al.* 2000), titania or organic polymers (Terol *et al.* 2011). For some of these stationary phases, the temperatures can be as high as 200 °C without degradation of the stationary phase. In spite of the fact that refractory stationary phases are used, several studies have pointed out degradation and bleeding problems when high temperatures are applied (Marin *et al.* 2004).

25.3.2.2 Preheating

A critical component in HTLC is the eluent preheating system because temperature variations of the mobile phase along the column can counteract the possible benefits of the work at high temperatures (Yan *et al.* 2000). The role of the preheating tube, placed between the injection valve and the column inlet (Figure 25.1), is to limit temperature variations inside the column. The optimal length of the preheating tube has been intensively studied (Yan *et al.* 2000).

25.3.2.3 Heating Systems

The column temperature must be controlled correctly because small temperature variations have a large effect on compound retention and/or selectivity. Nowadays, virtually all the HPLC systems have a temperature control up to 100 °C. As the heating systems should keep constant the oven temperature and because the effective column temperature can differ according to the oven design, it is essential to correctly specify the oven type in every investigation carried out at high temperatures.

Isothermal separations can be carried out with conventional ovens. However, if the goal is to work with temperature programming, gas chromatography

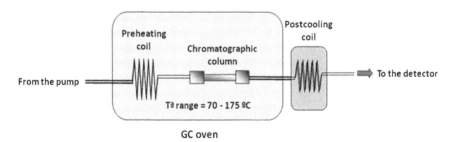

Figure 25.1 Basic set-up for High-temperature Liquid Chromatography. Original figure. An oven is employed to heat the mobile phase through the separation process. A preheating coil must be used to limit temperature variations of the mobile phase inside the column. After passing through the column, the mobile phase reaches a cooling coil to adapt its temperature to the detector requirements.

ovens, electric tapes, circulating fluid jackets, block heating systems, or homemade ovens can be used.

25.3.2.4   Postcolumn Cooling

In the coupling of HTLC with conventional detectors it is necessary to reduce the mobile phase temperature before it reaches the detection system without causing a significant peak broadening. There are different possibilities; an approach is to use an efficient heat exchanger to lower the mobile phase temperature below the upper limit of the detector (*e.g.*, 80 °C for many UV absorbance detectors) (McNeff *et al.* 2007). Note that if the temperature of the mobile phase at the detector is not closely controlled, an unstable baseline and increased noise can be obtained (Smith 2008). Sometimes the use of an extra capillary at the exit of the oven is employed to cool the mobile phase (Terol *et al.* 2010). It should be pointed out that the postcolumn cooling can be avoided with detectors that are compatible with hot mobile phases (*e.g.*, Inductively Coupled Plasma Atomic Emission Spectrometer, FID, *etc.*).

25.4   Inductively Coupled Plasma Atomic Emission Spectrometry for the Determination of Sugars in Food

Inductively Coupled Plasma Atomic Emission Spectrometry (ICP-AES) is an elemental technique routinely employed in the analysis laboratories. Trace metals and some non metals can be quickly determined in samples of a wide nature by means of this technique. Generally speaking, the analysis of liquid samples through ICP-AES involves the following steps:

(i)   Aerosol generation. The liquid sample should be transformed into an aerosol as this medium is compatible with high-temperature plasmas. Normally, a pneumatic nebulizer is employed in which the aerosol is generated through the exposure of the liquid solution to a high-velocity gas stream.

(ii)   Aerosol filtering. Only the finest droplets must be introduced into the high temperature plasma. The nebulizer is adapted to a spray chamber so as to eliminate the coarsest droplets. Many different designs of spray chambers are available. Typically, after the spray chamber those droplets with diameters higher than 20 µm are eliminated and they are lost as they impact against the inner walls of the spray chamber. Figure 25.2 shows a scheme of a conventional liquid sample introduction system for ICP-AES.

(iii)   The droplets at the exit of the chamber reach a high temperature argon plasma, they are completely evaporated, the generated solid particles are thus vaporized and the element, or analyte, is atomized, the atoms being further excited and/or ionized.

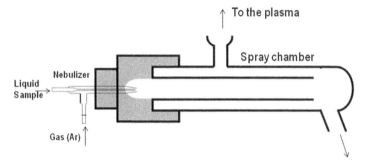

Figure 25.2 Scheme of a conventional sample introduction system for ICP-AES consisting of a pneumatic nebulizer adapted to a double pass spray chamber. Original figure. The pneumatic nebulizer commonly used has two different inlets: one for the liquid sample and the other one for the gas flow. An aerosol is generated as a result of the liquid and high velocity streams interaction. The spray chamber eliminates the coarsest droplets and only the finest ones are introduced into the high temperature plasma.

(iv) As the excited species deactivate, they emit light at characteristic wavelengths and a monochromator selects the wavelength, whereas a detector measures the intensity emitted at one (sequential instruments) or various (simultaneous instruments) wavelengths.

The measured emission intensity is directly proportional to the analyte concentration in the sample (Montaser and Golightly 1987). In ICP-AES, the sample introduction system precludes the measured analytical signal. In general terms, it can be stated that the finer the aerosol generated by the nebulizer, the higher the mass of analyte delivered to the plasma and hence the higher the sensitivity.

ICP-AES is seldom used for the determination of organic compounds. To determine the concentration of sugars by ICP-AES, the carbon emission signal is measured. This detector has some advantages over conventional chromatographic detectors such as: *(i)* lower limits of detection for some organic compounds; *(ii)* wider dynamic ranges; *(iii)* universality for non-volatile compounds; and *(iv)* cleaner backgrounds. A drawback observed is that if retention times are too long, the cost of the determination becomes too high. Furthermore, the carbon emission background is usually high, because this element can diffuse from the air surrounding the plasma as well as from the carbon dioxide dissolved into the mobile phase. Furthermore, carbon is present as an impurity in argon. Finally, organic solvents should be avoided in the mobile phase to minimize background signals.

There are some studies that demonstrate the suitability of an ICP-AES as HPLC detector for the determination of organic compounds. The HPLC-ICP-AES hyphenation has been applied to the determination of aminoacids (Peters *et al.* 2004), saccharides (Jinno *et al.* 1984), carboxylic acids (Paredes *et al.*

2006) and alcohols (Jinno *et al.* 1985). Because of the high argon consumption and the sometimes long retention times (*i.e.*, longer than 20 min) the development of rapid separation methods, such as HTLC, is crucial to make this detector interesting in terms of analytical cost.

25.5   HTLC-ICP Hyphenation

Recently, HTLC has been applied to the determination of maltose using an ICP-AES as the detector. Figure 25.3 shows a scheme of this coupling. In this case a cyclonic (jacketed) spray chamber is depicted. HTLC can be considered as a good approach for coupling with ICP-AES for several reasons: *(i)* this chromatographic technique is fast, thus shortening the analysis time as compared to HPLC and reducing significantly the argon consumption; *(ii)* as the effluent of the chromatographic column emerges at a high temperature, once the aerosol is generated, the solvent evaporation inside the chamber is

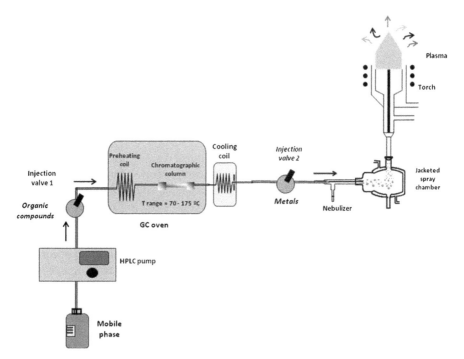

Figure 25.3   Experimental set-up employed to determine maltose through HTLC-ICP-AES. Original figure. Two valves were used in the determination of maltose. Valve 1 was used to inject the sample (or standards) for separating the organic compounds in the column. Valve 2 in turn was used to inject the sample (or standards) after the column for the determination of metals. In this way, peaks obtained for the metals present in the sample were registered before the first organic compound left the column.

promoted, thus providing higher sensitivities than at room temperature; *(iii)* it is possible to remove the postcolumn cooling capillary and, therefore, to reduce the peak dispersion; *(iv)* under some circumstances, the aerosol is generated at the exit of the column without the use of a nebulizer, thus making the interface simple; *(v)* many applications of HTLC are carried out using only water as mobile phase; *(vi)* as ICP-AES is an elemental technique, it is possible to obtain information about the content of organic compounds as well as elements (*i.e.*, metals) from a single chromatographic run. Among the problems found, we can mention the fact that the background emission intensity increases as the column temperature goes up. This is due to the enhancement in the release of dissolved carbon dioxide from the mobile phase as it is heated.

Maltose shows an especially interesting performance when using HTLC. Thus, the signal to background ratio for the maltose peak increases by a factor close to two as the flow rate goes from 0.25 to 1.00 mL/min. This is obviously a direct consequence of the decrease in peak dispersion along the chromatographic column. Additionally, the column temperature is also critical. For example, for graphite-based stationary phase (Hypercarb column) at 100 °C, the maltose peak signal to background ratio is 25, whereas at 150 °C this parameter increases up to 200. Note that maltose does not leave the column at temperatures below 70 °C. Therefore, for this stationary phase, HTLC is required if maltose is to be determined. Concerning retention times, at 75 °C this parameter is 7 min, whereas at 150 °C retention times close to 2 minutes are measured. For other sugars, such as sucrose, retention times shorten only by a factor of 2.

The interface between HTLC and ICP-AES must provide a high analyte mass transport efficiency. Moreover the dead volumes must be kept at minimum levels to avoid peak dispersion. To reach these goals the so-called micronebulizers can be employed (Todolí and Mermet 2008). As the nebulizer, the spray chamber also plays a very important role. Low inner volume chambers are preferred when HTLC is coupled to ICP-AES. Thus, for example, if a single pass small spray chamber is used, the sensitivity for maltose is three times higher than when using a cyclonic design (such as that depicted in Figure 25.2). This can be assigned to the higher mass of analyte transported towards the plasma and the lower peak dispersion produced for the former device.

As mentioned before, in HTLC-ICP-AES, a hot liquid stream is continuously nebulized. The solvent evaporation is obviously intense and it may cause degradation of the plasma thermal state. In order to prevent this effect, a thermostated spray chamber can be used. A cyclonic design is shown in Figure 25.3. The solvent evaporated from the aerosol generated by the nebulizer condenses and does not reach the plasma. In a different approach, the nebulizer can be removed and the aerosol is directly produced as the hot effluent leaves the column. In order to better control the temperature of the mobile phase, a heating system (*e.g.*, a tape) can be wound around a stainless steel capillary (Figure 25.4). In this case, it has been observed that the higher the temperature of the tape, the stronger the sensitivity. Thus on increasing this variable from

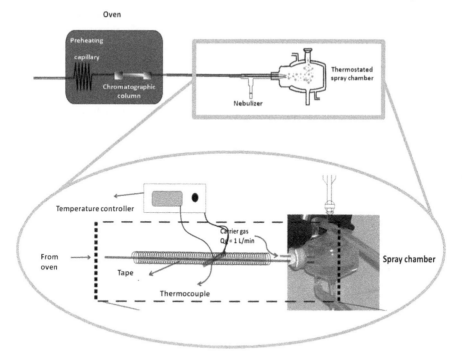

Figure 25.4   Scheme of the interface employed to couple HTLC with ICP-AES (up)
and interface without pneumatic nebulizer (down). By thermostating the
capillary with the help of a tape, it is possible to generate an aerosol at
the exit of the stainless steel capillary thus avoiding the use of a nebulizer.
The solution is thus directly introduced into the spray chamber and the
peak broadening caused by the dead volume of the pneumatic nebulizer
is reduced.

170 °C to 220 °C maltose peak height grow by a factor close to three. Furthermore, LODs for maltose were up to three times lower with this new
approach as compared to the conventional system.

25.6   Comparison of HTLC-ICP-AES Hyphenation
with the Techniques Commonly used for the
Determination of Maltose

Comparatively speaking, the use of HTLC-ICP-AES for the determination of
maltose in food samples has several advantages against other detectors such the
ELSD. Among them is the extended dynamic range. The ICP-AES maltose
dynamic range goes from about 10 to 7,000 mg/L (Terol *et al.* 2010). HTLC-
ICP-AES coupling can be applied for the determination of both metals and
maltose (as well as other organic compounds) in a single chromatographic
run. To carry out this procedure a second valve can be used (see valve 2 in

Figure 25.3). This capability, not shown by chromatographic detectors, is especially possible when a simultaneous ICP-AES system is employed. Finally, as regards limits of detection for maltose, this parameter takes values of about 5 and 10 mg/L for the ELSD and ICP-AES, respectively. This is a clear consequence of the high background found for the latter design.

25.7 Conclusion

The determination of maltose in food samples has been traditionally performed by applying HPLC. The drawbacks encountered (*e.g.*, long retention times) together with the need for determining additional sugars has promoted the application of additional separation techniques. HTLC is an appropriate alternative to conventional HPLC. The most important advantage is the shortening in the retention times. Besides, high sensitivities can be obtained when the mobile phase is heated at an appropriate temperature. However, HTLC also suffers from several drawbacks such as the need for a stable stationary phase and the requirement of a dedicated instrumental setup. Furthermore, HTLC is an appropriate technique when few compounds are to be determined because peak overlapping can be a problem otherwise.

The use of HTLC for maltose determination together with an ICP-AES detector is a promising coupling because of several reasons: *(i)* as the retention time is shortened at high temperatures, the argon consumption is reduced; *(ii)* this elemental detector can be used for the determination of several carbohydrates; and *(iii)* the detector can be employed to determine heavy metal concentration. In this way inorganic as well as organic information can be obtained from a single chromatographic run.

Within this filed, several advances can be predicted such as the development of temperature programming to better separate sugars. Another point is the use of microbore columns that allow for a quick separation with a decrease in the mobile phase flow rate. Additionally, ICP-AES can be used to determine some anions containing elements such as sulfur or phosphorous.

Summary Points

- This chapter focuses on the determination of maltose in food samples by High-temperature Liquid Chromatography coupled to Inductively Coupled Plasma Atomic Emission Spectrometer (ICP-AES).
- Maltose is a disaccharide composed of two glucose units associated by an α-1,4 glycosidic linkage.
- Maltose has limited application as a sweetener in foods. However, its determination (together with glucose) is of great importance in fermentation processes.
- The most common analytical method used for the determination of maltose in food samples has been High Performance Liquid Chromatography (HPLC) coupled to different detectors.

- High-temperature Liquid Chromatography (HTLC) has emerged as an alternative chromatographic technique that has shown good results for the analysis of carbohydrates, especially with maltose.
- Inductively Coupled Plasma Atomic Emission Spectrometer (ICP-AES) is popular in elemental analysis but it is seldom used for the determination of organic compounds such as sugars.
- The use of High-temperature Liquid Chromatography Inductively Coupled Plasma Atomic Emission Spectrometer (HTLC-ICP-AES) for the determination of maltose in food samples has shown several advantages against other detectors.
- The interface between High-temperature Liquid Chromatography and Inductively Coupled Plasma Atomic Emission Spectrometer (ICP-AES) is a key point that dictates the quality of the results obtained both in terms of sensitivity and resolution.

Key Facts of Maltose in Beer

1. Maltose is not common naturally in foods. It is found in malt beverages such as beer because it is derived from starch by treatment with β amylase.
2. Maltose plays a principal and important role in the fermentation process of beer because it is the most abundant sugar in the first phase of beer production.
3. The alcoholic fermentation is the necessary process to convert sugars into alcohol. To carry out this process the presence of yeast is necessary.
4. As maltose is the most abundant sugar in wort, it requires long time to be completely fermented.
5. The evaluation of the carbohydrate profile in brewing technology is an important issue, because any in it may influence the sensory characteristics of beer.
6. Some studies have researched an appropriate method for the determination of maltose in beer.

Definition of Words and Terms

Glycosidic linkage. It is a type of covalent bond that joins a saccharide or molecule derived from a saccharide (hemiacetal group) to another group (hydroxyl), which may or may not be another carbohydrate.

Amylases. Refers to a group of enzymes that have the function to digest glycogen and starch to form sugars.

Brewing process. The brewing process is the production of beer by cooking a source of starch, usually a cereal grain, adding bitterness and additional flavoring, and then fermenting it with yeast.

Wort. It is the liquid obtained from the mashing process during the brewing of beer. Wort contains the sugars that will be fermented by the brewing yeast to produce alcohol.

Maillard reaction. It is the reaction between carbohydrates (reducing sugar) and amino acids, and is responsible for changes in color, flavor and nutritive in food, usually required heat.

Derivatization. It is a technique used in chemistry which transforms an analyte into a product with similar chemical structure but that it is easier to analyze. For example, in GC the derivatization process is used to obtain more volatile species.

Chromatographic techniques. It is a term that includes laboratory techniques for the separation of mixtures.

van Deemter equation. It is applied in chromatography and allows the determination of the optimum mobile phase velocity.

Nebulizer. The nebulizer is an instrument for converting a liquid into a fine spray.

Spray chamber. The spray chamber is used in techniques such as ICP-AES and ICP-MS and is the responsible to eliminate the coarsest droplets and to select only the finest ones to introduce them into the high temperature plasma.

List of Abbreviations

CE	Capillary Electrophoresis.
ELSD	Evaporative Light Scattering Detection.
ESI-MS	Electrospray Ionization Mass Spectrometry.
FID	Flame Ionization Detection.
HPLC	High Performance Liquid Chromatography.
HTLC	High Temperature Liquid Chromatography.
ICP-AES	Inductively Coupled Plasma Atomic Emission Spectrometry.
MS	Mass Spectrometry.
NMR	Nuclear Magnetic Resonance.
PAD	Pulsed Amperometric Detection.
RI	Refractive Index.
visible-UV	Visible Ultraviolet Spectroscopy.

Acknowledgement

The authors wish to thank to the Spanish Science and Innovation Ministry (CTQ2009-14063) for the financial support. Amanda Terol also thanks the Generalitat Valenciana for the grant (FPI/2008/126).

References

Del Pozo-Insfran, D., Urias-Lugo, D., Hernandez-Brenes, C., and Saldivar, S.O.S., 2004. Effect of amyloglucosidase on wort composition and fermentable carbohydrate depletion in sorghum lager beers. *Journal of the Institute of Brewing*. 110: 124–132.

Fields, S.M., Ye, C.Q., Zhang, D.D., Branch, B.R., Zhang, X.J., and Okafo, N., 2001. Superheated water as eluent in high-temperature high-performance liquid chromatographic separations of steroids on a polymer-coated zirconia column. *Journal of Chromatography A*. 913: 197–204.

Gagliardi, L.G., Castells, C.B., Rafols, C., Roses, M., and Bosch, E., 2005. Effect of temperature on the chromatographic retention of ionizable compounds - II. Acetonitrile-water mobile phases. *Journal of Chromatography A*. 1077: 159–169.

Gotsick, J.T., and Benson, R.F. 1991. HPLC analysis of carbohydrates important to beer brewing using an aminopropyl stationary phase. *Journal of Liquid Chromatography*. 14: 1887–1901.

Greenway, G.M., and Kometa, N., 1995. The determination of sugars in beverages and medicines using online dialysis for sample preparation. *Food Chemistry*. 53: 105–110.

Guillarme, D., and Heinisch, S., 2005. Detection Modes with High Temperature Liquid Chromatography—A Review. *Separation & Purification Reviews*. 34: 181–216.

Guillarme, D., Heinisch, S., Gauvrit, J.Y., Lanteri, P., and Rocca, J.L., 2005. Optimization of the coupling of high-temperature liquid chromatography and flame ionization detection. Application to the separations of alcohols. *Journal of Chromatography A*. 1078: 22–27.

Jinno, K., Nakanishi, S., and Fujimoto, C., 1985. Direct Sample Introduction System for Inductively Coupled Plasma Emission Spectrometric Detection in Microcolumn Liquid Chromatography. *Analytical Chemistry*. 57: 2229–2235.

Jinno, K., Nakanishi, S., and Nagoshi, T., 1984. Microcolumn Gel Permeation Chromatography with Inductively Coupled Plasma Emission Spectrometric Detection. *Analytical Chemistry*. 56: 1977–1979.

Kaine, L.A., and Wolnik, K.A. 1998. Detection of counterfeit and relabeled infant formulas by high-pH anion-exchange chromatography pulsed amperometric detection for the determination of sugar profiles. *Journal of Chromatography A*. 804: 279–287.

Koizumi, K., 1996. High-performance liquid chromatographic separation of carbohydrates on graphitized carbon columns. *Journal of Chromatography A*. 720: 119–126.

Lee, Y.C., 1996. Carbohydrate analyses with high-performance anion-exchange chromatography. *Journal of Chromatography A*. 720: 137–149.

Marin, S.J., Jones, B.A., Felix, W.D., and Clark, J., 2004. Effect of high-temperature on high-performance liquid chromatography column stability and performance under temperature-programmed conditions. *Journal of Chromatography A*. 1030: 255–262.

McNeff, C.V., Yan, B., Stoll, D.R., and Henry, R.A., 2007. Practice and theory of high temperature liquid chromatography. *Journal of Separation Science*. 30: 1672–1685.

Montaser, A., and Golightly, D.W., 1987. *Inductively Coupled Plasmas in Analytical Atomic Spectrometry*. VCH Publishers, New York, 1017 pp.

Muir, J.G., Rose, R., Rosella, O., Liels, K., Barrett, J.S. Shepherd, S.J., and Gibson, P.R., 2009. Measurement of Short-Chain Carbohydrates in Common Australian Vegetables and Fruits by High-Performance Liquid Chromatography (HPLC). *Journal of Agricultural and Food Chemistry*. 57: 554–565.

Nollet, L., 2004. *Handbook of Food Analysis*, 2nd Ed., CRC Press, New York, 2296 pp.

Ouchemoukh, S., Schweitzer, P., Bey, M.B., Djoudad-Kadji, H., and Louaileche, H. HPLC sugar profiles of Algerian honeys. *Food Chemistry*. 121: 561–568.

Paredes, E., Maestre, S.E., Prats, S., and Todolí, J.L., 2006. Simultaneous Determination of Carbohydrates, Carboxylic Acids, Alcohols, and Metals in Foods by High-Performance Liquid Chromatography Inductively Coupled Plasma Atomic Emission Spectrometry. *Analytical Chemistry*. 78: 6774–6782.

Peters, H.L., Davis, A.C., and Jones, T.B., 2004. Enantiomeric separations of amino acids with inductively coupled plasma carbon emission detection. *Microchemical Journal*. 76: 85–89.

Smith, R.M., 2008. Superheated water chromatography – A green technology for the future. *Journal of Chromatography A*. 1184: 441–455.

Terol, A., Paredes, E., Maestre, S.E., Prats, S., and Todoli, J.L., 2010. High-Temperature Liquid Chromatography Inductively Coupled Plasma Atomic Emission Spectrometry hyphenation for the combined organic and inorganic analysis of foodstuffs. *Journal of Chromatography A*. 1217: 6195–6202.

Terol, A., Paredes, E., Maestre, S.E., Prats, S., and Todoli, J.L., 2011. Alcohol and metal determination in alcoholic beverages through high-temperature liquid-chromatography coupled to an inductively coupled plasma atomic emission spectrometer. *Journal of Chromatography A*. 1218: 3439–3446.

Teutenberg, T., 2010. *High-Temperature Liquid Chromatography. A User's guide for Method Development*. RSC Chromatography Monographs. Cambridge, 210 pp.

Vanhoenacker, G., and Sandra, P., 2008. High temperature and temperature programmed HPLC: possibilities and limitations. *Anal. Bioanal. Chem*. 390: 245–248.

Verardo, G., Duse, I., and Callea, A., 2009. Analysis of underivatized oligosaccharides by liquid chromatography/electrospray ionization tandem mass spectrometry with post-column addition of formic acid. *Rapid Communications in Mass Spectrometry*. 23: 1607–1618.

Wei, Y., and Ding, M.Y., 2000. Analysis of carbohydrates in drinks by high-performance liquid chromatography with a dynamically modified amino column and evaporative light scattering detection. *Journal of Chromatography A*. 904: 113–117.

Yan, B., Zhao, J., Brown, J.S., Blackwell, J., and Carr, P.W., 2000. High-Temperature Ultrafast Liquid Chromatography. *Analytical Chemistry*. 72: 1253–1262.

Yan, Z., Zhang, X.D., and Niu, W.J., 1997. Simultaneous determination of carbohydrates and organic acids in beer and wine by ion chromatography. *Mikrochimica Acta*. 127: 189–194.

CHAPTER 26

Analysis of Maltose and Lactose by U-HPLC-ESI-MS/MS

LUCA VALGIMIGLI,*[a] SIMONE GABBANINI[b] AND RICCARDO MATERA[b]

[a] University of Bologna, Department of Organic Chemistry "A. Mangini", Via San Giacomo 11, 40126 Bologna, Italy; [b] R&D division, BeC s.r.l., Via C. Monteverdi 49, 47100 Forlì, Italy
*Email: luca.valgimigli@unibo.it

26.1 Introduction

Analysis of maltose and lactose in food and diet supplements is of high relevance for quality and safety assurance. Both carbohydrates contribute to the caloric content of food and such contribution should be considered to establish the nutritional value and compose the corresponding table, which has to be exposed in external packaging. The widespread diffusion of lactose intolerance among the population (Buller and Grand 1990), and the necessity for intolerant customers to comply with a lactose-free diet to avoid gastrointestinal symptoms associated with malabsorbed lactose in the colon (Grand *et al.* 2008), require the content of lactose in food and diet supplements to be clearly indicated by appropriate labelling. This necessity has actually the form of a strict obligation in the European Community (Directives 2003/89/EC and 2007/68/EC, and Commission Regulation No 575/2006).

Extension of the obligation to indicate lactose content also for diet supplements and nutraceutical preparations poses an extremely challenging analytical task, since both stereoisomeric disaccharides, lactose and maltose, are

Food and Nutritional Components in Focus No. 3
Dietary Sugars: Chemistry, Analysis, Function and Effects
Edited by Victor R Preedy
© The Royal Society of Chemistry 2012
Published by the Royal Society of Chemistry, www.rsc.org

commonly employed as inert support for the preparation of dry herbal extracts, either individually, in combination, or mixed with other oligosaccharides (maltodextrins). Even lactose-free preparations could actually contain lactose, due to cross-contamination in extraction or food industries. Therefore, food products often contain at the same time lactose, maltose and other mono-, di-, or oligo-saccharides. Beside the difficulties originated from the complex matrix in food products, analysis of carbohydrates is further complicated by isomerism and lack of significant UV-absorbance above 200 nm.

Mass spectrometry has gained growing importance for the analysis of underivatized oligosaccharides (El Firdoussi *et al.* 2007; Guignard *et al.* 2005; Wuhrer *et al.* 2004; Zaia 2004) and both positive (Hofmeister *et al.* 1991; Liu *et al.* 2005) and negative (Ikegami *et al.* 2008; Zaia 2004) ESI-MS methods have been documented. The similarity of in-source and CID fragmentation patterns has however represented a major obstacle toward the identification of isomeric carbohydrates. Lactose and maltose differ solely for the stereochemistry at C1 and C4 positions in the non-reducing pyranose moiety (see Figure 26.1), and both comprise a mixture of α and β anomers, originating isobaric ions both in positive and negative ESI, as well as isobaric in-source and CID fragment ions. Recent efforts to identify and rationalize distinctive fragmentation patterns for oligosaccharides differing for linkage stereochemistry and position, has contributed a significant body of knowledge in the field (El Firdoussi *et al.* 2007; Fang *et al.* 2007; Zhang *et al.* 2008). While this knowledge allows identification of a single disaccharide from the relative abundance of fragment ions, it does not allow determining the actual content of each disaccharide when they are present in a mixture, in the absence of good chromatographic resolution.

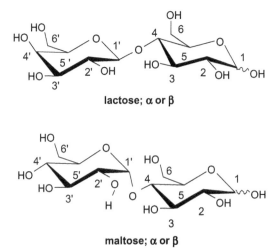

Figure 26.1   Structures of maltose and lactose. Maltose and lactose are formed by condensation of a molecule of glucose with a second glucose (linked α 1'-4) or galactose molecule (linked β 1'-4), respectively.

In this chapter we will discuss and summarize our experience with a recently developed U-HPLC-ESI-MS/MS method (Gabbanini *et al.* 2010) that allows the rapid screening of underivatized disaccharides in dietary vegetable matrices and food products, including the discrimination of maltose and lactose α- and β-anomeric forms.

26.2 Method Description and Discussion

26.2.1 Chromatography

Separation of disaccharides by liquid chromatography is certainly not a trivial task and co-elution of maltose and lactose, or that of one of the two with other disaccharides (*e.g.* cellobiose or sucrose), or with oligosaccharides, is often encountered with conventional reverse-phase HPLC approaches, using C18 stationary phases. Improved chromatographic methods have been discussed in the literature and they often involve the use of aminic-phase columns. For instance Nikolov *et al.* (1984) reported the separation of seven disaccharides (including maltose, but not lactose) present in natural honey, corn syrup and other food products, on Zorbax NH$_2$, an amine modified silica column. More recently, the aminic-phase Luna NH$_2$ eluted with acetonitrile/water was successfully employed to analyze sugars in commercial fruit-juice drinks (Verardo *et al.* 2009), however, co-elution of lactose with sucrose and that of maltose with trehalose were reported. Moreover, aminic-phase columns do not allow discrimination of anomeric forms and suffer from important pH effects; hence acidic reagents for ESI (*e.g.* formic acid) need to be injected post-column. Strong anion exchange columns eluted with potassium hydroxide solutions (high-performance anion exchange, HPAE, chromatography) have also been used to separate carbohydrates from vegetable matrices and food products (Guignard *et al.* 2005; Stroop *et al.* 2002), however, some co-elution is still encountered and subsequent coupling with mass spectrometry poses some difficulties. High salinity of the eluate and limited volatility of KOH require post-column manipulation of the eluate (to avoid crystallization and capillary clogging), such as by addition of water (resulting in decreased sensitivity), or by so called electrochemical desalting (Guignard *et al.* 2005), which requires dedicated equipment. Ion-exclusion chromatography (IEC) is another LC approach used in the separation of carbohydrates, which has been coupled with electrospray ionization mass spectrometry to determine sugars (glucose, fructose and sucrose) in soft drinks using acidic eluent (Chen *et al.* 2009); however, this chromatographic approach commonly results in poor separation and detection limits for carbohydrates.

A cation exchange phase in Ca^{2+} form, specifically designed for the analysis of sugars (Sugar Pack column, Waters), was reported to afford good resolution for glucose, galactose, fructose, mannitol, and sorbitol, when eluted with water; however maltose, lactose and sucrose co-eluted (Muir *et al.* 2009).

Specific cyclodextrins-based stationary phases offer good resolution of carbohydrates, including separation of anomeric forms of reducing sugars, and could be the chromatographic approach of choice. However, they often require complex elution techniques that might be difficult to couple to electrospray mass spectrometry. For instance, Shumacher and Kroh (1995) reported excellent separation of anomeric saccharides using a chiral β-cyclodextrin based stationary phase (Nucleodex β-OH, Macherey Nagel), with refractometric detection. Lactose and maltose had resolution (R) between α and β anomers of 0.6 and 0.8, respectively; however, this performance was achieved only by using 80% ethyl acetate (in mixture with methanol and water) as mobile phase, while it was noted that any methanol/water or acetonitrile/water eluting mixture was unable to afford similar performance, or to allow any separation of anomers (Shumacher and Kroh 1995). More recently, Liu *et al.* (2005) described a successful LC-MS method for the analysis of oligosaccharides using cyclo-dextrin based stationary phases (Cyclobond I 2000) and acetonitrile/water/ formic acid as mobile phase, however, no anomeric resolution was reported and maltose was co-eluted with cellobiose.

High chromatographic resolution in LC-MS analysis of underivatized oli-gosaccharides in mixture has been reported with direct-phase elution on amide column, although long analysis time (1 h) was required (Wuhrer *et al.* 2004).

In our own experience, the use of columns based on Porous Graphitic Carbon (PGC) offers the most interesting approach for the analysis of dis-accharides. Although it is a versatile general-purpose stationary phase that can be eluted both in direct- or in reverse-phase (Li *et al.* 1997; Michel and Buszewski 2009), PGC is relatively uncommon in the scientific literature. It was successfully applied to the analysis of carbohydrates in food products, working in conditions of reverse phase HPLC with post-column derivatization and fluorescence detection (Komuir *et al.* 1996). Under such settings, a 7 μm par-ticle-size column (Shandon Hypercarb 100 mm) was shown to afford good separation of disaccharides, including their anomeric forms, although it was lamented by the authors that separation of anomers resulted in decreased sen-sitivity. Furthermore, using a microbore graphitized carbon column, Itoh *et al.* (2002) were able to analyze N-linked oligosaccharides in a glycoprotein. These results prompted our research group to implement a U-HPLC approach based on PGC stationary phase for the analysis of underivatized disaccharides, sui-table for coupling with electrospray mass spectrometry (Gabbanini *et al.* 2010).

We chose a 3 μm particle-size column (Thermo Hypercarb 100 × 2.1 mm). Although such particle size is larger than 2 microns, normally encountered in U-HPLC columns, when eluted with aqueous mixtures it offers a backpressure of 350–400 bars already at flows of 250 μL/min.

By gradient programming of a water (+0.1% formic acid)/acetonitrile binary mixture, it allowed an excellent separation of disaccharides from vegetable matrices with resolution $R > 2.0$ for all peaks, affording also an unprecedented resolution between α and β anomers of the reducing dis-accharides, with coefficient of selectivity α of 1.1 and 1.2 for maltose and lactose respectively (*Method A*, see Table 26.1). A typical chromatogram obtained by *Method A* is presented in Figure 26.2.

Such an amazing performance was somewhat unexpected and was inter-preted on the basis of the unusual mode of interaction of the stationary phase

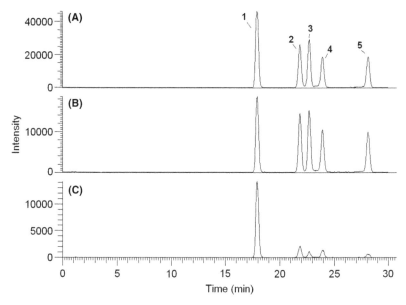

Figure 26.2 U-HPLC-ESI-MS analysis of a mixture of disaccharides (lactose, maltose and sucrose), on PGC column with Method A. Chromatograms, obtained with *Method A*, of a disaccharides mixture (lactose, maltose and sucrose) in formate buffer, generated in negative ion mode by monitoring: (A) the $[M+HCOO]^-$ adduct at m/z 387.0 ± 1.5 in MS; (B) the extracted $[M-H]^-$ ion at m/z 341.0 ± 1.5 in MS^2 $387 \rightarrow$; (C) the extracted ions at m/z 161.0 ± 1.5 and 179 ± 1.5 in MS^3 $387 \rightarrow 341 \rightarrow$. Numbers indicate sucrose (**1**), β-maltose (**2**), β-lactose (**3**), α-maltose (**4**), α-lactose (**5**). The figure is redrawn with changes after Gabbanini *et al.* (2010).

with analytes. The extended and polarisable π-system at the graphite surface is capable of establishing donor-acceptor interactions with electron dense or charged moieties of the analyte. Such interactions are sensitive both to the unique pattern of electron-donor/-withdrawing substituents in the analytes and to the spatial distribution of charges or electron-dense functions (Li *et al.* 1997). This results in a unique selectivity of PGC stationary phases, capable of finely discriminating analytes differing both for the pattern and for the spatial arrangement of dipoles or charges. In other words, it will be sensitive to the stereochemistry of solutes. Under our experimental settings, based on a mobile phase containing a significant amount of formic acid, we hypothesized that the complex between the carbohydrates and formate anion (which are monitored in negative-mode ESI, *vide infra*) is produced directly in-column, further enhancing the interaction with the graphite surface by contributing [complex charge]–[induced dipole at graphite] interactions (Gabbanini *et al.* 2010). Despite its performance *Method A* also had drawbacks; particularly the relatively long analysis time (approximately 30 minutes) compared to typical U-HPLC methods. In the search for a faster chromatographic approach that could take full advantage of U-HPLC on short low dead-volume columns, the best compromise between resolution and analysis time was represented by *Method B*, based on gradient elution with 2 mM ammonium formate buffer (pH 5.6)/ methanol binary mixtures, which had overall analysis time of 12 minutes including column reconditioning. Although resolution was lower (*R* 1.2–1.6 for all peaks) it allowed complete separation of mixtures of disaccharides (maltose, lactose and sucrose), including anomeric forms of reducing sugars (selectivity $\alpha \geq 1.1$), in less than 7 minutes (see Figure 26.3), being sufficiently fast to allow investigation of the kinetics of mutarotation (Gabbanini *et al.* 2010). A summary of the chromatographic conditions herein discussed is presented in Table 26.1.

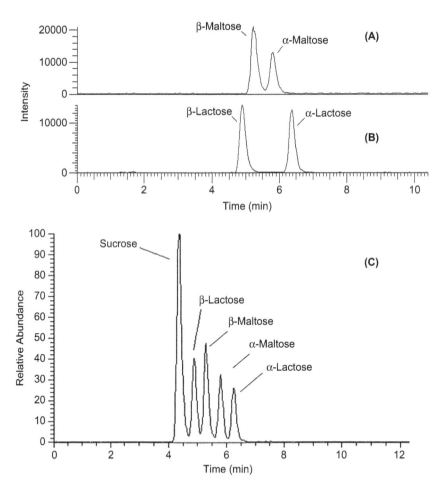

Figure 26.3   U-HPLC-ESI-MS analysis of disaccharides (lactose, maltose and sucrose), on PGC column with Method B. Representative chromatograms of standard disaccharides solutions in formate buffer: maltose (A); lactose (B); or a mixture of maltose, lactose and sucrose (C). Analyses were performed using *Method B* and monitoring in negative ion mode the $[M + HCOO]^-$ adduct at m/z 387.0 ± 1.5 in MS. The figure is modified after Gabbanini *et al.* (2010).

Table 26.1 Instrumental settings and operating conditions for U-HPLC separation of disaccharides on PGC. The table summarizes the chromatographic conditions optimized to separate mixtures of disaccharides (maltose, lactose and sucrose), including complete resolution of α and β anomers of maltose and lactose, prior to ESI-MS detection with instrumental settings described in Table 26.2. Typical chromatograms obtained with *Methods A* and *B* are illustrated in Figures 26.2 and 26.3, respectively.

System: U-HPLC Accela® (Thermo Scientific) with mass spectrometer LCQ Fleet® (Thermo Scientific) and ESI interface.
Column: HYPERCARB 100 × 2.1 mm, particle size 3 µm (Thermo Scientific)

Method A

Features: Eluent A: Formic acid 0.1% w/v	Run time: 30 min	Injection volume: 10 µL		Column Temp.: 5 °C Eluent B: Acetonitrile	R ≥ 2.0
Gradient programming Step	Time [min]	A % [v/v]	B % [v/v]	Flow-rate [µL/min]	
1	0.00	100.0	0.0	250	
2	0.50	100.0	0.0	250	
3	28.00	97.0	3.0	250	
4	28.50	100.0	0.0	250	
5	30.00	100.0	0.0	250	

Method B

Features: Eluent A: 2 mM amonium formate in formic acid 0.1% w/v (pH 5.6)	Run time: 12 min	Injection volume: 10 µL		Column Temp.: 20 °C Eluent B: Methanol	R ≥ 1.2
Gradient programming Step	Time [min]	A % [v/v]	B % [v/v]	Flow-rate [µL/min]	
1	0.00	80.0	20.0	200	
2	1.00	80.0	20.0	200	
3	10.00	65.0	35.0	200	
4	12.00	80.0	20.0	200	

26.2.1.1 Role of Temperature

Temperature control during the chromatographic run by thermostatting the column has become a standard procedure to increase chromatographic performance, both in terms of resolution and stability of retention times. Increase in resolution is commonly observed at higher temperatures, partly due to a decrease in the viscosity of the mobile phase. Contrary to normal behaviour, in our chromatographic approach (with both Methods A and B) the resolution increased with *decreasing* the temperature, and the best results were obtained at 5 °C (lower temperatures were not investigated). Variation of performance as a function of temperature is illustrated in Figure 26.4 for *Method B*. Although the reasons for this reversed temperature dependence are not fully understood, it can be postulated that they are related to decreased solubility of analytes in the mobile phase and their enhanced exothermic interactions with the graphite surface. Enhanced interaction with the stationary phase is visible from the increase of retention time on decreasing the temperature. Aiming at maintaining the highest possible resolution with *Method A*, and the shortest analysis time compatible with sufficient resolution with *Method B*, the two methods were normally run at 5 °C and 20 °C, respectively.

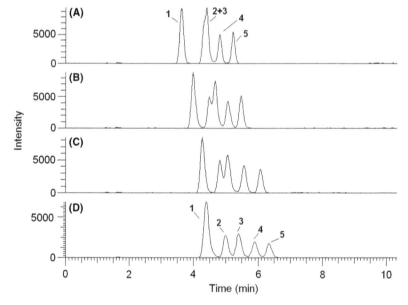

Figure 26.4 Role of temperature on the chromatographic performance of PGC column in the analysis of a mixture of disaccharides. The figure (unpublished) illustrates the dependence of chromatographic resolution on the temperature. Chromatograms were recorded under identical settings using *Method B* and monitoring the $[M+HCOO]^-$ adduct at m/z 387.0 ± 1.5 in MS. Column temperature was: 30 °C (A); 25 °C (B); 22.5 °C (C) and 20 °C (D). Numbers indicate sucrose (**1**), β-lactose (**2**), β-maltose (**3**), α-maltose (**4**), α-lactose (**5**).

26.2.1.2 Loss of Chromatographic Performance

In the course of our ongoing practice with the described LC-MS methods, we experienced sudden and irreversible losses of chromatographic performance, consisting in a significant decrease in resolution associated with sensibly increased backpressure. Loss of performance is attributed to fragmentation of the first layers of graphite particles, resulting in variation of particle size and distribution, and in "clogging" of the column. Indeed, while PGC stationary phases are chemically very stable and can manage an exceptionally broad range of solvent polarity and pH, they are comparably less robust from a mechanical perspective. Therefore, it is recommended to avoid pressure jumps and crystallization of solutes in column. The use of a pre-column appears very important.

26.2.2 Mass Spectrometry

Several methods for the ESI-MS analysis of underivatized oligosaccharides have been described in the literature, comprising both positive and negative ionization modes. Positive-ion modes include formation of protonated ions $[M + H]^+$ (Liu *et al.* 2005; Zhu and Sato 2007), alkali and alkaline earth adducted ions such as Na^+, Li^+, K^+, Cs^+, NH_4^+; Ca^{2+}, Pb^{2+} (Asam *et al.* 1997; El Firdoussi *et al.* 2007; Liu *et al.* 2005; Zhu and Sato 2007; Zaia 2004), while negative modes include formation of deprotonated ions $[M-H]^-$ (Chen *et al.* 2009; Guignard *et al.* 2005), or adducts with small anions such as formate, acetate, or chloride (Chen *et al.* 2009; Verardo *et al.* 2009; Zaia 2004). With both positive and negative ionization modes, nano-electrospray has been shown to yield improved sensitivity, as compared to conventional micro-ESI sources, in the analysis of neutral carbohydrates, due to improved desolvation of the highly hydrophilic analytes (Bahr *et al.* 1997; Zaia 2004). Indeed, in a comparative study on an ion-trap mass spectrometer, nano-ESI outperformed the sensitivity of micro-ESI by orders of magnitude (Bahr *et al.* 1997). Nonetheless, in a recent work very good sensitivity in the LC-MS analysis of underivatized mono-, di-, and oligo-saccharides in food products was obtained in already with conventional micro-ESI in negative ion mode, by monitoring the formate adduct $[M + HCOO]^-$ obtained by post column addition of formic acid/ammonium formate (Varardo *et al.* 2009). Clearly such an approach was ideally suited to our chromatographic methods, in which formic acid (or formate buffer) was already contained in the mobile phase, allowing us to avoid post-column additions.

26.2.2.1 Mass Spectra of Lactose and Maltose Formate Adducts $[M + HCOO]^-$

Under the experimental setting summarized in Table 26.2 the $[M + HCOO]^-$ ion at m/z 387 was the most abundant in full MS spectra of lactose or maltose.

It also underwent facile fragmentation allowing investigating the MS^n spectra of the analytes. Fragmentation of formate adduct occurred mainly by loss of HCOOH, leading to the deprotonated molecule [M-H]$^-$ at m/z 341 (MS^2), accompanied by minor amounts of ions at *m/z* 179 (deprotonated monosaccharide) and *m/z* at 161, due to neutral loss of a monosaccharide molecule. Extraction and subsequent fragmentation of *m/z* 341 (MS^3) resulted mainly in product ions at *m/z* 179 and 161 for any disaccharide, although their relative abundance and/or the collision energy required for fragmentation differed slightly for lactose and maltose (Figures 26.5 and 26.6), and largely between reducing and non-reducing disaccharides, as can be inferred from comparison with sucrose under identical experimental settings (Figure 26.7).

March and Stadey (2005) using ESI-Q-TOF high-resolution spectrometry showed that the relative abundance of fragment ions in product ion mass spectra of [M-H]$^-$ from disaccharides depends both on stereochemistry and on collision energy. In line with this study, we subjected to statistical analysis normalized (relative) abundance of major fragment ions in MS^2 and MS^3 spectra, obtained in a Paul trap, from CID of formate adducts [M + HCOO]$^-$, recorded both during LC/MS runs and by direct infusion of standard solutions. At collision energy of 17 a.u. the abundance of precursor ion [M + HCOO]$^-$ at *m/z* 387 was $4.2 \pm 0.5\%$ (Mean $\pm$ SD) in MS^2 spectrum of lactose, which was significantly higher ($p < 0.0001$, two-tailed) than the value $1.1 \pm 0.1\%$ recorded for maltose (Figure 26.5), and significantly lower ($p < 0.0001$, two-tailed) than the value of $18.9 \pm 0.8\%$ for sucrose used for comparison (Figure 26.7). This indicates major differences in adduct ions gas-phase stability. In MS^2 spectra, base peak (100%) was [M-H]$^-$ at *m/z* 341 for all three disaccharides: while almost no fragment ion was visible for sucrose (ions at *m/z* 179 and 161 were $< 0.1\%$), lactose had distinctive fragments at *m/z* 179, corresponding to the monosaccharide anion, and *m/z* 161 (due to loss of one monosaccharide), with respective intensities of $1.6 \pm 0.3\%$ and $9.4 \pm 0.9\%$, both significantly higher ($p < 0.0001$, two-tailed) than corresponding signals for maltose of $0.4 \pm 0.2\%$ and $2.1 \pm 0.4\%$ respectively. MS^3 spectra of both lactose and maltose obtained by CID of [M-H]$^-$ at *m/z* 341 at collision energy 14 a.u. had base peak at *m/z* 161, however, they significantly ($p < 0.0001$, two-tailed) differed for the abundance of precursor *m/z* 341 ($11.5 \pm 0.6\%$ for lactose *vs.* $24.6 \pm 1.9\%$ for maltose) and fragment ion at *m/z* 179 ($11.4 \pm 2.0\%$ for lactose *vs.* $27.74 \pm 2.2\%$ for maltose). Highly diagnostic ($p < 0.0001$) was also fragment ion at *m/z* 281 due to neutral loss of 2-hydroxyacetaldehyde (60 Da), whose abundance changed from $4.1 \pm 0.6\%$ in lactose to $10.1 \pm 0.8\%$ in maltose. At higher or lower collision energies, the relative abundance of fragment ions in MS^3 spectra was visibly different; however, significative differences between lactose and maltose were conserved. MS^3 spectrum of sucrose (Figure 26.7) largely differed from the reducing disaccharides, fragment ion at *m/z* 179 being the base peak while, at collision energy of 22 a.u., *m/z* 161 was only $19.7 \pm 1.1\%$. Fragment ion at *m/z* 221, due to one intact non-reducing sugar glycosidically linked to a C2 aglycone (2-hydroxyacetaldehyde), was $7.4 \pm 1.2\%$, while ions at *m/z* 119 and 113, derived respectively from *m/z*

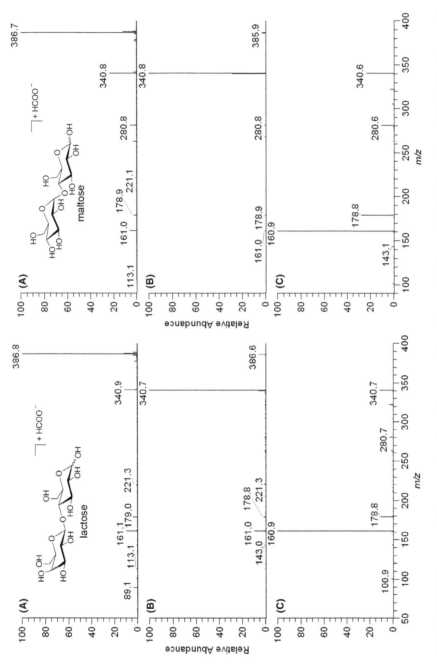

Figure 26.5 ESI-MS[n] spectra of lactose and maltose formate adducts in negative ion mode. The figure (unpublished) depicts negative ESI-MS[n] spectra of lactose (left) and maltose (right) by direct infusion of a solution (1 mg/L) in formate buffer. (A) full scan (50–400 Da) MS; (B) MS[2] $387 \rightarrow$ obtained at collision energy 14 a.u.; (C) MS[3] $387 \rightarrow 341 \rightarrow$ obtained at collision energy 17 a.u.

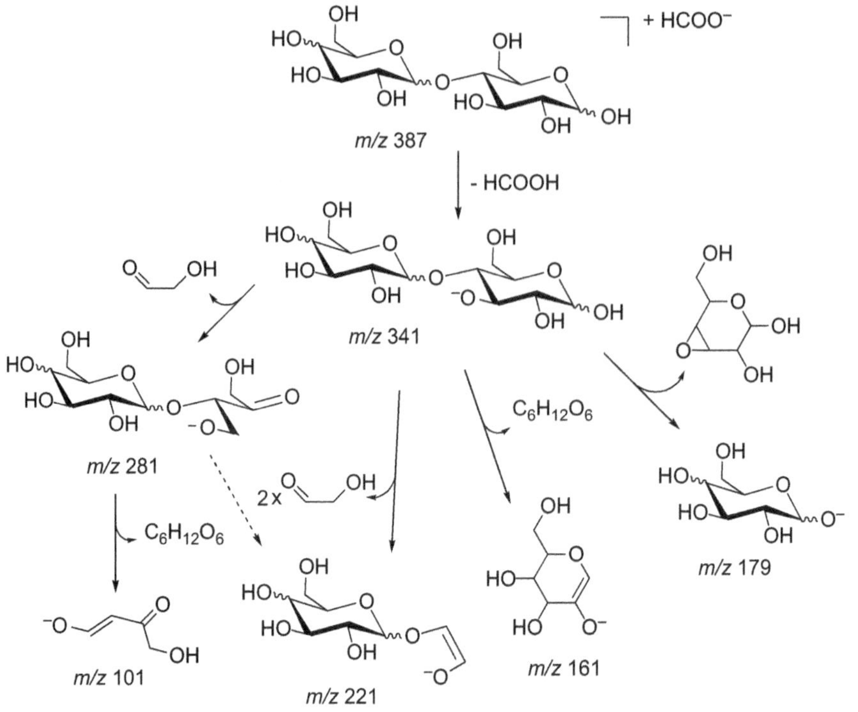

Figure 26.6 Fragmentation pattern of lactose and maltose formate adducts in negative ion mode. The figure (unpublished) illustrates the main fragmentations observed in ESI-MS^n spectra of lactose and maltose formate adducts, in negative ion mode.

179 and 161 by neutral loss of 2-hydroxyacetaldehyde (60 Da) and formaldehyde + water (48 Da), had abundance of $15.0 \pm 1.1\%$ and $16.3 \pm 2.0\%$ respectively. Main fragmentations of the reducing disaccharides are summarized in Figure 26.6. Interestingly, no significant difference in the MS^2 and MS^3 spectra could be found, at any collision energy, between α and β anomers of reducing disaccharides (*i.e.* maltose α *vs.* β or lactose α *vs.* β).

26.2.2.2 Coupling U-HPLC with ESI-MS

Best analytical performance in our experience was obtained by direct infusion of column eluate (at 250 µL/min) into the ESI source, without splitting or post-column reagent additions.

We monitored and extracted for instrument calibration both adduct [M + HCOO]$^-$ at *m/z* 387 in full MS and the [M-H]$^-$ ion at *m/z* 341 (MS^2), while MS^3 was used only to confirm the identity of the analytes. Typical chromatograms obtained with a standard disaccharides aqueous mixture are shown in Figure 26.2 (*Method A*) or Figure 26.3 (*Method B*).

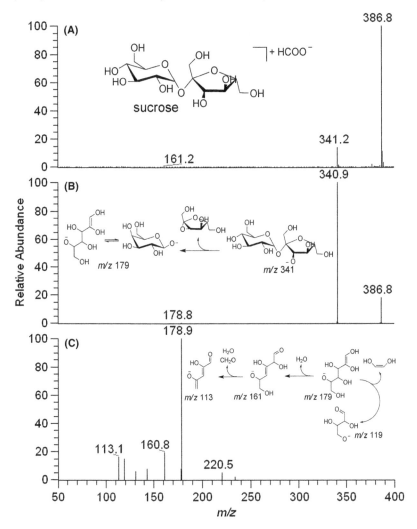

Figure 26.7 ESI-MS^n spectra and fragmentation pattern of sucrose formate adduct in negative ion mode. The figure (unpublished) depicts negative ESI-MS^n spectra of sucrose by direct infusion of a solution (1 mg/L) in formate buffer. (A) full scan (50–400 Da) MS; (B) MS^2 387→ obtained at collision energy 14 a.u.; (C) MS^3 387→341→ obtained at collision energy 17 a.u. Main fragmentations are interpreted.

With both ions at *m/z* 387 and 341, a six levels calibration resulted in good linearity ($r^2 > 0.99$), with limits of detection (LOD) for lactose of 3.2 and 4.5 µg/L (ppb) respectively, corresponding respectively to 32 and 45 pg of injected lactose, while LOD values for maltose were 4.8 and 6.6 µg/L (at *m/z* 387 and 341) respectively. A summary of the optimized ESI-MS operating conditions is given in Table 26.2.

Table 26.2   ESI-MS settings and performance for the analysis of lactose and maltose. The table summarizes the instrumental settings and the expected sensitivity for ESI-MS analysis of lactose and maltose, eluted from a PGC column as described in Table 26.1. Examples of chromatograms recorded by monitoring events 1, 2 or 3 are displayed in Figure 26.2. LOD (limit of detection) and LOQ (limit of quantitation) are evaluated by repeated injections at the lowest analyte concentration giving a Signal/Noise ratio of 3 or 10, respectively. CID: collision induced dissociation.

System: Mass spectrometer LCQ Fleet® (Thermo Scientific) with ESI interface. Mass analyzer: Paul-type (tridimensional) ion trap (IT).

Electrospray ionization (ESI) settings

Parameter	Instrumental setting	Unit	Value (measured)	Unit
Sheat Gas Flow Rate	40	Arbitrary (a.u.)	8.0	L/min
Auxiliary Gas Flow Rate	5	Arbitrary (a.u.)	1.0	L/min
Sweep Gas Flow Rate	5	Arbitrary (a.u.)	1.0	L/min
Transfer Line Capillary Temperature	250	°C	-	-
Ion Spray Voltage	3.00	kV	-	-

Mass spectrometry settings and performance

Collision gas: He (1 mTorr)	Normalized collision energy: 12–22 (a.u.)	Activation Q: 0.25	Activation time: 30 ms	Isolation width: 3.0 m/z (±1.5)

Event 1: adduct monitoring [M+HCOO]⁻ in full MS negative mode

Disaccharide	MS^n: Ions used for analysis [m/z]	Linear range [$\mu g/L$]	LOD [$\mu g/L$]	LOQ [$\mu g/L$]
Lactose	MS: 387	11–900	3.2	10.7
Maltose	MS: 387	16–900	4.8	16.0

Event 2: monitoring [M-H]⁻ in MS/MS of ion 387 $m/z \rightarrow$ @ CID 14 a.u.

	MS^n: Ions used for analysis [m/z]	Linear range [mg/L]	LOD [mg/L]	LOQ [$\mu g/L$]
Lactose	MS²: 341	15–900	4.5	15.0
Maltose	MS²: 341	22–900	6.6	21.8

Event 3: monitoring of fragment ions in MS/MS/MS of ion 387 $m/z \rightarrow$ 341 $m/z \rightarrow$ @ CID 17 a.u.

	MS^n: Ions used for analysis [m/z]	Linear range [mg/L]	LOD [mg/L]	LOQ [$\mu g/L$]
Lactose	MS³: 179; 161		Only for confirmation, not for quantitative analysis	
Maltose	MS³: 179; 161		Only for confirmation, not for quantitative analysis	

26.2.3 Method Validation

Analysis of 3-level standard solutions of disaccharides' mixture during three days (n = 15), resulted in good accuracy, with % recovery between 95.6% and 103.1%, and good precision, with coefficients of variation (CVs) in the range 1.9%–3.4% intra-day and 2.7–4.2% inter-day.

In order to evaluate potential interference from the very complex vegetable matrices that are the actual target of application for our LC-MS approach, we validated our method by preparing 3 reconstructed green tea dry extracts, each containing a known mixture of the three disaccharides, in different ratios and 3 concentration levels (Gabbanini *et al.* 2010). Analysis of the extracts resulted in recoveries ranging from 92.1% to 96.5% with CVs (intra-day) from 2.6% to 4.8%, indicating the good accuracy, precision and selectivity of our analytical approach.

26.3 Applications

26.3.1 Analysis of Nutraceutical Preparations

Our experience with the described LC-MS method is mostly focused on the analysis of lactose and maltose (as well as sucrose) in plant dry extracts used for the manufacture of diet supplements, and on the analysis of commercial nutraceutical formulations, where the method showed to be particularly practical and reliable (see Figure 26.8). Indeed it always showed to be suffi-ciently specific to avoid interference from the complex matrix, with only minimal sample preparation. For specimens in the form of a powder or a capsule (after capsule removal), sample preparation consisted simply in dis-solving or suspending the powder in aqueous formate buffer (2 mM ammonium formate dissolved in 0.1% w/v formic acid solution; pH 5.6), sonicate for 1 min, stir for 2 min, then filter with 0.45-mm PTFE syringe filters, followed by dilu-tion to the desired level using the same solvent composition. Solid specimens

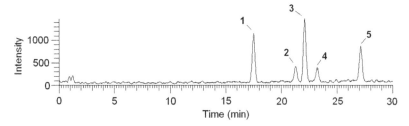

Figure 26.8 Analysis of a green tea dietary extract containing disaccharides. The figure (unpublished) depicts the analysis with *Method A* of a green tea dry extract, containing 5.0% w/w sucrose, 2.0% w/w maltose and 10.0% w/w lactose. Chromatogram was generated monitoring the extracted [M-H]⁻ ion at *m/z* 341.0 ± 1.5 in MS² 387→. Numbers indicate sucrose (**1**), β-maltose (**2**), β-lactose (**3**), α-maltose (**4**), α-lactose (**5**).

(*e.g.* compresses) simply required fine mechanical triturating as preliminary step. With aqueous or hydro-alcoholic solutions (*e.g.* tinctures or syrups), simple dilution in the formate buffer (and filtration) was effective.

26.3.2 Analysis of Food Products

With food products the approach for sample preparation may vary considerably, due to the large variety in the form, composition and texture of food products. While simple dilution/filtration, previously described for syrups, is a convenient method for fruit juices and other drinks, other food products might require preliminary steps, such as triturating and extracting in a stomacher processor. Several valuable reviews on sample preparation in food analysis are available in the literature (Buldini *et al.* 2002).

26.4 Conclusions

The described U-HPLC-MS/MS approach for the analysis of underivatized lactose and maltose is practical and high-performing, favourably comparing with the most valuable approaches available in the literature. Ultra-high performance liquid chromatography on (general purpose) porous graphitic carbon columns allows excellent resolution of disaccharide mixtures and of the anomeric forms of reducing disaccharides. At the same time, it is sufficiently robust to allow gradient elution with formic acids mixtures, with no need for post-column regent additions. Negative ESI with double monitoring of $[M + HCOO]^-$ formate adducts (in MS) and deprotonated disaccharides $[M-H]^-$ (in MS^2) yields good linearity, precision and accuracy, with limits of detection in the low pg range already in a Paul-ion-trap mass spectrometer. The described method is particularly suited for food analysis, as well as the selective quantitation of disaccharides' anomeric forms, such as in studies of mutarotation, where it can replace more conventional approaches like optical rotation or NMR spectroscopy (Gabbanini *et al.* 2010).

Summary Points

- This chapter focuses on the analysis of lactose and maltose in food products.
- Analysis of disaccharides in mixture is complicated by stereoisomerism and absence of UV absorbance.
- LC-ESI-MS analysis of underivatized saccharides in mixture is a valuable approach only if sufficient chromatographic resolution is achieved.
- U-HPLC-ESI-MS/MS using Porous Graphitized Carbon columns provides a rapid, specific and sensitive approach.
- A simple and practical method is discussed in detail and examples of applications to the analysis of food products are provided.
- The method allows discrimination of α and β anomers of lactose and maltose and can be used also for mutarotation analysis.

Key Facts of HPLC and U-HPLC

- High pressure (or high performance) liquid chromatography, HPLC is a reference separation technique used both in analytical or preparative scale.
- In analytical scale liquid chromatography can be associated with electrospray ionization (ESI) mass spectrometry (MS) to obtain LC-ESI-MS (or HPLC-ESI-MS) systems.
- In normal analytical operations HPLC uses columns 100 to 300 mm long, with 3.2 to 4.6 mm diameter, packed with silica particles or silica-coated polymers with a particle size of 3.5–10 μm.
- Silica columns (stationary phase) are eluted with mixtures of organic solvents (mobile phase) being less polar that silica itself in the so-called, direct-phase chromatography.
- In reverse-phase chromatography, silica particles are derivatized with relatively apolar organic compounds (*e.g.* C18 aliphatic chains in a, so called, C18 column), and the column is eluted with aqueous mixtures, more polar than the stationary phase.
- Lower particle size increases both the chromatographic resolution and the column backpressure.
- Typical backpressure in a conventional HPLC system, operating in reverse-phase with a 100–300 mm × 4.6 mm × 5 μm column, ranges from 50 to 250 bar with a flow-rate of 0.5–1.0 mL/min (maximum pressure 400 bar).
- Ultra-high pressure liquid chromatography (U-HPLC or UPLC) was first introduced in 2003 and became commercially available in 2004.
- U-HPLC systems can reach backpressures of 1000 bar and work routinely at 300–600 bar; hence they can use columns of lower particle size.
- In U-HPLC columns typical particle size is 1.7–2.1 μm and the resolution is much higher, requiring shorter columns to achieve separation.
- Typical U-HPLC columns are 50–150 mm long, with 2–2.5 mm diameter. Due the to lower volume they require much less solvent for elution. Normal flow rate is 200–400 μL/min.
- Due to higher resolution U-HPLC allows shorter analysis time, typically 5–20 min compared to typical 20–60 min for HPLC.

Definitions of Words and Terms

PGC. PGC is the abbreviation of Porous Graphitized Carbon, a material used for separation and adsorption of organic material, such as for packing of chromatographic columns, for analytical or preparative purposes. It is manufactured by impregnating chromatography silica particles with a phenol-formaldehyde resin, then carbonizing at 1000 °C in inert atmosphere and dissolving the silica in alkali. Finally, it is heated to above 2000 °C to graphitize the carbon.

Nutraceutical. Term coming from the crasis of the words *nutrition* and *pharmaceutical,* which indicates a special type of food product having health-oriented scope, beside basic nutrition, *i.e.* aiming at improving or maintaining the health of the consumer. It can be in the form of a diet supplement (*e.g.* a pill) or a fortified food, *i.e.* a food product deliberately enriched with vitamins, minerals or other nutrients with health functions.

Electrospray. A ionization technique consisting in the fine nebulisation of a solution in a highly dielectric solvent (*e.g.* water or methanol) through a heated, electrically conducting capillary, held at a few kVolt potential more negative or more positive with respect of a surrounding chamber, where the solvent from the nebulised solution is rapidly evaporated under an inert gas flow (nitrogen).

kVolt. Kilo-Volt corresponding to 1000 Volt, the international unit for electrical potential or electromotive force.

Carbohydrate. An organic molecule with a general formula $C_n(H_2O)_n$, *i.e.* erroneously resembling an hydrated form or carbon. Carbohydrates or saccharides are distinguished in monosaccharides (*e.g.* glucose), disaccharides ($=2$ saccharides condensed by loss of a water molecule), oligosaccharides (3, 4 or a small number of condensed monosaccharides), or polysaccharides ($=$ many condensed monosaccharide units).

Disaccharide. See carbohydrate.

Anomer. The term is used only in carbohydrate chemistry to indicate epimers at the hemiacetalic carbon, formed in carbohydrates' cyclic structure, *i.e.* C1 in aldose or C2 in chetose compounds. In aldohexose like glucose, anomers are indicated as α (alpha) or β (beta) according to the position, down or up respectively in Haworth projections, of the –OH group at C1 with respect of main molecular plane.

Epimer. A stereoisomer that differ solely for the stereochemistry at one carbon atom. If there is only one chiral centre in the molecule, epimers are enantiomers, *i.e.* non-superimposable mirror images.

Mutarotation. The conversion of one carbohydrate anomer in the other until an equilibrium mixture is reached, having composition specific for any carbohydrate, and a specific optical rotation of plane-polarized light, different form that of any of the two anomers. Mutarotation literally means change in rotation.

Plane-polarized light. An electromagnetic radiation having the electrical vector component (or the orthogonal magnetic vector) that oscillates only in one plane, orthogonal to the direction of propagation.

List of Abbreviations

U-HPLC	ultra high pressure liquid chromatography
LC	liquid chromatography
ESI	electro-spray ionization
MS	mass spectrometry
CID	collision induced dissociation
m/z	mass/charge ratio in gas-phase ions
HPAE	high performance anion exchange chromatography
a.u.	arbitrary units

Da	Dalton = atomic mass unit, corresponding to 1/12 of ^{12}C atomic weight or to $1,660565 \times 10^{-27}$ kg in I.S. units
LOD	limit of detection
LOQ	limit of quantitation
CV	coefficient of variability = % relative standard deviation.

References

Asam, M.R., and Glish, G.L., 1997. Tandem Mass Spectrometry of Alkali Cationized Polysaccharides in a Quadrupole Ion Trap. *Journal of the American Society for Mass Spectrometry.* 8: 987–995.

Bahr, U., Pfenninger, A., Karas, M., and Stahl, B., 1997. High-sensitivity analysis of neutral underivatized oligosaccharides by nanoelectrospray mass spectrometry. *Analytical Chemistry.* 69:4530–4535.

Buldini, P.L., Ricci, L., and Sharma, J.L., 2002. Recent applications of sample preparation techniques in food analysis. *Journal of Chromatography A.* 975: 47–70.

Buller, H.A., and Grand, R.J., 1990. Lactose intolerance. *Annual Review of Medicine.* 41: 141–148.

Chen, Z., Jin, X., Wang, Q., Lin, Y., and Gan, L., 2009. Confirmation and Determination of Sugars in Soft Drink Products by IEC with ESI-MS. *Chromatographia.* 69: 761–764.

El Firdoussi, A., Lafitte, M., Tortajada, J., Kone, O., and Salpin, J.-Y., 2007. Characterization of the glycosidic linkage of underivatized disaccharides by interaction with Pb2+ ions. *Journal of Mass Spectrometry.* 42: 999–1011.

Fang, T.T., and Bendiak, B., 2007. The Stereochemical Dependence of Unimolecular Dissociation of Monosaccharide-Glycolaldehyde Anions in the Gas Phase: A Basis for Assignment of the Stereochemistry and Anomeric Configuration of Monosaccharides in Oligosaccharides by Mass Spectrometry via a Key Discriminatory Product Ion of Disaccharide Fragmentation, m/z 221. *Journal of the American Chemical Society.* 129: 9721–9736.

Gabbanini, S., Lucchi, E., Guidugli, F., Matera, R., and Valgimigli, L., 2010. Anomeric discrimination and rapid analysis of underivatized lactose, maltose, and sucrose in vegetable matrices by U-HPLC–ESI-MS/MS using porous graphitic carbon. *Journal of Mass Spectrometry.* 45: 1012–1018.

Grand, R.J., and Montgomery, R.K., 2008. Lactose Malabsorption. *Current Treatment Options in Gastroenterology.* 11: 19–25.

Guignard, C., Jouve, L., Bogéat-Triboulot, M.B., Dreyer, E., Hausman, J.-F., and Hoffmann, L., 2005. Analysis of carbohydrates in plants by high-performance anion-exchange chromatography coupled with electrospray mass spectrometry. *Journal of Chromatography A.* 1085: 137–142.

Hofmeister, G.E., Zhou, Z., and Leary, J.A., 1991. Linkage position determination in lithium-cationized disaccharides: tandem mass spectrometry and semiempirical calculations. *Journal of the American Chemical Society.* 113: 5964–5970.

Ikegami, T., Horie, K., Saad, N., Hosoya, K., Fiehn, O., and Tanaka, N., 2008. Highly efficient analysis of underivatized carbohydratesusing monolithic-silica-based capillary hydrophilic interaction (HILIC) HPLC. *Analytical and Bioanalytical Chemistry.* 391: 2533–2542.

Itoh, S., Kawasaki, N., Ohta, M., Hyuga, M., Hyuga, S., and Hayakawa, T., 2002. Simultaneous microanalysis of *N*-linked oligosaccharides in a glyco-protein using microbore graphitized carbon column liquid chromatography–mass spectrometry. *Journal of Chromatography A.* 968: 89–100.

Koimur, M., Lu, B., and Westerlund. D., 1996. High performance liquid chromatography of disaccharides on a porous graphitic carbon column applying post-column derivatization with benzamidine. *Chromatographia.* 43: 256–260.

Li, C.-H., Low, P.M.N., Li, S., Lee, H.K. and Hor, T.S.A., 1997. Porous Graphitic Carbon (PGC) High-Performance Liquid Chromatography of Diphosphine-Bridged Complexes with Heteronuclear Au-M (M = Mn, Re) Bonds. *Chromatographia.* 44: 381–385.

Liu, Y., Urgaonkar, S., Verkade, J.G., Daniel W. and Armstrong, D.W., 2005. Separation and characterization of underivatized oligosaccharides using liquid chromatography and liquid chromatography–electrospray ionization mass spectrometry. *Journal of Chromatography A.* 1079: 146–152.

March, R.E., and Stadey. C.J., 2005. A tandem mass spectrometric study of saccharides at high mass resolution. *Rapid Communications in Mass Spectrometry.* 19: 805–812.

Michel, M., and Buszewski, B., 2009. Porous graphitic carbon sorbents in biomedical and environmental applications. *Adsorption.* 15: 193–202.

Muir, J.G., Rose, R., Rosella, O., Liels, K., Barrett, J., Shepherd, S.J., and Gibson, P., 2009. Measurement of Short-Chain Carbohydrates in Common Australian Vegetables and Fruits by High-Performance Liquid Chromatography (HPLC). *Journal of Agricultural and Food Chemistry.* 57: 554–565.

Nikolov, Ž.L., Jakovljević, J.B., and Boškov, Ž.M., 1984. High Performance Liquid Chromatographic Separation of Oligosaccharides Using Amine Modified Silica Columns. *Starch.* 36: 97–100.

Shumacher, D., and Kroh, L.W., 1995. A rapid method for separation of anomeric saccharides using a cyclodextrin bonded phase and for investiga-tion of mutarotation. *Food Chemistry.* 54: 353–356.

Verardo, C., Duse, I., and Callea, A., 2009. Analysis of underivatized oligo-saccharides by liquid chromatography/electrospray ionization tandem mass spectrometry with post-column addition of formic acid. *Rapid Communica-tion in Mass Spectrometry.* 23: 1607–1618.

Wuhrer, M., Koeleman, C.A.M., Deelder, A.M., and Hokke, C.H., 2004. Normal-Phase Nanoscale Liquid Chromatography – Mass Spectrometry of Underivatized Oligosaccharides at Low-Femtomole Sensitivity. *Analitical Chemistry.* 76: 833–838.

Xingyu Zhu X., and Sato, T., 2007. The distinction of underivatized mono-saccharides using electrospray ionization ion trap mass spectrometry Rapid Communications in Mass Spectrometry 221: 191–198.

Zaia, J., 2004. Mass spectrometry of oligosaccharides. *Mass Spectrometry Reviews*. 23: 161–227.

Zhang, H., Brokman, S.M., Fang, N., Pohl, N.L., and Yeung. E.S., 2008. Linkage position and residue identification of disaccharides by tandem mass spectrometry and linear discriminant analysis. *Rapid Communications in Mass Spectrometry*. 22: 1579–1586.

CHAPTER 27

Assays of Fructose in Experimental Nutrition

KEIICHIRO SUGIMOTO,*[a] HIROSHI INUI[b] AND
TOSHIKAZU YAMANOUCHI[c]

[a] Research and Development Center, Nagaoka Perfumery Co. Ltd., 1-3-30
Itsukaichi, Ibaraki, Osaka 567-0005, Japan; [b] Department of Applied
Biological Chemistry, Graduate School of Life and Environmental Sciences,
Osaka Prefecture University, 1-1 Gakuen-cho, Naka-ku, Sakai, Osaka
599-8531, Japan, Email: inui@biochem.osakafu-u.ac.jp; [c] Department of
Internal Medicine, Teikyo University School of Medicine, 2-11-1 Kaga,
Itabashi-ku, Tokyo 173-8605, Japan
*Email: ksugimoto@npc-nagaoka.co.jp

27.1 Introduction

This chapter describes the methods for determination of fructose level in body
fluids. Blood glucose concentrations are routinely measured and examined as
key factors in diagnosis of diabetes. In contrast, fructose concentration in
blood has been barely studied as it is a minor component. Thus, the kinetics
and physiological implications of fructose level in the body are poorly understood. Fructose is a major component of nutrition, and excessive fructose
ingestion is a serious human health problem. Studies of the behavior of fructose
in the body will become important in the field of clinical nutrition.

In humans, fructose is derived from exogenous dietary intake in
significant quantities. However, postprandial fructose concentration in systemic vessels is low because most dietary fructose is rapidly metabolized in the

Food and Nutritional Components in Focus No. 3
Dietary Sugars: Chemistry, Analysis, Function and Effects
Edited by Victor R Preedy
© The Royal Society of Chemistry 2012
Published by the Royal Society of Chemistry, www.rsc.org

Table 27.1 Determination methods for fructose concentration in body fluids.

Category	Sensitivity	Operability	Expenditure
Chemical methods	low	simple	low (only reagents)
Enzymatic methods	middle	simple	low (only reagents)
HPLC methods	high	complicated	needs expensive apparatus
GC/MS methods	very high	very complicated	needs expensive apparatus

liver by ketohexokinase, which metabolic pathway bypasses the main regulation at 6-phosphofructokinase in glycolysis. In the fasting state, peripheral fructose concentration (10–100 μM, measured by GC/MS) is markedly lower than the glucose concentration (5–8 mM). On the other hand, fructose is endogenously synthesized in minute amounts by the polyol pathway and is detectable in other body fluids, such as sperm, urine, and cerebrospinal fluid.

Since fructose levels are generally very low in body fluids, it is necessary to examine fructose concentration using highly sensitive and selective measurement methods. However, selective detection of fructose is very difficult because coexisting substances in the specimen mostly interfere with the sensitivity and accuracy of fructose assays. Reducing substances also disrupt the assays that leverage the reducing character of fructose. In particular, in the case of blood, glucose markedly disturbs the fructose assays. Therefore, there are no highly sensitive methods for microdetermination of fructose in body fluids that can be performed easily at low cost. Various methods have been applied for fructose determination. Simple, quick, and inexpensive methods have been developed, including chemical methods such as colorimetry by reaction of fructose with chemical reagents and enzymatic methods using fructose-specific enzymes, but the thresholds of these methods are high. HPLC methods are more sensitive than chemical and enzymatic methods and can measure many polyols simultaneously, but they require HPLC apparatus. GC/MS methods are the most sensitive and accurate but they require expensive and sophisticated instrumentation. Table 27.1 shows representative microdetermination methods for fructose in body fluids.

27.2 Physiological Significance

Fructose levels in blood and several tissues have been examined for the purposes outlined below. In addition, changes in fructose concentration after fructose-loading may be helpful to determine individual variability of susceptibility to potential adverse effects of fructose.

- To examine digestion, absorption, and metabolism of dietary fructose and sucrose, fructose levels in blood and various tissues have been measured (Crossley and Macdonald 1970).
- To investigate the effects of fructose consumption on endocrine function and metabolism, postprandial blood fructose concentrations have been

measured (Adams *et al.* 2008; Yamamoto *et al.* 1999). There have been many studies in which the effects of fructose were compared with glucose.

- To investigate the correlations of diabetic complications, blood fructose level has been measured in the fasting state or after fructose consumption; myocardial infarction (Gul *et al.* 2009), pancreatic cancer (Hui *et al.* 2009), retinopathy (Kawasaki *et al.* 2004).
- To study the correlations of diabetes complications including the polyol pathway, the concentrations of related polyols including fructose have been measured in plasma and heart (Noh *et al.* 2009), urine (Yoshii *et al.* 2001), lens (Lal *et al.* 1995), and sciatic nerve (Hotta *et al.* 1985).
- The inhibitory effects of drugs or natural products on intestinal fructose absorption have been examined by measurement of postprandial hyperfructosemia for the development of anti-obesity and/or anti-diabetic agents (Sugimoto *et al.* 2005; 2010).
- As a marker of soft drink ketosis, which is caused by frequent long-term consumption of high-fructose-containing beverages, blood fructose level has been measured in subjects with atypical diabetes (Kawasaki *et al.* 2010).
- To examine the effects of fructose infusion on the body during parenteral nutrition, the changes in blood fructose level have been measured (Adams *et al.* 2008; Shiota *et al.* 2005; Yamamoto *et al.* 1999).
- To examine seminal vesicle activity, fructose concentrations in seminal plasma have been measured (Anderson *et al.* 1979; Al-Daghistani *et al.* 2010). Concentrations of polyols in the conceptus have been examined (Jauniaux *et al.* 2005; Regnault *et al.* 2010).
- The amount of cellular fructose released by oxidized low-density lipoproteins was measured (Takeda *et al.* 1996).

27.3 Sample Treatment

When fluid is preserved inadequately, the sugar content gradually decreases due to the action of bacteria. Especially, minute amounts of sugars in blood disappear overnight despite keeping at 4 °C. Fluid samples should be kept at least under –20 °C or a preservative should be added. When urine is stored at 4 °C, 5 g/person of sodium benzoate is added to the specimen. Some assays need deproteinization and/or desalting. To remove protein, centrifuge sample solution at 10,000–16,000 *g* for 10–20 min and collect the supernatant after adding ethanol, acetonitrile, acetone, trifluoroacetic acid, an equivalent mixture of 0.3 N $Ba(OH)_2$ and 0.3 N $ZnSO_4$, etc. For desalting, solid-phase extraction cartridges packed with ion-exchange resin are available.

Most assays can be treated with heparin for blood specimens. Hematocytes in blood samples should be quickly separated by centrifugation just after collecting because cythemolysis seriously interferes with the sensitivity of fructose assays. Treatment of sperm specimens was described in the World Health Organization manual for semen analysis (Al-Daghistani *et al.* 2010).

27.4 Determination of Fructose Concentration

27.4.1 Chemical Method

Chemical methods include colorimetry by color-forming reaction of fructose with chemical reagents and can be carried out at low cost and with simple procedures. These methods can be applied to samples in which the fructose concentration is high (over 100 μM). However, coexisting reducing substances, especially glucose, complicate the interpretation of fructose analysis because many assays utilize the reductive activity of fructose. Enzymatic methods are recommended rather than these methods. The following methods have been developed.

- Resorcinol method (Al-Daghistani *et al.* 2010; Crossley and Macdonald 1970; Roe *et al.* 1934)
  This is the most commonly used method for clinical and/or research applications among the chemical methods. Truswell *et al.* (1988) developed Roe's method by previous removal of glucose using glucose oxidase.
- Anthrone method (Nixon 1969).
- Cysteine hydrochloride-tryptophan method (Messineo and Musarra 1972).
- Skatole-hydrochloric acid method (Carvalho and Pogell 1957; Gul *et al.* 2009).
- Thiobarbituric acid method (Zender and Falbriard 1966).

27.4.2 Enzymatic Method

27.4.2.1 Background

As enzymatic methods selectively change fructose into detectable substance(s) using a specific enzyme for fructose, they detect fructose with greater sensitivity than the chemical methods, and can also be performed without sophisticated and expensive instrumentation. These methods are applicable to samples containing high concentrations of fructose (over 100 μM) such as testis fluid, plasma or serum after fructose-loading, and culture solution containing fructose. However, their accuracy is lower than HPLC and GC/MS methods because complete exclusion of coexisting interfering substances is difficult. As these methods can measure fructose only or few polyols, it is impossible to measure many polyols concurrently.

Methods generally used for fructose determination are glucose assays using PGI which converts F6P to G6P after phosphorylation of D-fructose by hexokinase as shown in the following equations. The formation of NADPH is measured spectrophotometrically through oxidation of G6P to 6-phosphogluconate by G6PDH. The fructose concentration is calculated as the remainder of subtraction of glucose content before adding PGI (Adams *et al.* 2008; Hotta *et al.* 1985; Takeda *et al.* 1996; Yamamoto *et al.* 1999). This method is very convenient. However, for test samples consisting of blood, a large amount of coexisting glucose negatively affects the quantitative performance of fructose. Many assay kits are commercially available such as F-kit Glucose/Fructose (Roche Diagnostics, Basel, Switzerland), FA-20 (Sigma-Aldrich, St. Louis, MO).

$$\text{Fructose} \xrightarrow[\text{hexokinase}]{} \text{F6P} \xrightarrow[\text{PGI}]{} \text{G6P} + \text{NADP}^+ \text{ (as an indicator)} \xrightarrow[\text{G6PDH}]{}$$

$$\text{6-phosphogluconate} + \text{NADPH} \, (\lambda_{\text{max}} \, 340 \, \text{nm})$$

There are enzymatic assays in which the enzymatic reaction is directly proportional to the fructose concentration, *i.e.*, the assays using sorbitol dehydrogenase (Anderson *et al.* 1979) and FDH (Hui *et al.* 2009; Noh *et al.* 2009; Sugimoto *et al.* 2005) as key enzymes. This section describes the method using FDH (Sugimoto *et al.* 2005).

27.4.2.2 *Principle*

FDH specifically catalyses oxidation of D-fructose to 5-keto-D-fructose as shown in the following equations. The concomitant reduction of WST-1 in the presence of 1-methoxy PMS as an electron carrier provides the water-soluble pigment formazan with which the progression of the reaction can be monitored by absorbance spectroscopy.

$$\text{D-fructose} + \text{WST-1(as an electron accepter)} \xrightarrow[\text{1-methoxy PMS (as an electron carrier)}]{\text{FDH}}$$

$$\text{5-keto-D-fructose} + \text{formazan}(\lambda_{\text{max}} \, 438 \, \text{nm})$$

The assay mixture consists of 1% Triton X-100, 0.2 mM WST-1, 8 µM 1-methoxy PMS, 5 U/ml fructose dehydrogenase, and a test sample in 0.1 M KPB (pH 6.0).

27.4.2.3 *Sample Solution*

Dilute the test sample with water to 702 µl. For example, 10 µl of plasma from the portal vein is diluted with 692 µl of water.

To make 0, 10, 20, 30, and 50 nmol/tube fructose solutions for calibration curve preparation, dilute the corresponding amounts (0, 10, 20, 30, and 50 µl) of 1 mM fructose with water to 702 µl (always use freshly prepared solution).

27.4.2.4 *Reagents*

- 100 U/ml FDH solution
  Dissolve FDH powder (from *Gluconobacter* sp.; EC 1.1.99.11) in McIlvaine buffer (pH 4.5) prepared by mixing of 0.1 M citric acid and 0.2 M Na$_2$HPO$_4$. Always use freshly prepared solution, and keep it under 4 °C on ice by adding it to the assay mixture.

- McIlvaine buffer (pH 4.5), 1 M KPB (pH 6.0), and 1 mM 1-methoxy PMS can be stored for a long time at 4 °C. Five mM WST-1 kept at 4°C should be used within 3 months.

27.4.2.5 Procedure

Add 702 μl of the sample solution to 248 μl of a mixture prepared with 100 μl of 1 M KPB (pH 6.0), 100 μl of 10% (v/v) Triton X-100, 40 μl of 5 mM WST-1, and 8 μl of 1 mM 1-methoxy PMS. After preincubation for 5 min, add 50 μl of 100 U/ml FDH solution and vortex mix rapidly. Leave to stand at 30 °C for 3 h in the dark. During the incubation, the microplate or test tube containing the assay solution should be kept under a wrap to exclude oxygen as much as possible. Measure the optical density at 438 nm against water.

27.4.3 HPLC Methods

27.4.3.1 Background

The methods using HPLC can measure many sugars and other polyols simultaneously. The sensitivity and specificity for fructose is superior to those of enzymatic methods. Some HPLC methods can measure the peripheral blood collected in the fasting state because HPLC columns have large numbers of theoretical stages. The threshold is 20–50 μM fructose. Although an HPLC apparatus is needed, most biochemistry-related laboratories already possess such equipment.

In general, monosaccharides including fructose have been separated on a column in normal phase mode and measured by RI detector due to the lack of characteristic UV absorbance or fluorescence. The sensitivity of RI detectors is very low, and gradient mode is inapplicable. Various techniques have been developed to achieve high sensitivity for detecting sugars. A postcolumn labeling method using phenylhydrazine, a type of labeling reagent, selectively reacts with reducing sugars and allows fluorescence detection of sugars the sensitivity of which is two orders of magnitude higher than that of RI detector, although a reaction unit and additional pump are necessary. This section describes the HPLC method using postcolumn labeling with phenylhydrazine after separation through an amino column (Abe 1991; Suzuki *et al.* 2009).

On the other hand, methods using an anion exchange column and a pulse amperometric detector have been applied for fructose determination (Jauniaux *et al.* 2004; Lal *et al.* 1995; Regnault *et al.* 2010).

27.4.3.2 Principle

- Analysis system
  After separation using a polymer-based amino column with a linear gradient in normal phase mode, fructose reacts with a reagent containing phenylhydrazine sent from another tube to produce fluorescent adducts, and is analyzed by fluorescence detector (Figure 27.1).

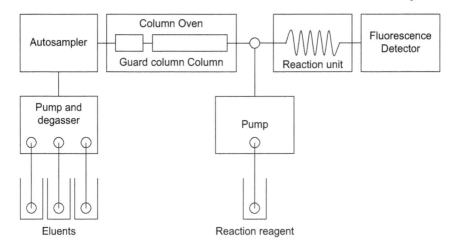

Figure 27.1 Schematic representation of the HPLC system for sugar analysis by postcolumn labeling with phenylhydrazine.

- Separation of fructose using an amino column

  The separation of fructose is carried out using an amino column, which incorporates an amino group in the base gel. As an amino column, Shodex Asahipak NH2P-50 series (Showa Denko, Tokyo, Japan) is recommended due to its high durability.

  The amino groups have high polarity, forming a hydrophilic environment around the surface of the column as a solid phase, and the partition and equilibrium of solute (sugars) forms between such solid phase and low-polar mobile phase. As the partition of solute verges to the mobile phase with increasing water-to-eluent ratio, the retention time of sugars decreases.

  The amino column, which is alkaline internally, separates sugars without causing anomer separation at low temperature (40 °C) because reducing sugars do not form anomers under alkaline conditions in contrast to amide columns in which the acrylamide groups are not alkaline.

  For ease of pretreatment of body fluids, a polymer-based amino column, which incorporates a pentaethylenehexamine-bonded vinyl alcohol copolymer gel, is recommended. This is superior to silica-based amino columns with regard to reproducibility and durability.
- Labeling fructose by phenylhydrazine

  Reducing sugars containing fructose react with phenylhydrazine to produce Schiff base, which shows fluorescence.

27.4.3.3 Sample Solution

- Pretreatment of body fluid

  Collect the supernatant by centrifugation of blood at 10 000 *g* for 1 min under 4 °C. Other body fluids are also pretreated appropriately. Add an equivalent volume of acetonitrile to the specimen. After vortexing, centrifuge the mixture and collect the supernatant again for deproteinization and preservation. The supernatant is stored as a sample solution under −20°C. Filter the sample solution through a 0.45 μm membrane filter just before loading onto the HPLC column.

As pretreatment of body fluids for extraction of fructose, a spin column packed with monolithic ODS-silica disk is convenient. Monoliths, which are structured hierarchical pores, have a high-surface area. The elution of target compounds can be achieved by centrifugation for a short time compared with silica gel. In the case of MonoSpin C18 (GL Sciences, Tokyo, Japan), the extraction can be performed by centrifugation at $10\,000\,g$ for 1 min from serum.

- Preparation of fructose solution for calibration curve

    Dilute fructose with water to 3600 μg/ml. Then, add the equivalent of acetonitrile to the fructose solution (The fructose concentration is 1800 μg/ml). Furthermore, make serial dilutions with the fructose solutions and 50% acetonitrile.

    When fructose is dissolved in a solvent with a ratio of acetonitrile lower than the eluent, fructose deposits in the injection loop and slowly dissolves. As it causes the peak to be broad, fructose should be dissolved in a solvent the composition of which is as close as that of the eluent as possible. However, fructose in over 65% acetonitrile becomes cloudy; 50% acetonitrile or higher may be a better composition in this experiment.

27.4.3.4 Reagents

- Reaction reagent

    Mix 220 ml of acetic acid (GR grade) and 6 ml of phenylhydrazine. Next, add 180 ml of phosphoric acid (GR grade) to the mixture and stir well. Do not mix 3 reagents together to avoid the solution becoming cloudy. The reaction reagent, which has high viscosity, in a brown glass bottle must be degassed by sonication under vacuum pressure for at least 30 min. The reaction reagent should be used within 2 weeks.

- Eluent

    As the mobile phase, prepare the following 3 solvents. The GR grade (not HPLC grade) acetonitrile is available for preparation of these solvents.

    Solvent   A, acetonitrile/H_2O/H_3PO_4 90 : 9.5 : 0.5(v/v/v)
    　　　　　B, acetonitrile/H_2O/H_3PO_4 75 : 24.5 : 0.5(v/v/v)
    　　　　　C, acetonitrile/H_2O/H_3PO_4 65 : 34.5 : 0.5(v/v/v)

    Prepare each solvent without H_3PO_4 for washing the column.

- Other reagents

    Prepare 100 mM ammonium acetate for conditioning pretreatment of the amino column.

27.4.3.5 HPLC Apparatus

The HPLC system shown in Figure 27.1 is necessary for these studies. GL-7400 series (GL Sciences) or LaChrom Elite series (Hitachi High-Technologies, Tokyo, Japan) are recommended because they are customized systems built for the analysis of sugars by postcolumn labeling with phenylhydrazine. For efficient reaction with phenylhydrazine, the length of the reaction coil, which is a heat-resistant tube, such as Peek Tubing (GL Sciences) is set as 0.5 mm ID × 7 m in the reaction unit.

Our system consisted of MIDAS as an autosampler (Spark Holland B.V., Emmen, Netherlands), 2 units of GL-7410 (GL Sciences) as pumps with a built-in degasser, CO-705 (GL Sciences) as a column oven, P.C.S. Pump Model 022 (Flom, Tokyo, Japan) as a pump for reaction reagent, Reactor 522 (Flom) installed with the Peek Tubing as a reaction unit, and GL-7453 (GL Sciences) as a fluorescence detector. Data were analyzed using EZChrom Elite ver. 3.1.5J software (Agilent Technologies, Santa Clara, CA).

27.4.3.6 Procedure

HPLC conditions are shown in Table 27.2. When the peaks in a chromatogram cause deformation and/or the retention times are abnormal, 100 mM ammonium acetate is passed at 1 ml/min for 30 min through the column for activation, and then change the 100 mM ammonium acetate to an eluent without H_3PO_4 and flow for at least 1 h. Change all eluents without H_3PO_4 for the corresponding eluents containing H_3PO_4. In parallel with the above operation, flow water from the pump for reaction reagent and set the temperature of the reactor at 95 °C. When the temperature reaches 95 °C, replace water with the reaction reagent and set the reactor temperature at 150 °C. The analysis can be started approximately 30 min later.

Table 27.2 Gradient program and conditions of HPLC for fructose analysis by postcolumn labeling with phenylhydrazine.

Time (min)	Eluent A (%)	Eluent B (%)	Eluent C (%)
0	100	0	0
20	100	0	0
25	0	100	0
30	0	0	100
40	0	0	100
40.1	100	0	0
60	100	0	0

Flow rate of mobile phase, 1.0 ml/min
Column, Shodex Asahipak NH2P-50 4E (4.6 mmID × 250 mm) with a guard column.
Column Temperature, 40 °C
Flow rate of the reagent, 0.4 ml/min
Reaction temperature, 150 °C
Fluorescence Detection: *Excitation* 330 nm, *Emission* 470 nm
Injection volume, 20 µl.

27.4.4 GC/MS Methods

27.4.4.1 Background

GC/MS-based methods can measure fructose concentrations in body fluid with high sensitivity and specificity because capillary columns have exceptional separation ability and mass spectroscopy can be used to analyze the chemical structures of minute amounts of samples. The threshold is of the order of approximately 1 μM by 0.1 unit. GC/MS methods can analyze other polyols simultaneously with high sensitivity. However, the procedure is very cumbersome and expensive instruments are necessary.

Sugars in samples must be reacted to volatile derivatives. Most examinations have employed derivatization of the carbonyl group as *O*-alkyloximes and the hydroxyl groups as trimethylsilyl ethers or acetate esters. The derivatives are usually separated using a low-polar capillary column in which the solid phase consists of 5% phenyl and 95% methylpolysiloxane or the equivalent, such as Agilent J&W HP-5MS (Agilent Technologies) or Inertcap 5MS (GL Sciences). As an internal standard, $^{13}C_6$-fructose is added to a specimen and fructose concentration is calculated by comparison of the intensity of characteristic fragment ion peaks of sample fructose with those of $^{13}C_6$-fructose.

This section describes a GC/MS method for accurate microdetermination of blood fructose concentration by exclusion of glucose (Kawasaki *et al.* 2004). The threshold is approximately 1 μM. However, the instrumentation is very complicated and needs proficiency for operations of HPLC and GC/MS. Recently, Wahjudi *et al.* (2010) and Preston *et al.* (2010) have developed more simple methods without exclusion of glucose, respectively.

27.4.4.2 Principle

- Analysis procedure

  Addition of internal standard → deionization → Fractionation of fructose by preparative HPLC → Derivatization to oximes → Fractionation of oximes by preparative HPLC → Peracetylation → GC/MS analysis.

- Exclusion of glucose from blood by HPLC.

  An amido column is used for fractionation of fructose in normal phase mode because the separation is performed under neutral condition. As an amido column, Amido-80 HR (Tosoh, Tokyo, Japan) is recommended due to high number of theoretical stages. Column size is for analysis scale. The elution time of fructose must be expected by preliminary experiments (not retention time; because retention time indicates the time a sample just goes through the detector).

- Derivatization of fructose

  Reaction with *O*-ethyl hydroxylamine affords to *O*-ethyloxime derivatives as a mixture of geometric isomer, *syn-* and *anti-*form. Therefore, fructose *O*-ethyloxime appears as 2 peaks on a chromatogram.

- Purification of fructose *O*-ethyloximes by HPLC

   Since fructose *O*-ethyloximes are more low-polar than original fructose, ODS column is available for fractionation in reversed phase mode. As an ODS column, Kromosil 100-$_5$C$_{18}$ (Eka Chemicals AB, Bohus, Sweden) is recommended because it permits alkali cleaning. The elution time of fructose *O*-ethyloximes must be expected by preliminary experiments.

- GC/MS analysis

   After peracetylation of hydroxyl groups, the solution containing fructose *O*-ethyloximes is subjected to the injector of GC/MS and separated through a low-polar capillary column. In the fragment ion peaks of fructose, m/z 271 is observed as fructose in the original sample. The characteristic peak of internal standard, ^{13}C$_6$-fructose is presented as m/z 277.

27.4.4.3   *Sample Solution*

When the concentration of the internal standard is 100 μM, prepare 500 μM ^{13}C$_6$-fructose solution with water. Add 20 μl of 500 μM ^{13}C$_6$-fructose to 100-μl sample. After vortexing, apply the mixture to a two-layer column containing (from the bottom): 250 μl of the acetated form of an anion exchanger AG1-X8 (Bio-Rad, Richmond, CA) and 250 μl of the H form of the cation exchanger AG50W-X8 (Bio-Rad) for deionization. When preparation of two-layer column is impossible, pass the sample through each column. Next, wash the column with 1 ml of water and collect all of the eluate. After evaporation in the centrifugal evaporator, the dry residue is dissolved in 40 μl of water and diluted with 160 μl of acetonitrile.

27.4.4.4   *Separation of Fructose by HPLC*

An HPLC system equipped an autosampler, pump with degasser, column oven, RI detector, and fraction collector is used. All solvents for elution must be of HPLC grade. As an isocratic eluent, acetonitrile-water (80:20, v/v) is delivered through Amido-80 (4.6 mm ID × 250 mm) with a guard column at a flow rate of 0.8 ml/min at 80 °C. Apply the portion onto the column and monitor the elution by RI detector. Collect the peak corresponding to fructose using a fraction collector and evaporate.

27.4.4.5   *Derivatization of Fructose*

The dry residue containing purified fructose is dissolved in 150 μl of water containing 2% *O*-ethylhydroxylamine hydrochloride. Leave the mixture to stand at 110 °C in the dark for 1 h.

27.4.4.6   *Purification of Fructose O-ethyloximes by HPLC*

The HPLC system equipped with a UV detector is used. All solvents for elution must be of HPLC grade. The purification is performed through Kromosil 100-$_5$C$_{18}$ (4.6 mm ID × 150 mm) using a guard column with isocratic 100% water in 8 min at a flow rate of 1 ml/min at 40°C, followed by cleaning with acetonitrile-water (80:20, v/v) for 5 min and conditioning with 100% water for 7 min. Apply the portion onto the column and monitor the elution by measuring UV absorbance at 210 nm. Collect the 2 peaks corresponding to fructose *O*-ethyloximes and evaporate.

27.4.4.7 *Peracetylation of Fructose O-ethyloximes*

Add 75 μl of acetic anhydride-pyridine (1:2, v/v) into the tube containing the residue and leave the mixture to stand at 80 °C for 20 min. After evaporation, the residue is dissolved in 200 μl of ethyl acetate.

27.4.4.8 *GC/MS Analysis*

A 1-μl aliquot is injected onto a fused silica capillary column HP-5MS (0.25 mm ID × 30 m, 0.25 μm film thickness) in a GC/MS at 130 °C and the final separation is performed at 250 °C. The rate of increase in temperature is 5 °C/min from 180 °C to 230 °C and 20 °C/min among the other ranges. Record the intensities of the fragment ion peak corresponding to fructose at $m/z = 271$ and the peak corresponding to $^{13}C_6$-fructose at $m/z = 277$ in the mass spectrum, respectively. The amount of fructose in the original sample is calculated from the area ratio of the peaks of m/z 271 to m/z 277.

27.5 Applications

27.5.1 Sucrose Tolerance Test in Rats

27.5.1.1 *Materials and Methods*

Male Wistar rats (8 weeks old; body weight, 241–300 g; $n = 21$) were treated according to the previously described method (Sugimoto *et al.* 2005; 2010) (Table 27.3). Serum fructose concentration was measured using the enzymatic method described in section 27.4.2 and the GC/MS method described in section 27.4.4.

Table 27.3 Comparison of fructose concentrations in portal sera in rats after sucrose ingestion between the enzymatic method and the GC/MS method. Rats ($n = 21$) starved for 16 h were orally administered sucrose solution (2 g/kg body weight), and blood was sampled from the portal vein under diethyl ether anesthesia before and after fructose ingestion. Serum fructose concentration was analyzed by the enzymatic method and the GC/MS method. Values are the means of 7 rats with their SD.

Time after ingestion (min)	Fructose concentration (μM)	
	Enzymatic method	*GC/MS method*
0	77 ± 40	112.2 ± 71.0
30	916 ± 483	902.2 ± 441.8
60	630 ± 370	683.1 ± 369.6

27.5.1.2   Results

Table 27.3 shows the changes in postprandial fructose concentrations. Values were different between the enzymatic and GC/MS methods. However, they showed the same tendencies.

27.5.2   Fructose Tolerance Test in Mice

27.5.2.1   Materials and Methods

The experimental procedure with female Balb/c mice (10 weeks old; body weight, 20–22 g; $n = 8$) is described in Table 27.4. Serum fructose concentration was measured using the enzymatic method described in section 27.4.2 and the HPLC method described in section 27.4.3.

27.5.2.2   Results

Figure 27.2 illustrates the chromatograms of fructose analysis using HPLC. The changes in postprandial fructose concentrations are shown Table 27.4. The data analyzed by HPLC distinctly showed the differences in fructose concentration between mice under fasting conditions and after fructose ingestion.

27.5.3   Sucrose Tolerance Test in Humans

27.5.3.1   Materials and Methods

The experimental procedure with healthy Japanese men ($n = 3$) is described in Table 27.5. Serum fructose concentration was measured using the GC/MS method described in section 27.4.4.

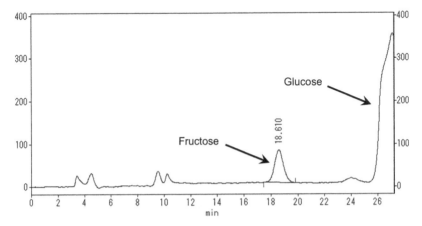

Figure 27.2   HPLC chromatograms of peripheral blood in mice. A mouse starved for 4 h was orally administered fructose solution (4 g/kg body weight), and blood was sampled from the postcaval vein under diethyl ether anesthesia 20 min after the ingestion of fructose. In the chromatogram, fructose was detected as 370 µM using a fluorescence detector at Excitation 330 nm and Emission 470 nm after postcolumn labeling with phenylhydrazine.

Table 27.4 Comparison of the fructose concentrations in portal sera from mice after fructose ingestion between the enzymatic method and the HPLC method. Mice starved for 4 h were orally administered fructose solution (4 g/kg body weight). Blood was sampled from the postcaval vein under diethyl ether anesthesia at fasting and 20 min after the ingestion of fructose. Serum fructose concentration was analyzed by the enzymatic method and the HPLC method. Values are presented as means ± SD.

	Fructose concentration (µM)	
	Enzymatic method	HPLC method
Fasting (n = 3)	49 ± 36	11 ± 0.2
Fructose ingestion (n = 5)	108 ± 28	350 ± 52

Table 27.5 Changes in fructose concentration in peripheral sera in human subjects after sucrose ingestion using the GC/MS method. Human subjects (n = 3) starved overnight were orally administered sucrose solution (25 g in 200 ml of water), and blood was sampled from the antecubital vein before and after sucrose ingestion. Serum fructose concentration was analyzed by the GC/MS method. Values are presented as means ± SD.

		After ingestion			
Time (min)	Before ingestion	30	60	90	120
Fructose (µM)	11.2 ± 2.1	186.6 ± 23.5	106.9 ± 15.5	57.4 ± 20.5	27.32 ± 9.8

27.5.3.2 Results

Table 27.5 shows the changes in peripheral serum fructose concentrations. The fasting serum fructose concentration was accurately measured and its mean value was 11.2 µM.

27.5.4 Evaluation of Inhibitive Activity on Fructose Absorption in the Caco-2 Cell Line

27.5.4.1 Materials and Pretreatment of Cells

BD Falcon cell culture inserts and 6-well cluster plates (BD Biosciences, Franklin Lakes, NJ; Catalogue No. 353090) were used. As a test sample, EGCG (Sigma-Aldrich) was dissolved in 10% dimethylsulfoxide.

Human intestinal epithelial Caco-2 cells obtained from Dainippon Sumitomo Pharma (Osaka, Japan) were cultured as described previously (Aspenström-Fagerlund *et al.* 2009). For absorption experiments, subcultured cells were allowed to reach confluence in DMEM (Sigma-Aldrich; Catalogue

Table 27.6   Inhibitory effect of EGCG on fructose absorption in Caco-2 cell line. Caco-2 cells cultured on membrane filters in an insert were exposed to 1 mM fructose with or without EGCG for 4 h. Fructose concentration in the medium on the basolateral side was analyzed by the enzymatic method.

	Dose (µg/ml)	Fructose permeation (nM)	Inhibition (%)
Blank (10% dimethylsulfoxide)		155	
EGCG	5.0	111	28

No. D5796) containing 10% fetal calf serum (Biowest, Nuaillé, France) and 1% non-essential amino acids (Sigma-Aldrich; Catalogue No. M7145), and seeded on the inside of the insert at a density of 2×10^5 cells/insert. The DMEM was changed both apically (inside the insert) and basolaterally (the plate) every 3–4 days. Cells were allowed to differentiate for 3 weeks.

27.5.4.2   Absorption Experiment

Both inside and outside of the inserts were washed with PBS(-) (pH 7.4). As a medium for absorption experiment, 1 ml of D-PBS (pH 7.4; Gibco, Langley, OK; Catalogue No. 21300-025), which is glucose-free, was added to the plate and the inside of the insert. The cells were incubated for 30 min at 37 °C in a humidified incubator containing 5% CO_2. The formation of a monolayer by Caco-2 cells was confirmed by measuring TEER of the medium between the plate and in the inside of the insert using Millicell-ERS (Millipore, Bedford, MA). After addition of 10 µl of sample solution to the medium on the inside of the insert followed by preincubation for 5 min, 10 µl of 100 mM fructose solution (final concentration, 1 mM) was added to the medium on the inside of the insert, and cells were incubated for an additional 4 h. After measurement of TEER again, the medium in the plate was collected and stored at –20 °C in a sterilized tube. Fructose concentration in the medium was analyzed using the enzymatic method described in section 27.4.2.

27.5.4.3   Results

TEER of cell monolayers did not differ between the different culture plates (data not shown). The results are shown in Table 27.6.

Summary Points

- Fructose concentrations in body fluids have been measured for examination of diabetes, semen, and the absorption and metabolism of fructose.
- Determination of fructose concentration in body fluids is difficult because coexisting substances interfere with the sensitivity of the assay systems.
- In blood, the large amounts of coexisting glucose interrupts the fructose assays.

- Chemical methods using colorimetric reagents and enzymatic methods have low sensitivity (adaptable over 100 µM) but are simple and inexpensive to perform.
- High-performance liquid chromatography (HPLC) and gas chromatography/mass spectrometry (GC/MS) methods can concurrently measure fructose and other polyols with high sensitivity and accuracy (over 20 µM and 1 µM, respectively), but these methods are expensive and require sophisticated instrumentation.

Key Facts of Diabetes and Blood Fructose Levels

1. Fructose is much more reactive in glycation than glucose *in vitro* and high-toxicity precursors of advanced glycation end-products (AGEs), such as methylglyoxal, are yielded *via* the fructose metabolism pathway as by-products.
2. "Fructation" means glycation by fructose, and fructated AGEs may be responsible for microvascular diabetic complications.
3. Fructose level in circulating blood is very low compared with that of glucose, but some studies have shown that dietary fructose consumption increases fructated AGEs in blood and various tissues.
4. The polyol pathway also supplies fructose endogenously.
5. It has been reported that inhibitors of aldose reductase, a key enzyme in this pathway, reduced the elevation of plasma fructose concentrations.
6. Although the relationships between fructose levels in the blood and/or tissues and diabetic complications have yet to be investigated in sufficient detail, these may be important risk factors that should be taken into consideration in investigations and treatment of diabetes.

Definitions of Words and Terms

AGEs. AGEs are yielded by glycation in tissues and blood in the chronic hyperglycemic state. They damage peripheral vessels and nerves, and are responsible for diabetic complications.

Glycation. Glycation means reaction with amino group and reducing sugar. Under conditions of hyperglycemia, proteins react with a reducing sugar to lose its function and yield toxic substances, namely AGEs.

EGCG. This is a major component of green tea (*Camellia sinensis*). Many physiological functions of EGCG have been reported, such as anticancer activity, antioxidative activity, and inhibitory effects on digestive enzymes.

Oxime. Hydroxylamine (NH_2OH) and alkyl hydroxylamine (NH_2OR^1) react with carbonyl groups ($>C=O$) in organic compound to form oximes, $R^2R^3C=N-OH$, $R^2R^3C=N-O-R^1$, respectively. These products are used for sugar analysis using GC/MS after derivatization of hydroxyl groups.

Phenylhydrazine. This compound has a $C_6H_5NHNH_2$ structure and reacts with carbonyl groups in organic compounds to form phenylhydrazones,

$R^1R^2C=NNHC_6H_5$ and osazone adducts, in which α-hydroxy ketone (or aldehyde) are substituted with 2 phenylhydrazine residues. These compounds are fluorescent.

Polyol pathway. In cells showing insulin-independent glucose uptake, hyperglycemia causes activation of enzymatic activity in the polyol pathway, such as aldose reductase and sorbitol dehydrogenase. This metabolic pathway converts glucose to fructose through sorbitol. Aldose reductase inhibitors have been used to treat diabetic complications.

Retinopathy. Series hyperglycemia causes microvascular damage and finally leads to vision loss through retinopathy, *i.e.*, retina hemorrhage and retinal detachment.

Schiff base. A type of imine expressed in $R^1R^2C=N\text{-}R^3$ (R^3 is not H). In tissues, reducing sugars react with protein to form unstable Schiff bases. The Schiff bases rapidly convert to precursors of AGEs.

TEER. When Caco-2 cells form a monolayer, electrical resistance occurs between the apical and basolateral sides. TEER is measured to confirm whether Caco-2 cells tightly form a monolayer.

List of Abbreviations

1-methoxy PMS	1-methoxy-5-methylphenazinium methyl sulfate
AGEs	advanced glycation end-products
DMEM	Dulbecco's modified Eagle's medium
D-PBS	Dulbecco's phosphate-buffered saline
EGCG	(-)-epigallocatechin 3-*O*-gallate
F6P	fructose-6-phosphate
FDH	fructose dehydrogenase
G6P	glucose-6-phosphate
G6PDH	glucose 6-phosphate dehydrogenase
GC/MS	gas chromatography/mass spectrometry
GR	guaranteed reagent
HPLC	high performance liquid chromatography
ID	internal diameter
KPB	potassium phosphate buffer
M	mol/l
NADP$^+$	nicotinamide adenine dinucleotide phosphate (oxidized form)
NADPH	nicotinamide adenine dinucleotide phosphate (reduced form)
ODS	octa decyl silyl
PBS(-)	phosphate-buffered saline without magnesium and calcium
PGI	phosphoglucose isomerase
RI	refractive index
SD	standard deviations
TEER	transepithelial electrical resistance

UV   ultraviolet
WST-1   2-(4-Iodophenyl)-3-(4-nitrophenyl)-5-(2,4-disulfophenyl)-
               2*H*-tetrazolium, monosodium salt

References

Abe, S., 1991. Sugar analysis in blood plasma using phosphoric acid/phenyl-hydrazine by HPLC. *Hitachi Scientific Instrument News*. 34: 3464–3468.

Adams, S.H., Stanhope, K.L., Grant R.W., Cummings, B.P., and Havel, P.J., 2008. Metabolic and endocrine profiles in response to systemic infusion of fructose and glucose in rhesus macaques. *Endocrinology*. 149: 3002–3008.

Al-Daghistani, H.I, Hamad, A.W., Abdel-Dayem, M., Al-Swaifi, M., and Abu Zaid, M., 2010. Evaluation of serum testosterone, progesterone, seminal antisperm antibody, and fructose levels among Jordanian males with a history of infertility. *Biochemistry Research International*. 2010: Article ID 409640. DOI: 10.1155/2010/409640.

Anderson, R.A., Reddy, J.M., Oswald, C., and Zaneveld, L.J., 1979. Enzymic determination of fructose in seminal plasma by initial rate analysis. *Clinical Chemistry*. 25: 1780–1782.

Aspenström-Fagerlund, B., Sundström, B., Tallkvist, J., Ilbäck, N.G., and Glynn, A.W., 2009. Fatty acids increase paracellular absorption of aluminium across Caco-2 cell monolayers. *Chemico-Biological Interactions*. 181: 272–278.

de Carvalho, C.A., and Pogell, B.M., 1957. Modified skatole method for microdetermination of fructose and inulin. *Biochimica et Biophysica Acta*. 26: 206.

Crossley, J.N., and Macdonald, I., 1970. The influence in male baboons, of a high sucrose diet on the portal and arterial levels of glucose and fructose following a sucrose meal. *Nutrition and Metabolism*. 12: 171–178.

Gul, A., Rahman, M.A., and Hasnain, S.N., 2009. Influence of fructose concentration on myocardial infarction in senile diabetic and non-diabetic patients. *Experimental and Clinical Endocrinology and Diabetes*. 117: 605–609.

Hui, H., Huang, D., McArthur, D., Nissen, N., Boros, L.G., and Heaney, A.P., 2009. Direct spectrophotometric determination of serum fructose in pancreatic cancer patients. *Pancreas*. 38: 706–712.

Jauniaux, E., Hempstock, J., Teng, C., Battaglia, F.C., and Burton, G.J., 2005. Polyol concentrations in the fluid compartments of the human conceptus during the first trimester of pregnancy: maintenance of redox potential in a low oxygen environment. *The Journal of Clinical Endocrinology and Metabolism*. 90: 1171–1175.

Kawasaki, T., Igarashi, K., Ogata, N., Oka, Y., Ichiyanagi, K., and Yamanouchi, T., 2010. Markedly increased serum and urinary fructose concentrations in diabetic patients with ketoacidosis or ketosis. *Acta Diabetologica*. DOI: 10.1007/s00592-010-0179-3.

Kawasaki, T., Ogata, N., Akanuma, H., Sakai, T., Watanabe, H., Ichiyanagi, K., and Yamanouchi, T., 2004. Postprandial plasma fructose level is associated with retinopathy in patients with type 2 diabetes. *Metabolism.* 53: 583–588.

Lal, S., Szwergold, B.S., Taylor, A.H., Randall, W.C., Kappler, F., and Brown, T.R., 1995. Production of fructose and fructose-3-phosphate in maturing rat lenses. *Association for Research in Vision and Ophthalmology.* 36: 969–973.

Messineo, L., and Musarra, E., 1972. Sensitive spectrophotometric determination of fructose, sucrose, and inulin without interference from aldohexoses, aldopentoses, and ketopentoses. *International Journal of Biochemistry.* 3: 691–699.

Nixon, D.A., 1969. The determination of fructose in biological fluids using anthrone. *Clinica Chimica Acta.* 26: 167–169.

Noh, H.L., Hu, Y., Park, T.S., DiCioccio, T., Nichols, A.J., Okajima, K., Homma, S., and Goldberg, I.J., 2009. Regulation of plasma fructose and mortality in mice by the aldose reductase inhibitor lidorestat. *The Journal of Pharmacology and Experimental Therapeutics.* 328: 496–503.

Regnault, T.R., Teng, C., de Vrijer, B., Galan, H.L., Wilkening, R.B., and Battaglia, F.C., 2010. The tissue and plasma concentration of polyols and sugars in sheep intrauterine growth retardation. *Experimental Biology and Medicine.* 235: 999–1006.

Roe, J.H., 1934. A colorimetric method for the determination of fructose in blood and urine. *The Journal of Biological Chemistry.* 107: 15–22.

Shiota, M., Galassetti, P., Igawa, K., Neal, D.W., and Cherrington, A.D., 2005. Inclusion of low amounts of fructose with an intraportal glucose load increases net hepatic glucose uptake in the presence of relative insulin deficiency in dog. *American Journal of Physiology. Endocrinology and Metabolism.* 288: E1160–E1167.

Sugmimoto, K., Hosotani, T., Kawasaki, T., Nakagawa, K., Hayashi, S., Nakano, Y., Inui, H., and Yamanouchi, T., 2010. Eucalyptus leaf extract suppresses the postprandial elevation of portal, cardiac and peripheral fructose concentrations after sucrose ingestion in rats. *Journal of Clinical Biochemistry and Nutrition.* 46: 205–211.

Sugmimoto, K., Suzuki, J., Nakagawa, K., Hayashi, S., Fujita, T., Yamaji, R., Inui, H., and Nakano, Y., 2005. Eucalyptus leaf extract inhibits intestinal fructose absorption, and suppresses adiposity due to dietary sucrose in rats. *British Journal of Nutrition.* 93: 957–963.

Suzuki, H., Kato, E., Matsuzai, A., Ishikawa, M., Harada, Y., Tanikawa, K., and Nakagawa, H., 2009. Analysis of saccharides possessing post-translational protein modifications by phenylhydrazine labeling using high-performance liquid chromatography. *Analytical Sciences.* 25: 1039–1042.

Takeda, H., Higashi, T., Nishikawa, T., Sato, Y., Anami, Y., Yano, T., Kasho, M., Kobori, S., and Shichiri, M., 1996. Release of fructose and hexose phosphates from perivascular cells induced by low density lipoprotein and acceleration of protein glycation *in vitro*. *Diabetes Research and Clinical Practice.* 31: 1–8.

Truswell, A.S., Seach, J.M., and Thorburn, A.W., 1988. Incomplete absorption of pure fructose in healthy subjects and the facilitating effect of glucose. *The American Journal of Clinical Nutrition.* 48: 1424–1430.

Wahjudi, P.N., Patterson, M.E., Lim, S., Yee, J.K., Mao, C.S., and Lee, W.N., 2010. Measurement of glucose and fructose in clinical samples using gas chromatography/mass spectrometry. *Clinical Biochemistry.* 43: 198–207.

Yamamoto, T., Moriwaki, Y., Takahashi, S., Tsutsumi, Z., Yamakita, J., and Higashino, K., 1999. Effects of fructose and xylitol on the urinary excretion of adenosine, uridine, and purine bases. *Metabolism.* 48: 520–524.

Yoshii, H., Uchino, H., Ohmura, C., Watanabe, K., Tanaka, Y., and Kawamori, R., 2001. Clinical usefulness of measuring urinary polyol excretion by gas-chromatography/mass-spectrometry in type 2 diabetes to assess polyol pathway activity. *Diabetes Research and Clinical Practice.* 51: 115–123.

Zender, R., and Falbriard, A., 1966. Colorimetric analysis of keto-hexoses and inulin by means of the thiobarbituric acid reaction. *Clinica Chimica Acta.* 13: 246–250.

CHAPTER 28
Amperometric Detection for Simultaneous Assays of Glucose and Fructose

MITHRAN SOMASUNDRUM[a] AND
WERASAK SURAREUNGCHAI*[b]

[a] Biochemical Engineering and Pilot Plant Research and Development Unit,
National Center for Genetic Engineering and Biotechnology,
Email: mithran.somasundrum@gmail.com; [b] School of Bioresources and
Technology, Biological Engineering Graduate Program, King Mongkut's
University of Technology Thonburi, Bangkhuntien, Bangkok 10150,
Thailand
*Email: werasak.sur@kmutt.ac.th

28.1 Introduction

Voltammetry is an electrochemical technique in which the potential of one electrode (denoted the working electrode, WE) is controlled relative to a second electrode (denoted the reference electrode, RE). The potential difference between the two electrodes is set such that a redox reaction of interest is driven at the WE. The current, I, resulting from this reaction can provide information about the quantity of the reacting species, as originally described by Faraday:

$$\int_0^t I \mathrm{d}t = Q = mnF \tag{1}$$

Food and Nutritional Components in Focus No. 3
Dietary Sugars: Chemistry, Analysis, Function and Effects
Edited by Victor R Preedy
© The Royal Society of Chemistry 2012
Published by the Royal Society of Chemistry, www.rsc.org

where Q is the charge passed over the duration t of the current flow, m is the molar quantity of reactant, n is the stoichiometric number of electrons transferred in the reaction and F is the Faraday constant (96485 C mol^{-1}). Voltametric detection has often been applied to analytes in the food industry (Gulaboski and Pereira 2008). The advantages of voltammetry are that it can provide relatively high sensitivities, while remaining unaffected by the optical state of the sample matrix, and that because voltammetry produces an electrical signal without needing a further transduction step, voltammetric detectors are relatively cheap to construct. Also, electrochemical equipment can be readily coupled to existing micro-fabrication technologies to produce microchip/microfluidic based detection (Manz *et al.* 1990). This allows measurements to be made in extremely small volumes, requiring only small quantities of sample and reagents, and thus producing less waste. The disadvantages of voltammetry are that problems in selectivity can occur, depending on the sample matrix and the potential being used, and that not all compounds react "cleanly" at the electrode, which can result in the electrode surface becoming blocked over time. Both these issues arise in the measurement of glucose and fructose, and both can be resolved. In this chapter we will describe the methods for doing so, and will outline the theory and practical details necessary to carry out the voltammetric determination of glucose and fructose.

28.2   Principles of Voltammetry

If two electrodes are placed in an electrolyte and allowed to reach equilibrium, a potential difference (E), measureable by a voltmeter, will develop between them. The equilibrium reached will be a dynamic one, in which the redox species in the solution, and/or the solvent itself, are interconverted between their oxidised and reduced states at each electrode. For a generalised conversion between reactants and products of the form:

$$aA + bB \leftrightarrow cC + dD$$

the measured potential difference can be related to the activities of the species by the Nernst equation:

$$E = E^{\circ} - \frac{RT}{nF} \ln \frac{a_C^c \, a_D^d}{a_A^a \, a_B^b} \tag{2}$$

where E° is the potential difference under thermodynamic standard conditions (all activities are unity, liquids are in a pure state, *etc.*) R is the gas constant (8.314 J K^{-1} mol^{-1}) and T is the temperature. In most electrochemical experiments the concentrations used are low enough to replace activities in eq. (2).

If the voltmeter is substituted by a voltage source then the potential of the cell can be driven away from its equilibrium value. Driving the potential high enough can cause the analyte to be oxidised or reduced at the WE. To complete the circuit, electron transfer must also occur at the RE and ions must pass

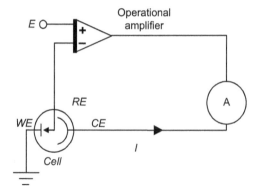

Figure 28.1 Usual configuration for voltammetric experiment. "A" denotes a current measuring device.

through the solution between the two electrodes. It is often the case that passing a current through the RE causes a variation in the reference potential. For this reason voltammetry is usually performed using a third electrode (denoted the counter electrode, CE), in the configuration shown in Figure 28.1.

Hence, the experimenter controls the potential difference between the WE and RE, while measuring the current flow between the WE and CE. The current due to a redox reaction is termed the Faradaic current and can be related to the applied potential by the Butler-Volmer equation:

$$I = nFAk^\circ \left[C_R^s \exp\left\{ \frac{(1-\alpha)nF(E-E^\circ)}{RT} \right\} - C_O^s \exp\left\{ \frac{-\alpha nF(E-E^\circ)}{RT} \right\} \right] \quad (3)$$

where C_O^s and C_R^s are the concentrations of the oxidised (O) and reduced (R) forms of the reactant near the surface of the electrode, k° is the standard rate constant for the reaction (*i.e.* the rate constant at the standard potential, E°), and α is the transfer coefficient, which describes the symmetry of the energy barrier between O and R. For a symmetrical barrier (often assumed to be the case for convenience) $\alpha = 0.5$. Eq. (3) applies for as long as the rate of electron transfer is slower than the rate of mass transport. The quantity $E - E^\circ$ is referred to as the overpotential. When the overpotential is zero ($E = E^\circ$) the rates of electron flow into and out of the electrode will be equal. The rate of the flow at $E = E^\circ$ is termed the exchange current, I_o, and as described in section 28.3.2, can be an important parameter in experimental design. Note that I_o cannot be measured directly; since it is the same magnitude in both directions the overall *measured* current will be zero.

28.3   Experimental Details

28.3.1   Electrochemical Cells

For electroanalysis in non-corrosive media, electrochemical cells are usually made of glass. A typical cell configuration is shown in Figure 28.2. If the

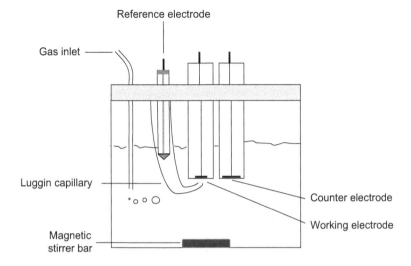

Figure 28.2 Typical configuration of an electrochemical cell.

intended operating potential is going to be negative enough to reduce dissolved oxygen, then oxygen needs to be removed from the cell by purging with an inert gas such as nitrogen. A further consideration is that the volume of solution between RE and WE will have a resistance, R_U, and therefore from Ohm's law will have a voltage drop IR_U across it. Hence, the actual potential experienced at the WE will be $E - IR_U$. To minimise this drop in potential it is necessary to place the RE and WE as close together as possible. This can be done by positioning the RE itself, by using an RE inside a glass tube which tapers to a narrow end (called a Luggin capillary), which can be positioned near the WE, or by using a salt bridge (typically a glass tube containing agar and NaCl, tapering to a fine point) in the reference electrode solution, and again positioning it close to the WE surface.

Often the cell will contain a magnetic stirrer bar so that convection can be used to increase the rate of mass transport to the WE. Other ways of increasing the rate of mass transport are to rotate the WE, or embed the electrodes in a flow cell, as described in section 28.4.2.

28.3.2 Electrodes

Although carbon and mercury WEs are used in other types of electrochemistry, for the determination of glucose and fructose either Pt or Au WEs are required. This is because oxide formation on the WE is an integral part of the experimental technique, as described in section 28.7.

The RE is a particular half-cell. The main requirement is that the potential across the RE should remain stable for different values of potential difference between WE and RE, as applied by the potentiostat. To achieve this the half-cell needs to have a high value of I_o, so that the half-cell equilibrium is maintained when the applied potential changes. The most common reference half-cells used are Hg contacting Hg_2Cl_2 in saturated KCl, called a saturated calomel electrode (SCE), and Ag metal coated with a layer of AgCl placed in saturated KCl or in 3.5 M NaCl. Both types are commercially available.

The requirements of the CE are that it should remain chemically inert during the measurement, should have good conductivity and should have a surface area at least roughly equal to the area of the WE. This is to ensure that current flow through the CE does not become the rate limiting factor. Typically the CE is a platinum wire or a platinum disk.

28.3.3 Electrolyte

The electrolyte solution should contain a relatively high concentration of an inert, ionized salt. Typically, in aqueous solutions KCl or NaCl is used at a value of at least 0.1 M. In organic solutions tetrabutyl ammonium salts are usually used instead. The purpose of these ions, referred to as a supporting electrolyte, is to increase the conductivity of the solution. Conductivity between the WE and CE is necessary to avoid Joule heating and to maintain a uniform distribution of potential. Conductivity between the WE and RE is necessary to reduce the value of IR_U, as noted in section 28.3.1. Also, every electrode/solution interface has a capacitance, due to the fact that it consists of opposite layers of charge. This can affect the output of the voltammetric experiment, as will be described later. At low ionic concentrations the value of the interfacial capacitance is dependent on ion concentration, which means experimental results can change as the ion content changes.

28.4   Types of Voltammetry

There are a number of different types of voltammetry possible, based on different methods of applying a potential to the WE under different hydrodynamic conditions (Bard and Faulkner 2001, Greef et al. 1993). In this chapter we will confine ourselves to the voltammetric techniques necessary to understand how glucose and fructose can be measured.

28.4.1   Fixed Potential – Potential Step

In a potential step experiment the WE is held in a quiescent solution and the potential is stepped from a value where no reaction occurs, to a value where a solution species is either oxidised or reduced at a diffusion-controlled rate. That is, where the rate of electron transfer is much faster than the rate of diffusion to the electrode. (The potential where this occurs can be determined from a potential ramp experiment, as described in section 28.4.3). Two types of current will flow during this process. Due to the capacitance of the electrode-solution interface, noted earlier, there will be a capacitive current, I_C. This is equivalent to the current that flows when a voltage is applied across a capacitor and a resistor connected in series. There will also be a Faradaic current, I_F, due to the electrode reaction. The two currents have different time dependencies, of the forms $I_C \propto \exp(-t)$ and $I_F \propto 1/\sqrt{t}$. Hence, the measured current, $I (= I_C + I_F)$ falls with time, and at longer times is composed almost entirely of I_F.

28.4.2   Fixed Potential – Amperometry

In amperometry convective mass transport to the WE is provided, either by stirring the solution or by rotating the WE. Initially, the solution does not contain analyte and so the measured current falls according to the decay of I_C. Eventually, a stable baseline current is reached, representing the residual redox processes of the solvent. If the electrode potential has been set to a value where the analyte reacts, then when analyte is injected into the solution a Faradaic current will flow. Instead of decaying, the current will reach a steady state, due to the convective mass transport. Injection of further analyte will cause the current to rise to a second steady state. A typical output from the process is shown in Figure 28.3. Usually, the working potential is chosen such that the redox reaction is diffusion-controlled. This ensures that small changes in the

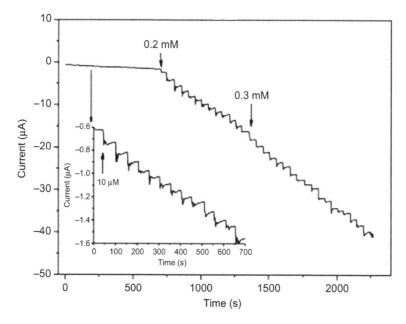

Figure 28.3   Example of the output from an amperometric experiment. Current response is measured as the height of each step from the current baseline. (Reprinted with permission from Li, Z., Wang, X., Wen, G., Shuang, S., Dong, C., Paau, M.C., and Choi, M.M.F., 2011. *Biosensors and Bioelectronics*. 26: 4619–4623. © Elsevier Ltd)

reference potential do not change the current response. Under this condition the steady state current, I_{SS}, can be related to the analyte concentration by:

$$I_{SS} = nFAC_jD_j/\delta \qquad (4)$$

where δ is a layer of unstirred solution at the electrode surface, often referred to as the Nernst diffusion layer. Increasing the stirrer speed makes δ thinner, thereby increasing the concentration gradient across δ. For this reason, increasing the stirrer speed increases the response. However, it can also increase the electrical noise.

Another way to provide convective mass transport to the WE is to place the electrodes in a flowing stream. Analyte is injected into the stream and so a peaked response is observed as the segment of analyte solution passes the WE surface. A typical example of the response is shown in Figure 28.4. The flowing stream can be sent through a liquid chromatography column before detection, thus allowing for separation of different analytes, as discussed elsewhere in this book.

The peak height will be proportional to the analyte concentration. The background current has the same origin as for amperometry and, as with amperometry, higher working potentials will cause a higher background. Also as with amperometry, a working potential is chosen so that the current response is diffusion-limited. Generally, the flow cell configuration can either be that of thin layer or wall jet. In the thin layer arrangement the electrodes are located in a section of flow channel between the solution inlet and outlet. The order of the electrodes is such that the WE is upstream of the CE and RE. This is so that reaction products at the CE, or leakage from the RE, do not affect the WE reaction. In the wall jet system the inlet is set so that the flowing stream impacts perpendicularly at the centre of the WE surface and then flows to

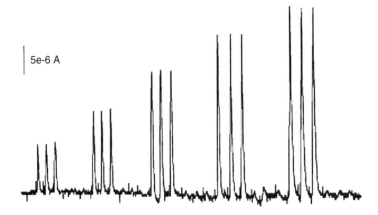

5e-6 A

Figure 28.4   Example of the output from a flow analysis experiment. Current response is measured as either the peak height or peak area.
(Reprinted with permission from Surareungchai, W., Deepunya, W., and Tasakorn, P., 2001. *Analytica Chimica Acta.* 2001: 215–220. © Elsevier Ltd)

outlets on either side. The inlet diameter is significantly less than the WE diameter. Because relatively large inlet-WE separations are possible, relatively large volume wall jet detectors can be made, which makes the fabrication process simple.

28.4.3 Potential Ramp

The simplest method of potential variation is a linear potential ramp. The start and end points of the ramp are chosen such that the redox reaction of interest takes place between them. The ramp can either be applied in one direction only (linear sweep voltammetry, LSV) or can be reversed to return to the starting potential (cyclic voltammetry, CV). As described in eq. (3), once the redox reaction begins, the current at the working electrode will increase with the potential applied, first linearly and then exponentially. This is because the rate of the electrode reaction is increasing. The electrolyte solution for the potential ramp is usually quiescent and so the rate of electron transfer will eventually become faster than the rate at which the reacting species diffuses to the electrode surface. Hence, a thin solution volume adjacent to the electrode will progressively empty of reactant. As this happens the electrode current will decrease. Thus, the redox reaction is observed as a peak in the voltammogram. The peak height is proportional to the concentration of analyte. Any potential beyond the voltammetric peak (*i.e.* positive of a reduction peak or negative of a reduction peak) will produce a diffusion-limited response when used in the fixed potential techniques described in section 28.4.2.

If the product of the electrode reaction is chemically stable, when the direction of the voltage ramp is reversed a second peak will be observed in the opposite direction to the first. This signifies the conversion of the product back to reactant. An ideal CV is shown in Figure 28.5. In the case of glucose and

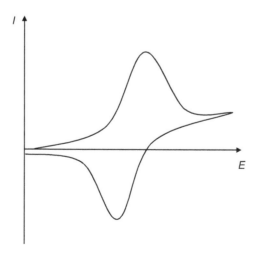

Figure 28.5 Example of an ideal cyclic voltammogram for a redox couple in solution.

fructose voltammetry this ideal behaviour is not seen, due to passivation, as described below.

28.5   Passivation in Voltammetry

For a CV to exhibit the ideal behaviour illustrated in Figure 28.5, in addition to chemical stability, the concentrations of the oxidised and reduced species at the electrode must change fast enough with E for eq. (3) to be followed, even though the cell is no longer at equilibrium. The characteristics of such a system (described as reversible) are that at 25 °C the peak-to-peak separation will be $59/n$ mV, the ratio of the oxidation to the reduction peak will be 1:1, and the peak potentials will be independent of the scan rate. For this to happen $k°$ must have a high value, as can be understood by re-arranging eq. (3) to give

$$C_R^S \exp\left\{\frac{nF(E-E°)}{RT}\right\} - C_O^S = \frac{I}{nFAk°} \exp\left\{\frac{\alpha nF(E-E°)}{RT}\right\} \tag{5}$$

If $k°$ is so high that the right-hand side of the equation is virtually zero, then eq. (5) reduces to eq (2). Inherent in this requirement is the assumption that the species at the electrode react "cleanly", meaning that after electron transfer the reaction products diffuse back into bulk solution. In organic electrode reactions this is often not the case, due to reaction products remaining adsorbed on the electrode. This means the available area for reaction decreases, and thus in CV or LSV the peak height decreases with subsequent scans as the electrode becomes blocked, or passivated. In amperometry passivation is observed as a current which gradually decreases to zero instead of reaching a steady state.

28.6   Electrochemistry of Glucose and Fructose

The electrochemistry of aldoses (such as glucose) and ketoses (such as fructose) are quite similar, and both can be detected at Au and Pt electrodes in alkaline media (pH > 12). We will use CV studies of glucose to illustrate the general characteristics of the electrochemistry of both analytes. Figure 28.6, curve A, shows a CV of a rotated Au electrode in 0.10 M NaOH. The peak in the positive direction (about $+0.2$ V) is due to the formation of AuO on the electrode and the peak in the negative direction (about $+0.1$ V) is due to its reduction back to Au metal. Curve B shows a CV of the same electrode in the same electrolyte containing 0.5 M glucose. The direction of the scan is again from negative to positive. At around -0.6 V (point a) the current increases in a positive direction. This is due to a fast 2-electron oxidation of the glucose aldehyde group to the corresponding carboxylic acid (Johnson and LaCourse 1990).

$$CH_2OH-[CHOH]_4-CHO + H_2O \rightarrow CH_2OH-[CHOH]_4-COO^- + 3H^+ + 2e^-$$

$$\tag{6}$$

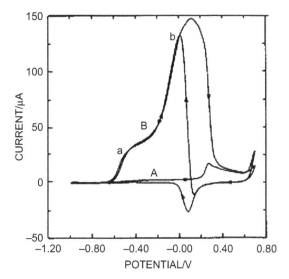

Figure 28.6 Cyclic voltammogram of rotated Au electrode in 0.10 M NaOH (A) and in 0.10 M NaOH + 5 mM glucose (B). Point (a) is the oxidation of the glucose aldehyde group as described in eq (6). Point (b) is the oxidation of the glucose alcohol groups as described in eqs (7) and (8).
(Reprinted with permission from Johnson, D.C., Dobberpuhl, D., Roberts, R., and Vandeberg, P., 1993. *Journal of Chromatography*. 640: 79–96. © Elsevier Ltd).

and occurs concurrently with the adsorption of a submonolayer of $^{\bullet}OH$ on the Au surface. The $^{\bullet}OH_{ads}$ is thought to be the intermediate in the transfer of oxygen from H_2O to the glucose molecule. Moving the potential farther positive, the large oxidation peak at about $+0.15$ V (point b) is seen at all carbohydrates and in the case of glucose comes first from the fast cleavage of the C_1–C_2 bond:

$$CH_2OH-[CHOH]_4-COO^- + H_2O \rightarrow {}^-OOC-[CHOH]_3-COO^- \\ + HCO_2H + 6e^- \tag{7}$$

followed by the slower cleavage of the C_5–C_6 bond:

$$^-OOC-[CHOH]_3-COO^- + H_2O \rightarrow {}^-OOC-[CHOH]_2-COO^- \\ + HCO_2H + 2H^+ + 2e^- \tag{8}$$

It is thought that the oxidation of all carbohydrates proceeds from the terminal carbons. Moving the potential farther positive still, it can be seen that at around $+0.2$ V the current begins to fall rapidly. This is due to the formation of the insulating oxide layer indicated in curve A. Hence, when the scan direction is reversed, reduction of the oxide layer back to metal (about $+0.1$ V) causes the oxidation of glucose to be resumed, and so we see a second oxidation peak. This process is observed for all carbohydrates, although obviously in the case of ketoses (such as fructose) the current comes only from the oxidation of the alcohol groups, since there is no aldehyde present. Similar characteristics are found when reacting carbohydrates at Pt electrodes, with the difference that at Pt carbohydrate oxidation occurs at potentials where dissolved oxygen is reduced. In principle, amperometry, as described above, should be suitable for measuring carbohydrates by electro-oxidation. However, a complicating factor is that passivation occurs due to the adsorption of reaction products (LaCourse and Johnson 1991). This problem can be solved by using the technique of pulsed amperometric detection, as described below.

28.7   Pulsed Amperometric Detection (PAD)

The concept of applying a pulsed potential to a noble metal electrode as a method of cleaning the electrode can be traced back to the work of Hammett in 1924, who used a pulsed waveform to reactivate a Pt electrode for H_2 oxidation. Armstrong and co-workers used a similar method in 1933 to study oxygen reduction at Pt, and in 1954 Kolthoff and Tanaka used pulsed methods to look at Pt electrodes in different supporting electrolytes. Following on from these studies, pulsed cleaning techniques were applied to Pt for the oxidation of ethylene (Clark *et al.* 1972) and *p*-aminophenol (MacDonald and Duke 1973). A form of electrode cleaning was first used for glucose oxidation by Marinčić *et al.* (1979), who applied successive positive and negative potential sweeps prior to recording the glucose $I - E$ curve. A problem with this approach is that it requires around 6 min for each measurement, and obviously cannot be applied to flow analysis. To achieve a quicker method of measurement, in the late 1970s a research group led by Denis C. Johnson at Iowa State University developed a general method for conducting electroanalysis at noble metal electrodes based on triple-pulsed waveforms, and applied the method to the detection of alcohols (Hughes *et al.* 1981a) and carbohydrates (Hughes *et al.* 1981b). The technique became known as *pulsed amperometric detection* (PAD), and has since been applied to a large number of passivating analytes, including aldehydes, amines and organo-sulphurcompounds. A monograph by LaCourse (1997) provides a good overview of the technique.

The basic waveform used for PAD is illustrated in Figure 28.7(a). The potential E_1 is referred to as the detection potential and is set to a value where the relevant analyte oxidation occurs. This value can be identified from a CV of the analyte. For example, to detect glucose by PAD at a Au electrode, a potential of about $+0.15$ V would be used, based on the CV shown in Figure 28.6. The first part of the E_1 period is a delay time (t_d) of around 240 ms during which no current is measured. The purpose of this is to allow the capacitive component of the current to decay to virtually zero, as explained in section 28.4.1. The remainder of the period of E_1 is a sampling time (t_s) of about 200 ms, during which the current is measured. This current will be almost

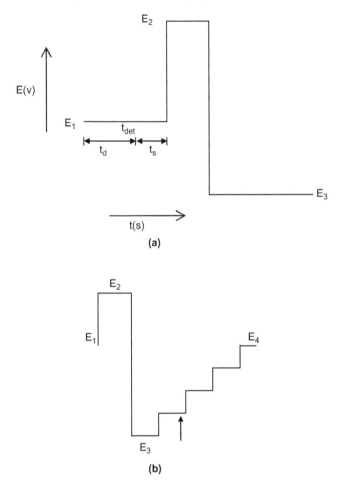

Figure 28.7 (a) Waveform used in pulsed amperometric detection (PAD) at a fixed potential. Measurement is conducted at E_1 during t_s. E_2 and E_3 are used to clean the electrode. Suitable for batch or flow analysis. (b) Waveform used in PAD at a ramped potential. Measurement is conducted from E_3 to E_4 with the current sampled close to the end of each step (see arrow for example). Axis as in (a).

wholly Faradaic in nature and therefore can be related to the concentration of the analyte. After the measurement period, the current is stepped to a value (E_2) of around $+0.8\,V$ for about 180 ms, followed by a step to a negative potential (E_3) of about $-0.3\,V$ for about 360 ms. During E_2 an insulating oxide layer is formed on the electrode and during E_3 it is stripped off to return the bare metal. This process has the effect of cleaning the electrode of adsorbed reaction products, either because the high value of E_2 causes desorption after further oxidation, or because when the oxide film is removed at E_3 the fouling species are removed with it. Hence, the next application of E_1 occurs at a clean surface.

The application of PAD requires the formation of a metal oxide film and so will be favoured by an alkaline pH, due to the fact that oxide formation generates protons:

$$M + H_2O \rightarrow M^{\bullet}OH_{ads} + H^+ \rightarrow MO + H^+ + 2e^- \tag{9}$$

In the case of carbohydrate oxidation the electrode reactions themselves also supply protons, as illustrated in equations Eqs (6)–(8). This is the reason that an alkaline media is necessary for glucose and fructose oxidation at Au. However, with Pt WEs oxidation can even be achieved in acidic media. This is thought to be due to the better adsorptive properties of Pt, due to the metal having a greater number of unpaired d-orbitals (Johnson and LaCourse 1990).

28.8   Simultaneous Detection of Glucose and Fructose

Using the PAD technique described above, either glucose or fructose can be detected either by batch measurement (stirred solution/rotated electrode) or by electrodes in a flow cell. Since PAD lacks any inherent selectivity it is often applied in flow cell measurement downstream from a liquid chromatography unit. The application of liquid chromatography to the resolution of glucose and fructose is described elsewhere in this chapter.

In the absence of chromatographic separation, glucose and fructose can be detected by extending the PAD technique to a potential ramp, as illustrated in Figure 28.7(b). After the cleaning step the potential is scanned from E_3 to E_4. However, instead of a conventional ramp, the value of E changes in a series of steps of typically 5–10 mV amplitude and 20–50 ms duration. The current is sampled shortly before the end of each step, so that, as with the delay time function in Figure 28.7(a), the capacitive part of the current is suppressed. The experiment is performed in a quiescent solution and hence the analyte reaction is observed as a peak. In the case of glucose and fructose both compounds exhibit a common oxidation peak at approx. $+0.15$ V which, as noted in eqs (7) and (8), is due to oxidation of the alcohol groups. However, only glucose has an aldehyde group, and so the aldehyde oxidation in eq. (6) (approx. -0.6 V) is observed at glucose only. This fact was employed by Fung and Mo (1995) to detect the two compounds simultaneously. The aldehyde peak was used to determine the glucose concentration and then this value was subtracted from the "total sugar" reading, determined from the alcohol peak, to reveal the fructose concentration. The authors examined the effect of a number of potential interferents including sweeteners, colours, preservatives, amino acids, thickening agents, stabilisers and vitamins. They found some sweeteners (sorbital, mannitol and sacharin) could be present at up to 0.1% before causing interference, preservatives such as sodium metabisulphide, sorbic acid, benzoic acid and salicyclic acid, could be tolerated up to 100 ppm, and all other compounds tested could be present at 1000 ppm before interfering. Some real samples, including fruit juices, honey and a glucose energy drink, were analysed by both PAD and HPLC, and the results were found to agree closely.

Bessant and Saini (1999) noted that the peak currents produced by glucose and fructose mixtures were not always simple summations of the individual responses, and so developed an artificial neural network to recognise the separate current contributions.

A problem with using the Figure 28.7(b) waveform is that it is not applicable to flow analysis, which is the technique of choice when a high sample throughput is required. To adapt simultaneous glucose and fructose PAD detection to a flow cell, Surareungchai *et al.* (2001) split E_1 in Figure 28.7(a) into two separate detection potentials, $E_{det1} = -0.45\,\mathrm{V}$ for aldehyde detection and $E_{det2} = +0.20\,\mathrm{V}$ for alcohol detection. Analogous to the potential scan method of Fung and Mo (1995), the flow peak at E_{det1} gives the glucose concentration and the peak at E_{det2} gives the glucose + fructose concentration. Hence, the fructose concentration can be extracted. In examining interferences the authors found mannose gave a very similar response to glucose at both E_{det1} and E_{det2}, which is reasonable since mannose is sterioisometric with glucose. The non-reducing sugar sucrose gave a lower response than glucose, which is probably due to one of the steps of the sucrose electrode reaction being much slower. In these experiments a Pt WE was used, coated with a layer of Nafion. This is a negatively charged ionomer and hence the interference caused by the oxidation of organic acids and some amino acids could be reduced, as charge repulsion prevented the interfering anions from reaching the electrode. Analysis of five fresh fruit samples (tomato, watermelon, cantaloupe, apple and grape) by PAD and HPLC gave a good agreement, with the closest agreement found in the samples free of sucrose (tomato and grape).

Summary Points

- If an analyte can be made to react at an electrode, then the current from that reaction can be used to determine the analyte concentration.
- Both glucose and fructose can be oxidised at noble metal electrodes.
- To oxidise glucose and fructose at Au an alkaline media (pH > 12) must be used, but Pt electrodes can achieve oxidation even in acidic media.
- Glucose and fructose oxidation causes passivation, or electrode fouling, due to the strong adsorption of oxidation products.
- Because of passivation, conventional voltammetry is not suitable for glucose and fructose measurement.
- The technique known as pulsed amperometric detection (PAD) can be used to oxidise glucose and fructose without passivation occurring.
- PAD functions by applying a high positive and then high negative potential to the electrode (both for ~1 s) before measuring the current from the analyte oxidation.
- During PAD the high positive potential breaks down adsorbed reaction products further while depositing an oxide layer on the metal surface; the

high negative potential strips off the oxide layer, taking off the adsorbed material in the process, thus returning the electrode to its clean state.
- Glucose and fructose can be measured by adding the PAD waveform to the application of either a fixed potential (suitable for batch or flow analysis) or a potential scan (batch measurement only).
- Glucose and fructose share a common oxidation potential, due to alcohol oxidation, but glucose also has a second less positive oxidation potential due to oxidation of the aldehyde group.
- In a mixture of glucose and fructose oxidation of the aldehyde group can be used to determine the glucose concentration, while oxidation of the alcohol group can be used to determine the total glucose + fructose concentration; subtraction of the former from the latter hence gives the fructose concentration.
- Interference from organic acids can be minimised by coating the electrode with a negatively charged polymer such as Nafion.
- Other oxidisable sugars such as sucrose and mannose will interfere and therefore, if present in the sample matrix, must be removed by liquid chromatography or other means.

Key Facts of Voltammetry

- When a species in solution reacts at an electrode, the reaction is the product of two separate processes: mass transport from bulk solution to the electrode surface and electron transfer at the electrode surface.
- Mass transport can be driven by a gradient in: concentration (the phenomena of diffusion), electrical potential (the phenomena of migration), pressure (the phenomena of convection).
- The concentration of the reacting species is always lower at the electrode surface than in bulk and therefore diffusion always occurs.
- The electrode potential acts on all charged species in solution and since the supporting electrolyte (section 28.3.3) is present at a much higher concentration than the analyte, it will carry most of the migrational charge. Hence, the effect of migration can be ignored when supporting electrolyte is present.
- A gradient of pressure – leading to convection – is generated by stirring the solution, rotating the electrode or using a flowing stream.
- Out of electron transfer and mass transport, the slower rate (known as the rate determining step) will control the value of the electrode current.
- If electron transfer is the rate determining step, then the current will change exponentially with applied potential (eq. (3)), which means a small change in the reference electrode will give a large change in response.
- Small variations in reference potential may be expected in practice. In contrast, mass transport effects, such as stirrer speed or pump flow rate, are much easier to reproduce. For this reason, the usual working potential in analytical voltammetry is one where diffusion is the rate determining step.

- The potential range providing a diffusion-limited reaction can be identified from a potential ramp experiment (section 28.4.3, a value beyond the peak potential), or by repeating an amperometric experiment (section 28.4.2) at a series of applied potentials and choosing the potential at which the response no longer changes.

Definitions of Words and Terms

Electrode: Phase through which charge is transported by the movement of electrons.

Electrolyte: Phase through which charge is transported by the movement of ions.

Voltammetry: Measurement of current through an electrode under conditions of controlled potential.

Working electrode (WE): Electrode at which the redox reaction of interest takes place.

Reference electrode (RE): Reference electrode/solution interface across which the potential is controlled relative to the WE.

Counter electrode (CE): Chemically inert electrode at which reaction occurs to complete the current path from the WE.

Amperometry: Voltammetric experiment performed at a fixed potential under conditions of convective mass transport (see section 28.4.2)

Cyclic voltammetry: Voltammetric experiment in which the potential is ramped linearly from one potential to another, and then back to the starting point (see section 28.4.3)

Flow analysis: Voltammetric technique where analyte is injected into a mobile phase flowing past or onto the WE. Usually used in conjunction with amperometry or pulsed amperometric detection.

Passivation: Phenomena whereby reaction products remain adsorbed on the electrode surface, leading to a deterioration in the electrode response (see section 28.5). Also called electrode fouling.

Pulsed amperometric detection: Voltammetric technique in which the potential is applied to the WE in a pulsed form, as a way of cleaning the electrode surface during measurement (see section 28.7).

List of Abbreviations

a_j	Activity of species j
A	Electrode area (cm^2)
C_j	Concentration of species j in bulk solution ($mol\ cm^{-3}$)
C_j^s	Concentration of species j at electrode surface ($mol\ cm^{-3}$)
CE	Counter electrode
CV	Cyclic voltammetry

D_j	Diffusion coefficient of species j $(cm^2\ s^{-1})$
E	Electrode potential (V)
E_1	Detection potential in pulsed amperometric detection (V)
E_2	Anodic cleaning potential in pulsed amperometric detection (V)
E_3	Cathodic cleaning potential in pulsed amperometric detection (V) *or* start of potential ramp in pulsed amperometric detection (V)
E_4	End of potential ramp in pulsed amperometric detection (V)
E_{det1}	First detection potential in bi-analyte pulsed amperometric detection (V)
E_{det2}	Second detection potential in bi-analyte pulsed amperometric detection (V)
E^o	Standard electrode potential (V)
F	Faraday constant (96485 C mol^{-1})
HPLC	High performance liquid chromatography
I	Current (A)
I_o	Exchange current (A)
I_C	Capacitive current (A)
I_F	Faradaic current (A)
I_{SS}	Steady state current (A)
k^o	Standard rate constant for electron transfer $(cm\ s^{-1})$
Q	Charge (C)
LSV	Linear sweep voltammetry
m	Molar quantity of species converted at electrode (mol)
n	Stoichiometric number of electrons in overall electrode reaction (dimensionless)
PAD	Pulsed amperometric detection
R	Gas constant (8.314 J $K^{-1}\ mol^{-1}$)
R_U	Solution resistance between working electrode and reference electrode (Ω)
RE	Reference electrode
SCE	Saturated calomel electrode
t	Time (s)
t_d	Delay time in pulsed amperometric detection before current is measured (s)
t_s	Sampling time in pulsed amperometric detection during which current is measured (s)
T	Temperature (K)
WE	Working electrode
α	Transfer coefficient (dimensionless)
δ	Width of Nernst diffusion layer (cm)

Acknowledgements

The authors are grateful to Chatuporn Phanthong for assistance in producing the figures in this chapter.

References

Armstrong, G., Himsworth, F.R., and Butler, J.A.V., 1933. The kinetics of electrode processes. Part III. The behaviour of platinum and gold electrodes in sulphuric acid and alkaline solutions containing oxygen. Proceeding of the Royal Society London A. 143: 89–103.

Bard, A.J., and Faulkner, L.R., 2001. *Electrochemical methods fundamentals and applications*, 2nd edn. John Wiley & Sons, Inc. New York, USA.

Bessant, C., and Saini, S., 1999. Simultaneous determination of ethanol, fructose, and glucose at an unmodified platinum electrode using artificial neural networks. *Analytical Chemistry*. 71: 2806–2813.

Clark, D., Fleischman, M., and Pletcher, D., 1972. The partial anodic oxidation of propylene in acetonitrile. *Journal of Electroanalytical Chemistry*. 36: 137–146.

Fung, Y.S., and Mo, S.Y., 1995. Application of dual-pulse staircase voltammetry for simultaneous determination of glucose and fructose. *Electroanalysis*. 7: 160–165.

Greef, R., Peat, R., Peter, L.M., Pletcher, D., and Robinson, J., 1993. *Instrumental Methods in Electrochemistry*. Ellis Horwood, New York, USA.

Gulaboski, R., and Pereira, C.M., 2008. Electroanalytical techniques and instrumentation in food analysis. In: Otles, S. (ed.) *Handbook of Food Analysis Instruments*. CRC Press, Boca Raton, USA, pp. 379–402.

Hammett, L.P., 1924. The velocity of the hydrogen electrode reaction on platinum catalysts. *Journal of the American Chemical Society*. 46: 7–19.

Hughes, S., Meschi, P.S., and Johnson, D.C., 1981a. Amperometric detection of simple alcohols in aqueous solutions by application of a triple-pulse potential waveform at platinum electrodes. *Analytica Chimica Acta*. 132: 1–10.

Hughes, S., and Johnson, D.C., 1981b. Amperometric detection of simple carbohydrates at platinum electrodes in alkaline solutions by application of a triple-pulse potential waveform. *Analytica Chimica Acta*. 132: 11–22.

Johnson, D.C., and LaCourse, W.R., 1990. Liquid chromatography with pulsed amperometric detection. *Analytical Chemistry*. 62: 589A–597A.

Johnson, D.C., Dobberpuhl, D., Roberts, R., and Vandeberg, P., 1993. Pulsed amperometric detection of carbohydrates, amines and sulfur species in ion chromatography - the current state of research. *Journal of Chromatography*. 640: 79–96.

Kolthoff, I.M., and Tanaka, N., 1954. Rotated and stationary platinum wire electrodes. Analytical Chemistry. 26: 632–636.

LaCourse, W.R., and Johnson, D.C., 1991. Optimization of waveforms for pulsed amperometric detection (p.a.d.) of carbohydrates following separation by liquid chromatography. *Carbohydrate Research*. 215: 159–178.

LaCourse, W.R., 1997. *Pulsed Ampeometric Detection in High Performance Liquid Chromatography*. Wiley-Interscience, New York, USA.

MacDonald, A., and Duke, P.D., 1973. Small-volume solid-electrode flowthrough electrochemical cells: Preliminary evaluation using pulse polarographic techniques. *Journal of Chromatography A*. 83: 331–342.

Manz, A., Graber, N., and Widmer, H.M., 1990 Miniaturized total chemical analysis systems: a novel concept for chemical sensing. *Sensors and Actuators B: Chemical.* 1: 244–248.

Marinčić, L., Soeldner, J.S., Colton, C.K., Giner, J., and Morris, S., 1979. Electrochemical glucose oxidation on a platinized platinum electrode in Krebs-Ringer solution. *Journal of the Electrochemical Society.* 126: 43–49.

Surareungchai, W., Deepunya, W., and Tasakorn, P., 2001. Quadruple-pulsed amperometric detection for simultaneous flow injection determination of glucose and fructose. *Analytica Chimica Acta.* 2001: 215–220.

CHAPTER 29

Sucrose Determination by Raman Spectroscopy

LEONARDO M. MOREIRA,*[a] FABIO V. SANTOS,[b]
JULIANA P. LYON,[c] PATRÍCIA LIMA,[a] VANESSA
J. S. V. SANTOS,[b] PEDRO C. G. DE MORAES,[a] JOSÉ
PAULO R. F. MENDONÇA,[a] VALMAR C. BARBOSA,[d]
CARLOS J. DE LIMA,[e] FABRÍCIO L. SILVEIRA[e] AND
LANDULFO SILVEIRA JR.[e]

[a] Universidade Federal de São João Del Rei, Praça Dom Helvécio, 74,
Fábricas, São João del Rei, MG, Brazil; [b] Center of Health Science, Federal
University at São João del Rei, 400 Sebastião Gonçalves Coelho Street,
Divinópolis, MG, Brazil; [c] Department of Natural Sciences, Federal
University at São João del Rei, Praça Dom Helvécio, 74, Fábricas, São João
del Rei, MG, Brazil; [d] Universidade Federal do Rio de Janeiro (UFRJ),
Instituto de Física (IF), Departamento de Física Nuclear, Av. Athos da
Silveira Ramos, 149, Centro de Tecnologia, Bloco A, Cidade Universitária,
Ilha do Fundão, Rio de Janeiro, RJ; [e] Universidade Camilo Castelo Branco,
São José dos Campos, São Paulo, Brazil
*Email: leonardomarmo@gmail.com

29.1 Introduction

Raman spectroscopy is one of the most versatile instrumental techniques currently used for structural characterisation. This technique has been applied to several biological systems for the chemical characterisation of these respective

Food and Nutritional Components in Focus No. 3
Dietary Sugars: Chemistry, Analysis, Function and Effects
Edited by Victor R Preedy
© The Royal Society of Chemistry 2012
Published by the Royal Society of Chemistry, www.rsc.org

biochemical environments (Moreira *et al.* 2008). The application of Raman spectroscopy in the characterisation of nanomaterials is also an important area for the employment of this vibrational technique. For instance, studies focused on graphitic carbon nanocoils with a high degree of crystallinity have been synthesised using hydrothermally carbonised samples as intermediates, which are known as hydrochar (Sevilla and Fuertes 2009). These structures are obtained from three representative carbohydrates, glucose, sucrose and starch. According to Sevilla and co-workers (Sevilla and Fuertes 2009), this synthesis scheme allows for the fabrication of uniform nanostructures of graphitic nanocoils, which have been well characterised by Raman spectroscopy among other techniques.

In fact, carbohydrates are considered to be versatile compounds and interesting precursors in the nanoscience field. For example, sucrose esters, which have a wide range of hydrophilic-lipophilic balance values, can, consequently, can be used as surface active agents (surfactants) or as solubility or penetration enhancers (Szűts *et al.* 2008).

In a biological context, the physiological implications of the chemical composition of various materials with or without pathological alterations have been successfully evaluated using several spectroscopic tools. Some instrumental analysis techniques have already been tested *in situ*, such as ultraviolet-visible electronic absorption, infrared, fluorescence and Raman spectroscopies, which have been proposed and tested for biochemical and clinical analysis with significant success (Pons *et al.* 2004). These techniques are relatively rapid compared to the conventional clinical methodologies and can be developed for *in vivo* screening of disease, thus avoiding the need for a patient to undergo a painful biopsy. In this way, these minimally invasive techniques can be exploited for population screening, early diagnosis, prognosis, monitoring of therapy and subsequent aspects (Krishna *et al.* 2006). Furthermore, the increasing application of instrumental analysis as a result of the development of the biomedical engineering field has introduced new perspectives on the use of spectroscopic techniques in medicine. In this context, the spectroscopy of molecular vibrations has been experiencing a great advancement due to the substantial technical improvements in experimental methods, the increased computational capabilities and the growing analytical demands (Rohleder *et al.* 2005). In addition, the decreased cost of acquisition of Raman spectrometers has made this instrumental analytical tool more accessible to many laboratories.

Because of their heterogeneous characteristics, biological samples present significant issues that need to be addressed. The versatility of Raman spectroscopy allows for the characterisation of complex mixtures, which is a property that is not found in many other instrumental tools. Furthermore, although most components are present in low physiological concentrations in biological fluids (on the order of mmol L^{-1} to nmol L^{-1}) (Reyes-Goddard *et al.* 2005), Raman spectroscopy can still be used to obtain a significant amount of physico-chemical information from these samples. Despite these difficulties, which are inherent to the biological medium, some spectroscopic techniques, especially the vibrational spectroscopies, have been considered as a basis for minimally invasive and non-destructive measuring systems. This characteristic is a relevant pre-requisite in the biomedical engineering field for improving the quality of life for great number of patients (Zeng *et al.* 2004), and this non-destructive property is also highly important in food analysis where it permits the systematic methodology of chemical analysis.

The notion that vibrational spectroscopies might play a role in clinical diagnostics and food analysis emerged through the work of early pioneers, which showed that even complex biomolecules such as proteins, lipids, and nucleic acids have distinct vibrational signatures that reflect the structure and conformation of tissues and biochemical systems. In fact, the biological medium contains several types of molecules that can mutually interact, which results in perturbations of their respective "pure" spectra.Infrared (FT-IR) and Raman spectroscopy (RS) have demonstrated high sensitivity that has improved immensely. Several methods have been developed to obtain high-quality spectra for various complex samples from virtually any type of biological materials. Thus, vibrational spectroscopists are working to address the challenges inherent to the study of cells, tissues, biological fluids and even intact human beings (Shaw and Mantsch 1999). Moreover, these vibrational spectroscopic techniques are important analytical tools with several different applications in the areas of pharmacy and chemistry (Wartewig and Neubert 2005).

29.2   Vibrational Spectroscopy

Several types of spectroscopies involve vibrational transitions. Two of the most widespread vibrational techniques are Fourier transform infrared spectroscopy (FT-IR) and Raman scattering spectroscopy, or, simply, Raman spectroscopy. Infrared spectroscopy is well established as a great technique for structural characterisation. This instrumental analytical method probes the transitions between vibrational energy levels of a molecule that are induced by the absorption of infrared (IR) electromagnetic radiation. This technique is usually more sensitive than Raman spectroscopy but often results in various spectral lines related to the vibrational and rotational transitions of water, a major component of biological samples. Alternatively, Raman spectroscopy probes the vibrational transitions that occur as a result of light scattering by molecules, which is frequently associated with a lower sensitivity. However, the influence of water peaks is not a significant limitation to Raman analysis because Raman spectroscopy is related to the change in polarisability of the molecules, whereas infrared spectroscopy is associated with the dipole moment variation of the molecules. Consequently, some normal modes of vibration can be detected by both Raman and infrared spectroscopies, whereas other vibrational transitions are detectable by only one technique.

29.2.1 Raman Spectroscopy

Raman spectroscopy relies on the inelastic scattering of monochromatic light, which is known as Raman scattering. This process frequently involves laser light in the visible, near-infrared, or near-ultraviolet ranges. The laser light interacts with molecular vibrations, phonons or other excitations in the system, resulting in a shift in the energy of the laser photons. This shift in energy provides information about the phonon modes in the system. Thus, Raman spectroscopy furnishes a significant amount of structural information by investigating the vibrational normal modes.

It is important to note that infrared spectroscopy yields similar, but complementary, information with respect to Raman spectroscopy. In fact, there are some active vibrational modes in Raman spectroscopy that are not active in infrared spectroscopy, and, conversely, there are some active vibrational modes in infrared spectroscopy that are inactive in Raman spectroscopy. Thus, depending on the molecules under evaluation and the relevant spectral range, one type of spectroscopy (*i.e.*, infrared or Raman) may be more suitable for the desired analysis. For example, water's influence on the spectral lines, which is more significant in infrared than in Raman spectroscopy, can negatively affect some spectral analyses if the spectral regions containing the water peaks coincide with those containing the spectral lines of the sample. Alternatively, because infrared spectroscopy is not based on scattering, whereas Raman spectroscopy is, some low-intensity peaks are more easily detected in infrared spectroscopy than in Raman spectroscopy.

A schematic representation of a Raman spectrometer is presented in Figure 29.1. The instrument consists of a laser diode that emits light at a wavelength of 830 nm and lenses that focus the laser photons onto the sample, resulting in elastic and inelastic light scattering. The photons generated by the light

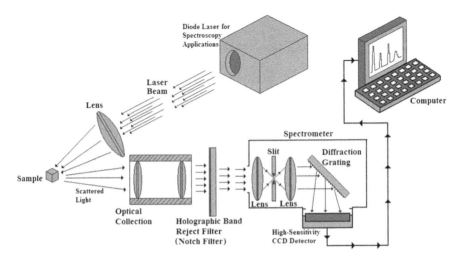

Figure 29.1 Schematic representation of a Raman spectroscopic device.

scattering are collected and filtered by an optical system, which allows for the maximum attenuation of the light from the laser. This process permits the transmission of the spectroscopic radiation from the sample. Then, the radiation passes through a structure composed of lenses to arrive at a reflexive diffraction net, which allows the radiation to be collected until it is transmitted to a photosensor with high sensitivity. Finally, this photosensor is connected to an electronic system for signal treatment that produces a readout of the Raman spectrum corresponding to the sample.

Raman spectroscopy can be used in conjunction with other analytical tools to obtain a more consistent physico-chemical evaluation of the chemical mixtures that are found in biological systems. For instance, the Raman microprobe, which combines the analytical properties of Raman spectroscopy and the visualisation capabilities of a high-quality microscope, allows for both the localisation and the identification of species and phases in time intervals of seconds to a few minutes and with a spatial resolution down to 1 μm (Celedon and Aguilera 2002). This spectroscopic association is very promising for future applications of Raman spectroscopy aimed at obtaining physico-chemical information in biochemical systems. For example, this instrumental association can be applied to food chemistry for the *in situ* differentiation of sucrose and lactose crystals and of amorphous and crystalline phases in sucrose as well as for the detection of semi-crystalline and gelatinised zones in starch granules (Celedon and Aguilera 2002).

The suitable employment of a Raman spectrometer depends directly on adequate instrumental calibration. Several organic compounds have been utilised for this objective. For instance, naphthalene is a well-known organic compound that presents very characteristic spectral peaks. These spectral "fingerprints" permit precise calibration of a Raman spectrometer, which is a fundamental pre-requisite to the subsequent application of this technique to the characterisation of several materials. This calibration is necessary because biological materials are usually very complex samples that contain different types of compounds. The quantitative determination of carbohydrates in a biological medium, for instance, is only a precise analytical procedure after a careful calibration of the Raman spectrometer (Figure 29.2).

29.2.2 Raman Spectroscopy Applied to the Characterisation of Foods and Drugs in Biological Systems

The use of Raman spectroscopy to evaluate different types of products has advanced significantly and has been applied in several areas of study, such as cellular biology, pharmaceutical analysis (Larmour *et al.* 2010) and food chemistry (Salameh *et al.* 2006).

Raman spectroscopy has evolved rapidly and is currently considered to be a direct and non-destructive technique in pharmaceutical analysis (de Veij *et al.* 2009). Generally, the focus in this field is on the active ingredients and not on the excipients present in the drugs. However, this technique has the potential to

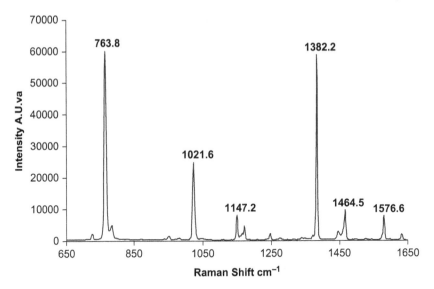

Figure 29.2 Raman spectrum of the organic compound naphthalene, which is commonly employed to calibrate the Raman spectrometer.

investigate several components of a given medicine without previously isolating these respective chemical components.A collection of the Raman spectra of widely used pharmaceutical excipients was amassed by de Veij and co-workers (de Veij *et al.* 2009), which can serve as a reference for the interpretation of Raman spectra during drug analysis. In this excellent work, de Veij and co-workers (de Veij *et al.* 2009) performed the analysis of 43 excipients that can be classified into seven categories: mono- and disaccharides (dextrose, lactose and sucrose), polysaccharides (microcrystalline cellulose, methylcellulose, carboxy-methylcellulose, hydroxypropyl cellulose, hydroxypropyl methylcellulose, wheat starch, maltodextrin, primojel, tragacanth and pectin), polyalcohols (propylene glycol, lactitol, maltitol, erythritol, xylitol, mannitol and sorbitol), carboxylic acids and salts (alginic acid, glycine, magnesium stearate, sodium acetate and sodium benzoate), esters (arachis oil, Lubritab, dibutyl sebacate, triacetin, Eudragit E100 and Eudragit RL100), inorganic compounds (calcium phosphate, talc, titanium dioxide (anatase and rutile), calcium carbonate, magnesium carbonate, sodium bicarbonate and calcium sulphate) and some unclassified products (gelatin, macrogol 4000 (polyethylene glycol), polyvinyl pyrrolidone and sodium lauryl sulphate) (de Veij *et al.* 2009).

As an example of the use of reference compounds to interpret Raman data, the various polymorphic forms of mannitol were characterised and quantified based on the Raman spectra of the existing pure forms, and these results are consistent with the X-ray powder diffraction data for the same compounds (Xie *et al.* 2008). This Raman method (a novel multivariate curve resolution (MCR)-based Raman spectroscopic methodology) has been demonstrated to be very suitable for both monitoring and controlling the mannitol polymorphic forms in the lyophilised drugs during formulation and process development, allowing the quality of the drug product to be evaluated and improved (Xie *et al.* 2008).

Bulk Raman investigations of food and their constituents have been carried out, but this subject has not been extensively explored in the literature addressing the application of Raman mapping capabilities to food. The applicability of Raman spectroscopic mapping to the analysis of several types of food, such as white and milk chocolate samples, has been demonstrated, with promising results (Larmour *et al.* 2010). It is important to note that the economic impact of the advancement of food chemical analysis is very high, suggesting that the scientific and technological contribution of Raman spectroscopy to food chemical analysis has a very pronounced economic relevance.

Many common food ingredients (crystalline sugars, inorganic salts, and organic acids) exhibit deliquescence, a moisture-induced solid-to-solution phase transition that occurs at a characteristic relative humidity. Studies that include Raman spectroscopy analysis demonstrate that lowering the deliquescence of a mixture of ingredients, such as sucrose and citric acid mixtures, has the potential to impact both the chemical and physical stability (caking) of powdered ingredient mixtures (Salameh *et al.* 2008).

de Oliveira and co-workers (de Oliveira *et al.* 2002) observed several vibrational bands in the 500–1800 cm^{-1} region that they identified as fingerprints of the two major components in honey: fructose and glucose, and at least one vibrational band characteristic of sucrose. Moreover, the relative intensities of the vibrational bands in the C-H stretching region of the FT-Raman spectra were found to be sensitive to the observed physical states of the specimens (crystallised and fluid samples) (de Oliveira *et al.* 2002). To characterise the changes that occur in honey during the crystallisation process, near-infrared spectroscopy (NIRS) has been used with great success (Bakier 2009). This methodology provided an analysis of the same type of honey that was used in other methods and involved the subtraction of the spectrum of the crystallised species from that of the liquid species, *i.e.*, the isolation of these two main components of honey. Interestingly, comparing the difference spectra allowed for the determination of the specific water-bonding processes that were affected by the crystallisation process. The peak that appeared during the absorption of water was observed between 5330 and 4965 cm^{-1}, with a maximum at 5155 cm^{-1}. The area under the peak was found to be directly related to the water activity changes that occurred in the honey during the crystallisation process (Bakier 2009).

Other vibrational spectroscopic techniques have also been employed to evaluate the carbohydrate concentration in honey, such as Fourier transform infrared (FT-IR) spectroscopy in conjunction with a microattenuated total reflectance (mATR) sampling accessory and chemometrics (partial least squares and principal component regression). This method was applied to the simultaneous determination of saccharides, such as fructose, glucose, sucrose, and maltose. The spectral data obtained from FTIR-mATR were corroborated by HPLC data, demonstrating the accuracy of this methodology (Tewari and Irudayaraj 2004).

Polyacetylenes, carotenoids, and polysaccharides that are present in carrots (*Daucus carota* L.) have also been well characterised by Raman spectroscopy. Interestingly, the components were measured *in situ* in the plant tissue without any preliminary sample preparation. The molecular structures of the main carrot polyacetylenes, falcarinol and falcarindiol, are similar, but their Raman spectra exhibit specific differences including a shift of their fingerprint bands, which are observed at 2258 and 2252 cm^{-1}, respectively. Carotenoids can be identified by the -C=C- stretching vibrations (approximately 1520 and 1155 cm^{-1}) of the conjugated system of their polyene chain, whereas the characteristic Raman band at 478 cm^{-1} is due to the skeletal vibration mode of the starch molecule. The other polysaccharide, pectin, can be identified by its characteristic band at 854 cm^{-1}, which is due to the -C-O-C- skeletal vibration mode of carbohydrates. The Raman technique that was applied in the present case has revealed detailed structural information regarding the relative distribution of polyacetylenes, carotenoids, starch, and pectin in the investigated plant tissues, demonstrating the ability of this tool to differentiate peaks associated with several organic functions.

Raman spectroscopy has also been used to determine the ethanol content of fermentation. Specifically, a regression equation that was obtained from the Raman intensity ratios of standard ethanol solutions (relative to an internal standard) was used to quantify the ethanol concentration (Li *et al.* 2010). The processes of strain screening for ethanol fermentation were measured, and the results obtained with this method had an accuracy equivalent to those obtained with gas chromatography (Li *et al.* 2010). This method requires only between 10 and 15 seconds to collect Raman spectra because the results can be read out simultaneously, which indicates that this method is very rapid (real-time) and high-throughput for monitoring the ethanol content of fermentation for both the ethanol industry and strain screening (Li *et al.* 2010).

29.2.3   Raman Spectroscopy Applied to the Characterisation of Saccharides

Carbohydrates are widespread in nature, and some saccharides occur in a practically pure form, such as sucrose, glucose fructose, amide and cellulose. Interestingly, cellulose is considered to be one of the most available and relevant molecules in nature, occurring in cotton, wood and paper; thus, it is considered one of the most important carbohydrate resources on Earth (Ferreira *et al.* 2001).

The renewable biomass on Earth consists of several products of low and high molecular masses, such as carbohydrates, amino acids, and lipids. These chemical compounds are utilised for nutrition, the production of a significant variety of fuels, and chemical and energetic products (Ferreira *et al.* 2009). The great molecular mass of these biological macromolecules makes this type of chemical compound highly relevant in several areas, such as nutrition and ecological equilibrium. Interestingly, these carbohydrates constitute the main source of energy in nature.

The management of the renewable biomass includes biological methods, chemical and thermal transformations, and mechanical treatment to obtain solid, liquid and gas fuels or chemical products of higher aggregated value (Ferreira *et al.* 2009). Recently, the rational use of natural resources has constituted one of the most relevant discussions in the biological context. Thus, significant attention has been focused on analysing the degree of purity of important macromolecules, such as various carbohydrates. The necessity of improving the analytical methodologies employed to characterise these relevant biochemical compounds has motivated several research groups to develop new procedures of physico-chemical evaluation, with special attention to the spectroscopic tools.

In terms of the volume of world production, the saccharides and the vegetal oleos are considered extraordinarily relevant biochemical compounds. In fact, approximately 95% of the biomass produced in nature is formed from carbohydrates (200 billion tons). Despite this great quantity, only 3% of carbohydrates are used by human beings (Ferreira *et al.* 2001). The remainder of the carbohydrates are recycled by nature. In the industrial utilisation of carbohydrates, excluding nutrition, this percentage decreases significantly, and its use is restricted to some mono- and disaccharides of low molecular mass (Ferreira *et al.* 2001). Sucrose is a disaccharide that has been known since 200 B.C. and is the most available crystalline carbohydrate in nature. It is produced on a large scale in several countries (100 000 ton/year), mainly Brazil, where its cost is approximately R$ 0.50/Kg, which means that Brazilian sugar is a very accessible and economically competitive product. In fact, Brazilian cane sugar is very competitive in comparison with European beet sugar, which has led to governmental discussions regarding free trade. Interestingly, the hydrolysis of sucrose in an acid medium produces an equimolar mixture of D-Glucose (aldose) and D-fructose (ketose), which is known as inverted sugar because of its inverted optical rotation. Figure 29.3 presents the Raman spectrum of "crystal sugar", *i.e.*, sucrose in its less refined commercial form.

Figure 29.4 presents the Raman spectrum of "refined sugar", *i.e.*, sucrose in its more refined commercial form. It is interesting to note that the spectral profiles of the "crystal sugar" and "refined sugar" are very similar despite the industrial process of sugar refinement that is used to produce the "refined sugar".

Figure 29.5 presents "pulverised refined sugar", which refers to refined sucrose after submission to a physical process of pulverisation for application in the food industry. In fact, the production of many types of foods uses this form of sucrose. Interestingly, the Raman spectrum of the "pulverised refined sugar" exhibits a peculiar spectral line above 1300 cm^{-1} that is not observed in the spectra of "crystal sugar" and "refined sugar". It is possible that this spectral peak is associated with an intermolecular interaction originating from the pression used in the pulverisation process. In this way, the different aggregation could generate the distinct interactions that would slightly modify the profile of the resulting Raman spectrum. Indeed, a similar phenomenon occurs in some infrared (IR) measurements when salts are used as a support in

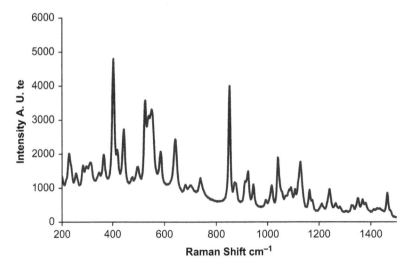

Figure 29.3   Raman spectrum of sucrose in its less refined commercial form ("crystal sugar").

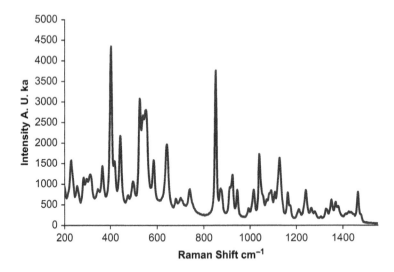

Figure 29.4   Raman spectrum of sucrose in its refined commercial form ("refined sugar").

the sample analysis. KBr, for example, is a salt that is widely used in IR measurements. However, in some cases, the excessive pression in the formation of the KBr pellet can lead to significant interactions between the ions of the salt and the compound being detected by IR. Figure 29.5 presents a general spectral profile, which contains both the characteristic spectral lines of sucrose and a line that can be used as a "fingerprint" peak for the "pulverised refined sugar".

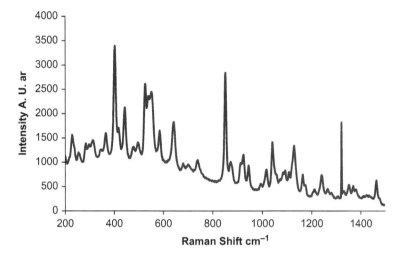

Figure 29.5 Raman spectrum of sucrose in its refined commercial form that has been pulverised for application in the food industry ("pulverised refined sugar").

Carbohydrates are compounds that, despite their organic character, present significant polarity, implying that they have a great ability to form hydrogen bonds. In an aqueous medium, the physico-chemical properties of carbohydrates are associated with this capability, which is also related to a significant interaction with their crystal structures. This interesting property has been evaluated in studies of the anhydrous crystal forms of several types of sugars (Sussich *et al.* 2009).

Inelastic neutron scattering (INS), Raman scattering, and infrared spectroscopy experiments have been used together to generate significant information about the structural properties of carbohydrates, especially in an aqueous medium. In fact, aqueous solutions of homologous disaccharides, such as trehalose and sucrose, have been extensively studied to analyse the structural modifications induced by water on trehalose and sucrose and to identify the different structural arrangements that can account for variations in their effectiveness as bioprotectors (Branca *et al.* 2003). These studies involve the analysis of intramolecular stretching, bending and intermolecular vibrational modes.

The sugar-induced thermostabilisation of lysozyme was analysed by Raman scattering and modulated differential scanning calorimetry investigations for three disaccharides (maltose, sucrose, and trehalose) that are characterised by the same chemical formula ($C_{12}H_{22}O_{11}$) (Hedoux *et al.* 2006). Indeed, it is well established that sugars are known to stabilise proteins (Wright *et al.* 2003). In this context, it is important to note the action of trehalose, which is the most effective sugar for stabilising the folded secondary structure of the protein. The bioprotective properties of trehalose, sucrose and maltose on lysozyme protein have been studied using Raman spectroscopy and molecular dynamics simulations because it has been found that the temperature dependencies of the alpha-helical unfolding processes can be determined in the presence of different sugars and at various concentrations using the amide I band. Indeed, thermal unfolding is responsible for the loss of the natural spatial arrangement of several biological molecules, which has prompted significant attention from several researchers with respect to the methods that can inhibit or, at least, decrease the unfolding mechanisms. Ionov and co-workers have evaluated the energies of stabilisation and the acting forces successfully with Raman spectroscopy (Ionov *et al.* 2006). The influence of sugars on the mechanism of thermal denaturation was carefully investigated by Raman scattering experiments carried out both in the low-frequency range and in the amide I band region. It was determined that the thermal stability of the hydrogen-bond network of water, which is highly dependent on the presence of sugars, contributes to the stabilisation of the native tertiary structure and inhibits the first stage of denaturation, which includes the transformation of the tertiary structure into a highly flexible state with an intact secondary structure (Hedoux *et al.* 2006). Studies focused on the structure-activity relationship of proteins require a consideration of the influence of the water molecules as well as the stabilising role of the sugars in several aspects of the structure-function relation, such as spatial arrangement, unfolding and denaturation.

The binding between quantum dots (QDs) and mannitol was analysed with a combination of instrumental techniques, including UV-vis, Fourier transform infrared and Raman spectroscopies (Ghosh *et al.* 2010). Near-infrared (NIR) spectroscopic analysis of noncrystalline polyols and saccharides (*e.g.*, glycerol, sorbitol, maltitol, glucose, sucrose and maltose) was performed at different temperatures to elucidate the effect of the glass transition on molecular interactions. Heating the samples decreased the intensity of an intermolecular hydrogen-bonding OH vibration band (6200–6500 cm^{-1}) with a concomitant increase in the intensity of a free and intramolecular hydrogen-bonding OH group band (6600–7100 cm^{-1}). A significant reduction in the intensity of the intermolecular hydrogen-bonding band at temperatures above the glass transition of the individual solids may explain the higher molecular mobility and lower viscosity observed in the rubber state (Izutsu *et al.* 2009).

In addition to Raman spectroscopy, it is important to note that other vibrational spectroscopies have been used extensively to study sugars. For instance, infrared spectroscopy is an interesting alternative to other methods such as polarimetry, which, despite being a destructive method, has been the reference procedure used for the identification of beet sugar since 1964 (Maalouly *et al.* 2004). Infrared spectroscopy is a non-invasive, non-destructive technique requiring minimal or no sample preparation (Blanco and Villarroya 2002), and it can also be used to simultaneously determine parameters other than the percentage of sugar.

Similarly, near-infrared (NIR) spectroscopy has been widely used for sucrose identification in aqueous solutions and fruit juices (Maalouly *et al.* 2004). Although it has been less widely used, mid-infrared (MIR) spectroscopy is also an interesting alternative for the determination of sucrose in beets and sugarcane. NIR spectroscopy has also been used to investigate the water–sugar interactions at different temperatures and physical states.This vibrational technique has been used to investigate the spectroscopic response in the 1100–2400 nm range of the spectra of solutions of glucose, fructose, and sucrose with significant sensitivity, including samples with concentrations as low as 5% (Giangiacomo 2006). It has been revealed that the shifts in the frequencies of the bands arising from mono- and disaccharides in the H_2O and D_2O solutions can be used to indicate the level of interaction between the saccharides and water (Kanou *et al.* 2005). Kanou and co-workers also demonstrated that the glycosidic linkage position of the constituent monosaccharides has a significant influence on the infrared spectroscopic characterisation of disaccharides in an aqueous solution (Kanou *et al.* 2005). Moreover, other data indicated that increasing the sugar concentration results in a more symmetric water band and a shift of the absorption maximum towards longer wavelengths. Interestingly, it was demonstrated by Giangiacomo (2006) that sugars at low concentrations behave as structure breakers of water clusters, whereas at higher concentrations they act as structure makers. It is likely that the mechanism of hydrogen bond formation is significantly different in these two cases, implying that the relative distance between the molecules as well as their spatial arrangement, such as aggregation phenomena, has a significant influence upon the stability of the sugar-sugar and sugar-water hydrogen bonds. Indeed, Mathlouthi and co-workers (Mathlouthi *et al.* 1980) noted a "structure-maker" effect in sugar solutions, suggesting that at concentrations of approximately 30% (w/w) the sugar–sugar association is reinforced and that this phenomenon is followed by the water–sugar dissociation.

Infrared spectral characteristics that were derived from disaccharides and the interaction between disaccharides and water were evaluated by Kanou and co-workers (Kanou *et al.* 2003), including six types of sugar-H_2O solutions (trehalose, kojibiose, nigerose, maltose, isomaltose and sucrose) and three types of sugar-D_2O solutions (trehalose, maltose and sucrose). The results demonstrated that the solvent affected the absorption intensity rather than frequency shift, which was caused by the interaction between disaccharides and solvents. The absorbance of the glycosidic linkage and C-OH vibration bands in H_2O and D_2O exhibited good linearity in the trehalose samples, whereas non-linearity was observed in the sucrose samples because the bands due to trehalose shifted only slightly in both H_2O and D_2O, but the shifts of the bands due to sucrose were significant. Judging from the spectral patterns, trehalose seems to be stable in both H_2O and D_2O (Kanou *et al.* 2003).

29.2.4 Raman Spectroscopy Applied to the Characterisation of Sucrose

The mechanisms of protein thermostabilisation by sugars have been analysed by Raman scattering for three disaccharides (maltose, sucrose and trehalose) that are characterised by the same chemical formula ($C_{12}H_{22}O_{11}$). These studies have demonstrated that the principal effect of sugars is to stabilise the tertiary structure, in which the biomolecule preserves its native conformation, by strengthening the hydroxyl group (OH^-) interactions. This effect is associated with the bioprotective properties of sugars that are mainly based on interactions between water and sugar (Hedoux *et al.* 2007).

In polycrystalline sugars, it is possible to observe a series of distinct absorption lines originating from the lowest intermolecular vibrational modes, whereas in amorphous sugars a broad, featureless absorption spectrum is observed. Thus, a less distinct spectrum with broader spectral peaks is indicative of the amorphous character of the carbohydrate. It is also interesting to note that an anomalous temperature dependence of the absorption line position of sucrose indicates that the effective potential of the weakest intermolecular vibrational modes in sucrose is determined by a balance between the hydrogen bond strength and the van der Waals forces (Walther *et al.* 2003).

Disaccharides, such as sucrose and trehalose, stabilise biological membranes and proteins during drying, which plays a relevant biochemical role in several biological systems, such as anhydrobiotic organisms and dehydration-resistant seeds, that contain large quantities of these disaccharides (Crowe *et al.* 1988; Crowe *et al.* 1996; Kiselev *et al.* 2005). In fact, disaccharides can replace water molecules in dry lipid and protein systems. The membrane microdomains remain intact in platelets freeze-dried with trehalose (Crowe *et al.* 2003; Kiselev *et al.* 2005). The interaction of sucrose and trehalose with phospholipid molecules remains incompletely understood, but it can serve as a model system for the elucidation of other relevant biological interactions.

The interaction between the vanadate ion (VO_3^-) and the carbohydrates sucrose, glucose and fructose in aqueous solutions (pH <6, 298.15 K) has been studied with several instrumental techniques, such as Raman spectroscopy. A significant number of hydroxyl groups are present in these compounds that can act as Lewis bases, *i.e.*, they can act as ligands to several metallic ions. These metal ions can, in turn, act as Lewis acids at the centre of a metal complex. Raman spectroscopy has been able to identify several physico-chemical aspects of the interaction between carbohydrates and metal ions. In the case of sucrose and glucose, anion hydrolysis in the absence of the sugars was indicated by a decrease in the diffusion coefficient with an increasing concentration of the anion. Significant effects on the diffusion coefficients were observed in the presence of sucrose and glucose, suggesting interactions between the carbohydrates and vanadate ion (Ribeiro *et al.* 2004). Confirmation of oligomer formation was obtained by Raman spectroscopy, where it was possible to identify the species present and to analyse their dependence on the specific carbohydrate evaluated. It is interesting to note that a vanadyl (IV) complex of the disaccharide lactose was obtained in aqueous solution at pH $= 13$ in an interesting work by Etcheverry and co-workers (Etcheverry *et al.* 2001). The inhibitory effect on alkaline phosphatase activity was tested for this compound as well as for the vanadyl (IV) complexes with maltose, sucrose, glucose, fructose and galactose. For comparative purposes, the free ligands and the vanadyl (IV) cation were also studied independently. The free sugars and the sucrose/VO complex exhibited the lowest inhibitory effect. Lactose-VO, maltose-VO, and the free VO^{2+} cation showed an intermediate inhibition potential, whereas the monosaccharide/VO complexes appeared to be the most potent inhibitory agents (Etcheverry *et al.* 2001).

It is important to note that sucrose is a very relevant stabilising agent of microheterogeneous systems, such as vesicles. Indeed, it has been found that the addition of sucrose decreases the polydispersity of the vesicle population (Kiselev *et al.* 2003).

Infrared (IR) studies of samples in the crystalline phase and in other solid phases are also among the earliest spectroscopic approaches to studying saccharides (Brauer *et al.* 2011). The spectra of ordered and crystalline samples present several nicely resolved bands, but other bands can be influenced by intermolecular interactions involving hydrogen bonds with neighbouring molecules in the solids, which are typically quite strong. Consequently, the lowest energy conformer in the crystal is not the same as that in the gas phase. Alternatively, sugars in a rare gas matrix only experience a weakly interacting environment. The vibrational frequency shifts are typically modest, and the sugar conformers in matrices are expected to correspond to those populated in the gas phase at low temperatures. Unfortunately, relatively few spectroscopic studies of sugars in matrices have been reported so far (Brauer *et al.* 2011).

29.2.5 Quantitative Analysis of Sucrose by Raman Spectroscopy

Generally, the application of Raman spectroscopy to quantitative analysis requires the use of a calibration curve as a reference in relation to a very pure standard. In a previous work, a calibration curve using sucrose from refined sugar (sugar cane) diluted in spring water was constructed in the range between 0 and 15.0 g/100 mL, and a quantification model based on a partial least-squares (PLS) regression was developed to correlate the Raman spectra to the amount of sucrose in each dilution (Silveira Jr. *et al.* 2009). The sucrose in each sample, which was a lemon soft drink, was then predicted from the calibration curve. The mean error of calibration for the PLS method was 0.30 g/100 mL (3.0%) because the results indicated that the samples have sugar contents ranging from 8.1 to 10.9 g/100 mL, and the error of the predicted values compared to those obtained from the nutritional table ranged from 1.1% to 5.5% (Silveira Jr. *et al.* 2009). Therefore, Raman spectroscopy in association with a PLS regression was an effective method for quantifying the sucrose with a small prediction error.

A quantification procedure for sugars in aqueous solutions has been successfully developed for other vibrational spectroscopies. For example, the use of Fourier transform infrared (FTIR) spectroscopy coupled with an attenuated total reflectance (ATR) accessory was applied by Kameoka and co-workers (Kameoka *et al.* 1998) to quantitatively evaluate seven types of sugar solutions (glucose, maltose, sucrose, fructose, lactose, galactose and mannose) that were prepared in different concentrations as food models. These authors showed that the shape of the spectrum in the fingerprint region for each sugar was completely different. Specifically, for non-ideal solutions like maltose, sucrose, fructose and lactose, all peak areas other than that of the C-OH vibration band exhibited excellent linearity when plotted against the concentration of the sugar solution, whereas all peak areas showed excellent linearity in an ideal solution like glucose (Kameoka *et al.* 1998). A simple analytical procedure using FT-NIR and multivariate techniques for the rapid determination of individual sugars in fruit juices was evaluated, and aqueous solutions of sugar mixtures (glucose, fructose, and sucrose; 0–8% w/v) have been used to develop a calibration model. In summary, FT-NIR spectroscopy allowed for a rapid (approximately 3 min analysis time), accurate and non-destructive analysis of sugars in juices and could be used for the quality control of beverages or to monitor for their adulteration or contamination (Rodriguez-Saona *et al.* 2001).

29.2.6 Physico-Chemical Analysis of Sucrose by Raman Spectroscopy

Raffinose pentahydrate, trehalose dihydrate and sucrose have been used as model compounds in conjunction with their amorphous counterparts to analyse the physico-chemical aspects of the structure-function relationship of these carbohydrates. Following exposure to D_2O vapour, the exchange of water among hydration and/or hydroxyl groups has been successfully monitored by Raman spectroscopy (Ahlqvist and Taylor 2002). For the amorphous materials, all of the sugar hydroxyl groups were found to exchange upon exposure to D_2O, providing evidence that water has no fixed site in these amorphous materials, nor is access to different parts of the molecule restricted. For raffinose pentahydrate and trehalose dihydrate, exchange of both hydrate waters and hydroxyl groups was incomplete, suggesting that there are specific pathways for diffusion into and within the crystal structure (Ahlqvist and Taylor 2002).

FT-Raman spectroscopy in conjunction with rheological tests indicated that protein denaturation and its eventual texture changes were minimised in the presence of cryoprotectants (solutions of a mixture of sucrose and sorbitol), as well as in the presence of antioxidants with citrate; antioxidants alone; or a mixture of antioxidants, citrate, and cryoprotectants (Badii and Howell 2002). Similarly, surimi and natural actomyosin (NAM) from ling cod (*Ophiodon elongatus*) were subjected to frozen storage in the absence or presence of cryoprotectants (sorbitol, sucrose, lactitol, and Litesse, either individually or in combination) (Sultanabawa and Li-Chan 2001). In this work, surimi or NAM frozen in the absence of cryoprotectants or in a 4% solution with an individual cryoprotectant showed an increased percentage of alpha-helical content by Raman analysis. Additionally, in samples without cryoprotectants, increased disulphide content was observed by the Raman S-S stretching band and by chemical determination.

Studies focused on the Raman spectra of glucose demonstrate that spectral changes in the O-H stretching region indicate a weakening of the hydrogen bond network during the glass transition (Soderholm *et al.* 2000). There are other spectral changes associated with a weakening of the van der Waals interactions, which demonstrates the complexity of the function-activity relationship of carbohydrates. Other interesting approaches to the study of the molecular structure of sugar crystallinity have been evaluated for carbohydrates, such as sucrose, and involve spectroscopic techniques, such as near-infrared spectroscopy (NI-RS), in which the method error analysis demonstrates a comparable accuracy between NI-RS and X-ray powder diffraction (XRPD), with NI-RS showing slightly better precision in repeated crystallinity determinations for a 50% crystalline sucrose sample (Seyer *et al.* 2000).

A typical application of infrared reflection-absorption (RA) consists of the study of thin films, such as monolayer Langmuir-Blodgett (LB) films of a phospholipid with sucrose. In fact, the use of sugars in Langmuir films has been reported in several works. The LB film consists of distinct associations between phospholipid/sucrose mixtures and sucrose layers. The residual minute water molecules bound to the head group of the phospholipid were readily detected by this method (Hasegawa 1999).

The hydrogen bonding interactions between sugars and polymers, such as poly(vinylpyrrolidone) (PVP), were monitored using FT-Raman spectroscopy in an interesting work by Taylor and Zografi (Taylor *et al.* 1995). Sucrose was found to hydrogen bond with PVP at a particular sugar:polymer ratio to a greater extent than did the other disaccharides studied including trehalose and the trisaccharide raffinose. This approach can be interesting considering the significant number of applications of carbohydrates in the areas of new materials and nanoscience. Indeed, the extent of hydrogen bonding was found to correlate inversely with the glass transition temperature of the sugar, demonstrating the importance of the hydrogen bonding interactions to the thermodynamics of mixing in amorphous solids.

The shift of the vibrational peak frequencies with the temperature of hemoproteins, such as myoglobin, in sucrose/water and glycerol/water solutions has been used to probe the expansion of the hydrogen bond network (Demmel *et al.* 1997), which is a promising area for biophysical evaluations because of the relevance of carbohydrates to the stabilisation of several three-dimensional structures of proteins.

It is important to note that resonance Raman spectroscopy is a powerful technique to investigate transient species with short lifetimes. Resonance Raman spectroscopy can selectively detect a short-lived transient species whose absorption is in resonance with the probing wavelength, thus favouring interesting studies of transient species that can be found in a variety of compounds of biological relevance, such as sugars. For example, the work developed by Limantara and coworkers (Limantara *et al.* 1998) demonstrates that a pair of transient species with different lifetimes can be selectively detected by transient Raman spectroscopy using pulses whose durations differ by 2 orders of magnitude.

29.3 Conclusions

The vibrational spectroscopies, together with nuclear magnetic resonance (NMR), are considered to be some of the most important instrumental tools for the characterisation of chemical compounds. It is important to note that the vibrational spectrometers frequently involve a lower cost relative to the modern NMR spectrometers. Moreover, the vibrational spectroscopies are more versatile for samples in different physical states. Among the vibrational tools, Raman spectroscopy (RS) has a very relevant role due to its high applicability and accessibility to the analysis of the most varied and complex chemical systems. In this context, biochemical systems receive great attention due to their structural complexity and relevance. Organic compounds such as carbohydrates are not easily characterised by many techniques, but RS has provided a significant amount of very precise information regarding the various physico-chemical properties of these compounds.Considering the instrumental advances of RS, this technique must be considered to be an extraordinary alternative for the characterisation of complex biological macromolecules, such as carbohydrates.

Summary Points

- This chapter is focused on the detection of sucrose by Raman spectroscopy.
- Raman spectroscopy is a vibrational technique that relies on the inelastic scattering of monochromatic light.
- This technique has been used for the chemical characterisation of a series of compounds in foods, drugs and biological systems.
- Raman spectroscopy can be a powerful tool for the detection of sucrose and other saccharides. In this chapter, examples of Raman spectra of refined sugar, crystal sugar and pulverised sugar are provided.

Key Facts of Raman Spectroscopy

- Raman spectroscopy is a vibrational technique based on the inelastic scattering of monochromatic light, which is generally provided by a laser source.
- As a result of the inelastic scattering, the frequency of the photons present in monochromatic light is changed by their interaction with the sample.
- The photons of the laser light are absorbed by the sample and then re-emitted.
- The Raman effect consists of the up- or down-shift of the energy of the photons relative to the incident monochromatic frequency.
- Raman spectroscopy can be used to study solid, liquid and gaseous samples.
- The Indian scientist C.V. Raman won the 1930 Nobel Prize in Physics for his observation of the Raman effect by means of sunlight.

Definitions of Words and Terms

"Fingerprint": A characteristic spectral peak, which, even in isolation, is useful for identifying and quantifying well-defined compounds present in a sample.
Head group: The polar site of surfactant compounds.

HPLC: High-performance liquid chromatography is one the most employed chromatographic techniques to the preparation of samples and their qualitative and quantitative analysis.

Lewis base: Molecule or ion that can furnish one or more electron pairs to form a coordinated bond with a receptor compound, which is called a Lewis acid.

Ligand: Lewis base that is one of the fundamental components of coordination compounds, such as metallic complexes, together with the coordination centre, which is a Lewis acid.

Metal complex: Coordination compound that contains a metal ion or atom as its coordination centre. The metal centre is formed by the coordination centre and its ligands. However, when the metal complex has a non-zero oxidation state, a third component is necessary to form the complex salt to isolate the complex from the solution as a solid compound. This third component of the metal complexes is known as a counter-ion or ionic pair.

R$: Symbol of the Brazilian monetary unit, which is called "Real".

Surfactants: Surface active agents, which characterise the amphipathic compounds that decrease the surface tension in solvents, such as water, as a function of the disruption of hydrogen bonds. Frequently, these compounds consist of a head group and a structural moiety.

Thin films: Very small layers, on a molecular scale, involving some chemical compounds with defined function in the relevant chemical system. The formation of thin films, which can reach the level of only one layer (molecular monolayers), can be developed through different methodologies, such as self-assembled films and Langmuir and Langmuir-Blodgett films. The surface modification of thin films is a relevant resource to obtain new materials in several areas, such as the formation of biosensors and heterogeneous catalysts.

Vibrational spectroscopies: Several types of spectroscopic techniques, in which the optical phenomenon is a vibrational transition that occurs in the infrared region of the electromagnetic spectrum.

List of Abbreviations

RS   Raman Spectroscopy
FT   Fourier Transformed
IR   Infrared
mATR   microattenuated total reflectance
HPLC   High Performance Liquid Chromatography
INS   Inelastic Neutron Scattering
QD   Quantum Dot
NIR   Near Infrared
MIR   Mid Infrared
PLS   Partial Least Squares
ATR   Attenuated Total Reflectance
NAM   Natural Actomyosin
LB   Langmuir Bloodge
PVP   Polyvinilpyrrolidone

References

Ahlqvist, M.U.A., and Taylor, L.S., 2002. Water diffusion in hydrated crystalline and amorphous sugars monitored using H/D exchange. *Journal of Pharmaceutical Sciences*. 91: 690–698.

Badii, F., and Howell, N.K., 2002. Effect of antioxidants, citrate, and cryoprotectants on protein denaturation and texture of frozen cod (Gadus morhua). *Journal of Agricultural and Food Chemistry*. 50: 2053–2061.

Bakier, S., 2009. Capabilities of near-infrared spectroscopy to analyse changes in water bonding during honey crystallisation process. *International Journal of Food Science and Technology*. 44: 519–524.

Blanco, M., and Villarroya, I., 2002. NIR Spectroscopy: A rapid response analytical tool. *Trends in Analytical Chemistry*. 21: 240–250.

Branca, C., Magazu, S., Maisano, G., Bennington, S.M., and Fak, B., 2003. Vibrational studies on disaccharide/H2O systems by inelastic neutron scattering, Raman, and IR Spectroscopy. *Journal of Physical Chemistry B*. 107: 1444–1451.

Brauer, B., Pincu, M., Buch, V., Bar, I., Simons, J.P., and Gerber, R. B., 2011. Vibrational Spectra of r-Glucose, β-Glucose, and Sucrose: Anharmonic Calculations and Experiment. *Journal of Physical Chemistry A*. 115: 5859–5872.

Celedon, A., and Aguilera, J.M., 2002. Applications of microprobe Raman spectroscopy in food science. *Food Science and Technology International*. 8:101–108.

Crowe, J.H., Crowe, L.M., Carpenter, J.F., Rudolph, A.S., Wistrom, A.A., Spargo, B.J., and Anchordogy, T.J., 1988. *Biochimica et Biophysica Acta*. 947: 367–383.

Crowe, L.M., Reid, D.S., and Crowe, J.H., 1996. Is trehalose special for preserving dry biomaterials? *Biophysical Journal*. 71: 2087–2093.

Crowe, J.H., Tablin, F., Wolkers, W.F., Gousset, K., Tsvetkova, N.M., and Ricker, J., 2003. Stabilization of membranes in human platelets freeze-dried with trehalose. *Chemistry and Physics of Lipids*. 122: 41–52.

de Oliveira, L.F.C., Colombara, R., and Edwards, H.G.M., 2002. Fourier transform Raman spectroscopy of honey. *Applied Spectroscopy*. 56: 306–311.

de Veij, M., Vandenabeele, P., De Beer, T., Remonc, J.P., and Moens, L., 2009. Reference database of Raman spectra of pharmaceutical excipients. *Journal of Raman Spectroscopy*. 40: 297–307.

Demmel, F., Doster, W., Petry, W., and Schulte, A., 1997. Vibrational frequency shifts as a probe of hydrogen bonds: thermal expansion and glass transition of myoglobin in mixed solvents. *European Biophysics Journal with Biophysics Letters*. 26: 327–335.

Etcheverry, S.B., Barrio, D.A., Williams, P.A.M., and Baran, E.J., 2001. On the interaction of the vanadyl (IV) cation with lactose - Inhibition effects of vanadyl (IV)/ monosaccharide and disaccharide complexes upon alkaline phosphatase activity. *Biological Trace Element Research*. 84: 227–238.

Ferreira, V.F., Silva, F.C., and Perrone, C.C., 2001. Sucrose in undergraduate organic chemistry laboratory. *Quimica Nova*. 24: 905–907.

Ferreira, V.F., Rocha, D.R., and Silva, F.C., 2009. Potentiality and opportunity in the chemistry of sucrose and other sugars. *Quimica Nova*. 32: 623–638.

Giangiacomo, R., 2006. Study of water-sugar interactions at increasing sugar concentration by NIR spectroscopy. *Food Chemistry*. 96: 371–379.

Ghosh, D., Ghosh, S., and Saha, A., 2010. Quantum dot based probing of mannitol: An implication in clinical diagnostic. *Analytica Chimica Acta*. 675: 165–169.

Hasegawa, T., 1999. Detection of minute chemical species by principal component analysis. *Analytical Chemistry*. 71: 3085–3090.

Hedoux, A., Willart, J.F., Ionov, R., Affouard, F., Guinet, Y., Paccou, L., Lerbret, A., and Descamps, M., 2006. Analysis of sugar bioprotective mechanisms on the thermal denaturation of lysozyme from Raman scattering and differential scanning calorimetry investigations. *Journal of Physical Chemistry B*. 110: 22886–22893.

Hedoux, A., Affouard, F., Descamps, M., Guinet, Y., and Paccou, L., 2007. Microscopic description of protein thermostabilization mechanisms with disaccharides from Raman spectroscopy investigations. *Journal of Physics Condensed Matter*. 19: 205142.

Ionov, R., Hédoux, A., Guinet, Y., Bordat, P., Lerbret, A., Affouard, F., Prevost, D., and Descamps, M., 2006. Sugar bioprotective effects on thermal denaturation of lysozyme: Insights from Raman scattering experiments and molecular dynamics simulations. *Journal of Non-Crystalline Solids*. 352: 4430–4436.

Izutsu, K., Hiyama, Y., Yomota, C., and Kawanishi, T., 2009. Near-Infrared Analysis of Hydrogen-Bonding in Glass- and Rubber-State Amorphous Saccharide Solids. *AAPS Pharmscitech*. 10: 524–529.

Kameoka, T., Okuda, T., Hashimoto, A., Noro, A., Shiinoki, Y., and Ito, K., 1998. FT-IR analysis of sugars in aqueous solution using ATR method. *Journal of the Japanese Society for Food Science And Technology*. 45: 192–198.

Kanou, M., Nakanishi, K., Hashimoto, A., and Kameoka, T., 2005. Influences of monosaccharides and its glycosidic linkage on infrared spectral characteristics of disaccharides in aqueous solutions. *Applied Spectroscopy*. 59: 885–892.

Kanou, M., Nakanishi, K., Hashimoto, A., and Kameoka, T., 2003. Infrared spectroscopic analysis of disaccharides in aqueous solutions. *Journal of the Japanese Society for Food Science and Technology*. 50: 57–62.

Kiselev, M.A., Wartewig, S., Janich, M., Lesieur, P., Kiselev, A.M., Ollivon, M., and Neubert, R., 2003. Does sucrose influence the properties of DMPC vesicles? *Chemistry and Physics of Lipids*. 123: 31–44.

Kiselev, M.A., Zbytovska, J., Matveev, D., Wartewig, S., Gapienko, I.V., Perez, J., Lesieur, P., Hoell, A., and Neubert, R., 2005. Influence of trehalose on the structure of unilamellar DMPC vesicles. *Colloids and Surfaces A*. 256: 1–7.

Krishna, C.M., Prathima, N.B., Malini, R., Vadhiraja, B.M., Bhatt, R.A., Fernandes, D.J., Kushtagi, P., Vidyasagar, M.S., and Kartha, V.B., 2006. Raman spectroscopy studies for diagnosis of cancers in human uterine cervix. *Vibrational Spectroscopy*. 41: 136–141.

Larmour, I.A., Faulds, K., and Graham, D., 2010. Rapid Raman mapping for chocolate analysis. *Analytical Methods.* 2: 1230–1232.

Li, Z.D., Shen, N.K., Lai, J.Z., Qin, Y., Liu, J.X., and Wang, G.W., 2010. Determination of Ethanol in Fermentation Broth Using 96-well Plate and Raman Spectroscopy. *Chinese Journal of Analitical Chemistry.* 38:1267–1271.

Limantara, L., Fujii, R., Zangh, J.P., Kakuno, T., Hara, H., Kawamori, A., Yagura, T., Coqdell, R.J., and Koyama, Y., 1998. Generation of triplet and cation-radical bacteriochlorophyll a in carotenoidless LH1 and LH2 antenna complexes from Rhodobacter sphaeroides. *Biochemistry.* 37: 17469–17486.

Maalouly, J., Eveleigh, L., Rutledge, D.N., and Ducauze, C.J., 2004. Application of 2D correlation spectroscopy and outer product analysis to infrared spectra of sugar beets. *Vibrational Spectroscopy.* 36: 279–285.

Mathlouthi, M., Luu, C., Meffroy-Biget, A.M., and Luu, D.V., 1980. Laser-Raman study of solute-solvent interactions in aqueous solutions of D-fructose, D-glucose, and sucrose. *Carbohydrate Research.* 81: 213–233.

Moreira, L.M., Silveira Jr., L., Santos, F.V., Lyon, J.P., Rocha, R., Zangaro, R.A., Villaverde, A.B., and Pacheco, M.T.T., 2008. Raman spectroscopy: a powerful technique for biochemical analysis and diagnosis. *Spectroscopy.* 22: 1–19.

Pons, M., Le Bonté, S., and Potier, O., 2004. Spectral analysis and fingerprinting for biomedia characterization. *Journal of Biotechnology.* 113: 211–230.

Reyes-Goddard, J.M., Barr H., and Stone, N., 2005. Photodiagnosis using Raman and surface enhanced Raman scattering of bodily fluids. *Photodiagnosis and Photodynamic Therapy.* 2: 223–233.

Ribeiro, A.C.F., Valente, A.J.M, Lobo, V.M.M., Azevedo, E.F.G., Amado, A.M., da Costa, A.M.A., Ramos, M.L., and Burrows, H.D., 2004. Interactions of vanadates with carbohydrates in aqueous solutions. *Journal of Molecular Structure.* 703: 93–101.

Rodriguez-Saona, L.E., Fry, F.S., McLaughlin, M.A., and Calvey, E.M., 2001. Rapid analysis of sugars in fruit juices by FT-NIR spectroscopy. *Carbohydrate Research.* 336: 63–74.

Rohleder, D., Kocherscheidt, G., Gerber, K., Kiefer, W., Köhler, W., Möcks, J., and Petrich, W., 2005. Comparison of mid-infrared and Raman spectroscopy in the quantitative analysis of serum. *Journal of Biomedical Optics.* 10: 031108.

Salameh A.K., Mauer, L.J., and Taylor, L.S., 2006. Deliquescence lowering in food ingredient mixtures. *Journal of Food Science.* 71: E10–E16.

Sevilla, M., and Fuertes, A.B., 2009. Easy synthesis of graphitic carbon nanocoils from saccharides. *Materials Chemistry and Physics.* 113: 208–214.

Seyer, J.J., Luner, P.E., and Kemper, M.S., 2000. Application of diffuse reflectance near-infrared spectroscopy for determination of crystallinity. *Journal of Pharmaceutical Sciences.* 89: 1305–1310.

Shaw, R.A., and Mantsch, H.H., 1999. Vibrational biospectroscopy: from plants to animals to humans. A historical perspective. *Journal of Molecular Structure.* 480/481: 1–13.

Silveira Jr., L., Moreira, L.M., Conceição, V.G.B., Casalechi, H.L., Munoz, I.S., da Silva, F.F., Silva, M.A.S.R., de Souza, R.A., and Pacheco, M.T.T., 2009. Determination of sucrose concentration in lemon-type soft drinks by dispersive Raman spectroscopy. *Spectroscopy.* 23: 217–226.

Soderholm, S., Roos, Y.H., and Meinander, N., 2000. Temperature dependence of the Raman spectra of amorphous glucose in the glassy and supercooled liquid states. *Journal of Raman Spectroscopy.* 31: 995–1003.

Sultanabawa, Y., and Li-Chan, E.C.Y., 2001. Structural changes in natural actomyosin and surimi from ling cod (*Ophiodon elongatus*) during-frozen storage in the absence or presence of cryoprotectants. *Journal of Agricultural and Food Chemistry.* 49: 4716–4725.

Sussich, F., Princivalle, F., and Cesaro, A., 2009. The interplay of the rate of water removal in the dehydration of alpha, alpha-trehalose. *Carbohydrate Research.* 322: 113–119.

Szűts, A., Makai, Z., Rajkó, R., and Szabó-Révész P., 2008. Study of the effects of drugs on the structures of sucrose esters and the effects of solid-state interactions on drug release. *Journal of Pharmaceutical and Biomedical Analysis.* 48: 1136–1142.

Taylor, L.S., York, P.; Williams, A.C., Edwards, H.G.M., Mehta, V., Jackson, G.S., Badcoe, I.G., and Clarke, A.R., 1995. Sucrose reduces the efficiency of protein denaturation by a chaotropic agent. *Biochimica et biophysica Acta.* 1253: 39–46.

Tewari, J., and Irudayaraj, J., 2004. Quantification of saccharides in multiple floral honeys using Fourier transform infrared microattenuated total reflectance spectroscopy. *Journal of Agricultural and Food Chemistry.* 52: 3237–3243.

Xie, Y., Cao, W.J., Krishnan, S., Lin, H., and Cauchon, N., 2008. Characterization of mannitol polymorphic forms in lyophilized protein formulations using a multivariate curve resolution (MCR)-based Raman spectroscopic method. *Pharmaceutical Research.* 25: 2292–2301.

Walther, M., Fischer, B.M., and Jepsen, P.U., 2003. Noncovalent intermolecular forces in polycrystalline and amorphous saccharides in the far infrared. *Chemical Physics.* 288: 261–268.

Wartewig, S., and Neubert, R.R.H., 2005. Pharmaceutical applications of Mid-IR and Raman spectroscopy. *Advanced Drug Delivery Reviews.* 57: 1144–1170.

Wright, W.W., Guffanti, G.T., and Vanderkooi, J.M., 2003. Protein in sugar films and in glycerol/water as examined by infrared spectroscopy and by the fluorescence and phosphorescence of tryptophan. *Biophysical Journal.* 85: 1980–1995.

Zeng, H., McWilliams, A., and Lam, S., 2004. Optical spectroscopy and imaging for early lung cancer detection: a review. *Photodiagnosis and Photodynamic Therapy.* 1: 111–122.

CHAPTER 30

Analysis of Sucrose from Sugar Beet

J. MITCHELL McGRATH*[a] AND KAREN K. FUGATE[b]

[a] Research Geneticist, U.S. Department of Agriculture – Agricultural Research Service, Sugarbeet and Bean Research Unit, 494 Plant and Soil Science Building, Michigan State University, East Lansing, Michigan 48824-1325, USA; [b] Research Molecular Biologist, U.S. Department of Agriculture – Agricultural Research Service, Sugarbeet and Potato Research, 1307 18th Street North, Fargo, North Dakota 58102-2765, USA
*Email: mitchmcg@msu.edu

30.1 Introduction

Sucrose (saccharose), commonly known as table sugar or sugar, is perhaps the most abundant, chemically pure, renewable compound produced, with 168.5 billion metric tons produced worldwide in the 2011/2012 campaign year (U.S. Dept. of Agriculture Economic Research Service 2011). It occurs naturally in most plants as a product of photosynthesis and serves the plant as a transportable source of stored energy. Primary sources of crystalized sucrose are sugar cane (*Saccharum* spp.) and sugar beet (*Beta vulgaris*), and other economic sources of sugar include maples (*Acer saccharum*), sorghum (*Sorghum bicolor*), and a number of species in the palm family. Analyses of sucrose derived and manufactured from these sources are methodologically similar, differing in the preparation of the source material for analysis and the standards of purity required at any one stage of the purification process.

Food and Nutritional Components in Focus No. 3
Dietary Sugars: Chemistry, Analysis, Function and Effects
Edited by Victor R Preedy
© The Royal Society of Chemistry 2012
Published by the Royal Society of Chemistry, www.rsc.org

Sucrose (β-D-fructofuranosyl-(2→1)-α-D-glucopyranoside; CAS number 57-50-1; $C_{12}H_{22}O_{11}$, physical properties in Tables 30.1 and 30.2) is a disaccharide composed of D-glucose and D-fructose monosaccharides linked *via* a glycosidic bond (Figure 30.1), where the anomeric carbons of glucose and fructose are linked. Further polymerization of sucrose is, therefore, inhibited under physiological conditions, although chemical conversion of sucrose for renewable chemical feed-stocks is possible (Peters *et al.* 2010; Godshall 2010). Sucrose decomposes to a complex carbohydrate mixture known as caramelization upon heating above 135 °C (Lee et al. 2011), but is otherwise stable as crystals at room temperature. Color results from impurities co-purifying with sucrose during processing and is generally considered undesirable except in cases of brown sugar products or in unrefined cane sugar (Asadi 2007). Sucrose also participates in reactions with other biological compounds, such as amino acids (*e.g.* Maillard reactions), and the analysis of sucrose in prepared foods and its dietary effects is confounded by its degradation to glucose and fructose and reactions of these with other compounds.

Sucrose is not generally consumed as a pure food, however, it is widely used as a sweetener and for calories in the home, food and beverage industries, and in the production of pharmaceuticals and other fermentation reactions (*e.g.* biofuels). Crystalline sucrose is used in baking for its bulking and texture modifying properties as well as for its sweetening properties, and sucrose dissolved in solutions serves as a binding agent due to its stickiness. High concentrations of

Table 30.1   Physical properties of sucrose.

Molecular Formula	$C_{12}H_{22}O_{11}$
Molecular Weight	342.3
Melting Point (decomposes at lower temps)	185.5 °C
Density (17 °C)	1.5805 g/mL
Specific Gravity (10% solution)	1.0381
Refractive Index (10% solution)	1.34782
pKa	12.62
pH	neutral
solubilty: 1 g pure sucrose in:	
water	0.5 ml
methanol	100 ml
ethanol	170 ml
Optical Rotation (in water @ 20 C)	+ 66.37
Index of Refraction	1.5376
Shelf Life	stable in air
Hazards	
combustion: dust	
inhalation: irritating fumes	
explosive: strong oxidizers	
carcinogenicity: none known	
Appearance	
color	white
form	crystalline
odor	odorless
taste	sweet

Table 30.2   Government sponsored web resources for physical, chemical, and biological information on sucrose.

Identifier	ID Code	Source (accessed November 2011)
CAS number	57-50-1	http://webbook.nist.gov/cgi/cbook.cgi?ID=57-50-1
PubChem	5988	http://pubchem.ncbi.nlm.nih.gov/summary/summary.cgi?cid=5988
ChemSpider	5768	http://www.chemspider.com/Chemical-Structure.5768.html
ChemIDplus	Sucrose	http://chem.sis.nlm.nih.gov/chemidplus/
UNII	C151H8M554	http://fdasis.nlm.nih.gov/srs/
CHEBI	17992	https://www.ebi.ac.uk/chebi/searchId.do?chebiId=17992
RTECS	WN6500000	http://www.cdc.gov/niosh-rtecs/WN632EA0.html
CHEMBL	253582	https://www.ebi.ac.uk/chembldb/index.php/compound/inspect/CHEMBL253582
KEGG	C00089	http://www.genome.jp/dbget-bin/www_bget?cpd:C00089
PDB	SUC	http://www.ebi.ac.uk/msd-srv/chempdb/cgi-bin/cgi.pl?FUNCTION=getByCode&CODE=SUC
IUPAC Standard InChIKey	CZMRCDWA GMRECN- UHFF- FAOYSA-N	http://webbook.nist.gov/cgi/inchi/InChI%3D1S/C12H22O11/c13-1-4-6(16)8(18)9(19)11(21-4)23-12(3-15)10(20)7(17)5(2-14)22-12/h4-11%2C13-20H%2C1-3H2
Food safety	Sucrose	http://www.fda.gov/Food/FoodIngredientsPackaging/GenerallyRecognizedasSafeGRAS/GRASSubstancesSCOGSDatabase/ucm260083.htm
Food safety	57-50-1	http://toxnet.nlm.nih.gov/cgi-bin/sis/search/r?dbs+hsdb:@term+@rn+@rel+57-50-1

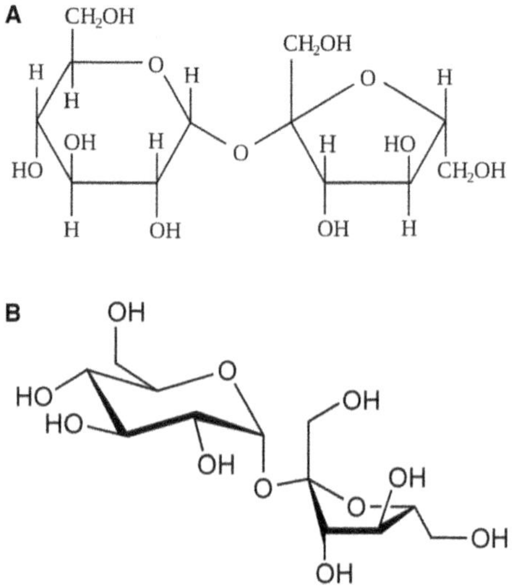

Figure 30.1   Structure of sucrose. A: Arrangement of molecular constituents. B: 2D conformation.

sucrose in solution inhibit microbiological growth due to high osmotic pressure. Medicinally, among other uses, sucrose in combination with sodium chloride in solution is an inexpensive and effective oral rehydration therapy in patients suffering from acute diarrhea (Vettorazzi and Macdonald 1988).

Sucrose's near ubiquity in raw and processed agricultural materials makes a uniform and straightforward set of analysis methods difficult, and methods still undergo revision after over 150 years of method development. Official methods are developed by ICUMSA (International Commission for Uniform Methods of Sugar Analysis) primarily for the cane and beet sugar industries and these methods are accepted as authoritative by national and international food safety organizations (*e.g.* Codex Alimentarius Commission). The Association of Analytical Communities (AOAC) promulgates dozens of standardized analytical methods for sucrose and sugar analyses, sometimes in conjunction with other official bodies such as ICUMSA. These methods include polarimetry, refractometry, gas chromatography, liquid chromatography, and several types of spectroscopic assays. The most commonly used method of sucrose analysis by the beet industry is polarimetry. Refractometry is commonly used in sugar content determinations for many other fruits and vegetables since it can be accomplished quickly and requires minimal equipment and training for approximate analyses. Of note, methods to directly measure sucrose on the basis of ELISA assays have not been possible since the development of antibodies to sucrose is practically impossible due its low immunogenicity.

30.2 Extraction for Measurement

Sucrose in natural products generally exists in solution with a large number of other solutes. A beet juice sample is most often obtained by macerating the root tissue, usually with a rasp or saw, and the juice is expressed from the resulting brei (puree) or separated from insoluble solids by filtration after dilution of the brei using prescribed methods (ICUMSA 1964, McGinnis 1982). Most analytical methods require a clarified juice sample, both to allow an unobstructed light path for polarization and to remove the majority of optically active nonsucrose impurities. Salting-out (*e.g.* defecation) is used to clarify samples, where salts compete for the water of hydration from the surface of solutes thus rendering them insoluble. Official ICUMSA standards use salts of lead or aluminum and changes in pH to selectively precipitate impurities (Chen and Chou 1993), and selectivity of the specific salts and conditions has been evaluated empirically over many years to arrive at these standard methodologies, which require exacting care for accurate analyses.

30.3 Analytical Methods

30.3.1 Historical Perspective

Several sucrose analytical methods in common use were codified in the late 1800s, with many handbooks for sugar laboratory analysts appearing in the early 1900s (see Browne 1912, Spencer 1910). Through the 1940s these methods were refined and expanded, and reference works by Bates and Associates (1942) and Browne and Zerban (1941) still rank among the most important and comprehensive works available for the sugar analyst today. It was not until the latter part of the 20th century that requirements for uniform analyses were implemented for standardizing national and international contractual trading agreements and food identity and labeling requirements. Currently, ICUMSA standardizes and validates official methods for cane and beet sugar analysis developed among cooperating institutions, and the AOAC reviews these and other methods, as they may be required to comply with governmental food safety regulations as may be specified by the U.S. Food and Drug Administration or the Customs and Border Protection arm of the U.S. Department of Homeland Security, for instance. These official methods are available by subscription, and their presentation is beyond the scope of this broad overview.

Two historically important methods for sugar analyses have been drying of sugar solutions and specific gravity (density) determination. Neither is practiced at an appreciable level in sugar houses now. Simply drying (in an oven or under vacuum) a known volume of sugar solution and calculating the percent sugar in relation to the mass of water lost to drying is a measure of concentration of sucrose in the original solution. This works well for pure sucrose solutions, but all soluble solids remaining after drying will be measured, thus this technique is not particularly sensitive for sucrose. Specific gravity (and refractometry, see below) also measures total dissolved solids and a great deal of method development has been devoted to assessing and correcting measured values to those of pure sucrose in complex sugar mixtures. Tables of specific gravity of sucrose in solution, either pure sucrose or in the presence of impurities, have been specifically developed for the sugar industry. Most sugars have specific gravities similar to sucrose and particular measurement of sucrose requires exacting methods, using instruments such as the pycnometer or hydrometer (Browne 1912). Various measurement systems (scales) have been applied for sugar solutions that are still in use today in sugar factories, such as Baumé and Brix (Asadi 2007). Interestingly, the first method applied to improving sucrose content in sugarbeet was that of density or specific gravity (Vilmorin 1859, cited in Galon and Zallen 1998), by weighing sugarbeets in air and comparing them to their weight in water. Mother roots with higher densities were thus identified, and seed produced from these denser mother roots rapidly led to sugarbeet varieties with higher sucrose content. Certainly, sucrose is not the only concern of sugarbeet processors, and a host of other attributes factor into sugarbeet quality that the sucrose analyst needs to consider (Dutton and Huijbregts 2006).

30.3.2   Refractometry

Light changes direction when it passes from air into a sugar solution, a phenomenon known as refraction. As with specific gravity, the refractive index of simple sugars is similar at the same temperature. The magnitude of refraction is proportional to the solute concentration and also to the specific gravity of the solution, and tables were developed relating these values to that of the most commonly used unit of Brix (°Bx) as derived from comparisons with the Brix specific gravity scale. Tables of refraction index for sucrose and other sugars have been developed, and the values are reported as "Refractometric Dry Substance" or RDS. The refractometer is a relatively inexpensive device and has found wide use in many industries, including the wine and beet industries, as well as for the determination of soluble solids, including sucrose, in fresh and processed foods, and in diagnostic medicine. Some modern electronic refractometers can also utilize additional wavelengths in the near infrared (specifically at 880 nm) that allow more specific quantification of sucrose in a mixture.

30.3.3 Polarimetry

Polarimetry is perhaps the most commonly employed method for industrial sugar quantification, and thus much work has been devoted to developing standardized methods (Bates and Associates 1942; Browne and Zerban 1941; ICUMSA 1964). Polarimetry takes advantage of the principle that linearly polarized light rotates as it passes through optically active solutions such as sucrose and other sugars. Light waves that have passed through a calcite prism (Nichol or Glan-Thompson) transmit light that oscillates in one direction, and the rotation can be measured as degrees of rotation, either right or left. Saccharimeters are polarimeters that have been calibrated in a 100-point system for degrees of rotation (°Z) calibrated to sugar solution of 26.000 grams of pure sucrose dissolved in water at 20.00 °C in a final volume of 100.000 milliliters in a 200 mm tube, which is defined as 100 °Z. Increased accuracy of modern analytical instruments demonstrated slight inaccuracies in the scale used prior to 1988, leading to a change in international units from °S to °Z (Lescure 1995). Degree of rotation is affected by temperature, wavelength of polarized light (standardized to 589 nm), the nature of substances in the sample, the light path length, and of course, concentration of optically active compounds in the sample. In addition to sucrose (specific rotation = +66.5°), glucose (+52.5°), fructose (−92.4°) and raffinose (+123.2°) are common in sugarbeet juice, albeit at just a few percent of sucrose in an undegraded sample. Sucrose can be hydrolyzed to give equimolar concentrations of glucose and fructose, and when this happens, the direction of polarization changes (inverts) due to the high negative specific rotation of fructose, hence the origin of the term "invert sugar". Although values are reported as degrees of polarization, or simply "pol", and do not directly correspond to percent sugar, "pol" closely mimics the percent of sugar in solution.

Polarimeters and saccharimeters are readily available commercially at moderate cost. Newer developments include automatic temperature compensation, options for different path lengths, continuous or batch processing, and software suites that allow data processing and collection by workers skilled in instrument operation. A large advantage of some newer instruments is the addition of a multispectral light source using infrared wavelengths. These instruments can accurately measure sucrose in dark and colored samples, *e.g.* molasses, and perhaps obviate the need for juice clarification (Schoonees 2003; Singleton *et al.* 2002).

30.3.4   Enzyme-based Spectroscopic Assays

Enzyme-based spectroscopic assays utilize a series of enzyme-catalyzed reactions to quantitatively couple sucrose reaction to the synthesis of a spectroscopically detectable compound (Figure 30.2). The most common methods use invertase to hydrolyze sucrose to glucose and fructose, and the glucose is converted to 6-phosphogluconate by the combined action of hexokinase and glucose 6-phosphate dehydrogenase with concomitant reduction of either NAD^+ or $NADP^+$ to NADH or NADPH (Figure 30.2A). The change in absorbance at 340 nm is proportional to the NADH (or NADPH) formed and the sucrose present in the original sample, since one molecule of NAD(P)H is produced from each molecule of sucrose. Standard sucrose solutions and control reactions containing all reactants except invertase are commonly included in the analysis to generate a calibration equation and correct for NAD(P)H produced from glucose present in samples. Since invertase and the glucose-converting enzymes differ substantially in their pH optima, sucrose hydrolysis and glucose conversion to 6-phosphogluconate are typically performed in separate reactions (Schoenrock and Costesso 1975), although a one-step protocol in which all enzymes are combined has been described (Spackman and Cobb 2001). In some descriptions of the method, phosphoglucose isomerase (PGI), which catalyzes the interconversion of fructose 6-phosphate and glucose 6-phosphate, is also included, allowing NAD(P)H to be produced by both fructose and glucose. PGI is added when simultaneous quantification of invert sugars or enhanced sensitivity is desired. With PGI, control reactions measure the combined concentration of fructose and glucose, and sensitivity in samples with low sucrose concentrations is improved since two molecules of NAD(P)H are produced from each sucrose molecule.

Other enzyme-based methods have been described. A fluorometric assay (Figure 30.2B) that couples the activities of invertase, glucose oxidase, and peroxidase to hydrolyze sucrose, oxidize the resulting glucose with coincident production of hydrogen peroxide, and react the hydrogen peroxide formed with Amplex Red (Molecular Probes, Carlsbad, CA) was developed as a tool for screening sugarbeet breeding materials (Trebbi and McGrath 2004). This method provides accurate and sensitive sucrose quantification of tissue from young sugarbeet roots, but is subject to inaccuracies with older, field-grown roots. A protocol that utilizes sucrose phosphorylase, phosphoglucomutase, and glucose 6-phosphate dehydrogenase and couples sucrose reaction with NAD(P)H production (Figure 30.2C) has also been described (Birnberg and Brenner 1984). This protocol combines all assay components into a single reaction since all enzymes have similar pH optima and is not subject to background absorbance from glucose or fructose since these compounds are not substrates for any of the method's enzymes.

In general, enzyme-based methods provide sensitive and accurate quantification of sucrose and are low-cost alternatives to other analytical techniques. In microplate format, the methods are easily automated and suitable for the rapid analysis of large numbers of samples. These methods can also be used for

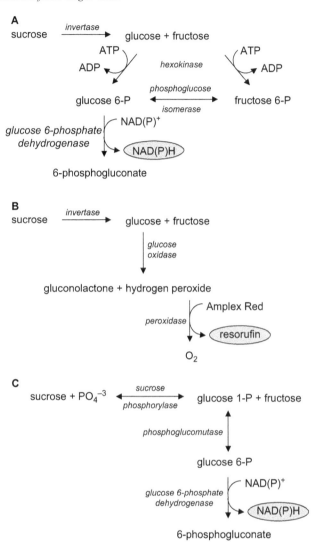

Figure 30.2 Reactions employed by enzyme-based spectroscopic assays. Schematic diagram of the enzymatic reactions used in three sucrose-quantifying assays. A: Reactions employed by a commonly used invertase/hexokinase/glucose 6-P dehydrogenase assay. B: Reactions employed by an invertase/glucose oxidase/peroxidase fluorometric assay. C: Reactions employed by a sucrose phosphorylase/phosphoglucomutase/glucose 6-P dehydrogenase assay. Compounds measured spectroscopically for each assay are highlighted.

quantification of glucose, fructose and invert sugars and can rapidly generate information regarding carbohydrate impurities. Enzymatic assays have also been adapted for flow-through operation coupled with near-infrared detection (Lendl and Kellner 1995).

30.3.5 Gas Chromatography (GC)

Gas chromatography has been used since the 1970s to determine sucrose concentration in sugarbeet roots (Karr and Norman 1974). Prior to GC analysis, sucrose is converted to a volatile derivative, most commonly by trimethylsilylation of the eight hydroxyl moieties of sucrose. Trehalose, a nonreducing sugar comprised of two glucose residues linked via an α,α-1,1 bond, is typically added as an internal standard to correct for inaccuracies caused by incomplete reaction or losses during derivatization. GC analyses have been conducted with both packed and capillary columns, and analytes have been detected most commonly with flame ionization detectors, although thermal conductivity detectors and mass spectrometers have also been employed.

GC is an accurate and sensitive method for sucrose quantification and can be used for simultaneous quantification of other sugars including the invert sugars, glucose, and fructose (Long and Chism 1987). GC requires specialized, relatively expensive equipment, uses harmful chemicals, and is slower than other methods since samples must be derivatized prior to analysis, and analyzed sequentially. With the development of HPLC protocols utilizing stable and sensitive detectors, *i.e.*, electrochemical and evaporative light scattering detectors, the use of GC methodology has declined.

30.3.6 High-performance Liquid Chromatography (HPLC)

HPLC protocols have been available since the 1970s but were not widely used until the 1990s when methodology changes dramatically improved the sensitivity, accuracy, reliability and ruggedness of the technique. The earliest HPLC protocols separated sample components using amino-bonded silica or sodium- or calcium-based cation exchange columns and detected sucrose by refractive index or low wavelength UV (190 to 200 nm). While these protocols are still used, they are expensive, slow, and prone to inaccuracies, since amino-bonded silica columns prematurely deteriorate and lose resolving power under the conditions used to separate carbohydrates. Cation exchange columns require elevated temperatures, low flow rates, and long equilibration times, and refractive index and UV detectors operated at low wavelength are insensitive, non-selective, subject to baseline drift and limited to isocratic conditions (Day-Lewis and Schäffler 1990; Kort 1980).

In the 1980s, HPLC protocols utilizing high performance anion exchange chromatography coupled with pulsed amperometric detection (HPAE-PAD) and reverse phase columns coupled with evaporative light scattering detectors (ELSD) were developed (Macrae *et al.* 1982; Rocklin and Pohl 1983). These methods overcome many of the shortcomings of earlier HPLC protocols. Anion exchange chromatography is possible because carbohydrates, including sucrose, are weakly acidic and are ionized under strongly alkaline conditions. HPAE-PAD methods, therefore, utilize strongly alkaline mobile phases to ionize carbohydrates and pH-stable anion exchange columns to effect their separation. The alkaline conditions necessitate a metal-free flow path but facilitate carbohydrate oxidation, and an electrochemical detector equipped with a gold electrode is used for quantification. The detector is operated in pulsed amperometric mode by rapidly pulsing the electrode between three potentials to sequentially oxidize carbohydrates, remove oxidized products, and reactivate the electrode. HPLC methods utilizing evaporative light scattering detectors, typically coupled with reverse phase chromatography, have also been developed (Herbreteau *et al.* 1990). ELSDs are universal detectors that quantify any compound less volatile than the mobile phase and function by nebulizing the HPLC column outflow, evaporating the solvent, and measuring the light scattered by sample components. Calibration curves are required with ELSDs, since detector response is nonlinear over wide concentration ranges. Both HPAE-PAD and ELSD-utilizing methods accurately and sensitively quantify sucrose and can be used to quantify other carbohydrates that are common in sugarbeet root, including fructose, glucose, and raffinose. They are compatible with gradients and, therefore, more rapid than earlier-developed HPLC methods. Like all HPLC methods, however, they require specialized and expensive equipment and are relatively slow since samples are analyzed sequentially (Mulcock *et al.* 1985).

30.3.7 Near-infrared Spectroscopy (NIR)

Reflection or transmission of light in the near-infrared wavelengths (780 to 2500 nm) has more recently been shown to have robust information that correlates well with polarimetry and other wet chemistries, with the proviso that proper illumination and instrumentation, the calibration model, and specific quality attributes of the sample are adequately taken into account (Huang *et al.* 2008; Nicolaï *et al.* 2007). Absorption and scattering of NIR is affected by the angle of illumination and reflection, chemical constituents of the sample, the heterogeneity of the sample (*e.g.* presence of cell walls and cellular organelles), which along with strong water absorption in NIR, results in broad and diffuse peaks that are detected over a wide range of wavelengths. Statistical multivariate analyses, generally multivariate techniques such as partial least squares and principle component analyses, are required to extract useful information from these peaks, and these statistical treatments are used to fit wet chemistry observations in such a way as to approximate a best fit between empirical and NIR observations (Cozzolino *et al.* 2011). Calibration is the process of creating the best fit and the resulting model is used in place of wet chemistry in additional samples. NIR spectroscopy is rapid, generally non-destructive, can be performed on a wide range of unprocessed and processed materials, and can be relatively inexpensive depending on which of the wide variety of available instruments is used (Figure 30.3). The most important aspect to keep in mind is the calibration of the predictive model for the particular form of sample presented for analysis, and this is where the majority of work is required in developing new official methods of sucrose analysis (Figure 30.4).

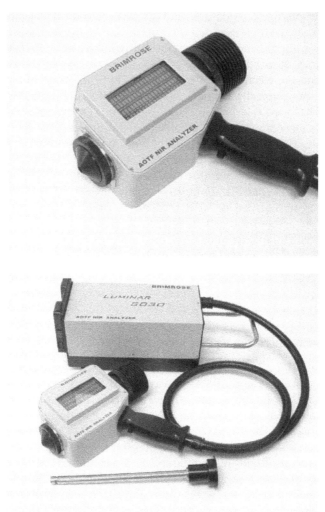

Figure 30.3   A handheld NIR spectrophotometer.

For sugar beet, Roggo *et al.* (2003) tested eight different calibration strategies and showed large differences between each of them, although three related strategies based on principle components or partial least squares were able to more accurately classify direct observations (*e.g.* disease resistance, harvest location, and harvest date) than other tested strategies. These workers followed up their calibration and model robustness work to demonstrate the potential of NIR spectroscopy to provide equivalent information as that obtained from polarimetry for the purpose of grower payment calculations of factory-delivered beets (Roggo *et al.* 2004). Using a modified partial least squares calibration strategy on more than 2700 brei samples from 15 different factories with NIR and polarimetry readings performed almost simultaneously, a standard error of prediction for sucrose content of 0.11% was obtained, indicating high confidence. Further, these results were compared with other work reported in the literature regarding the individual NIR wavelengths contributing the higher discriminant power for sucrose or total solids determination in various matrices (fresh tissue or aqueous solutions, reflectance or transmittance modes), demonstrating the necessity to evaluate each matrix independently. For NIR spectroscopy to replace polarimetry for grower payment and routine factory sucrose evaluation, influences of instrumentation variability, results over years of operation due to certain environmental variability, changes in agronomic practices, and new varieties need to be ascertained. Some guidance may suggest that brei spectral databases can be updated yearly with new samples derived from the current campaign and for similar instruments that calibration models can be transferred between factories (Roggo *et al.* 2002). Much of current research and development is being done within the context of industry and instrument manufacturers.

30.3.8 Other Methods

An almost bewildering array of methods has been applied to sucrose analysis over the years, alone and in combination, but they have not persisted due to their complexity, toxicity, requirements for specialized apparatus, sensitivity, or development of better methods. These are discussed in Bates and Associates (1942), Browne and Zerban (1941), and Mathlouthi and Reiser (1995). Sugar methods included in these books are based on measurements of viscosity, calorimetry, osmotic pressure, freezing or boiling points, colorimetry, isotopic methods and others, along with their applicability to certain sample types and starting materials. Nuclear magnetic resonance (NMR) spectroscopy for sucrose analysis in sugar beet juice has shown good precision (Lowman and Maciel 1979) and could be useful for *in situ* imaging of sucrose distribution in plants if further development of the method was pursued.

Comparison of methods for sucrose determination.

Method	Sensitivity	Precision	Handling	Time consuming	Cost for equipment	Running cost
Refractometry	C	C	A	A	A	A
Polarimetry	A	A	C	C	C	B
Enzymatic assay	B	B	B	B	B	B
GC	B	B	C	C	D	C
HPLC	A	A	C	C	D	C
NIR	A–C	A–C	A	A	A–C	A

A: excellent, B: good, C: below average, D: poor.

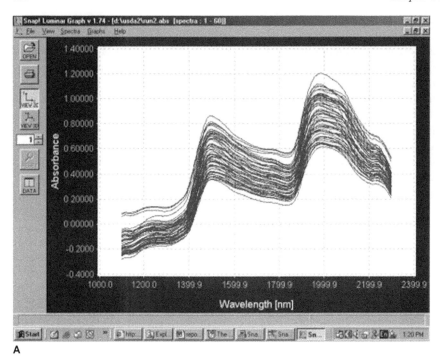

A

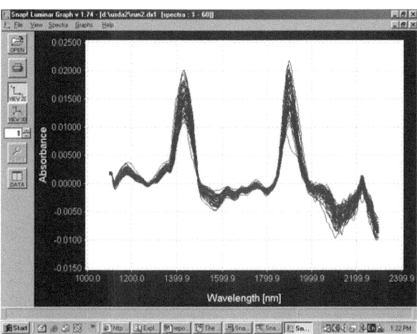

B

30.4 Future Trends

Sucrose analysis has a long history, and the importance of sucrose to the human condition ensures that chemistry of sucrose and sugars will remain an important topic for scientific investigation. Refinement of sucrose analytical methods as well as combining methodologies for unique applications can be anticipated. Perhaps the strongest trend is the adoption of NIR spectroscopy for beet and cane sugar industry and trade applications. One could envision replacement of current polarimetry-based payment schedules with one based on NIR technologies, with the long-trusted polarimeter taking a venerable position in the sugar laboratory as the arbiter of NIR calibration models to augment their validation over time.

Summary Points

- This chapter focuses on analytical methods for sucrose, primarily those used in beet sugar manufacture.
- In the beet sugar industry, polarimetry and refractometry are commonly used for routine sucrose quantification.
- Polarimetry and refractometry provide rapid and accurate quantification of sucrose in healthy roots at harvest, but are subject to inaccuracies when used to quantify sucrose in diseased roots or roots after prolonged post-harvest storage.
- Enzyme-based spectroscopic assays that couple sucrose reaction with production of a spectroscopically detectable compound provide accurate quantification of sucrose from healthy, diseased or stored roots and are relatively rapid and inexpensive when used in a microplate format.
- Gas chromatography methods are accurate and sensitive, but are not commonly used due to expense and slow analysis times.
- Many high performance liquid chromatography (HPLC) methods for sucrose analysis have been developed, although the most commonly used method today utilizes high performance anion exchange chromatography with pulsed amperometric detection (HPAE-PAD).
- HPAE-PAD provides accurate and sensitive detection and quantification of sucrose and other carbohydrates present in sugarbeet roots but is

---

Figure 30.4 NIR absorbance spectra of 148 frozen sugar beet brei samples (McGrath and Trivedi, unpublished). A: Original absorbance spectra screenshot. B: First derivative spectra screenshot. Samples were scanned three times using the wavelength range from 1100 nm to 2300 nm, in 2 nm increments. 200 scans were collected per reading and averaged into one spectrum. Wavelengths beyond 1900 nm were noisy and these wavelengths were not used when building the calibration model. PLS regression models for Brix were created using the absorbance and first derivative spectra at different wavelength ranges with the best modeling results coming from the first derivative spectra. The PLS regression model showed a correlation coefficient of 0.939 for calibration and 0.910 for validation. The standard error of prediction was 0.872, indicating reasonable predictive power.

expensive, relatively slow and of limited use when large numbers of samples must be analyzed in a limited time.
- NIR spectroscopy is rapid, non-destructive, utilizes inexpensive instrumentation, and is a promising analytical technique for sucrose quantification, but multivariate analyses models of NIR spectra must be developed with consideration to the wet chemistry results and be validated robustly.

Key Facts

Key Facts of AOAC INTERNATIONAL

1. AOAC INTERNATIONAL was formally called the Association of Official Analytical Chemists and the Association of Agricultural Chemists.
2. It is a globally recognized, independent, not-for-profit association founded in 1884.
3. The organization's web address is http://www.aoac.org/ (accessed November 2011).
4. AOAC serves the analytical science community by providing a forum to develop technical standards through consensus and working groups where methods and their performance criteria are established and documented.
5. AOAC Official Methods are accepted by many regulatory entities and recognized worldwide.

Key Facts of ICUMSA

1. The International Commission for Uniform Methods of Sugar Analysis (ICUMSA) is a worldwide body which brings together the activities of the National Committees for Sugar Analysis in more than thirty member countries.
2. ICUMSA held its first meeting in Hamburg, Germany in 1897.
3. Methods are validated by cooperating members and recommended for Tentative (T) approval by ICUMSA in the first instance. Upon meeting all the Commission's requirements, methods are accorded Official (O) status. Methods which are demonstrably useful and have found an established application, or which do not lend themselves to collaborative testing, are given an Accepted (A) status.
4. The organization's web address is http://www.icumsa.org/ (accessed November 2011).

Key Facts of the Codex Alimentarius Commission

1. The Codex Alimentarius (food code) is a global reference point for consumers, food producers and processors, national food control agencies and the international food trade.

2. The Codex Alimentarius generates authoritative publications for food safety management and consumer protection based on the work of the best informed individuals and organizations in food and related fields.
3. The Codex Alimentarius Commission was established by FAO and WHO in 1963 to harmonize international food standards, guidelines and codes of practice to protect the health of the consumers and ensure fair trade practices in the food trade. The Commission also promotes co-ordination of all food standards work undertaken by international governmental and non-governmental organizations.
4. The organization's web address is http://www.codexalimentarius.net (accessed November 2011).

Definitions of Words and Terms

Gas chromatography (GC). GC is a chromatographic method that separates volatilized compounds based on their partioning with a liquid or solid stationary phase and is used for separating compounds that can be vaporized without thermal decomposition. Vaporized samples are transported over the stationary phase *via* the flow of an inert gas, and sample components are separated based on the extent of their interaction with the stationary phase. A detector at the terminus of the stationary phase senses and quantifies compounds.

High performance liquid chromatography (HPLC). HPLC is a chromatographic method that separates compounds based on their partitioning between the solvent in which they are dissolved and a solid phase over which they are passed, such that compounds emerge from the solid support at a rate that is inversely related to the extent that they chemically or physically interact with it. High pressure is used to increase flow rate over the solid support and increase the separation rate of sample components, and a detector located at the terminus of the solid support is used to sense and quantify compounds based on their physical or chemical properties.

Mass spectrometry (MS). MS is an analytical technique used to separate and detect an ionized molecule by its mass-to-charge ratio. MS may be combined with liquid- (LC) or gas- (GC) chromatography. The advantages of LC/MS and GC/MS are very high selectivity and sensitivity, and thus these procedures are considered to offer the most precise analytical methods.

Near-infrared spectroscopy (NIR). NIR spectroscopy irradiates samples with light in the 780–2500 nm wavelength range, and measures reflected or transmitted wavelengths with a charged-coupled device such as silicon or indium gallium arsenide (InGaAs) semiconductors. The sample absorbs NIR, usually in a broad and irregular pattern, and the reflected or transmitted spectrum is processed using multivariate statistics to identify major signal peaks that best predict the analyte measured by an accepted reference method.

Polarimetry. Polarimetry utilizes the property of optically active molecules to rotate the plane of polarized light in solution. The degree of rotation is calibrated against known solution concentrations. Saccharimeters are calibrated specifically for the measurement of pure sucrose solutions.

Refractometry. Refractometry takes advantage of the principle that light changes speed as it passes through different mediums. The angle of refraction is used to calibrate the instrument to a standard scale, most often reported as degrees Brix, named after Adolph Brix, a German mathematician and civil servant who developed precise tables of refraction for sucrose in solution.

Spectroscopy. Spectroscopy is an analytical technique used to quantify compounds based on their interactions with radiated energy at specified wavelengths. Interactions typically measured include absorbance, emission, reflectance or transmission. These interactions form the basis for colorimetric assays (absorbance), fluorometric assays (emission), and near-infrared assay methods (reflectance or transmission).

List of Abbreviations

AOAC	Association of Analytical Communities
ELISA	enzyme-linked immunosorbent assay
ELSD	evaporative light scattering detector
FAO	Food and Agriculture Organization of the United Nations
GC	gas chromatography
HPAE-PAD	high performance anion exchange chromatography with pulsed amperometric detection
HPLC	high-performance liquid chromatography
ICUMSA	International Commission for Uniform Methods of Sugar Analysis
InGaAs	indium gallium arsenide
LC	liquid chromatography
MS	mass spectrometry
NADH	Nicotinamide adenine dinucleotide
NADPH	Nicotinamide adenine dinucleotide phosphate
NAD(P)H	denotes either NADH or NADPH
NIR	Near-infrared
PGI	phosphoglucose isomerase
PLS	Partial Least Squares
RDS	refractometric dry substance
UV	ultraviolet
WHO	World Health Organization

Acknowledgements

The authors thank Kyle H. McGrath, Department of Chemistry, Kalamazoo College for critically reading the manuscript and providing valuable feedback. Mention of a proprietary product does not constitute an endorsement for its use by the USDA.

References

Asadi, M., 2007. *Beet-sugar handbook*. John Wiley & Sons. Hoboken, NJ, USA, 866 pp.

Bates, F.J., and Associates, 1942. *Polarimetry, saccarimetry, and the sugars*. Circular of the National Bureau of Standards C440.

Birnberg, P.R., and Brenner, M.L., 1984. A one-step enzymatic assay for sucrose with sucrose phosphorylase. *Analytical Biochemistry*. 142: 556–561.

Browne, C.A., 1912. *A handbook of sugar analysis: a practical and descriptive treatise for use in research, technical and control laboratories*. John Wiley and Sons, New York.

Browne, C.A., and Zerban, F.W., 1941. *Physical and chemical methods of sugar analysis: a practical and descriptive treatise for use in research, technical and control laboratories*. John Wiley and Sons, New York. 1353 pp.

Chen, J.C.P., and Chou, C.-C., 1993. *Cane sugar handbook: A manual for cane sugar manufacturers and their chemists*, 12th Edition. John Wiley and Sons, Hoboken, NJ, USA. 1120 pp.

Cozzolino, D., Cynkar, W.U., Shah, N., and Smith, P., 2011. Multivariate data analysis applied to spectroscopy: Potential application to juice and fruit quality. *Food Research International*, 44: 1888–1896.

Day-Lewis, C.M.J., and Schäffler, K.J., 1990. HPLC of sugars: the analytical technique for the 1990s? *Proceedings of the Fourth African Sugar Technologists Association*, Durban, Natal, South Africa, 64: 174–178.

Dutton, J., and Huijbregts, T., 2006. Root quality and processing. In: Draycott, A.P. (ed) *Sugar Beet*. Blackwell Publishing, Oxford, UK. pp. 409–442.

Galon, J., and Zallen, D.T., 1998. The role of the Vilmorin Company in the promotion and diffusion of the experimental science of heredity in France, 1840–1920. *J. Hist. Biol*. 31: 241–262.

Godshall, M.A., 2010. Value-added products for a sustainable sugar industry. In: Eggleston, G. (ed.) *Sustainability of the Sugar and Sugar Ethanol Industries*. ACS Symposium Series, American Chemical Society, Washington, DC. pp. 253–268.

Herbreteau, B., Lafosse, M., Morin-Allory, L., and Dreux, M., 1990. Automatic sugar analysis in the beet industry. Part 1: Parameter optimization of a reversed phase HPLC carbohydrate determination. *Journal of High Resolution Chromatography*. 13: 239–243.

Huang, H., Yu, H., Xu, H., and Ying, Y., 2008. Near infrared spectroscopy for on/in-line monitoring of quality in foods and beverages: A review. *Journal of Food Engineering*. 87: 303–313.

ICUMSA, 1964. *ICUMSA methods of sugar analysis*. de Whalley, H.C.S. (ed) Elsevier Publishing Co., New York. 153 p.

Karr, J., and Norman, L.W., 1974. The determination of sucrose in concentrated Steffen filtrate by G.L.C. *Journal of the American Society of Sugar Beet Technologists*. 18: 53–59.

Kort, M.J., 1980. High-Pressure Liquid Chromatography. In: *Sugar Milling Research Institute Annual Report for 1979–1980*. pp. 13–14.

Lee, J.W., Thomas, L.C., and Schmidt, S.J., 2011. Investigation of the heating rate dependency associated with the loss of crystalline structure in sucrose, glucose, and fructose using a thermal analysis approach (Part I). *J. Agric. Food Chem.* 59: 684–701.

Lendl, B., and Kellner, R., 1995. Determination of sucrose by flow injection analysis with fourier transform infrared detection. *Mikrochim. Acta*, 119: 73–79.

Lescure, J.P., 1995. Analysis of sucrose solutions. In Mathlouthi, M., and Reiser, P. (eds) *Sucrose: Properties and Applications.* Blackie Academic and Professional, London, UK. pp. 155–185.

Long, A.R., and Chism, G.W., III. 1987. A rapid direct extraction-derivatization method for determining sugars in fruit tissue. *Journal of Food Science.* 52: 150–154.

Lowman, D.W., and Maciel, G.E., 1979. Determination of sucrose in sugar beet juices by nuclear magnetic resonance spectrometry. *Analytical Chemistry.* 51: 85–90.

Macrae, R., Trugo, L.C., and Dick, J., 1982. The mass detector: a new detection system for carbohydrate and lipid analyses. *Chromatographia.* 15: 476–478.

Mathlouthi, M., and Reiser, P., 1995. *Sucrose: Properties and Applications.* Blackie Academic and Professional, London, UK. 294 pp.

McGinnis, R.A., 1982. *Beet Sugar Technology* (3rd edition). Beet Sugar development Foundation, Fort Collins, CO, USA. 855 pp.

Mulcock, A.P., Moore, S., Barnes, F., and Hickey, B., 1985. The analysis of sugars in beet part III: HPLC analysis. *International Sugar Journal.* 87: 203–207.

Nicolaï, B.M., Katrien Beullens, K., Bobelyn, E., Peir, A., Saeys, W., Theron, K.I., and Lammertyn, J., 2007. Nondestructive measurement of fruit and vegetable quality by means of NIR spectroscopy: A review. *Postharvest Biology and Technology.* 46: 99–118.

Peters, S., Rose, T., Moser M., 2010. Sucrose: A prospering and sustainable organic raw material. *Topics Curr. Chem.* 294: 1–23.

Rocklin, R.D., and Pohl, C.A., 1983. Determination of carbohydrates by anion exchange chromatography with pulsed amperometric detection. *Journal of Liquid Chromatography.* 6: 1577–1590.

Roggo, Y., Duponchel, L., and Huvenne, J.-P., 2003. Comparison of supervised pattern recognition methods with McNemar's statistical test Application to qualitative analysis of sugar beet by near-infrared spectroscopy. *Analytica Chimica Acta*, 477: 187–200.

Roggo, Y., Duponchel, L., and Huvenne, J.-P., 2004. Quality evaluation of sugar beet (*Beta vulgaris*) by near-infrared spectroscopy. *J. Agric. Food Chem.*, 52: 1055–1061.

Roggo, Y., Duponchel, L., Noe, B., and Huvenne, J.-P., 2002. Sucrose content determination of sugar beets by near infrared reflectance spectroscopy. Comparison of calibration methods and calibration transfer. *J. Near Infrared Spectroscopy.* 10: 137–150.

Schoenrock, K.W.R., and Costesso, D., 1975. The spectral photometric determination of sucrose in sugar beets and sugar beet products via specific enzyme systems. *Journal of the American Society of Sugar Beet Technologists.* 18: 349–359.

Schoonees, B.M., 2003. Transition from leaded pol to NIR pol in the South African sugar industry. *Proc. S. Afr. Sug. Technol. Assn.* 77: 404–413.

Singleton, V., Horn, J., Bucke, C., and Adlard, M., 2002. A new polarimetric method for the analysis of dextran and sucrose. *Journal American Society of Sugarcane Technologists.* 22: 112–119.

Spackman, V.M.T., and Cobb, A.H., 2001. An enzyme-based method for the rapid determination of sucrose, glucose and fructose in sugar beet roots and the effects of impact damage and postharvest storage in clamps. *Journal of the Science of Food and Agriculture.* 82: 80–86.

Spencer, G.L., 1910. *A hand-book for chemists of beet-sugar houses and seed-culture farms.* John Wiley and Sons, New York.

Trebbi, D., and McGrath, J.M., 2004. Fluorometric sucrose evaluation for sugar beet. *Journal of Agricultural and Food Chemistry.* 52: 6862–6867.

U.S. Dept. of Agriculture Economic Research Service. 2011. Briefing Rooms / Sugar and Sweeteners / Recommended Data Table 1 http://www.ers.usda.gov/Briefing/Sugar/Data.htm (accessed November 29, 2011).

Vettorazzi, G., and Macdonald, I., 1988. *Sucrose: Nutritional and safety aspects. ILSI Human Nutrition Reviews*, Springer-Verlag, London. 192 pp.

Vilmorin, P.-P.-A.L., 1859. Notice sur l'amélioration de la Carotte sauvage. In L. Vilmorin. *Notices sur l'amélioration des plantes par le semis et considérations sur l'hérédité dans les végétaux.* 1st edn. Librairie Agricole, Paris.

Function and Effects

CHAPTER 31

Lactose in Milk and Dairy Products: A Focus on Biosensors

FELIPE CONZUELO, A. JULIO REVIEJO
AND JOSÉ M. PINGARRÓN*

Electroanalysis and (bio)Sensors group, Dpt. Analytical Chemistry,
Complutense University of Madrid, 28040-Madrid, Spain,
*Email: pingarro@quim.ucm.es

31.1 Introduction

Milk and dairy products are major constituents of humans' daily diet. Milk is an economical and readily available source of nutrition all over the world, even among poor populations. It contains calcium and magnesium phosphates, high-quality proteins, a variety of vitamins, and lactose, as a characteristic carbohydrate.

Lactose is a disaccharide formed by glucose and galactose units joined by a β-(1–4) linkage. It is found as a major sugar in milk from almost all mammal species, therefore it is present in dairy products. It is present in a large number of food products as well from milk and its derivatives; namely, soups, processed cereals, candies, salad dressings, bakery products such as bread and cookies, and even in beer. Moreover, lactose is also present in medicines and prescription drugs.

Lactose is a basic parameter in wastewater control, and also its determination is clinically relevant because an excessive amount of lactose in blood indicates the presence of gastrointestinal affections (Göktuğ et al. 2005). In addition, milk from cows suffering from mastitis has low lactose levels; lactose

Food and Nutritional Components in Focus No. 3
Dietary Sugars: Chemistry, Analysis, Function and Effects
Edited by Victor R Preedy
© The Royal Society of Chemistry 2012
Published by the Royal Society of Chemistry, www.rsc.org

content is an important factor in veterinary medicine as it is a basic indicator for evaluating milk quality and detection of abnormal milk.

For those people suffering from lactose intolerance, some milks and dairy products are produced by previously hydrolysing lactose by means of enzymatic reactors, making such products more edible for people who are unable to digest this sugar.

For all these reasons, the quality assurance of milk and dairy products requires lactose determination during the production steps and also in the final products. This kind of analysis is carried out in the dairy industry as a product control, and in this sense new methodologies capable of carrying out direct on-line monitoring with rapid assay time, low cost, enough accuracy and unskilled personnel operation, are highly demanded.

31.2   Lactose Analysis

There are many methods available in the literature for the analysis of lactose in milk and dairy products. Although they demonstrated some usefulness, a lot of them exhibit a rather high complexity, are time consuming, tedious and difficult to automate or require expensive instruments and skilled personnel. Many of them demand an intricate sample treatment, including lactose extraction and filtration and, therefore, they are not adequate to implement real-time measurements and setting up in automated systems thought for the milking parlour and dairy industry.

Table 31.1 summarises various methods developed for lactose determination. Although some of them were extensively used in the past, they exhibit important disadvantages, such as time-consuming and complex sample treatment steps and interference from other carbohydrates present in the samples.

Biosensors are versatile analytical devices that have been extensively applied in a wide range of fields such as biotechnology, medicine, agriculture, environmental monitoring and food control, due to their well known advantages in terms of high degree of selectivity, high sensitivity, simplicity, low cost, easy miniaturisation, and the possibility of coupling in an automated system for on-line measurements. Moreover, they can be used for in field analysis since the high selectivity of the bio-recognition elements allows the analysis time to be reduced minimising sample treatments and allowing, in most cases, low-cost analyses.

Biosensors can be used for lactose determination and, due to the above-mentioned advantages, can be envisaged as useful tools for the food industry. In this chapter, the role of biosensors for this particular purpose is critically reviewed, featuring not only the already existing work but also the future perspectives in this field.

31.2.1   Enzyme Reactions Involved in Lactose Determination

The reactions involved in lactose determination by using biosensors are depicted in Figure 31.1. The main reaction entails the cleavage of lactose

Table 31.1 Non-electrochemical methods for lactose determination. The table summarises some classical methods that were used for lactose determination as well as the fundamentals on which they are based. Many of these methods are now unusual, displaced by more accurate and simple methodologies. The principal drawbacks of these methods are long and complex sample preparation steps and the interference from other carbohydrates in the sample. The use of biosensors for lactose determination can overcome these problems and therefore be useful for the dairy industries. *Unpublished.*

Method	Principle	Drawback	Reference
Redox titration	Iodimetric titration	Cumbersome, gives erratic results owing to under- and over-oxidation in the presence of certain ions in sample	(Cheng and Christian 1977)
Colorimetric (phenol-sulphuric acid method)	Coloured aromatic complex formed between phenol and carbohydrate	Non-specific colour absorption and reaction, use of highly corrosive reagents	(Cheng and Christian 1977)
High-Performance Liquid Chromatography	Chromatographic determination of lactose	Tedious sample pre-treatment, long time of analysis, expensive, skilled personnel	(Tkáč *et al.* 2000)
Enzymatic assay	Coupled enzymatic reaction and measurement of NADH absorbance	Long and elaborate procedures	(Cheng and Christian 1977)
Infrared spectrophotometry	measurement of the absorption of the hydroxyl group	Only fluid samples, interference from other carbohydrates	(Yang *et al.* 2007)
Gravimetric analysis	Precipitation of lactose with $CuSO_4$	Interference from all reducing carbohydrates	(Webb *et al.* 1974)
Polarimetry	Measurement of the specific rotation of polarised light due to asymmetric carbon in lactose	Interference from other carbohydrates and optically active substances in sample	(Webb *et al.* 1974)

disaccharide into glucose and galactose moieties. The different possible strategies to perform lactose determination are outlined in this figure as well as the final species monitored, to be related with the lactose content in different samples.

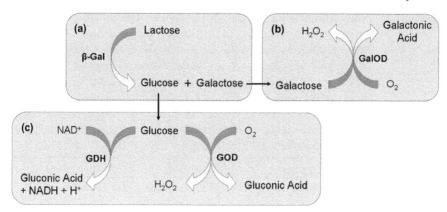

Figure 31.1 Enzyme reactions involved in lactose determination. This figure summarises the main enzyme reactions involved in the different biosensor
designs reported in the literature for lactose determination. (a) The
enzyme β-galactosidase (β-Gal) hydrolyses lactose to produce glucose
and galactose. (b) Galactose is selectively oxidised by galactose oxidase
(GalOD) to form hydrogen peroxide and galactonic acid. (c) Glucose can
be either catalytically oxidised by glucose oxidase (GOD), producing
gluconic acid and H_2O_2 or by glucose dehydrogenase (GDH) in the
presence of NAD^+, to produce gluconic acid, NADH and hydrogen ion.
Unpublished figure.

31.2.2   Biosensors for Lactose Determination

This section is devoted to comment on some of the biosensors found in the
literature for the determination of lactose. Most of them correspond to enzyme-
based amperometric biosensors, which, due to their low cost, simplicity and the
possibility of them being operated by unskilled personnel can be envisaged as
appropriate analytical tools for such purpose.

31.2.2.1   Amperometric Biosensors – Early Configurations

The first approaches to the determination of lactose involving the use of
enzymes and electrodes were accomplished by measuring the rate of oxygen
depletion after lactose hydrolysis and subsequent glucose oxidation with a
Clark-type oxygen electrode (Cheng and Christian 1977). Solutions of the
enzymes β-galactosidase (β-Gal) and glucose oxidase (GOD) were pipetted into
a cell with the sample and the rate of oxygen depletion, directly related with the
lactose content in sample, was monitored. Although, in fact this methodology
cannot be considered as based on the use of biosensors, the whole determination took less than 5 min, but it was necessary to add iodide to reduce the
hydrogen peroxide generated in order to avoid the interference of O_2 produced
by disproportionation. A negative aspect of this methodology is that every
single determination requires the addition of enzyme solutions with the subsequent increase in the cost per analysis even if small volumes are employed.

On the contrary, the immobilisation of the enzymes involved in lactose analysis on a transducer substrate, that is, using a true biosensor system, allows the reusing of the biological catalysts leading to more economical and accurate determinations.

Pioneer electrochemical determination of lactose by means of a biosensor was reported by Pilloton and Mascini (Pilloton and Mascini 1990). They developed a biosensor composed of a dialysis membrane with immobilised β-Gal, and another membrane with immobilised GOD. Both membranes were held on a Pt electrode polarised at $+650\,mV$ vs. Ag/AgCl. As it was schematized in Figure 31.1, the amount of lactose in the sample was related with the measured hydrogen peroxide generated in the hydrolysis of lactose by β-Gal, and the consecutive oxidation of released glucose by GOD. This was the first semi-automated method described for the determination of lactose, and also permitted the determination of glucose and lactose in the same system by removing β-Gal from the biosensor and its further incorporation as an in-line enzymatic reactor.

More or less at the same time, Watanabe *et al.* reported another biosensor for the determination of lactose and glucose. It consisted of a Clark-type oxygen electrode coated with dialysis membranes where β-Gal and GOD were immobilised (Watanabe *et al.* 1991). The analytical signal corresponded to the measurement of oxygen depletion in the oxidation reaction of glucose by GOD, and it was the first biosensing system that allowed a simultaneous determination of lactose and glucose using a flow injection system without previous separation or pre-hydrolysis steps. This system differentiated among carbohydrates by monitoring the time lag for the initiation of enzyme reactions and diffusion to the electrode surface, according to the multiple enzyme membrane setting fixed on the oxygen electrode. The assay took approx. 10 min and was successful when the amount of glucose was in the range 2.8–8.3% of lactose.

Another biosensor based on the measurement of oxygen depletion as a consequence of the glucose release in the lactose hydrolysis by β-Gal and its subsequent oxidation by GOD, was developed by immobilising these two enzymes with gelatin and glutaraldehyde on a mercury thin film electrode. This electrode was constructed by deposition of Hg on the surface of a glassy carbon electrode (Göktuğ *et al.* 2005). Lactose detection was achieved at $-200\,mV$ (vs. Ag/AgCl) as the difference in the measured current values before and after the addition of lactose. The assay was completed in 8 minutes but it suffered from a low stability of the mercury thin film allowing only 40 measurements to be performed before the biosensor response started to decrease.

Screen-printed electrodes were employed firstly for lactose determination by co-immobilising the enzymes GOD and β-Gal on a Pt working electrode by cross-linking with glutaraldehyde (Jäger and Bilitewski 1994). The sensor was stable for nearly 3 months without noticeable loss of activity.

An extended linear range for the determination of lactose was attained with a biosensor incorporating the enzymes β-Gal and GOD into a poly(carbamoyl)sulfonate gel and using screen-printed platinum electrodes. The H_2O_2 produced in the enzyme reactions was monitored at $+600\,mV$

(vs. Ag/AgCl). The extension of the linear range was achieved by employing membranes and spray-coated aqueous polymer dispersions as diffusion barriers (Loechel *et al.* 1998). However, the initial response time between 20 and 25 s increased up to several minutes with the extended range of linearity as a consequence of slowing down the diffusion process.

31.2.2.2 More Recent Enzyme Immobilisation Strategies

More recently, some more complex strategies have been used to achieve an efficient immobilisation of the enzymes involved in lactose determination on electrode substrates. Langmuir-Blodgett films of poly(3-hexyl thiophene)/ stearic acid with immobilised β-galactosidase and galactose oxidase (GalOD) were transferred onto ITO-coated glass plates to construct a lactose biosensor which exhibited a lifetime of more than 120 days at 4 °C (Sharma *et al.* 2004). It was based on the hydrolysis of lactose and the subsequent oxidation of galactose monitoring the generated H_2O_2 (see Figure 31.1). The biosensor maintained 97% of the original response after 10 uses and decreased until 50% after 30 uses.

An asymmetrical alternating current electrophoretic deposition was used by Ammam and Fransaer for the deposition of thick and highly active layers of β-Gal and GOD on a platinum electrode (Ammam and Fransaer 2010). This strategy allowed a high reproducibility in the biosensor construction to be reached. After polarization of the Pt working electrode at $+0.65\,V$ (vs. Ag/ AgCl) the oxidation of the H_2O_2 released from the enzyme's chain reaction permitted lactose determination with a fast response of approximately 8 s. It is interesting to remark that the high reproducibility in the biosensor construction, avoiding manual immobilisation procedures, was translated in a high reproducibility for lactose determination. Nevertheless, the immobilisation methodology would imply an elevated cost of the fabrication of the biosensors and, therefore, in the analysis costs.

31.2.2.3 Second Generation Biosensors

As it has been commented above, after lactose cleavage by the enzyme β-galactosidase, a direct measurement of H_2O_2 formation or the consumption of oxygen involved in the oxidation of lactose hydrolysis products can be accomplished. The so-called first generation biosensors have as their main disadvantage the use of an extreme potential applied to the working electrode, as it has been evident for the biosensor designs commented on above, which increased the possibility of interference from electroactive species potentially present in the samples to be analysed. Improved second generation biosensors were developed by introducing low molecular weight artificial mediators, which shuttle electrons between the active site of the enzyme and the electrode, thus allowing lower overpotentials to be proceeded (Stoica *et al.* 2006).

As an example, an amperometric sensor sensitive for lactose and glucose employing ferrocene as a mediator was constructed by cross-linking the enzymes β-Gal, GOD and mutarotase in a β-cyclodextrin polymer whose cavities included the ferrocene trough a host-guest chemical interaction (Liu *et al.* 1998). The coupled mutarotase enzyme in the enzymatic chain reactions permitted an enhanced signal for lactose to be achieved because of the isomerisation of the α-D-glucose liberated by the hydrolysis of lactose in β-D-glucose which is the specific substrate of GOD. The biosensor exhibited a fast lactose determination within 40 s at an applied potential of $+0.35$ V (vs. Ag/AgCl) enabling the electrochemical oxidation of ferrocene. An important drawback of this design is the long time required, more than 34 h, for its construction.

The incorporation of the enzyme peroxidase leads to a decrease in the detection potential. For instance, a three-cascaded-enzymes biosensor was prepared involving the use of a glassy carbon electrode modified with the enzymes, β-Gal, GOD and horseradish peroxidise (HRP) and the mediator 5-aminosalycilic acid. Upon hydrolysis of lactose by β-Gal, and glucose oxidation by GOD, HRP catalytically oxidises the generated H_2O_2, and the oxidised form of the mediator is then reduced at the electrode surface at an applied potential of 0.0 V (vs. SCE), thus minimising potential interference from other electro-active species (Eshkenazi *et al.* 2000).

Because the coupled enzyme cascade reaction mostly comprises the use of GOD, the interference of glucose is a major concern. Therefore, biosensors constructed with immobilised GalOD, instead of GOD, have been developed to avoid this kind of interference. The enzyme GalOD is able to oxidise oligo-saccharides with D-galactose in terminal position, one of them being lactose. In this context, Tkáč and co-workers prepared biosensors for the determination of lactose either by direct oxidation of lactose by GalOD or incorporating also β-Gal for previous lactose hydrolysis (Tkáč *et al.* 2000). Both configurations implied incorporation of HRP and the mediator ferrocene. As it could be expected, the main interferences for lactose determination were free galactose and lactulose (another disaccharide with D-galactose in terminal position which is found in milk after thermal treatment). It is important to remark that galactose interference is particularly important in the analysis of foodstuffs with low lactose content due to the fact that these products are industrially fabricated by hydrolysing lactose with the subsequent formation of glucose and galactose as reaction products.

Recently, our research group developed an amperometric biosensor for lactose involving co-immobilisation of the enzymes β-Gal, GOD, and HRP as well as the mediator tetrathiafulvalene (TTF) by physical entrapment with a dialysis membrane on a gold electrode, (Conzuelo *et al.* 2010). Figure 31.2 shows the enzyme and electrode reactions involved in the determination of lactose with the developed biosensor.

25 commercial milk and dairy products were analysed with this electro-chemical biosensor and the achieved results correlated fairly well with those obtained using a commercial enzyme test kit. Also, the biosensor was validated

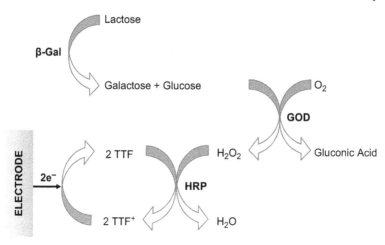

Figure 31.2 Enzyme and electrode reactions involved in lactose determination with the amperometric biosensor developed by Conzuelo *et al.* Lactose is hydrolysed by β-galactosidase (β-Gal) and the formed glucose is oxidised by glucose oxidase (GOD) to produce hydrogen peroxide and δ-gluconolactone which is spontaneously hydrolysed to gluconic acid. H_2O_2 is reduced in the presence of peroxidase (HRP) whose reduced form regeneration is mediated by tetrathiafulvalene (TTF). The electrochemical reduction of the oxidised form of the mediator, TTF^+, at an applied potential of 0.0 V (vs. Ag/AgCl), is the analytical signal employed to quantify lactose. *Unpublished figure.*

by analysing two certified reference materials with satisfactory results. As a particular analytical figure of merit, this lactose biosensor exhibited one of the lowest limits of detection (0.46 μM) when compared with data reported using other biosensors. This allowed the determination of this disaccharide in foodstuffs containing low lactose contents either by performing the subtraction of the signal corresponding to the glucose present in the sample, if any, or by removing it previously. The biosensor was employed as a detector in a flow-injection system, and applied to the analysis of different milk samples in order to evaluate its applicability in an automated system. All the obtained results demonstrated that the developed biosensor fulfilled the essential requirements to be considered as a useful analytical tool for the dairy industries.

Table 31.2 summarises the main compositional and analytical characteristics of the amperometric biosensors reported in the literature for the determination of lactose. The Table allows visualization of the evolution of the analytical strategies employed to improve the performance of lactose quantification in real samples, such as the use of redox mediators and peroxydase enzyme in order to minimise potential interferences by decreasing the overpotentials needed to carry out the amperometric detection, and the search for sensitivities in agreement with those required in real industrial applications.

Table 31.2 Amperometric enzyme-based biosensors for lactose determination. This table summarises the main characteristics for the amperometric electrochemical biosensors for lactose reported in the literature. Early biosensors were based on direct measurement of the hydrogen peroxide generated from the enzyme reactions, thus requiring large overpotentials and increasing the number of potential interfering species. The use of redox mediators allowed lower detection potentials to be applied to obtain the electroanalytical signal. Several electrode materials and immobilisation strategies have been used. Nevertheless, simple and low-cost biosensor fabrication procedures must be followed in order to construct economical and easy-to-operate systems for industrial applications, while maintaining the accuracy and reproducibility of the determinations. *Unpublished.*

Electrode	Enzyme/s (Redox mediator)	Immobilisation type	E_{appl}/V	Linear Range/M	Sensitivity	LOD	Stability	Reference
Pt	GOD, β-Gal (—)	Immobilisation with polyazetidine in cellulose acetate membranes on electrode surface	+0.65 vs. Ag/AgCl	—	—	—	—	(Pilloton and Mascini 1990)
Pt	GOD, β-Gal (—)	Immobilisation on polyurethane by cross-linking with polyvinylisocyanate	+0.60 vs. Ag/AgCl	—	—	10 μM*	More than 15 days	(Pfeiffer et al. 1990)
Clark-type oxygen electrode	GOD, β-Gal, mutarotase (—)	Immobilisation on cellulose membranes by cross-linking with GA	—	up to 180*	—	—	More than 100 assays. Enzyme membranes stable more than 10 days	(Watanabe et al. 1991)
Pt	GOD, β-Gal (—)	Cross-linking with GA	+0.60 vs. Ag/AgCl	$2 \times 10^{-6} - 2.5 \times 10^{-3}$	$250 \, \text{nA} \, \text{mM}^{-1}$	—	Nearly 3 months without noticeable loss of activity	(Jäger and Bilitewski 1994)
Carbon paste	GOD, β-Gal (—)	Incorporation of enzymes in a carbon paste matrix (60% graphite powder, 40% mineral oil)	+0.90 vs. Ag/AgCl	up to 2.5×10^{-3}	$38 \, \text{nA} \, \text{mM}^{-1}$	0.1 mM	—	(Katsu et al. 1994)

Table 31.2 (*Continued*)

Electrode	Enzyme/s (Redox mediator)	Immobilisation type	E_{appl}/V	Linear Range/M	Sensitivity	LOD	Stability	Reference
GCE	GOD, β-Gal, mutarotase (ferrocene)	Cross-linking with GA and a β-cyclodextrin polymer	+0.35 vs. Ag/AgCl	$50 \times 10^{-6} - 13.5 \times 10^{-3}$	—	10 μM	Maintains 85.7% of initial sensitivity after two months storage	(Liu et al. 1998)
Pt	GOD, β-Gal (—)	Immobilisation with poly(carbamoyl) sulphonate gel	+0.60 vs. Ag/AgCl	$3.5 \times 10^{-6} - 2.0 \times 10^{-3}$ (and up to 40×10^{-3} with diffusion barriers)	$1 \, \mu A \, mM^{-1}$	3.5 μM	—	(Loechel et al. 1998)
GCE	HRP, GOD, β-Gal (5-aminosalicylic acid)	Cross-linking with GA	0.0 vs. SCE	$2.9 \times 10^{-5} - 9.9 \times 10^{-4}$	—	—	Maintains 40% of initial sensitivity after 140 days of extensive use	(Eshkenazi et al. 2000)
Graphite rod	HRP, GalOD, β-Gal (ferrocene)	Entrapment with a dialysis membrane	+0.16 vs. Ag/AgCl	$0.09 \times 10^{-3} - 3.6 \times 10^{-3}$	$445.2 \, nA \, mM^{-1}$	44 μM	Half-life of 47 days	(Tkáč et al. 2000)
Pt	GalOD, β-Gal (—)	Immobilisation on polyethersulphone membranes	+0.70 vs. Ag/AgCl	$4.3 \times 10^{-3} - 3.1 \times 10^{-2}$	$6.81 \, nA \, M^{-1}$	—	Enzyme membranes show stability for 20 days	(Lourenço et al. 2003)
ITO-coated glass plates	GalOD, β-Gal (—)	Immobilization in LB films of P3HT/SA	+0.40 vs. Pt	0.029–0.175	—	—	Has a shelf-life above 120 days	(Sharma et al. 2004)
GCE	GOD, β-Gal (—)	Immobilisation on a mercury thin film with gelatin and GA	-0.20 vs. Ag/AgCl	$1.0 \times 10^{-4} - 3.5 \times 10^{-3}$	$52.1 \, nA \, M^{-1}$	1.0×10^{-4} M	40 measurements	(Göktuğ et al. 2005)

Electrode	Enzyme	Immobilisation method	Potential	Linear range	Sensitivity	LOD	Stability	Reference
Pt	GOD, β-Gal (—)	Immobilisation using gelatin as a carrier system and GA for chemical cross-linking	+0.70 vs. Ag/AgCl	$(0.1 - 15) \times 10^{-3}$	—	0.1 mM	Shelf-life of 30 days	(Loğoğlu et al. 2006)
Solid spectrographic graphite electrode	CDH (—)	Chemo-physical adsorption	+0.30 vs. Ag/AgCl	$(1 - 100) \times 10^{-6}$*	$17800\,nA\,mM^{-1}cm^{-2}$*	$1\,\mu M$*	Maintains 98% of its initial signal after 11h of continuous use	(Stoica et al. 2006)
Pt	GOD, β-Gal (—)	Deposition of enzymes by an alternating current electrophoretic deposition	+0.65 vs. Ag/AgCl	up to 14×10^{-3}	$111\,nA\,mM^{-1}mm^{-2}$	0.1 mM	One week with a quasi-stable response (augmented with a polyurethane thin layer)	(Ammam et al. 2010)
Gold disk electrode modified with a MPA-SAM	β-Gal, GOD, HRP (TTF)	Entrapment with a dialysis membrane	0.0 vs. Ag/AgCl	$(1.8 - 120) \times 10^{-6}$ $(5.0 - 1000) \times 10^{-6}$	$(6.04) \times 10^{5}\,nA\,M^{-1}$ $(1.11) \times 10^{4}\,nA\,M^{-1}$*	$0.46\,\mu M$ $3.81\,\mu M$*	28 days of continuous measurements	(Conzuelo et al. 2010)
Screen-printed carbon electrode	CDH (—)	Chemo-physical adsorption	+0.10 vs. Ag/AgCl	$(0.5 - 200) \times 10^{-6}$*	—	$250\,nM$*	Maintains 86–96% of its initial activity after 8 days	(Safina et al. 2010)

*Flow injection analysis mode.

Abbreviations: β-Gal: β-galactosidase; CDH: cellobiose dehydrogenase; GA: glutaraldehyde; GalOD: Galactose oxidase; GCE: glassy carbon electrode; GDH: glucose dehydrogenase; GOD: Glucose oxidase; HRP: horseradish peroxidase; ITO: indium-tin oxide; LB: Langmuir-Blodgett; LOD: limit of detection; M: molar concentration; MPA: 3-mercaptopropionic acid; P3HT/SA: poly (3-hexyl thiophene)/stearic acid; SAM: self-assembled monolayer; SCE: saturated calomel electrode; TTF: tetrathiafulvalene.

31.2.2.4 Potentiometric and Conductometric Biosensors

Other strategies for the determination of lactose are those using potentiometric and conductometric sensors. Using a field effect transistor (FET) whose response was sensitive to pH changes produced in solution by the enzyme reactions, lactose was determined (Sevilla III *et al.* 1994). The biosensor was constructed by immobilising the enzymes β-Gal and glucose dehydrogenase (GDH) on the pH sensitive area of the FET. After lactose cleavage by β-Gal, the generated glucose was oxidised by GDH in the presence of NAD^+ (see Figure 31.1 for the sequence of reactions). The produced gluconic acid and the hydrogen ion formed were measured as the difference between the voltages in this bioFET and a pH-sensitive FET and this related with lactose concentration. Owing to the use of GDH, the biosensor suffered from the interference of glucose.

Amárita and co-workers developed a hybrid potentiometric biosensor for the determination of lactose in milk (Amárita *et al.* 1997). The biosensor was based on the hydrolysis of lactose by β-Gal and the subsequent fermentation of the sugars liberated (glucose and galactose) by *Saccharomyces cerevisiae*, producing CO_2 and ethanol. A CO_2 electrode was used to monitor the formation of CO_2. An important drawback of this biosensor is the very long response time, 1 h, as well as the time needed to recover the baseline after one analysis of lactose in milk (4.28%), 2 h, which restricts largely the real applicability of this approach.

To our knowledge, only one conductometric biosensor for lactose has been described in the literature. It makes use of the enzymes β-Gal and GOD and was applied to the determination of lactose in milk avoiding the interference of glucose (Marrakchi *et al.* 2008). The biosensor measured the resistance of a thin layer in the immediate surface of the electrode; the signal was measured as the difference between the working electrode and a reference electrode. The suppression of the glucose interference was achieved by immobilising GOD on both working and reference electrodes while β-Gal was only immobilised on the former one. Therefore, the conductance variation was only due to lactose hydrolysis and the effect of glucose in the sample was cancelled. This biosensor showed a linear range comprised between 30 and 600 µM lactose.

31.2.2.5 Cell-based Biosensors

Apart from enzyme biosensors, other biological recognition elements have been employed to develop biosensors for the determination of lactose. In this context, the transport protein lactose permease from *Escherichia coli* was used as the lactose receptor for the development of a fluorescence biosensor (Klee *et al.* 1992). This protein catalyses the transport of lactose through a supported bilayer in a 1:1 stoichiometry with protons. Upon changes in lactose concentration, lactose and protons were transported through the membrane into a space where a pH-sensitive fluorescent dye was entrapped. The change in the relative fluorescence was then related to lactose concentration.

A flow-injection analysis system with mutant strains of *Escherichia coli* immobilised on a multi-sensor array was developed for the determination of glucose and lactose (Fritzen *et al.* 1996). Two different strains of *Escherichia coli* K12 with defects in their carbohydrate transport systems were used. Cells of *E. coli* K12 JWL184, able to interact with glucose but not with lactose, and *E. coli* K12 LR2-168, able to take up lactose but not glucose, were immobilised by gel formation with κ-carrageenan on an electrode array. The biosensing system measured the oxygen partial pressure by electrochemical reduction of O_2 at $-600\,mV$ (vs. Ag/AgCl) on platinised working electrodes. When a carbohydrate is added, either glucose or lactose, the corresponding cells metabolised it, thus producing a depletion of oxygen in the working medium and, accordingly, giving rise to a decrease in the recorded current, which can be correlated with the respective carbohydrate concentration.

31.2.2.6   Other Sensors

Other sensing strategies, different to those commented on above, were also found in the literature. For instance, a microdialysis-coupled flow injection amperometric sensor was developed for the determination of glucose, galactose and lactose in milk, (Rajendran and Irudayaraj 2002). The detection of glucose and galactose was accomplished by sequential injection of their corresponding oxidase enzymes, while a mixture of β-Gal and GOD was injected into the system for the determination of lactose. The sensor was a Pt electrode held at $+500\,mV$ vs. an Ag/AgCl reference electrode and exhibited a linear response for lactose between 0.2 and 20 mM. Although the amount of enzyme used was only 4 μL, this methodology implied a continuous use of enzymes solution which makes the determination more expensive and complex.

A manometric biosensor was also described for the determination of glucose and lactose (Jenkins and Delwiche 2003). It was based on the measurements of pressure changes due to oxygen depletion in the enzymatic oxidation of glucose and their correlation with lactose concentration. The sensitivity of this device was adjustable by changing the volume of gas relative to the sample volume.

Immobilisation of GOD and β-Gal on calcium alginate fibres and amine-modified nanosized mesoporous silica was employed to develop a flow-injection chemiluminescence biosensor for lactose (Yang *et al.* 2010). The biosensor involved the hydrolysis of lactose by β-Gal, the subsequent oxidation of glucose by GOD, and the reaction of the generated H_2O_2 with luminol in the presence of diperiodatonickelate to produce luminescence with a maximum absorbance at 425 nm.

Table 31.3 summarises the main characteristics of these strategies available for the determination of lactose.

31.2.2.7   New Trends

Lately, new perspectives regarding the development of novel biosensing approaches for the determination of lactose have emerged. For example,

Table 31.3 Biosensors for lactose determination based on other different measurement principles. This table summarises the main characteristics of non-amperometric biosensors reported in the literature for the determination of lactose. Unpublished.

Type of measurement	Electrode	Biological components	Immobilisation type	Linear Range / M	Stability	Reference
Fluorescence	—	Protein lactose permease	Receptor embedded in an E. coli phospholipid membrane	—	—	(Klee et al. 1992)
Potentiometric	Field-effect transistor	GDH, β-Gal	Cross-linking with GA	—	20 days of continuous measurements	(Sevilla III et al. 1994)
Potentiometric	CO$_2$ electrode	β-Gal, S. cerevisiae	Entrapment with a dialysis membrane	0.015–0.292	2 months	(Amárita et al. 1997)
Manometric	—	HRP, GOD, β-Gal	none (enzymes in solution)	adjustable	—	(Jenkins and Delwiche 2003)
Conductometric	Gold interdigitated electrodes	GOD, β-Gal	Cross-linking with GA	$(30-600) \times 10^{-6}$	28.5% loss of signal after 5 days of measurements	(Marrakchi et al. 2008)
Chemiluminescence	—	GOD, β-Gal	Immobilisation on amine modified nanosized mesoporous silica	2.33×10^{-7} -1.88×10^{-5}	2 months	(Yang et al. 2010)

Abbreviations: β-Gal: β-galactosidase; GA: glutaraldehyde; GDH: glucose dehydrogenase; HRP: horseradish peroxidase; M: molar concentration.

a lactose nano-probe involving the immobilisation of β-Gal on gold nano-particles (AuNps) has been reported (Dwevedi *et al.* 2009). This approach combines the catalysed lactose hydrolysis by β-Gal from *Pisum sativum* at AuNps with the oxidation of glucose by means of an enzyme kit at a test strip, thus providing a simple method for lactose determination in foodstuffs with contents between 0.1% and 2.0% lactose.

Another innovative strategy consisted of a wireless biosensor using a magnetoelastic ribbon as transducer (Yang *et al.* 2007). This magnetoelastic sensor was coated with a layer of a pH sensitive polymer, β-Gal, GOD, and catalase. The enzymatic reactions that take place in the presence of lactose lead to the formation of gluconic acid resulting in a pH decrease. The pH sensitive film then underwent a reversible shrink resulting in a shift in the resonance frequency measured which can be related to the lactose content in the sample. The sensor was applied successfully for lactose analysis in milk samples, with results in good agreement with an infrared methodology.

One of the most promising advances in the construction of biosensors for lactose determination implies the use of the enzyme Cellobiose Dehydrogenase (CDH). This enzyme allowed the development of a third-generation biosensor based on direct electron communication between CDH and the electrode substrate (Stoica *et al.* 2006). Lactose was oxidised by the FAD domain of the enzyme, and an internal electron transfer occurred between the FAD and the heme cofactor. By application of a potential value of $+300\,mV$ (vs. Ag/AgCl) to a graphite working electrode, a direct electron transfer (DET) from the heme cofactor to the electrode was produced as a consequence of the close location of this cofactor at the surface of the protein. Figure 31.3 shows schematically the reactions involved in the DET from CDH to the electrode surface. The biosensor was employed in a flow-injection system achieving a detection limit of $1\,\mu M$ as well as a high sample throughput and a good operational stability.

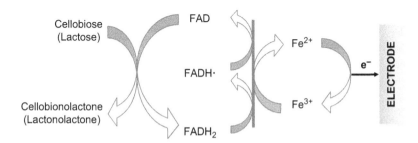

Figure 31.3 Scheme of reactions occurring at a cellobiose dehydrogenase enzyme electrode. Reactions involved in the direct electron transfer (DET) between the enzyme cellobiose dehydrogenase and an electrode surface making possible the development of a third generation biosensor for lactose. The iron ions correspond to the heme cofactor in the enzyme. *Unpublished figure.*

Moreover, no interference from other sugars found in milk like glucose and galactose were apparent. Although the main substrates for CDH, apart from lactose, are cellodextrins (cellobiose among them), their possible interference in the determination of lactose is negligible due to their absence in milk (Stoica *et al.* 2006).

Another application involving the use of CDH was reported by employing screen-printed electrodes in a flow-injection system (Safina *et al.* 2010). The enzyme was immobilised on the electrodes by simple adsorption, although cross-linking of CDH with two different compounds (glutaraldehyde and poly(ethyleneglycol)diglycidyl ether) was studied in order to stabilise the enzyme. The measurements were carried out at $+100\,mV$ (vs. pseudo Ag/AgCl) without any sample treatment.

However, the relative elevated overpotentials employed in the amperometric biosensors based on CDH for lactose determination could produce significant interference from electroactive species present in milk and dairy products when the analysis of low-content lactose samples is performed. In this kind of analysis, the sample dilution is much smaller and therefore the concentration of that species becomes important. Another limiting factor for CDH biosensors is that this enzyme is not commercially available yet, which largely prevents its wide applicability in the dairy industry.

31.3 Concluding Remarks

Biosensors offer many advantages in the development of efficient analytical tools for the analysis of lactose in milk and dairy products, as well as in other foodstuffs containing lactose. Early methods for lactose analysis were time-consuming, tedious, much more expensive and exhibited remarkable interference from other carbohydrates present in the samples to be analysed. However, late developments in biosensor technology lead to the possibility of implementing easy and accurate routine analytical methodologies for lactose quantification with immediate application in the food industry.

Summary Points

- This chapter focuses on biosensors developed for lactose determination.
- Direct on-line monitoring permits quality assurance in industry.
- Lactose analysis is highly relevant for people suffering from lactose intolerance.
- The use of enzymatic reactions contributes to the high selectivity achieved by biosensors.
- Electrochemical methods are advantageous over other methods in terms of time and costs.
- The use of biosensors in lactose analysis permits an accurate determination in short times, in an economical way and can be done by unskilled personnel.

Key Facts

Key Facts of Biosensors

1. Biosensors are devices capable of providing analytical information.
2. They use a biological recognition element that interacts with the analyte to be measured and contributes to selectivity of the biosensor.
3. This bio-recognition element is in close contact with a transducer, which is responsible for the biosensor sensitivity.
4. Enzyme biosensors use enzymes as bioreceptors.
5. Electrochemical biosensors use well-established electrochemical techniques that make possible the transduction of the biochemical event into an amplified electronic signal.
6. Biosensors provide simple, rapid and sensitive methods for the measurement of many substances of analytical significance.

Key Facts of Lactose Intolerance

1. More than 50% of the world population suffers from this affection.
2. Ingested lactose cannot be directly assimilated by the body, and must be previously hydrolysed in the small intestine by the enzyme lactase.
3. When lactose is transported undigested to the colon, colonic bacteria ferment this sugar producing short-chain fatty acids and gas.
4. Abdominal distension and diarrhea can then occur.
5. Lactose content determination is critical for those people suffering from lactose intolerance.

Definitions of Words and Terms

Amperometry. Electrochemical technique based on monitoring an electric current, as a consequence of an electrochemical reaction from electroactive species, at an applied potential regarding to a reference electrode. The response observed as a current increase can be related to the analyte concentration.

Bio-recognition element. Element of a biosensor that is capable of a selective recognition interaction with the analyte to be measured. It is the biological component of a biosensor.

Clark-type oxygen electrode. Electrode based on the enzyme electrode developed by L. C. Clark in 1962 for oxygen measurement.

Conductometry. Technique based on the measurement of conductance, reciprocal of the solution resistance, in the surrounding area between two electrode surfaces.

Indium-tin-oxide (ITO). A conducting oxide that consists of indium(III) oxide and tin(IV) oxide used as substrate in the construction of biosensors.

Langmuir-Blodgett (LB) films. Films formed by molecules in an air-water interface. These molecules are densely packed and highly oriented, approximately perpendicular to the surface.

Mastitis. This affection consists of inflammation of the mammary gland caused by several bacteria. It is the most frequent and costly disease associated with dairying.

On-line monitoring. Direct measurement of sample composition in real time during the production processes; the measurement is made in the same production line. It can be done *in situ* where the sensor is taken to the sample, or *ex situ* when the sample is taken to the sensor.

Potentiometry. Technique based on the measurement of the potential difference between an indicator and a reference electrode, when there is no significant current flowing between them. These potential differences are proportional to the logarithm of the analyte activity (or concentration).

Screen-printed electrodes. Low cost and disposable electrodes made by screen printing technology on strips coated with several inks according to the materials to be deposited.

List of Abbreviations

β-Gal	β-galactosidase
Ag/AgCl	silver/silver chloride reference electrode
AuNps	gold nanoparticles
CDH	cellobiose dehydrogenase
DET	direct electron transfer
FAD	flavin adenine dinucleotide
FET	field effect transistor
GA	glutaraldehyde
GalOD	galactose oxidase
GCE	glassy carbon electrode
GDH	glucose dehydrogenase
GOD	glucose oxidase
HRP	horseradish peroxidase
ITO	indium-tin oxide
LB	Langmuir-Blodgett
LOD	limit of detection
M	molar concentration
MPA	3-mercaptopropionic acid
mV	millivolts
NAD^+	nicotinamide adenine dinucleotide
P3HT/SA	poly(3-hexyl tiophene)/stearic acid
SAM	self-assembled monolayer
SCE	saturated calomel electrode
TTF	tetrathiafulvalene

References

Amárita, F., Rodríguez Fernández, C., and Alkorta, F., 1997. Hybrid Biosensors to Estimate Lactose in Milk. *Analytica Chimica Acta*. 349: 153–158.

Ammam, M., and Fansaer, J., 2010. Two-Enzyme Lactose Biosensor Based on β-Galactosidase and Glucose Oxidase Deposited by AC-Electrophoresis: Characteristics and Performance for Lactose Determination in Milk. *Sensors and Actuators B: Chemical*. 148: 583–589.

Cheng, F.S., and Christian, G.D., 1977. Rapid Enzymatic Determination of Lactose in Food Products Using Amperometric Measurement of the Rate of Depletion of Oxygen. *Analyst*. 102: 124–131.

Conzuelo, F., Gamella, M., Campuzano, S., Ruiz, M.A., Reviejo, A.J., and Pingarrón, J.M., 2010. An Integrated Amperometric Biosensor for the Determination of Lactose in Milk and Dairy Products. *Journal of Agricultural and Food Chemistry*. 58: 7141–7148.

Dwevedi, A., Singh, A.K., Singh, D.P., Srivastava, O.N., and Kayastha, A.M., 2009. Lactose nano-Probe Optimized Using Response Surface Methodology. *Biosensors and Bioelectronics*. 25: 784–790.

Eshkenazi, I., Maltz, E., Zion, B., and Rishpon, J., 2000. A Three-Cascaded-Enzymes Biosensor to Determine Lactose Concentration in Raw Milk. *Journal of Dairy Science*. 83: 1939–1945.

Fritzen, M., Schuhmann, W., Lengeler, J.W., and Schmidt, H.-L., 1996. Immobilized Transport Mutants of Bacterial Cells in Biosensor Arrays. Improved Selectivity for the Simultaneous Determination of Glucose and Lactose. *Progress in Biotechnology*. 11: 821–827.

Göktuğ, T., Sezgintürk, M.K., and Dinçkaya, E., 2005. Glucose Oxidase- β-Galactosidase Hybrid Biosensor Based on Glassy Carbon Electrode Modified with Mercury for Lactose Determination. *Analytica Chimica Acta*. 551: 51–56.

Jäger, A., and Bilitewski, U., 1994. Screen-printed Enzyme Eletrode for the Determination of Lactose. *Analyst*. 119: 1251–1255.

Jenkins, D.M., and Delwiche, M.J., 2003. Adaptation of a Manometric Biosensor to Measure Glucose and Lactose. *Biosensors and Bioelectronics*. 18: 101–107.

Katsu, T., Zhang, X., and Rechnitz, G.A., 1994. Simultaneous Determination of Lactose and Glucose in Milk Using two Working Enzyme Electrodes. *Talanta*. 41: 843–848.

Klee, B., John, E., and Jähnig, F., 1992. A Biosensor based on the Membrane Protein Lactose Permease. *Sensors and Actuators B*. 7: 376–379.

Liu, H., Li, H., Ying, T., Sun, K., Qin, Y., and Qi, D., 1998. Amperometric Biosensor Sensitive to Glucose and Lactose Based on co-Immobilization of Ferrocene, Glucose Oxidase, β-Galactosidase and Mutarotase in β-Cyclodextrin Polymer. *Analytica Chimica Acta*. 358: 137–144.

Loechel, C., Chemnitius, G.-C., Borchardt, M., and Cammann, K., 1998. Amperometric bi-Enzyme Based Biosensor for the Determination of Lactose with an Extended Linear Range. *European Food Research and Technology*. 207: 381–385.

Loğoğlu, E., Sungur, S., and Yildiz, Y., 2006. Development of Lactose Biosensor Based on β-Galactosidase and Glucose Oxidase Immobilized into Gelatin. *Journal of Macromolecular Science – Part A: Pure and Applied Chemistry*. 43: 525–533.

Lourenço, R.J.M., Serralheiro, M.L.M., and Rebelo, M.J.F., 2003. Development of a New Amperometric Biosensor for Lactose Determination. *Portugaliae Electrochimica Acta*. 21: 171–177.

Marrakchi, M., Dzyadevych, S.V., Lagarde, F., Martelet, C., and Jaffrezic-Renault, N., 2008. Conductometric Biosensor Based on Glucose Oxidase and beta-Galactosidase for Specific Lactose Determination in Milk. *Materials Science and Engineering C*. 28: 872–875.

Pfeiffer, D., Ralis, E.V., Makower, A., and Scheller, F.W., 1990. Amperometric Bi-enzyme Based Biosensor for the Detection of Lactose-Characterization and Application. *Journal of Chemical Technology & Biotechnology*. 49: 225–265.

Pilloton, R., and Mascini, M., 1990. Flow Analysis of Lactose and Glucose in Milk with an Improved Electrochemical Biosensor. *Food Chemistry*. 36: 213–222.

Rajendran, V., and Irudayaraj, J., 2002. Detection of Glucose, Galactose, and Lactose in Milk with a Microdialysis-Coupled Flow Injection Amperometric Sensor. *Journal of Dairy Science*. 85: 1357–1361.

Safina, G., Ludwig, R., and Gorton, L., 2010. A Simple and Sensitive Method for Lactose Detection Based on Direct Electron Transfer between Immobilised Cellobiose Dehydrogenase and Screen-Printed Carbon Electrodes. *Electrochimica Acta*. 55: 7690–7695.

Sevilla, III, F., Kullick, T., and Scheper, T., 1994. A bio-FET Sensor for Lactose based on co-Immobilized β-Galactosidase/Glucose Dehydrogenase. *Biosensors and Bioelectronics*. 9: 275–281.

Sharma, S.K., Singhal, R., Malhotra, B.D., Sehgal, N., and Kumar, A., 2004. Lactose Biosensor based on Langmuir-Blodgett films of Poly(3-hexyl thiophene). *Biosensors and Bioelectronics*. 20: 651–657.

Stoica, L., Ludwig, R., Haltrich, D., and Gorton, L., 2006. Third-Generation Biosensor for Lactose Based on Newly Discovered Cellobiose Dehydrogenase. *Analytical Chemistry*. 78: 393–398.

Tkáč, J., Šturdík, E., and Gemeiner, P., 2000. Novel Glucose non-Interference Biosensor for Lactose Detection Based on Galactose Oxidase-Peroxidase with and without co-Immobilised β-Galactosidase. *Analyst*. 125: 1285–1289.

Watanabe, E., Takagi, M., Takei, S., Hoshi, M., and Shu-gui, C., 1991. Development of Biosensors for the Simultaneous Determination of Sucrose and Glucose, Lactose and Glucose, and Starch and Glucose. *Biotechnology and Bioengineering*. 38: 99–103.

Webb, B.H., Johnson, A.H., and Alford, J.A. (eds.), 1974. *Fundamentals of Dairy Chemistry*. Avi Pub. Co., Westport, CT, 929 pp.

Yang, W., Pang, P., Gao, X., Cai, Q., Zeng, K., and Grimes, C.A., 2007. Detection of Lactose in Milk Samples Using a Wireless Multi-Enzyme Biosensor. *Sensor Letters*. 5: 405–410.

Yang, C., Zhang, Z., Shi, Z., Xue, P., Chang, P., and Yan, R., 2010. Application of a Novel co-Enzyme Reactor in Chemiluminescence Flow-Through Biosensor for Determination of Lactose. *Talanta*. 82: 319–324.

CHAPTER 32

Analysis of Human Milk Lactose

DAVID S NEWBURG,*,† CENG CHEN† AND
GHERMAN WIEDERSCHAIN†

Program in Glycobiology, Department of Biology, Boston College,
140 Commonwealth Avenue, Chestnut Hill, MA 02467-3961, USA
*Email: david.newburg@bc.edu

32.1 Introduction

32.1.1 Biological Importance of Lactose Analysis in Human Milk

The typical milk sugar is lactose, a disaccharide of glucose and galactose. It is formally named lactobiose, whose chemical structure is 4-O-β-D-galactopyranosyl-D-glucopyranose (Scheme 32.1). This disaccharide can be digested into its monosaccharides by lactase (β-galactosidase), an enzyme that is found in intestinal mucosa of infants and many adults, and in some bacteria. The digestion products of lactose are glucose and galactose. Glucose is the major source of energy in most cells, and galactose can be converted to glucose in the liver. We hypothesize that galactose could also be used directly by the infant to make galactose-containing molecules, such as the molecules essential for myelin formation during brain development. Galactosylceramide (cerebroside)

†Supported by grants from the National Institutes of Health HD013021.

Food and Nutritional Components in Focus No. 3
Dietary Sugars: Chemistry, Analysis, Function and Effects
Edited by Victor R Preedy
© The Royal Society of Chemistry 2012
Published by the Royal Society of Chemistry, www.rsc.org

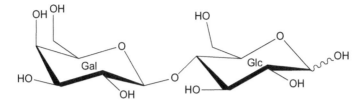

Scheme 32.1 Chemical structure of lactose.

and its sulfated analog, sulfatide, are predominant molecules in myelin, and the peak of myelination coincides with the nursing period of the breastfed infant. This hypothesis suggests that the synthesis of key glycosphingolipids in the infant brain could depend in part on galactose released from human milk lactose.

The amount of lactose in milk is distinct for different species, and in some species lactose is known to change over the course of lactation (Newburg and Neubauer 1995). Human milk was thought to have constant and consistent levels of lactose (68 g/L) once the milk matured (~ 4 weeks), but more recent reports suggest that mature human milk lactose levels can exhibit pronounced variation (Coppa *et al.* 1993). Additional variation is apparent between fore-milk and hindmilk. Lactose, which is primarily in the aqueous phase of milk, decreases as nursing progresses, because the concentration of fat in milk rises as the breast empties. Therefore, different milk acquisition protocols can intro-duce a great deal of variation in the reported values of lactose in milk (Allen *et al.* 1991). Thus, the variation, and even the precise concentration, of lactose in milk are still not certain.

Precise measurement of lactose concentrations is also important for studying lactogenesis. High among the manifestations of abrupt and extensive hormonal changes in mothers during parturition is the induction of milk synthesis in her mammary glands, *i.e.*, lactogenesis (Neville 1995). The first production just prior to or in the first several days after parturition is colostrum, a low-volume fluid that is high in secretory antibodies and oligosaccharides and low in casein and lactose. As the volume increases, the mounting levels of casein and lactose define the fluid as milk (Kent 2007). This process of lactogenesis is of great clinical and basic research interest, and can be better defined through reliable measurement of low concentrations of lactose in small volumes of multiple samples of milk.

Lactose levels in human milk are also clinically relevant when human milk is fed to premature infants; it is important to maximize the amount of calories that the infant is fed, but not to overfeed the infant. Overfeeding the premature infant is associated with intestinal overgrowth, inflammation and, sometimes, necrotizing enterocolitis (Okada *et al.* 2010). Thus, for many very low and extremely low birth weight infants, the amount of fat, protein, and lactose provided through human milk is carefully monitored (Sauer and Kim 2011).

Lactose was the first of the milk components to be isolated (Bartoletti 1628), and was defined as a sugar by Carl Scheele in 1780 (Linko 1982); by the 1880s human milk lactose was being quantified by direct chemical methods in multiple human milk samples (Leeds 1884). Despite this longstanding interest, comparisons of the lactose in milk within individuals, across lactation, across individuals, and across ethnic groups have not been definitive. Part of the reason is that human milk contains many other carbohydrates that can interfere with the most common types of analysis. Milk oligosaccharides contain lactose at their reducing end, and for many types of analysis the oligosaccharides will make the lactose values appear considerably higher than warranted. Milk also contains monosaccharides, glycoproteins, glycolipids, glycopeptides, mucins and glycosaminoglycans, all of which are potential confounders of lactose measurement for some techniques of lactose analysis. Another cause of uncertainty regarding milk lactose concentrations is the variation in how milk samples are obtained. Because foremilk and hindmilk vary in their lactose concentration secondary to their variation in fat content, different sampling techniques can lead to variation in lactose content. Hand-pumped milk, where the mother decides when the pumping is finished, would give more variable lactose values than milk that is obtained in a reliable electric pump set to a fixed end point. It is essential that the entire milk of one full breast be pooled, and that neither fore- nor hindmilk be considered representative of an individual's milk. Finally, milk *per se* may vary in the amount of lactose that it contains at a given time. This point is somewhat controversial: on the one hand, lactose is thought to be an osmoregulator that defines how much milk is made, and thus the lactose concentration should always be similar. On the other hand, production of most milk constituents is variable enough that lactose should also be expected to vary, at least within a range of values. To define the variation in milk lactose, a method whose precision and accuracy is less than the biologic variation intrinsic to human milk needs to be used to measure lactose in meticulously collected representative human milk samples.

Herein, we discuss the methods currently available to measure lactose levels in milk. Many earlier published analyses of milk lactose used these methods on pooled milk to deliberately suppress individual variation and to suppress variation due to any sampling inconsistencies. We now describe a new LC-MS method with unprecedented precision and accuracy. This LC-MS method could be applied to a large number of individual samples to define lactose levels and variability in human milk.

32.1.2   Techniques for Lactose Analysis in Human Milk

The earliest assessment of human milk lactose was by crystallization, but this process is not inherently quantitative. A more common early method is subtractive gravimetric analysis: the dry weight of the milk is determined, from that is subtracted the deduced weight of protein, the weight of the lipids, and the weight of the ash, leaving a value that is assumed to be the lactose. This proximate analysis is not suitable for human milk because of the high amount of glycosylation in milk ($\sim 20\,\text{g/L}$) that leads to excessive error in calculation of its lactose content (68 g/L on average). An improvement over this method used direct chemical reactions with the reducing end of sugars to produce specific chromophores that could be measured by spectrophotometry. But these chemical techniques assume that the sugars in milk are all lactose without correcting for the increased values due to reducing ends of other carbohydrates, or for depressed values from the action of compounds in milk that may interfere with chromophore production. A more specific technique is the measurement of free monosaccharide before and after treatment of milk with β-galactosidase (lactase), an enzyme that specifically cleaves lactose into free glucose and galactose. However, under some conditions, this enzyme can also catalyze side reactions that transfer the galactose from one lactose molecule onto other sugars, which would lead to underestimation of the amount of lactose present. Several protocols for instrumental analysis of lactose in milk have been employed. In one approach, the above enzymes can be attached to a solid matrix around a chemical probe. A technique with increasing usage is infrared detection of milk molecules, but this measurement of lactose may be susceptible to massive interference by the high amounts of other glycans of human milk. Chromatographic separation of the lactose followed by various types of detection is intrinsically the most accurate but, again, there are strengths and weaknesses to each analytical strategy, which will be discussed in more detail below.

32.2   Methods of Analysis

32.2.1   Gravimetric and Chemical

Assessing lactose in milk indirectly by gravimetry is one of the earliest techniques devised and is still in use today. The dry weight of a milk sample is determined gravimetrically. The protein content is generally calculated from a Kjeldahl analysis of total nitrogen through multiplication by a constant. The nitrogen from amino sugars, which are plentiful in human milk, would cause an overestimation of protein. The extensive glycosylation of human milk proteins by mixed sugars would result in an underestimation of protein, as the calculation of protein weight from the nitrogen of amino acids assumes a typical polypeptide without accounting for its extensive post-translational modifications. The lipid content is based on gravimetric analysis of the materials that are extracted by organic solvent, and this value is considered the total fat. The ash is the weight of the residue of the milk solids after complete combustion and represents the total minerals in the milk. The dry weight of milk minus the total of fat, protein, and ash is assumed to be the lactose. The imprecision intrinsic to this method is the sum of the errors for each of the component measures. Moreover, the method does not account for oligosaccharides or the glycan moieties of the glycoproteins, glycolipids, and other glycans that are especially prevalent in human milk, further confounding the analysis.

Another classic technique still in use is measuring the reducing sugars in milk, and assuming that they all represent lactose. There are ways to try to make such measurements more specific, as summarized in *The Handbook of Milk Composition* (Newburg and Neubauer 1995). However, the oligosaccharides in human milk, which contain a lactose moiety at the reducing end, are found in unusually high amounts ($\sim 10\text{--}20\,\text{g/L}$), and are measured as lactose even with these corrections and modifications.

These chemical techniques can be quite useful for measuring lactose in bovine milk ($\sim 40\,\text{g/L}$ lactose, $\sim 1\,\text{g/L}$ oligosaccharides) and other foods, but are not as suitable for analysis of lactose ($\sim 68\,\text{g/L}$ lactose, $\sim 10\text{--}20\,\text{g/L}$ oligosaccharides) in human milk.

32.2.2 Enzymatic

The specificity of enzymes can be utilized to make the quantification of lactose more accurate. The enzyme β-galactosidase cleaves the glycosidic bond of lactose, thereby releasing one molecule of free glucose and one molecule of galactose from each molecule of lactose (Scheme 32.2). Glucose levels in milk are measured before and after treatment of the milk sample with β-galactosidase. The difference in glucose levels is deemed to reflect the lactose content, and this assumption is valid if the β-galactosidase only releases stoichiometric amounts of glucose and galactose from lactose, and the reaction goes to completion. The glucose levels are most often measured by the glucose oxidase method, as described in Schemes 32.3 and 32.4. Alternatively, the free glucose can be oxidized to glucose-6-P with stoichiometric reduction of $NADP^+$; the NADPH produced is detected by spectrophotometry. Yet another option is to measure the released galactose by galactose oxidase, as galactose is also released stoichiometrically from lactose. The accuracy of these techniques depends upon the specificity of the enzymes. Thus, a potential source of error would be if β-galactosidase were to transfer the galactose of hydrolyzed lactose to the terminal galactose of another lactose molecule to form artificial oligosaccharides, thereby decreasing the measured lactose value, as discussed in section 32.1.2. Such side-reactions have been reported under some conditions (Prenosil and Stuker 1987; Zárate and López-Leiva 1990). Some of these limitations were reportedly obviated through the use of another enzyme, lactose oxidase, whose substrate specificity was considered desirable. The peroxide formed by the action of lactose oxidase on lactose was measured electrolytically. To the extent that peroxide detection is independent of the milk matrix, this enzymatic technique showed promise, but this technique has not gained widespread popularity.

Each of these enzymatic techniques described above depends on the differences between two measurements. Thus the overall error of the technique is the sum of the error of each individual measurement, which is intrinsically higher than the error of a single measure.

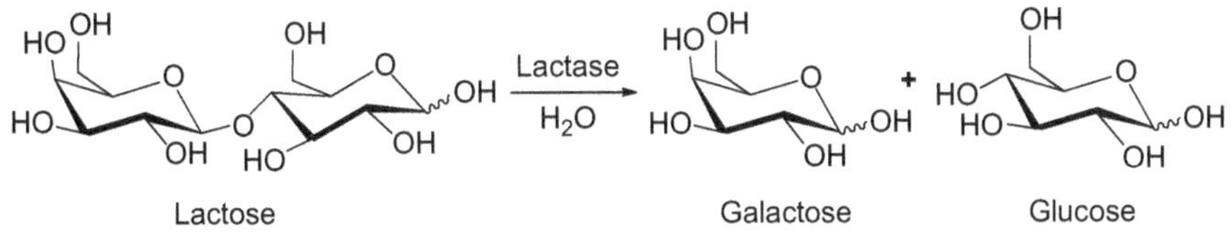

Scheme 32.2 Lactose hydrolysis by lactase (β-galactosidase).

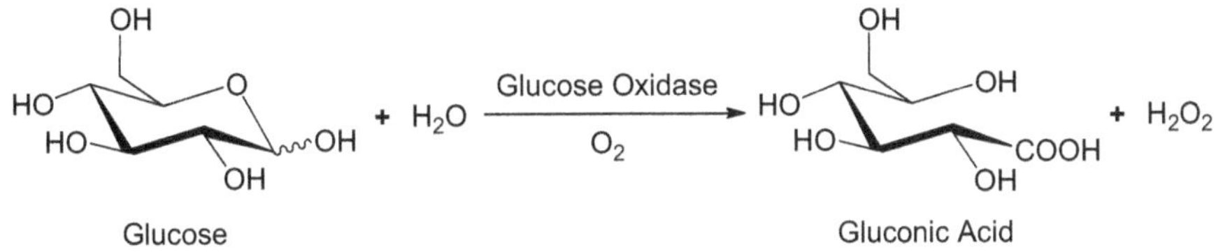

Scheme 32.3 Glucose oxidase reaction.

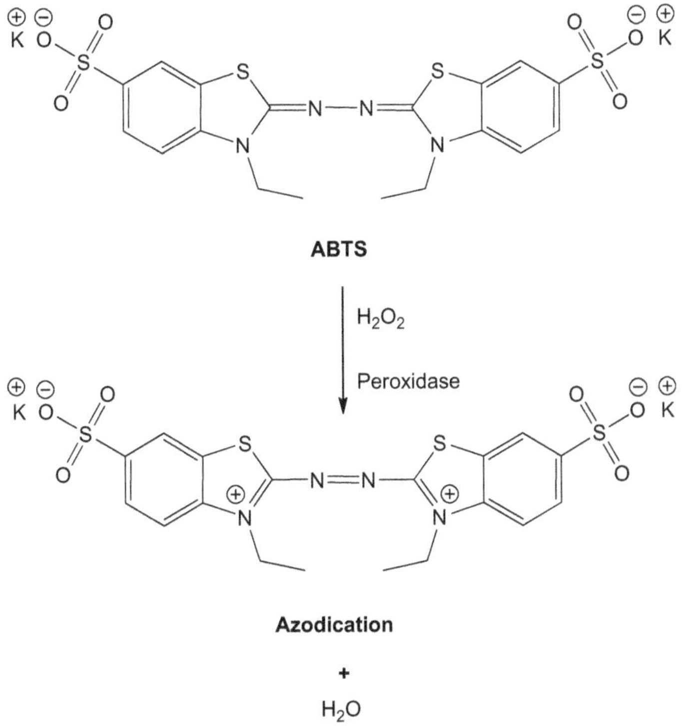

ABTS

H_2O_2

Peroxidase

Azodication

+

H_2O

Scheme 32.4 ABTS oxidation by peroxide.

32.2.2.1   Protocols

Modified glucose oxidase method of Kuhn (Kuhn and Lowenstein 1967):
human milk sample (2 mL) is skimmed at 3000 × g, 45 min, 4 °C. Incubate two
sets of skimmed milk aliquots of 0.3 mL for 2 h at 30 °C; to one is added 0.1 mL
of 0.1 M sodium phosphate buffer, pH 7.4, and to the other is added the same
buffer containing 0.7 mg of yeast β-galactosidase (Roche Diagnostics, Mannheim, Germany). Glucose is measured in the aliquot by addition of 0.1 mL of
glucose oxidase incubation mixture. The released glucose is the difference in the
average measured glucose between the two sets of aliquots. For each batch of
samples, a standard lactose curve is run. Quantitative cleavage of lactose to
glucose is linear to at least 250 μg of lactose. In a modification of this procedure, Henderson and coauthors (Henderson *et al.* 2008) reported an average
recovery ± SD of 101.9 ± 3.5%, as determined by adding known amounts of
authentic lactose to milk samples.

Micro glucose oxidase method of Arthur and Smith: the above assay was
modified to measure lactose in a microtiter plate. After defatting, 5 μL aliquots of
diluted (1:150 v/v) milk are mixed with lactase reagent (50 μL) and incubated at
37 °C for 1 h in a microtiter plate. The lactase reagent contains 8 U/mL β-galactosidase (Sigma, St. Louis, MO), and 0.1 mM magnesium chloride in 100 mM
potassium phosphate, pH 7.2. After the hydrolysis of lactose, the released glucose
is measured by addition of 200 μL of glucose assay reagent [4.8 U/mL glucose
oxidase, 1.25 U/mL peroxidase (Sigma, St. Louis, MO), and 500 μg/mL reduced
azinodiethylbenzothiazoline sulphonate (Boehringer Mannheim, Germany) in
100 mM potassium phosphate buffer, pH 7.2]. After incubation at 20 °C for
45 min, absorbance at 405 nm is measured. Even without correcting for the glucose
that was present before the hydrolysis of lactose, the maximum reported overestimation for lactose was <1.5%. Intra- and inter-assay coefficients of variation
are reported to be 4% and 6%, respectively. The detection limit (3 standard
deviations above the mean blank value) of this assay was 8 mM. According to the
authors, the recovery of the method is very close to 100% (Arthur and Smith 1989).

Biosensor analysis of lactose: enzymes are immobilized on a solid matrix to
create biosensors for routine determination of lactose in large numbers of bovine
milk and milk products. Electrochemical sensors coated with β-galactosidase,
galactose oxidase and glucose oxidase allow rapid, inexpensive lactose assays
(Adanyi and Szabo 1999). Integrated amperometric biosensors produced a linear
response from 1.5×10^{-6} to 1.2×10^{-4} M lactose, with a limit of detection of 4.6×10^{-7} M. Glucose, galactose and ascorbic acid can produce spurious amperometric
responses, but removing glucose or subtracting the glucose content made this
technique more specific for lactose in bovine milk and dairy products (Conzuelo
et al. 2010). A polyurethane spray stabilizes a similar system for at least 20 days
(Ammam 2010). Biosensor analysis could be applied to human milk.

32.2.3   Infrared Analysis

Infrared (IR) analysis may be among the least accurate of the instrumental
methods for determining levels of lactose in human milk. Most macromolecules

in human milk can be evaluated based on the IR absorption of their characteristic functional groups. However, lactose content is deduced by the IR absorption of hydroxyl (OH) groups. Thus, all oligosaccharides, other glycans, and any other molecules that carry OH groups will be measured as lactose. Correction constants are built into the algorithm to try to compensate, but the variation in glycan expression in human milk nonetheless can introduce much imprecision and inaccuracy into the measurement.

The mean value of the lactose content determined by a Milko-scan 104 IR-based analyzer (74 g/L) was statistically higher than the mean value (65 g/L) generated by a conventional enzymatic method. Furthermore, only 60% of the variation around the mean could be attributed to variation intrinsic to the IR analytical method (Michaelsen *et al.* 1988). Therefore, the authors attribute the overestimated results to interfering absorption by other carbohydrate moieties of milk.

Lactose levels measured by a Miris® IR Human Milk Analyzer were positively correlated to HPLC values, but the values were significantly lower than those measured by high performance liquid chromatography (HPLC) with refractive index (RI) detection. The more dilute the milk sample, the greater the observed discrepancy (Menjo *et al.* 2009).

The Miris® AB Human Milk Analyzer provides lactose results (56 g/L) that are lower than those of an enzymatic method (60 g/L). Dilution of milk samples reduced the variation about the mean values (Casadio *et al.* 2010).

These and more current versions of IR-based analyzers are facile for generating a macronutrient content database of a great number of donor milk samples simultaneously, but poor for providing actual lactose concentrations in human milk that are reliably close to what is present as a nutrient in each sample. Thus, IR quantification may be unsuitable for assessing the caloric content of milk destined for premature infants. For highly accurate results, critical lactose concentrations may be measured by more reliable methods, including those utilizing HPLC described below.

32.2.4 HPLC

HPLC can resolve mixtures of closely related compounds. The ability to resolve lactose from all of the other glycans of human milk makes this one of the most intrinsically accurate methods for determining lactose in human milk. Indeed, HPLC is often the standard measurement for assessing the accuracy of other types of measurement.

HPLC was first adopted for quantitative analysis of carbohydrates in bovine milk. Its high sensitivity, linearity, and low limits of detection are increasingly popular for analyzing lactose in dairy and other products in the food industry. With a refractive index detector, neither pre- nor post-column derivatization is required.

Using HPLC, Butte and Calloway (1981) found lactose levels that were slightly less than those obtained by other procedures. A popular HPLC procedure

(Coppa *et al.* 1991) allows simultaneous measurement of monosaccharides, lactose and total oligosaccharides in samples of human milk and colostrum.

32.2.4.1   HPLC Measurement of Human Milk Lactose (Coppa)

To each human milk sample, an equal volume of acetonitrile is added (1:1 vol/vol) for deproteination. After centrifuging at 4000 rpm for 10 min, the supernatant is filtered through a 0.22 μm membrane (Millipore, Billerica, MA). The clarified sample (20 μL) is injected into the HPLC with HPLC-grade water as the mobile phase (0.6 mL/min, 85 °C). The HPLC system consists of a Kontron 420 dual-piston pump, a refractive index detector, an amino column (300 × 7.8 mm Aminex HPX 87 C, Bio-Rad, Richmond, CA). A Carbohydrate De-ashing System (Bio-Rad, Richmond, CA) precolumn may be added to protect the more expensive main column (Coppa *et al.* 1991; Coppa *et al.* 1993).

This straightforward technique allowed measurement of the carbohydrate compositions of milk samples from multiple subjects over the first 4 months of lactation. Lactose levels varied within and among individuals, consistent with some earlier reports using other techniques. The same analyses allowed the total oligosaccharides to be defined over the first 4 months of lactation.

Limitations of this technique include its inability to baseline resolve lactose, or to resolve oligosaccharides from one another; the sensitivity is limited to that of the refractive index detector, and the cost of the HPLC instrumentation is significant (but much less than that of a mass spectrometer). With a range of literature values from 30–80 g/L, human milk lactose analysis does not require the sensitivity of HPLC. When only lactose levels within a given range are of interest, many research laboratories are satisfied with the lower sensitivity and accuracy of less expensive techniques.

HPAEC-PAD (High-Performance Anion-Exchange Chromatography with Pulsed Amperometric Detection) is a specialized form of HPLC. HPAEC-PAD detection is based on the property of neutral sugars to display weak acidity at high pH. At high pH the sugars are ionized sufficiently to resolve by anion-exchange chromatography. Classical silica-based columns degrade at high pH, so Dionex Corporation (Sunnyvale, CA, USA) has developed base-stable polymer anion exchange columns to be used with PAD detectors. Carbohydrates are detected by measuring the electrical current generated by their oxidation at the surface of a gold electrode. HPAEC-PAD does not require preliminary derivatization or separation (Cataldi and Campa 2000; Cataldi *et al.* 2003).

HPAEC-PAD has been used for identification and quantification of lactose in bovine milk and bovine milk products. When the focus is lactose analysis *per se* in bovine milk, the CarboPac PA1, PA10, SA10 and especially the PA20 columns, which were designed to separate mono- and disaccharides, are recommended.

Because the concentration of lactose in human milk is much greater than the concentrations of other glycans, simultaneous measurement presents an analytical challenge. This is exacerbated by the complexity of the human milk matrix. HPAEC-PAD allows separation and quantification not only of lactose, the major component of human milk, but of other carbohydrates, with some degree of sensitivity and specificity.

A comprehensive analysis of the sugars and sugar alcohols of human milk was conducted using HPAEC-PAD (Cavalli *et al.* 2006). Many of the peaks were symmetrical and baseline resolved for the multiple sugars and sugar alcohols of milk, but the lactose peaks were not. The values for lactose concentrations in mature milk were 15% below the consensus of published values (Newburg and Neubauer 1995). It is worth noting that for human milk glycan analysis, HPAEC-PAD has been applied mainly to the analysis of the oligosaccharides.

Another form of separation that can utilize the property of lactose to ionize at extremely high pH is capillary electrophoresis (CE). CE separates mixtures of compounds in a buffer-filled capillary tube with high voltages applied across the ends of the tube. CE could resolve lactose at pH 9.3, 30–60 °C, but could only detect lactose at relatively high concentrations of approximately 20 mg/mL. If lactose is complexed with borate, enhanced photometric detection at 195 nm can be obtained (Hoffstetter-Kuhn 1991). However, quantification of lactose was not attempted. Note that CE has proved to be a powerful instrument for analysis of human milk acidic oligosaccharides (Bao and Newburg 2008).

The data in Table 32.1 make evident that the definitive concentration of lactose in human milk is still uncertain. Each of the above methods has strengths and weaknesses that bias the apparent lactose concentration. Limitations include interference by other glycans, error intrinsic to using the difference of two indirect measures, lack of full resolution from other milk molecules, lack of specificity by the detector, lack of sensitivity, and other interference by the human milk matrix. Therefore, a method that addressed each of these limitations was devised, developed, tested and, most importantly, validated for use in human milk *per se*.

32.2.4.2 LC-MS

HPLC is currently the most powerful technique for separation of the carbohydrates of milk. Mass spectrometry (MS) is the most sensitive and specific method for the detection of individual molecules in a mixture. When these two techniques are combined, tandem HPLC-MS, known as LC-MS, becomes a method of choice for analytical instrumentation. In our experience, a porous graphitic carbon column provides excellent HPLC separation of glycans, including lactose. Even the two lactose anomers are baseline resolved (Gabbanini *et al.* 2010). Mass spectrometers can selectively detect molecular ions of a specific mass to charge ratio, allowing highly specific lactose detection without any form of derivatization. The potential for improved detection of human milk lactose by LC-MS led us to devise, test, and validate the protocol described below.

Reagents and Samples Lactose is from Phanstiehl Laboratories, Inc. (Waukegan, IL); Acetic Acid is from Sigma-Aldrich (Missouri, MO). Optima®

Table 32.1 Published values of lactose in human milk. Values for concentrations of lactose in human milk hover around 68 g/L in mature milk, and are lower in immature milk and milk from mothers of premature infants. The degree of inter-individual and intraindividual variation of lactose in human milk is not yet fully defined.

Method	Sampling	Lactose (g/L)	Reference
Miris AB mid-Infrared Analyzer vs. HPLC	Term (13) and preterm (7); various stages of lactation n = 30	64 ± 6 (22–75) 69 ± 7 (24–85)	Casadio et al. 2010
Milko-scan 104 Infrared Analyzer vs. Enzymatic assay	n = 30	68–79 (total carbohydrates) 54–72	Michaelsen et al. 1988
SpectraStar 2400 Near Infrared Analyzer vs. HPLC	Study day 1, one morning sample one evening sample Study day 4, separate foremilk and hindmilk Study day 7, 14, 21, single morning sample n = 52	56–66 57–71	Sauer and Kim 2011
Enzymatic analysis	Preterm (<34 weeks) mother: Gestational age <28 wks n = 13 Gestational age ≥ 28 wks n = 37	Day 2: 22 (<28 wks); 41 (≥ 28 wks) Day 3: 32 (<28 wks); 47 (≥ 28 wks) Day 4: 46 (<28 wks); 49 (≥ 28 wks) Day 6–8: 58 (<28 wks); 58 (≥ 28 wks)	Henderson et al. 2008
Enzymatic analysis	Electric pump n = 75	Day 21: 63 ± 1 (59–68) Day 45: 66 ± 1 (62–68) Day 90: 68 ± 1 (62–72) Day 180: 69 ± 1 (57–77)	Allen et al. 1991
HPLC	Electric pump 7 am after term infants suckled for 5 minutes n = 46	Day 4: 56 ± 6 Day 10: 63 ± 6 Day 30: 64 ± 6 Day 60: 66 ± 7 Day 90: 66 ± 7 Day 120: 69 ± 8	Coppa et al. 1993
Meta Analysis	Values from literature	68	Newburg and Neubauer 1995

acetonitrile and Optima® methanol are from Thermo Fisher Scientific (Waltham, MA). HPLC-grade water is generated by Q-POD® of Millipore (Billerica, MA). Milk samples are from breastfeeding mothers of term and preterm babies (Boston, MA) who were at various stages of lactation; all samples were stored at $-80\,°C$ until use.

Sample Preparation Milk samples (1 mL) are thawed at 37 °C for 2 min followed by vigorous mixing (vortex) for 5 min or until all visible precipitate is dissolved and the samples appear homogeneous. The samples are centrifuged at $4000 \times g$ for 10 min at room temperature. From the supernatant under a creamy layer of fat, 100 µL clear liquid is transferred to a test tube and mixed with 200 µL Optima® methanol (final methanol concentration 66.7%) to precipitate proteins. This mixture is left at 4 °C for 1 h and centrifuged at $10,000 \times g$ for 10 min to remove the precipitate. The supernatant contains carbohydrates of human milk, including monosaccharides, lactose, oligosaccharides, and some other glycans in relatively high concentrations. The supernatant is diluted 3000 fold with mobile phase (0.02% acetic acid and 5% acetonitrile in water) to bring the lactose concentrations into the linear range of the calibration curve. The diluted sample is filtered through a 0.22 µm PTFE membrane syringe filter (4 mm, Millipore, Billerica, MA) before LC-MS analysis.

Instrumentation A 1200 series Agilent HPLC with a 6460 triple-quadrupole mass spectrometer is equipped with a Hypercarb porous graphite column (3 µm, 100 × 2.1 mm, Thermo Scientific, Waltham, MA) set for 25 °C. For each analysis, 2 µL of diluted filtered sample is injected and eluted with 5% aqueous acetonitrile containing 0.02% acetic acid at 0.3 mL/min over 5 min. An Agilent Jet Stream (AJS) electrospray ionization source with the negative target ion $[M + Cl]^-$ set at m/z 377 allowed detection of each lactose anomer, with a typical spectrum shown in Figure 32.1.

Calibration Curves Lactose standard is diluted in mobile phase (made with deionized Milli-Q water, 5% MS grade acetonitrile and 0.02% acetic acid) to concentrations of 31.25, 15.6, 7.8, 3.9, 1.95, 0.97, and 0.49 µg/mL. Each lactose standard resolves into two anomer peaks. The sum of the areas of these two peaks (area/3000 = **Y**) is regressed against the nominal concentrations (**X**, µg/mL).

Method Validation Technique Three nominal concentrations of lactose, 15.6, 3.9, 0.97 µg/mL, typical milk values after dilution, are prepared as above to verify the precision and accuracy of this method. Each (n = 5) nominal working solution of each concentration is subjected independently to the milk preparation process. Measured values are calculated from the above calibration curve and expressed as the mean of the 5 resulting values.

The precision is the standard deviation divided by the mean value of a repeated concentration, which, when multiplied by 100%, is known as the coefficient of variation (CV). (CV = SD/Mean × 100%) The accuracy is the ratio between a measured value and its nominal value multiplied by 100%. (Accuracy = measured/actual × 100%).

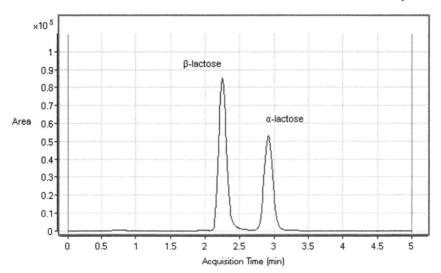

Figure 32.1   LC-MS area (counts %) of lactose anomers resolved by porous graphitic
column. The α and β lactose anomers are baseline resolved and their
ratio is consistent across various concentrations (unpublished data).

To determine whether the milk matrix (*i.e.*, the other milk components)
affects the analysis of lactose, lactose standard solutions from 0.001 to
0.01 mg/mL were added to aliquots of a pooled human milk sample. Each
concentration was analyzed in 5 replicates. The measured increase in lactose
concentration was compared with the amount of authentic lactose added to the
pooled milk sample. The mean increases and the standard deviation of these
mean increases are plotted against the concentrations of the authentic standards added to the pooled milk to determine the linearity of response in the
presence of milk. Precision of measurement in the presence of milk matrix is
calculated as above. The recovery is the increase in measured lactose relative to
its nominal value of added lactose multiplied by 100%. (Recovery = measured/
actual × 100%).

Results and Discussion   Each lactose standard gives two baseline resolved
anomer peaks, β-lactose (t = 2.3 min) and α-lactose (t = 2.9 min) (Figure 32.1).
The relationship between the concentration (**X**), and the sum of the areas of the
peaks (**Y**) divided by 3000, is described by Y = 106771 X − 26.885 with a
coefficient of determination (R^2) of 0.997 (Table 32.2). This relationship was
determined within the above calibration curve whose range was 0.49–31.25 µg/
mL, as this covered the range of all anticipated milk samples. However, the
limits of quantitation can extend well beyond this range: 0.039–50 µg/mL
(0.1–139 nmol/mL), which is 78–100,000 pg/2 µL injection (0.2–278 pmol/2 µL
injection), and these extended ranges would allow this technique to be useful
for analyzing samples whose lactose concentrations differ greatly from those
typical of human milk and colostrum.

Table 32.2 Calibration curve for lactose analysis. The analysis of lactose by LC-MS is highly linear over a wide range of concentrations (unpublished data).

Regression Equation ($\times$ 3000)	Y = 106771 X − 26.885
Correlation Coefficient (r)	0.9986
Dynamic Range (μg/mL)	0.49–31.25 μg/mL
Limit of Quantitation	0.039 μg/mL (S/N > 10)
Sensitivity	0.2 pmol/2 μL
Ratio of β- to α-lactose anomers (%)	$60 \pm 1\% : 40 \pm 1\%$

Table 32.3 Precision and accuracy of the method. The analysis of lactose by LC-MS is more precise, accurate, and sensitive than all previously reported methods (unpublished data).

Nominal Concentration (μg/mL)	Measured Concentration Mean ± SD (μg/mL)	Precision (CV) n = 5	Accuracy (%) n = 5
15.6	15.5 ± 0.3	2	99
3.9	3.8 ± 0.1	3	98
0.97	0.93 ± 0.02	2	95

Table 32.4 Precision and accuracy in the presence of milk matrix. The presence of milk matrix does not interfere with the linearity, precision, accuracy, or sensitivity of this method (unpublished data).

Nominal Concentration (μg/mL)	Measured Concentration Mean ± SD (μg/mL)	Precision (CV) n = 5	Accuracy (%) n = 5
1	0.97 ± 0.02	2	97
2	1.96 ± 0.08	4	98
3	2.73 ± 0.06	2	91
4	3.80 ± 0.08	2	95
5	5.1 ± 0.2	3	102
6	6.2 ± 0.1	2	104
7	6.79 ± 0.07	1	97
8	8.24 ± 0.08	1	103
9	8.9 ± 0.4	4	99
10	9.8 ± 0.3	3	98

The calibration curve was validated at three concentrations of lactose with 5 replicates (n = 5) per concentration (Table 32.3). The precision was 2% to 3% and the accuracy was 95% to 99%.

The above data validated the ability of this method to reliably measure lactose in pure aqueous solution, but did not address its ability to measure lactose in the presence of potentially interfering non-lactose substances that might be present in human milk. From 1 to 10 μg/mL of authentic lactose standard was added to pooled human milk samples. The increase in measured lactose with increased amounts of added lactose was linear (R = 0.999). With repeated measures at a given concentration (n = 5) the precision (CV) was 1% to 4% and the accuracy (recovery) was 91% to 103% (Table 32.4). Thus, the other molecules in human milk do not interfere with the ability of this technique to accurately and reproducibly quantify lactose in human milk.

In this study, negative mode MS allowed both the chloride adduct $[M + Cl]^-$ and the deprotonated $[M\text{-}H]^-$ to be observed. We chose to use the chloride adduct, $[M + Cl]^-$ rather than $[M\text{-}H]^-$, for technical reasons relating to our instrument. But we believe the $[M\text{-}H]^-$ ion may also be useful. Even with using only the chloride adduct ion, our ratio of β- to α-lactose anomers (60%:40%) agreed closely with the values reported in the literature (Gabbanini *et al.* 2010). Note that this method was carefully validated for analysis of lactose in human milk, but has not yet been validated for analysis of lactose in other types of samples that may contain a different matrix, including other disaccharides.

32.3   Conclusions

The most challenging aspect of lactose analysis in human milk is differentiating lactose from the large number and amounts of other glycans uniquely present in the matrix. The ability to discern lactose varies in the methods described herein. At one extreme, absorption in the hydroxyl fingerprint region of the IR spectrum would provide a reasonable measure of pure lactose except that human milk glycans also absorb in this region. This confounded absorption measurement is automatically adjusted to produce a more accurate value, but introduces further imprecision in lactose values. This is unfortunate, as these lactose values are critical to calculating the energy content of milk fed to premature infants. In contrast, enzymatic techniques can improve the specificity of lactose analysis: the difference in monosaccharide content (glucose or galactose) before and after β-galactosidase hydrolysis is measured to calculate total lactose. This relatively inexpensive technique is popular for accurate lactose analysis and the popularity seems apt. However, some concerns were raised regarding potential errors introduced by side reactions and the multiple steps of sequential reactions. Another approach toward differentiating the signal of lactose from that of the other glycans is chromatographic instrumental analysis. This depends upon finding appropriate conditions to separate lactose from the other human milk glycans, and upon the availability of efficient detectors, such as refractive index, spectrophotometric detection of UV absorbent or fluorescent lactose derivatives, and electrochemical (amperometric) detection. The most sensitive of the detectors is a mass spectrometer. LC-MS directly measures lactose without any prior separation or derivatization, and MS has much higher sensitivity than most detectors. Although the initial instrumentation costs can be formidable, and the instruments need regular maintenance, sample preparation is simple, and thus, incremental preparation cost per sample and time per analysis is minimized. Instrumentation costs may be justified if high throughput, high accuracy, and broad capability can be fully utilized. Also, simple types of MS instruments are suitable. The LC-MS technique allows measurements with enough precision and accuracy to define variation in individual expression of lactose in milk. These can be followed over the course of lactation, and can be compared within individuals and across populations. The data made possible by LC-MS may resolve longstanding discrepancies in reported lactose values of human milk. Furthermore, it could help elucidate genetic and environment influences on human milk composition and advance our understanding of lactogenesis.

Summary Points

- Earlier non-specific methods of lactose analysis in human milk are being replaced by more specific methods.
- Enzymatic, HPLC and LC-MS techniques remain the most useful for human milk lactose analysis.
- A new method for analysis of lactose in human milk by LC-MS is described.
- The LC-MS technique is accurate and reliable.
- Sample preparation for LC-MS is facile.
- LC-MS can provide data to study individual variation in lactose expression in milk as a function of genetic and environmental determinants.

Key Facts

1. Lactose is the principal carbohydrate in human milk.
2. Human milk contains more non-lactose glycans than milk from cows.
3. Methods for measuring lactose in bovine milk can overestimate lactose in human milk.
4. LC-MS analysis of lactose in human milk is not confounded by other milk glycans.
5. LC-MS analysis of lactose requires minimal sample preparation.
6. LC-MS instruments are expensive, but high throughput can attenuate cost per analysis.
7. Accurate analysis in human milk allows variation in lactose expression to be studied.

Definition of Words and Terms

Carbohydrates. Molecules composed of sugars $(CH_2O)_n$ and their derivatives, including monosaccharides, disaccharides, oligosaccharides, and polysaccharides. Among the disaccharides is lactose.

Glycans. Any complex molecule that contains carbohydrates.

Glycopeptides. A peptide molecule having one or more covalently bound glycans.

Glycoproteins. A protein molecule with covalently bound carbohydrate.

Glycolipids. Molecules containing a hydrophobic lipid region covalently bound with a glycan moiety.

Glycosaminoglycans (GAGs). Complex polysaccharides composed of linear disaccharide repeating units of a hexosamine and a hexose or hexuronic acid. Major GAGs are heparin, heparin sulfate, chondroitin sulfate, dermatan sulfate, and hyaluronan.

Lactogenesis. Initiation of secretion of milk, a mammary gland fluid that includes substantial lactose and casein.

Monosaccharides. The basic molecular unit of carbohydrates.

Mucins. Very large glycoproteins occurring in glycocalyx and some exocrine secretions.

Parturition. The process or action of giving birth to offspring.

List of Abbreviations

ABTS	2,2′-azino-di-(3-ethyl-benzthiazoline-6-sulphonic acid)
CE	Capillary Electrophoresis
HPAEC	High-Performance Anion-Exchange Chromatography
HPLC	High Performance Liquid Chromatography
IR	Infrared
LC-MS	Liquid Chromatography with tandem Mass Spectrometry
MS	Mass Spectrometry
PAD	Pulsed Amperometric Detection
PTFE	Polytetrafluoroethylene
RI	Refractive Index
UV	Ultraviolet

References

Adanyi, N., and Szabo, E., 1999. Multi-enzyme biosensors with amperometric detection for determination of lactose in milk and dairy products. *European Food Research and Technology*. 209: 220–226.

Allen, J., Keller, R., Archer, P., and Neville, M., 1991. Studies in human lactation: milk composition and daily secretion rates of macronutrients in the first year of lactation. *American Journal of Clinical Nutrition*. 54: 69–80.

Ammam, M., 2010. Two-enzyme lactose biosensor based on β-galactosidase and glucose oxidase deposited by AC-electrophoresis: Characteristics and performance for lactose determination in milk. *Sensors and Actuators B: Chemical*. 148: 583–589.

Arthur, P., and Smith, M., 1989. Milk lactose, citrate, and glucose as markers of lactogenesis in normal and diabetic women. *Journal of Pediatric Gastroenterology & Nutrition*. 9: 488–496.

Bao, Y., and Newburg, D., 2008. Capillary electrophoresis of acidic oligosaccharides from human milk. *Electrophoresis*. 29: 2508–2515.

Bartoletti, F., 1628. *Methodus in Dyspnoeam; seu, de Respirarionibus*. N. Tebaldini, Bomoniae, Bologna. pp. 561.

Butte, N., and Calloway, D., 1981. Evaluation of lactational performance of Navajo women. *The American Journal of Clinical Nutrition*. 34: 2210–2215.

Cavalli, C., Teng, C., Battaglia, F., and Bevilacqua, G., 2006. Free sugar and sugar alcohol concentrations in human breast milk. *Journal of Pediatric Gastroenterology and Nutrition*. 42: 215–221.

Casadio, Y., Williams, T., Lai, C., Olsson, S., Hepworth, A., and Hartmann, P., 2010. Evaluation of a mid-infrared analyzer for the determination of the macronutrient composition of human milk. *Journal of Human Lactation*. 26: 376–383.

Cataldi, T., Angelotti, M., D'Erchia, L., and Altieri, G., 2003. Ion-exchange chromatographic analysis of soluble cations, anions and sugars in milk whey. *European Food Research and Technology*. 216: 75–82.

Cataldi, T., Campa, C., and Benedetto, G., 2000. Carbohydrate analysis by high-performance anion-exchange chromatography with pulsed amperometric detection: The potential is still growing. *Fresenius Journal of Analytical Chemistry*. 368: 739–758.

Conzuelo, F., Gamella, M., Campuzano, S., Ruiz, M., Reviejo, A., and Pingarrón, J., 2010. An integrated amperometric biosensor for the determination of lactose in milk and dairy products. *Journal of Agricultural and Food Chemistry*. 58: 7141–7148.

Coppa, G., Gabrielli, O., Pierani, P., Zampini, L., Rottoli, A., Carlucci, A., and Giorgi, P., 1991. Qualitative and quantitative studies of carbohydrates of human colostrum and mature milk. *Rivista Italiana Di Pediatria-italian*. 17: 303–307.

Coppa, G., Gabrielli, O., Pierani, P., and Catassi, C., 1993. Changes in carbohydrate composition in human milk over 4 months of lactation. *Pediatrics*. 91: 637–641.

Gabbanini, S., Lucchi, E., Guidugli, F., Matera, R., and Valgimigli, L., 2010. Anomeric discrimination and rapid analysis of underivatized lactose, maltose, and sucrose in vegetable matrices by U-HPLC–ESI-MS/MS using porous graphitic carbon. *Journal of Mass Spectrometry*. 45: 1012–1018.

Henderson, J., Hartmann, P.E., Newnham, J.P., and Simmer, K., 2008. Effect of preterm birth and antenatal corticosteroid treatment on lactogenesis II in women. *Pediatrics*. 121: e92–100.

Hoffstetter-Kuhn, S., Paulus, A., Gassmann, E., and Widmer, H., 1991. Influence of borate complexation on the electrophoretic behavior of carbohydrates in capillary electrophoresis. *Analytical Chemistry*. 63: 1541–1547.

Kent, J., 2007. How breastfeeding works. *Journal of Midwifery & Women's Health*. 52: 564–570.

Kuhn, N., and Lowenstein, J., 1967. Lactogenesis in the rat. Changes in metabolic parameters at parturition. *Biochemical Journal*. 105: 995–1002.

Leeds, A., 1884. The composition and methods of analysis of human milk. *Journal of American Chemical Society*. 6: 252–279.

Linko, P., 1982. Lactose and Lactitol, in Birch, G.G. & Parker, K.J, *Nutritive Sweeteners*, London & New Jersey: Applied Science Publishers, USA. pp. 109–132.

Menjo, A., Mizuno, K., Murase, M., Nishida, Y., Taki, M., Itabashi, K., Shimono, T., and Namba, K., 2009. Bedside analysis of human milk for adjustable nutrition strategy. *Acta Paediatrica*. 98: 380–384.

Michaelsen, K., Pedersen, S., Skafte, L., Jæger, P., and Peiterson, B., 1988. Infrared analysis for determining macronutrients in human milk. *Journal of Pediatric Gastroenterology & Nutrition*. 7: 229–235.

Neville, M., 1995. Determinants of Milk Volume and Composition. Lactogenesis in Women: A Cascade of Events Revealed by Milk Composition. *Handbook of Milk Composition*. Food science and technology international series. Academic Press, San Diego, USA. pp. 87–98.

Newburg, D., and Neubauer, S., 1995. Carbohydrates in Milks: Analysis, Quantities, and Significance. *Handbook of Milk Composition*. Food science and technology international series. Academic Press, San Diego, USA. pp. 273–349.

Okada, K., Fujii, T., Ohtsuka, Y., and Yamakawa, Y., 2010. Overfeeding can cause NEC-like enterocolitis in premature rat pups. *Neonatology*. 97: 218–224.

Prenosil, J.E., Stuker, E., and Bourne, J., 1987. Formation of oligosaccharides during enzymatic lactose: Part I: State of art. *Biotechnology and Bioengineering*. 30: 1019–1025.

Sauer, C., and Kim, J., 2011. Human milk macronutrient analysis using point-of-care near-infrared spectrophotometry. *Journal of Perinatology*. 31: 339–343.

Zárate, S., and López-Leiva, M., 1990. Oligosaccharide formation during enzymatic lactose hydrolysis: a literature review. *Journal of Food Protection*. 53: 262–268.

CHAPTER 33

Sweetened Beverages and Added Sugars in Obesity

ODILIA I. BERMUDEZ

Department of Public Health and Community Medicine, Tufts University
School of Medicine, 136 Harrison Avenue, Boston, MA 02111, USA
Email: Odilia.Bermudez@tufts.edu

33.1 Introduction

Carbohydrates (CHO), specifically sugars and starches, are essential components in our diets. Carbohydrates provide dietary energy to the body. One gram of CHO provides 17 kilojoules (kJ), or 4 kilocalories (kcal).

Both the World Health Organization (WHO), in their nutrition goals for the prevention of chronic diseases and the US Food and Nutrition Board (FNB) at the Institute of Medicine, in their nutrient recommendations for the United States (US) and Canada (FAO/WHO 2001; FNB 2002), had set recommendations for the proportion of food energy that should be provided by the major macronutrients; CHO, protein and fats. WHO recommends that, out of the total food energy intake, carbohydrates should provide 55–75% of food energy, but with no more than 10% coming from simple sugars. The rest of the needed dietary energy should come from fats (15–30%) and proteins (10–15%) (WHO, 2001).

Many foods and beverages contain simple sugars in their natural states. However, additional sugars may be added to them to increase sweetness, add palatability and enhance appearance. Several publications related to the consumption of foods and beverages rich in simple sugars, either present naturally

Food and Nutritional Components in Focus No. 3
Dietary Sugars: Chemistry, Analysis, Function and Effects
Edited by Victor R Preedy
© The Royal Society of Chemistry 2012
Published by the Royal Society of Chemistry, www.rsc.org

or as added sugars, have reported the rapid increases in consumption of simple sugars that have occurred during the last 40 years (Bray 2008; French *et al.* 2003; Guthrie *et al.* 2000). This rapid increase in the consumption of food and beverage sources rich in simple or added sugars seemed to go along towards the dramatic increases in rates of obesity observed across the world (CDC 2004). However, analysis of the existing data regarding the association between obesity and intakes of sweetened beverages and added sugars has been inconclusive (Allport 2004; Ginevan 2004).

This chapter presents details about trends in intakes of simple and added sugars, particularly those present in sweetened beverages, and discusses the links observed between this eating practice and obesity. In this chapter we have focused only on sugar-sweetened beverages, with the exclusion of those beverages that are sweetened with products that provide zero or reduced amounts of food energy.

33.2   Simple Sugars as Dietary Components

Simple sugars and complex carbohydrates provide dietary energy for body function. However, basic sugars, because of their simple chemical structure, only provide energy and small or zero amounts of nutrients. Furthermore, more complex carbohydrates, in addition to being sources of dietary energy, are also rich in several nutrients and other dietary components (*e.g.* fiber, water and phytochemicals) needed for proper nutrition and health.

Once ingested, simple sugars and starches are broken down by the digestive system into glucose, which is the simple sugar that circulates in the body. Glucose supplies cells with the energy needed for vital functions. Two of the most common disaccharides, important sources of glucose, are: (a) sucrose, or table sugar, which contains one molecule of glucose and one of fructose and (b) lactose, or milk sugar, composed of glucose and galactose. Extra glucose is accumulated in the liver and muscles as glycogen. However, when the storage capacity for glycogen is exceeded, then glucose is metabolized and stored as fat, which is the reserve fuel for the body.

Some complex carbohydrates are processed to make them more attractive and palatable for consumers. The processing could include the separation of some of the complex carbohydrates (*e.g.* fiber) and nutrients (*e.g.* vitamins and minerals) from naturally nutrient-rich foods, converting them into refined, energy-dense, but nutrient-depleted products. For example, this is the case for refined cereals and grains. Another interesting example of a refined, highly processed product is the sugar extracted from corn, which is known as high-fructose corn syrup (HFCS). This corn syrup is composed of fructose and glucose, the component monosaccharides that make up the disaccharide sucrose (table sugar). HFCS has become the most widely used sweetener in processed foods and beverages.

Simple sugars are nutrients that contain one or two saccharides. Sugars are sweet: in general, the simpler the sugar, the sweeter it is. Simple sugars are dense in energy, but are mainly depleted of nutrients. Keeping the contribution of simple sugars below 10% of the total energy consumed is recommended (WHO

2001). In addition, the contribution of added sugars to the total intake of energy should stay below 5%.

33.3 Sugar-sweetened Beverages

33.3.1 Sugar-sweetened Beverages

Some of the simple sugars commonly used in the formulation of home-based or industry-based drinks and beverages include sugar (brown or white), HFCS, maple syrup, other food-based syrups, honey, and molasses. Beverages to which simple sugars are added during formulation could also be dense in energy, but poor in nutrients. Among them are carbonated beverages (*e.g.* soda), sport drinks, energy drinks, fruit drinks and artificially flavored drinks. Table 33.1 details the amount of CHO, total sugars and added sugars for some

Table 33.1 Total and added sugars in one cup (8 fluid ounces or 250 ml) in sweetened beverages. *Source:* Values estimated with information from the USDA Database for the Added Sugars – Content of Selected Foods (USDA 2006).

Food Description	CHO 250 ml	Total Sugars 250 ml	Added Sugars 250 ml
Grape juice, concentrate, sweetened, diluted with water	31.9	31.6	26.1
Orange and apricot juice drink, canned	31.8	30.2	21.1
Cranberry juice cocktail, bottled	33.8	29.7	20.3
Orange drink, canned	30.9	27.6	23.3
Root beer soda	26.5	26.5	26.5
Lemon-lime soda	26.1	25.5	25.5
Orange juice drink	33.5	23.4	3.5
Soda, cola type	23.9	22.4	22.4
Tonic water, carbonated	22.0	22.0	22.0
Ginger ale soda	21.9	21.8	21.8
Orange breakfast drink, ready-to-drink	27.0	16.4	10.1
Sports drink, fruit-flavored, ready-to-drink	16.9	13.9	13.8
Pineapple juice, frozen concentrate, unsweetened, diluted with 3 volume water	31.9	31.4	0.0
Grapefruit juice, white, frozen concentrate, unsweetened, diluted with 3 volume water	24.3	24.1	0.0
Orange juice, frozen concentrate, unsweetened, diluted with 3 volume water	27.0	21.0	0.0
Blackberry juice, canned	19.5	19.3	0.0
Baby food, juice, apple	29.3	26.8	0.0
Baby food, juice, apple - cherry	28.0	26.3	0.0
Baby food, juice, apple and plum	30.8	28.9	0.0
Baby food, juice, apple-sweet potato	28.5	24.0	0.0
Baby food, juice, fruit punch, with calcium	31.8	26.5	0.0
Baby food, juice, mixed fruit	29.0	21.2	0.0
Baby food, juice, orange	25.5	20.7	0.0
Baby food, juice, orange-carrot	24.8	21.0	0.0

examples of common sweetened beverages. The Table also includes beverages without added sugars that contain important amounts of naturally occurring sugars, such as fruit juices and beverages formulated as baby foods. Data from the USDA Database for the Added Sugars Content of Selected Foods (USDA, 2006) were used to estimate the values in Table 33.1.

33.4   Trends in the Intake of Sweetened Beverages

Sugar-sweetened beverages (SSB) have become the main energy provider in the diets of various population groups, particularly those from the US. Research data from the 1999–2000 National Health and Examination Survey (NHANES) suggested that milk and other more nutritious foods and drinks may have been displaced by SSB as the main contributor of energy to the diets of American people (Block 2004). It was reported that among young adults aged 20–39 years, SSB were consumed by 84% of the population, with those located in the highest quartile of intake consuming more than 6 servings/day (Bermudez *et al.* 2010). This represented about 50 teaspoons of added sugars/day. One teaspoon contains about 5 g, therefore the contribution of energy from added sugar reached 4184 kJ (1000 kcal). While individuals in the first quartile of intake consumed 8972 kJ per day, those in the fourth quartile consumed 11 000 kJ, from which 38% were provided by added sugars.

There is also evidence that the levels of SSB consumption among adults have increased significantly over the years. A comparison of SSB consumption among US adults, > 20 years of age, between the periods 1988–1994 and 1999–2004, revealed that the proportion of adults drinking SSB went from 58% in the first period to 62% in the second. Also, the amount of SSB consumed increased by 6 oz in the same period of time. And while the energy contribution of SSB to the total energy intake was 9%, in 1999–2004, it exceeded (12%) the recommendation from WHO of no more than 10% of energy from added sugars (Bleich *et al.* 2009).

Use of energy-rich sweeteners has increased substantially during the last 50 years. In the American population, the per capita consumption of caloric sweeteners between 1950 to 2000 rose by 43 lbs, or 39% (USDA 2006), as reflected in Figure 33.1. However, the type of caloric sweetener added to food and beverages also changed. As seen in Figure 33.1, the total amount of caloric sweetener grew steadily due to the dramatic increase in the use of corn sugars, despite the decrease in consumption of cane and beet sugar.

33.5   Obesity as a World Epidemic

33.5.1   Concepts and Definitions of Obesity

The WHO defines overweight and obesity as an excessive fat accumulation that could impair health. The basic cause of overweight and obesity is the imbalance that occurs between consumed and expended dietary energy. Precise and

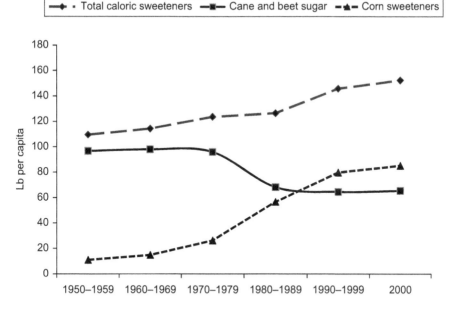

Figure 33.1 Trends in the consumption of caloric sweeteners in the US between 1950–2000.
Source: Data taken from *Agriculture Fact Book*, Chapter 2 (USDA 2006).

complex body mechanisms control the balance that takes place when the consumed dietary energy is utilized to provide the body with the required energy for the basic metabolic rate, the energy for physical activity and the energy for food thermogenesis. Only modest amounts of energy are stored as a reserve, mainly as glycogen and less as fat. Excess accumulation of body fat has been associated with health problems. Several reports about the health-related consequences of obesity have shown that as weight increases, the risk for multiple health conditions also increases (Mokdad *et al.* 2003; Must *et al.* 1999).

Several risk factors are associated with weight gain, including genetic, socio-economic and environmental factors. For example, genetic characteristics influence the efficiency and rate of fat accumulation and distribution in the body, and some cultural groups consider obesity a desirable attribute. However, environmental factors, mainly dietary patterns and physical activity, are strongly associated with overweight and obesity.

WHO attributes the dramatic rise in obesity to the increased consumption of foods dense in dietary energy (*e.g.* from fat- or sugar-rich foods or beverages), but low in nutrients, which is not compensated by adequate levels of physical activity (WHO 2011). The lack of sufficient physical activity has been partially explained by the current trends of people becoming more sedentary, in part due to changes associated with the modernization in living styles, including modes of transportation, communication and social interactions.

33.5.2   The Worldwide Epidemic of Obesity

The world is in the middle of an obesity epidemic. This epidemic is affecting children and adults. In 2010, the International Obesity Task Force (IOTF) estimated that, worldwide, approximately 200 million school-aged children were overweight or obese, with about one-quarter of them considered obese. Furthermore, IOTF also reported that over one billion adults were either overweight or obese, with almost half of them included in the obesity category (IOTF 2011).

33.5.2.1   *Childhood Obesity*

Childhood obesity has increased dramatically in the last 30 years. In the US, childhood obesity was relatively stable between 1960 and 1989 (CDC 2011). However, as seen in Figure 33.2, the rate of obesity for the 6–11 y age group went from 7% in 1976–1980 to 19% in 2003–2004 (269% increase), while for the 12–19 y age group it increased from 5% in 1976–1980 to 17% in 2003–2004, (348% increase). The IOTF estimated that among children 5–17 y, worldwide, prevalence reached 10% for overweight and approximately 3% for obesity (IOTF 2011; Wang *et al.* 2006). However, these figures are much higher in several countries around the globe, particularly those from Europe and the Americas, as can be appreciated from Table 33.2.

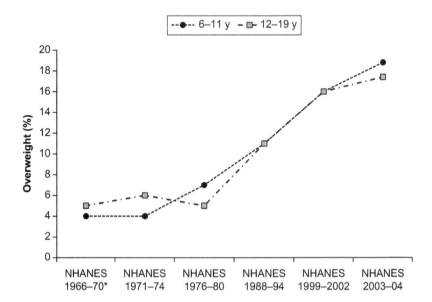

Figure 33.2   Trends in overweight among US children (6–11 y) and adolescents (12–17 y), 1966–2004.
Notes: Data for 1966–1970 are for adolescents 12–17 y.
Source: (CDC 2011).

Table 33.2 Countries with elevated prevalence of childhood obesity among boys and girls, according to data from the International Obesity Task Force. *Source:* Data, and year of survey, taken from the International Obesity Task Force Electronic Database for children (IOTF 2011).

Countries with Prevalence of 20–29%, Plus Year of the Survey	*Countries with Prevalence of 30% or more, Plus Year of the Survey*
Australia, 2007	Greece, 2005–2006
Bolivia, 2003 (only girls)	Italy, 2001–2002
Brazil, 2002	Spain, 1998–2000 (only boys)
Canada, 2004	Unites States, 2003–2004
Chile, 2002	
Czech Republic, 2005 (only boys)	
Finland, 2006 (only boys)	
Germany, 2008 (only boys)	
Hungary, 2005	
India, 2007–2008 (only boys)	
Mexico, 2006	
New Zealand, 2006–2007	
Portugal, 2008	
Russian Federation, 1992	
Spain, 1998–2000 (only girls)	
United Kingdom, 2007	

Childhood obesity can be assessed with different indicators. However, the most widely accepted methods are based on anthropometric measurements of weight and height or length of children. The most widely used anthropometric indicators are the weight for height (or length) index (W/H, W/L) and the body mass index (BMI, weight in kg/height in m^2) (WHO 2008).

There is no universal consensus on the standards and reference values to define obesity in children. The US Centers for Disease Control and Prevention (CDC) has established overweight- and obesity-specific cutoffs and labels. For example, they define "at risk for overweight" as being between the 85th and 95th percentiles of BMI for age and sex, and "overweight" as being at or above the 95th percentile of the BMI adjusted for age and sex. They avoid the use of the term "obesity" for children below two years of age.

WHO, using data from the Multicentre Growth Reference Study (WHO 2008), developed child growth standards, including those for the evaluation of obesity with the W/H or W/L index or the BMI. W/H and W/L are usually expressed in Z-scores, with those above 2 standard deviations (SD) as overweight and above 3 SD as obese. With BMI as the indicator, WHO applied the same cutoffs for overweight and obesity as the US CDC, 85th and 95th percentiles, respectively. The IOTF has developed a reference based on sex-age-specific BMI using cutoffs that correspond to BMIs of 25 for overweight and 30 for obesity at age 18.

Recently, the US standards are in the process of being adjusted, particularly those for infants and children under 2 years of age. The Institute of Medicine

(IoM), in their report for Early Childhood Obesity Prevention Policies, recommends that children under 24 months of age are measured by the WHO standards (IoM 2011).

33.5.2.2   Adult Obesity

The BMI is also used to measure overweight and obesity in adults. WHO has developed definitions for nutritional status of individuals based on the BMI. These definitions are the same as those used in the US. Overweight is defined as BMI between 25.0 and 29.9 kg/m^2 and obesity as BMI equal to or above 30.0 kg/m^2. Obesity is further defined by subcategories of severity: obesity class I (30–34 kg/m^2), class II (35–39 kg/m^2) and class III (≥ 40 kg/m^2). Furthermore, WHO has established additional country-specific criteria for the definition of overweight and obesity that are applicable in countries where the risk for mortality and morbidity appears to increase at different BMI cutoffs.

Although BMI is the internationally accepted indicator to evaluate obesity, the regional distribution of body fat is an important characteristic that has been related to the risk for chronic conditions in adults. Excessive abdominal fat has been associated with higher risk for mortality and morbidity of several diseases, including type 2 diabetes and cardiovascular disease. Measurement of waist circumference (WC) is the indicator used for the evaluation of abdominal obesity in the US (DHHS 1998).

The global epidemic of obesity, also called "globesity", is rapidly affecting more and more population groups from developing and least developed countries. Numbers reported by WHO reveal the dramatic situation that currently affects most nations. By 2008, overweight affected a 1.5 billion adults 20 y and older; 65% of the world population is from nations where overweight and obesity surpass underweight as a direct or associated cause of mortality (WHO).

Worldwide, adults are increasing in BMI. We used data from 192 countries, obtained at the WHO Global InfoBase (WHO 2011), to perform an analysis of the world situation. We observed increases in BMI that occurred in adults, both males and females, in less than a decade. Similar increases were also observed in the growing proportion of adults with overweight and obesity (BMI $\geq$ 25.0) as seen in Figure 33.3. Those changes in larger BMI values and higher proportions of overweight or obese adults explain the shifts observed among the countries analyzed for this chapter. Among adults, more countries are moving to the overweight and obese categories. In the year 2002, there were 13 nations in which the female population, 30 y and older, had a mean BMI of 30 kg/m^2 or higher. By 2010, the number of nations with females at the cutoff for obesity increased to 21. More worrisome is the list of countries with higher than desirable BMI (Table 33.3) and with 50% or more of those populations affected by the current "globesity" epidemic (Table 33.4). With the exception of the United States, the lists presented in Tables 33.3 and 33.4 almost exclusively contain developing nations, which most likely lack the enormous resources and efforts needed to combat this epidemic.

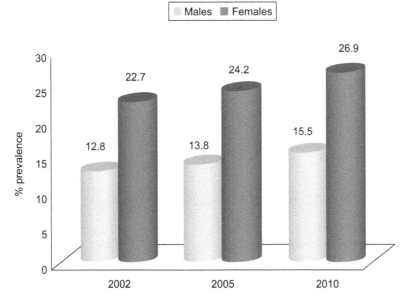

Figure 33.3 Time changes in worldwide prevalence of overweight and obesity in males and females 30 y and older.
Source: Estimates based on data from 192 countries with available data in the WHO Global InfoBase (WHO 2011).

33.6 Sweetened Beverages and Added Sugars and Obesity

As discussed in the previous sections, both childhood and adult obesity have reached epidemic proportions globally. The root causes, causal effects and precipitating events interconnected in the evolution and magnitude of obesity are complex. The influence of dietary patterns in protecting or exposing individuals to obesity is strong. Along with the increases in the rates of obesity, there have been similar increases in the consumption of SSB and added sugars (Bray *et al.* 2004).

Although several studies have found an association between the consumption of sweetened beverages and obesity (Allison *et al.* 2009; Dhingra *et al.* 2007; Malik *et al.* 2006), other studies have found weak or no association between consumption of SSB and obesity (Allison *et al.* 2009; Pereira 2006). However, independent of the food sources, and the potential increases in energy intake if the consumption of those foods exceeds the caloric needs, weight accumulation, and therefore obesity, could increase. The association of excessive consumption of SSB with obesity is partially explained by the provision of unnecessary food energy, which, along with insufficient levels of physical activity, provokes the accumulation of body fat. In addition, there are indications of differences in the activation of satiety mechanisms when solid or

Table 33.3   Countries with adult populations (30 y and older) with mean BMI
at 30 kg/m^2 or higher, according to sex. *Source:* Data for 192
countries obtained at the WHO Global InfoBase (WHO 2011).

	2002	*2005*	*2010*
Country Name		*Mean BMI*	
Males			
Nauru	34.9	35.1	35.5
Micronesia, Federated States of	32.6	32.9	33.3
Cook Islands	31.8	32.1	32.6
Tonga	31.8	32.1	32.5
Samoa			30.3
United States			30.0
Females			
Nauru	36.8	37.2	37.8
Tonga	36.3	36.7	37.3
Micronesia, Federated States of	35.3	35.7	36.4
Cook Islands	33.5	33.9	34.6
Samoa	33.4	33.8	34.5
Niue	32.7	33.2	33.9
Kuwait	32.3	32.9	33.4
Barbados	30.5	31.1	32.2
Palau	30.9	31.3	32.1
Trinidad and Tobago	30.1	30.8	31.8
Egypt	30.2	31.3	31.7
Dominica	30.0	30.7	31.7
United States			30.8
United Arab Emirates		30.0	30.5
Jordan	30.9	30.0	30.5
Jamaica			30.4
Malta		30.0	30.3
Mexico			30.2
Nicaragua			30.2
Seychelles			30.1
Saint Lucia			30.1

liquid foods are ingested (DiMeglio *et al.* 2000). The evidence points to
decreases in satiety when SSB are consumed, which leads to overeating.

Excessive consumption of SSB is also linked to decreases in the intake of
other, more nutritious beverages. For example, from 1977–1978 and 1994–
1998, consumption of milk decreased by 36% among US female adolescents
(Bowman 2002). During the same time period, French *et al.* (2003) reported
increases in the prevalence of consumption of SSB among girls aged 6–17 y,
with increases from a prevalence of 37% in 1977–1978 to 56% in 1994–1998.
Also, mean daily intakes of SSB more than doubled, from 5 fl oz to 12 fl oz
(French *et al.* 2003).

Consumption of SSB was inversely associated with intakes of milk and
orange juice in US adults, 20–39 y (Bermudez *et al.* 2010). Milk, particularly
low-fat milk, and 100% fruit juices are sources of several nutrients. Moreover,
intakes of fruit and low-fat milk have been associated with lower BMI and

Table 33.4 Countries with adult populations (30 y and older) with prevalence of overweight and obesity 50% or higher, according to sex. *Source:* Data for 192 countries obtained at the WHO Global InfoBase (WHO 2011).

	2002	*2005*	*2010*
Country Name	*% Prevalence of Overweight & Obesity*		
Males			
Nauru	79.2	80.2	81.9
Micronesia, Federated States of	69.0	70.7	73.2
Tonga	64.0	65.9	68.8
Cook Islands	63.2	65.0	67.9
Samoa			52.2
United States			50.0
Females			
Nauru	79.6	80.5	81.9
Tonga	78.2	79.3	80.8
Micronesia, Federated States of	75.3	76.6	78.5
Cook Islands	68.0	69.8	72.5
Samoa	67.8	69.6	72.4
Niue	65.1	67.1	70.1
Kuwait	61.5	64.7	66.6
Barbados	53.2	57.0	62.7
Palau	55.2	57.9	61.9
Trinidad and Tobago	50.8	54.7	60.7
Dominica		54.0	60.0
Egypt	50.9	57.0	59.4
United States			54.8
Jamaica			52.6
United Arab Emirates			52.4
Jordan	54.8		52.2
Nicaragua			51.2
Mexico			51.1
Malta			50.6
Seychelles			50.4
Saint Lucia			50.4

reduced risk for obesity-related health conditions such as type 2 diabetes and hypertension (Moore *et al.* 1999; Sargeant *et al.* 2000).

The association between intake of SSB and obesity is also explained by the greater intake of added sugars included in the formulation of those SSB. Among US adults, it was observed that 10 additional teaspoons of added sugars per day represented a 52% higher risk for obesity (Bermudez *et al.* 2010). Most of the added sugars consumed are incorporated into the food supply as sweeteners and enhancers for drinks and beverages.

33.7 Final Remarks and Future Perspectives

As documented in this chapter, both obesity and inadequate dietary intakes, particularly with the inclusion of large servings of SSB, are significant health concerns across the globe. While the complex mechanisms behind the association between obesity and large consumption of added sugars, specially from SSB, are still being elucidated, it seems prudent to assist all levels of society to form alliances and design strategies that could lead to the modification or improvement of the two direct causal factors of obesity: increases in intake of food energy and decreases in expenditure of energy. Although consumption of SSB is not the only factor responsible for the dramatic rise in rates of overweight and obesity across the globe, it is a part of the problem. Therefore, it should remain in the equations developed to achieve favorable changes, which will halt the rise of overweight and obesity.

Summary Points

- In this chapter, we review types of sugars added to sweetened beverages and evaluate the association of sugar sweetened beverage (SSB) consumption with obesity.
- The basic cause of overweight and obesity is the imbalance that occurs between consumed and expended dietary energy.
- This worldwide epidemic is affecting children and adults, and individuals from different cultures, societies and ethnicities.
- The association of excessive consumption of SSB with obesity is partially explained by the provision of unnecessary food energy, which, along with insufficient levels of physical activity, provokes the accumulation of body fat.
- In countries affected by the obesity epidemic, it is imperative to design and implement strategies for the modification or improvement of the two direct causal factors of obesity: increases in intake of food energy and decreases in expenditure of energy.

Key Facts of Obesity and Intake of Sugar Sweetened Beverages

- The World Health Organization (WHO), the directing and coordinating authority for health within the United Nations, defines obesity as an excessive fat accumulation that could impair health.
- Obesity, a global epidemic also called globesity, was considered a health problem in high-income countries, but it is now on a rapid rise in low- and middle-income countries.
- Obesity is a major risk factor for chronic diseases such as diabetes, musculoskeletal disorders (especially osteoarthritis), cardiovascular diseases, particularly heart disease and stroke, and some types of cancers.
- According to the WHO, obesity in children increases the chances for obesity, premature death and disability in adulthood.
- Both obesity and inadequate dietary intakes, particularly with the inclusion of large servings of sugar sweetened beverages (SSB), are significant health concerns across the globe.

- The consumption of SSB has increased substantially in most countries around the world.
- The increased intakes have occurred in most age groups, but particularly in children and adolescents.
- The association between intake of SSB and obesity is also explained by the greater intake of added sugars included in the formulation of those SSB.

Key Facts of the US National Health and Nutrition Examination Survey (NHANES)

- The NHANES is a program of the National Center for Health Statistics (NCHS), which is part of Centers for Disease Control and Prevention (CDC).
- NHANES is designed to assess the health and nutritional status of the US population.
- Every year, NHANES examines a nationally representative sample of about 5000 persons.
- Findings from NHANES are used to assess the prevalence of major diseases and risk factors for diseases affecting the US population.
- Dietary, anthropometric (weight and height) and physical activity data collected regularly by the NHANES, have been used to document the prevalence of overweight and obesity in the US population, as well as the changes in diet and physical activity patterns linked to this epidemic.

Definitions of Words and Terms

Abdominal obesity: Also known as central obesity, **this** is an excessive accumulation of fat in the abdominal area. The waist circumference is measured to assess abdominal obesity.

Added sugars: Sugars added to food and beverages when they are processed or prepared or ready to eat or drink.

Body Mass Index: A measure used as an indicator of overweight and obesity, defined as the relation between body weight in kg divided by the height in squared meters.

Globesity: Combination of global and obesity. A relatively new term coined to convey the message that obesity is a global epidemic.

Kilocalories: Traditional (English) unit of measure for food energy. One kilocalorie (kcal) equals 4.186 kJ.

Kilojoules: Metric unit of measure for food energy. One kilojoule (kJ) equals 0.24 kcal.

Obesity: An excessive fat accumulation that could impair health. For adults, obesity is defined as BMI equal or higher than $30 \, kg/m^2$.

Overweight: High fat accumulation that increases the risk for several health conditions. For adults, overweight is defined as BMI equal or higher than $25\,kg/m^2$.

Simple sugars: Nutrients that contain one or two saccharides. Sugars are sweet: in general, the simpler the sugar, the sweeter it is.

Table sugar: Obtained mainly from sugar cane and beets. It is a sweet crystalline or powdered substance, white when pure. Table sugar is used in many foods, beverages and medicines as a flavor enhancer and for its properties as a sweetener.

List of Abbreviations

BMI   Body mass index
CDC   Centers for Disease Control and Prevention
CHO   Carbohydrates
ECOG   European Childhood Obesity Group
FNB   Food and Nutrition Board
HFCS   High-fructose corn syrup
IoM   Institute of Medicine
IOTF   International Obesity Task Force
kcal   Kilocalories
kJ   Kilojoules
SSB   Sugar-sweetened beverages
WC   Waist circumference
WHO   World Health Organization

References

Allison, D.B., and Mattes, R.D., 2009. Nutritively sweetened beverage consumption and obesity: the need for solid evidence on a fluid issue. *JAMA* 301(3): 318–20.

Allport, J.H., 2004. Soft drinks and obesity. *J Pediatr* 144(4): 554–5.

Bermudez, O.I., and Gao, X., 2010. Greater consumption of sweetened beverages and added sugars is associated with obesity among US young adults. *Ann Nutr Metab* 57(3–4): 211–8.

Bleich, S.N., Wang, Y.C., Wang, Y., and Gortmaker, S.L., 2009. Increasing consumption of sugar-sweetened beverages among US adults: 1988–1994 to 1999–2004. *Am J Clin Nutr* 89(1): 372–81.

Block, G., 2004. Foods contributing to energy intake in the US: data from NHANES III and NHANES 1999–2000. *J Food Comp Anal* 17(3/4): 439–47.

Bowman, S.A., 2002. Beverage choices of young females: changes and impact on nutrient intakes. *J Am Diet Assoc* 102(9): 1234–9.

Bray, G.A., 2008. Fructose: should we worry? *Int J Obes (Lond)* 32 Suppl 7: S127–31.

Bray, G.A., Nielsen, S.J., and Popkin, B.M., 2004. Consumption of high-fructose corn syrup in beverages may play a role in the epidemic of obesity. *Am J Clin Nutr* 79(4): 537–43.

CDC, 2004. Overweight and Obesity. An overview. URL: http://www.cdc.gov/nccdphp/dnpa/obesity/index.htm. (accessed 6-22-11).

CDC, 2011. Prevalence of Overweight among Children and Adolescents: United States, 1999–2002. NCHS Health E-Stat. URL: http://www.cdc.gov/nchs/data/hestat/overweight/overweight99.htm. (accessed 7-14-11).

DHHS, 1998. Clinical guidelines on the identification, evaluation and treatment of overweight and obesity in adults: executive summary. *Am J Clin Nutr* 68: 899–917.

Dhingra, R., Sullivan, L., Jacques, P.F., Wang, T.J., Fox, C.S., Meigs, J.B., D'Agostino, R.B., Gaziano, J.M., and Vasan, R.S., 2007. Soft drink consumption and risk of developing cardiometabolic risk factors and the metabolic syndrome in middle-aged adults in the community. *Circulation* 116(5): 480–8.

DiMeglio, D.P., and Mattes, R.D., 2000. Liquid versus solid carbohydrate: effects on food intake and body weight. *Int J Obes Relat Metab Disord* 24(6): 794–800.

FAO/WHO, 2001. Human Vitamin and Mineral Requirements. Technical Report Series No. 724. URL: www.fao.org/es/esn/vitrni/pdf/TOTAL.PDF. (accessed 2-24-02).

FNB, 2002. *Dietary Reference Intakes for Energy, Carbohydrates, Fiber, Fat, Protein and Amino Acids.* Washington, DC, National Academy Press.

French, S.A., Lin, B.H., and Guthrie, J.F., 2003. National trends in soft drink consumption among children and adolescents age 6 to 17 years: prevalence, amounts, and sources, 1977/1978 to 1994/1998. *J Am Diet Assoc* 103(10): 1326–31.

Ginevan, M.E., 2004. Soft drinks and obesity. *J Pediatr* 144(4): 555–6.

Guthrie, J.F., and Morton, J.F., 2000. Food sources of added sweeteners in the diets of Americans. *J Am Diet Assoc* 100(1): 43–51, quiz 49–50.

IoM, 2011. *Early Childhood Obesity Prevention Policies.* The National Academies Press. Washington, DC.

IOTF, 2011. The Global Epidemic. URL: http://www.iaso.org/iotf/obesity/obesitytheglobalepidemic/. (accessed 7-14-11).

IOTF, 2011. Overweight children around the world. Electronic database. URL: http://www.iaso.org/iotf/obesity/?map = children. (accessed 7-16-11).

Malik, V.S., Schulze, M.B., and Hu, F.B., 2006. Intake of sugar-sweetened beverages and weight gain: a systematic review. *Am J Clin Nutr* 84(2): 274–88.

Mokdad, A.H., Ford, E.S., Bowman, B.A., Dietz, W.H., Vinicor, F., Bales, V.S., and Marks, J.S., 2003. Prevalence of obesity, diabetes, and obesity-related health risk factors. *JAMA* 289(1): 76–9.

Moore, T.J., Vollmer, W.M., Appel, L.J., Sacks, F.M., Svetkey, L.P., Vogt, T.M., Conlin, P.R., Simons-Morton, D.G., Carter-Edwards, L., and Harsha, D.W., 1999. Effect of dietary patterns on ambulatory blood pressure: results from the Dietary Approaches to Stop Hypertension (DASH) Trial. DASH Collaborative Research Group. *Hypertension* 34(3): 472–7.

Must, A., Spadano, J., Coakley, E.H., Field, A.E., Colditz, G.A., and Dietz, W.H., 1999. The disease burden associated with overweight and obesity. *JAMA* 282(16): 1523–9.

Pereira, M.A., 2006. The possible role of sugar-sweetened beverages in obesity etiology: a review of the evidence. *Int J Obes* 30: S28–S36.

Sargeant, L., Wareham, N., Bingham, S., Day, N., Luben, R., Oakes, S., Welch, A., and Khaw, K., 2000. Vitamin C and hyperglycemia in the European Prospective Investigation into Cancer--Norfolk (EPIC-Norfolk) study: a population-based study. *Diabetes Care* 23(6): 726–32.

USDA, 2006. *Profiling Food Consumption in America Agriculture Fact Book*, Chapter 2. USDA Office of Communications.

USDA, 2006. USDA Database for the Added Sugars - Content of Selected Foods, Release 1. URL: http://www.ars.usda.gov/nutrientdata. (accessed 7-15-11).

Wang, Y., and Lobstein, T., 2006. Worldwide trends in childhood overweight and obesity. *Int J Pediatr Obes* 1(1): 11–25.

WHO, 2001. *Diet, Nutrition and the Prevention of Chronic Diseases*. WHO Technical Report Series 916. Geneva.

WHO, 2008. The WHO Child Growth Standards. Electronic database. URL: http://www.who.int/childgrowth/en/. (accessed 6-22-11).

WHO, 2011. Global InfoBase. Global Comparable Estimates and Maps. Electronic database. URL: https://apps.who.int/infobase/. (accessed 7-8-11).

WHO, 2011. Obesity and overweight. Fact sheet N°311. URL: http://www.who.int/mediacentre/factsheets/fs311/en/index.html. (accessed 7-14-11).

CHAPTER 34

Maternal Glucose and Offspring Child BMI

ANDREA DEIERLEIN

Department of Preventive Medicine, Mount Sinai School of Medicine, 17 East 102nd St. D3-125, New York, NY 10029
Email: andrea.deierlein@mssm.edu

34.1 Introduction

Glucose metabolism is altered throughout pregnancy to meet the energy needs of the mother and the growing fetus. During a normal pregnancy, metabolic changes include a decrease in fasting blood glucose and increases in hepatic glucose production and fasting insulin. These changes lead to an increase in insulin resistance, which allows for greater glucose and other fuel availability to the fetus (Lain and Catalano 2007). In some women, insulin resistance and hyperglycemia are exacerbated, resulting in abnormal glucose tolerance (AGT). AGT is usually diagnosed in the late 2nd/early 3rd trimester and includes a range of blood glucose concentrations from mild hyperglycemia to the more extreme gestational diabetes mellitus (GDM). GDM can be defined as "the presence of glucose concentrations that are at the upper end of the population distribution for glucose in pregnant women and are first detected during pregnancy" (Lain and Catalano 2007). Impaired glucose tolerance (IGT, hyperglycemia below the range of a GDM diagnosis) and GDM are linked to several adverse pregnancy outcomes for the mother and the infant and are thought to negatively influence fetal and postnatal development.

Food and Nutritional Components in Focus No. 3
Dietary Sugars: Chemistry, Analysis, Function and Effects
Edited by Victor R Preedy
© The Royal Society of Chemistry 2012
Published by the Royal Society of Chemistry, www.rsc.org

Nearly 60 years ago, Pedersen hypothesized that fetal exposure to maternal hyperglycemia *via* diabetes resulted in fetal hyperinsulinemia, excessive fetal growth, and greater adiposity at birth (Pedersen 1954). Since then, results from animal and human studies support these observations and the role of fuel-mediated teratogenesis in the development of obesity and related conditions in the offspring (Freinkel 1980). Fuel-mediated teratogenesis refers to the negative influence of excess nutrients, such as glucose, in the intrauterine environment on fetal growth and organ development. Experiments in animals provide substantial evidence that fetal exposure to hyperglycemia permanently alters the programming of appetite and body weight regulation systems leading to increased adiposity as well as type 2 diabetes and the metabolic syndrome later in life (Gluckman and Hanson 2008; Warner and Ozanne 2009). Observational studies in pregnant women also support an adverse effect of maternal hyper-glycemia on offspring anthropometric outcomes, including BMI and fat mass, from birth through childhood (Philipps *et al.* 2011). This chapter will highlight and review results from studies that examined the association between maternal glucose concentrations and offspring anthropometric outcomes, focusing on the first several years of life.

34.2 Anthropometric Outcomes in Early Infancy

IGT and GDM are associated with adverse anthropometric outcomes in early infancy. Maternal blood glucose concentrations, in the absence of pre-gestational diabetes or GDM, are positively correlated with birth weight, length, body mass index (BMI), and ponderal index (Knight *et al.* 2007). It is estimated that each 1 milligram per deciliter (mg/dl) increase in blood glucose concentration is associated with a 1.5–2.0 gram increase in birth weight (Scholl *et al.* 2002). Although other observational studies report similar results, a large amount of current knowledge on the effects of blood glucose concentrations on infant outcomes, including birth weight and adiposity, was obtained from the Hyperglycemia and Adverse Pregnancy Outcomes (HAPO) Study (HAPO Study Cooperative Research Group 2002).

The HAPO Study was initiated in 1999 as a five-year prospective, multinational observational study designed "to clarify the risks of adverse outcomes associated with various degrees of maternal glucose intolerance less severe than that in overt diabetes mellitus" (HAPO Study Cooperative Research Group 2002). Therefore, it did not include women with pre-gestational diabetes or GDM. The study improved on shortcomings in the previous literature by collecting data on potential confounders, such as maternal BMI, age, and medical complications, and blinding clinicians to participants' glucose test results to reduce bias in their medical care. The results showed positive, continuous associations for maternal blood glucose concentrations and infant outcomes of birth weight $>90^{th}$ percentile, sum of skinfold thicknesses (based on triceps, subscapular, and flank skinfold thicknesses) $>90^{th}$ percentile, and percent body fat $>90^{th}$ percentile. After adjustment for potential confounders, each one standard deviation increase in maternal blood glucose concentration was associated with a 35% to 46% increased odds of these infant outcomes. The mean birth weights of infants of women with the highest blood glucose concentrations were 240 to 300 grams greater than the mean birth weights of infants of women with the lowest concentrations. The odds of having an infant with a birth weight $>90^{th}$ percentile, sum of skinfold thicknesses $>90^{th}$ percentile, or percent body fat $>90^{th}$ percentile were approximately 3.5- to 5-fold greater among women with the highest blood glucose concentrations compared to those with the lowest concentrations. These findings suggest that hyperglycemia that is below the threshold for a GDM diagnosis negatively influences infant anthropometric outcomes (HAPO Study Cooperative Research Group 2008; 2009).

Adverse infant anthropometric outcomes related to GDM, including high birth weight and adiposity are well documented in the literature. Among women in the Avon Longitudinal Study of Parents and Children, a large, population-based birth cohort in England, infants of women with GDM had a 0.92 higher mean birth weight z-score and approximate 5-fold greater odds of macrosomia and large for gestational age (LGA) compared to infants of women without a diabetes diagnosis during pregnancy (Lawlor *et al.* 2006). Higher birth weights of infants exposed to GDM are likely due to differences in body composition. Among a population of women in the United States (US), Catalano and colleagues found that blood glucose concentration in women with GDM was the strongest predictor of infant fat mass (Catalano *et al.* 2003). Infants of women with GDM had a higher percentage of body fat and greater fat mass but lower fat-free mass compared to infants of women with normal glucose tolerance during pregnancy. These anthropometric differences persisted when the sample population was restricted to infants born average-for-gestational age (AGA), suggesting that GDM influences infant adiposity independent of birth weight (Catalano *et al.* 2003).

34.3 Anthropometric Outcomes in Early Childhood

Relatively fewer studies have examined the effects of fetal exposures to AGT, especially IGT, on offspring anthropometric outcomes beyond birth and throughout early childhood. Collectively, results from these studies suggest that the effects are long-term; however, they also reveal a pattern such that the observed associations of AGT and infant adiposity resolve at approximately 1–2 years but then may recur later in childhood (Pettitt *et al.* 2010; Silverman *et al.* 1998).

34.3.1 Anthropometric Outcomes at 1–2 Years

Studies of children measured within the ages of 1–2 years are not indicative of an association between AGT and offspring anthropometric outcomes. Using data from the Belfast, United Kingdom (UK) center of the HAPO study, Pettitt and colleagues found no association between maternal glucose concentrations and child overweight/obesity (defined as BMI $> = 85^{th}$ percentile) and obesity (defined as $BMI > = 95^{th}$ percentile) at 2 years. The only exception was a significant increase in the prevalence of child overweight/obesity across strata of one measure of maternal blood glucose concentrations. Weak correlations were observed for maternal glucose concentration and sum of skinfold thicknesses (based on triceps, subscapular, and suprailiac skinfold thicknesses) and suprailiac skinfold thickness but were more notable in boys than in girls (Pettitt *et al.* 2010). Similarly, in a non-diabetic Caucasian population in Exeter, UK, significant correlations between maternal glucose concentrations and infant weight, length, and BMI at birth were not observed when these measurements were repeated at ages 12 weeks, 1 year, and 2 years (Knight *et al.* 2007).

34.3.2   Anthropometric Outcomes in Early Childhood

Although there is limited evidence to support an association of maternal glucose concentrations and offspring anthropometric outcomes at 1–2 years, longitudinal studies and studies in children at later ages suggest a positive association. Longitudinal studies that compare children of women with and without GDM during pregnancy reveal significant differences in their patterns of growth. Among a community of Pima Indians in Arizona (US), infants of women with diabetes (defined as pre-gestational diabetes or GDM) were heavier at birth compared to infants of women without diabetes; however, at 1.5 years, children of women with diabetes were shorter but their weights and adjusted weights (weight adjusted for age, sex, and height) were similar to those of children of women without diabetes. At 7.7 years, children of women with diabetes had significantly greater weights but similar heights than those of women without diabetes (Touger *et al.* 2005). In the German GDM Offspring Study, the prevalence of overweight (defined as $BMI > = 90^{th}$ percentile) was significantly higher and increased from 2 to 11 years among children of women with GDM compared to those of women without GDM. The prevalence of overweight was 17.2%, 20.2%, and 31.1% at 2, 8, and 11 years, respectively, in children of women with GDM compared to 11.4%, 10.3%, and 15.5%, respectively, in children of women without GDM. Maternal obesity at early pregnancy was found to be the most important risk factor (compared to birth weight, GDM treatment type, maternal smoking status during pregnancy, and duration of breast feeding) for overweight status among children of women with GDM (Boerschmann *et al.* 2010).

Differences in anthropometric outcomes of offspring exposed to AGT may first be detectable during the preschool years. At age 3 years, children exposed to elevated maternal glucose concentrations have an increased risk of being overweight/obese (defined as $BMI > = 85^{th}$ percentile) (Deierlein *et al.* 2011) and children exposed to GDM have greater sum of skinfold thicknesses (based on triceps and subscapular skinfold thicknesses) (Wright *et al.* 2009) compared to children of women with normal glucose tolerance. These observed differences are consistent with results from studies of children at later ages. At ages 4–9 years, mean BMI and BMI z-scores are significantly higher among children of women with GDM compared to those of women without GDM or with

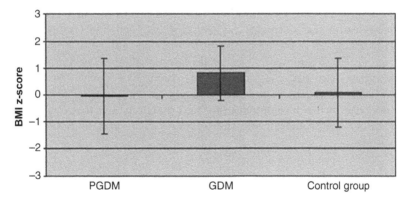

Figure 34.1 Mean values of BMI z-scores among children of women with pre-gestational diabetes mellitus (PGDM), gestational mellitus, and without these conditions (control group). The mean values of BMI z-scores in children ages 4–9 years were significantly higher ($p < 0.05$) among women with GDM compared to PGDM or the control group. There was no difference in the mean values of child BMI z-scores between the control and PGDM groups.
(From Wroblewska-Seniuk *et al.* 2009; with permission).

pre-gestational diabetes mellitus (Figure 34.1) (Wroblewska-Seniuk *et al.* 2009). In a multiethnic US population of women with normal glucose tolerance, increasing glucose concentrations (mg/dl) were associated with child over-weight/obesity (defined as BMI $>85^{\text{th}}$ percentile) and obesity (defined as BMI $>95^{\text{th}}$ percentile) at 5–7 years. The odds of child overweight/obesity and obesity were approximately 25% higher among children of women in the highest category of glucose concentrations compared to those of women in the lowest category. There were nearly 2-fold greater odds of overweight/obesity and obesity at 5–7 years among children of women with GDM compared to those of women with normal glucose tolerance (Hillier *et al.* 2007). The higher BMI of children who are exposed to elevated glucose concentrations or GDM is likely due to greater adiposity in these children. Among children aged 6–13 years, those exposed to GDM have higher BMI, larger waist circumferences, greater amounts of visceral and subcutaneous adipose tissue, as well as a more centralized distribution of fat (based on the ratio of subscapular to triceps skin fold thicknesses) compared to children of women who did not have GDM; however, many of these associations were attenuated after inclusion of maternal pre-pregnancy BMI and infant birth weight in the statistical models (Crume *et al.* 2011).

34.3.3 Other Outcomes in Childhood

In addition to associations with child anthropometric outcomes, there is evidence that fetal exposure to increasing maternal glucose concentrations are associated other unfavorable health outcomes, including IGT and

poor intellectual and psychomotor development (Silverman *et al.* 1998).
Reduced insulin sensitivity and higher static beta cell response are observed
in children at 5–10 years, independent of their percentage of body fat (Bush
et al. 2011). These findings suggest that maternal glucose concentrations play
a role in the fetal programming of the pancreas, as well as insulin target
tissues, such as the liver and skeletal muscle. They may also explain the
increased risk of the development of type 2 diabetes and other characteristics
of metabolic syndrome among the offspring of women with GDM (Bush
et al. 2011).

34.4 Limitations of Studies and Gaps in our Understanding

The current literature suggests that AGT has long-term consequences for offspring anthropometric outcomes at birth and into early childhood; however,
there are short-comings of previous studies that limit the ability to discern the
true nature of this association. These include, but are not limited to, issues
related to the role of maternal prepregnancy obesity, measurement of offspring
anthropometric outcomes, and assessment of AGT.

34.4.1 Role of Maternal Prepregnancy Obesity

It is not clear how much of the association of AGT and offspring anthropometric outcomes can be attributed to shared genetic and environmental factors,
specifically those related to maternal pre-pregnancy BMI. Women who develop
AGT are more likely to be obese prior to pregnancy (Torloni *et al.* 2009).
Figure 34.2 displays the prevalence of GDM according to maternal pre-
pregnancy BMI. Independent of GDM, maternal obesity is associated with
enhanced insulin resistance and increased delivery of glucose and other fuel
sources to the fetus; increased risk of high infant birth weight outcomes and
overweight/obesity later in childhood; as well as an obesogenic postnatal
environment with respect to infant feeding, diet, and physical activity-related
behaviors (King 2006; Oken 2009). Results from a recent meta-analysis of six
studies reported a 0.28 (95% CI 0.05, 0.51) higher mean BMI z-score of children of women with GDM compared to those of women without GDM. There
were only three studies included in the meta-analysis that adjusted for maternal
pre-pregnancy BMI, among these studies the predicted difference in child mean
BMI z-score was attenuated to 0.07 (95% CI − 0.15, 0.28) (Phillips *et al.* 2011).
These estimates suggest that maternal pre-pregnancy BMI has an influential
role in the development of offspring obesity independent of maternal glucose
tolerance but do not reject an independent effect of maternal glucose tolerance
(Dabelea *et al.* 2000; Deierlein *et al.* 2011; HAPO Study Cooperative Research
Group 2009; Vohr *et al.* 1999). The independent and possible joint effects of
pre-pregnancy obesity and AGT on fetal and post-natal growth and development remain to be elucidated.

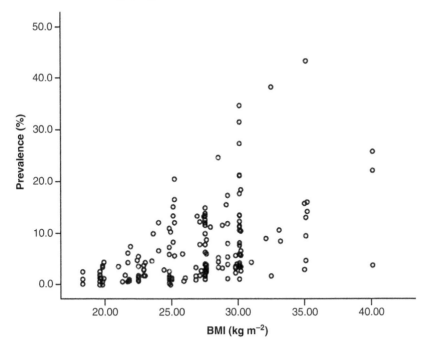

Figure 34.2 Prevalence of gestational diabetes mellitus according to initial maternal body mass index (BMI). The prevalence of gestational diabetes mellitus according to increasing prepregnancy body mass index based on the results of a meta-analysis of 56 cohort studies that included 631 763 women. It was estimated that for every one unit increase in prepregnancy BMI, the risk of GDM increased by 0.92% (95%CI, 0.73–1.10).
(From Torloni *et al.* 2011; with permission).

34.4.2 Measurement of Offspring Anthropometric Outcomes

The majority of studies defined child anthropometric outcomes using BMI and/or skinfold thickness measurements (measured at various sites on the body). Although BMI is an easy and inexpensive measurement that is used in most large cohort studies, it is a crude assessment of adiposity and does not provide measurements of fat and fat-free mass or body fat patterning. Similarly, skinfold thicknesses are a measure of subcutaneous fat that can be used to estimate percentage of body fat but tend to be measured in a limited number of body regions and are poorly reproduced and less reliable than BMI (Mei *et al.* 2007). Future studies that assess child anthropometric outcomes using sophisticated body composition techniques that provide accurate measurements of fat and fat-free mass, percent body fat, and body fat patterns, including visceral and subcutaneous adiposity are necessary to understand how AGT influences offspring adiposity and other adverse health outcomes.

34.4.3 Assessment of Abnormal Glucose Tolerance

There is a lack of consistency across studies in methods used for measuring maternal blood glucose concentration. There have been no internationally accepted standards for screening methods or diagnostic thresholds and consequently they have differed across studies (See Key Facts of Screening for GDM). The use of different methods and cut-points results in varying levels of sensitivity and specificity of tests and possible misclassification of cases of GDM within studies. Misclassification of women based on disease status (GDM or no GDM), has the potential to bias effect estimates upwards or downwards, depending on the study. Additionally, not all studies account for the effects of treatment of GDM in statistical analyses. This may also bias effect estimates since the influence of GDM on offspring anthropometric outcomes may be less severe when women receive either diet-related or medical treatment (Hillier *et al.* 2007; Whitaker *et al.* 1998). The possibility of bias in studies as a result of misclassification or lack of adjustment for GDM treatment status needs to be considered when attempting to determine the strength of the association between AGT and offspring anthropometric outcomes.

Summary Points

- Impaired glucose tolerance and gestational diabetes mellitus (GDM) are associated with adverse offspring anthropometric outcomes in early infancy.
- A large amount of data on the effects of maternal hyperglycemia on infant outcomes was obtained from the Hyperglycemia and Adverse Pregnancy Outcomes Study. The results showed positive, continuous associations for maternal glucose concentrations and infant birth weight $>90^{th}$ percentile, sum of skinfold thicknesses $>90^{th}$ percentile, and percent body fat $>90^{th}$ percentile.
- Adverse infant anthropometric outcomes related to GDM, including high birth weight and adiposity are well documented in the literature.
- Results from studies suggest the effects of fetal exposures to abnormal glucose tolerance (AGT) on offspring anthropometric outcomes in early childhood are long-term. However, they also reveal a pattern such that the observed associations of AGT and infant adiposity resolve at approximately 1–2 years but then may recur later in childhood.
- Differences in anthropometric outcomes of offspring exposed to AGT are observed in children aged 3 years and older.
- Fetal exposure to increasing maternal glucose concentrations is also associated with reduced insulin sensitivity and higher static beta cell response in children at 5–10 years.
- There are several short-comings of previous studies that limit the ability to discern the true nature of the association of abnormal glucose tolerance

and offspring anthropometric outcomes at birth and into early childhood. These include, but are not limited to, issues related to the role of pre-pregnancy obesity, measurement of offspring anthropometric outcomes, and assessment of AGT.

Key Facts of Screening for GDM (Virally *et al.* 2010)

- Abnormal glucose tolerance is often diagnosed with an oral glucose tolerance test (OGTT), administered at approximately 24–28 weeks gestation after an overnight fast. The test requires drinking a glucose solution and measuring venous blood glucose concentrations at specified times thereafter.
- The two most common screening methods are: (1) the 50 g glucose challenge test (GCT) followed by a 100 g OGTT and (2) the 75 g OGTT.
- The 50 g GCT followed by a 100 g OGTT is a two-step process that involves drinking 50 g of glucose with a blood measurement at 1 hour followed by drinking 100 g of glucose with blood measurements at 0, 1, 2, and 3 hours (the 100 g OGTT is only administered among women with abnormal values on the 50 g GCT).
- The 75 g OGTT is a one-step process that involves drinking 75 g of glucose with blood measurements at 0 and 2 hours.
- The first diagnostic criteria for GDM were proposed in 1964 by O'Sullivan and Mahan. These criteria were derived from non-pregnancy values and based on predicting the risk of future development of maternal type 2 diabetes. They involved the use of the two-step OGTT method. Though widely used, these criteria were controversial because they were not based on the risks of any short-term outcomes of maternal, fetal, or perinatal complications or mortality.
- In 1980, the World Health Organization (WHO) proposed the use of the one-step 75 g OGTT in place of the two-step method.
- From 1980 through 2008 various European and United States-based organizations recommended the use of either the one-step or two-step method with varying diagnostic thresholds, culminating in 2008 with the National Institute for Health and Clinical Excellence (UK) recommendation of the one-step 75 g OGTT and the United States Preventive Services Task recommendation of the two-step method.
- Most recently in 2010, the International Association of Diabetes and Pregnancy Study Groups issued recommendations for the diagnosis of hyperglycemia and GDM during pregnancy based on findings from the Hyperglycemia and Adverse Pregnancy Outcomes Study. The recommendation included the use of the one-step 75 g OGTT.
- Screening and diagnostic methods have varied across research studies due to the lack of internationally accepted standards. This limits the ability to compare results from studies in a consistent manner.

Key Facts of Diet and GDM

- Predictors of GDM include prepregnancy BMI, physical activity, and smoking status (Radesky *et al.* 2009).
- There is inconsistent evidence to support an association between dietary factors, such as sugar intake, and the development of GDM. Some studies have shown that consumption of sugar-sweetened beverages (Chen *et al.* 2009; Western type diets, which are characterized by high intakes of red meat, processed foods, and refined sugars (Zhang *et al.* 2006a); and diets with high glycemic loads (a measure of dietary carbohydrate) (Zhang *et al.* 2006b) are associated with an increased risk of GDM. However, other studies have not observed an association of dietary carbohydrates and GDM (Bo *et al.* 2001; Radesky *et al.* 2009).
- The effects of prepregnancy and early pregnancy diet on the development of GDM are not well-established; prepregnancy BMI, an indicator of nutritional status prior to pregnancy, is likely to be a more important risk factor for GDM than diet (Radesky *et al.* 2008).

Definitions of Word and Terms

Abnormal Glucose Tolerance: A condition of elevated blood glucose concentrations during pregnancy, which includes IGT as well as GDM.

Adiposity: Fatness

Body Mass Index: A proxy measurement for body fatness derived from weight (kilograms) divided by height (meters) squared (kg/m^2).

Gestational Diabetes Mellitus: A condition of high blood glucose concentrations during pregnancy with first onset during pregnancy and usually resolves after pregnancy.

Hyperglycemia: A condition of high concentrations of glucose in the blood.

Hyperinsulinemia: A condition of high concentrations of insulin in the blood.

Impaired Glucose Tolerance: A condition of elevated blood glucose concentrations during pregnancy that are below the diagnostic threshold of GDM.

Insulin Resistance: A condition in which the body produces insulin but is unable to use it efficiently. As a result, glucose is not adequately absorbed from the blood by liver, muscle, and fat cells (there is a reduction in glucose uptake by these cells). The pancreas continues to produce and secrete insulin as a response to the increased concentrations of glucose; therefore, a common consequence of insulin resistance is elevated circulating levels of blood glucose and insulin.

Insulin Sensitivity: A measure of the body's (muscle, liver, and fat cells) ability to absorb glucose from the blood in response to insulin. When an individual has optimal insulin sensitivity, insulin levels peak after a meal resulting in glucose uptake from the blood into cells and insulin levels decline thereafter. Poor insulin sensitivity results when cells do not properly respond to insulin and do not efficiently absorb glucose from the blood. Low insulin sensitivity is synonymous with the condition of insulin resistance.

Large-for-Gestational Age: Birth weight greater than the 90th percentile for gestational age.

Macrosomia: Birth weight greater than 4000g.

Normal Glucose Tolerance: A condition of blood glucose concentrations within a normal range during pregnancy.

Ponderal Index: A proxy measurement for infant body fatness derived from weight (kilograms) divided by length (meters) cubed (kg/m^3).

Skinfold Thicknesses: A method of measuring subcutaneous fat in various regions of the body, which can be used to estimate percent body fat by incorporating them into prediction equations. Common regions include the triceps, biceps, subscapular (below the shoulder blade), and suprailiac or flank (on side of body between rib and hip). The process involves pinching the skin and measuring the thickness of the fold using calipers.

Subcutaneous Adiposity: Fat located beneath the skin.

Visceral Adiposity: Fat located in the intra-abdominal region between organs in this region, such as the stomach, liver, and kidneys. This type of fat has been linked to increased risk of certain diseases such as type 2 diabetes and cardiovascular disease.

z-score: A measure used in statistics that indicates how many standard deviations an observation lies above or below the mean (0).

List of Abbreviations

AGA	Average for Gestational Age
AGT	Abnormal Glucose Tolerance
ALSPAC	Avon Longitudinal Study of Parents and Children
BMI	Body Mass Index
GCT	Glucose Challenge Test
GDM	Gestational Diabetes Mellitus
HAPO	Hyperglycemia and Adverse Pregnancy Outcomes Study
IGT	Impaired Glucose Tolerance
LGA	Large for Gestational Age
OGTT	Oral Glucose Tolerance Test
UK	United Kingdom
US	United States

References

Bo, S., Menato, G., Lezo, A., Signorile, A., Bardelli, C., De Michieli, F., Massobrio, M., and Pagano, G., 2001. Dietary fat and gestational hyperglycemia. *Diabetologia*. 44: 972–978.

Boerschmann, H., Pfluger, M., Henneberger, L., Ziegler, A-G., and Hummel, S., 2010. Prevalence and predictors of overweight and insulin resistance in offspring of mothers with gestational diabetes mellitus. *Diabetes Care*. 33: 1845–1849.

Bush, N.C., Chandler-Laney, P.C., Rouse, D.J., Granger, W.M., Oster, R.A., and Gower, B.A., 2011. Higher maternal gestational glucose concentration is associated with lower offspring insulin sensitivity and altered b-cell function. *Journal of Clinical Endocrinology and Metabolism.* 96: E803–809.

Catalano, P.M., Thomas, A., Huston-Presley, L., and Amini, S.B., 2003. Increased fetal adiposity: A very sensitive marker of abnormal in utero development. *Obstetrics and Gynecology.* 189: 1698–1704.

Chen, L., Frank, F.B., Yeung, E., Willett, W., and Zhang, C., 2009. Prospective study of pre-gravid sugar-sweetened beverage consumption and the risk of gestational diabetes mellitus. *Diabetes Care.* 32: 2236–2241.

Crume, T.L., Ogden, L., West, N.A., Vehik, K.S., Scherzinger, A., Daniels, S., McDuffie, R., Bischoff, K., Hamman, R.F., Norris, J.M., and Dabelea, D., 2011. Association of exposure to diabetes in utero with adiposity and fat distribution in a multi-ethnic population of youth: the Exploring Perinatal Outcomes among Children (EPOCH) Study. *Diabetologia.* 54: 87–92.

Dabelea, D., Hanson, R.L., Lindsay, R.S., Pettitt, D.J., Imperatore, G., Gabir, M.M., Roumain, J., Bennett, P.H., and Knowler, W.C., 2000. Intrauterine exposure to diabetes conveys risks for type 2 diabetes and obesity: a study of discordant sibships. *Diabetes.* 49: 2208–2211.

Deierlein, A.L., Siega-Riz, A.M., Chantala, K., and Herring, A.H., 2011. The association between maternal glucose concentration and child BMI at age 3 years. *Diabetes Care.* 34: 480–484.

Freinkel, N. 1980. Banting Lecture, 1980. Of pregnancy and progeny. *Diabetes.* 29: 1023–1035.

Gluckman, P.D. and Hanson, M.A., 2008. Developmental and epigenetic pathways to obesity: an evolutionary–developmental perspective. *International Journal of Obesity.* 32: S62–S71.

HAPO Study Cooperative Research Group, 2002. The hyperglycemia and adverse pregnancy outcome (HAPO) study. *International Journal of Gynecology and Obstetrics.* 78: 69–77.

HAPO Study Cooperative Research Group, 2008. Hyperglycemia and adverse pregnancy outcomes. *New England Journal of Medicine.* 358: 1991–2002.

HAPO Study Cooperative Research Group, 2009. Hyperglycemia and adverse pregnancy outcome (HAPO) study: Associations with neonatal anthropometrics. *Diabetes.* 59: 453–459.

Hillier, T.A., Pedula, K.L., Schmidt, M.M., Mullen, J.A., Charles, M.A., and Pettitt, D.J., 2007. Childhood obesity and metabolic imprinting. *Diabetes Care.* 30: 2287–2292.

King, J.C., 2006. Maternal obesity, metabolism, and pregnancy outcomes. *Annual Review of Nutrition.* 26: 271–291.

Knight, B., Shields, B.M., Hill, A., Powell, R.J., Wright, D., and Hattersley, A.T., 2007. The impact of maternal glycemia and obesity on early postnatal growth in a non-diabetic Caucasian population. *Diabetes Care.* 30: 777–783.

Lain, K.Y. and Catalano, P.M., 2007. Metabolic changes in pregnancy. *Clinical Obstetrics and Gynecology.* 50: 938–943.

Lawlor, D.A., Smith, G.D., O'Callaghan, M., Alati, R., Mamun, A.A., Williams, G.M., and Najman, J.M., 2006. Epidemiologic evidence for the fetal overnutrition hypothesis: Findings from the mater-university study of pregnancy and its outcomes. *American Journal of Epidemiology*. 53: 89–97.

Mei, Z., Grummer-Strawn, L.M., Wang, J., Thornton, J.C., Freedman, D.S., Pierson, R.N., Dietz, W.H., and Horlick, M., 2007. *Pediatrics*. 119: e1306–e1313.

Oken, E. 2009. Maternal and child obesity: the causal link. *Obstetrics and Gynecology Clinics of North America*. 36: 361–377.

Pedersen, J. 1954. Weight and length at birth of infants of diabetic mothers. *Acta Endocrinologica*. 16: 330–342.

Pettitt, D.J., McKenna, S., McLaughlin, C., Patterson, C.C., Hadden, D.R., and McCance, D.R., 2010. Maternal glucose at 28 weeks of gestation is not associated with obesity in 2-year-old offspring. *Diabetes Care*. 33: 1219–1223.

Philipps, L.H., Santhakumaran, S., Gale, C., Prior, E., Logan, K.M., Hyde, M.J., and Modi, N., 2011. The diabetic pregnancy and offspring BMI in childhood: a systematic review and meta-analysis. *Diabetologia*. [May 31 Epub ahead of print]

Radesky, J.S., Oken, E., Rifas-Shiman, S.L., Kleinman, K.P., Rich-Edwards, J.W., and Gillman, M.W., 2008. Diet during early pregnancy and development of gestational diabetes. *Paediatric Perinatal Epidemiology*. 22: 47–59.

Scholl, T.O., Chen, X., Gaughan, C., and Smith, W.K., 2002. Influence of maternal glucose level on ethnic differences in birth weight and pregnancy outcome. *American Journal of Epidemiology*. 156: 498–506.

Silverman, B.L., Rizzo, T.A., Cho, N.H., and Metzger, B.E., 1998. Long-term effects of the intrauterine environment. The Northwestern University Diabetes in Pregnancy Center. *Diabetes Care*. 21: B142–149.

Torloni, M.R., Betran, A.P., Horta, B.L., Nakamura, M.U., Atallah, A.N., Moron, A.F., and Valente, O. Prepregnancy BMI and the risk of gestational diabetes: a systematic review of the literature with meta-analysis. 2009. *Obesity Reviews*. 10: 194–203.

Touger, L., Looker, H.C., Krakoff, J., Lindsay, R.S., Cook, V., and Knowler, W.C., 2005. Early growth in offspring of diabetic mothers. *Diabetes Care*. 28: 585–589.

Virally, M., and Laloi-Michelin, M., 2010. Methods for the screening and diagnosis of gestational diabetes mellitus between 24 and 28 weeks of pregnancy. *Diabetes and Metabolism*. 36: 549–565.

Vohr, B.R., McGarvey, S.T., and Tucker, R., 1999. Effects of maternal gestational diabetes on offspring adiposity at 4-7 years of age. *Diabetes Care*. 22: 1284–1291.

Warner, M.J. and Ozanne, S.E., 2009. Mechanisms involved in the developmental programming of adulthood disease. *Biochemistry Journal*. 427: 333–347.

Whitaker, R.C., Pepe, M.S., Seidel, K.D., Wright, J.A., and Knopp, R.H., 1998. Gestational diabetes and the risk of offspring obesity. *Pediatrics*. 101: e9–316.

Wright, C.S., Rifas-Shiman, S.L., Rich-Edwards, J.W., Taveras, E.M., Gillman, M.W., and Oken, E., 2009. Intrauterine exposure to gestational diabetes, child adiposity, and blood pressure. *American Journal of Hypertension*. 22: 215–220.

Wroblewska-Seniuk, K., Wender-Ozegowska, E., and Szczapa, J., 2009. Long-term effects of diabetes during pregnancy on the offspring. *Pediatric Diabetes*. 10: 432–440.

Zhang, C., Schulze, M.B., Solomon, C.G., and Hu, F.B., 2006a. A prospective study of dietary patterns, meal intake, and the risk of gestational diabetes mellitus. *Diabetologia*. 49: 2604–2613.

Zhang, C., Lui, S., Solomon, C.G., and Hu, F.B., 2006b. Dietary fiber intake, dietary glycemic load, and the risk for gestational diabetes mellitus. *Diabetes Care*. 29: 2223–2230.

CHAPTER 35

Dextrose in Total Parenteral Nutrition

KAREN C. McCOWEN

Harvard Vanguard Medical Associates and Harvard Medical School, 133 Brookline Avenue, Medical Specialties, Boston, MA 02215
Email: kmccowen@bidmc.harvard.edu

35.1 Introduction

Successful delivery of hyperconcentrated parenteral nutrition by a central vein was established in the 1960s to allow survival of patients with intestinal failure. Dextrose (a name usually applied to glucose in solution or used synonymously with glucose) and protein hydrolysates formed the mainstay of therapy initially; subsequently, amino acids became the source of protein. Lipid emulsions were also added to prevent essential fatty acid deficiency initially, and then as a source of energy. Patients with gastrointestinal fistulae and with short bowel syndrome were the primary recipients. Now total parenteral nutrition (TPN) is a common source of feeding in malnourished hospitalized patients who have temporary intestinal failure and is less commonly used in outpatients with chronic bowel failure (in which cases, patients administer the treatment at home) (Table 35.1).

TPN is infused through a large central vein, and is composed of an appropriate mixture of amino acids, dextrose, lipids, electrolytes, trace minerals and vitamins, although dextrose is the main energy source in most cases. The combination of dextrose and amino acids has been shown to be nitrogen-sparing in the intensive care unit (ICU) but does not completely suppress

Food and Nutritional Components in Focus No. 3
Dietary Sugars: Chemistry, Analysis, Function and Effects
Edited by Victor R Preedy
© The Royal Society of Chemistry 2012
Published by the Royal Society of Chemistry, www.rsc.org

Table 35.1 Indications for total parenteral nutrition. In general, use of the gut to provide nutrition is preferred. However, in certain patients, TPN should be used to prevent or treat malnutrition.

- Patient unable to tolerate enteral nutrition because of intestinal obstruction, acute pancreatitis, high output fistulae or other condition
- Anticipated duration of the failure of the gut > 1 week
- Patient with short-term intestinal failure with baseline malnutrition
- Patient has accessible central venous access to the circulation (small peripheral vein is inadequate)
- Absence of relative contra-indications such as hyperglycemia, azotemia, encephalopathy, hyperosmolarity, severe fluid and electrolyte disturbances – such problems should be controlled prior to initiation of TPN

either endogenous muscle catabolism or glucose production. The application of TPN must be monitored closely, because of the potential for side effects from any of the components. The most common side effect over the years has been hyperglycemia from the hypertonic dextrose. Dextrose is by far the most commonly used carbohydrate energy source in TPN, and it provides 3.4 kcal/gram when hydrated. Commercially available concentrations range from 2.5% up to 70%. The concentration of dextrose infused is limited by the percent amino acids and lipids in the mixture.

35.2   Metabolic Response to Injury

Hyperglycemia can be considered as a normal component of the response to injury and illness and results from release of endogenous mediators of insulin resistance such as counter-regulatory hormones (catecholamines, cortisol, glucagon and growth hormone) and compounds such as TNF alpha and IL6 (McCowen *et al.* 2001). Insulin resistance leads to hyperglycemia if limited pancreatic reserve for insulin production or release is present. The metabolic milieu allows ongoing gluconeogenesis (in liver and kidney) even in the face of euglycemia or hyperglycemia (Figure 35.1). The biochemical mechanisms that promote such insulin resistance remain somewhat poorly understood, but likely relate to abnormalities in the intracellular insulin signaling cascade (Figure 35.2). The major risk factors for insulin resistance in intensive care unit (ICU) patients are shown in Table 35.2, along with risk factors that also promote hyperglycemia in such patients.

Infusion of concentrated dextrose loads in this situation can compound the hyperglycemia and lead to substantial toxicity (Figure 35.1). This occurs partly because intravenous dextrose feeding during illness does not restrain endogenous glucose production. Protein catabolism in critical illness has a role to produce substrate for the immune system and acute phase proteins, although the end result is depletion of host lean mass. In addition, proteolysis allows a source for gluconeogenesis. Lipolysis is also accelerated as a result of the circulating catabolic hormones and results in production of free fatty acids and

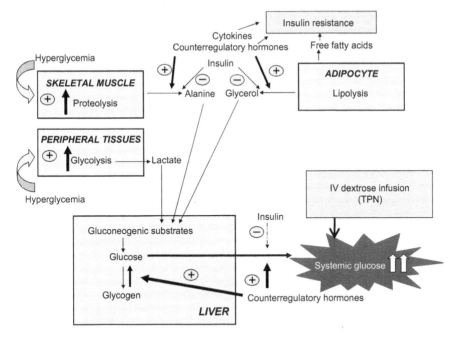

Figure 35.1 Factors that lead to hyperglycemia in critical illness. The patient has ongoing glucose production in liver (and kidney) in association with impaired glucose utilization, so is primed for hyperglycemia when intravenous infusion of dextrose is added. Sources of carbon for the new glucose production include breakdown products from lipids and protein.

glycerol. Glycerol is funneled towards gluconeogenesis. However, ketogenesis (from the fatty acids) in ICU patients occurs to a lesser degree than in simple starvation (due to elevated insulin concentrations) such that the protein sparing effect that settles in during sustained starvation (without illness) does not occur in ICU patients. Feeding through the gut is at least temporarily impossible in many ICU patients because of their underlying illnesses, and hence TPN is a reasonable alternative method of nutrition support in these circumstances. Complete absence of nutritional supplementation would be associated with worse survival, so whether or not to feed the ICU patient is not a debate at this time, rather route of feeding is controversial.

35.3 Route of Feeding in ICU Patients – TPN Versus Enteral

Classical ICU teaching is that nutrition *via* the gastrointestinal tract is always superior to intravenous feeding. Early studies of TPN showed high rates of infectious complications, probably linked to hyperglycemia (The Veterans

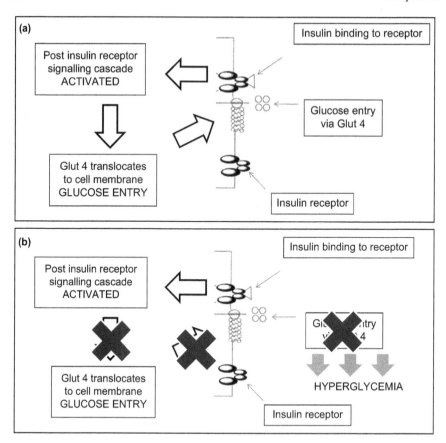

Figure 35.2 Glucose transport into insulin-sensitive tissues. Under normal conditions, insulin activates its cell surface receptor, permits a chain of
intracellular signaling events, which allows movement of specialized
glucose transporters to the cell membrane, such that glucose can be taken
up from the bloodstream (a). However, under conditions of insulin
resistance (b) such as occur in patients that are critically ill receiving
TPN, the post-receptor signaling is in some way impaired, so that glucose
uptake is reduced, contributing to hyperglycemia.

Affairs Total Parenteral Nutrition Study Group 1991). The ease with which
TPN can be delivered allows infusion of large amounts of concentrated dextrose, leading to florid hyperglycemia in critically ill, insulin resistant patients
unless appropriate insulin therapy is instituted. Thus, many studies in which
enteral and parenteral nutrition have been compared were flawed in that
unequal amounts ("overfeeding" in the case of TPN) of nutrients were delivered, and different rates of hyperglycemia surely contributed to the differences
in infectious complications.

Therefore, TPN has fallen into some disrepute. Most ICU nutrition guidelines recommend the use of enteral nutrition wherever possible in preference to

Table 35.2 Risk factors for developing hyperglycemia in critically ill patients. There are many risk factors for the development of high blood glucose in sick patients. Most factors operate either through inhibition of pancreatic insulin production, or through the development of insulin resistance.

Factor	Main Mechanism of Hyperglycemia
Pre-existing diabetes mellitus	Insulin deficiency and insulin resistance
Hyperglycemia*	Insulin deficiency and insulin resistance
Catecholamine infusion	Insulin resistance
Illness severity	Higher counterregulatory hormones leading to insulin resistance
Glucocorticoid therapy	Insulin resistance
Obesity	Insulin resistance
Ageing	Relative insulin deficiency
Dextrose infusion	Glucose removal rates overwhelmed
Pancreatitis	Insulin deficiency
Sepsis	Insulin resistance
Uremia	Insulin resistance

*Hyperglycemia alone can inhibit pancreatic beta cell function as well as resulting in reduced insulin action at multiple sites.

parenteral (Kondrup *et al.* 2003; Martindale *et al.* 2009). Interestingly, however, a meta-analysis of studies comparing TPN *vs.* enteral nutrition, which excluded poor quality studies, showed a mortality benefit for TPN over enteral (Simpson and Doig 2005). Clearly, in this current era where the deleterious effects of elevations in plasma glucose concentrations have been established, the ability to attain normoglycemia might allow some of the other benefits of TPN to be appreciated. In the most nutritionally depleted patients, achieving goal feeding rates is easier with TPN than with enteral, allowing an advantage for the former. Interestingly, all of the mortality advantages of TPN were found in comparison to "late" enteral feeding. These findings might suggest, at least in patients at highest nutritional risk, that the ability to feed early in the illness is key, and that any way it can be done is acceptable. Delaying feeding is associated with poorer outcomes in these patients, although not in patients who do not have baseline malnutrition (Casaer *et al.* 2011). Meta-analyses of early versus delayed enteral nutrition have generally concurred and showed better results in terms of infectious complications and length of hospital stay from early feeding, but no effects on non-infectious complications or mortality.

There is some early evidence that even very low rates of enteral nutrition in combination with reasonable TPN infusion rates might be associated with less hyperglycemia. In animal models such practice improved intestinal integrity and promoted incretin hormone release. In a randomized trial of patients undergoing oesophagectomy for adenocarcinoma, the combination of TPN with enteral feeding resulted in lower glucose concentrations, improved insulin sensitivity and increased plasma concentrations of certain incretin hormones when compared with TPN alone (Lidder *et al.* 2010).

35.4   Hyperglycemia

35.4.1   Dextrose Infusion Rates

Only very high rates of dextrose infusion reduce gluconeogenesis in critical illness, and then only when accompanied by marked hyperinsulinemia. However, such infusion generally allows overfeeding, with the resultant complications of *de novo* lipogenesis (leading to fatty liver) and excessive production of CO_2 (which might cause ventilatory disturbances), both discussed below. Glucose oxidation occurs in ICU patients in glucose – dependent organs as well as in wounds and in inflammatory tissues. However, rates of glucose oxidation are lower in critical illness than in the healthy state for a given rate of dextrose infusion (Shaw and Wolfe 1986).

Glucose oxidation increases in parallel with increasing dextrose infusion rates, although a plateau is reached at approximately 4 mg/kg/min infusion rate (Wolfe 1996). In healthy individuals, endogenous glucose production rates approximate 2 mg/kg/min, and this rises to 3 mg/kg/min during illness. Hence, there is enormous potential for TPN to result in additional disturbances in glucose homeostasis. In a retrospective study of patients without diabetes receiving TPN, 18 of 37 who received dextrose infusions at > 5 mg/kg/min demonstrated hyperglycemia, in contrast to 0 of 19 patients given dextrose at 4 mg/kg/min or less. The dextrose infusion rate was independently correlated with plasma glucose concentration in regression analysis (Rosmarin *et al.* 1996).

35.4.2   Effects of Hyperglycemia Induced by TPN on Clinical Outcome

The degree of hyperglycemia induced by TPN in one study was correlated with the clinical outcome. Each 10 mg/dL increase in mean blood glucose level was associated with an increased risk of infection, cardiovascular dysfunction, acute renal failure and respiratory failure. The risk of adverse outcomes increased with hyperglycemia, independent of confounding variables including body weight, prior diagnosis of diabetes, insulin therapy, blood sugar readings before TPN treatment, and frequency of blood sugar measurements (Lin *et al.* 2007). Clearly, patients in whom hyperglycemia develops on TPN are sicker, and it is likely that the presence and degree of hyperglycemia reflect illness severity. However, hyperglycemia itself is potentially contributing to a worse outcome. In a similar study with comparable results, when the data were examined by quartiles of blood glucose levels, the mortality of patients in the highest quartile was 11-fold that of those in the lowest quartile, and the risk of developing any complication was over 4-fold increased. These effects were independent of age, sex, or prior diabetes status (Cheung *et al.* 2005).

35.4.3   Effect of Hyperglycemia to Promote Infections

There are multiple mechanisms whereby hyperglycemia itself promotes further infectious complications. Abnormalities in polymorphonuclear function, immunoglobulin inactivation by glycosylation, reduction in oxidative burst and phagocytosis in macrophages have all been described (McCowen *et al.* 2001). Observational studies have consistently shown that hyperglycemia predicts subsequent infectious complications. Prospective randomized clinical trials have shown that treatment of even mild hyperglycemia in a SICU population leads to a reduction in new infectious complications, although this treatment cannot be globally recommended as an anti-infectious therapy due to the side effects of intensive insulin therapy (discussed in detail below).

35.5   Hyperglycemia-induced Muscle Catabolism

Aside from the propensity for hyperglycemia to increase infectious complications of TPN, uncontrolled hyperglycemia may also worsen catabolism. In a series of elegant studies, Wolfe and co-workers studied alanine efflux from the leg muscles of burn victims. In these critically injured patients, acute hyperglycemia (up to 400 mg/dl) increased muscle catabolism despite an endogenous insulin response. In contrast, exogenous insulin given in sufficient amount impeded muscle protein loss (Gore *et al.* 2002a). Subsequent patients were stratified by plasma glucose values at the time of metabolic measurements (*i.e.*, normal, glucose at $</=130$ mg/dL; mild hyperglycemia, glucose at 130–200 mg/dL; severe hyperglycemia, glucose at >200 mg/dL). All had similar caloric nutrition support and simultaneous measurements of indirect calorimetry and leg net balance of phenylalanine (as an index of muscle protein catabolism) were performed. Severe hyperglycemia was associated with significantly higher efflux of phenylalanine from the leg. Similarly, in a study of healthy humans, during hyperinsulinemia, acute elevations of plasma glucose to two times basal levels (191 mg/dl vs 94 mg/dl) result in a marked stimulation of whole body proteolysis despite hyperinsulinemia (Flakoll *et al.* 1993). These findings demonstrate an association between hyperglycemia and an increased rate of muscle protein catabolism in severely burned patients. This suggests a possible link between resistance of muscle to the action of insulin for both glucose clearance and muscle protein anabolism (Gore *et al.* 2002b).

35.5.1   Intensive Insulin Therapy in Patients Receiving Nutrition Support in the ICU

Prior to 2001, apathy reigned with respect to ICU-related hyperglycemia. TPN dextrose rates were often excessive. In 2001, a landmark study from Leuven, Belgium, demonstrated that targeting normal blood sugars in ICU patients (80–110 mg/dl) versus a more lax approach was associated with a mortality benefit and fewer infectious complications (van den Berghe *et al.* 2001). Much of the nutrition support in this trial was *via* the parenteral route, and most of the patients had undergone cardio-thoracic surgery. Many practitioners then instituted a policy of strict glucose control following this, while awaiting further data. However, none of the subsequent randomized clinical trials demonstrated this mortality benefit, and the largest of them (Finfer *et al.* 2009) showed higher mortality with intensive insulin therapy. Most of the trials that did not show tight glucose control to be beneficial were performed mainly under conditions of enteral nutrition in which case much higher rates of hypoglycemia would be expected, which might explain the higher mortality (Kavanagh and McCowen 2010). However, the one institution in which tight glucose control was associated with clinical benefit gave a substantial portion of the nutrition prescription parenterally (van den Berghe *et al.* 2001, 2006). On ICU admission, all patients were given IV dextrose for 24 hours, 200–300 g per day. Then nutrition support with TPN and or enteral nutrition was initiated, providing an average of 200–260 gram of dextrose per 24 hours. The aim was to switch to 100% enteral nutrition, but most patients continued to receive TPN. When intensive insulin therapy is co-administered with primarily enteral nutrition, rates of severe hypoglycemia (defined as <40 mg/dl) in prospective clinical trials have been staggeringly high – up to 28% in some (reviewed in Kavanagh and McCowen 2010). Under intensive insulin therapy, even when TPN is a primary source of dextrose energy, hypoglycemia is a major risk factor for death (Meyfroidt *et al.* 2010). Enteral nutrition is far more frequently interrupted than is parenteral, and this is likely a major contributor to hypoglycemia in real-world setting. However, in situations where TPN is needed, and leads to rise in blood sugars, intensive insulin infusions might be beneficial although the precise glycemic target remains unclear (Malhotra 2006). Expert recommendation is currently targeting glucose concentrations of 150–180 mg/dL (Kavanagh and McCowen 2010) (Table 35.3).

Countering this last hypothesis were two intriguing studies recently published. In the first of these, the authors hypothesized that TPN-related complications would be ameliorated by tight glucose control and be equivalent to

Table 35.3 Guidelines from professional organizations for blood sugar control in ICU patients. Most professional organizations no longer recommend strict glycemic control in ICU patients, however not all have updated their published guidelines.

Year	Professional Organization	Treatment Threshold (mg/dL)	Target Glucose Range (mg/dL)	Definition of hypoglycemia (mg/dL)
2009	*American Association of Clinical Endocrinologists-American Diabetes Association (*joint statement*)	180	140–180	<70
2009	†Surviving Sepsis Campaign	180	150	Not stated
2009	‡Institute for Healthcare Improvement	180	<180	<40
2007	††European Society of Cardiology	Not Stated	'strict'	Not stated

*http://www.aace.com/pub/pdf/guidelines/InpatientGlycemicControlConsensusStatement.pdf
†http://www.survivingsepsis.org/About_the_Campaign/Documents/
SSC%20Statement%20on%20Glucose%20Control%20in%20Severe%20Sepsis.pdf
‡http://www.ihi.org/IHI/Topics/CriticalCare/IntensiveCare/Changes/
ImplementEffectiveGlucoseControl.htm

those in enteral nutrition patients. A prospective cohort study of 155 patients admitted to a surgical ICU was conducted, comparing TPN –fed and enterally fed patients. Glycemic target in both groups was 80 to 110 mg/dL. Mean daily glucose values were lower for the TPN patients than for the enteral (118 *vs.* 126 mg/dL). Nonetheless, the incidence of bloodstream infection and catheter-related bloodstream infection was significantly higher in the TPN group. In a multivariate logistic regression model, TPN (compared with enteral nutrition) was associated with a >4-fold increase in the risk of having a catheter-related bloodstream infection (Matsushima 2010). Similarly a large study from Belgium, 4600 ICU patients were randomized to receive TPN either 2 days or 8 days after admission in addition to early enteral feeding as tolerated. TPN was used to bring the nutritional intake up to goal, and large disparities were seen by design between the two groups in nutritional intake between day 2 and day 8. The groups randomized to receiving "late TPN" had significantly fewer infections and an earlier discharge from the ICU than the group randomized to "early TPN". Normal glucose values were targeted (and presumably achieved) in both groups, eliminating hyperglycemia as a potential causative factor in the difference (Casaer *et al.* 2011). Clearly, more trials on this topic need to be conducted to determine whether it is the high intravenous dextrose infusion rates, the attained plasma glucose or some other factor altogether that permits higher infectious complication rates in TPN patients.

35.5.2 Glycemic Lability

Another concern about the use of intensive insulin therapy during nutrition support with TPN is the issue of glycemic variability, which has been defined in several ways, but primarily relates to glucose fluctuations. Variability may be calculated using the standard deviation around the mean of the measured glucose values, or is often defined as "mean amplitude of glycemic excursions". The importance of labile glycemic control is that such fluctuations are potentially more deleterious to the host defense than a stable glucose value with the same mean concentration. Markers of oxidative stress have been correlated with glycemic variability in patients with hyperglycemia, although there are limited studies in ICU patients at this time (Monnier *et al.* 2006). In an analysis of the studies from Leuven of intensive insulin therapy in which TPN was a major component of the nutrition prescription, glucose variability was increased by the insulin infusions (Meyfroidt *et al.* 2010).

35.5.3 Hypocaloric TPN

Hypocaloric dextrose infusions as part of TPN are increasingly recognized as a potentially beneficial strategy. Such therapy may be useful in obesity, or other situations where the risk of hyperglycemia is elevated (Patiño *et al.* 1999). Alternatively, in clinical situations where fluid restriction is imperative, only a small volume is allowed for nutrition support, thus hypoenergetic infusions are preferred. In prospective, randomized trials in hospitalized obese patients, a 50% reduction in dextrose (to provide 50% of measured energy expenditure as nonprotein calories) (versus 100%) did not show any negative outcomes. Protein was provided as 1.2 g/kg/day; and nitrogen balance was similar between the groups (Burge *et al*. 1994). Nitrogen balance was improved by a hypocaloric regimen in ventilated, medical ICU patients in a small study of 20 adults where a variety of TPN regimens were compared (Müller *et al*. 1995). However, in a small study of 40 hospitalized adults (only some in the ICU), the hypocaloric TPN regimen (1 L fluid, containing at goal 70 gram protein and 210 g dextrose) was associated with a more negative nitrogen balance than a weight based TPN regimen (25 kcal/kg), although insulin requirements were less in the experimental group (McCowen *et al*. 2000).

35.5.4   Insulin in TPN

In a patient known to have diabetes, or who has demonstrated hyperglycemia prior to initiation of TPN, insulin can simply be added to the TPN in the pharmacy. Additional insulin can be supplemented by subcutaneous injection, or by a continuous infusion if severe hyperglycemia develops. The usual initial regimen is 0.1 units of insulin per gram of dextrose, up to 0.2 units per gram in a hyperglycemic patient. Two-thirds of the previous day's subcutaneous need can be added to the TPN infusion on the following day. In patients with type 1 diabetes who are prone to wide glycemic excursions, an additional insulin drip is likely to produce smoother glycemic control than a subcutaneous supplement. Dextrose infusion rates from TPN should not be increased until some degree of control of the blood sugars has been achieved.

35.6   Complications Associated with Dextrose Infusions in Parenteral Nutrition

Excessive dextrose infusion rates as part of TPN have been associated with problems such as increased work of breathing, and of *de novo* lipogenesis, which leads to fat accumulation in the liver. In one representative study, 16 surgical ICU patients were randomized to receive isocaloric isonitrogenous TPN) containing either 75% (TPN-dextrose) or 15% (TPN-lipid) dextrose over 5 days (Tappy *et al*. 1998) (Table 35.4). Glucose metabolism and *de novo* lipogenesis were assessed using concurrent stable isotope infusions. Compared with TPN-lipid, TPN-dextrose increased plasma glucose more (by 26% *vs*. 7%), and increased total CO_2 more (by 15% *vs*. 0%). Fractional *de novo* lipogenesis was markedly increased by TPN-dextrose to 17% in comparison with 3% for TPN-lipid.

35.6.1   Liver Dysfunction

Hepatic dysfunction is relatively common among patients receiving TPN, especially for chronic bowel failure. Simple steatosis likely relates to

Table 35.4 Major clinical effects of acute hyperglycemia on different organ systems in critical illness. Hyperglycemia has multiple effects on a variety of organ systems that can compound critical illness.

Organ	Effect and Mechanism
Kidney	Diuresis, related to glycosuria. Leads to electrolyte losses
Immune system	Immune depression, increased chance of infection
Brain	Confusion, caused by hyperosmolarity and fluid shifts
Gut	Reduced intestinal motility related to neural dysfunction from glucose
Muscle	Proteolysis
Pancreas	Beta cell failure secondary to hyperglycemia
Liver	Steatosis
Heart	Myocardial depression (hyperglycemia), arrhythmias (potassium homeostasis)
Adipocytes	Lipolysis unchecked, allows ketogenesis and ketoacidosis if severe
Eyes	Visual disturbance related to hyperosmolarity
Lungs	Tachypnea if ketoacidosis develops

carbohydrate provision in excess. Energy excess plus an abundance of insulin promotes lipogenesis and inhibits fatty acid oxidation (similar to overfeeding through the enteral route). Such steatosis has become less common now that the risks of overfeeding and hyperglycemia have become more appreciated, but may eventually result in cirrhosis, just as can occur in the obese ambulatory patient. More serious is TPN-associated cholestasis, which is more common in children receiving TPN, but can cause fatalities in adults with short bowel syndrome on chronic TPN. Cholestasis typically presents with elevations in alkaline phosphatase and bilirubin and may progress to cirrhosis. The underlying mechanism is poorly understood and is not clearly related to dextrose infusion rates. Some studies suggest that infusion of lipids at rates $> 1 \, g/kg/d$ for prolonged periods is a risk factor, others that the degree of severity of the short bowel is the critical factor, or perhaps that underlying inflammation is a promotor (Chan *et al.* 1999) but most studies are too small for strong conclusions. In children, infusion of n-3 containing fatty acids has been associated with clinical improvements, but adults have been less well studied (Gura *et al.* 2006). In rats, either hypocaloric feeding as dextrose-based TPN or TPN provided at maintenance energy levels with the addition of fish oil reduces liver fat accumulation and limits evidence of hepatic and systemic injury found with higher dextrose infusion rates or TPN containing both dextrose and soybean oil (Ling *et al.* 2011). Patients with end-stage liver disease related to TPN and intestinal failure have very high mortality, with dual liver and intestine transplantation being the main therapy when cirrhosis is established.

35.6.2 Respiratory Insufficiency

Catabolism associated with critical illness results in muscle wasting, including respiratory muscles, such that malnutrition may hinder ventilation. However, injudicious feeding can cause further ventilator compromise. Infusion of dextrose in excess leads to increases in CO_2 production; glucose oxidation produces more CO_2 than fat oxidation (Tappy *et al.* 1998). This rise in CO_2 production requires increased minute ventilation, which might result in ventilator dependence. Use of slow lipid infusions to replace some of the dextrose, while keeping the total energy allotment low, is recommended under these circumstances.

35.6.3   Essential Fatty Acid Deficiency (EFAD)

EFAD is a potential complication of dextrose-containing TPN that is infused continuously without any fat emulsions for a period of > 1–2 weeks. Linoleic and alpha-linolenic acid are both essential in that they cannot be synthesized *in vivo* (McCowen and Bistrian 2005). Scaly dermatitis, bone marrow dysfunction and fatty liver can result. Usually with simple starvation or malnutrition, EFAD does not occur when TPN is not infused, because even if there are no sources of fat in the diet, lipolysis of endogenous fat stores can provide sufficient essential fatty acids. However, with ongoing infusion of dextrose, circulating plasma insulin concentrations are such that lipolysis may be relatively inhibited. EFAD, when suspected, can be confirmed by measurement of triene/tetraene ratio. Prevention of EFAD in a patient receiving TPN can be easily prevented by giving 1% of energy as linoleic acid, and 0.5% as alpha-linolenic acid. As little as 500 ml of a 20% lipid infusion per week will suffice for this purpose.

35.6.4   Refeeding Syndrome

The delivery of carbohydrate energy may result in "refeeding syndrome" in malnourished patients. This is particularly likely in cases of cachexia where initial dextrose infusion exceeds 200 g/24 hours. A pronounced insulin response to the dextrose can occur, which leads to electrolyte shifts – hypophosphatemia, hypokalemia and hypomagnesemia; this can be fatal. Phosphate, potassium and magnesium are all used up intracellularly in the anabolic process that is permitted with feeding, and the shift from extra- to intracellular space under insulin stimulation. Insulin also has potent anti-natiuretic properties such that salt and water retention promote congestive heart failure in susceptible individuals.

35.7   Conclusions and Recommendations

While TPN remains a controversial form of nutrition, undoubtedly it is life-saving for many patients. In expert hands, such infusions can be administered safely. The major principles of TPN therapy to minimize occurrence of hyperglycemia (which itself can reduce many of the benefits of nutrition support) are summarized in Table 35.5. A dedicated nutrition support team is likely to allow beneficial effects of nutrition support to be realized, while complications can be avoided.

Table 35.5 The major principles of dextrose administration as part of TPN therapy to minimize occurrence of hyperglycemia. TPN can be safely administered without hyperglycemia if the nutrition support team uses sensible dextrose infusion rates and adequate protein.

- Calculate energy goal as 25 kcal/kg usual weight (use ideal weight if obese)
- Protein 1.5 g/kg usual weight (2 g/kg ideal weight if obese)
- Lipid emulsions can be considered as an alternative energy source to dextrose (up to 30% of total kcal), particularly if complications arise with infusion of concentrated dextrose
- Initiate TPN with maximum of 100–150 g dextrose/24 h
- Add regular insulin to TPN if already hyperglycemic (suggest 50% of previous day's insulin requirement) and plan for q4–6 h subcutaneous insulin in addition to treat established hyperglycemia.
- Advance dextrose infusion rates by ∼50 g per 24 h if euglycemia is present
- Maximum dextrose infusion rates of 4 mg/kg/min

Summary Points

- Dextrose (a name usually applied to glucose in solution or used synonymously with glucose) is the major component of TPN, providing calories in patients with critical illness and intestinal failure, or patients with chronic malabsorption.
- A large amount of dextrose infused in a susceptible host may lead to severe hyperglycemia requiring insulin therapy. Blood sugars should always be monitored and frequently in patients newly started on TPN.
- In critically ill persons, even infusion of dextrose and amino acids as TPN cannot completely prevent the catabolic sequelae of the illness, although the degree of wasting is blunted by TPN.
- Other side effects of TPN infusion include fatty liver, which might contribute to cirrhosis in patients receiving long-term TPN, and increased work of breathing, which might increase ventilator dependency in ICU patients.
- TPN can be safe and effective therapy if initiated with low dextrose infusion rates, and careful monitoring by experienced clinicians, but the potential for harm is great.

Key Facts

- Concentrated dextrose infusion as a component of TPN can be life-saving in many clinical situations.
- Enteral feeding is generally preferred over TPN, however intestinal failure is common in critical illness.
- Hyperglycemia should be prevented during TPN, both by careful infusion of dextrose initially, as well as addition of insulin. Goal plasma glucose values are 150–180 mg/dL.
- Complications of dextrose infusion include fatty liver and ventilatory dysfunction.

Definitions of Words and Terms

Total parenteral nutrition. This is the provision of all of a person's nutrition through a vein, rather than by the gastro-intestinal tract. There are 2 main patient populations, the usual is a critically ill hospitalized patient, with temporary intestinal failure. A smaller patient group has chronic intestinal failure, usually related to bowel resection or other form of severe malabsorption. These patients can eat but do not absorb the nutrition, and they usually infuse their nutrition solution overnight during sleep. In all cases, a large central vein is used.

Enteral nutrition. This is feeding of nutrients using the gastrointestinal tract, but usually suggests use of a feeding tube that leads directly into the stomach, or a more distal area in the small intestine. Theoretically, this route of feeding results in presentation of nutrients in a more physiological manner, although many patients have intolerance related to their underlying illnesses.

Incretins. Incretins are hormones released from the gastrointestinal tract that have multiple effects to minimize perturbations in blood glucose after feeding. They include Glucagon-like peptide 1 and gastric inhibitory peptide and together they slow gastric emptying, reduce glucagon release and enhance glucose-induced insulin release from the beta cells of the pancreas.

Liver steatosis. This refers to fat deposition in the form of triglyceride in the liver and can be associated with liver dysfunction. It is common in obesity, type 2 diabetes, and other situations of either energy excess or insulin resistance. There are multiple pathways leading to this condition. Uninterrupted, end-stage liver disease can result.

List of Abbreviations

TPN   total parenteral nutrition
ICU   intensive care unit
EFAD   essential fatty acid deficiency

References

Burge, J.C., Goon, A., Choban, P.S., and Flancbaum, L., 1994. Efficacy of hypocaloric total parenteral nutrition in hospitalized obese patients: a prospective, double-blind randomized trial. *JPEN J Parenter Enteral Nutr*. 18: 203–7.

Casaerm, M.P., Mesotten, D., Hermans, G., Wouters, P.J., Schetz, M., Meyfroidt, G., Van Cromphaut, S., Ingels, C., Meersseman, P., Muller, J., Vlasselaers, D., Debaveye, Y., Desmet, L., Dubois, J., Van Assche, A., Vanderheyden, S., Wilmer, A., and Van den Berghe, G., 2011. Early versus Late Parenteral Nutrition in Critically Ill Adults. *N Engl J Med*. 365: 506–17.

Chan, S., McCowen, K.C., Bistrian, B.R., Thibault, A., Keane-Ellison, M., Forse, R.A., Babineau, T., and Burke, P., 1999. Incidence, prognosis, and etiology of end-stage liver disease in patients receiving home total parenteral nutrition. *Surgery.* 126: 28–34.

Cheungm, N.W., Napier, B., Zaccaria, C., and Fletcher, J.P., 2005. Hyperglycemia is associated with adverse outcomes in patients receiving total parenteral nutrition. *Diabetes Care.* 28: 2367–71.

Finfer, S., Chittock, D.R., Su, S.Y., Blair, D., Foster, D., Dhingra, V., Bellomo, R., Cook, D., Dodek, P., Henderson, W.R., Hébert, P.C., Heritier, S., Heyland, D.K., McArthur, C., McDonald, E., Mitchell, I., Myburgh, J.A., Norton, R., Potter, J., Robinson, B.G., and Ronco, J.J., 2009. Intensive versus conventional glucose control in critically ill patients. *N Engl J Med.* 360: 1283–97.

Flakoll, P.J., Hill, J.O., and Abumrad, N.N., 1993. Acute hyperglycemia enhances proteolysis in normal man. *Am J Physiol.* 265(5 Pt 1): E715–21.

Gore, D.C., Wolf, S.E., Herndon, D.N., and Wolfe, R.R., 2002a. Relative influence of glucose and insulin on peripheral amino acid metabolism in severely burned patients. *JPEN J Parenter Enteral Nutr.* 26: 271–7.

Gore, D.C., Chinkes, D.L., Hart, D.W., Wolf, S.E., Herndon, D.N., and Sanford, A.P., 2002b. Hyperglycemia exacerbates muscle protein catabolism in burn-injured patients. *Crit Care Med.* 30: 2438–42.

Gura, K.M., Duggan, C.P., Collier, S.B., Jennings, R.W., Folkman, J., Bistrian, B.R., and Puder, M., 2006. Reversal of parenteral nutrition-associated liver disease in two infants with short bowel syndrome using parenteral fish oil: implications for futuremanagement. *Pediatrics.* 118: e197–201.

Kavanagh, B.P., and McCowen, K.C., 2010. Clinical practice. Glycemic control in the ICU. *N Engl J Med.* 363: 2540–6.

Kondrup, J., Allison, S.P., Elia, M., Vellas, B., and Plauth, M., 2003. ESPEN guidelines for nutrition screening 2002. *Clin Nutr.* 22: 415–21.

Lidder, P., Flanagan, D., Fleming, S., Russell, M., Morgan, N., Wheatley, T., Rahamin, J., Shaw, S., and Lewis, S., 2010. Combining enteral with parenteral nutrition to improve postoperative glucose control. *Br J Nutr.* 103: 1635–41.

Lin, L.Y., Lin, H.C., Lee, P.C., Ma, W.Y., and Lin, H.D., 2007. Hyperglycemia correlates with outcomes in patients receiving total parenteral nutrition. *Am J Med Sci.* 333: 261–5.

Ling, P.R,, Andersson, C., Strijbosch, R., Lee, S., Silvestri, A., Gura, K.M., Puder, M., and Bistrian, B.R., 2011. Effects of glucose or fat calories in total parenteral nutrition on fat metabolism and systemic inflammation in rats. *Metabolism.* 60: 195–205.

Malhotra, A., 2006. Intensive insulin in intensive care. *N Engl J Med.* 354: 516–8.

Martindale, R.G., McClave, S.A., Vanek, V.W., McCarthy, M., Roberts, P., Taylor, B., Ochoa, J.B., Napolitano, L., and Cresci, G., 2009. Guidelines for the provision and assessment of nutrition support therapy in the adult critically ill patient: Society of Critical Care Medicine and American Society for Parenteral and Enteral Nutrition: Executive Summary. *Crit Care Med.* 37: 1757–61.

Matsushima, K., Cook, A., Tyner, T., Tollack, L., Williams, R., Lemaire, S., Friese, R., and Frankel, H., 2010. Parenteral nutrition: a clear and present danger unabated by tight glucose control. *Am J Surg.* 200: 386–90.

McCowen, K.C., Friel, C., Sternberg, J., Chan, S., Forse, R.A., Burke, P.A., and Bistrian, B.R., 2000. Hypocaloric total parenteral nutrition: effectiveness in prevention of hyperglycemia and infectious complications--a randomized clinical trial. *Crit Care Med.* 28: 3606–11.

McCowen, K.C., Malhotra, A., and Bistrian, B.R., 2001. Stress-induced hyperglycemia. *Crit Care Clin.* 17: 107–24.

McCowen, K.C., and Bistrian, B.R., 2005. Essential fatty acids and their derivatives. *Curr Opin Gastroenterol.* 21: 207–15.

Meyfroidt, G., Keenan, D.M., Wang, X., Wouters, P.J., Veldhuis, J.D., and Van den Berghe, G., 2010. Dynamic characteristics of blood glucose time series during the course of critical illness: effects of intensive insulin therapy and relative association with mortality. *Crit Care Med.* 38: 1021–9.

Monnier, L., Mas, E., Ginet, C., Michel, F., Villon, L., Cristol, J.P., and Colette, C.. 2006. Activation of oxidative stress by acute glucose fluctuations compared with sustained chronic hyperglycemia in patients with type 2 diabetes. *JAMA.* 295: 1681–7.

Müller, T.F., Müller, A., Bachem, M.G., and Lange, H., 1995. Immediate metabolic effects of different nutritional regimens in critically ill medical patients. *Intensive Care Med.* 21: 561–6.

Patiño, J.F., de Pimiento, S.E., Vergara, A., Savino, P., Rodríguez, M., and Escallón, J., 1999. Hypocaloric support in the critically ill. *World J Surg.* 23: 553–9.

Rosmarin, D.K., Wardlaw, G.M., and Mirtallo, J., 1996. Hyperglycemia associated with high, continuous infusion rates of total parenteral nutrition dextrose. *Nutr ClinPract.* 11: 151–6.

Shaw, J.H., and Wolfe, R.R., 1986. Glucose, fatty acid, and urea kinetics in patients with severe pancreatitis. The response to substrate infusion and total parenteral nutrition. *Ann Surg.* 204: 665–72.

Simpson, F., and Doig, G.S., 2005. Parenteral vs. enteral nutrition in the critically ill patient: a meta-analysis of trials using the intention to treat principle. *Intensive Care Med.* 31: 12–23.

Tappy, L., Schwarz, J.M., Schneiter, P., Cayeux, C., Revelly, J.P., Fagerquist, C.K., Jéquier, E., and Chioléro, R., 1998. Effects of isoenergetic glucose-based or lipid-based parenteral nutrition on glucose metabolism, de novo lipogenesis, and respiratory gas exchanges in critically ill patients. *Crit Care Med.* 26: 860–7.

The Veterans Affairs Total Parenteral Nutrition Cooperative Study Group. 1991. Perioperative total parenteral nutrition in surgical patients. *N Engl J Med.* 325: 525–32.

van den Berghe, G., Wouters, P., Weekers, F., Verwaest, C., Bruyninckx, F., Schetz, M., Vlasselaers, D., Ferdinande, P., Lauwers, P., and Bouillon, R., 2001. Intensive insulin therapy in the critically ill patients. *N Engl J Med.* 345: 1359–67.

Van den Berghe, G., Wilmer, A., Hermans, G., Meersseman, W., Wouters, P.J., Milants, I., Van Wijngaerden, E., Bobbaers, H., and Bouillon, R., 2006. Intensive insulin therapy in the medical ICU. *N Engl J Med.* 354: 449–61.

Wolfe, R.R., 1996. Herman Award Lecture, 1996: relation of metabolic studies to clinical nutrition--the example of burn injury. *Am J Clin Nutr.* 64: 800–8.

CHAPTER 36

The Intestinal Transport of Galactase

MARÍA JESÚS RODRÍGUEZ YOLDI

Physiology Unit, Department of Pharmacology and Physiology, Veterinary Faculty, University of Zaragoza, Miguel Server 177, E.-50013, Zaragoza (Spain)
Email: mjrodyol@unizar.es

36.1 Galactose

Dietary carbohydrate in humans and omnivorous animals is a major nutrient and the alimentary tract is well adapted for its digestion and subsequent absorption (Figure 36.1).

The carbohydrates that we ingest range from simple monosaccharides (glucose, fructose and galactose) to disaccharides (lactose, sucrose) to complex polysaccharides. Most carbohydrates are digested by salivary and pancreatic amylases, and are further broken down into monosaccharides by enzymes in the brush border membrane (BBM) of enteroctes. Once monosaccharides are presented to the BBM, mature enterocytes expressing transporters to transport the sugars into the enterocytes (Drozdowski and Thomson 2006).

Dietary molecules as large as disaccharides do not cross the small intestinal epithelium, although larger molecules such as polyethylene glycols can be absorbed and excreted into the urine. It is the high concentration and hydrolytic efficiency of the disaccharidase in the brush border that effectively hydrolyze all the disaccharides, leaving none to pass across intact. The hydrolysis of lactose into monosaccharides, glucose and galactose, is crucial for

Food and Nutritional Components in Focus No. 3
Dietary Sugars: Chemistry, Analysis, Function and Effects
Edited by Victor R Preedy
© The Royal Society of Chemistry 2012
Published by the Royal Society of Chemistry, www.rsc.org

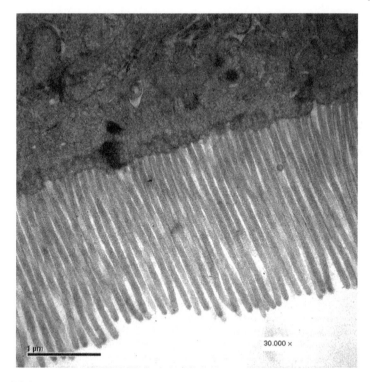

Figure 36.1   Electron microscopy image of the small intestine taken from rabbits.
Jejunum microvilli at × 30.000 magnification. Bars corresponds to 1 μm.
The image is from our group (unpublished).

the nutrition of humans and they are the fuel that provides energy for normal
activity in humans. Lactose is found primarily in milk and milk products.
Glucose is the major form of absorbed carbohydrate, and glucose also serves as
a signal for activation of numerous regulatory events (Freeman *et al.* 2006).

Galactose metabolismo, which coverts galactose into glucose, is carried out
by the three principal enzymes: galactokinase (GALK), galactose-1-phosphate
uridyltransferase (GALT), and UDP-galactose-4'-epimerase (GALE).

Galactose is also found in dairy products, sugar beets, and other gums and
mucilages. It is synthesized by the body, where it forms part of glycolipids and
glycoproteins in several tisues (Figure 36.2).

36.2   SGLT1 and GLUT2, D-Galactose Intestinal Transporters

Intestinal sugar delivery depends on the levels of expression of dissacharidases
(*i.e.*, sucrase-isomaltase, lactase) and sugar transporters (Le Gall *et al.* 2007).

The rate of absorption and the intestinal region where sugars are absorbed
affects the time course of appearance of sugars in the blood, and the availability

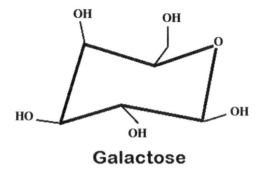

Galactose

Figure 36.2 Molecule of galactose. Cyclic form of hexose D-galactose. The image is from our group (unpublished).

Table 36.1 Intestinal transporters. Transporters of sugars in the small intestine on brush-border (BBM) or basolateral membrane (BLM).

	BBM	*BLM*
D-Galactose	SGLT1	GLUT2
D-Glucose	SGLT1	GLUT2
D-Fructose	GLUT5	GLUT2

of those sugars to other parts of the body. Absorptive systems in the small intestine therefore influence plasma sugar concentrations and play a vital role in nutrition.

Intestinal sugar transporters are responsible for transporting the monosaccharides glucose, galactose and fructose from the intestinal lumen to the blood. SGLT1 is located in the brush-border or apical membrane, and transports glucose and galactose from the intestinal lumen into the cytosol. GLUT5 is also apical, a unique member of the ubiquitous facilitative glucose transporter family, and trasnports fructose from the lumen into the cytosol. GLUT2 is basolateral transports all three monosaccharides from the cytosol to the blood (Table 36.1, Figure 36.3) (Ferraris 2001).

SGLT1 is a member of a very large solute carrier family (SLC5) which transports various solutes into cells using the Na^+ electrochemical potential gradient across the plasma membrane. SGLT1 strongly discriminates among monosaccharides, D-glucose and D-galactose being the natural substrates (Wright and Turk 2004).

Intestinal sugar absorption in many species is Na^+-dependent and crucially mediated by SGLT1. Rabbit SGLT1 was first cloned in 1987 (Hediger *et al.* 1987) and since then has been studied extensively. It has been proposed that SGLT1 contains 14 transmembrane α-helices with both the N- and C-termini facing the extracellular compartment (Lin *et al.* 1999). With regard to the structure-function relationship, the N-terminal half of SGLT1 participates in Na^+ binding while the C-terminal domain, particulary helices 10–13 of the

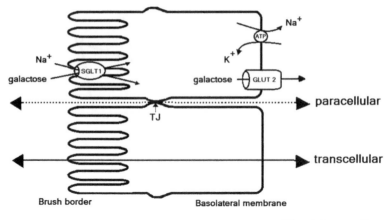

Figure 36.3 Classical model of intestinal sugar transport. The hexose galactose has
entry across the brush-border membrane (BBM) mediated by the
sodium-dependent sugar transporter (SGLT1). Exit of hexose across the
basolateral membrane (BLM) into the blood stream is mediated by
facilitated transporter, GLUT2. The Na^+/K^+-ATPase on the BLM
maintains the gradient necessary for the functioning of SGLT1. The
image is from our group (unpublished).

protein, participates in sugar transport (Nagata and Hata 2006) (Figure 36.4).
Cotransport is supposed to be initiated when two external Na^+ ions bind to the
SGLT1 and induce structural alterations in the protein, which allow sugar
binding, followed by the simultaneous translocation of sodium and sugar
across the membrane (Wright 2001). The energy stored in the sodium electrochemical potential gradient across the brush-border membrane provides the
energy to drive sugar accumulation in the enterocyte against its concentration
gradient. Sodium that enters the cell along with glucose or galactose is then
transported out into blood by the Na^+/K^+ pump in the basolateral membrane
(BLM), thereby maintaining the driving force for sugar transport. As the sugar
accumulates within enterocytes, this sets up a driving force to transport glucose
or galactose out of the cells into the blood via GLUT2 espressed in basolateral
membranes. The net result is that one mole of sugar and two moles of sodium
are transported across the enterocyte from gut lumen into blood, and this is
followed by two moles of anions to ensure electroneutrality, and water. The
energy for the overall process comes from the ATP consumed by the basolateral Na^+/K^+ pump (Wright *et al.* 2007). The Na^+/K^+-ATPase in the BLM is
responsible for maintaining the Na^+ and K^+ electrochemical gradient across
the cell membrane. This ATPase is up-regulated in experimental diabetes (Wild
et al. 1999) and experimental ileitis (Wild and Thomson 1995), with post-
translational modifications playing an important role in its regulation. This up-
regulation may influence the functioning of SGLT1 and subsequently alter
intestinal sugar uptake in these conditions.

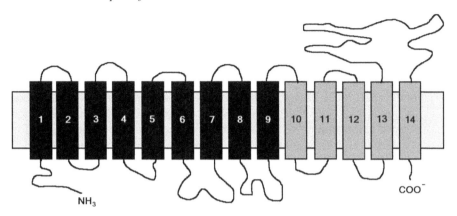

Figure 36.4 Image of SGLT1: a secondary structure model for SGLT1. The protein has 14 transmembrane helices with extracellular N-and C-terminal. The image is from our group (unpublished).

Na^+ and sugar cotransport by SGLT1 is referred to as secondary active transport because the driving forces – Na gradients – are maintained by the primary active Na^+/K^+-pump, or Na^+/K^+-ATPase. Quite simply, the direction and rate of sugar transport by SGLT1 are a function of the direction and magnitude of the Na-gradients across the plasma membrane. In normal cells, it is the Na^+/K^+-pump that sets the direction and magnitude of the sodium gradients. The beta-glucoside phlorizin is a potent competitive inhibitor of $Na^+/$glucose transport (Wright *et al.* 2007).

Protons can substitute equally well for Na^+ to drive glucose or galactose transport but the affinity for sugar is about an order of magnitude lower than in Na^+. This may be of physiological significance in the proximal duodenum, where the chyme has an acid pH and high sugar concentration D-glucose and D-galactose are transported equally well by SGLT1 but fructose is not transported (Wright *et al.* 2007).

GLUT2 is a low-affinity, high-capacity facilitative transporter in the basolateral membrane (BLM) that transports glucose, fructose and galactose. It has 12 trnasmembrane domains, with intracellular N and C terminals (Cheeseman 1993).

In the case of a sugar rich meal, GLUT2 can be recruited into the apical membrane where it complements SGLT1 and GLUT5 transport capacities (Kellet and Brot-Laroche 2005). Furthermore, the small intestine adapts to repeated sugar ingestion by increasing transporter expression over the course of a few days (Goda 2000).

Hence, GLUT2, depending on its relative abundance in the apical and basolateral membranes, is likely to trigger sugar signals from the blood or the intestinal lumen content. Whereas GLUT2 may be transiently expressed in the apical enterocytes membrane, it is permanently high in the basolateral membrane, and may thereby determine the polarity of sugar detection.

On the other hand, there is evidence to support the hypothesis that SGLT3, a member of the SLC5 family, is involved in glucose-sensing in the wall of the gut

(Freeman *et al.* 2006). Sensing of glucose at the brush border membrane of the intestine is important to mount appropriate reflex control of gastrointestinal secretion and motility in the post-prandial period. Glucose-mediated inhibition of gastric motility is mediated via a vago-vagal reflex pathway involving 5-HT3Rs located on vagal afferent nerve terminals in the intestinal mucosa (Raybould *et al.* 2003). Galactose is ineffective in stimulating vagal afferent fiber activity and it is less effective than glucose in releasing 5-HT. Because galactose has a much lower affinity for SGLT3 than glucose, yet the same affinity for SGLT1, these results suggest that the mechanism involved in detection of glucose in the intestinal mucosa involves SGLT3 rather than SGLT1.

36.2.1 Functional Disorders of SGLT1

Na^+/sugar co-transporters are prototypes of secondary active transporters that drive the accumulation of molecules into cells. These transporters have critical roles in human physiology, where mutations in their genes are responsible for severe congenital diseases (Wright *et al.* 2007) and are the molecular targets for drugs to treat diabetes and obesity (Isaji 2007).

Glucose-galactose malabsorption (GGM) is a very rare autosomal recesive disease characterized by severe life-threatening diarrhea in the neonate that resolves when the offending sugars (glucose, galactose and lactose) are removed from the diet (Wright *et al.* 2002). Normal intestinal mucosal histology is observed, while phlorizin binding studies show reductions in SGLT1 protein in the BBM. Electrophysiological studies and freeze fracture electron microscopy have shown that this disease is due to a failure of the SGLT1 protein to traffic normally to the BBM (Martin *et al.* 1997).

SGLT1 is also used as a potential therapeutic target for obesity and associated type 2 diabetes mellitus, highlighted by emerging evidence for dysregulation of SGLT1 in these disease states (Dyer *et al.*, 2002; Osswald *et al.*, 2005).

36.3 Factors Involved in Galactose Intestinal Transport

A number of factors are involved in the regulation of intestinal sugar transport. These factors may modify galactose transport by altering the abundance of sugar transporters in the intestine. Alternatively, sugar transport may be regulated at an entirely different level. The intrinsic activity of the transporters (amount of substrate transported per unit of transporter protein) may be altered, in the absence of detectable changes in transporter abundance. The post-translational mechanism by which intrinsic activity is regulated may involve phosphorylation or dephosphorylation of the transporter or the activation or inhibition of the transporter by a regulatory protein.

36.3.1 Proteins

The role of SGLT1 phosphorylation was studied. Protein kinase A (PKA) activation increased glucose transport by approximately 30%, while protein kinase C (PKC) activation reduced transport by 60%. The change in maximal transport rates was accompanied by alterations in the number of transporters in the plasma membrane, as well as changes in the surface area of the membrane. Since endocytosis and exocytosis alter the membrane surface area, the findings of the effects of PKA and PKC on SGLT1 suggest that these proteins may be involved in the regulation of active sugar transport (Wright *et al.* 1997).

RS1 is an intracellular protein that is involved in the regulation of the Na^+-sugar cotransporter SGLT1. RS1 contains consensus sequences for protein kinase C and casein kinase II and an ubiquitin-associated domain that is conserved between different species. The RS1 protein is localized intracellularly and associated with the plasma membrane. Human RS1 is involved in post-transcriptional down-regulation of SGLT1 and increased by activation of PKC. RS1 participates in short-term regulation of SGLT1 by inhibiting the exocytotic pathway and modulates dynamin-dependent trafficking to the BBM of intracellular vesicles containing SGLT1 (Veyhl *et al.* 2006, Osswald *et al.* 2005).

These researchers speculated that therapeutic strategies aimed at reducing glucose uptake by increasing RS1 might potentially be used to treat obesity.

The transcription factors hepatocyte nuclear factor-1 (HNF-1) and Sp1-multigene member may also regulate SGLT1. These factors have enhancing the basal level of its transcription in Caco-2 cells (Martin *et al.* 2000).

On the other hand, *Foxl1* is a winged-helix transcription factor expressed in the mesenchymal cells bordering the crypts in the small intestine. The effect of the loss of Foxl1 on SGLT1 was specific due to decreased production of SGLT1 protein in the small intestine (Katz *et al.* 2004).

Kellet's working hypothesis proposes that before a meal, when luminal concentrations of glucose are low, GLUT2 levels in the BBM are also low, which would minimize the escape of glucose from the cell. Once a meal is ingested and BBM enzymes hydrolyse disaccharides, luminal glucose concentrations increase. Glucose uptake *via* SGLT1 causes increases in enterocyte volume due to a rise in osmolarity (and the co-transport of water molecules by SGLT1), and may trigger the entry of Ca^{2+}, activating PKCβII and promoting the insertion of GLUT2 in the BBM (Kellett and Helliwell 2000).

Helliwell *et al.* (2003) established a role for the phosphoinositol 3-kinase (PI-3K) and the mammalian target of Rapamycin (mTOR) pathways in the phosphorylation, turnover and degradation of PKCβII. Using an *in vivo* perfusion model, they showed that inhibitors of these pathways block GLUT2 trafficking to the BBM and inhibit sugar absorption. In their model, they suggest that as sugar absorption increases, the plasma sugar concentration increases, stimulating the release of insulin, which activates PI 3-K, resulting in the phosphorylation of PKCβII. They also proposed a model by which amino acids promote the formation of competent PKCβII by activating the mTOR pathway, which prevents dephosphorylation of PKCβII. Thus, the dynamic control of intestinal sugar absorption may be achieved by the rapid turnover and degradation of PKCβII.

The trafficking of GLUT2 to the BBM may represent a mechanism by which sugar absorptive capacity is matched to dietary intake.

On the other hand, Walker *et al.* (2005) demonstrated that the activation of AMPK resulted in the recruitment of GLUT2 to the BBM and a down-regulation of the energy-requiring SGLT1-mediated sugar uptake.

36.3.2   Hormones

Intestinal galactose is actively transported by the Na^+/sugar cotransporter (SGLT1) and passively by GLUT2. Moreover, it is also becoming increasingly evident that the gut is not just site of nutrient absorption but is also an active endocrine organ (Kellett *et al.* 2008). A paracrine regulation of hexose absorption by intestinal hormones such as glucagon-like peptide 2 (GLP-2) in relation to GLUT2 insertion on intestinal brush-border membrane has been shown (Au *et al.* 2002)

Cholecystokinin (CCK) is a gut hormone secreted from cells of the intestinal mucosa and is known to play a significant role in many physiological processes.

Cholecystokinin-8 (CCK-8) rapidly inhibits SGLT1 by reducing its abundance at the apical membrane which trafficking is regulated on a time scale of minutes by glucose. However, CCK-8 does not seem to be involved in apical GLUT2 translocation (Hirst and Cheeseman, 1998).

Vasoactive intestinal peptide (VIP) is widely distributed through out the body but is most highly concentrated in the nervous system and the gut. Studies, *in vivo* and *in vitro*, in rabbit jejunum suggest that the VIP inhibitory action on D-galactose intestinal absorption is due to the binding of VIP to receptors on enterocyte membranes and to a consequent increase in intracellular cAMP (Arruebo *et al.* 1990; Rodriguez-Yoldi *et al.* 1988,). In the same way, serotonin (5-hydroxytryptamine, 5-HT) present in the enterochromaffin cells, neurons, and mast cells in the lamina propia, inhibits D-galactose absorption in rabbit jejunum but no affects the activity of Na^+/K^+-APTase (Salvador *et al.* 2000).

Indeed, certain gastro-intestinal peptides such as leptin, have a mucosal effect on hexose transport. This peptide, released from adipose tissue and stomach in response to nutrient ingestion, is delivered to the circulation and to the small intestinal lumen. Leptin receptors are expressed at the apical and basolateral membranes of enterocytes and Iñigo *et al.* (2007) found that luminal leptin inhibited intestinal sugar absorption at low galactose concentration, which indicates that leptin regulates SGLT1 activity *in vivo*. The inhibition was reversed in the absence of hormone in the intestinal lumen, suggesting that it was produced by post-translational regulation processes.

36.3.3   Stress and Glucocorticoids

A clasical response to chronic stress is elevated plasma corticosterol, which is associated with gastrointestinal disorders of various intensities. Environmental stress created by perturbed housing conditions inhibited the apical GLUT2 component of glucose absorption while preserving the SGLT1 component (Shepherd *et al.*, 2004).

36.3.4 Diet

The absorption of hexoses by the small intestine in mammalian omnivores is a very efficient process which has a safety factor to ensure that hexoses do not spill over into the colon. The arrival of sugars in the lower bowel can produce diarrhoea resulting from an osmotically induced back-flux of water into the lumen or as the consequence of the increased nutrients driving a bacterial over-growth. One mechanism designed to avoid this situation is the ability of the small intestine to increase its absorptive capacity for hexoses when their dietary load increases. In this way, alterations in dietary carbohydrate are sustained for three or more days, modified the expression of the sugar transport proteins in parallel and it was independent of differences in villus morphology (Ferraris 2001).

The time course of regulation by dietary Na^+ is similar to that of regulation by dietery carbohydrate. Sodium depletion reduces intestinal glucose and galactose transport within a day after consumption of a low-salt diet, and the decrease reaches a maximum within 2 days (De la Horra *et al.* 2001).

Adenosine is a conditionally essential nutrient for the intestine but also functions as an ubiquitous host signalling molecule that regulates the intestinal functions. Kimura *et al.* (2005) indicate that luminal adenosine causes a rapid increase in SGLT1 that is of clinical relevance and acts via receptors linked to a signalling pathway that involves intracellular cAMP production.

As we have shown above, Kellet and Helliwell (2000) indicate that the expression of GLUT2 in the apical membrane is highly regulated through the activation of PKCβII when SGLT1 transports glucose. GLUT2 in the BBM would provide a high capacity, low affinity pathway for the entry of glucose, galactose and fructose into the epithelium. Both SGLT1 and GLUT2 are highly regulated by the insertion or removal of transporter protein from the BBM in response to the high presence of hexoses in the lumen.

36.3.5 Heavy Metals

Heavy metals are absorbed in variable proportion across the intestine by different mechanisms. Retention in the intestinal mucosa is very usual, by fixation to a wide spectrum of cell constituents including proteins at the membrane level or cytosol. Thiol groups of proteins seem to be the prevailing chemical groups involved in the binding of heavy metals.

Heavy metals (cadmium, zinc, copper and mercury) appear to inhibit D-galactose intestinal absorption by their binding to proteins (prevailing to thiol groups) of the luminal membrane of enterocytes, which pertain or is functionally related to the Na^+-sugar cotransport system (Rodriguez-Yoldi *et al.* 1989, 1992).

36.3.6 Infection

The main role of the intestinal epithelium is the absorption of nutrients from the gut lumen to the circulation. In addition, it acts as a barrier, preventing the passage of pathogens from the lumen to the bloodstream. Effectively, inflammatory and infectious diseases that affect the gastrointestinal tract can induce changes in the intestinal absorption of nutrients.

The family of proteins called RELMs (resistin-like molecules) has been reported to be involved with insulin resistance, diabetes, and inflammatory processes. RELM-β is a protein homologous to resistin that is localized mainly within the digestive tract and acts as a hormone. RELM-β is highly expressed in goblet cells and is secreted in response to bacterial colonization. Glucose reduces the enterocyte expression of RELM-β, while insulin and tumor necrosis

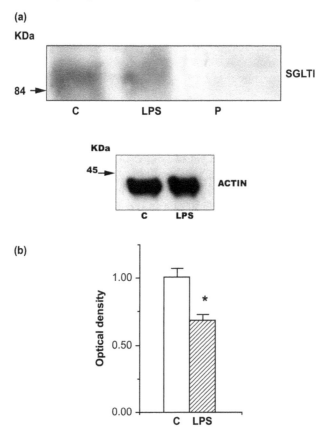

Figure 36.5 Effect of Lipopolysaccharide (LPS) on SGLT1 protein expression in brush-border membrane (BBM). (a) The antibody recognized an immunoreactive protein of about 84 kDa in the lanes from control (C) and treated (LPS) animals. When the antibody was previously adsorbed with the antigenic peptide, no signal was detected (P). Actin was used as a loading control of total protein onto the electrophoresis gels. (b) Relative abundance of SGLT1 protein measured by optical density (counts/mm^2). Values represent percentage means recorded in five separate experiments for each group (control and LPS-treated). *p < 0.05 with respect to control animals.

Reproduced with permission from Amador *et al.* (2007a), *Journal of Membrane Biology*. 215, 125–133. Springer.

factor-alpha (TNFα) can upregulate its expression (Fujio *et al.* 2008). This suggests that intestinal RELM-β may not only be associated with inflammation but can also be a regulator of energy homeostasis.

Krimi *et al.* (2009) show that luminal RELM-β inhibited SGLT1 activity in line with a diminished SGLT1 abundance in BBMs. Further, the potentiating effect of RELM-β on jejunal glucose uptake was associated with an increased

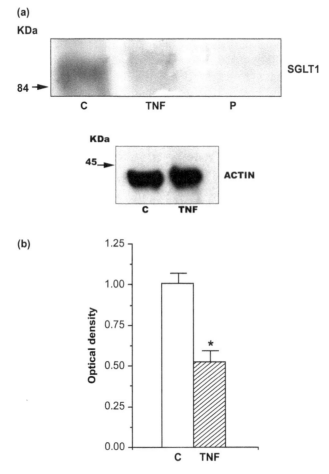

Figure 36.6 Effect of Tumor Necrosis Factor Alpha (TNFα) on SGLT1 protein expression in brush-border membrane (BBM). (a) The SGLT1 antibody recognized specifically an immunoreactive protein of about 84 kDa in control (C) and treated (TNF) animals. When the antibody was previously adsorbed with the antigenic peptide, no signal was detected (P). Actin was used as a loading control of total protein onto the electrophoresis gels. (b) Relative abundance of SGLT1 protein measured by optical density (counts/mm²). Values represent percentage means recorded in five separate experiments for each group (control and TNF-treated). *p < 0.05 with respect to control animals.

Reproduced with permission from Amador *et al.* (2007b). The image is from our group (unpublished).

abundance of GLUT2 at BBMs. The effects of RELM-β were associated with an increased amount of PKCβII in BBMs and an increased phosphorylation of AMP-activated protein kinase (AMPK).

Sepsis is a systemic response to infection and, in this state, the presence of toxins such as lipopolysaccharide (LPS) produced by bacteria stimulates the production of inflammatory mediators like the cytokine TNFα. Amador *et al.* (2007a, 2007b) have shown that intravenous administration of LPS or TNFα inhibits D-galactose intestinal absorption in rabbit by reducing the number of SGLT1 transporters at the brush-border membrane (Figures 36.5 and 36.6). Intracellular signaling pathways associated with protein kinase C (PKC), protein kinase A (PKA), p38 mitogen-activated protein kinase (p38MAPK), Jun N-terminal kinase (JNK), MAPK/extracellular signal-regulated kinases 1 and 2 (MEK1/2) and proteasome were found to be involved in this effect. On the other hand, in experiments *in vitro* when LPS is additioned to tissue, the endotoxin is able to diminish the Na^+/K^+-ATPase activity and the Na^+-dependent D-galactose intestinal absorption by decreasing SGLT1 sugar affinity without reduction in the transporter expression level (Figure 36.7). The

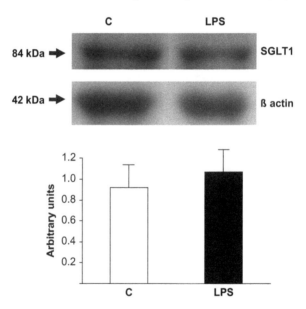

Figure 36.7   Effect of Lipopolysaccharide (LPS) on SGLT1 protein expression in brush-border membrane (BBM) of jejunum. Representative Western blot analysis of BBM SGLT1 prepared from control or LPS treated intestinal tissue after 15 min. The immunoreactive protein weighs around 84 kDa. The results represent data obtained by densitometric analysis of immunoblotted signals for protein normalized to those of β-actin on the same gels. Representative blots and data expressed as arbitrary units values (means ± SEM) are given. The preparations of intestinal vesicles per animal of each group (n = 5) were prepared and analysed in duplicate. Reproduced with permission from Amador *et al.* (2008), *Cellular of Physiology and Biochemistry*. 22, 715-724. S. Karger AG, Basel.

PKC activation and other intracellular messengers such as MAPKs seem to be involved in the endotoxin effect. Thus, the final effect would be the consequence of the cross talking between all of them (Amador *et al.* 2008). Finally, the different modifications on SGLT1 to reduce galactose absorption by LPS, when it was intravenously administered, indicates the importance of the administration route (local or systemic).

Summary Points

- Carbohydrates are an important component of the diet. The hydrolysis of lactose into glucose and galactose is crucial for the nutrition of human in facts.
- Galactose is a monosaccharide present at dairy products: milk and milk products, sugar beets, and other gums and mucilages.
- The galactose absorption across intestinal brush-border membrane is mediated by the sodium-dependent transporter SGLT1.
- The exit of galactose across the enterocyte basolateral membrane into the blood stream is mediated by the facilitated transporter GLUT2.
- Several factors (proteins, hormones, stress, diet, heavy metals, infection *etc.*) are involved in the regulation of intrinsic activity and/or the abundance of SGLT1 and GLUT2 in the gut. These factors have effect on the galactose intestinal absorption.
- The activation of several interrelated signaling cascades like kinases can be related to effect of these factors on sugar transporters. Study, in depth, the action of several kinases could be the aim a future research.

Key Facts

- The primary function of the small intestine is the absorption of nutrients.
- Dietary molecules as large as disaccharides or proteins do not cross the small intestine.
- It is the high concentration and efficiency of the hydrolytic enzymes in the brush border membrane of enterocyte that effectively hydrolyze all the molecules leaving none to pass across intact.
- The polysaccharides are digested by intestinal enzymes in the brush border and are further broken down into monosaccharides which can be absorbed from intestinal lumen to blood.
- The most important monosaccharides of the diet are galactose, glucose and fructose. These sugars are absorbed by specific transporters.
- When the sugars are in the blood, the cells take these molecules that are metabolized to obtain energy.
- The energy is necessary to carry out vital organism functions.
- A number of factors are involved in the regulation of intestinal sugar transport.
- These factors can modify the cellular permeability or alter the activity of specifics sugar transporters.

Definitions of Words and Terms

Gut: is a part of the digestive system by which the animals transfer the nutrients to blood. The small intestine is well adapted for digestion and subsequent absorption of nutrients.

Enterocyte: is a polarized cell in the gut.

Microvilli: Are microscopic finger-like projections that increase the surface area of enterocytes and are involved in absorption of nutrients.

Brush-border membrane (BBM): is a specialized portion of apical surface of enterocyte on luminal intestine. This surface contains absorptive microvilli and glycocalys which is rich in hydrolytic enzymes.

Basolateral membrane (BLM): is the part of enterocyte that forms its basal and lateral surface. This surface contains Na^+-K^+ pump.

Na^+-K^+ ATPase or Na^+-K^+ pump: is an enzyme responsible to moves 3 sodium ions out by hydrolysing ATP and allows 2 potassium ions in through active transport. The sodium exports from the enterocyte provides the driving force for several secondary active transporters membrane which import glucose/galactose and other nutrients into the cell by use of the sodium gradient.

Galactose: is a sugar presents in the diet of humans and omnivorous animals.

SGLT1: Sodium-dependent sugar cotransporters are a family of proteins that transport glucose-galactose from the intestinal lumen across the brush-border membrane into the cells. These proteins use the energy from a downhill sodium gradient to transport glucose-galactose across the uphill sugar gradient. Sodium and sugar cotransport by SGLT1 is referred to as secondary active transport because where Na gradient is maintained by Na^+/K^+-pump.

GLUT2: Facilitated glucose transporter is a protein that has a low affinity transporter, and it is responsible for transporting glucose, galactose and fructose across basolateral membrane out of the cells into the blood, via facilitative diffusion. In the case of a sugar rich meal, GLUT2 can be recruited into the apical membrane where it complements SGLT1 transport capacity.

GLUT5: is a fructose transporter expressed on the apical boder of enterocyte. GLUT5 allows for fructose to be transported from the intestinal lumen into the enterocyte by facilitated diffusion.

Protein kinases: are enzymes implicated in numerous intracellular signaling cascades that can provoke changes in the intestinal transport of sugars.

List of Abbreviations

BBM	Brush border membrane
BLM	Basolateral membrane
GALK	Galactokinase
GALT	Galactose-1-phosphate uridyltransferase
GALE	UDP-galactose-4'-epimerase
GGM	Glucose-galactose malabsorption

PKA	Protein kinase A
PKC	Protein kinase C
PI-3K	Phosphoinositol 3-kinase
p38MAPK	p38 Mitogen-activated protein kinase
JNK	Jun N-terminal kinase
MEK1/2	MAPK/extracellular signal-regulated kinases 1 and 2
mTOR	Mammalian target of Rapamycin
cAMP	Cyclic adenosine monophosphate
Na^+-K^+ATPase or Na^+-K^+pump	Sodium-potasium adenosine triphosphatase
RELMs	Resistin-like molecules
GLP-2	Glucagon-like peptide 2
CCK	Cholecystokinin
CCK-8	Cholecystokinin-8
VIP	Vasoactive intestinal peptide
HNF-1	Hepatocyte nuclear factor-1
Fox11	Winged-helix transcription factor
TNF-α	Tumor necrosis factor-alpha
LPS	Lipopolysaccharide

Acknowledgements

We thank Sonia Gascón Mesa for help with illustrations. The laboratory is member of "Intestinal Transport" group supported by grants from the Departamento de Ciencia, Tecnología y Universidad del Gobierno de Aragón (A-32), Spain.

References

Amador, P., García-Herrera, J., Marca, M.C., de la Osada, J., Acín, S., Navarro, M.A., Salvador, M.T., Lostao, M.P., and Rodríguez-Yoldi, M.J., 2007a. Intestinal D-galactose transport in an endotoxemia model in the rabbit. *Journal of Membrane Biology*. 215: 125–133..

Amador, P., García-Herrera, J., Marca, M.C., de la Osada, J., Acín, S., Navarro, M.A., Salvador, M.T., Lostao, M.P., and Rodríguez-Yoldi, M.J., 2007b. Inhibitpry effect of TNF-α on the intestinal absorption of galactose. *Journal of Cellular Biochemistry*. 101: 99–111.

Amador, P., Marca, M.C., García-Herrera, J., Lostao, M.P., Guillén, N., de la Osada, J., and Rodríguez-Yoldi, M.J., 2008. Lipopolysaccharide induces inhibition of galactose intestinal transport in rabbits *in vitro*. *Cellular of Physiology and Biochemistry*. 22: 715–724.

Arruebo, M.P., Sorribas, V., Rodríguez-Yoldi, M.J., Murillo, M.D., and Alcalde, A.I., 1990. Effect of VIP on sugar transport in rabbit small intestne *in vitro*. *Journal of Veterinary Medicine, Series A*. 37: 123–129.

Au, A., Gupta, A., Schembri, P., and Cheeseman, C.I., 2002. Rapid insertion of GLUT2 into the rat jejunal brush-border membrane promoted by glucagon-like-peptide 2. *Biochemical Journal*. 367: 247–254.

Cheeseman, C.I., 1993. GLUT2 is the transporter for fructose across the rat intestinal basolateral membrane. *Gastroenterology*. 105: 1050–1056.

De la Horra, M.C., Cano, M., Peral, M.J., Calonge, M.L., and Ilundain A.A., 2001. Hormonal regulation of chicken intestinal NHE and SGLT-1 activities. *American Journal of Physiology*. 280: R655–R660.

Drozdowski, L.A., and Thomson, A.B.R., 2006. Intestinal sugar transport. *World Journal of Gastroenterology*. 12 (11): 1657–1670.

Dyer, J., Wood, I.S., Palejwala, A., Ellis, A., and Shirazi-Beechey, S.P., 2002. Expression of monosaccharide transporters in intestine of diabetic humans. *American Journal of Physiology*. 282 (2): G241–G248.

Ferraris, R.P., 2001. Dietary and developmental regulation of intestinal sugar transport. *Biochemical Journal*. 360: 265–276.

Freemman, S.L., Bohan, D., Darcel, N., and Raybould H.E., 2006. Luminal glucose sensing in the rat intestine has characteristics of a sodium-glucose cotransporter. *American Journal of Physiology*. 291: G439–G445.

Fujio, J., Kushiyama, A., Sakoda, H., Fujishiro, M., Ogihara, T., Fukushima, Y., Anai, M., Horike, N., Kamata, H., Uchijima, Y., Kurihara, H., and Asano, T., 2008. Regulation of gut-derived resistin-like molecule β expression by nutrients. *Diabetes Research and Clinical Practice*. 79: 2–10.

Goda, T., 2000. Regulation of the expression of carbohydrate digestion/absorption-related genes. *British Journal of Nutrition*. 84: S245–S248.

Hediger, M.A., Coady, M.J., Ikeda, T.S., and Wright, E.M., (1987). Expression cloning and cDNA sequencing of the Na+/glucose co-transporter. *Nature*. 330: 379–381.

Helliwell, P.A., Rumsby, M.G., and Kellet, G.L., 2003. Intestinal sugar absorption is regulated by phosphorylation and turnover of protein kinase C betaII mediated by phosphatidylinositol 3-kinase- and mammalian target of rapamycin-dependent pathways. *Journal of Biological Chemistry*. 278: 28644–28650.

Hirst, A.J., and Cheeseman, C.I., 1998. Cholecystokinin decreases intestinal hexose absorption by a parallel reduction in SGLT1 abundance in the brush-border membrane. *Journal of Biological Chemistry*. 273: 14545–14549.

Iñigo, C., Patel, N., Kellet, G.L., Barber, A., and Lostao. M.P., 2007. Luminal leptin inhibits intestinal sugar absorption *in vivo*. *Acta Physiologica*. 190 (4): 303–310.

Isaji, M., 2007. Sodium-glucose cotransporter inhibitors for diabetes. *Current Opinion in Investigational Drugs*. 8: 285–292.

Katz, J.P., Perreault, N., Goldstein, B.G., Chao, H.H., Ferraris, R.P., and Kaestner, K.H., 2004. Foxl1 null mice have abnormal intestinal epithelia, postnatal growth retardation, and defective intestinal glucose uptake. *American Journal of Physiology*. 287: G856–G864.

Kellett, G.L., and Helliwell, P.A., 2000. The diffusive component of intestinal glucose absorption is mediated by the glucose-induced recruitment of GLUT2 to the brush-border membrane. *Biochemical Journal.* 350 (1): 155–162.

Kellet, G.L., and Brot-Laroche, E., 2005. Apical GLUT2: A major pathway of intestinal sugar absorption. *Diabetes.* 54: 3056–3062.

Kellett, G.L., Brot-Laroche, E., Mace, O.J., and Leturque, A., 2008. Sugar absorption in the intestine: the role of GLUT2. *Annual Review of Nutrition.* 28: 35–54.

Kimura, Y., Turner, J.R., Braasch, D.A., and Buddington, R.K., 2005. Lumenal adenosine and AMP rapidly increase glucose transport. *American Journal of Physiology.* 289: G1007–G1014.

Krimi, R.B., Letteron, P., Chedid, P., Nazaret, C., Ducroc, R., and Marie, J.C., 2009. Resistin-like molecule-β inhibits SGLT-1 activity and enhances GLUT2-dependent jejunal glucose transport. *Diabetes.* 58: 2032–2038.

Le Gall, M., Tobin, V., Stolarczyk, E., Dalet, V., Leturque, A., and Brot-Laroche, E., 2007. Sugar sensing by enterocytes combines polarity, membrane bound detectors and sugar metabolism. *Journal of Cellular Physiology.* 213: 834–843.

Lin, J., Kormanec, J., Homerova, D., and Kinne, R.K., 1999. Probing trans-membrane topology of the high-affinity sodium/glucose cotransporter (SGLT1) with histidine-tagged mutants. *Journal of Membrane Biology.* 170: 243–252.

Martin, M.G., Lostao, M.P., Turk, E., Lam, J., Kreman, M., and Wright, E.M., 1997. Compound missense mutations in the sodium/D-glucose cotransporter result in trafficking defects. *Gastroenterology.* 112: 1206–1212.

Martin, M.G., Wang, J., Solorzano-Vargas, R.S., Lam, J.T., Turk, E., and Wright, E.M., 2000. Regulation of the human Na^+-glucose cotransporter gene, SGLT1, by HNF-1 and Sp1. *American Journal of Physiology.* 278: G591–G603.

Nagata, K., and Hata, Y., 2006. Substrate specificity of a chimera made from *Xenopus* SGLT1-like protein and rabbit SGLT1. *Biochimica et Biophysica Acta.* 1758: 747–754.

Osswald, C., Baumgarten, K., Stumpel, F., Gorboulev, V., Akimjanova, M., Knobeloch, K.P., Horak, I., Kluge, R., Joost, H.G., and Koepsell, H., 2005. Mice without the regulate gene Rsc1A1 exhibit increased Na^+-D-glucose cotransport in small intestine and develop obesity. *Molecular of Cellular Biology.* 25 (1): 78–87.

Raybould, H.E., Glatzle, J., Robin, C., Meyer, J.H., Phan, T., Wong, H., and Sternini, C., 2003. Expression of 5-HT$_3$ receptors by extrinsic duodenal afferents contribute to intestinal inhibition of gastric emptying. *American Journal of Physiology.* 284: G367–G372.

Rodríguez Yoldi, M.C., Mesonero, J.E., and Rodríguez Yoldi, M.J., 1992. Inhibition of D-galactose transport across the small intestine of rabbit by zinc. Journal of Veterinary Medicine Series A 39: 687–695.

Rodríguez Yoldi, M.J., Arruebo, M.P., Alcalde, A.I., and Murillo, M.D., 1988. Influence of VIP on D-galactose transport across rabbit jejunum *in vivo*. *Revista española de Fisiología*. 44 (82): 127–130.

Rodríguez Yoldi, M.J., Lugea, A., Barber, A., Lluch, M., and Ponz, F., 1989. Inhibition of sugar and amino acid transporta across rat jejunum by cadmium, copper and mercury. *Revista española de Fisiología*. 45: 207–214.

Salvador, M.T., Murillo, M.D., Rodríguez-Yoldi, M.C., Alcalde, A.I., Mesonero, J.E., and Rodríguez-Yoldi, M.J., 2000. Effects of serotonin on the physiology of the rabbit small intestine. *Canadian Journal of Physiolgy and Pharmacology*. 78: 359–366.

Shepherd, E.J., Helliwell, P.A., Lister, N., Mace, O.J., Morgan, E.L., Patel, N., and Kellet, G.L., 2004. Stress and glucocorticoid inhibit apical GLUT2-trafficking and intestinal glucose absorption in rat small intestine. *Journal of Physiology*. 560: 281–290.

Veyhl, M., Keller, T., Gorboulev, V., Vernaleken, A., and Koepsell, H., 2006. RS1 (RSC1A1) regulates the exocytotic pathway of Na^+-D-glucose cotransporter SGLT1. *American Journal of Physiology*. 291: F1213–F1223.

Walker, J., Jijon, H.B., Diaz, H., Salehi, P., Churchill, T., and Madsen, K.L., 2005. 5-aminoimidazole-4-carboxamide riboside (AICAR) enhances GLUT2-dependent jejunal glucose transport: a possible role for AMPK. *Biochemical Journal*. 385: 485–491.

Wild, G.E., and Thomson, A.B., 1995. Na^+/K^+-ATPase alpha 1 and beta 1 mRNA and protein levels in rat small intestine in experimental ileitis. *American Journal of Physiology*. 269: G666–G675.

Wild, G.E., Thomson, J.A., Serles, L., Turner, R., Hasan, J., and Thomson, A.B., 1999. Small intestinal Na^+/K^+-adenosine triphosphatase activity and gene expression in experimental diabetes mellitus. *Digestive Diseases and Sciences*. 44: 407–414.

Wright, E.M., 2001. Renal Na^+-glucose cotransporters. *American Journal of Physiology*. 280. F10–F18.

Wright, E.M., Hirsch, J.R., Loo, D.D., and Zampighi, G.A., 1997. Regulation of Na^+/glucose cotransporters. *Journal of Experimental Biology*. 200: 287–293.

Wright, E.M., Turk, E., and Martin, M.G., 2002. Molecular basis for glucose-galactose malabsorption. *Cell Biochemistry and Biophysics*. 36: 115–121.

Wright, E.M., and Turk, E., 2004. The sodium/glucose cotransport family SLC5. *Pfluegers Archiv-European Journal of Physiology*. 447: 510–518.

Wright, E.M., Hirayama, B.A., and Loo, D.F., 2007. Active sugar transport in health and disease. *Journal of Internal Medicine*. 261: 32–43.

CHAPTER 37

Galactose and Galactose Tracers in Metabolic Studies

ANA FRANCISCA SOARES[†] AND
JOHN GRIFFITH JONES*

CNC - Center for Neurosciences and Cell Biology, University of Coimbra,
Largo Marquês de Pombal, 3004-517 Coimbra, Portugal
*Email: jones@cnc.uc.pt

37.1 Overview of Galactose in Nutrition and Health

After glucose and fructose, galactose is the most abundant hexose in adult human nutrition and is a major nutritional component for all mammals in their early stages of life when they are wholly dependent on milk for nutrition. Galactose represents one-half of milk carbohydrate and accounts for ~20% of total food calories for suckling mammals and infants. For adult humans that ingest typical amounts of dairy products during a meal, *i.e.* a 200 ml glass of milk with breakfast, galactose constitutes a small fraction of total carbohydrate intake, but since it is efficiently captured and metabolized by the liver (Cohn and Segal 1973), it can represent a significant fraction of postprandial hepatic carbohydrate metabolism after such a meal.

For subjects with a normal complement of galactose metabolizing enzymes, the selective metabolism of galactose by the liver provides a basis for assessing hepatic function (Tarantino 2009). The galactose elimination capacity (GEC)

[†]Current Address: CIBM - Center for Biomedical Imaging, EPFL SB IPSB LIFMET, CH F1 592 (Bâtiment CH), Station 6, CH-1015 Lausanne, Switzerland. Email: francisca.soares@epfl.ch

Food and Nutritional Components in Focus No. 3
Dietary Sugars: Chemistry, Analysis, Function and Effects
Edited by Victor R Preedy
© The Royal Society of Chemistry 2012
Published by the Royal Society of Chemistry, www.rsc.org

test is used in the clinical setting to determine the rate of galactose disappearance from the blood after the intravenous injection of a galactose load sufficient to saturate galactokinase. In pig models, PET imaging of the galactose analog 2-[^{18}F]fluoro-2-deoxy-D-galactose has been developed to determine regional differences in hepatic liver function (Sørensen 2011). The galactose single point (GSP) method, where a single measurement of blood galactose is performed 1 hour after its injection, has been proposed as a more practical alternative to the GEC and shows good correlation with GEC determinations and the severity of liver disease (Young *et al.* 2007).

37.1.1   Principal Metabolic Fate of Galactose

Galactose is metabolized in the liver, as shown in Figure 37.1. Galactose uptake by the hepatocyte is mediated by the main hepatic glucose transporter, GLUT2. Accordingly, hypergalactosemia is present, along with impaired glucose metabolism, in patients with "Fanconi-Bickel syndrome", or congenital GLUT2 deficiency (Santer *et al.* 1998). After uptake, galactose is rapidly metabolized to common intermediates of glucose metabolism *via* the Leloir pathway. This involves the sequential activity of four enzymes: galactose mutarotase, galactokinase, galactose-1-phosphate uridyltransferase (GALT), and UDP-galactose 4-epimerase (GALE) (Holden *et al.* 2003). Uridine diphosphate-glucose (UDPG) is formally the first common intermediate of hepatic galactose and glucose metabolism. Genetic defects in the latter three enzymes of the Leloir pathway hamper galactose disposal, resulting in galactosemia, and clinical manifestations that include cirrhosis, cataracts, or mental retardation (Cohn and Segal 1973).

   As depicted in Figure 37.1, galactokinase phosphorylates position 1 of α-D-galactose yielding gal-1-P, then GALT catalyses the transference of the uridine monophosphate moiety from UDPG to Gal-1-P, producing uridine diphosphate-galactose (UDPgal) and glucose-1-phosphate (G1P). UDPgal can also be produced from gal-1-P by UDP-galactose pyrophosphorylase activity, but the activity of this enzyme is normally very low compared with that of GALT (Leslie *et al.* 2005). By GALE activity, UDPgal is converted to UDPG which can either accept a new molecule of gal-1-P to perpetuate the Leloir pathway (whereby the initial galactose moiety now becomes G1P) or alternatively donates its glucosyl units to other acceptors such as glycogen. The extensive equilibration between G1P produced by the Leloir pathway and glucose-6-phosphate (G6P) coupled to glucose-6-phosphatase activity allows the net conversion of galactose to glucose that can be subsequently utilized by the whole body. Whether the carbons of galactose are converted to glycogen or glucose is largely dependent on the status of glycogen synthase and glycogen phosphorylase activities that in turn are sensitive to nutritional status and insulin levels. In the feeding phase, glycogen synthase is active while glycogen phosphorylase is inactive, hence the glycosyl of UDPG is incorporated into glycogen. During fasting, glycogen synthase is inactive,

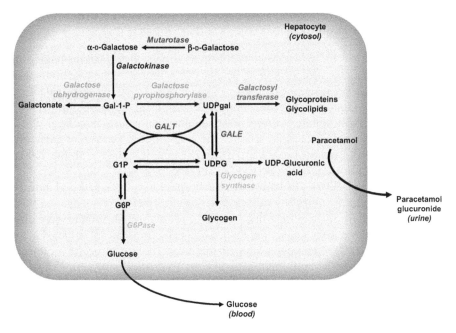

Figure 37.1 Principal pathways and enzymes of hepatic galactose metabolism. Galactose is converted to uridine diphosphate glucose (UDPG) *via* the sequential activities of Leloir pathway enzymes. This links galactose to glucose metabolism, and the subsequent activity of enzymes depicted by red font allows for the production of glucose, which can be released to the blood stream, or glycogen that is stored in the liver. The UDPG pool can be non-invasively sampled *via* the administration of paracetamol that is conjugated with glucuronic acid in the liver prior to being excreted in the urine. In green font is depicted the use of galactose in the galactosylation of glycoproteins and glycolipids. Alternative pathways of galactose metabolism (enzymes shown in light blue font) may operate in the liver when the activity of Leloir pathway enzymes is absent or saturated. G1P, glucose-1-phosphate; G6P, glucose-6-phosphate; G6Pase, glucose-6-phosphatase; Gal-1-P, galactose-1-phosphate; GALE, uridine diphosphate-galactose 4-epimerase; GALT, galactose-1-phosphate uridyltransferase; UDPgal, uridine diphosphate-galactose; UDP-glucuronic acid, uridine diphosphate glucuronic acid.

hence UDPG recycling *via* the Leloir pathway is more prevalent resulting in the net synthesis of G1P from galactose. Since G1P is also a common product of glycogenolysis, the provision of galactose under these conditions can be used as a surrogate of glycogenolytic flux (Staehr *et al.* 2007). In addition to being converted to UDPG by GALE, UDPgal formed in the Leloir pathway can also serve as a galactose donor in the galactosylation of proteins and lipids. In certain tissues, UDPgal for galactosylation can also be produced from UDPG via GALE, for example, in the developing brain for myelin formation (Cohn and Segal 1973).

37.1.2 Minor Metabolic Fates of Galactose: Galactitol and Galactonate

Two alternate fates for galactose metabolism have been identified: the reduction of galactose to galactitol catalyzed by NADPH-dependent aldose reductase (Cohn and Segal 1973) and the oxidation of galactose to galactonate by galactose dehydrogenase, followed by decarboxylation to form xylulose-5-phosphate (X5P), a pentose phosphate pathway intermediate (Cuatrecasas and Segal 1966). These alternate routes normally account for a very minor portion of galactose metabolism, but become prominent when the activity of any of the Leloir pathway enzymes is deficient (Berry *et al.* 1995b; Ning *et al.* 2001; Segal *et al.* 2006; Wehrli *et al.* 2007). Galactitol and galactonate have been detected in trace amounts in the urine of healthy subjects after administration of a galactose load (Segal *et al.* 2006). In wild-type mice, small amounts of galactonate were also detected in the liver after a galactose load, while galactitol was not detected at all (Wehrli *et al.* 2007). The appearance of these metabolites in healthy subjects and animals, albeit in small amounts, suggests that there is some overflow of galactose metabolism even with fully functional Leloir pathway enzymes. This characteristic is exploited in experimental animal models of galactosemia that are created by feeding a high galactose diet (Berry 1995; Mackic *et al.* 1994).

37.1.3 Endogenous Galactose Production

Endogenous production of galactose became evident with the observation that even with a galactose-free diet, galactosemic patients still developed long-term clinical and biochemical manifestations consistent with the accumulation of galactose side products (Schweitzer *et al.* 1993). Galactose appearance rates have been quantified by isotope dilution methods and demonstrate that both healthy and galactosemic subjects produce gram amounts of galactose *de novo* over a 24-hour period (Berry *et al.* 1995a; Huidekoper *et al.* 2005; Ning *et al.* 2000). Moreover, this process does not appear to be acutely influenced by galactose intake (Huidekoper *et al.* 2005). Endogenous galactose production may be important for assuring adequate galactosylation of lipids and proteins in the absence of dietary galactose. For the study of galactose as a macronutrient in subjects with functional Leloir pathway activity, galactolysation reactions, the minor oxidation and reduction pathways, and endogenous production rates are assumed to be negligible relative to galactose disposal via the Leloir pathway. This assumption also applies to isotope dilution studies of UDPG fluxes that involve the use of galactose tracers, *i.e.* the rate of appearance of the label in UDPG is directly related to the rate of galactose tracer infusion.

37.2 Galactose Tracers as Probes of Hepatic Metabolic Fluxes

Galactose tracers have two distinctive applications in the study of mammalian carbohydrate metabolism. Firstly, they inform galactose carbon fluxes through the various pathways described, culminating in oxidative and non-oxidative disposal of galactose carbons. Secondly, galactose tracers are used for determining hepatic glucose fluxes between G6P and glycogen. In this setting, the Leloir pathway serves solely as a means of delivering a hexose tracer at a known rate into the hepatic UDPG pool. The use of radioactive ^3H and ^{14}C galactose tracers has been largely superseded by stable ^2H and ^{13}C galactose isotopes both for reasons of safety and for exploiting positional ^2H- and ^{13}C-enrichment measurements to more precisely describe the metabolic fate of the tracer. Hepatic UDPG can be non-invasively sampled in both humans and animals *via* the glucuronidation of commonly used medications such as Paracetamol. The glucuronide moiety originates from the glucosyl of UDPG and the glycoconjugate is rapidly cleared into urine where it can be sampled for tracer enrichment or specific activity (Hellerstein *et al.* 1986). A diversity of galactose stable isotope ^2H- and ^{13}C-tracers are now commercially available for studying specific aspects of hepatic carbohydrate metabolism and their applications will be outlined in this chapter.

37.2.1 Oxidative and Non-oxidative Disposal of Galactose

As with glucose, disposal of galactose carbons into oxidative and non-oxidative pathways can be determined by measuring $^{13}CO_2$ enrichment of expired breath after the administration of ^{13}C-enriched galactose tracers. The fraction of CO_2 derived from galactose oxidation is thus derived and multiplying this by the total CO_2 output (estimated by indirect calorimetry) yields the total amount of galactose oxidized. The difference between total ingested and total oxidized is attributed to non-oxidative disposal. The ^{13}C enrichment of CO_2 can be measured by gas chromatography coupled with isotopic ratio mass spectrometry (IRMS) after the administration of [1-^{13}C]- or [2-^{13}C]galactose (Berry *et al.* 1995b; Berry *et al.* 2004; Mion *et al.* 1999; Segal *et al.* 2006). IRMS can precisely quantify low levels of excess $^{13}CO_2$ enrichment, hence the galactose precursor can be enriched in a single carbon at relatively low enrichment levels, thereby reducing the cost of the tracer.

With this methodology, studies have demonstrated that the rate of CO_2 production from galactose increases rapidly in the first minutes to few hours after administration of a galactose load both in healthy humans (Segal *et al.* 2006) and wild-type rodents (Mion *et al.* 1999; Ning *et al.* 2001). Moreover, CO_2 production accounts for up to 75% of the administered galactose dose by 24 hours, in humans (Segal *et al.* 2006). It was also shown that galactosemic subjects with confirmed GALT deficiency can oxidize both carbons 1 (Berry *et al.* 1995b) and 2 (Segal *et al.* 2006) of galactose, demonstrating galactose metabolism to X5P and CO_2 via galactonate (where carbon 1 only is liberated as CO_2) is not the sole oxidation pathway for galactose in these subjects.

37.2.2 Galactose Flux into UDPG

Galactose is converted to UDPG by the Leloir pathway and UDPG enrichment can be non-invasively assayed in both humans and animals by glucuronide analysis as described (Hellerstein *et al.* 1986). Thus, enrichment of urinary Paracetamol [2-^{13}C]glucuronide was observed following administration of an oral [2-^{13}C]galactose load and paracetamol in healthy individuals, confirming the Leloir metabolic pathway. With galactosemic subjects, significant [2-^{13}C]glucuronide enrichment – albeit with slower appearance kinetics – was also found despite a confirmed deficiency of GALT (Segal *et al.* 2006). The explanation for this observation is that residual GALT activity is still sufficient for Leloir pathway metabolism of galactose to UDPG but at a slower rate than normal, and/or that UDPgal was generated from gal-1-P by UDP-galactose pyrophosphorylase activity, thereby bypassing GALT (Segal *et al.* 2006). These findings are consistent with the appearance of both [1-^{13}C]- and [2-^{13}C]galactose labels in plasma glucose in GALT-deficient subjects as well as the observed whole-body oxidation of [2-^{13}C]galactose.

37.2.3 Measuring UDPG Flux with Galactose Tracers

UDPG is the immediate precursor for glycogen hence UDPG flux informs the rate of hepatic glycogen synthesis from all hexose precursors. This can be measured by infusion of a galactose tracer and quantifying its dilution at the level of UDPG by measuring the ratio of UDPG enrichment to that of the infused galactose (Hellerstein *et al.* 1997). UDPG enrichment is assessed from urinary glucuronide analysis as described and UDPG rate of appearance is calculated as the product of the dilution and tracer infusion rate (Hellerstein *et al.* 1997). The principal assumptions of this measurement are that the galactose tracer is quantitatively converted to UDPG and that the labeled UDPG is subsequently cleared by net conversion into glycogen or glucose. In some situations, there may be futile cycling of hexose between UDPG, glycogen and hexose phosphates, known as glycogen cycling (Landau 2001). Glycogen cycling *per se* should not contribute to the disappearance of the galactose tracer, and with carbon isotopes this is assured since the hexose skeleton remains intact during cycling. However, with certain ^{2}H-galactose tracers, exchanges at the level of G6P may result in the removal of the ^{2}H-label thereby contributing to the disappearance of the tracer and resulting in an over-estimation of net UDPG flux rates.

As a result of extensive equilibration between G6P and fructose-6-phosphate (F6P) the position 2 hydrogen label is extensively exchanged with body water hydrogen if the UDPG hexose moiety is cycled *via* glycogen and hexose phosphates. On this basis, the fraction of UDPG undergoing futile cycling *versus* its net rate of appearance from G6P can be estimated by administration of a galactose tracer where a carbon and the position 2 hydrogen are both labeled. The enrichment or specific activity of UDPG carbon and position 2 hydrogen is then measured. If there is no cycling, then both hydrogen and carbon labels are diluted to an equivalent degree and the ratio of carbon to hydrogen enrichment or specific activity in UDPG is the same as for the galactose precursor. If cycling is active, UDPG hydrogen enrichment or specific activity is reduced relative to that of the carbon tracer, and the fraction of cycled UDPG relative to its overall rate of appearance is determined by the ratio of hydrogen to carbon enrichments or specific activities. Thus, in overnight fasted type 2 diabetics that were administered with [2-^{3}H, 6-^{14}C]galactose with the same specific activity for both isotopes (*i.e.* ^{3}H/^{14}C specific activity = 1.0), the UDPG ^{3}H/^{14}C specific activity ratio determined from analysis of glucuronide was found to range between 0.88 and 0.98, indicating that futile cycling accounted for 2–12% of UDPG appearance (Wajngot *et al.* 1991). Glycogen cycling also results in a partial loss of [1-^{2}H]galactose label as a result of F6P and mannose-6-phosphate exchange (Chandramouli *et al.* 1999).

37.2.4 Measuring Transaldolase Exchange Activity with Galactose Tracers

In addition to its role in the pentose phosphate pathway, transaldolase also mediates a simple exchange of F6P and glyceraldehyde-3-phosphate (G3P), with carbons 4, 5 and 6 of fructose being replaced by carbons 1, 2 and 3 of G3P. This process has important consequence for tracer measurements of gluconeogenesis and glycogenolysis, since unlabeled hexose phosphates assumed to be derived from glycogenolysis become labeled with the gluconeogenic tracer resulting in an overestimation of gluconeogenic activity and a corresponding underestimation of glycogenolytic fluxes (Basu *et al.* 2009). Since transaldolase exchange involves only the lower half of F6P, its activity can be assessed by providing a galactose tracer that is enriched in both triose halves and monitoring the production of labeled glucose. Transaldolase exchange causes selective loss of the label in the bottom triose moiety resulting in plasma glucose with depleted labeling of positions 4, 5 and 6 relative to positions 1, 2 and 3 (see Figure 37.2). Thus, Basu *et al.* infused both healthy and type 2 diabetic subjects with [3,5-^{2}H$_2$]galactose and measured a plasma glucose ^{2}H5/^{2}H3 enrichment ratio of $\sim$0.7, indicating that $\sim$30% of the hexose phosphate intermediates

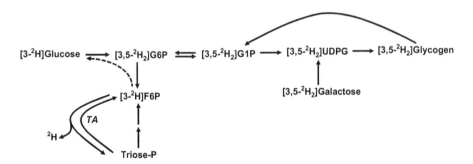

Figure 37.2 Metabolism of [3,5-^{2}H$_2$]galactose to glucose in the presence of transaldolase (TA)-mediated exchange of F6P and triose-P moieties. The position 3 label is retained regardless of TA activity but the position 5 label is removed from those hexose molecules that participate in TA, resulting in the generation of [3-^{2}H]glucose.

had undergone transaldolase-mediated exchange (Basu *et al.* 2009). Transaldolase exchange may also explain the observations of Coss-Bu *et al.*, where [U-^{13}C]galactose was infused and a mixture of plasma [U-^{13}C]glucose and [^{13}C$_3$]glucose (67% to 33%) were detected by GC-MS (Coss-Bu *et al.* 2009). The GC-MS analysis did not resolve the [^{13}C$_3$]glucose into [1,2,3-^{13}C$_3$]- and [4,5,6-^{13}C$_3$]glucose isotopomers, but assuming no re-incorporation of the labeled G3P into glucose, transaldolase exchange should only generate the [1,2,3-^{13}C$_3$]glucose isotopomer.

37.3 Integration of Galactose and Glucose Metabolic Flux Measurements

Dietary galactose is potentially an important precursor of postprandial hepatic glucose and glycogen under normal life conditions but it is not metabolized in isolation from other carbohydrates. Lactose is the principal dietary source of galactose, therefore each equivalent of galactose is accompanied by at least one of glucose. Since both sugars have UDPG as a common metabolic intermediate, it is necessary to resolve their individual contributions to UDPG and subsequent metabolites (*i.e.* hepatic glycogen and glucose). This has been recently achieved by ^{13}C-isotopomer analysis of ^{13}C-enriched glucose and galactose tracers (Soares *et al.* 2010) and with deuterated water (^{2}H$_2$O) (Barosa *et al.* 2012; Soares *et al.* 2010).

37.3.1 ^{13}C-isotopomer Analysis of Galactose and Glucose Metabolism

Exogenous glucose and galactose metabolism may be simultaneously monitored using ^{13}C-enriched galactose and glucose tracers coupled with ^{13}C NMR isotopomer analysis of glycogen, as shown in Figure 37.3. This approach needs to account for direct and indirect pathway metabolism of glucose, in particular the rearrangement of the ^{13}C-label by indirect pathway fluxes. With [U-^{13}C]glucose, indirect pathway metabolism generates a mixture of partially labeled glycogen isotopomers, but these do not include significant levels of [1-^{13}C]glycogen. Therefore, the formation of [1-^{13}C]glycogen from a mixture of [1-^{13}C]galactose and [U-^{13}C]glucose can be attributed to the [1-^{13}C]galactose precursor. Glycogen isotopomers generated from glucose and galactose metabolism can be resolved and quantified by ^{13}C NMR, as shown in Figure 37.3. This analysis can also be applied to urinary glucuronide (Mendes *et al.* 2006; Segal *et al.* 2006). The ^{13}C-isotopomer analysis was applied to study hepatic glycogen repletion from exogenous glucose or exogenous glucose plus galactose loads in 24-hour fasted rats. Substituting 10% of the glucose load with galactose did not significantly alter net hepatic glycogen synthesis. However, the minority galactose component was preferentially utilized for glycogen synthesis and displaced the direct pathway contribution from glucose (Soares *et al.* 2010). Nevertheless, the absolute contributions of galactose and glucose to hepatic

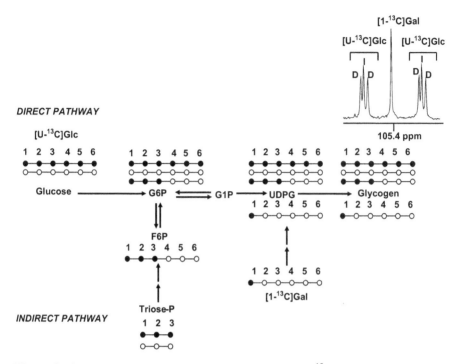

Figure 37.3 Resolution of galactose and glucose fluxes by ^{13}C-isotopomer analysis of [1-^{13}C]galactose ([1-^{13}C]Gal) and [U-^{13}C]glucose ([U-^{13}C]Glc) tracers. The ^{13}C atoms are represented by filled circles and ^{12}C by unfilled circles. Direct pathway metabolism of [U-^{13}C]Glc yields [U-^{13}C]UDPG and [U-^{13}C]glycogen. Indirect pathway metabolism of [U-^{13}C]Glc yields partially-labeled G6P and UDPG isotopomers such as the [1,2,3-^{13}C$_3$]G6P shown, but does not generate significant amounts of [1-^{13}C]G6P or [1-^{13}C]UDPG. Therefore, [1-^{13}C]Gal is the only source of [1-^{13}C]UDPG or [1-^{13}C]Glycogen isotopomers. As a result of ^{13}C-^{13}C-spin-spin coupling, glycogen ^{13}C-isotopomers derived from direct pathway (**D**) and indirect pathway (**I**) metabolism of [U-^{13}C]glucose and from [1-^{13}C]Gal are resolved in the ^{13}C NMR carbon 1 resonance of the monoacetone glucose derivative (Soares *et al.* 2010), shown above the glycogen isotopomers.

glycogen synthesis were surprisingly small, and the bulk of hepatic glycogen synthesis (~70%) was instead derived from endogenous gluconeogenic precursors *via* the indirect pathway. These studies demonstrate the need for specific tracers of exogenous glucose and galactose to determine their real contributions to hepatic glucose and glycogen synthesis.

37.3.2   Galactose and Glucose Metabolism by ^{2}H$_2$O

The contribution of galactose to hepatic glycogen synthesis can be estimated with ^{2}H$_2$O as shown in Figure 37.4. The basis of this measurement is that G6P

DIRECT PATHWAY

Figure 37.4 Quantifying glucose-6-phosphate (G6P) and galactose contributions to hepatic glycogen synthesis by analysis of UDPG position 2 enrichment form deuterated water (^2H_2O). Exchange of G6P and fructose-6-phosphate (F6P) is extensive and results in the equilibration of G6P position 2 and body water hydrogens regardless of whether G6P was derived *via* direct or indirect pathways. As a result, [2-^{2}H]G6P enrichment can be assumed to be equivalent to that of body water following ingestion of ^2H_2O and isotopic equilibration with bulk body water. UDPG that is synthesized from this source is also enriched to the same level as body water. Metabolism of galactose results in the generation of unlabeled UDPG and a dilution of [2-^{2}H]UDPG enrichment relative to body water.

position 2 enrichment is equivalent to that of body water regardless of whether it originated *via* direct or indirect pathways. UDPG that originated from G6P is also enriched in position 2 while UDPG derived from galactose is unlabeled, thereby reducing the overall UDPG position 2 enrichment compared to that of body water. The fractional contribution of galactose to UDPG flux is thus estimated from the ratio of UDPG position 2 and body water enrichments (Barosa *et al.* 2012). The main advantage of ^2H_2O over carbon tracers is that the contribution of unlabeled dietary galactose to hepatic glycogen synthesis can be directly evaluated (Barosa *et al.* 2012). The critical assumption of this method is that exchange of hepatic G6P position 2 and body water is complete under the study conditions. This can be verified in a separate experiment where the meal is labeled with [U-^{2}H$_7$]glucose and depletion of UDPG position 2 enrichment is measured (Barosa *et al.* 2012). Thus, for healthy individuals that ingested a mixed breakfast meal that included 200 ml of skimmed milk, hepatic G6P exchange with body water was essentially complete, and the galactose component of the milk was estimated to contribute ∼20% of the total postprandial glycogen synthesis flux (Barosa *et al.* 2012).

Summary Points

- Galactose represents a small fraction of the daily caloric intake but a significant fraction of postprandial hepatic metabolism.
- The major metabolic fate of galactose is to be converted to glucose intermediates *via* the Leloir pathway.
- Galactose tracers are used to follow galactose disposal.
- Galactose tracers also report the hepatic UDPG allowing monitoring fluxes to glycogen and glucose-6-phosphate.

- Galactose disposal depends on the nutritional, hormonal and health status of the organism.
- Novel tracer methods for assessing metabolism of dietary galactose, isolated galactose, and mixtures of galactose and other carbohydrates will provide new insights into its metabolism in both humans and animal models of nutrition and disease.

Key Facts of Hepatic Galactose Metabolism

1. Galactose is metabolized mainly in the liver
2. Galactose loads are used in the clinical setting to assess hepatic function
3. Defects in galactose-metabolizing enzymes are related with liver disease

Definitions of Words and Terms

Metabolic tracer: substance used to follow the biological transformation of an endogenous substrate (tracee). The tracer must be metabolically indistinguishable from the tracee but at the same time possess a unique property allowing for its detection. This is the case, for example, of substances composed of rare isotopes such as ^{13}C or ^{2}H that can be detected by NMR spectroscopy methods.

Cirrhosis: scarring of the liver and deficient liver function as a consequence of chronic liver disease.

Galactosemia: literally meaning galactose in the blood, this results from the inability to adequately metabolize galactose due to genetic defects on the galactose-metabolizing enzymes.

Leloir Pathway: the main route for galactose metabolism in the liver named after the Argentinian biochemist Luis Federico Leloir.

Paracetamol: also known as acetaminophen, is an analgesic drug metabolized in the liver by glucuronidation and excreted in the urine.

Metabolic flux: Rate of conversion of a precursor metabolite to a product, *via* a single enzyme or a specific sequence of enzymes that constitute a metabolic pathway. Units are µmol of metabolite per unit tissue or body mass per unit time.

Metabolite rate of appearance/disappearance: Also referred to as metabolite turnover or lifetime, this means the rate at which new molecules enter or leave a specific metabolite pool. Unlike metabolic flux, metabolite appearance or disposal rate is not necessarily defined by a single enzyme or metabolic pathway. For example, the rate of appearance of plasma glucose can involve both gluconeogenic and glycogenolytic pathway fluxes.Appearance/disappearance rates can also reflect metabolite transport between different tissues or compartments independently of conversion to another metabolic intermediate. Units are µmol of metabolite per unit tissue or body mass per unit time.

Pentose phosphate pathway: series of reactions involving the oxidation of glucose-6-phosphate to pentose phosphates and subsequent carbon skeleton rearrangements to glucose-6-phosphate.

Isotopomer: Contraction of the words isotope and isomer, refers to molecules that differ one from the other solely by their isotopic composition. For example, considering the glucose molecule and the ^{13}C isotope, each carbon may be ^{12}C or ^{13}C. Since there are 6 carbons in glucose, there are $26 = 64$ possible isotopomers.

NMR: Abbreviation of Nuclear Magnetic Resonance. In an NMR experiment, nuclei of atoms with a positive spin are placed in an external magnetic field generated by the NMR spectrometer. The net magnetization vector of the nuclei is aligned with the external field but is displaced when a second magnetic field perpendicular to the first one and generated by a radio-frequency pulse that processes at the same frequency as the spinning nuclei is applied. After this excitation pulse the magnetization decays with a characteristic time. This decay originates a current that is detected by the receiver coil. NMR-visible nuclei of interest in the metabolic field are: ^{13}C, ^{1}H, ^{2}H, or ^{31}P.

List of Abbreviations

F6P	fructose-6-phosphate
G1P	glucose-1-phosphate
G3P	glyceraldehyde-3-phosphate
G6P	glucose-6-phosphate
Gal-1-P	galactose-1-phosphate
GALE	uridine diphosphate-galactose 4-epimerase
GALT	galactose-1-phosphate uridyltransferase
GEC	galactose elimination capacity
GSP	galactose single point
^{2}H$_2$O	deuterated water
IRMS	isotopic ratio mass spectrometry
UDPG	uridine diphosphate-glucose
UDPgal	uridine diphosphate-galactose
X5P	xylulose-5-phospate.

References

Barosa, C., Silva, C., Fagulha, A., Barros, L., Caldeira, M.M., Carvalheiro, M., and Jones, J.G., 2012. Sources of hepatic glycogen synthesis following a milk-containing breakfast meal in healthy subjects. *Metabolism: Clinical and Experimental.* 61: 250–254.

Basu, R., Chandramouli, V., Schumann, W., Basu, A., Landau, B.R., and Rizza, R.A., 2009. Additional Evidence That Transaldolase Exchange,

Isotope Discrimination During the Triose-Isomerase Reaction, or Both Occur in Humans Effects of Type 2 Diabetes. *Diabetes.* 58: 1539–1543.

Berry, G.T., 1995. The role of polyols in the pathophysiology of hypergalactosemia. *European Journal of Pediatrics.* 154: S53–S64.

Berry, G.T., Nissim, I., Lin, Z., Mazur, A.T., Gibson, J.B., and Segal, S., 1995a. Endogenous synthesis of galactose in normal men and patients with hereditary galactosaemia. *The Lancet.* 346: 1073–1074.

Berry, G.T., Nissim, I., Mazur, A.T., Elsas, L.J., Singh, R.H., Klein, P.D., Gibson, J.B., Lin, Z.P., and Segal, S., 1995b. In Vivo Oxidation of [13C]Galactose in Patients with Galactose-1-Phosphate Uridyltransferase Deficiency. *Biochemical and Molecular Medicine.* 56: 158–165.

Berry, G.T., Reynolds, R.A., Yager, C.T., and Segal, S., 2004. Extended [13C]galactose oxidation studies in patients with galactosemia. *Molecular Genetics and Metabolism.* 82: 130–136.

Chandramouli, V., Ekberg, K., Schumann, W.C., Wahren, J., and Landau, B.R., 1999. Origins of the hydrogen bound to carbon 1 of glucose in fasting: significance in gluconeogenesis quantitation. *American Journal of Physiology-Endocrinology and Metabolism.* 277: E717–723.

Cohn, R.M., and Segal, S., 1973. Galactose metabolism and its regulation. *Metabolism.* 22: 627–642.

Coss-Bu, J.A., Sunehag, A.L., and Haymond, M.W., 2009. Contribution of galactose and fructose to glucose homeostasis. *Metabolism.* 58: 1050–1058.

Cuatrecasas, P., and Segal, S., 1966. Galactose conversion to D-xylulose: an alternate route of galactose metabolism. *Science.* 153: 549–551.

Hellerstein, M.K., Greenblatt, D.J., and Munro, H.N., 1986. Glycoconjugates as Noninvasive Probes of Intrahepatic Metabolism: Pathways of Glucose Entry into Compartmentalized Hepatic UDP-glucose Pools during Glycogen Accumulation. *Proceedings of the National Academy of Sciences of the United States of America.* 83: 7044–7048.

Hellerstein, M.K., Letscher, A., Schwarz, J.M., Cesar, D., Shackleton, C.H., Turner, S., Neese, R., Wu, K., Bock, S., and Kaempfer, S., 1997. Measurement of hepatic Ra UDP-glucose in vivo in rats: relation to glycogen deposition and labeling patterns. *Am J Physiol Endocrinol Metab.* 272: E155–162.

Holden, H.M., Rayment, I., and Thoden, J.B., 2003. Structure and Function of Enzymes of the Leloir Pathway for Galactose Metabolism. *Journal of Biological Chemistry.* 278: 43885–43888.

Huidekoper, H.H., Bosch, A.M., van der Crabben, S.N., Sauerwein, H.P., Ackermans, M.T., and Wijburg, F.A., 2005. Short-term exogenous galactose supplementation does not influence rate of appearance of galactose in patients with classical galactosemia. *Molecular Genetics and Metabolism.* 84: 265–272.

Landau, B.R., 2001. Methods for measuring glycogen cycling. *American Journal of Physiology-Endocrinology and Metabolism.* 281: E413–E419.

Leslie, N., Yager, C., Reynolds, R., and Segal, S., 2005. UDP-galactose pyrophosphorylase in mice with galactose-1-phosphate uridyltransferase deficiency. *Molecular Genetics and Metabolism.* 85: 21–27.

Mackic, J.B., Ross-Cisneros, F.N., McComb, J.G., Bekhor, I., Weiss, M.H., Kannan, R., and Zlokovic, B.V., 1994. Galactose-induced cataract formation in guinea pigs: morphologic changes and accumulation of galactitol. *Investigative Ophthalmology & Visual Science.* 35: 804–810.

Mendes, A.C., Caldeira, M.M., Silva, C., Burgess, S.C., Merritt, M.E., Gomes, F., Barosa, C., Delgado, T.C., Franco, F., Monteiro, P., Providencia, L., and Jones, J.G., 2006. Hepatic UDP-glucose ^{13}C isotopomers from [U-^{13}C]glucose: A simple analysis by ^{13}C NMR of urinary menthol glucuronide. *Magnetic Resonance in Medicine.* 56: 1121–1125.

Mion, F., Géloën, A., and Minaire, Y., 1999. Effects of ethanol and diabetes on galactose oxidative metabolism and elimination in rats. *Can. J. Physiol. Pharmacol.* 77: 182–187.

Ning, C., Fenn, P.T., Blair, I.A., Berry, G.T., and Segal, S., 2000. Apparent Galactose Appearance Rate in Human Galactosemia Based on Plasma [^{13}C]Galactose Isotopic Enrichment. *Molecular Genetics and Metabolism.* 70: 261–271.

Ning, C., Reynolds, R., Chen, J., Yager, C., Berry, G.T., Leslie, N., and Segal, S., 2001. Galactose Metabolism in Mice with Galactose-1-Phosphate Uridyltransferase Deficiency: Sucklings and 7-Week-Old Animals Fed a High-Galactose Diet. *Molecular Genetics and Metabolism.* 72: 306–315.

Santer, R., Schneppenheim, R., Dombrowski, A., Gotze, H., Steinmann, B., and Schaub, J., 1998. Fanconi-Bickel syndrome - A congenital defect of the liver-type facilitative glucose transporter. *Journal of Inherited Metabolic Disease.* 21: 191–194.

Schweitzer, S., Shin, Y., Jakobs, C., and Brodehl, J., 1993. Long-term outcome in 134 patients with galactosaemia. *European Journal of Pediatrics.* 152: 36–43.

Segal, S., Wehrli, S., Yager, C., and Reynolds, R., 2006. Pathways of galactose metabolism by galactosemics: Evidence for galactose conversion to hepatic UDPglucose. *Molecular Genetics and Metabolism.* 87: 92–101.

Soares, A.F., Carvalho, R.A., Veiga, F.J., and Jones, J.G., 2010. Effects of galactose on direct and indirect pathway estimates of hepatic glycogen synthesis. *Metab Eng.* 12: 552–560.

Sørensen, M., 2011. Determination of hepatic galactose elimination capacity using 2-[^{18}F]fluoro-2-deoxy-d-galactose PET/CT: reproducibility of the method and metabolic heterogeneity in a normal pig liver model. *Scandinavian Journal of Gastroenterology.* 46: 98–103.

Staehr, P., Hother-Nielsen, O., Beck-Nielsen, H., Roden, M., Stingl, H., Holst, J.J., Jones, P.K., Chandramouli, V., and Landau, B.R., 2007. Hepatic autoregulation: response of glucose production and gluconeogenesis to increased glycogenolysis. *American Journal of Physiology-Endocrinology and Metabolism.* 292: E1265–E1269.

Tarantino, G., 2009. Could quantitative liver function tests gain wide acceptance among hepatologists? *World Journal of Gastroenterology.* 15: 3457–3461.

Wajngot, A., Chandramouli, V., Schumann, W.C., Efendic, S., and Landau, B.R., 1991. Quantitation of glycogen/glucose-1-P cycling in liver. *Metabolism*. 40: 877–881.

Wehrli, S., Reynolds, R., and Segal, S., 2007. Metabolic fate of administered [^{13}C]galactose in tissues of galactose-1-phosphate uridyl transferase deficient mice determined by nuclear magnetic resonance. *Mol Genet Metab*. 90: 42–48.

Young, T.-H., Tang, H.-S., Lee, H.-S., Hsiong, C.-H., and Hu, O.Y.-P., 2007. Effects of hyperglycemia on quantitative liver functions by the galactose load test in diabetic rats. *Metabolism*. 56: 1265–1269.

CHAPTER 38

D-Galactose, Dietary Sugars and Modeling Neurological Aging

KODEESWARAN PARAMESHWARAN,[a] MICHAEL H. IRWIN,[b] KOSTA STELIOU[c] AND CARL A. PINKERT*[d]

[a] 240D Greene Hall, Auburn University College of Veterinary Medicine, Auburn, AL 36849, USA; [b] 240B Greene Hall, Auburn University College of Veterinary Medicine, Auburn, AL 36849, USA; [c] PhenoMatriX, Inc., 9 Hawthorne Place, Suite 4R, Boston, MA 02114 and Cancer Research Center, Boston University School of Medicine, 715 Albany Street, Room K-701, Boston, MA 02118; [d] 202 Samford Hall, Auburn University, Auburn, AL 36849-5112, USA
*Email: cap@auburn.edu

38.1   Introduction

38.1.1   Human Intake of Dietary Sugars

Dietary carbohydrates provide a major source of metabolic energy (45–70%) in developed countries. Dietary sugars refer to both disaccharides (*e.g.*, sucrose, lactose, and maltose) and monosaccharides (*e.g.*, glucose, fructose, galactose, and sorbitol). Though fructose is the most commonly occurring monosaccharide in nature, glucose is the main sugar used for metabolic energy in tissues. Along with short-chain ketoacids, glucose is the primary fuel source for brain and nerve tissues. Diet-induced effects on the brain and aging nervous system are correlated with advancements in food technology and increased

Food and Nutritional Components in Focus No. 3
Dietary Sugars: Chemistry, Analysis, Function and Effects
Edited by Victor R Preedy
© The Royal Society of Chemistry 2012
Published by the Royal Society of Chemistry, www.rsc.org

Table 38.1   Dietary sugars and neurological function.

Sugar	Neurological Relevance	Mouse Models	References
Glucose	Principle energy source for brain and nervous tissue	SAMP, TgAPP, TgABAD Tg601	Kurokawa *et al.* 1996; Ohta *et al.* 1996; Takuma *et al.* 2005; Kambe *et al.* 2011;
		STZ-induced diabetic	Chan *et al.* 2003; Magarinos and McEwen 2000; Revsin *et al.* 2008; Stranahan *et al.* 2008a.
Fructose	Altered energy metabolism; impairment of cognitive function	Chronic administration	Stranahan *et al.* 2008b.
Galactose	Accelerated aging	Chronic administration	Cui *et al.* 2004; Xu and Zhao 2002; Wei *et al.* 2005; Zhang *et al.* 2011; Lei *et al.* 2008.

consumption of refined foods and sweetened drinks. In this chapter, we provide an overview of our work on D-galactose-induced aging in mice and the role of other dietary sugars in neurological aging in rodent models (see Table 38.1).

38.1.2 Brain Energy Metabolism: Energy Requirements and Glucose Metabolism in Health, Disease, and Aging

Even though the brain represents about 2.0–2.3% of adult body weight in humans, its energy requirements are much higher – about 20–23% of the total energy expended or consumed by the body (Sokoloff 1999). Factors that regulate energy metabolism in the brain are cerebral blood flow, oxygen consumption, and glucose metabolism. Studies demonstrated that a major proportion of glucose is utilized to maintain synaptic ion gradients required for glutamate neurotransmission while the rest is used to maintain the resting potential of neurons (Attwell and Iadecola 2002; Shulman *et al.* 2004). Uptake of glucose in the brain is controlled by GLUTs, of which, a 55 kDa isoform of GLUT1 mediates uptake through the endothelium of the blood-brain barrier, the 45 KDa isoform aids transport into astrocytes, and GLUT3 transfers glucose into neurons (Duelli and Kuschinsky 2001). Glucose metabolism in the brain is regulated by insulin and cortisol. The source of brain insulin is mainly from pancreatic β-cells with contributions from pyramidal neurons in the hippocampus, cortex, entorhinal cortex, and the olfactory bulb. Consistently high expression of insulin receptors in the brain is found in these regions and hypothalamus.

Metabolic enzyme activity and the subsequent utilization of glucose in the brain are critically dependent upon mitochondrial respiration. PDH and COX are two important enzymes involved in mitochondrial energy production. The majority ($>95\%$) of acetyl-coA is oxidized in the tricarboxylic acid cycle to

ATP. About 1–2% is used for the synthesis of acetylcholine (Gibson *et al.* 1975) and cholesterol synthesis in the 3-hydroxy-3-methylglutaryl-CoA cycle (Michikawa and Yanagisawa 1999).

A critical variable in metabolism is aging; decreases in metabolic function in the brain are correlated with age. Enzymes for the anaerobic catabolism of glucose are more abundant in the neonatal brain. Aerobic enzymes, for the most part, develop postnatally. Accordingly, compared to adults, neonatal mammals are highly resistant to hypoxia induced injury suggesting enzymes for the anaerobic catabolism of glucose are more abundant in the neonatal brain (Duffy *et al.* 1975; Kabat 1940). Aging increases vulnerability to experimental oxygen and glucose deprivation (Baltan *et al.* 2008). Though some studies reported no change in CMRg with age, CMRg and aging correlations revealed that CMRg decreases with age in humans (Cunnane *et al.* 2011) and both regional cerebral blood flow and glucose metabolism were reduced with aging, supporting the theory that aging is associated with decreased glucose metabolism in the brain (Bentourkia *et al.* 2000).

Brains of senescence accelerated (SAMP) mice consumed less glucose than those of younger mice (Borras *et al.* 2009; Takeda *et al.* 1994; see Figure 38.1). During aging, control over neuronal glucose metabolism is reduced, likely a result of a decline of neuronal insulin signal transduction involving tyrosine kinases, tyrosine phosphatases, serine kinases, and cAMP-dependent kinases. Increased concentrations of cortisol with advanced aging and elevated levels due to stress may play a role in the attenuation of insulin receptor function (Lupien *et al.* 1994; Sapolsky *et al.* 1986). A decline in neuronal insulin receptor function may also be due to increased levels of noradrenaline in the brain with aging (Ida *et al.* 1982; Harik and McCracken, 1986). Therefore, aging-induced changes, which encompass stress and deficient insulin signal transduction, may play a vital role in reduced brain glucose metabolism.

Apart from the natural aging processes, age dependent and metabolic diseases such as AD and diabetes disrupt brain glucose metabolism at an early stage and more intensively. Mice overexpressing mutant forms of APP and ABAD, a mitochondrial enzyme, showed decreased oxygen consumption in the brain (Takuma *et al.* 2005). Analysis of postmortem brain tissue from AD patients revealed decreased expression and activity of COX and PDH (Blass *et al.* 2000). In the 3XTg-AD mouse model, reduced activities of PDH and COX were found in the forebrain. Additionally, an association of reproductive senescence with declines in PDH and COX activities was noted. This was indicative of diminished glucose metabolism in ATP production (Yao *et al.* 2009; Yao *et al.* 2010). About a 45% reduction in cerebral glucose utilization in AD patients indicates that there is a generalized shift from glycolytic energy production towards use of an alternative fuel, ketone bodies, in the AD brain (Ishii *et al.* 1997). Decreased glucose metabolism in AD brains was also supported by decreased expression of two major brain GLUTs; GLUT1 and GLUT3, and decreased expression of hypoxia inducible transcription factor-1 (the main regulator of GLUT1 and GLUT3). Furthermore, decreased GLUT1 and GLUT3 also correlated with increased tau hyper-phosphorylation and an

(a)

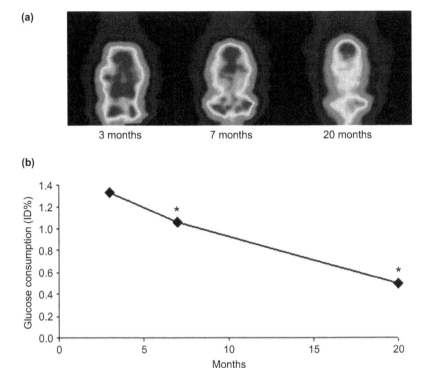

Figure 38.1 Effect of aging on glucose consumption *in vivo* by the brain of SAMP mice. (a) Brain glucose consumption *in vivo* measured by positron emission tomography. (b) Quantification of brain glucose consumption. Mean ± S.E.M.* P<0.05. n = 3.
(Reproduced with permission from Borras *et al.* 2009).

increased density of NFTs, indicating involvement of GLUTs in AD neuro-degeneration (Liu *et al.* 2008; Ogunshola and Antoniou 2009). A direct role of Aβ on GLUT function was also demonstrated. In rodent neuronal cultures Aβ oligomers produced hydrogen peroxide and hydroxyl radicals with subsequent deleterious effects that included lipid peroxidation, impairment of the function of Ca^{2+} and Na^{+}/K^{+} ATPases and GLUTs, finally resulting in alterations in cellular Ca^{2+} and energy homeostasis, and impaired synaptic function (Mark *et al.* 1997). Thus, alterations in GLUT expression and function, along with disrupted mitochondrial function, represent progressive events in AD.

AD is closely linked to diabetes and obesity and altered energy metabolism is increasingly recognized as a critical determinant of disease mechanisms in AD. Risk factors for aging and AD are high-energy diets and diabetes. In contrast, dietary energy restriction and exercise are associated with increased cognitive abilities and reduced aging effects. In non-human primates, a negative correlation between metabolic syndrome severity and oxidative function of mito-chondria as well as aging-associated decreases in the number and activity of functional mitochondria were also shown in the hippocampus (Blalock *et al.* 2010). Similarly, rodents given high sugar and/or high fat diets revealed poor learning and memory (Winocur and Greenwood 1999).

In young animals, excessive body weight impairs some cognitive domains, indicating that age alone may not be limiting in disease progression (Mielke *et al.* 2006). In AD transgenic mice, high energy diet consumption exacerbated disease pathology while dietary energy restriction prevented or alleviated the disease symptoms (Cao *et al.* 2007; Halagappa *et al.* 2007). Mild metabolic stress, such as mitochondrial uncoupling and caloric restriction, enhanced glucose utilization and increased expression of GLUT1 in mice (Liu *et al.* 2008). Other studies also demonstrated age-related cognitive deficits were reduced by caloric restriction, attenuating age-related deficits in learning and memory (Komatsu *et al.* 2008; Pitsikas and Algeri 1992). Therefore, elevated oxidative stress in the brain, inflammatory processes, and deficient adaptive responses to stress are major mechanisms that contribute to the cognitive deficits associated with high energy diet consumption.

38.1.3 Mouse Models of Glucose Metabolism and Neurological Aging

One of the early models of aging is the Senescence-Accelerated Mouse (SAM); developed through the selective inbreeding of the AKR/J strain of mice and based on a graded score for senescence, life span, and pathologic phenotype. Nine senescence-prone inbred strains (SAMP) and three senescence-resistant inbred strains (SAMR) were developed. The SAMP mice show decreased glucose metabolism in the brain linked to learning deficits (Kurokawa *et al.* 1996; Ohta *et al.* 1996). Aged transgenic mice expressing wild-type human tau (Tg601) showed decreased glucose metabolism in the nucleus accumbens region of the brain (Kambe *et al.* 2011). More conventional models include mice lacking a functional leptin receptor. The Stat3-C^{POMC} mouse model expresses a constitutively active version of Stat3 (Stat3-C) in POMC-expressing neurons of the hypothalamus. These mice exhibited obesity, increased food intake and leptin resistance. Glucose homeostasis was not altered, indicating that constitutive activation of Stat3-dependent signaling in POMC-expressing neurons results in mild obesity with increases in body weight and fat mass, without overtly affecting whole-body insulin sensitivity (Ernst *et al.* 2009).

Evidence for a causal link between metabolic diseases like diabetes and harmful neurological effects is well established. Diabetes mellitus is often accompanied by reduced cognitive function in humans. Patients with altered metabolic homeostasis are at a higher risk of affective mental disorders, dementia and AD (Brismar *et al.* 2007; Northam *et al.* 2006). Altered glucose metabolism and hyperglycemia taken together represent a strong risk factor for many other problems including brain and vascular disorders.

Type 1 diabetes caused by insufficient insulin production is usually induced in rodents by injection of STZ, which results in several changes in the brain including increased expression of hypothalamic hormones (Saravia *et al.* 2001), hyperactivity of the hypothalamo-pituitary-adrenal axis (Chan *et al.* 2003), high vulnerability to stress (Magarinos and McEwen 2000), and elevated blood glucocorticoid levels (Revsin *et al.* 2008; Stranahan *et al.* 2008a). In addition, type 1 diabetes also results in hippocampal astrogliosis and reduced neurogenesis (Beauquis *et al.* 2008; Revsin *et al.* 2005; Saravia *et al.* 2004). Many of these effects are also found in encephalopathy associated with aging, suggesting that the diabetic brain could be considered similar to an aged brain (Saravia *et al.* 2007). A comprehensive list of mouse models available for the study of metabolic syndrome was provided in a recent report (Kennedy *et al.* 2010).

HD is an inherited neurodegenerative disorder (Landles and Bates, 2004) with the expansion of CAG trinucleotide repeats in exon 1 of the HD gene. This alteration in the gene results in polyglutamine expansions in the htt protein (Andrew *et al.* 1993) leading to its atypical processing and intracellular aggregation, particularly in degenerating neurons in the striatum and cerebral cortex (Bates 2003). Although this disease is considered mainly a neurological disorder, other symptoms include progressive weight loss, appetite dysfunction, and poor glycemic control (Aziz *et al.* 2007). In particular, glucose metabolism is impaired in the brain and peripheral nervous system (Lodi *et al.* 2000; Powers *et al.* 2007). Rodent models include mice expressing N-terminal human htt with various CAG repeats, knock-in mice with pathogenic CAG repeats inserted into the existing CAG expansion in murine Hdh, mice that express the full-length human HD gene (plus murine Hdh), and toxin induced models (mainly excitotoxins and 3-NP) (Ferrante 2009). Depending on the range of CAG repeats and the type of mutation involved, several HD mouse models were developed and the role of dietary sugar consumption in the disease progression and severity in these models remains to be explored.

Analysis of preclinical and clinical reports highlights the critical role of dietary sugar metabolism in regulating nervous system physiology, cell survival, and cell genesis as well as promoting neurological aging. In this context, generation of animal models that better mimic human disease conditions and transgenic research that advances cost-effective production of pharmaceutical and nutra-ceutical compounds gains added significance (Dunn *et al.* 2005; Pinkert 2002; Pinkert and Trounce 2002). Recent developments in the generation of xenomi-tochondrial mouse models and models with mitochondrial mutations (both nuclear and mtDNA) could be very useful in unraveling the mechanisms that contribute to dietary sugar induced accelerated aging (Cannon *et al.* 2010; Ingraham *et al.* 2009; McKenzie *et al.* 2004; Trounce *et al.* 2004; Trounce and Pinkert 2007; Pogozelski *et al.* 2008). Mouse models of metabolic diseases, particularly obese and diabetic models, are now being tested for their susceptibility to effects on the brain, with emphasis on AD, gustatory mechanisms, and cerebrovascular pathology. Overall, transgenic mouse models exhibiting a spectrum of metabolic phenotypes have proven essential in our understanding of the impact of dietary sugars and nutrition on nervous system health.

38.2   D-Galactose

38.2.1   D-Galactose as Dietary Sugar and Modifications in Metabolism

Dietary sources of galactose include mainly milk products but significant amounts are also found in sugar beets and some plant products. Four enzymes (galactose mutarotase, galactokinase, galactose-1-phosphate uridyltransferase, and UDP-galactose 4-epimerase) that constitute the Leloir pathway (Holden *et al.* 2003) are involved in D-galactose metabolism to produce glucose 1-phosphate. Mutations in the genes encoding these enzymes result in a defective metabolic disease state known as galactosemia with symptoms including intellectual retardation (Novelli and Reichardt 2000), and can also be lethal. Oversupply of D-Galactose alters the normal metabolism of D-galactose, which is a reducing sugar, resulting in accumulation of galactitol, osmotic stress and production of reactive oxygen species in cells (Hsieh *et al.* 2009). Production of AGEs and disruption in mitochondrial function are potential mechanisms underlying chronic D-galactose induced accelerated aging in mice (Kumar *et al.* 2009; Song *et al.* 1999). Hence, chronic administration of reducing sugars, including D-galactose to mice, represents an intriguing model for neurological aging.

When rats were given D-galactose along with green tea polyphenols, changes in plasma metabolites associated with the disruptions of lecithin metabolism, amino acid metabolism and phospholipid metabolism were observed (Fu *et al.* 2011). D-Galactose administration was reported to cause oxidative stress, and contributed to the generation of cytotoxic lysophosphatidylcholine through free radical induced phospholipase A2 mediated hydrolysis (Lee and Charlton 2001).

A variation of the equilibrium between nutrition supply and consumption during accelerated aging could result in decreases in tryptophan; a dietary requirement (not endogenously synthesized) and serotonin precursor. In concert, dihydrosphingosine plasma levels were increased with D-galactose treatment. Formation of dihydrosphingosine by condensation of serine and palmitoyl-CoA and subsequent generation of ceramide may be a factor in cellular aging processes (Venable *et al.* 1995). D-Galactose administration induced increases in plasma dihydrosphingosine and phytosphingosine levels in rats, suggestive that sphingolipids might be among the first cell elements to face cellular stress (Hannun and Obeid 2002). These reports correlate with multiple pathways through which chronic D-galactose administration may modify energy metabolism in the brain and other tissues.

38.2.2   Chronic D-galactose Administration of Oxidative Stress in Rodents

Chronic D-galactose administration was shown to generate ROS and result in raised lipid peroxidation, nitrite concentration, depletion of reduced glutathione, catalase activity, and apoptosis in the mouse brain (Kumar *et al.* 2011; Li *et al.* 2011). D-Galactose treatment in mice also reduced T-SOD, GSH-Px activity and high levels of MDA in serum (He *et al.* 2009). D-Galactose-induced mimetic aging is associated with an increase in oxidative stress and consequent damage of hippocampal neurons through apoptotic cascades leading to neuronal loss and cognitive dysfunction. Chronic D-galactose treatment was also shown to induce formation of AGEs, ultra-structural changes of retinal pigment epithelium, and changes of transcriptome response similar to generalized aging in the retinal pigment epithelium-choroid of mice (Tian *et al.* 2005). D-Galactose-induced senescence acceleration was widely used as a model for studying aging mechanisms and screening drugs (Song *et al.* 1999). Different mechanisms, including the hypothesis of increasing oxidant generation and consequent oxidative damage, were proposed (Cui *et al.* 2004). However, the molecular mechanisms underlying D-galactose-induced aging and neurodegeneration remain unclear.

38.2.3   D-Galactose and Neurological Aging

Many reports have illustrated that chronic administration of D-galactose results in accelerated aging and changes that resemble natural aging in animals, such as a shortened lifespan (Cui *et al.* 2004), cognitive dysfunction (Xu and Zhao 2002; Wei *et al.* 2005), neurodegeneration (Cui *et al.* 2006; see Figure 38.2), oxidative stress (Li *et al.* 2011), decreased immune response (Lei *et al.* 2003), AGE formation (Song *et al.* 1999) and gene transcriptional changes (Tian *et al.* 2005) Chronic administration of D-galactose also resulted in decreased acetylcholine and choline acetyl transferase in the visual cortex of rats along with memory impairment (Zhang *et al.* 2011). Furthermore, cholinergic terminals in the hippocampus and medial septal region were reduced in D-galactose treated rats (Lei *et al.* 2008).

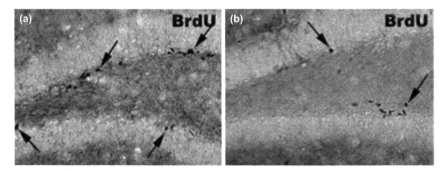

Figure 38.2   Neurogenesis in the subgranular zone of hippocampal dentate gyrus region is reduced in D-galactose administered mice. (a) 5-bromodeoxy-uridine positive neurons in control mice treated with saline. (b) D-galactose treatment resulted in decreased 5-bromodeoxyuridine positive neurons indicating reduced neurogenesis.
(Reproduced with permission from Cui *et al.* 2006).

In a recent study, D-galactose administration in rats resulted in an increase in the prevalence of the mtDNA common deletion and decreased mitochondrial base excision repair capacity in the inner ear (Zhong *et al.* 2011). This finding supports the hypothesis that impairment in mitochondrial base excision repair capacity and an increase in mtDNA replication resulting from mitochondrial transcription factor A overexpression contributed to the accumulation of mtDNA deletions in the inner ear during aging.

Curiously, several studies reported cognitive deficits in mice chronically treated with D-galactose and subjected to water maze (Kumar *et al.* 2009) and Y-maze testing (Wang *et al.* 2009). However, in our recently published study, no impairment in memory was detected when mice chronically administered D-galactose were tested using a Y-maze novelty seeking paradigm (Parameshwaran *et al.* 2010).

Corresponding to earlier testing strategies, neurobehavioral effects were reported to accompany aging including reduced performance in the open field test (Lu *et al.* 2006); yet we detected neither impairment in open field exploration nor any decline in sensory motor coordination when D-galactose treated mice were subjected to a Rota-rod test (Parameshwaran *et al.* 2010).

Combined administration of D-galactose and aluminum chloride and effects on memory (Luo *et al.* 2009; Xiao *et al.* 2011) were recently confirmed. Curiously, most of the studies were performed with Kunming strain of mice suggesting choice of strain could be an influential factor. In agreement with our results, one study reported that novelty-seeking behavior was not affected by D-galactose administration (Huang *et al.* 2009). These data again illustrate that D-galactose administration may not be suitable for modeling experimental aging-induced memory deficits. This also holds in experimental paradigms that do not involve reward/penalty stimuli, as in the case of forced swimming in a Morris water maze (Parameshwaran *et al.* 2010).

38.3   Fructose Metabolism and Brain Aging

Since fructose is metabolized more slowly than sucrose, it is frequently used in dietary products for people with diabetes or similar metabolic deficits. GLUT5 is the high affinity GLUT for fructose (Payne *et al.* 1997; Rand *et al.* 1993). Fructose feeding enhanced GLUT5 expression in the jejunum (Inukai *et al.* 1993) and in the brains of rats (Shu *et al.* 2006). Importantly, the GLUT5 increase was also found in the hippocampus and in cerebellar Purkinje neurons, suggesting that these brain regions are able to metabolize fructose (Funari *et al.* 2005).

Fructose is a reducing sugar capable of producing AGEs. Food products containing HFCS have been linked to several health problems in humans. Rats fed HFCS in drinking water along with a high-fat, high-glucose diet showed alterations in energy metabolism, high glucose levels and elevated cholesterol and triglycerides. When fed for 8 months, reductions were observed in learning abilities, spine density, synaptic plasticity and the levels of BDNF, suggesting that high-energy diets with HFCS reduce hippocampal synaptic plasticity and impair cognitive function, possibly through BDNF-mediated effects on dendritic spines (Stranahan *et al.* 2008b). Advances in research on energy metabolism in the brain (and other tissues) have provided evidence that food rich in fructose can lead to accelerated aging induction in the brain.

Summary Points

- Dietary carbohydrates constitute a major source of metabolic energy – as much as 45–70% for populations in developed countries.
- Glucose is the main metabolic fuel in the brain and nerve tissues – which require a continuous supply of energy to support synaptic activity.
- Glucose metabolism in the brain is impaired in several age-dependent neurological diseases such as AD and HD and in diseases such as diabetes.
- Reducing sugars such as D-galactose react with proteins to produce AGEs that promote aging.
- Experimental chronic administration of D-galactose results in advanced brain aging and the strain of mouse may be a determining factor in this process.
- Advances in transgenic animal modeling technology would be desirable to enhance our understanding of dietary sugar and its nutritional impact in neurological aging.

Key Facts

Key facts of energy metabolism: this term broadly refers to biochemical reactions that take place primarily in mitochondria of living cells to produce energy. In all living organisms, ATP serves as the common bio-molecule of energy exchange and is thus an integral part of energy metabolism.

Key facts of Alzheimer's disease: AD is an age-related disease of the brain with a progressive loss of memory and other intellectual abilities serious enough to interfere with daily life. Risk factors include age and family history as well as metabolic diseases including diabetes.

Key facts regarding the aging brain: brain aging is a major risk factor for many diseases in which memory decline as well as reduction in brain volume, thinning of the cortex and loss of neurons are common symptoms.

Key facts on insulin resistance: insulin resistance refers to a physiological situation in which insulin becomes inefficient at lowering blood sugars within normal limits.

Key facts regarding oxidative phosphorylation: OXPHOS is a metabolic process that occurs within mitochondria that utilizes energy released by the oxidation of nutrients to produce adenosine triphosphate (ATP).

Key facts on mitochondrial DNA: mtDNA is present in the mitochondria and is maternally inherited. The two strands of mtDNA are distinguished by the content of their nucleotides which are part of the building blocks of DNA. The two strands are the guanine-rich strand (heavy strand) and the cytosine-rich strand (light strand). Together, these two strands contain 37 genes, 13 of which encode proteins involved in OXPHOS.

Definitions of Words and Terms

Monosaccharides: the simplest form of carbohydrates; cannot be further broken down by enzymatic hydrolysis.

Amyloid beta: peptides of 33–43 amino acids produced from enzymatic cleavage of amyloid precursor protein.

Huntingtin: gene located in short arm of chromosome 4 at position 16.3, from base pair 3,113,411 to base pair 3,282,655; this gene encodes huntingtin protein produced in the brain of HD patients.

Glucose transporters: membrane proteins with 12 membrane spanning domains with both termini facing the cytoplasm. These proteins play the physiological role of transporting glucose across membranes.

Glycation: the process of enzymatic binding of proteins with reducing sugars such as D-galactose.

Oxidative phosphorylation: a metabolic process that occurs within mitochondria and results in the production of ATP from nutrients.

Kinases: a group of enzymes that add a phosphate group to molecules, such as proteins, from high energy donors resulting in phosphorylation of recipient molecules.

BDNF: brain derived neurotrophic factor is a class of neurotrophins, which are secreted growth factors, and is involved in many pro-survival, differentiation, and synaptic functions.

Transcription factors: proteins that bind to DNA at specific sequences and control the transcription of DNA to mRNA.

Oxidative stress: is a harmful condition in which altered mitochondrial function leads to an excessive generation of reactive oxygen species and inadequate antioxidant activity.

List of Abbreviations

ABAD	Aβ-binding alcohol dehydrogenase
Aβ	amyloid beta
AGE	advanced glycation end products
APP	amyloid precursor protein
AD	Alzheimer's disease
BDNF	brain derived neurotrophic factor
CMRg	cerebral metabolic rate of glucose
COX	cytochrome c oxidase
GLUT	glucose transporter
GSH-Px	glutathione peroxidase
HD	Huntington's disease
HFCS	high fructose corn syrup
Htt	huntingtin
MDA	malondialdehyde
NFT	neurofibrillary tangles
OXPHOS	oxidative phosphorylation

PDH   pyruvate dehydrogenase
SAMP   senescence accelerated mice prone
STZ   streptozotocin
T-SOD   total superoxide dismutase

References

Andrew, S.E., Goldberg, Y.P., Kremer, B., Telenius, H., Theilmann, J., Adam, S., Starr, E., Squitieri, F., Lin, B., Kalchman, M.A., Graham, R.K., and Hayden, M.R. 1993. The relationship between trinucleotide (CAG) repeat length and clinical features of Huntington's disease. *Nature Genetics* 4: 398–403.

Attwell D., and Iadecola, C., 2002. The neural basis of functional brain imaging signals. *Trends Neurosci.* 25: 621–625.

Aziz, N.A., Swaab, D.F., Pijl, H., and Roos, R.A., 2007. Hypothalamic dysfunction and neuroendocrine and metabolic alterations in Huntington's disease: clinical consequences and therapeutic implications. *Reviews in Neuroscience* 18: 223–251.

Baltan, S., Besancon, E.F., Mbow, B., Ye, Z., Hamner, M.A., and Ransom, B.R., 2008. White matter vulnerability to ischemic injury increases with age because of enhanced excitotoxicity. *J. Neuroscience* 28: 1479–1489.

Bates, G., 2003. Huntingtin aggregation and toxicity in Huntington's disease. *Lancet* 361: 1642–1644.

Beauquis, J., Saravia, F., Coulaud, J., Roig, P., Dardenne, M., Homo-Delarche, F., and De Nicola, A., 2008. Prominently decreased hippocampal neurogenesis in a spontaneous model of type 1 diabetes, the nonobese diabetic mouse. *Exp. Neurology* 210: 359–367.

Bentourkia, M., Bol, A., Ivanoiu, A., Labar, D., Sibomana, M., Coppens, A,, Michel, C., Cosnard, G., and De Volder, A.G., 2000. Comparison of regional cerebral blood flow and glucose metabolism in the normal brain: effect of aging. *J. Neurological Sci.* 181: 19–28.

Blalock, E.M., Grondin, R., Chen, K.C., Thibault, O., Thibault, V., Pandya, J.D., Dowling, A., Zhang, Z., Sullivan, P., Porter, N.M,, and Landfield, P.W., 2010. Aging-related gene expression in hippocampus proper compared with dentate gyrus is selectively associated with metabolic syndrome variables in rhesus monkeys. *J. Neuroscience* 30: 6058–6071.

Blass, J.P., Sheu, R.K., and Gibson, G.E., 2000. Inherent abnormalities in energy metabolism in Alzheimer disease. Interaction with cerebrovascular compromise. *Annals New York Acad. Sci.* 903: 204–221.

Borras, C., Stvolinsky, S., Lopez-Grueso, R., Fedorova, T., Gambini, J., Boldyrev, A., and Vina, J., 2009. Low in vivo brain glucose consumption and high oxidative stress in accelerated aging. *FEBS Letters* 583: 2287–2293.

Brismar, T., Maurex, L., Cooray, G., Juntti-Berggren, L., Lindstrom, P., Ekberg, K., Adner, N., and Andersson, S. 2007. Predictors of cognitive impairment in type 1 diabetes. *Psychoneuroendocrinology* 32: 1041–1051.

Cannon, M.V., Dunn, D.A., Irwin, M.H., Brooks, A.I., Bartol, F.F., Trounce, I.A., and Pinkert, C.A., 2010. Xenomitochondrial mice: investigation into mitochondrial compensatory mechanisms. *Mitochondrion* 11: 33–39.

Cao, D., Lu, H., Lewis, T.L., and Li, L., 2007. Intake of sucrose-sweetened water induces insulin resistance and exacerbates memory deficits and amyloidosis in a transgenic mouse model of Alzheimer disease. *J. Biol. Chem.* 282: 36275–36282.

Chan, O., Inouye, K., Riddell, M.C., Vranic, M., and Matthews, S.G., 2003. Diabetes and the hypothalamo-pituitary-adrenal (HPA) axis. *Minerva Endocrinol.* 28: 87–102.

Cui, X., Wang, L., Zuo, P., Han, Z., Fang, Z., Li, W., and Liu, J., 2004. D-galactose-caused life shortening in Drosophila melanogaster and Musca domestica is associated with oxidative stress. *Biogerontology* 5: 317–325.

Cui, X., Zuo, P., Zhang, Q., Li, X., Hu, Y., Long, J., Packer, L., and Liu, J., 2006. Chronic systemic D-galactose exposure induces memory loss, neurodegeneration, and oxidative damage in mice: protective effects of R-alphalipoic acid. *J. Neuroscience Res.* 84: 647–654.

Cunnane, S., Nugent, S., Roy, M., Courchesne-Loyer, A., Croteau, E., Tremblay, S., Castellano, A., Pifferi, F., Bocti, C., Paquet, N., Begdouri, H., Bentourkia, M., Turcotte, E., Allard, M., Barberger-Gateau, P., Fulop, T., and Rapoport, S.I., 2011. Brain fuel metabolism, aging, and Alzheimer's disease. *Nutrition* 27: 3–20.

Duelli, R., and Kuschinsky, W., 2001. Brain glucose transporters: relationship to local energy demand. *News Physiological Sci.* 16: 71–76.

Duffy, T.E., Kohle, S.J., and Vannucci, R.C., 1975. Carbohydrate and energy metabolism in perinatal rat brain: relation to survival in anoxia. *J. Neurochemistry* 24: 271–276.

Dunn, D.A., Pinkert, C.A., and Kooyman, D.L., 2005. Foundation Review: Transgenic animals and their impact on the drug discovery industry. *Drug Discovery Today* 10: 757–767.

Ernst, M.B., Wunderlich, C.M., Hess, S., Paehler, M., Mesaros, A., Koralov, S.B., Kleinridders, A., Husch, A., Munzberg, H., Hampel, B., Alber, J., Kloppenburg, P., Bruning, J.C., and Wunderlich, F.T., 2009. Enhanced Stat3 activation in POMC neurons provokes negative feedback inhibition of leptin and insulin signaling in obesity. *J. Neurosci.* 29: 11582–11593.

Ferrante, R.J., 2009. Mouse models of Huntington's disease and methodological considerations for therapeutic trials. *Biochim. Biophysica Acta* 1792: 506–520.

Fu, C., Wang, T., Wang, Y., Chen, X., Jiao, J., Ma, F., Zhong, M., and Bi, K., 2011. Metabonomics study of the protective effects of green tea polyphenols on aging rats induced by D-galactose. *J. Pharm. Biomed. Analysis* 55: 1067–1074.

Funari, V.A., Herrera, V.L., Freeman, D., and Tolan, D.R., 2005. Genes required for fructose metabolism are expressed in Purkinje cells in the cerebellum. *Brain Res. Mol. Brain Res.* 142: 115–122.

Gibson, G.E., Jope, R., and Blass, J.P., 1975. Decreased synthesis of acet-ylcholine accompanying impaired oxidation of pyruvic acid in rat brain minces. *Biochemical J.* 148: 17–23.

Halagappa, V.K., Guo, Z., Pearson, M., Matsuoka, Y., Cutler, R.G., Laferla, F.M., and Mattson, M.P., 2007. Intermittent fasting and caloric restriction ameliorate age-related behavioral deficits in the triple-transgenic mouse model of Alzheimer's disease. *Neurobiol. Disease* 26: 212–220.

Hannun, Y.A., and Obeid, L.M., 2002. The Ceramide-centric universe of lipid-mediated cell regulation: stress encounters of the lipid kind. *J. Biol. Chem.* 277: 25847–25850.

Harik, S.I., and McCracken, K.A., 1986. Age-related increase in presynaptic noradrenergic markers of the rat cerebral cortex. *Brain Res.* 381: 125–130.

He, M., Zhao, L., Wei, M.J., Yao, W.F., Zhao, H.S., and Chen, F.J., 2009. Neuroprotective effects of (-)-epigallocatechin-3-gallate on aging mice induced by D-galactose. *Biol. Pharm. Bulletin* 32: 55–60.

Holden, H.M., Rayment, I., and Thoden, J.B., 2003. Structure and function of enzymes of the Leloir pathway for galactose metabolism. *J. Biol. Chem.* 278: 43885–43888.

Hsieh, H.M., Wu, W.M., and Hu, M.L., 2009. Soy isoflavones attenuate oxi-dative stress and improve parameters related to aging and Alzheimer's dis-ease in C57BL/6J mice treated with D-galactose. *Food Chemical Toxicology* 47: 625–632.

Huang, Y., Su, Z., Li, Y., Zhang, Q., Cui, L., Su, Y., Ding, C., Zhang, M., Feng, C., Tan, Y., Feng, W., Li, X., and Cai, L., 2009. Expression and Purification of glutathione transferase-small ubiquitin-related modifier-metallothionein fusion protein and its neuronal and hepatic protection against D-galactose-induced oxidative damage in mouse model. *J. Pharm. Exp. Therapeutics* 329: 469–478.

Ida, Y., Tanaka, M., Kohno, Y., Nakagawa, R., Iimori, K., Tsuda, A., Hoaki, Y., and Nagasaki, N., 1982. Effects of age and stress on regional nora-drenaline metabolism in the rat brain. *Neurobiology Aging* 3: 233–236.

Ingraham, C.A,, Burwell, L.S., Skalska, J., Brookes, P.S., Howell, R.L., Sheu, S.S., and Pinkert, C.A., 2009. NDUFS4: creation of a mouse model mimicking a Complex I disorder. *Mitochondrion* 9: 204–210.

Inukai, K., Asano, T., Katagiri, H., Ishihara, H., Anai, M., Fukushima, Y., Tsukuda, K., Kikuchi, M., Yazaki, Y., and Oka, Y., 1993. Cloning and increased expression with fructose feeding of rat jejunal GLUT5. *Endocri-nology* 133: 2009–2014.

Ishii, K., Sasaki, M., Kitagaki, H., Yamaji, S., Sakamoto, S., Matsuda, K., and Mori, E., 1997. Reduction of cerebellar glucose metabolism in advanced Alzheimer's disease. *J. Nuclear Medicine* 38: 925–928.

Kabat, H., 1940. The greater resistance of very young animals to arrest of the brain circulation. *Am. J. Physiology* 130: 588–599.

Kambe, T., Motoi, Y., Inoue, R., Kojima, N., Tada, N., Kimura, T., Sahara, N., Yamashita, S., Mizoroki, T., Takashima, A., Shimada, K., Ishiguro, K., Mizuma, H., Onoe, H., Mizuno, Y., and Hattori, N., 2011. Differential regional distribution of phosphorylated tau and synapse loss in the nucleus accumbens in tauopathy model mice. *Neurobiology Disease* 42: 404–414.

Kennedy, A.J., Ellacott, K.L., King, V.L., and Hasty, A.H., 2010. Mouse models of the metabolic syndrome. *Disease Models Mech.* 3: 156–166.

Komatsu, T., Chiba, T., Yamaza, H., Yamashita, K., Shimada, A., Hoshiyama, Y., Henmi, T., Ohtani, H., Higami, Y., de Cabo, R., Ingram, D.K., and Shimokawa, I., 2008. Manipulation of caloric content but not diet composition, attenuates the deficit in learning and memory of senescence-accelerated mouse strain P8. Exp. *Gerontology* 43: 339–346.

Kumar, A., Dogra, S., and Prakash, A., 2009. Effect of carvedilol on behavioral, mitochondrial dysfunction, and oxidative damage against D-galactose induced senescence in mice. *Naunyn Schmiedebergs Archives Pharm.* 380: 431–441.

Kumar, A., Prakash, A., and Dogra, S., 2011. Protective effect of curcumin (Curcuma longa) against D-galactose-induced senescence in mice. *J. Asian Natural Products Res.* 13: 42–55.

Kurokawa, T., Sato, E., Inoue, A., and Ishibashi, S., 1996. Evidence that glucose metabolism is decreased in the cerebrum of aged female senescence-accelerated mouse; possible involvement of a low hexokinase activity. *Neurosci. Letters* 214: 45–48.

Landles, C., and Bates, G.P., 2004. Huntingtin and the molecular pathogenesis of Huntington's disease. Fourth in molecular medicine review series. *EMBO Reports* 5: 958–963.

Lee, E.S., and Charlton, C.G., 2001. 1-Methyl-4-phenyl-pyridinium increases S-adenosyl-L-methionine dependent phospholipid methylation. *Pharm. Biochem. Behavior* 70: 105–114.

Lei, H., Wang, B., Li, W.P., Yang, Y., Zhou, A.W., and Chen, M.Z., 2003. Anti-aging effect of astragalosides and its mechanism of action. *Acta Pharmacologica Sinica* 24: 230–234.

Lei, M., Su, Y., Hua, X., Ding, J., Han, Q., Hu, G., and Xiao, M., 2008. Chronic systemic injection of D-galactose impairs the septohippocampal cholinergic system in rats. *NeuroReport* 19: 1611–1615.

Li, W.J., Nie, S.P., Xie, M.Y., Yu, Q., Chen, Y., and He, M., 2011. Ganoderma atrum polysaccharide attenuates oxidative stress induced by D-galactose in mouse brain. *Life Sci.* 88: 713–718.

Liu, D., Pitta, M., and Mattson, M.P., 2008. Preventing NAD(+) depletion protects neurons against excitotoxicity: bioenergetic effects of mild mitochondrial uncoupling and caloric restriction. *Annals of New York Acad. Sci.* 1147: 275–282.

Lodi, R., Schapira, A.H., Manners, D., Styles, P., Wood, N.W., Taylor, D.J., and Warner, T.T., 2000. Abnormal in vivo skeletal muscle energy metabolism in Huntington's disease and dentatorubropallidoluysian atrophy. *Annals Neurology* 48: 72–76.

Luo, Y., Niu, F., Sun, Z., Cao, W., Zhang, X., Guan, D., Lv, Z., Zhang, B., and Xu, Y., 2009. Altered expression of Abeta metabolism-associated molecules from D-galactose/AlCl(3) induced mouse brain. *Mech. Ageing Devel.* 130: 248–252.

Lupien, S., Lecours, A.R., Lussier, I., Schwartz, G., Nair, N.P., and Meaney, M.J., 1994. Basal cortisol levels and cognitive deficits in human aging. *J. Neurosci.* 14: 2893–2903.

Magarinos, A.M., and McEwen, B.S., 2000. Experimental diabetes in rats causes hippocampal dendritic and synaptic reorganization and increased glucocorticoid reactivity to stress. *Proc. Natl. Acad. Sci. USA* 97: 11056–11061.

Mark, R.J., Pang, Z., Geddes, J.W., Uchida, K., and Mattson, M.P., 1997. Amyloid beta-peptide impairs glucose transport in hippocampal and cortical neurons: involvement of membrane lipid peroxidation. *J. Neuroscience* 17: 1046–1054.

McKenzie, M., Trounce, I.A., Cassar, C.A., and Pinkert, C.A., 2004. Production of homoplasmic xenomitochondrial mice. *Proc. Natl. Acad. Sci. USA* 101: 1685–1690.

Michikawa, M., and Yanagisawa, K., 1999. Inhibition of cholesterol production but not of nonsterol isoprenoid products induces neuronal cell death. *J. Neurochemistry* 72: 2278–2285.

Mielke, J.G., Nicolitch, K., Avellaneda, V., Earlam, K., Ahuja, T., Mealing, G., and Messier, C., 2006. Longitudinal study of the effects of a high-fat diet on glucose regulation, hippocampal function, and cerebral insulin sensitivity in C57BL/6 mice. *Behavioral Brain Res.* 175: 374–382.

Northam, E.A., Rankins, D., and Cameron, F.J., 2006. Therapy insight: the impact of type 1 diabetes on brain development and function. *Nature Clinical Practice Neurology* 2: 78–86.

Novelli, G., and Reichardt, J.K., 2000. Molecular basis of disorders of human galactose metabolism: past, present, and future. *Mol. Genetics Metab.* 71: 62–65.

Ogunshola, O.O., and Antoniou, X., 2009. Contribution of hypoxia to Alzheimer's disease: is HIF-1alpha a mediator of neurodegeneration? *Cellular and Molecular Life Sci.* 66: 3555–3563.

Ohta, H., Nishikawa, H., Hirai, K., Kato, K., and Miyamoto, M., 1996. Relationship of impaired brain glucose metabolism to learning deficit in the senescence-accelerated mouse. *Neurosci. Letters* 217: 37–40.

Parameshwaran, K., Irwin, M.H., Steliou, K., and Pinkert, C.A., 2010. D-Galactose effectiveness in modeling aging and therapeutic antioxidant treatment in mice. *Rejuvenation Res.* 13: 729–735.

Payne, J., Maher, F., Simpson, I., Mattice, L., and Davies, P., 1997. Glucose transporter Glut 5 expression in microglial cells. *Glia* 21: 327–331.

Pinkert, C.A., 2002. *Transgenic Animal Technology: A Laboratory Handbook*, 2nd Ed., Academic Press, San Diego.

Pinkert, C.A,, and Trounce, I.A., 2002. Production of transmitochondrial mice. *Methods* 26: 348–357.

Pitsikas, N., and Algeri, S., 1992. Deterioration of spatial and nonspatial reference and working memory in aged rats: protective effect of life-long calorie restriction. *Neurobiology Aging* 13: 369–373.

Pogozelski, W.K., Fletcher, L.D., Cassar, C.A., Dunn, D.A., Trounce, I.A., and Pinkert, C.A., 2008. The mitochondrial genome sequence of Mus terricolor: comparison with Mus musculus domesticus and implications for xenomitochondrial mouse modeling. *Gene* 418: 27–33.

Powers, W.J., Videen, T.O., Markham, J., McGee-Minnich, L., Antenor-Dorsey, J.V., Hershey, T., and Perlmutter, J.S., 2007. Selective defect of in vivo glycolysis in early Huntington's disease striatum. *Proc. Natl. Acad. Sci. USA* 104: 2945–2949.

Rand, E.B., Depaoli, A.M., Davidson, N.O., Bell, G.I., and Burant, C.F., 1993. Sequence, tissue distribution, and functional characterization of the rat fructose transporter GLUT5. *Am. J. Physiology* 264: G1169–1176.

Revsin, Y., van Wijk, D., Saravia, F.E., Oitzl, M.S., De Nicola, A.F., and de Kloet, E.R., 2008. Adrenal hypersensitivity precedes chronic hypercorticism in streptozotocin-induced diabetes mice. *Endocrinology* 149: 3531–3539.

Revsin, Y., Saravia, F., Roig, P., Lima, A., de Kloet, E.R., Homo-Delarche, F., and De Nicola, A.F., 2005. Neuronal and astroglial alterations in the hippocampus of a mouse model for type 1 diabetes. *Brain Res.* 1038: 22–31.

Sapolsky, R.M., Krey, L.C., and McEwen, B.S., 1986. The neuroendocrinolog of stress and aging: the glucocorticoid cascade hypothesis. *Endocrine Reviews* 7: 284–301.

Saravia, F., Beauquis, J., Pietranera, L., and De Nicola, A.F., 2007. Neuroprotective effects of estradiol in hippocampal neurons and glia of middle age mice. *Psychoneuroendocrinology* 32: 480–492.

Saravia, F.E., Gonzalez, S.L., Roig, P., Alves, V., Homo-Delarche, F., and De Nicola, A.F., 2001. Diabetes increases the expression of hypothalamic neuropeptides in a spontaneous model of type I diabetes, the nonobese diabetic (NOD) mouse. *Cell. Mol. Neurobiol.* 21: 15–27.

Saravia, F., Revsin, Y., Lux-Lantos, V., Beauquis, J., Homo-Delarche, F., and De Nicola, A.F., 2004. Oestradiol restores cell proliferation in dentate gyrus and subventricular zone of streptozotocin-diabetic mice. *J. Neuroendocrinology* 16: 704–710.

Shu, H.J., Isenberg, K., Cormier, R.J., Benz, A., and Zorumski, C.F., 2006. Expression of fructose sensitive glucose transporter in the brains of fructose-fed rats. *Neuroscience* 140: 889–895.

Shulman, R.G., Rothman, D.L., Behar, K.L., and Hyder, F., 2004. Energetic basis of brain activity: implications for neuroimaging. *Trends Neuroscience* 27: 489–495.

Sokoloff, L., 1999. Energetics of functional activation in neural tissues. *Neurochemical Res.* 24: 321–329.

Song, X., Bao, M., Li, D., and Li, Y.M., 1999. Advanced glycation in D-galactose induced mouse aging model. *Mech. Ageing Devel.* 108: 239–251.

Stranahan, A.M., Arumugam, T.V., Cutler, R.G., Lee, K., Egan, J.M., and Mattson, M.P., 2008a. Diabetes impairs hippocampal function through glucocorticoid-mediated effects on new and mature neurons. *Nature Neurosci.* 11: 309–317.

Stranahan, A.M., Norman, E.D., Lee, K., Cutler, R.G., Telljohann, R.S., Egan, J.M., and Mattson, M.P., 2008b. Diet-induced insulin resistance impairs hippocampal synaptic plasticity and cognition in middle-aged rats. *Hippocampus* 18: 1085–1088.

Takeda, T., Hosokawa, M., Higuchi, K., Hosono, M., Akiguchi, I., and Katoh, H., 1994. A novel murine model of aging, Senescence-Accelerated Mouse (SAM). *Arch. Gerontology Geriatrics* 19: 185–192.

Takuma, K., Yao, J., Huang, J., Xu, H., Chen, X., Luddy, J., Trillat, A.C., Stern, D.M., Arancio, O., and Yan, S.S., 2005. ABAD enhances Abeta-induced cell stress via mitochondrial dysfunction. *FASEB J.* 19: 597–598.

Tian, J., Ishibashi, K., Ishibashi, K., Reiser, K., Grebe, R., Biswal, S., Gehlbach, P., and Handa, J.T., 2005. Advanced glycation endproduct-induced aging of the retinal pigment epithelium and choroid: a comprehensive transcriptional response. *Proc. Natl. Acad. Sci. USA* 102: 11846–11851.

Trounce, I.A., and Pinkert, C.A., 2007. Cybrid models of mtDNA disease and transmission, from cells to mice. *Current Topics Developmental Biology* 77: 157–183.

Trounce, I.A., McKenzie, M., Cassar, C.A., Ingraham, C.A., Lerner, C.A., Dunn, D.A., Donegan, C.L., Takeda, K., Pogozelski, W.K., Howell, R.L., and Pinkert, C.A., 2004. Development and initial characterization of xeno-mitochondrial mice. *J. Bioenergetics Biomembranes* 36: 421–427.

Venable, M.E., Lee, J.Y., Smyth, M.J., Bielawska, A., and Obeid, L.M., 1995. Role of ceramide in cellular senescence. *J. Biol. Chem.* 270: 30701–30708.

Wang, W., Li, S., Dong, H.P., Lv, S., and Tang, Y.Y., 2009. Differential impairment of spatial and nonspatial cognition in a mouse model of brain aging. *Life Sciences* 85: 127–135.

Wei, H., Li, L., Song, Q., Ai, H., Chu, J., and Li, W., 2005. Behavioural study of the D-galactose induced aging model in C57BL/6J mice. *Behavioral Brain Res.* 157: 245–251.

Winocur, G., and Greenwood, C.E., 1999. The effects of high fat diets and environmental influences on cognitive performance in rats. *Behavioral Brain Res.* 101: 153–161.

Xiao, F., Li, X.G., Zhang, X.Y., Hou, J.D., Lin, L.F., Gao, Q., and Luo, H.M., 2011. Combined administration of D-galactose and aluminium induces Alzheimer-like lesions in brain. *Neurosci. Bulletin* 27: 143–155.

Xu, X.H., and Zhao, T.Q., 2002. Effects of puerarin on D-galactose-induced memory deficits in mice. *Acta Pharmacologica Sinica* 23: 587–590.

Yao, J., Hamilton, R.T., Cadenas, E., and Brinton, R.D., 2010. Decline in mitochondrial bioenergetics and shift to ketogenic profile in brain during reproductive senescence. *Biochimica Biophysica Acta* 1800: 1121–1126.

Yao, J., Irwin, R.W., Zhao, L., Nilsen, J., Hamilton, R.T., and Brinton, R.D., 2009. Mitochondrial bioenergetic deficit precedes Alzheimer's pathology in female mouse model of Alzheimer's disease. *Proc. Natl. Acad. Sci. USA* 106: 14670–14675.

Zhang, W.W., Sun, Q.X., Liu, Y.H., Gao, W., Li, Y.H., Lu, K., and Wang, Z., 2011. Chronic administration of Liu Wei Dihuang protects rat's brain against D-galactose-induced impairment of cholinergic system. *Sheng Li Xue Bao* 63: 245–255.

Zhong, Y., Hu, Y.J., Chen, B., Peng, W., Sun, Y., Yang, Y., Zhao, X.Y., Fan, G.R., Huang, X., and Kong, W.J., 2011. Mitochondrial transcription factor A overexpression and base excision repair deficiency in the inner ear of rats with D-galactose-induced aging. *FEBS J.* 278: 2500–2510.

CHAPTER 39

Maltose Preference: Studies in Outbreed Weanling Rats

YOKO HORIKAWA AND NANAYA TAMAKI*

Faculty of Nutrition, University of Kobe-Gakuin, 651-2180 Nishiku, Kobe, Japan, Email: horikawa@nutr.kobegakuin.ac.jp
*Email: tamaki@nutr.kobegakuin.ac.jp; tamaki@silver.ocn.ne.jp

39.1   Introduction

Maltose is used as a sweetner, is the main sugar in starch syrup, is contained in beverages, and is also used as a parenteral supplement sugar for diabetics and as a pharmaceutical dispensing reagent. Maltose, as well as sucrose, glucose and fructose, is a natural sweetner. However, maltose preference in the short and long terms is not well established.

In this chapter, we focus on maltose preference in rats compared with other sugars with use of a dextrin diet (Dex), maltose diet (Mal), sucrose diet (Suc), glucose diet (Glc) and fructose diet (Fru), and conducted the following experiments: (1) food intake of rats fed Mal was compared with that of Dex, Suc, Glc and Fru with a single diet method (Kubota *et al.* 2003); (2) maltose preference from the selection of Mal and Suc (Mal + Suc), compared with Mal + Glc, Mal + Fru, Suc + Glc, Suc + Fru and Glc + Fru, was analyzed with a two-choice method (Yamamoto *et al.* 2010); (3) maltose preference from the selection of Dex + Mal + Glc was studied by a three-choice method (Horikawa *et al.* 2008); (4) change of maltose preference under various concentrations of protein in the diet was studied with the selection of Mal + Suc (Horikawa *et al.* 2011); and (5) The maltose selection pattern of zinc (Zn)-adequate and

Food and Nutritional Components in Focus No. 3
Dietary Sugars: Chemistry, Analysis, Function and Effects
Edited by Victor R Preedy

Zn-deficient rats was compared with a two-choice method from Mal + Suc (Horikawa *et al.* 2009).

Male albino rats (Wistar/ST, 4 weeks) were used in the studies of maltose preference. Food intake and body weight were determined daily between 09:00 and 11:00, local time. The rats were given the experimental diet for 28 days. Food was prepared according to the composition AIN-93G. Egg albumin was used as a source of protein and comprised 20% of the diet, unless otherwise indicated. Dextrin, maltose, sucrose, glucose and fructose were used as a carbohydrate on the preparation of the Dex, Mal, Suc, Glc and Fru diets, respectively. The minerals in the Zn-deficient diets were omitted from AIN-93G mineral mixture.

The diets were in powder form and contained in a 9-cm-diameter glass jar covered with a stainless steel lip containing nine 1.2 cm-diameter holes.

39.2 Maltose Preference with a Single Diet Method

When dextrin, maltose, sucrose, glucose and fructose were used as a source of carbohydrate in the diet, the food intake was evaluated daily over 28 days (Figure 39.1). The food intake of the Dex group was the highest and that of the Fru group was the lowest. The food intake Mal group was the middle level and

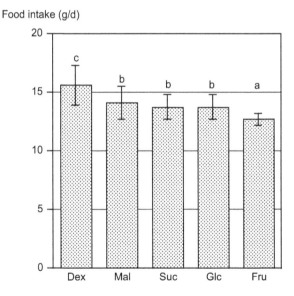

Figure 39.1 Food intake of rats fed diets containing various sugars. Daily food intake of rats is shown for diets with various sugar contents over a 28-day period. Dex, Mal, Suc, Glc and Fru, respectively, mean dextrin, maltose, sucrose, glucose or fructose used as the source of carbohydrate in the diet. Data are from 20 rats each and are shown as mean and SD (as bars). Note the highest is Dex, the lowest is Fru and the middle is Mal.

Data are from Kubota *et al.* (2003), with permission from the publishers.

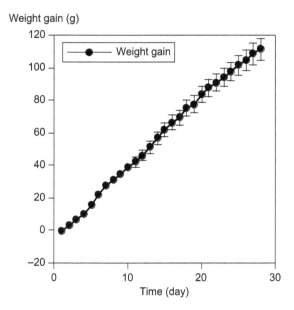

Figure 39.2 Weight gain over time in rats fed Mal. Rats were provided Mal prepared
with maltose as a carbohydrate in the diet. Each point is the mean with
SEM (as bars) from 20 rats. Note progressive weight gain of rats fed Mal.
Data are unpublished.

was not different from the Suc and Glc groups. The body weight gain of the Mal
group progressively increased (Figure 39.2). The body weight gains of Dex, Mal,
Suc, Glc, and Fru groups over 28 days were 142 ± 17^c, 111 ± 29^b, 112 ± 20^b,
105 ± 22^b, and 88 ± 10^a (mean $\pm$ SD, Values, not sharing a common superscript
letter are significantly different at $p < 0.05$) and followed their food intake.

An inverse relationship exists between the osmotic pressure exerted by a
dietary sugar and food intake (Harper and Spivey 1958). Therefore, Dex may
be more effective than disaccharide- and monosaccharide-diets on the food
intake and growth of rats. The molecular formula of fructose is the same as that
of glucose. The osmotic pressure in the digestive tracts after eating the diets
may not have any significant difference between the Glc and Fru groups. The
difference in catabolic pathway between glucose and fructose may have caused
the lower food intake and body weight gain of the Fru group.

39.3 Maltose Preference by Selection from Suc, Glc and Fru with a Two-choice Method

Various sugars in the diet affected the selection of foods on rats. Maltose,
sucrose, glucose and fructose were used as carbohydrate sources in the diet, and
the selection pattern was analyzed by a two-dish preference method. Rats were
separated individually in cages. Diets were provided as sets of two Mal + Suc,

Mal + Glc, Mal + Fru, Suc + Glc, Suc + Fru or Glc + Fru. Rats could freely select a diet from the two diets. Two diet jars were placed in a fixed position in the cage and were replaced daily. All rats were given the experimental diet and free access to water.

Over 28 days, the average food intake (g/d) of rat on groups was follows; Mal + Suc: 13.1 ± 1.0^a, Mal + Glc: $14.0 \pm 0.4^{a,b,c}$, Mal + Fru: $14.6 \pm 0.7^{b,c}$, Suc + Glc: $14.2 \pm 0.1^{b,c}$ and Glc $\pm$ Fru: 15.0 ± 0.8^c (Mean value $\pm$ SD. Values not sharing a common superscript letter are significantly different at $p < 0.05$). The weight gain of rats which simultaneously and continuously selected a diet from the two diets progressively increased each day. The body weight changes (g/d) of rats in each group were as follows; Mal + Suc, $3.6 \pm 0.4^{a,b}$: Mal + Glc, 3.2 ± 0.4^a: Mal + Fru, $3.5 \pm 0.3^{a,b}$: Suc + Glc, 3.9 ± 0.3^b: Suc + Fru, 4.0 ± 0.4^b: Glc + Fru, 3.7 ± 0.4^b (Mean value $\pm$ SD. Values not sharing a common superscript letter are significantly different at $p < 0.05$).

The value of the body weight change of Mal + Glc group was lower than those of Suc + Glc, Suc + Fru and Glc + Fru groups. Among the six groups there was no relationship between the food intake and the body weight change ($r^2 = 0.000$, $df = 4$, $p > 0.50$).

In the selection of Mal from Mal + Glc, rats preferred Mal at the same level as Glc on the first day after initiation of the experiment. The selection of Mal increased to 80% after three days and then gradually decreased to 50% (Figure 39.3A). In the case of selection of Mal from Mal + Fru, rats preferred Fru

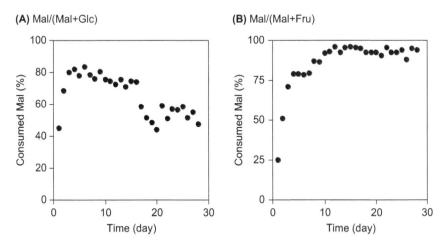

Figure 39.3 Percentage of Mal consumed by rats selecting from Mal + Glc, or Mal + Fru. Percentage of consumed Mal is shown over a 28-day period. The selected Mal in a two-dish preference test is shown for (A) Mal + Glc and (B) Mal + Fru. Each point is a mean value of 10 rats. Note preference for Mal increases with time and then gradually decrease in Mal + Glc (A), but maintain a high level in Mal from Mal + Fru.
Data are from Yamamoto *et al.* (2010), with permission from the publishers.

to Mal in the short term, and then exponentially changed their preference to Mal and sustained one-sided Mal preference through 28 days (Figure 39.3B).

The percentages of food selection in rats given a choice between two diets in the short and long term are summarized in Table 39.1. On the first and second day after initiation of the experiments, rats showed a preference for Suc = Fru > Mal = Glc, while through 28 days it was Suc > Mal > Glc > Fru. The order of increased response by the chorda tymani nerve is sucrose > fructose > maltose, glucose (Harada and Maeda 2004). Recently, the taste receptor T1R2 was identified as the sac gene product (Kitagawa *et al.* 2001; Nelson *et al.* 2001) and the coupled T1R2 and T1R3, which are sweet-responsive receptors, respond in the same order as the chorda tymani nerve (Damak *et al.* 2003; Harada *et al.* 2004; Nie *et al.* 2005). Therefore, rats may exclusively prefer Suc to Mal on the first and second days after the beginning of the selection of diets. When the taste preference of rats for solutions of

Table 39.1   Key Features of sugar preference of rats with a two-choice method.

1. Rats preferred Suc to Mal throughout 28 days.
2. Rats initially selected equally from Mal and Glc, and then preferred Mal to Glc.
3. Rats initially preferred Fru to Mal, but invariably selected Mal over Glc.
4. Rats invariably preferred Suc to Glc throughout 28 days.
5. Rats initially selected equally from Suc and Fru, but consumed dramatically more Suc over 28 days.
6. Rats initially selected equally from Glc and Fru, but invariably switched from Glc to Fru over 28 days.

This table lists the preference for sugars of rats fed for short and long terms with a two-choice method. Diets used were Mal, Suc, Glc and Fru. Values are the percentages of the indicated diet to the sum of two diets with SD after 1st day, 2nd day, and the average over 28 days. * and **, mean values are significantly different from the other diet, $p < 0.05$ and $p < 0.01$ respectively. ††, mean values are significantly different from the 1st day, $p < 0.01$. § and §§, mean values are significantly different from the 2nd day, $p < 0.05$ and $p < 0.01$, respectively. Data are from Yamamoto *et al.* (2010), with permission from the publishers.

	1st Day	2nd Day	Over 28 Days
	%	%	%
Mal/(Mal + Suc)	$14.9 \pm 9.0^{**}$	$1.5 \pm 2.0^{**}$	$28.0 \pm 19.3^{**,\S\S}$
Mal/(Mal + Glc)	45.0 ± 24.0	$68.7 \pm 20.2^{*}$	$65.0 \pm 14.5^{**,\dagger\dagger}$
Mal/(Mal + Fru)	$25.2 \pm 29.1^{*}$	51.4 ± 38.0	$87.9 \pm 1.6^{**,\dagger\dagger,\S\S}$
Suc/(Suc + Glc)	$89.6 \pm 6.1^{**}$	$89.8 \pm 18.4^{**}$	$95.4 \pm 5.1^{**}$
Suc/(Suc + Fru)	54.4 ± 30.9	48.3 ± 45.9	$95.4 \pm 5.1^{**,\dagger\dagger,\S\S}$
Glc/(Glc + Fru)	65.0 ± 29.8	$73.1 \pm 29.6^{*}$	$88.7 \pm 9.6^{**,\dagger\dagger,\S\S}$

different sugars was evaluated using brief two-bottle preference tests (Sclafani 1987; Sclafani and Mann 1987), the order of preference was maltose > sucrose > glucose = fructose at the low concentration of 0.03 M, and sucrose > maltose > glucose > fructose at the high concentration of 0.5 or 1.0 M. Preference for sugar in the long term with a two-diet method (Table 39.1) showed good agreement with that for high concentration solutions in the two-bottle test (Salafani and Mann 1987). The preference of Mal increased in the long term in all cases of Mal + Suc, Mal + Glc and Mal + Fru (Table 39.1 and Figure 39.3). Fructose feeding appears to interfere with glucose utilization *in vivo*, including an insulin-resistant state (Wei *et al.* 2007). Over-production of apo-lipoprotein is found after chronic fructose feeding for 3 weeks, but is not observed in short term (2 days) fructose feeding (Haidari *et al.* 2002). Hepatic *de novo* lipogenesis and post prandial triglyceride AUC (area under the blood concentration time curve) are increased during fructose consumption in humans (Stanhope *et al.* 2009). The post-ingestive effect may prevent Fru and Suc selection and increase maltose selection in the choice from Mal + Suc and Mal + Fru (Table 39.1).

39.4 Maltose Preference by Selection from Dex, Mal and Glc with a Three-choice Method

The chain length of glucose in the diet affects the selection of food. Dextrin, maltose and glucose were used as sources of carbohydrate in the diet and the selection patterns of rats were analyzed by a three-choice selection. The body weight of rats which continuously selected diets from among Dex, Mal and Glc increased linearly under the experimental conditions. The weight gain of rats was 102 ± 13 g for 28 days.

The percentages of daily selected diet from Dex, Mal and Glc with the three-choice method are shown in Figure 39.4. Although the three diet jars of Dex, Mal and Glc were replaced daily and rotated, all rats continuously maintained their preference by selection from the three diets. Therefore, rats could discriminate between Dex, Mal and Glc.

The feeding patterns and the preference of sugar in the three-choice method showed a characteristic phenomenon. At day 1 after onset of the experimental diets, rats equally selected Dex and Glc to 40% level, but selected Mal to 10% level, of the total food consumption. Dextrin preference increased to a 70% level at 2 and 3 days and then suddenly decreased. Maltose preference gradually increased, reaching a peak at day 9, maintained a high level until day 22 and then gradually decreased (Figure 39.4). Glucose preference remained at a low level through 28 days. From the two-choice method from Mal and Glc (Figure 39.3A) and the three-choice method, the post-ingestive effect shows a decrease after increase in the selection of Mal. These results show that the digestive tracts of rats adapts to catabolism of maltose and then the rat changes its utilization of maltose. Indeed, a single mutation of α-glucosidase from human gut microbiome changes substrate specificities (Tan *et al.* 2010).

Consumed diet (%)

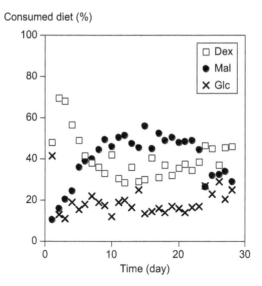

Time (day)

Figure 39.4   Percentages of consumed Dex, Mal and Glc in rats. Percentages of
consumed Dex, Mal and Glc are shown over a 28-day period. The
selections of Dex, Mal and Glc each in a three-dish preference test from
Dex + Mal + Glc are shown. Each point is the mean value of 10 rats.
Note gradual increase of Mal and decrease of Dex over time.
Data are unpublished.

39.5   Increased Maltose Preference in Rats Fed a Low-protein Diet

Protein concentrations in the diet affected the maltose preference. Maltose and
sucrose were used as sources of carbohydrate in the diet and the selection pattern
was analyzed by a two-choice method. Diets provided as a set of Mal and Suc
containing 10, 20 and 40% protein each were changed in position daily.

The increasing order of the average total food intake of rats was
$10\% > 40\% > 20\%$ protein, while the weight gain over 28 days showed no
significant difference among the groups. These results show that 20% protein
in the diet had a greater effect than 10% or 40% protein on the weight gain
of weanling rats.

The average values of daily selected Mal containing 10%, 20% and 40%
protein were 13.8 ± 2.8^c, 3.6 ± 2.7^b and 0.9 ± 1.3^a g/day, respectively (mean
value $\pm$ SD). Values not sharing a common superscript letter are significantly
different at $p < 0.05$, and those of Suc were 2.4 ± 1.8^a, 9.5 ± 2.6^b and
13.9 ± 1.7^c g/d, respectively (mean value $\pm$ SD). Values not sharing a common
superscript letter are significantly different at $p < 0.05$. The percentage of daily
consumed Mal from Mal + Suc, throughout 28 days, is illustrated in Figure 39.5.

On day 1, rats preferred Suc to Mal irrespective of the protein concentration in
the diet, and the rate of consumed Mal in the 10% protein group was larger than
those in the 20% and 40% groups. Mal preference increased with time in all the

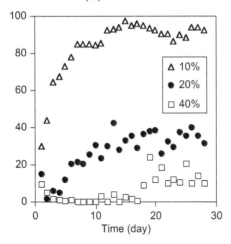

Figure 39.5   Effect of dietary proteins on maltose preference in a two-dish preference test from Mal + Suc. Percentage of consumed Mal is shown over a 28-day period. Rats were simultaneously and continuously provided with two dishes of Mal + Suc. △, ●, and □ indicate the percentages of the consumed Mal in 10%, 20%, and 40% protein groups, respectively. Each point is the mean value from 10 rats. Note exponential increase in preference of Mal in the 10% protein group over time, but predominant preference of Suc throughout in the 40% protein group.
Data are from Horikawa *et al.* (2011), with permission from the publishers.

groups. An exponential increase in the selection of Mal was found in the 10% protein group. Rats fed a high-protein diet preferred Suc to Mal through 28 days.

The food intake of rats fed Mal or Suc as a single diet was not different as shown in Section 39.1. Sucrose is a more potent stimulus than maltose for the responses from the chorida tymai nerve (Damak *et al.* 2003; Harada and Maeda 2004; Nie *et al.* 2005). Therefore, rats may have exclusively preferred Suc to Mal at day 1 after beginning of the selection of Mal and Suc, regardless of the protein concentrations in the diet. These results show that the taste effect of sucrose exceeds that of maltose. However, the preference for maltose increased with time. The increasing degree of maltose preference was affected by the concentration of protein in the diet; 10% > 20% > 40% protein. Glucose is a non-essential amino acid-related glycolysis (Waterhouse and Kwilson 1978; Pascual *et al.* 1997). Rats fed a low-protein diet may select maltose at an earlier stage to maintain their weight gain. The selection of Mal in the 10% protein diet group may reduce the disadvantage of sucrose induced by the post-ingestive effect.

39.6   Maltose Preference in Zn-deficient Rats

Zn deficiency changed the preference of foods as well as food intake in rats. Food intake of Zn-deficient rats falls to about 70% of that of control rats.

Daily food intake and body weight change data vary cyclically and can be analyzed by the Cosinor method (Tamaki *et al.* 1995; Horikawa *et al.* 2011). Food intake (*F*) and body weight change (*ΔB*) on day *t* were determined using the following equation:

$$F(\text{or } \Delta B) = M + A\cos(2\pi t/\tau + \phi) \tag{1}$$

Where *M*, *A*, *τ* and *φ* represent the mesor (the rhythm-adjusted mean), amplitude (maximum and minimum values of the adjusted mean), period (length of one complete cycle) and acrophase (phase of minimum value), respectively. The experimental data from Zn-deficient rats were fitted to the above equation by the nonlinear least-squared method, and the four parameters, *M*, *A*, *τ* and *φ*, were calculated using subroutine analysis.

All the Zn-deficient rats showed a cyclical variation in food intake and body weight change over 28 days (Horikawa *et al.* 2008, 2009, 2011; Kubota *et al.* 2003; Yamamoto *et al.* 2010).

We found a characteristic Mal selection from Mal + Suc by a two-choice method in the Zn-deficient rats. The total food intake and body weight change cycles in the rats fed a Zn-deficient diet showed a cyclical variation. Both cycles simulated from the average parameters of *M*, *A*, *τ* and *φ* are illustrated over 4 days in Figure 39.6. The level at the top of the food intake cycle of rats fed a Zn-deficient diet was the same as the average food intake in the Zn adequate control rats. When the food intake of Zn-deficient rats arrives at the top of the food intake cycle, they may avoid taking food. The top of the food intake cycle may reflect the average maximum energy requirement in all the cells. Therefore, the level of food intake decreases gradually to the bottom of the cycle, with a half period of 3.6 days. At the bottom of the food intake cycle, the Zn-deficient rats may fall to a negative energy balance and take food again. These feeding patterns were synchronized with the body weight change cycle. Weight change at the top of the weight change cycle in the Zn-deficient rats exceeds the average weight change in the control rats, reflecting the dry weight of the diet *versus* the wet weight of body mass.

The preference for maltose and sucrose in the Zn-deficient rats was different between the short and long terms. At the onset of the two-choice method selecting from Mal + Suc, the Zn-deficient rats exclusively preferred Suc, and then gradually changed their preference to Mal. The phase of the changing point was a trough in the food intake cycle. The food intake cycle followed the body weight change cycle. Zn-deficient rats may recognize their negative energy balance as being at the bottom of the body weight change cycle, and change their selection from sucrose to maltose. The change to a preference for Mal in the Zn-deficient rats thus may not merely depend on taste, but may be a sign of initiation of food intake. After changing their preference from Suc to Mal, the food intake cycle was maintained with a fixed period of 3.6 days.

Over 28 days, after beginning to select Mal, the control rats ate widely from Mal and Suc, while the Zn-deficient rats selected only Mal (Horikawa *et al.*

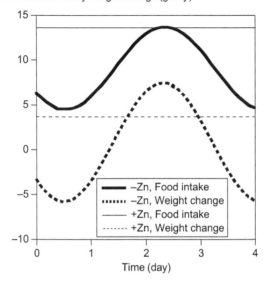

Figure 39.6 Simulation of food intake and body weight change of rats selectively fed Zn-deficient Mal + Suc or Zn-adequate Mal + Suc over time. Total food intake and body weight change of the Zn-deficient rats were simulated with a cosine curve. − Zn and + Zn mean rats fed Zn-deficient and Zn-adequate diets, respectively. Values of food intake and body weight change of the Zn-deficient and Zn-adequate diets were the average values of ten rats each, over a 28-day period. Food intake (F) and body weight change (ΔB) on day t of the Zn-deficient rats were simulated by $F = 9.1 + 4.6\cos(2\pi t/3.6 + 2.3)$ g/d and $\Delta B = 0.8 + 6.6\cos(2\pi t/3.6 + 2.3)$ g/d, respectively. The average total food intake and body weight change of rats fed a Zn-adequate diet were 13.1 ± 1.0 and 3.6 ± 0.4 g/d, respectively. The simulated values are illustrated for 4 days. Note the cyclic variation of total food intake and body-weight change of rats fed a Zn-deficient diet and the synchronization of the body-weight cycle with their food intake cycle.

Data are from Horikawa *et al.* (2009), with permission from the publishers.

2009). With a three-choice method for selection from Dex, Mal and Glc, Zn-adequate control rats selected the three diets uniformly, while Zn-deficient rats continued to select Dex or Dex and Glc (Horikawa *et al.* 2008). From these results, the preference for sugar as well as food selection and feeding pattern in Zn-deficient rats are different from those of Zn-adequate rats. Periods of nutrient deficiencies cause permanent changes to neurons caused by the nutritional deficiency itself (Kennedy *et al.* 1998). Zn-deficient rats may predominantly and continuously select Mal from Mal + Suc, after changing their Suc preference to Mal. Zn deficiency is a predominant factor underlying deviated food selection habits in animal models of anorexia, and should be useful for additional insight into its etiology in human populations.

Summary Points

- Rats could discriminate Mal from Dex, Suc, Glc and Fru on the 1st day after onset of the experimental diets.
- Increasing order of food intake is Dex > Mal = Suc = Glc > Fru with a single diet method.
- Increasing order of weight gain follows the food intake with a single diet method.
- Sugar preference in the short term is Suc = Fru > Mal = Glc with a two-choice method.
- Sugar preference in the long term is Suc > Mal > Glc > Fru with a two-choice method.
- Sugar preference among Dex, Mal and Glc, preference is Dex = Mal > Glc with a three-choice method.
- Mal preference to Suc with a two-choice method is exponentially increased with time under a low-protein diet.
- Mal preference to Suc with a two-choice method remains low level under a high-protein diet.
- Total food intake of rats fed a Zn-deficient diet from a two-choice method of Mal + Suc shows a variation with a cosine curve with a 3.6-day period.
- Zn-deficient rats predominantly and continuously selected Mal, after changing their Suc preference to Mal, with a two-choice method from Mal + Suc.
- The changing point of Suc preference to Mal is found in a trough of the food intake cycle in Zn-deficient rats.

Key Facts

- Inverse relationships between osmotic pressure and food intake are found in rats.
- Coupled proteins of T1R2 and T1R3 are isolated and found as a sweet-responsive receptor.
- The polycarbohydrate-taste receptor is different from those of other sugars.
- Food intake and weight change of rats fed a Zn-deficient diet shows a 3.5 ~ 4.0-day cycle.

Definitions of Words and Terms

Single diet method. Rat could take a diet from one dish.

Two-choice method. Rat could continuously select a diet from two dishes.

Three-choice method. Rat could continuously select a diet from three dishes.

Taste effect. Taste effect means an emotional desire for food, when animals touch the food with their mouth.

Post-ingestive effect. Post-ingestive effect means a physiological demand to select a food when animals take the food for a long time.

Preference for diet. Preference for diet is ordinated from the taste effect and the post-ingestive effect.

Weight gain. Weight gain is the increase in body-mass of the weanling rats over a 28-day term in this chapter.

Feeding pattern. Feeding pattern is the tendency to select a diet for several days.

Cosine curve in food intake or body weight change. Cosine curve is a pattern of food intake or body weight change of rats fed a Zn-deficient diet; the data of daily food and weight change can befit to a cosine formula over time.

Dermatitis. Dermatitis is a condition of abnormal cells in the epidermis induced by malnutrition such as Zn-deficiency.

List of Abbreviations

Zn   zinc
Dex dextrin diet
Mal maltose diet
Suc sucrose diet
Glc glucose diet
Fru fructose diet

References

Damak, S., Rong, M., Yasumatsu, K., Kokrashvili, Z., Varadarajan, V., Zou, S., Jang, P., Ninomiya, Y., and Margolskee, R.F., 2003. Detection of sweet and umami taste in the absence of taste receptor T1r3. *Science*. 301: 850–853.

Haidari, M., Leung, N., Mahbub, F., Uffelman, K.D., Kohen-Avramoglu, R., Lewis, G.F. and Adeli, K., 2002. Fasting and postprandial overproduction of intestinally derived lipoproteinins an animal model of insulin resistance. Evidence that chronic fructose feeding in the hamster is accompanied by enhanced intestinal de nove lipogenesis and ApoB48-containing lipoprotein overproduction. *The Journal of Biological Chemistry*. 277: 31646–31655.

Harada, S., and Maeda, S., 2004. Developmental changes in sugar responses of the chorda tympani nerve in preweanling rats. *Chemical Senses*. 29: 209–215.

Harper, A.E., and Spivey, H.E., 1958. Relationship between food intake and osmotic effect of dietary carbohydrate. *American Journal of Physiology*. 193: 483–487.

Horikawa, Y., Uehara, D., Matsuda, K., Fujimoto-Sakata, S., and Tamaki, N., 2008. Modulation of maltose preference by selection from dextrin, maltose and glucose diets in zinc-deficient rats. *Journal of Nutritional Science and Vitaminology*. 54: 203–209.

Horikawa, Y., Yamamoto, Y., Matsuda, K., Fujimoto-Sakata., Nakamura, M., and Tamaki, N., 2009. Change to a preference for maltose from sucrose over days in Zn-deficient rats selecting from maltose and sucrose diets. *Journal of Nutritional Science and Vitaminology*. 55: 353–360.

Horikawa, Y., Matsuda, K., Fujimoto-Sakata, S., and Tamaki, N., 2011. Decreased preference for sucrose in rats fed a low protein diet and its intensification by zinc-deficiency. *Journal of Nutritional Science and Vitaminology.* 57: 16–21.

Kennedy, K.J., Rains, T.M., and Shay, N.F., 1998. Zinc deficiency changes preferred macronutrient intake in subpopulations of Sprague-Dawley outbred rats and reduces hepatic pyruvate kinase gene expression. *Journal of Nutrition.* 128: 43–49.

Kitagawa, M., Kusakabe,Y., Miura, H., Ninomiya, Y., and Hino, A., 2001. Molecular genetic identification of a candidate receptor gene for sweet taste. *Biochemical and Biophysical Research Communications.* 283: 236–242.

Kubota, H., Ohyama, T., Horikawa, Y., Matsuda, K, Fujimoto-Sakata, S., and Tamaki, N., 2003. The influences of various carbohydrates on the feeding patterns of rats fed zinc-deficient diets. *Journal of Nutritional Science and Vitaminology.* 49: 228–233.

Nelson, G., Hoon, M.A., Chandrashekar, J., Zang, Y., Ryba, N.J., and Zuker, C.S., 2001. Mammalian sweet taste receptors. *Cell.* 106: 381–390.

Nie, Y., Vigues, S., Hobbs, J.R., Conn, G.L., and Munger S.D., 2005. Distinct contributions of T1R2 and T1R3 taste receptor subunits to the detection of sweet stimuli. *Current Biology.* 15: 1948–1952.

Pascual, M., Jahoor, F., and Reeds, P.J., 1997. Dietary glucose is extensively recycled in the splanchnic bed of fed adult mice. *Journal of Nutrition.* 127: 1480–1488.

Sclafani, A., 1987. Carbohydrate taste, appetite, and obesity: an overview. *Neuroscience and Biobehavioral Reviews.* 11: 131–153.

Sclafani, A., and Mann, S., 1987. Carbohydrate taste preferences in rats: glucose, sucrose, maltose, fructose and polycose compared. *Physiology and Behavior.* 40: 563–568.

Stanhope, K.L., Schwarz, J.M., Keim, N.L., Griffen, S.C., Bremer, A.A., Graham, J.L., Hatcher, B., Cox, C.L., Dyachenko, A., Zhang, W., McGahan, J.P., Seibert, A., Krauss, R.M., Chiu, S., Schaefer, E.J., Ai, M., Otokozawa, S., Nakajima, K., Nakano, T., Beysen, C., Hellerstein, M.K., Berglund, L., and Havel, P.J., 2009. Consumming fructose-sweetened, not glucose-sweetened, beverages increases visceral adiposity and lipids and decreases insulin sensitivity in overweight/obese humans. *The Journal of Clinical Investigation.* 119: 1322–1334.

Tamaki, N., Fujimoto-Sakata, S., Kikugawa, M., Kaneko, M., Onosaka, S., and Takagi, T., 1995. Analysis of cyclic feed intake in rats fed on a zinc-deficient diet and the level of dihydropyrimidinase (EC 3.5.2.2). *British Journal of Nutrition.* 73: 711–722.

Tan, K., Tesar, C., Wilton, R., Keigher, L., Babnigg, G., and Joachimiak, A., 2010. Novel α-glucosidase from human gut microbiome: substrate specificities and their switch. *The FASEB Journal.* 24: 3939–3949.

Waterhouse, C., and Keilson, J., 1978. The contribution of glucose to alanine metabolism in man. *Journal of Laboratory and Clinical Medicine.* 92: 803–812.

Wei, Y., Wang, D., Topczewski, F., and Pagliassotti, M.J., 2007. Fructose-mediated stress signaling in the liver: implications for hepatic insulin resistance. *The Journal of Nutritional Biochemistry*. 18: 1–9.

Yamamoto, Y., Horikawa, Y., Matsuda, K., Fujimoto-Sakata, S., and Tamaki, N., 2010. Preference of zinc-deficient rats for four common sugars. *Journal of Japanese Society of Nutrition and Food Science*. 63: 237–246 (in Japanese).

CHAPTER 40
Maltose and Other Sugars in Beer

GINÉS NAVARRO,[a] NURIA VELA[b] AND
SIMÓN NAVARRO*[c]

[a] Department of Agricultural Chemistry, Geology and Pedology. School of
Chemistry, University of Murcia, Campus de Espinardo, 30100 Murcia,
Spain; [b] School of Nursing, San Antonio Catholic University, Campus de los
Jerónimos, s/n, 30107 Murcia, Spain; [c] Department of Agricultural
Chemistry, Geology and Pedology, School of Chemistry, University of
Murcia, Campus de Espinardo, 30100 Murcia, Spain
*Email: snavarro@um.es

40.1   Introduction

Brewing is the art or science of making beer (Etym: From the Latin *bibere*,
meaning to drink). Four raw materials are required for beer production: barley,
hops, water and yeasts. The quality of these raw materials has a decisive
influence on the quality of the beverage. Barley is the main raw material for
beer production. Before use, the barley must first be converted into malt. Its use
depends on the fact that barley has a high starch content and the husk still
adheres to the grain, even after threshing and processing to malt. Unmalted
cereals, such as maize, rice, sorghum or wheat are often used as adjuncts. Hops
give beer its bitter taste and have an effect on the aroma. Water is quantitatively
the most important raw material. Finally, the alcoholic fermentation to form
beer depends on the activity of yeasts which consequently is essential for beer
production.

Food and Nutritional Components in Focus No. 3
Dietary Sugars: Chemistry, Analysis, Function and Effects
Edited by Victor R Preedy
© The Royal Society of Chemistry 2012
Published by the Royal Society of Chemistry, www.rsc.org

Commercial brewing is divided into six basic steps: malting (converting barley to malt), mashing (mixing grist with water), boiling (obtaining of brewer wort), fermentation (transformation of fermentable sugars in alcohol and CO_2 by yeasts), aging and finishing (flavour maturation and production of a brilliantly clear beverage), and packaging (beer filled into its final container). Figure 40.1 summarizes the main steps and sub-steps of the brewing process. More detailed descriptions can be found in Eaton (2006).

Beer comprises hundreds of different compounds (carbohydrates, inorganic salts, aldehydes, esters, *etc.*). Some of them are derived from raw materials and pass unchanged through the brewing process and others are produced during the process (Buiatti 2009). For years there has been a great deal of publicity highlighting the health benefits of a moderate consumption of beer. The influence of beer on health is related to the absence of negative attributes and the presence of positive attributes. Beer, a wholesome beverage that has been a staple part of our diet for many thousands of years, contains a number of components such as antioxidants, which can be beneficial to health. Furthermore, the nutritive aspects of beer include its low sugar content and significant amount of vitamins and minerals (Hughes and Baxter 2001).

40.2   Barley: Taxonomy, Cultivation and Uses

Barley is a grass belonging to the family Poaceae, the tribe Triticeae, and the genus *Hordeum*. Taxonomic description and details of morphology, ecology, and distribution of many species of *Hordeum*, included cultivated barley, have been described by Bothmer and Jacobsen (1985). *H. vulgare* comprises two subspecies and both cultivated and wild forms. The cultivated forms are now considered the subspecies *vulgare* of *H. vulgare,* whereas the wild forms are described as the subspecies *spontaneum* of *H. vulgare*. The former is both two-rowed and six-rowed with nonbrittle rachises, whereas the latter is two-rowed with a brittle rachis (Nilan and Ullrich, 1993).

Barley is a widely adaptable crop classified into two types and many varieties. It is currently popular in temperate areas where it is grown as a summer crop and tropical areas where it is sown as a winter crop. The winter type winter barleys are sown in about the middle of September and the summer type spring barleys in March and April. The two-row Barleys are preferably grown as spring barleys and combine all the desirable features for malt and beer production. They have a lower protein content than six-row barley and thus more fermentable sugar content.

40.2.1   Structure of the Barley Grain and Composition

Barley grain is mainly composed by husk, pericarp, testa, aleurone layer, starchy endosperm, and embryo (Palmer 2006) as can be seen in

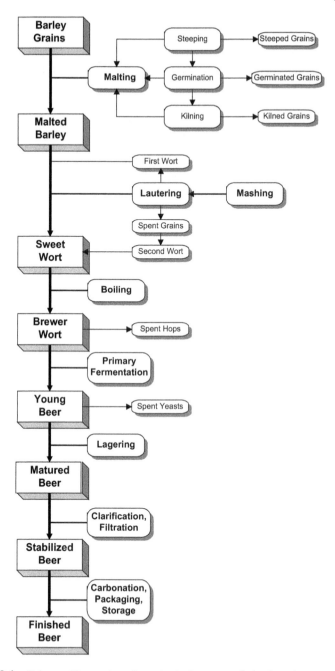

Figure 40.1 Scheme illustrating the principal stages of the brewing process. After malting, mashing, boiling and fermentation are the main steps in the brewing process.

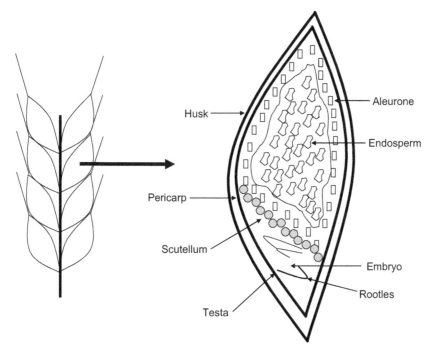

Figure 40.2   Internal structure of the barley grain. Starch is a carbohydrate mainly composed by amylose and amylopectin and it is deposited in the endosperm cells.

Figure 40.2. The husk is composed of two leaf-like structures, the ventral half (palea) and the dorsal half (lemma). The husk protects the underlaying structures of the grain, especially the embryo. The pericarp is the fruit wall of the grain. Like the husk, it contains a waxy cuticle. Below this semipermeable waxy layer is a compressed structure of cells. The testa comprises two lipid layers that enclose cellular material and it is permeable to gibberellic acid. The term endosperm comprises the pericarp, testa, aleurone, and the starchy endosperm. The starchy endosperm is the largest structure of the grain and it is made up of thousands of cells which contain the starch granules. During germination these endosperm cellules are slowly emptied in order to supply the embryo with the energy it needs for its development. The endosperm is surrounded by the aleurone layer consisting of protein-rich cells. This layer is the most important starting point for enzyme production during malting. The embryo (30–35% protein and 15% lipids) comprises two major tissues: the axis (shoot, node and roots) and the scutellum (a single cotyledon). During malting, gibberellic acid is sintethyzed and enzyme production initiated.

The moisture content of barley grain is about 14% on average although it can vary from 12–20% in dry and wet harvesting conditions, respectively. Table 40.1 shows its average chemical composition related to the dry weight.

Table 40.1 Chemical composition of barley grain. Starch is the main carbohydrate present in the barley grain. Data adapted from MacGregor and Fincher (1993).

Component	% (as dw)
Carbohydrates	78–83
Starch	63–65
Sucrose	1–2
Other sugars	1
WS polysaccharides	1–1.5
AS polysaccharides	8–10
Cellulose	4–5
Lipids	2–3
Protein	10–12
Albumins and globulins	3.5
Hordeins	3–4
Glutelins	3–4
Nucleic acids	0.2–0.3
Minerals	2
Other substances	5–6

40.2.1.1 Carbohydrates of the Barley Grain

Carbohydrates are quantitatively the most important class of compounds, but they differ considerably from one another with regard to their properties and therefore their importance in processing and the quality of the end product. In the malting and brewing industries, degradation products from both starch and wall polysaccharides are central in providing substrates for the fermentative phase of the brewing process.

Starch $(C_6H_{10}O_5)_n$ is a carbohydrate consisting of a large number of glucose units joined together by glycosidic bonds. It is formed in the slowly ripening barley grain by assimilation and subsequent condensation of glucose $(C_6H_{12}O_6)$ to form an energy reserve. Starch is deposited in granules in the endosperm cells. Most barley starches, like those from other cereal grains, contain two major components, amylose and amylopectin.

Amylose and amylopectine are built of glucose residues. However, they have very different structures and consequently differ in the ease with which they are broken down during malting and mashing. Amylose is the minor component (20–30%) of most cereal starches and consists of relatively long chains of α-(1→4)-linked D-glucose residues. Amylose is often referred to as the linear component of starch. However, although the extent of chain branching is much less than in amylopectin, it is not negligible, and so it would be more accurate to refer to amylose as having a low level of branching. Amylose can be hydrolysed completely to maltose by the combined actions of β-amylase and debranching enzymes such as isoamylase. On the other hand, amylopectin is the major component (70–80%) of most starches, and it, too, is composed of chains of α-(1→4)-linked glucose residues interconnected through α-(1→6) bonds. Unlike amylose, the unit chains of amylopectine are relatively small, and the 4–5% of interchain α-(1→6) linkages in the molecule leads to a highly branched. The extensive branching in amylopectin restricts the extent of hydrolysis by β-amylase, so the β-amylolysis limits achieved are significantly lower than those observed with amylose (MacGregor and Fincher 1993).

The major constituents of walls tissues of barley are (1→3),(1→4)- β-glucans and arabinoxylans. The (1→3),(1→4)-β-glucans from barley consists of linear chains of β-glucosyl residues polymerized through both (1→3) and (1→4) linkages. It is not possible to assign a single structure them because they are comprised of a family of polysaccharides that is heterogeneous with respect to size, solubility, and molecular structure (Woodward *et al.* 1988). The other noncellulosic polysaccharides of barley cell walls are the arabinoxylans. These polysaccharides consist predominantly of the pentoses, arabinose and xylose and are therefore often referred to as pentosans. The arabinoxylans of barley cells, like (1→3),(1→4)- β-glucans consists of a family of polysaccharides in which individual members differs in size, solubility, and molecular structure (Viëtor *et al.* 1993). Barley arabinoxylans generally have a (1→4)-β-xylopyranosyl backbone that carries single α-L-arabinofuranosyl residues.

Cellulose is a (1→4)-β- containing up to several thousand β-glucopyranosyl residues linked through (1→4)-β linkages to form very long, linear chains. The different spatial position of the glucose molecules relative one to another (β-position instead of the α-position in amylose) makes it insoluble and it cannot be broken down by malt enzymes. Therefore, it has no effect on the quality of beer.

In addition to the above polysaccharides, a number of simple sugars and oligosaccharides (glucose, fructose, sucrose, maltose, glucodifructose, raffinose and fruntans) have been found in barley kernels (Henry 1988). Most of these carbohydrates are present in low levels (<1.5%) and so are difficult to determine with high accuracy. The low levels of maltose (<0.1%) sometimes found in barley appear in endosperm regions adjacent to the embryo and aleurone. This suggests that maltose may be formed through limited starch degradation caused by preharvest sprouting of the barley. Raffinose and sucrose have the higher presence in the kernel (MacGregor and Fincher 1993).

40.3 Malting Process and Enzyme Development

Malt is germinated barley. Malt production from barley is the first step in beer production. Although malt can be made from other cereals (wheat, rye, sorghum or millet) barley malt has proved to be the most suitable malt for beer production. The purpose of malting is to produce enzymes in the barley kernel and to cause important changes in its chemical constituents.

Common malting operations involve four basic stages: barley intake, drying and storage; steeping; germination; and kilning (Bamforth and Barclay 1993). Drying enhances the storage properties of barley. It is necessary only in those climates where barley would otherwise be put into storage at moisture content in excess of 16%. After harvesting, the barley goes through a period of

dormancy (6–8 weeks). Only after this time does the barley reach its full germinative energy. A basic rule is that for malt to be made, the barley must be
capable of germination, so maltsters' source their barley with a minimum
germination of 98%.

The process commences with the steeping of barley in water to achieve a
moisture level sufficient to activate metabolism in the embryonic and aleurone
tissues, leading in turn to the development of hydrolytic enzymes. With regard
to water uptake it must be remembered that many barleys are water sensitive.
By water sensitivity, one means the situation in which the germinative energy is
greatly decreased when more water is available than is needed for germination.
In relation to water uptake, two basic barley types can be distinguished: barley
varieties showing strong rootlet growth and rapid germination and barley
varieties with a low germination vigour and weak rootlet growth. The water
content is of great importance for the germination process since enzyme formation, growth and metabolic transformations are decisively influenced by it.
The loss of oxygen from the steep is very rapid and, therefore, aeration of the
steep can improve germination. Air-resting reduces water sensitivity that
inhibits germination.

Germination is generally targeted to generate the maximum available
extractable material by promoting endosperm modification through the
development, distribution, and action of enzymes. An essential feature of
the germination process is that relative humidity (RH) of the airflow through
the grain bed should be as close to 100% as possible to avoid water loss, which
reduces the rate of modification of the endosperm. The levels of gibberellic acid
produced naturally by some grains are insufficient to produce the malting rates
required by many malsters. For this reason, malsters have been using gibberellic acid (0.2–0.3 ppm) since 1959 (Palmer 1989). However, although gibberellic acid can increase the malting rate, the asymmetric pattern of enzymic
modification of the endosperm remains the same. Aleurones from different
varieties may have different susceptibilities to gibberellins. In consequence, the
ability of barleys to develop hydrolytic enzymes may depend both on the extent
to which those barleys can synthesize gibberellin and on their responsiveness to
gibberellin (Kusaba *et al.* 1991). Also, other hormones like auxins and calcium
ions have a role to play in supporting the action of gibberellins and releasing of
α-amylase and other enzymes. With the notable exception of β-amylase, which
is formed or released in the endosperm, it is established that the aleurone cells
are capable of synthesizing most of the hydrolytic enzymes (endo $(1\rightarrow3),(1\rightarrow$
$4)$-β-glucanase, endo $(1\rightarrow4)$-β-xylanase, carboxypeptidase, α-amylases (I, II
and III) and limit dextrinase) necessary for dissolution of the starchy endosperm. All of them, with the exception of carboxypeptidase, are gibberellic acid
dependent (Bamforth and Barclay 1993). In contrast to α-amylase, which is not
detectable in ungerminated barley, β-amylase is already present in large
amounts in ungerminated barley. The amounts of α- and β-amylase formed
during germination depends on a number of factors such as varietal property,
kernel size, water content in the green malt, and temperature of germination
(Kunze 2010).

Finally, after a period of germination sufficient to achieve even modification, the "green malt" is kilned to arrest germination and stabilize the malt by lowering moisture levels, typically to less than 5%. This reduction in moisture permits long-term storage. During kilning, there is a development of colour, a reduction of enzyme activity, and an increase in acceptable flavours. Colour development results from reactions between sugars and aminoacids of the malt to form melanoidins *via* Maillard reaction (Palmer 2006).

In a well-modified malt, about 90% of the β-glucan is broken down. Although there is a positive correlation between the level of endo-β-glucanase and β-glucan breakdown during malting, the endo-β-glucanase levels of the individual grains of a barley sample do not correlate with β-glucan breakdown, suggesting that other enzymes may complement the action of β-glucanases during malting although evidence for these complementing enzymes has not yet been found (De Sa Marins and Palmer 2004).

During germination the starch content decreases and the sugar content increases. About 5–6% starch seems to have been broken down into sugar. The sugar in malt is mainly glucose, but fructose and sucrose are also present. There is hardly any maltose since it is immediately degraded (Kunze 2010).

40.4   Effect of Mashing and Boiling on the Sugar Content

Mashing is the most important process in wort production. During this process, malt grist, solid adjuncts, and water are mixed together at a suitable temperature for the malt enzymes to convert the cereal components into fermentable sugars and other nutrients (aminoacids, vitamins, *etc.*). This process occurs in three different stages (Kunze 2010): (i) gelatinisation (the swelling and bursting of starch granules in hot aqueous solution); (ii) liquefaction (the reduction of viscosity of the gelatinised starch by α-amylase); and (iii) saccharification (the complete degradation of starch to glucose, maltose, maltotriose and dextrins by amylases).

All the substances that go into solution are referred to as extract or wort. In the technical literature, the sugar content of a wort/beer is typically expressed as extract in degrees Plato (°P, specific gravity as the weight of extract in a 100 g solution at 17.5 °C).

The activity of enzymes mainly depends on temperature, pH, mashing time and mash concentration. By mashing at 62–65 °C the highest possible maltose content and highest attenuation limit is achieved. Maltose-rich worts ferment more quickly and hold the yeast longer in suspension. Temperatures higher than 72 °C for a long time produce a dextrin-rich beer with a low attenuation limit. The enzymes do not work uniformly during mashing (Guerra *et al.* 2009). The maximum activity is reached after 10–20 min and after 40–60 min enzyme activity decreases rapidly. Increasing mashing time at 62–63 °C the maltose content increases. The optimum pH for mashing is about 5.5 when more fermentable sugar is produced and the attenuation limit is higher. Finally, in thicker mashes the amount of fermentable sugars and consequently the attenuation limit is increased.

In the mashing step it is fundamental that the action of different enzymes, mainly α-amylase, β-amylase, limit dextrinase, and, α-glucosidase (Guerra *et al.* 2009). The main function of α-amylase is to break the links endo-1,4 (endoamylase) of amylose and amylopectin (Sogaard *et al.* 1993). As a result, maltose and dextrines are formed. Its optimum pH is in the range 5–7. It is resistant to the heat, conserving 70% of activity at 70 °C (Guerra *et al.* 2009). β-amylase releases maltose from the non-reducing ends of α-1,4-linked poly- and oliglucans and is therefore considered essential for starch degradation during mashing (Guerra *et al.* 2009; Ziegler 1999). This exo-enzyme acts on the unions α-1,4 of the lineal chain of amylopectin, and it stops its action at a distance of two glucose units before attacking the unions α-1,6. Its optimum pH is about 4.5 and it is less stable to the heat than α-amylase (Stenholm and Home 1999). Some investigations have verified that barley genotypes strongly differ in β-amylase thermostability (Kaneko *et al.* 2000). It is assumed that a higher degree of β-amylase thermostability enhances the formation of maltose during mashing, leading to a better fermentability (Kihara *et al.* 1998). Limit dextrinase (amylopectin-1,6-glucosidase) is the responsible enzyme of breaking dextrines in fermentable sugars. Its maximum activity is located from 67–70 °C (Stenholm 1997; Stenholm and Home 1999) in a pH ranging from 4.8 to 5.8. Finally, α-glucosidase acts breaking maltose and other sugar chains in molecules of glucose, although it is not significant in the final yield. This enzyme correctly acts between 60 and 70 °C and at pH 4.5–6.

As consequence of the enzymatic degradation of the starch, the major fraction of the extract produced in mashing consists of sugars (maltose, maltotriose and glucose) to which are added the preformed sugars present in barley (sucrose and fructose). Figures 40.3 and 40.4 show the schematic degradation of the starch (amylose and amylopectin) during mashing.

After mashing, the main objective of the wort boiling is to finalize the colour and sugar profile by the addition of hops to create the desired levels of hoppiness and bitterness (Keukeleire 2000). Enzyme activity after the normal mashing period can alter wort fermentation. Enzyme denaturation has normally taken place by the time the wort reaches boiling point (O'Rourke 2002). Table 40.2 shows the sugar content in the brewer wort after mashing and boiling.

40.5   Consumption of Sugars During Primary Fermentation

Wort is sterile prior to "pitching" with yeast. Oxygen is essential for the growth of yeast and therefore for proper fermentation. The amount of oxygen required depends on the yeast strain and its growth requirements for unsaturated fatty acids and sterols (Munroe 2006). Yeasts are classed into three categories: (i) bottom-fermenting yeast, or *Saccharomyces carlsbergensis*, reclassified

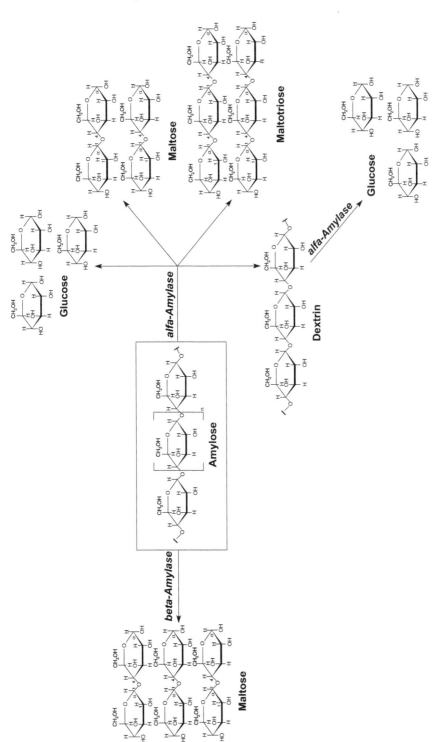

Figure 40.3 Scheme for enzymatic degradation of amylose during mashing. Main degradation carbohydrates produced during the enzymatic degradation of amylose.

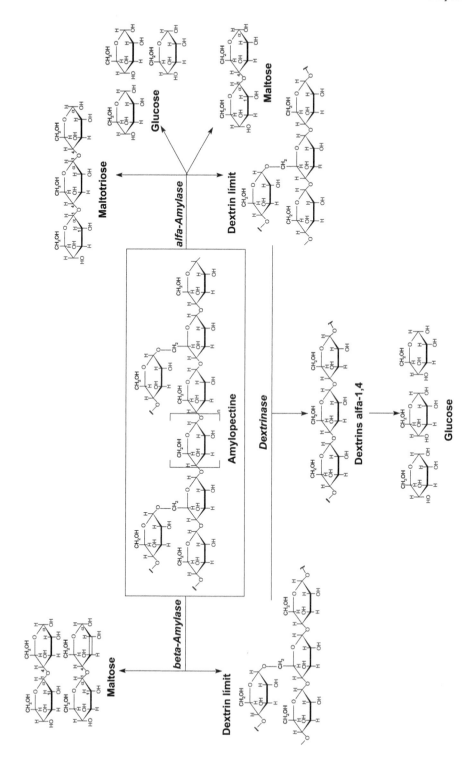

Figure 40.4 Scheme for enzymatic degradation of amylopectin during mashing. Main degradation carbohydrates produced during the enzymatic degradation of amylopectin.

Table 40.2 Sugar content in brewer wort, expressed as % of total carbohydrate. In the brewer wort, maltose has the higher content of all fermentable sugars. Data adapted from Palmer (1989).

Component	%
Monosaccharides (Hexoses)	
Fructose	1
Glucose	10
Disaccharides	
Maltose	46
Sucrose	5
Tri- and tetrasaccharides	
Maltotriose	15
Maltotetraose	10
Wort carbohydrates not fermentable	
Maltopentaose/Higher dextrins	13

Saccharomyces uvarum, (ii) top-fermenting yeast, or *Saccharomyces cerevisiae*, and (iii) wild yeasts, such as *Saccharomyces candida* and other species. Traditional lager fermentations (bottom-fermenting yeast) are conducting at temperatures ranging from about 7 to 14 °C while ale fermentations (top-fermenting yeast) make it at 15–25 °C.

During the fermentation process, yeast converts the sugars in the brewer wort into equal parts of alcohol and carbon dioxide. Sugars are fermented in accordance with the Gay-Lussac equation:

$$C_6H_{12}O_6 \rightarrow 2C_2H_5OH + 2CO_2 \, (AG = -230 \, KJ) \tag{1}$$

This process is exothermic and the energy released is stored and used by the yeast, *e.g.* in the form of ATP (adenosine triphosphate). However, this equation shows only the starting and end products. The Embden-Meyerhof-Parnas (EMP) pathway (the sequence of enzymatic reactions that produces pyruvate from glucose via the glycolysis pathway) provides more details of the reaction involved:

$$C_6H_{12}O_6 \rightarrow 2CH_3COCOO^- \, (EMP \, pathway)$$
$$2CH_3COCOO^- \rightarrow 2C_2H_5OH + 2CO_2 \, (Alcoholic \, fermentation) \tag{2}$$

In the absence of oxygen, the pyruvate is converted in ethanol and carbon dioxide primarily by way of acethaldehyde. Figure 40.5 schematizes the main compounds involved.

Figure 40.6 shows the assimilation of sugars by the yeast cell. Brewing yeast strains are capable of utilizing sucrose, glucose, fructose, maltose and maltotriose in this order, whereas the higher oligomers remain in solution. The

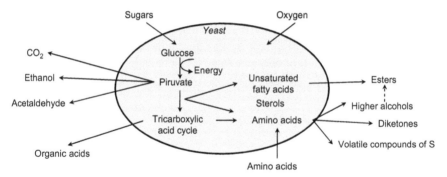

Figure 40.5   Yeast metabolism during alcoholic fermentation. Main compounds involved in the degradation of pyruvate during alcoholic fermentation.

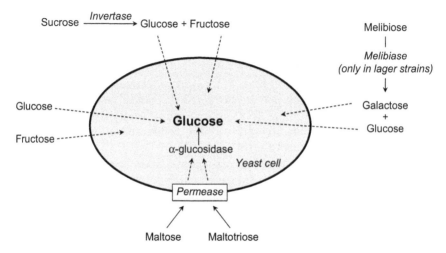

Figure 40.6   Assimilation of fermentable sugars by yeasts during fermentation. Metabolization of fermentable sugars by yeast cell during alcoholic fermentation.

sugars present in the brewer wort are not all fermented at the same time because yeast has to hydrolyse sugar polymers before it can use them (Figure 40.7). Glucose and fructose are consumed first; sucrose is easily split into fructose and glucose by yeast because the enzyme which decomposes it, invertase, is located in the cell wall; maltose (main fermentation sugar) and other disaccharides are easily and rapidly fermented by yeast after assimilation of hesoxes; Maltotriose is not fermented by yeast until the maltose has been fermented and dextrins are not fermented. The assimilation of maltose and maltotriose is produced trough of a specific permease whose synthesis does not begin until the concentration of glucose is residual. Finally it is necessary to consider the splitting possibility of melibiose in galactose and glucose by action of the enzyme melibiase although only in the case of lager yeast.

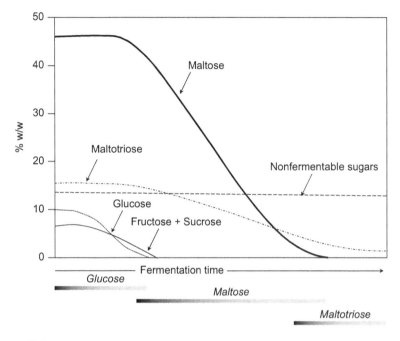

Figure 40.7 Consumption profile of fermentable sugars by yeasts during fermentation. Differences in the Assimilation rate of fermentable sugars by yeast during alcoholic fermentation.
(Adapted from Campbell 2003).

40.6 Factors Affecting Stuck and Sluggish Fermentation

By definition, a stuck fermentation is a fermentation that has stopped before all the available sugar in the wine has been converted to alcohol and CO_2 (Navarro and Navarro 2011). Symptoms of a stuck fermentation may be a long lag phase accompanied by a very slow fermentation rate and followed by no fermentation activity at all. In other cases, after a normal lag phase, active fermentation may simply stop before all fermentable carbohydrates are consumed (Munroe 2006). The serious dangers arising from the premature arrest of alcoholic fermentation are well known. Generally, residual sugar in beer is a dangerous and undesirable condition. If sugar is still present, bacteria may multiply and increase volatile acidity. Residual sugar in beer represents the major biological instability because fermentation may restart at anytime. However, retarded, sluggish or stuck fermentation of wort in beermaking not only has an unfavourable effect on beer quality but also on the economy of beer production.

It is well known that the flavour profile of beer is based on the yeast metabolism during fermentation. Consequently, all factors affecting performance and growth of yeast will also affect beer quality. To proceed properly, alcoholic fermentation needs certain environmental and chemical conditions, and these more or less coincide with those that yeasts need to grow. These conditions may change in the face of interference by other factors, some inherent in the vegetal material itself and others arising from modifications in the optimal conditions necessary for the fermentative biochemistry to function adequately. The main factors that may delay or even stop fermentation are: reduced supply of nutrients in the medium, substrate inhibition, dissolved oxygen, the presence of exogenous toxic substances (like pesticide residues), and alterations in pH and temperature.

A lack of extract and/or fermentable sugars in the wort can be due to a poor malt modification (Gunkel *et al.* 2002). If cell wall components and storage proteins are not sufficiently hydrolysed throughout the endosperm during malting, starch kernels will remain well embedded within the protein matrix and the interior of endosperm cell walls. In this state, the granules gelatinise incompletely during mashing and hence, are less accessible for amylolytic enzymes (Lauro *et al.* 1993).

The main cause of the stuck and sluggish fermentation is the restriction of nutrients in the must, since they are essential for the development and metabolic activity of the yeasts. In general, mineral elements (N and P), amino acids and vitamins are the main nutrients for the activity of the yeasts, although other compounds (survival factors) have a great influence on the development of the fermentation. Thus, zinc (as Zn^{2+}) is required (0.08–0.2 mg/L) as a cofactor in enzymatic reactions within the cells and therefore is a requirement for growth (Buiatti 2009; Munroe 2006). Also, it has been demonstrated that maltose and glucose fermentations by brewing yeast (*Saccharomyces cerevisiae*) are strongly affected by the structural complexity of nitrogen source being able to create conditions resembling those responsible for inducing sluggish/stuck fermentation (Batistote *et al.* 2006).

The fermentation process itself does not require oxygen, although the metabolism of the yeasts is dependent of the amount of dissolved oxygen to begin the fermentation. However, certain concentrations of this element favor fermentation since it encourages the direct oxidation of precursors in the biosynthesis of sterols (Henry 1982) and long chain unsaturated fatty acids (Kirsop 1982). For this reason, a decrease in the oxygen availability has as consequence the inhibition of the biosynthesis of fatty acids and sterols (Jackson 1994).

The pH decreases substantially during fermentation from 5.3–5.6 in the pitching wort to 4.3–4.6 in beer. In normal fermentations, during the fermentation phase the pH value slowly decreases and finally remains constant. An increase in pH indicates the beginning of yeast autolysis (Kunze 2010). On the other hand, a very important factor in fermentation is the control of temperature and fermentation time. Fermentation temperature control is necessary for consistent yeast growth. If a maximum fermentation temperature setpoint is specified, the excess heat must be carried away by cooling fermentor.

The presence of toxic substances in the wort (which negatively affect alcoholic fermentation) may be due to several reasons. They may originate through microbial activity in the wort (organic acids, mid-chain fatty acids and cis-fatty acids *etc.*) or they may be the consequence of *Saccharomyces* activity (Gutiérrez *et al.* 2001). Other compounds present formed during alcoholic fermentation such as acetic acid, acetaldehyde, higher alcohols or aldehydes are also considered toxic (Navarro and Navarro 2011). Some ions like carbonate (CO_3^{2-}) and bicarbonate (HCO_3^-) are not desirable as they take up hydrogen ions (H^+) from the wort to produce water and carbon dioxide (Buiatti 2009).

Also, it is very important to consider the presence of exogenous toxic substances in the barley grains like pesticide residues. As reported in the scientific literature, if barley contains pesticide residues they may remain in the malt obtained after common malting operations. Later, during mashing and boiling stages, the pesticides on the malted barley and hops can pass into the wort in different proportions. Also, if pesticide residues are present in the brewer wort they can alter the normal fermentative process being able to cause in certain cases sluggish and even stuck fermentation (Navarro and Vela 2009). Some authors have demonstrated that some triazole fungicides (SBIs), such as propiconazole or epoxiconazole, strongly affect the growth and fermentability of brewer's yeast, influencing the fermentative kinetic and, depending on the dose, causing stuck fermentation. As consequence, a higher amount of residual sugars (mainly maltose and maltotriose) has been found in beer (Navarro *et al.* 2005, 2011).

40.7 Sugar Content in Finished Beer

The analysis of sugars in beers is generally carried out by high-performance liquid chromatography (HPLC) which can provide nor only the qualitative analysis but also the quantitative determination. Refractive index (RI) measurement is the most popular detection method for sugars although other systems as evaporative light scattering detection (ELSD) have been proposed (Nogueira *et al.* 2005). Capillary electrophoresis with precolumn derivatization and UV detection has also been used for the analysis of carbohydrates in beverages (Cortacero-Ramírez *et al.* 2004). The total carbohydrates remaining in beer can be also estimated spectrophotometrically with anthrone in 85% sulphuric acid (EBC 1998).

Sugar content in beer may be different for each beer. In general, beers will contain only low levels of fermentable sugars (glucose, fructose, sucrose, maltose and maltotriose) other than those added as priming. Some yeast strains (mainly super-attenuating strains) will partially ferment maltotetraose against others (under-attenuating strains) that will not ferment maltotriose. Table 40.3 shows the content of unfermented sugars present in beer as reported by different authors.

The real extract of beer consists approximately of 75–80% of carbohydrates. The carbohydrates surviving into beer from wort are the non-fermentable dextrins (90%) and some polysaccharides (10%) coming from cell walls, in starchy endosperm of barley kernel (Buiatti 2009). Trace amounts of many other sugars including monosaccharides (D-ribose, L-arabinose, D-xylose, D-mannose and

Table 40.3 Fermentable sugar content (g/L) in finished beer. Sugar content in beer may be different for each beer. In general, beers will contain only low levels of fermentable sugars. BDL: Below Detection Limit.[1] Data adapted from Hughes and Baxter (2001).[2] Data adapted from Briggs *et al.* (2004) for ale beer.[3] Data adapted from Briggs *et al.* (2004) for lager beer.

Sugar	Range
Fructose	BDL–0.19[1]
	BDL–0.1[2]
	BDL–1.8[3]
Glucose	0.04–1.1[1]
	0.01–8.0[2]
	BDL–8.0[3]
Sucrose	0–3.3[1]
	BDL[2]
	BDL[3]
Maltose	0.7–3.0[1]
	BDL[1]–7.0[2]
	BDL–2.5[3]
Maltotriose	0.4–3.4[1]
	2.8–17.0[2]
	BDL–3.3[3]
Maltotetraose	0.4–4.0[2]
	0.9–2.0[3]

D-galactose), dissacharides (isomaltose, and cellobiose) and trisaccharides (panose and isopanose) have been detected in beer (Briggs *et al.* 2004).

Low-carbohydrate beers, initially brewed for diabetic patients, contain less carbohydrate because dextrins are digested and fermented in higher proportion than conventional beers. On the other hand, non-alcohol beers are brewed by restricting the ability of yeasts to ferment wort. These young beers have short fermentation and consequently, a great proportion of maltose, maltotriose and maltotetraose remain in them (Ferreira 2009).

Summary Points

- This chapter reports on fermentable sugars, mainly maltose, in beer.
- Carbohydrates, especially starch, are quantitatively the most important class of compounds in the barley grain.
- Starch (amylase 20–30% and amylopectine 70–80%) consists of a large number of glucose units joined together by glycosidic bonds.
- The purpose of malting is to produce enzymes in the barley kernel and to cause important changes in its chemical constituents.

- During germination the starch content decreases and the sugar content increases. The sugar in malt is mainly glucose, but fructose and sucrose are also present.
- Mashing is the most important process in wort production. During this process, the malt enzymes (mainly α- and β-amylases) transform the cereal components into fermentable sugars and other nutrients. As results, maltose and dextrines are formed.
- As consequence of the enzymatic degradation of the starch, the major fraction of the extract produced in mashing consists of sugars (maltose, maltotriose and glucose) to which are added the preformed sugars present in barley (sucrose and fructose).
- During the fermentation process, yeast (top- or bottom-yeast) converts the sugars in the brewer wort into equal parts of alcohol and carbon dioxide.
- The sugars present in the brewer wort are not all fermented at the same time because yeast has to hydrolyse sugar polymers before it can use them. Brewing yeast strains are capable of utilizing sucrose, glucose, fructose, maltose and maltotriose in this order, whereas the higher oligomers remain in solution.
- A stuck fermentation is a fermentation that has stopped before all the available sugar in the wine has been converted to alcohol and CO_2. Residual sugar in beer is a dangerous and undesirable condition. If sugar is still present, bacteria may multiply and increase volatile acidity. Residual sugar in beer represents the major biological instability because fermentation may restart at anytime.
- In general, beers will contain only low levels of fermentable sugars. The carbohydrates surviving into beer from wort are the non-fermentable dextrins (90%) and some polysaccharides (10%) although trace amounts of many other sugars may be detected in beer.

Key Facts

- The art of brewing is as old as civilization. Today, the brewing industry is a global business, consisting of several dominant multinational companies and many thousands of smaller producers.
- Beer is made from wholesome ingredients: malt, hops, yeast and water. All these materials have natural components which contribute to a "healthy balanced diet".
- Beer is a refreshing enjoyable beverage with relatively low alcoholic content which brings pleasure and social interaction to many people.
- Beer contains low sugar content and essential vitamins, minerals and antioxidants which can all contribute to a healthy diet.
- Many investigations have demonstrated that a moderate beer consumption has beneficial effects on many aspects of health such as reducing the risk of cardiovascular disease, diabetes, osteoporisis, *etc*.

Definitions of Words and Terms

Adjunct: Any unmalted cereal (wheat, corn, rice, sorghum) grain added to the mash to produce more fermentable sugars.

Aging: Maturation of beer.

Attenuation: The percentage of reduction in the wort's specific gravity caused by the transformation of sugars into alcohol and carbon dioxide through fermentation.

Barley: A cereal plant (*Hordeum vulgare*) used for malting purpose.

Brewing: Synonym of beermaking process.

Carbohydrates: Group of compounds composed of carbon, hydrogen, and oxygen, including sugars, starches, and celluloses.

Enzyme: An organic protein substance produced by living cells and that acts as a catalyst in biological and biochemical changes such as synthesis, hydrolysis, oxidative degradation, and isomerisation.

Extract: The dissolved materials in the sweet wort after mashing and lautering of malted barley and adjuncts.

Fermentation: The chemical conversion of fermentable sugars into approximately equal parts of ethyl alcohol and carbon dioxide by yeast.

Gibberellic acid: Material secreted by the barley embryo identical to a microbiological product having plant growth hormones extracted from a parasite mushroom (the fungus *Gibberella fujikuroi*).

Malt: Barley that has been steeped in water, germinated, and later dried for to convert the insoluble starch in barley in soluble substances and sugars.

Malting: The process to transform barley into malt.

Maltose: A fermentable sugar (dissacharide) consisting of two molecules of glucose obtained by the enzymatic hydrolysis of the starch.

Mashing: The process of mixing ground malt with water to extract the malt, degrade haze-forming proteins, and further convert grain starches to fermentable sugars and non-fermentable carbohydrates (dextrins).

Pitching: The addition of yeast to the wort.

Starch: Any of a group of carbohydrates or polysaccharides secreted in the form of granules by certain cereals, composed of amylose ($\approx 25\%$) amylopectin ($\approx 75\%$).

Sugars: A generic name for a class of carbohydrates including glucose, fructose, sucrose, maltose, maltotriose and others.

Wort: The bittersweet sugar solution obtained by mashing the malt and boiling in the hops before it is fermented into beer.

List of Abbreviations

AS	Alkaly soluble.
ELSD	Evaporative Light Scattering Detector.
EMP	Embden.Meyerhof-Parnas Pathway.
HPLC	High Performance Liquid Chromatography.
KJ	Kilojoule.
°P	Degree Plato.

RH   Relative Humidity.
RI   Refractive Index.
SBI   Sterol Biosynthesis-Inhibiting.
SG   Specific Gravity.
WS   Water soluble.

References

Bamforth, C.W., and Barclay, A.H.P., 1993. Malting technology and the uses of the malt. In: MacGregor, A. and Bhatty, R.S. (eds.). Barley: *Chemistry and Technology*. American Association of Cereal Chemists, Inc., St. Paul, MN, USA. pp. 297–354.

Batistote, M., Helena da Cruz, H., and Ernandes, J.R., 2006. Altered patterns of maltose and glucose fermentation by brewing and wine ywasts influenced by the complexity of nitrogen source. *Journal of Institute of Brewing*, 112: 84–91.

Bothmer, R., Von and Jacobsen, N., 1985. Origin, taxonomy, and related species. In: Barley, D.C. and Rasmusson (ed.). *American Society of Agronomy*. Madison, WI, USA. No. 26: 19–56.

Briggs, D.E., Boulton, C.A., Brookes, P.A., and Steven, R., 2004. *Brewing: Science and Practice*. Woodhead Publishing Limited. Abington Hall, Abington, Cambridge, GBR. pp. 359–360.

Buiatti, S., 2009. Beer composition. In: Preedy V.R. (ed.). *Beer in Health and Disease Prevention*. Elsevier Inc., Amsterdam, NED. pp. 213–225.

Campbell, I., 2003. Yeast and fermentation. In: Russell, I (ed.). *Whisky: Technology, Production and Marketing*. Academic Press, London, UK. pp. 115–150.

Cortacero-Ramírez, S., Segura-Carretero, A., Cruces-Blanco, C., Hernáinz-Bermúdez de Castro, M., and Fernández-Gutiérrez, A., 2004. Analysis of carbohydrates in beverages by capillary electrophoresis with precolumn derivatization and UV detection. *Food Chemistry*, 87: 471–476.

De Sa Marins, R., and Palmer, G.H., 2004. Assessment of malt modification by single grains analysis. *Journal of Institute of Brewing*, 110: 43–50.

Eaton, B., 2006. An overview of brewing. In: Priest, F.G. and Stewart G.G. (eds). *Handbook of brewing*. CRC Press. Boca Raton, FL, USA. pp. 77–89.

EBC., 1998. *Analytica European Brewery Convention*. 5th edn. Verlag Hans Carl Gentränke-Fachverlag. Grundwerk, GER.

Ferreira, I.M., 2009. Beer carbohydrates. In: Preedy V.R. (ed.). *Beer in Health and Disease Prevention*. Elsevier Inc., Amsterdam, NED. pp. 291–298.

Guerra, N.P., Torrado-Agrasar, A., López-Macías, C., Martínez-Carballo, E., García-Falcón, S., Simal-Gándara, J., and Pastrana-Castro, L.M., 2009. Use of amylolytic enzymes in brewing. In: Preedy, V. (ed.). *Beer in Health and Disease Prevention*. Academic Press, London, GBR. pp. 113–126.

Gunkel, J., Voetz, M., and Rath, F., 2002. Effect of the malting barley variety (*Hordeum vulgare* L.) on fermentability. *Journal of Institute of Brewing*, 108: 355–361.

Gutiérrez, A.R., Epifanio, S., Garijo, P., López, R., and Santamaría, P., 2001. Killer yeasts: Incidence in the ecology of spontaneous fermentation. *American Journal of Enology and Viticulture*, 52: 352–356.

Henry, S.A., 1982. The membrane lipids of yeasts: biochemical and genetic studies. In: Stratern, J.N., Jones, E.N. and Broach, J.R. (eds.). *The molecular biology of the yeast Saccharomyces: Metabolism and gene expression*. Cold Spring Harbor, New York: Cold Spring Harbor Laboratory, USA. pp. 101–158.

Henry, R.J., 1988. The carbohydrates of barley grains - a review. *Journal of Institute of Brewing*, 94: 71–78.

Hughes, P.S., and Baxter, E.D., 2001. *Beer: Quality, Safety and Nutritional Aspects*. Royal Society of Chemistry, Cambridge, GBR.

Jackson, R.S., 1994. *Wine science. Principles and applications*. Ed. Academic Press, New York, USA.

Kaneko, T., Kihara, M., and Ito, K., 2000. Genetic analysis of beta-amylase thermostability to develop a DNA marker for malt fermentability improvement in barley, Hordeum vulgare. Plant Breeding. 119: 197–201.

Keukeleire, D.D., 2000. Fundamentals of beer and hop chemistry. *Quimica Nova*, 23, 108–112.

Kihara, M., Kaneko, T., and Ito, K., 1998. Genetic variation of β-amylase thermostability among varieties of barley, Hordeum vulgare L., and relation to malting quality. *Plant Breeding*, 117: 425–428.

Kirsop, B.H., 1982. Developments in beer fermentation. *Topics in Enzyme and Fermentation Biotechnology*, 6: 79–131.

Kunze, W., 2010. *Technology of Brewing and Malting* (4th International English Edition). The Research and Teaching Institute for Brewing in Berlin (VLB). VLB's Publishing Department. Berlin, GER.

Kusaba, M., Kobayashi, O., Yamaguchi, I., and Takeda, C., 1991. Effects of giberellin on genetic variations in α-amylase production in germinating barley seeds. *Journal of Cereal Science*, 14: 151–160.

Lauro, M., Suortti, T., Autio, K., Linko, P., and Poutanen, K., 1993. Accessibility of barley starch granules to alpha-amylase during different phases of gelatinization. *Journal of Cereal Science*, 17: 125–136.

MacGregor, A.W., and Fincher, G.B., 1993. Carbohydrates of the barley grain. In: MacGregor, A. and Bhatty, R.S. (eds.). *Barley: Chemistry and Technology*. American Association of Cereal Chemists, Inc., St. Paul, MN, USA. pp. 73–130.

Munroe, J.H., 2006. Fermentation. In: Priest, F.G. and Stewart G.G. (eds.). *Handbook of brewing*. CRC Press, Boca Raton, FL, USA. pp. 487–524.

Navarro, S. and Vela, N. 2009. Fate of pesticide residues during brewing. In: Preedy, V.R. (ed.). *Beer in Health and Disease Prevention*. Elsevier Inc., Amsterdam, NED. pp. 415–428.

Navarro, S., Pérez, G., Vela, N., Mena, L., and Navarro, G., 2005. Behavior of myclobutanil, propiconazole and nuarimol residues during lager beer brewing. *Journal of Agricultural and Food Chemistry*, 53: 8572–8579.

Navarro, S., Vela, N., Pérez, G., and Navarro, G., 2011. Effect of sterol bio-synthesis-inhibiting (SBI) fungicides on the fermentation rate and quality of young ale beer. *Food Chemistry*, 126: 623–629.

Nilan, R.A., and Ullrich, S.E., 1993. Barley: Taxonomy, origin, distribution, production, genetics, and breeding. In: MacGregor, A. and Bhatty, R.S. (eds.). *Barley: Chemistry and Technology*. American Association of Cereal Chemists, Inc., St. Paul, MN, USA. pp. 1–29.

Nogueira, L.C., Silva, F., Ferreira, I.M.P., and Trugo, L.C., 2005. Separation and quantification of beer carbohydrates by high-performance liquid chromatography with evaporative light scattering detection. *Journal of Chromatography A*, 1065: 207–210.

O'Rourke, T., 2002. *The function of wort boiling*. Brewer International, 2: 17–19.

Palmer, G.H., 1989. *Cereal science and technology*. Aberdeen University Press. Aberdeen, GBR. 463 pp.

Palmer, G.H. 2006. Barley and malt. In: Priest, F.G., and Stewart G.G. (eds.). *Handbook of brewing*. CRC Press, Boca Raton, FL, USA. pp. 139–175.

Sogaard, M., Abe, J., Martin-Eauclaire, M.F., and Svensson, B., 1993. α-Amylases: Structure and function. *Carbohydrate Polymers*, 21: 137–146.

Stenholm, K., 1997. *Malt Limit Dextrinase and it's importance in Brewing*, Espoo: Technical Research Centre of Finland (VTT Publications 323).

Stenholm, K., and Home, S., 1999. A new approach to limit dextrinase and its role in mashing. *Journal of Institute of Brewing*, 105: 205–210.

Woodward, J.R., Phillips, D.R., and Fincher, G.B., 1988. Water soluble (1 → 3),(1 → 4)-β-D-glucans from barley (Hordeum vulgare) endosperm. IV. Comparison of 40 °C and 65 °C soluble fractions. *Carbohydrate Polymers*, 8: 85–97.

Ziegler, P., 1999. Cereal β-amylases. *Journal of Cereal Science*, 29: 195–204.

CHAPTER 41

Fructose Absorption and Enteric Metabolism

KATE WITKOWSKA AND CHRIS CHEESEMAN*

Dept of Physiology, 7-22 Medical Sciences Building, University of Alberta, Edmonton, Alberta, T6G 2H7, Canada
*Email: Chris.cheeseman@ualberta.ca

41.1 Sources of Fructose in the Human Diet

Fructose is a hexose with the same basic chemical formula as glucose ($C_6H_{12}O_6$). However, unlike glucose, it can exist in solution in both the furanose and the pyranose form, 30% with 70%, respectively (Figure 41.1). The majority of fructose in the diet occurs as the D-isomer, which is the only form transported or metabolized in humans. In today's Western diet, a large proportion of dietary fructose comes from sucrose or High Fructose Corn Syrup (HFCS). Sucrose is a dimer formed by the linkage between a glucose and a fructose molecule, whereas HFCS is a mixture of free fructose ($\sim 55\%$) and free glucose ($\sim 45\%$). Fruits and honey are also rich sources of fructose, but tend to make up a small proportion of the modern diet.

41.2 Intestinal Absorption Mechanisms

41.2.1 Site of Fructose Absorption

The small intestine is the site of carbohydrate absorption in mammals, including humans. Despite the complex structure of this tissue with its rich

---

Food and Nutritional Components in Focus No. 3
Dietary Sugars: Chemistry, Analysis, Function and Effects
Edited by Victor R Preedy
© The Royal Society of Chemistry 2012
Published by the Royal Society of Chemistry, www.rsc.org

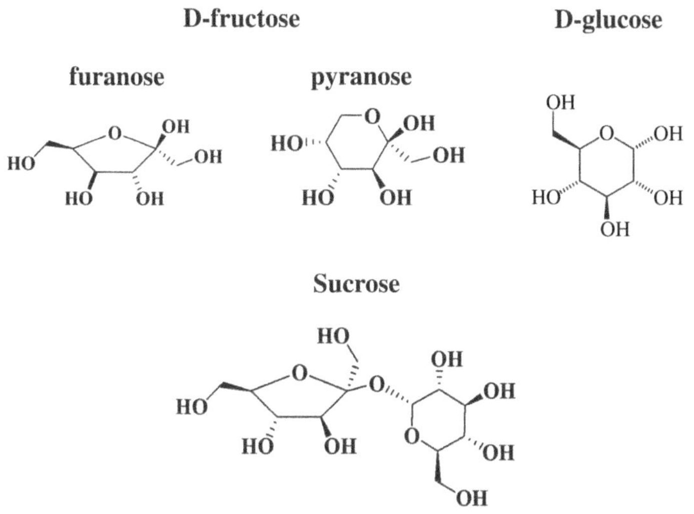

D-fructose   **D-glucose**

furanose   **pyranose**

Sucrose

Figure 41.1   Comparison of the structures of fructose in its furanose and pyranose forms with glucose in its pyranose form. Also shown is the disaccharide sucrose.

blood supply, abundant neural innervations, and lymphatic drainage, the lumen is separated from the blood by only a single-cell thick epithelium. This is comprised of multiple discrete cell types including sensory, endocrine, mucus secreting, immune and absorptive cells (enterocytes), which cover the finger-like villi. These appear to serve two functions: firstly, they increase the absorptive surface area and secondly, they provide a structural basis for rapid epithelial rejuvenation.

41.2.1.1   Regenerative Properties of Absorptive Epithelium

The epithelium is continually renewed by a process of stem cell division and migration from the crypts at the base of the villi, up to their tips. This migration takes approximately five days. Recent *vital* confocal microscopy has shown cell loss along the full length of the villi rather than just at the tips, as had been assumed to occur. The stimulus for this loss is not fully understood, but is clearly related to maintaining a fully integral barrier between the lumen and the blood stream (Watson *et al.* 2009). The death of these cells requires multiple steps, leaving the BLM in place until close to the final extrusion maintaining epithelial integrity (Wang *et al.* 2011). As the enterocytes migrate up the villi they acquire their transport and digestive capacity so that they are fully functional by two-thirds of the way up the villi.

41.2.2   Absorption of Free Fructose

Fructose uptake across the intestinal epithelium is mediated by specific transport proteins embedded in the apical (lumen) and basolateral (blood facing)

membranes of the enterocytes. Only free hexoses can be taken up across the
epithelial apical (brush border) membrane and so complete breakdown of complex carbohydrates to glucose, galactose and fructose is required for absorption.

41.2.3   GLUT5 (SLC2A5)

The major protein responsible for fructose uptake across the apical membrane
is the facilitated hexose transporter GLUT5. It is a member of the solute carrier
gene family SLC2A, of which there are 14 isoforms in humans. SLC2A5 has a
major preference for fructose and its K_m (concentration required to reach half
maximal transport rate, an apparent affinity measure) is about 5mM. The
human isoform has a very limited capacity to handle glucose (unlike the rat
isoform) and there is no evidence to suggest it can carry galactose (Joost
and Thorens 2001). The only energy to drive the entry of fructose into the
enterocytes is provided by the concentration gradient of the substrate between
the lumen and the cytoplasm (Manolescu *et al.* 2007). GLUT5 does not appear
to distinguish between the pyranose or furanose forms of fructose (Girniene *et al.*
2003). It is believed that upon entering the cells, fructose is rapidly phosphorylated providing a sink for further hexose uptake (see metabolism below).

Clearly, this route of entry differs significantly from that of glucose and
galactose, which can be taken up by SGLT1, a sodium-coupled active transporter. This contrasts with the uptake of fructose in being faster and likely
more efficient. This may explain why intake of high concentrations of fructose,
as found in some fruit juices, can overwhelm the sugar absorptive capacity,
manifesting in bloating and, in extreme cases, in osmotic diarrhea in some
individuals (Born 2007; Brian 2005).

41.2.4   GLUT2

In addition to GLUT5, there is now good evidence for the presence in the BBM
of a second facilitated hexose transporter with fructose-carrying capacity.
Human GLUT2 (SLC2A2) has a higher capacity for glucose than for fructose,
but with a lower affinity (Km $\sim$ 11 and $\sim$ 66 respectively, Colville *et al.*, 1993).
So this transporter functions best at relatively high luminal substrate concentrations (Joost and Thorens 2001). Furthermore, GLUT2 appears to reside
in the BBM transiently, and only when concentrations of carbohydrate within
the intestinal lumen are high (Mace *et al.* 2007).

41.2.5   Putative Fructose Transporters

GLUT7 is expressed in the apical membrane of the ileum and exhibits a very
high affinity, but low capacity, for fructose and glucose (Li *et al.* 2004). It is not
clear what role, if any, this transporter plays in the final stages of fructose
absorption. SGLT4, a member of the SGLT family of proteins, was cloned
recently from human small intestine (Tazawa *et al.* 2005). It can bind fructose,
but as yet, no role for this secondary active transporter in fructose absorption
has been demonstrated.

41.2.6 Intestinal Metabolism of Fructose

Fructose is rapidly phosphorylated to fructose-1-phosphate upon entry into the cytoplasm by either ketohexokinase (fructokinase), the enzyme which specifically attaches a phosphate to the one position of the carbon ring or hexokinase which can phosphorylate fructose as well as glucose at the 6 carbon position. Hexokinase has a far higher affinity for glucose than for fructose and so when both hexoses are being absorbed together there is little opportunity for fructose accessing this enzyme. Generation of either fructose 1- or 6-phosphate in the jejunal enterocytes results in the subsequent formation of lactate *via* glycolysis.

41.2.7 Release of Fructose and Metabolites into the Circulation

In most species of mammals, functional and biochemical studies indicate the presence of GLUT2 in the enterocyte BLM. However, recent immuno-gold studies using sections of human small intestine, indicate an abundant presence of GLUT5 in the BLM, as well as the BBM (Blakemore *et al.* 1995). In the absence of functional studies using basolateral membranes prepared from human jejunum, it seems reasonable to propose that GLUT5 could be a major route of exit for free fructose into the blood stream, while the similarly abundant GLUT2 would mediate primarily glucose fluxes.Studies using GLUT5 knockout mice indicate very poor absorption of fructose, which strongly supports the role of this transporter in fructose uptake and/or absorption into the bloodstream (Barone *et al.* 2009.).

However, fructose should be rapidly metabolised upon entry into the enterocyte cytoplasm resulting in the generation of lactate which could exit by the moncarboxylate transporter 1 (Kristl 2009). Both animal and human studies show that during a fructose meal the systemic concentration of the hexose increases significantly. However, it is always much lower than the concentration of glucose (50–500 µM fructose *versus* 5.5 mM glucose) (Neiwoehner 1986). Direct measurement of portal fructose concentrations in humans indicate that it reaches ∼ 1 mM maximum, which is much lower than that of glucose, which can exceed 5 mM. The liver expresses GLUT2 and has high hexokinase and fructokinase activity, suggesting that any fructose exiting the intestine would be cleared from the portal circulation, converted to glucose-6-phosphate, and stored as glycogen, or re-released as glucose.

41.2.8 Regulation of Fructose Uptake During a Meal

Recently, it has been shown that the intestine responds rapidly to the presence of carbohydates within the intestinal lumen by increasing the apical abundance of SGLT1 and GLUT2, but not GLUT5 (Helliwell *et al.* 2000). The sensing mechanisms appear to involve the expression of the taste receptors T1R2 & T1R3 within the enteroendocrine epithelial cells found all along the length of the small intestine (Le Gall *et al.* 2007). These two taste receptor proteins form a protein oligomer with the gustatory G-protein gustducin, which acts as a sweetness detector (Jang *et al.* 2007). Hexoses, sucrose, some amino acids, and artificial sweeteners can bind to this complex and initiate a second messenger cascade within these peptide releasing cells (Pepino and Bourne 2011; Renwick and Molinary 2010; Steinert *et al.* 2011). Animal knockout studies indicate that the presence of these molecules is detected through this taste receptor mechanism, resulting in the release of a number of enteric peptides, which have short- and long-term effects on hexose transporter expression. The endocrine K cells in the duodenum release glucose-dependent-insulintropic polypeptide, GIP into the circulation promoting the initial rapid release of insulin from the pancreas. This feed-forward control loop anticipates the entry of glucose (and fructose) into the circulation and primes the system to handle these key metabolites. GIP also promotes the release of glucagon-like peptides (GLP) 1 and 2 from L-type endocrine cells found predominantly in the ileum. GLP1 promotes insulin release from the pancreas, while GLP2 also causes the insertion of SGLT1 and GLUT2 into the intestinal BBM within minutes. GIP and GLP-2 are found on neuronal cells within the intestinal submucosa (Steinert and Beglinger 2011) which then act through one or more unknown pathways.

These complex control loops, which detect the presence of hexoses within the intestinal lumen, transiently increase the absorptive capacity for glucose and, to a lesser degree, fructose. The apical expression of GLUT5 appears to be unaffected by this mechanism, but is upregulated by the release of leptin into the lumen of the stomach. Leptin then reaches the jejunum and interacts with receptors on the BBM (Horio *et al.* 2010; Sakar *et al.* 2009.) (Figure 41.2). There is also good evidence that in the presence of luminal glucose, the absorption of fructose is increased. The absorption of fructose from sucrose is faster than fructose alone, which can be explained by the release of free glucose from sucrose, as it is hydrolyzed at the cell surface, triggering the insertion of both SGLT1 and GLUT2 through the enteric peptide signalling pathway (Jones *et al.* 2011; Mace *et al.* 2007; Stearns *et al.* 2010). Evidence supporting this possibility has recently been reported by Brot-Laroches' group who have found apical GLUT2 in the small intestine of diabetic patients (Ait-Omar *et al.* 2011).

41.2.9 Adaptation to Altered Dietary Intake of Hexoses

The small intestines of omnivores are able to alter their absorptive capacity in response to increased or decreased dietary carbohydrate content. This ensures an appropriate balance between energy expenditure to synthesize additional digestive enzymes and sustained reserve transport capacity. It has been argued by Diamond and co-workers that this provides a safety margin protecting the system from overload by a sudden surge in intake, which could lead to overspill into the colon (Lam *et al.* 2002). This would promote a bacterial overgrowth and also loss of potentially valuable nutrients, as carbohydrates represent a valuable source of carbon. The molecular mechanisms for this adaptation appear to involve detection of luminal hexose content with the induction, or repression, of promoters for transporter and

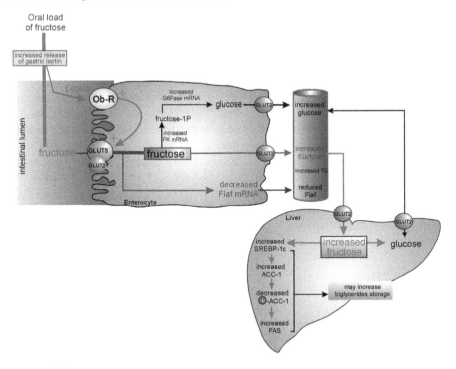

Figure 41.2   Routes of entry of fructose across the intestine and take up in the live. Interactions between transport, metabolism and their regulation by leptin (taken from Sakar *et al.* 2009). Permission to reproduce this figure was provided by A. Amar and is gratefully acknowledged.

hydrolase synthesis in the newly forming enterocytes within the crypts. The sensing mechanisms involve the same taste receptors described in the regulatory mechanisms for rapid increases in transporter functional activity. Because this increased expression starts in the cells within the crypt, it takes four to five days for these cells to reach the upper villus and be fully effective. Hence, full up-regulation of expression is congruent with this cell turnover rate, providing the higher intake is sustained over that time. This relatively slow response is presumably an ancestral feature of hunters and gatherers in order to match metabolic rates with changes in season and/or location of nutrient supply. In modern societies, the sourcing of food from all over the world, reducing seasonal variations in supply, and a surplus of foods rich in carbohydrate content, leads to a permanently elevated level of hydrolase and transporter expression in most individuals.

41.2.10   Fructose Malabsorption and Development

Fructose absorption is much slower than that of glucose and galactose and there is abundant evidence that it can be overwhelmed, resulting in an osmotically driven retention of fluid in the lumen and spillover into the colon. Both of these result in feelings of discomfort, bloating and cramps, which mimic the symptoms of irritable bowel disease (IBD). These symptoms appear to be quite common in children, but may also be more prevalent in adults than had been previously appreciated (Gibson *et al.* 2007; Kyaw and Mayberry 2011). When hexoses enter the colon, they are metabolized by the colonic bacteria to free fatty acids with a resulting release of free hydrogen, which passes rapidly across the wall of the colon to reach the circulating blood, and is then expired from the lungs. Thus, breath collection after the ingestion of a defined quantity of fructose, and the measurement of changes in free hydrogen (hydrogen breath test), can serve as an indirect determination of degree of hexose absorption. However, there is little agreement on the exact protocol needed to accurately assess fructose malabsorption. Gibson *et al.* (2007) have reviewed this literature thoroughly and provided an excellent critique of the method and how to interpret the clinical relevance of the literature. Measurements in adults have suggested that even a dose of 25 g of free fructose in a 10% solution produces a significant increase in breath hydrogen in 50% of adults tested, while 50 g shows a significantly higher peak in breath hydrogen in between 37% to 80% of adults.

Total fructose intake has increased significantly over the last 50 years in the United States and, more importantly, the form of intake has changed (Beyer *et al.* 2005). In the late '70s, the average fructose intake was about 37 g per day, of which two-thirds came from consumption of soft drinks rather than fruit, but sucrose was the main sweetener. However, by 1990, there had been a ten-fold increase in the consumption of HFCS, which is now used as the major soft drink additive. There is an active debate as to whether the use of HFCS in place of sucrose is playing a significant role in the rapidly increasing incidence of obesity and type II diabetes. To date, studies have found no difference in metabolism of free fructose and that which is bound as sucrose (Beyer *et al.* 2005). However, comprehensive studies looking at related fat metabolism have yet to be completed. In the absence of a full understanding of the regulation of hexose absorption and in particular of fructose uptake in the human small intestine, it is difficult to make a firm judgment. Also, the observation that fructose has a lower glycaemic index than glucose can lead to a naïve assumption that this does not represent a similar or equivalent caloric intake. Clearly, fructose intake does not result in the same increase in plasma glucose concentration as results from the same dose of glucose, since more than 50% of it is cleared from the portal circulation by the liver in a single pass and, subsequently, stored as glycogen in the liver and muscle. In addition, the uptake of fructose by the liver and its conversion to fructose-1-phosphate results in upregulation of hepatic glucose uptake and glycogen deposition. Thus, not only does the ingestion of glucose and fructose together promote the uptake of both hexoses, but it results in a more effective glucose clearance by the liver and a lower glycaemic index (McGuinness and Cherrington 2003). While these effects may be better for patients with type II diabetes, it is not clear what the metabolic consequences are for healthy individuals.

Summary Points

- This chapter reviews the processes by which dietary fructose is absorbed from the diet and enters the circulation.
- Recent findings indicate that this process is mediated by two facilitated hexose transporters (GLUT2 and GLUT5).
- The absorption of fructose does not appear to be as rapid and efficient as it is for glucose, which uses additional transport proteins.
- The ingestion of relatively large quantities of fructose leads to adverse consequences in many children, including bloating, diarrhea and flatulence.
- Recent evidence suggests that the poor tolerance to dietary fructose is not restricted to children and may be far more common in the adult population than previously appreciated.
- The significant increase in the fructose content of the Western diet appears to be related to the increased incidence of obesity and type II diabetes.

Key Facts of Fructose Absorption and Enteric Metabolism

- Fructose is present in the diet either as the free hexose or as a part of sucrose, a dimer of glucose and fructose.
- Fructose is absorbed by the epithelial cells of the small intestine as a monomer.
- Failure to fully absorb hexoses, including fructose, results in osmotic diarrhea, bloating and discomfort.
- Fructose absorption is mediated by two facilitated hexose transporters GLUT2 and GLUT5. GLUT5 is constitutively expressed on both sides of the epithelial cells. GLUT2 is constitutively expressed on the blood side but only transiently on the apical side during a meal.
- The transient expression of apical GLUT2 is mediated by a combination of taste receptors in the small intestinal epithelium, sensing by enteric endocrine cells and the release of enteric peptide hormones.
- Expression of GLUT2 and GLUT5 increases when the dietary carbohydrate content is elevated for more than five days.
- Fructose can be metabolised to lactate within the intestinal epithelium, but most appears in the portal circulation as free fructose.
- Fructose malabsorption is known to be common in children, but may also be quite prevalent in adults.

Definition of Words and Terms

Active transport: A mechanism by which energy is coupled to the movement of one or more substrates whose movement is mediated by that transporter. This may be direct hydrolysis of ATP as happens with the sodium/potassium ATPase or through indirect use of an existing energy gradient. Most cells have a sodium gradient across their plasma membranes and transporters, such as the sodium glucose transporter (SGLT1), and by coupling the sodium to the glucose movement the hexose can be driven into the cell against a concentration gradient.

Adaptation: This is a physiological process in which a controlled mechanism may change its set-point if a change in the environment is maintained for a prolonged period of time. This occurs to protect the system and allow for the maintenance of an appropriate safety margin. In the case of the intestine the capacity to absorb certain nutrients can be increased by expressing more transporter proteins or digestive enzymes if the intake of that nutrient is elevated for more than five days.

Enterocytes: Specialised epithelial cells making up about 80% of the intestinal epithelium. They are polarised with a very distinct apical surface folded into micro-projections or microvilli in which are embedded numerous transport proteins and on the surface of which are attached digestive enzymes. Their lateral surface also has numerous transport proteins embedded in it, which moves solutes into a space between the cells which connects to the blood stream.

Feed forward control loop: This is a control system designed to minimize variations of the controlled variable, such as the blood glucose concentration, by anticipating a change before it happens. In the case of insulin release, which will lower blood glucose, the intestine senses the presence of food and the subsequent entry of glucose into the circulation and promotes insulin release.

GLUTs or facilitative hexose transporters: These transport proteins are members of the solute carrier gene family 2A (SLC2A). They are expressed in almost all cell types and mediate the entry or exit of only free hexoses across the plasma membrane. The driving force for substrate movement is the concentration gradient of the substrate across the cell membrane.

Immuno-gold technique: This technique allows the detection of specific molecules within cellular structures using electron microscopy. Very thin sections of tissue are probed with antibodies which bind to a particular molecule like a transport protein. These in turn can then be found using a second antibody to which is attached a gold particle of a defined size. These then show up as black dots on the electron microscope image.

Intestinal villi: The epithelial surface of the small intestine is organized into finger-like projections which increase the surface area and provide a structural framework for cell turnover. Stem cells at their base divide to produce new cells which migrate towards the tips and mature as they progress along the villi. The cores of these villi contain rich blood, lymph and neural networks.

Kinases: These are proteins which catalyse the attachment of a phosphate group to a specific substrate. In the case of hexokinase this enzyme attaches a phosphate group to the carbon 6 of hexoses such as glucose or fructose.

Km or apparent affinity constant: The kinetic behaviour relating the rate of transport to substrate concentration of transport proteins can be described by constants which mimic those of enzyme kinetics. The substrate concentration at which the system reaches its half-maximal rate of transport is called the Km and is related to the affinity of the protein for its substrate/s.

Knock-out mice: In order to test the role of a particular gene or gene product within the physiology of an animal it is possible to genetically manipulate the animal such that the gene is silent. This can be done such that the animal never expresses that gene in any tissue, or only when a particular condition is induced or only in a specific tissue or tissues.

Taste receptors: This family of signalling molecules is found expressed on the surface of cells. They are made up of a complex of proteins which together can bind specific molecules on the outside of the cell and then signal through phosphorylation cascades inside the cell resulting in the detection of specific molecules. Originally they were found in special cells on the surface of the tongue and defined as conferring the sensation of taste.

List of Abbreviations

BBM	Brush border or apical membrane of the intestinal epithelial cells.
BLM	Basolateral membrane of the intestinal epithelial cells adjacent to the bloodstream.
GIP	Glucose-dependent-insulintropic polypeptide.
GLUT2	Human facilitated glucose/hexose transporter 2.
GLUT5	Human facilitated glucose/hexose transporter 5.
GLUT7	Human facilitated glucose/hexose transporter 7.
GLP1	Glucagon-like peptide 1.
GLP2	Glucagon-like peptide 2.
HFCS	High Fructose Corn Syrup.
K-cells	Endocrine cells found within the intestinal epithelium which sense the presence of nutrients within the lumen and release GIP into the circulation.
L-cells	Endocrine cells found within the intestinal epithelium which sense the presence of nutrients within the lumen and release GLP1 and GLP2 into the circulation.
SGLT1	Member of the sodium-coupled glucose transporter family, number 1.
SGLT4	Member of the sodium-coupled glucose transporter family, number 4.
T1R1	Taste receptor type one, member 1.
T1R3	Taste receptor type one, member 3.

References

Ait-Omar, A., Monteiro-Sepulveda, M., Poitou, C., Le Gall, M., Cotillard, A., Gilet, J., Garbin, K., Houllier, A., Château, D., Lacombe, A., Veyrie, N., Hugol, D., Tordjman, J., Magnan, C., Serradas, P., Clément, K., Leturque, A., and Brot-Laroche, E., 2011. GLUT2 Accumulation in Enterocyte Apical and Intracellular Membranes: A Study in Morbidly Obese Human Subjects and ob/ob and High Fat-Fed Mice. *Diabetes*. Aug 18. [Epub ahead of print].

Barone, S., Fussell, S.L., Singh, A.K., Lucas, F., Xu, J., Kim, C., Wu, X., Yu, Y., Amlal, H., Seidler, U., Zuo, J., and Soleimani, M., 2009. Slc2a5 (Glut5) is essential for the absorption of fructose in the intestine and generation of fructose-induced hypertension. *J Biol Chem.* 284(8): 5056–5066.

Beyer, P.L., Caviar, E.M., and McCallum, R.W., 2005. Fructose Intake at Current Levels in the United States May Cause Gastrointestinal Distress in Normal Adults. *Journal of the American Dietetic Association.* 105: 1559–1566.

Blakemore, S.J., Aledo, J.C., James, J., Campbell, F.C., Lucocq, J.M., and Hundal, H.S., 1995. The GLUT5 hexose transporter is also localized to the basolateral membrane of the human jejunum. *Biochemical Journal.* 309 (1): 7–12.

Born, P., 2007. Carbohydrate malabsorption in patients with non-specific abdominal complaints. *World Journal of Gastroenterology.* 13(43): 5687–5691.

Colville, C.A., Seatter, M.J., Jess, T.J., Gould, G.W., and Thomas, H.M., 1993. Kinetic analysis of the liver-type (GLUT2) and brain-type (GLUT3) glucose transporters in Xenopus oocytes: substrate specificities and effects of transport inhibitors. *The Biochemical Journal.* 290(3): 701–706.

Gibson, P.R., Newnham, E., Barrett, J.S., Shepherd, S.J., and Muir, J.G., 2007. Review article: fructose malabsorption and the bigger picture. *Alimentary Pharmacology Therapeutics.* 25(4): 349–363.

Girniene, J., Tatibouët, A., Sackus, A., Yang, J., Holman, G.D., and Rollin, P., 2003. Inhibition of the D-fructose transporter protein GLUT5 by fused-ring glyco-1,3-oxazolidin-2-thiones and -oxazolidin-2-ones. *Carbohydr Research.* 338(8): 711–719.

Helliwell, P.A., Richardson, M., Affleck, J., and Kellett, G.L., 2000. Regulation of GLUT5, GLUT2 and intestinal brush-border fructose absorption by the extracellular signal-regulated kinase, p38 mitogen-activated kinase and phosphatidylinositol 3-kinase intracellular signalling pathways: implications for adaptation to diabetes. *Biochemical Journal.* 350 (1): 163–169.

Horio, N., Jyotaki, M., Yoshida, R., Sanematsu, K., Shigemura, N., and Ninomiya, Y.J., 2010. New frontiers in gut nutrient sensor research: nutrient sensors in the gastrointestinal tract: modulation of sweet taste sensitivity by leptin. *Pharmacological Science.* 112(1): 8–12.

Jang, H.J., Kokrashvili, Z., Theodorakis, M.J., Carlson, O.D., Kim, B.J., Zhou, J., Kim, H.H., Xu, X., Chan, S.L., Juhaszova, M., Bernier, M., Mosinger, B., Margolskee, R.F., and Egan, J.M., 2007. Gut-expressed gustducin and taste receptors regulate secretion of glucagon-like peptide-1. *Procedings of the National Academy of Sciences, U S A.* 104(38): 15069–15074.

Jones, H.F., Butler, R.N., and Brooks, D.A., 2011. Intestinal fructose transport and malabsorption in humans. *American Journal of Physiology, Gastro-intestinal and Liver Physiology.* 300(2): G202–G206.

Joost, H.G., and Thorens, B., 2001. The extended GLUT-family of sugar/polyol transport facilitators: nomenclature, sequence characteristics, and potential function of its novel members (review). *Molecular Membrane Biology.* 18(4): 247–256.

Kristl, A., 2009. Membrane permeability in the gastrointestinal tract: the interplay between microclimate pH and transporters. *Chemical Biodiversity.* 6(11): 1923–1932.

Kyaw, M.H., and Mayberry, J.F., 2011. Fructose malabsorption: true condition or a variance from normality. *Journal of Clinical Gastroenterology.* 45(1): 16–21.

Lam, M.M., O'Connor, T.P., and Diamond, J., 2002. Loads, capacities and safety factors of maltase and the glucose transporter SGLT1 in mouse intestinal brush border. *Journal of Physiology.* 542(Pt 2): 493–500.

Le Gall, M., Tobin, M.V., Stolarczyk, V.E., Dalet, E.V., Leturque, V.A., and Brot-Laroche, E., 2007. Sugar sensing by enterocytes combines polarity, membrane bound detectors and sugar metabolism. *Journal of Cell Physiology.* 213(3): 834–843.

Li, Q., Manolescu, A., Ritzel, M., Yao, S., Slugoski, M., Young, J.D., Chen, X.Z., and Cheeseman, C.I., 2004. Cloning and functional characterization of the human GLUT7 isoform SLC2A7 from the small intestine. *American Journal of Physiology, Gastrointestinal and Liver Physiology.* 287(1): G236–242.

Mace, O.J., Affleck, J., Patel, N., and Kellett, G.L., 2007. Sweet taste receptors in rat small intestine stimulate glucose absorption through apical GLUT2. *Journal of Physiology.* 582(Pt 1): 379–392.

Manolescu, A.R., Witkowska, K., Kinnaird, A., Cessford, T., and Cheeseman C., 2007. Facilitated hexose transporters: new perspectives on form and function. *Physiology (Bethesda).* 22: 234–240.

McGuinness, O.P., and Cherrington, A.D., 2003. Effects of fructose on hepatic glucose metabolism. *Current Opinion in Clinical Nutrition and Metabolic Care.* 6(4): 441–448.

Niewoehner, C.B., 1986. Metabolic effects of dietary versus parenteral fructose. *Journal of the American College of Nutrition.* 5(5): 443–450.

Pepino, M.Y., and Bourne, C., 2011. Non-nutritive sweeteners, energy balance, and glucose homeostasis. *Current Opinion in Clinical Nutrition and Metabolism Care.* 14(4): 391–395.

Renwick, A.G., and Molinary, S.V., 2010. Sweet-taste receptors, low-energy sweeteners, glucose absorption and insulin release. *British Journal of Nutrition.* 104(10): 1415–1420.

Sakar, Y., Nazaret, C., Lettéron, P., Ait Omar, A., Avenati, M., Viollet, B., Ducroc, R., and Bado, A., 2009. Positive regulatory control loop between gut leptin and intestinal GLUT2/GLUT5 transporters links to hepatic metabolic functions in rodents. *PLoS One.* 4(11): e7935–79.

Steinert, R.E., and Beglinger, C., 2011. Nutrient sensing in the gut: Interactions between chemosensory cells, visceral afferents and the secretion of satiation peptides. *Physiology of Behaviour.* 30(4): 524–32.

Stearns, A.T., Balakrishnan, A., Rhoads, D.B., and Tavakkolizadeh, A., 2010. Rapid upregulation of sodium-glucose transporter SGLT1 in response to intestinal sweet taste stimulation. *Annals of Surgery.* 251(5): 865–871.

Tazawa, S., Yamato, T., Fujikura, H., Hiratochi, M., Itoh, F., Tomae, M., Takemura, Y., Maruyama, H., Sugiyama, T., Wakamatsu, A., Isogai, T., and Isaji, M., 2005. SLC5A9/SGLT4, a new Na + -dependent glucose transporter, is an essential transporter for mannose, 1,5-anhydro-D-glucitol, and fructose. *Life Sciences*. 76(9): 1039–1050.

Wang, F., Wang, F., Zou, Z., Liu, D., Wang, J., and Su, Y., 2011. Active deformation of apoptotic intestinal epithelial cells with adhesion-restricted polarity contributes to apoptotic clearance. *Laboratory Investigation*. 91(3): 462–471.

Watson, A.J., Duckworth, C.A., Guan, Y., and Montrose, M.H., 2009. Mechanisms of epithelial cell shedding in the Mammalian intestine and maintenance of barrier function. *Annals of the New York Academy of Science*. 1165: 135–142.

CHAPTER 42

Fructose and the Metabolic Syndrome

RAY ZHANG[a] AND MANAL ABDELMALEK*[b]

[a] Research Assistant, Division of Gastroenterology & Hepatology, Duke University, P.O. Box 3913, Durham, North Carolina 27710, United States; [b] Associate Professor of Medicine Division of Gastroenterology & Hepatology, Duke University
*Email: manal.abdelmalek@duke.edu

42.1 Introduction

Fructose is a monosaccharide that is widely available in natural food sources such as fruits and honey. However, in most countries the main source of fructose is from sucrose, a disaccharide composed of one glucose molecule linked through an α 1,4 glycoside bond to a fructose molecule. Sucrose remained the almost exclusive sweetener to be consumed, with only small amounts of glucose and fructose ingested essentially with fruits, until the 1960s, when the food industry developed and put into use technologies allowing for the isolation of starch from corn, hydrolysis into glucose and conversion to fructose through enzymatic isomerization (Marshall and Kooi 1957). The high sweetening power of the resulting mixture, known as high-fructose corn syrup (HFCS) (White 2008), its organoleptic properties, longer shelf-life, ability to maintain a long-lasting moisturization in industrial bakeries, together with its low cost, contributed to the very rapid increase in consumption at the expense of sucrose.

Food and Nutritional Components in Focus No. 3
Dietary Sugars: Chemistry, Analysis, Function and Effects
Edited by Victor R Preedy
© The Royal Society of Chemistry 2012
Published by the Royal Society of Chemistry, www.rsc.org

Fructose consumption has markedly increased over the past century and more than doubled in the past 30 years alone (Figure 42.1, complied from data from Electronic Outlook Report from the Economic Service Research of the USDA on Sugar and Sweetners) (USDA 2011). Soft drink consumption in the United States increased by 61% in adults from 1977 to 1997 (Putnam and Alshouse 1999) and more than doubled in children and adolescents from 1977–1978 to 1994–1998 (French 2003). Before 1900, US Americans consumed approximately 15 g of fructose per day (4% of total calories), mainly through fruits and vegetables. However by 1994, US Americans consumed approximately 55 g of fructose per day (10% of total calories) (Vos *et al.* 2008). Adolescents today consume over 72.8 g per day (12.1% of total calories) of fructose (Vos *et al.* 2008) with 20% of teenagers consuming 25% or more of their total calories as fructose (Bray 2007).

Of concern, the rise in fructose consumption has paralleled the increasing prevalence of obesity and diabetes mellitus (DM). A large body of evidence

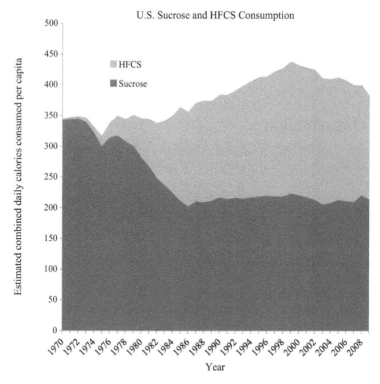

Figure 42.1 United States consumption of sucrose and HFCS since 1970. Daily caloric intake by year of sucrose and high fructose corn syrup (HFCS) in the United States from 1970–2009 per capita consumption of HFCS nearly doubled as that of sucrose halved.
(Data derived from the Electronic Outlook Report from the Economic Service Research of the USDA on Sugar and Sweetners, USDA, 2011).

shows that a high-fructose diet is associated with the development of obesity, DM, and dyslipidemia in rodents. In humans, fructose has long been known to increase plasma triglyceride concentrations. In addition, when ingested in large amounts as part of a hypercaloric diet, fructose can cause hepatic insulin resistance, increased total and visceral fat mass, and accumulation of ectopic fat in the liver and skeletal muscle. These early effects may be instrumental in causing, in the long run, the development of the metabolic syndrome. The major source of fructose in our diet is with sweetened beverages (*i.e.* soft drinks and fruit juices). Sweetened beverages have been associated with excess calorie intake, increased risk of DM and cardiovascular diseases. Thus, the association has led to the recommendation to limit daily intake of sugar calories. This chapter will discuss the effects of increased fructose consumption on health and its association with the metabolic syndrome.

42.2 Metabolic Effects of Fructose

42.2.1 Fructose, Weight Gain and Obesity

Increased fructose consumption may result in weight gain and obesity *via* several mechanisms.

Fructose does not acutely stimulate leptin or insulin release and hence may not trigger normal satiety responses (Teff *et al.* 2004). High fructose intake impairs leptin's actions thus inducing a state of a central leptin resistance leading to increased food intake and the development of visceral obesity (Shapiro *et al.* 2008). In addition, chronic consumption of diets high in fructose may increase ghrelin concentrations, a key signal to the central nervous system in the long-term regulation of energy balance, leading to increased caloric intake and ultimately contributing to weight gain and obesity (Teff *et al.* 2004). In humans, ghrelin is a major regulator of satiety; however, fructose feeding does not decrease ghrelin (Teff *et al.* 2004), and therefore, caloric intake is not suppressed. Further fructose consumption in the form of soft drinks increases the total calories consumed during meals (Havel 2005).

Linked to the rising prevalence of obesity is the increase in consumption of fructose-sweetened products, the majority of which being sucrose and HFCS. The prevalence of childhood and adulthood obesity in Americans compared to estimated daily caloric fructose consumption per capita of Americans by year from data recently reported by the Center of Disease Control (CDC 2011) is depicted in Figure 42.2. Several studies performed on children and adolescents showed a positive association between sugar-containing drink consumption and body weight (Berkey *et al.* 2004; Giammattei *et al.* 2003; Ludwig *et al.* 2001). Other cross-sectional studies even showed an inverse relation between total sucrose consumption (from all sources) and body weight (Bolton-Smith and Woodward 1994; Maillard *et al.* 2000), which certainly cannot be held as an indicator that sugar consumption promotes weight loss, but is rather explained by other uncontrolled variables. Meta-analyses linking body weight and soft

Correlation of Fructose Consumption and Obesity Prevalence

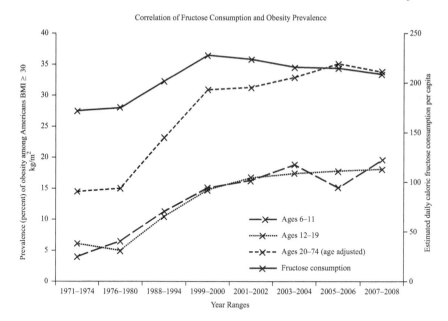

Figure 42.2 Cross-comparison of fructose consumption and obesity prevalence. Prevalence of childhood and adulthood obesity in Americans aged 6–11, 12–19, and 20–74 by year compared to estimated daily caloric fructose consumption per capita of Americans by year. 20–74 age group adjusted by the direct method to year 2000 by US Census Bureau.
Data suggest increasing fructose consumption parallels that of obesity in US Americans.

drink consumption also yield conflicting results. One meta-analysis of 88 published studies reported a significant positive association between soft drink consumption and body weight (Vartanian *et al.* 2007), while another meta-analysis of 12 studies showed no such association (Forshee and Storey 2003).

Intervention studies provide a clearer view of the relationship between sugar-containing beverages and body weight. In one study, addition of beverages sweetened with HFCS or aspartame, a non-calorie-containing sweetener, resulted in a significant weight gain with HFCS-sweetened beverages only (Tordoff and Alleva 1990). In another study, overweight subjects receiving sugar-containing beverages significantly increased their energy intake and gained weight, while subjects who received non-caloric-sweetened drinks as a control did not change weight (Raben *et al.* 2002). Conversely, several studies, mostly performed on children and adolescents, reduced the daily intake of sugar-sweetened beverages and showed a significant reduction in energy intake and/or body weight (Ditschuneit *et al.* 1999; Ebbeling *et al.* 2006). Consumption of fructose correlates not only with weight gain, but after adjusting for physical, demographic, dietary, and lifestyle differences, each additional serving of HFCS-sweetened soft drink increased BMI by 0.24 kg/m^2 and frequency of obesity by 60% in these children (Ludwig *et al.* 2001).

The incidence of obesity and excess fructose consumption has also been demonstrated in murine models. In a study of male and female rats, fructose ingestion resulted in increased body weight and abdominal fat pads. In male rats, the group with 24-hour access to HFCS with *ad libitum* chow gained more weight ($p < 0.05$) than the control group with *ad libitum* chow alone. After 6 months, *ad libitum* chow and HFCS resulted in final weight being 202% of baseline and 257% of baseline bodyweight, respectively. Also noted was the greater increase in abdominal fat pad weight in HFCS fed rats ($p < 0.05$) than in chow-only fed rats. Similar results procured for female rats with weight gain in control and HFCS fed rats being 177% and 200% of baseline, respectively. Abdominal fat pad weight was significantly heavier for HFCS fed rats compared to control rats ($p < 0.01$) (Bocarsly *et al.* 2010). Thus as evidence of ingested fructose and subsequent bodyweight and adipose tissue gain, scientists are considering increased fructose consumption to be one of the causal factors in the current obesity epidemic.

42.2.2 Fructose and Insulin Resistance

Although previously thought to be the favored sugar for patients with DM (Hawkins *et al.* 2002), recent studies point to a deleterious effect of fructose on glucose metabolism and insulin sensitivity. In rodent models, high-fructose or sucrose diets were associated with the development of insulin resistance and with disturbed glucose homeostasis. After 1 week of a high fructose diet, an impaired suppression of endogenous glucose production, indicating hepatic insulin resistance was observed (Pagliassotti *et al.* 1994; Pagliassotti *et al.* 1995; Pagliassotti *et al.* 1996). Between 2 and 5 weeks, fasting hyperinsulinemia developed, indicating whole body insulin resistance (Pagliassotti *et al.* 1996). The decrease in insulin's actions could indeed be documented by euglycemic, hyperinsulinemic clamps, showing a decreased insulin-mediated glucose disposal after 8 weeks (Pagliassotti *et al.* 1996). This sucrose-induced insulin resistance was independent of changes in body composition.

In healthy men a high-fructose diet led to hepatic insulin resistance (Faeh *et al.* 2005). In this study, fructose overfeeding increased hepatic *de novo* lipogenesis, increased plasma triglycerides and decreased hepatic insulin sensitivity. Under such conditions, supplementation with fish oil, which inhibited *de novo* lipogenesis, efficiently reduced plasma triglycerides but failed to normalize hepatic insulin sensitivity (Faeh *et al.* 2005). However, another study demonstrated that a high-fructose diet increased intrahepatic lipid deposition in humans, while hepatic insulin sensitivity remained unchanged (Le *et al.* 2009). Although in most studies, fructose elicited both hepatic insulin resistance and altered hepatic and extrahepatic lipid metabolism, it remains unclear whether these effects are distinct or causally related. In rats, a diet rich in fructose and trans fatty acid causes hepatic insulin resistance and hepatic steatosis; however, fructose appears more related to hepatic insulin resistance while trans fats were more involved in the development of steatohepatitis (Tetri *et al.* 2008).

It was further observed that sucrose elicited stress responses in hepatocytes. Changes in the redox state of the cells upon exposure to sucrose may be responsible for this activation of the c-Jun terminal kinase (JNK). Furthermore, normalization of JNK activity in hepatocytes isolated from sucrose-fed rats normalized insulin signaling. In addition, it was documented that the effects of sucrose on JNK activity and insulin sensitivity in the liver were essentially due to the fructose component of sucrose (Wei *et al.* 2005; Wei *et al.* 2007). Fructose administration was also shown to exert a marked oxidative stress on the organism (Busserolles *et al.* 2003). Providing fructose with honey, which is naturally rich in antioxidant substances, prevented both the oxidative stress induced by fructose and the reduction of insulin sensitivity (Busserolles *et al.* 2002).

In summary, there is evidence that high-fructose feeding can cause insulin resistance in rodents and possibly in humans. Interactions between fructose and fat or total energy intake remain to be assessed. Regarding the mechanisms possibly linking fructose to insulin resistance, altered lipid metabolism and lipotoxicity secondary to stimulation of *de novo* lipogenesis, or fructose-induced oxidative stress may be involved.

42.2.3   Fructose and Diabetes Mellitus

Few studies have specifically evaluated the relationship between fructose intake and the risk of developing DM. The association of fructose with increased risk of diabetes mellitus remains controversial. In a large prospective study of women (n = 39,345) older than 45 years of age, the relative risk of DM was not different when the lowest and highest quintiles of dietary sugar intake were compared. Furthermore, the absence of increased relative risk was also observed when the analysis was restricted to fructose intake (Janket *et al.* 2003). Another large prospective study of non-diabetic women aged 30–35 years (n = 71,346) demonstrated that fruit (and vegetable) intake was associated with a lower incidence of DM, while consumption of fruit juice tended to be associated with a higher incidence (Bazzano *et al.* 2008). Of those subjects with complete dietary assessment who were followed for up to 8 years, the risk of weight gain and developing DM was significantly increased in women who consumed one or more sugar-sweetened beverages per day (Schulze *et al.* 2004). In another large prospective study of non-diabetic middle-aged adults (n = 51,522), the combined intake of glucose and fructose increased the risk of DM, as did the consumption of sweetened fruit juices and soft drinks (Montonen *et al.* 2007). In African-American women (n = 59,000), the incidence of DM was significantly associated with sweetened beverage consumption, but this association was almost entirely mediated by effects of drink consumption on body weight (Palmer *et al.* 2008).

42.2.4   Fructose and Serum Lipids

Acute and long-term consumption of fructose can alter lipid homeostasis. In a clinical study consisting of 34 subjects, consumption of sucrose and HFCS significantly increased fasting triglyceride concentrations compared to baseline blood samples (Stanhope *et al.* 2008). Furthermore, postprandial triglyceride concentration elevation is significantly higher in subjects who consumed fructose compared to those who consumed glucose. This may be explained by increased hepatic *de novo* lipogenesis observed in fructose metabolism. Sucrose and HFCS postprandial triglyceride levels were comparable to that of pure fructose, suggesting a relationship between a threshold of fructose consumption and triglyceride elevation. A randomized, single blind, intervention trial asked 20 healthy participants to consume a total of either 150 g of fructose or 150 g of glucose 3 times daily for 4 weeks, while maintaining a balanced diet of 50% carbohydrates, 35% fat, and 15% protein. After 4 weeks, there was a significant increase in plasma triglyceride levels with the very high fructose diet by 350 mg/liter but no change in the very high glucose diet (Silbernagel *et al.* 2011). In addition to increased plasma triglyceride concentrations, fructose leads to a decreased level of blood high density lipoprotein (HDL), the lipoprotein which transport lipids such as triglycerides and cholesterol back to the liver for excretion or utilization. One randomized controlled trial of 74 healthy men showed that daily consumption of 200 g of fructose significantly decreased HDL cholesterol ($p < 0.001$) (Perez-Pozo *et al.* 2010). The increased triglycerides and decreased levels of blood HDL cholesterol associated with increased fructose consumption may contribute to metabolic syndrome and increased risk for cardiovascular disease and nonalcoholic fatty liver disease.

42.2.5 Fructose and Uric Acid

A number of studies have reported elevation of plasma uric acid after dietary consumption of fructose (Fiaschi *et al.* 1977; Nakagawa *et al.* 2005). Uric acid cannot be metabolized in humans and is excreted in the urine. Elevated uric acid is a risk factor for gout, cardiovascular disease, chronic kidney disease, hypertension and metabolic syndrome (Cirillo *et al.* 2006; Feig *et al.* 2006).

Fructose infusion studies have resulted in rapid increases in uric acid levels in the blood and urine of children, normal men, diabetics, and hypertensive and normotensive subjects (Raivio *et al.* 1975; Sestoft *et al.* 1985). There are few studies reporting uric acid levels after feeding dietary fructose, but a number of studies have reported increases after sucrose diets compared with glucose or starch in men and obese subjects, suggesting that it may be the fructose moiety of sucrose that is responsible for the increase (Reiser *et al.* 1989).

In a large prospective study of 78,906 women with no history of gout followed for nearly 22 years (1984–2006), sugar-sweetened beverages increased serum uric acid levels and the risk of gout (Choi *et al.* 2010). Multivariate relative risk (RR) of gout ratios was compared among consumption of less than 1 serving/month of sugar-sweetened beverages, 1 serving/day (RR 1.74) and 2 or more servings/day (RR 2.39) (Choi *et al.* 2010). In another prospective study of adults (n = 15,745) followed for up to 9 years to assess the association of sugar-sweetened beverages on the risk of hyperuricemia and chronic kidney disease (Bomback *et al.* 2010), patients who consumed more than 1 sugar-sweetened beverage per day showed serum uric acid levels over 9.0 mg/dl compared to normal serum uric acid levels of 4 mg/dl for women and 5 mg/dl for men. In the same patients, the odds ratio for chronic kidney disease significantly increased to 2.59 (Bomback *et al.* 2010).

42.2.6   Fructose and Hypertension

Increased fructose consumption has also been implicated in the risk of hypertension, renal and cardiovascular disease. In an experiment conducted on rats fed either a normal chow or a 66% fructose diet for 2 weeks, systolic blood pressure rose significantly ($p < 0.001$) in fructose fed rats compared to normal chow fed rats (Hwang *et al.* 1987). Similar effect of fructose consumption on blood pressure has been noted in humans. A recent study of 2,696 subjects, aged 40–59 years, noted that that drinking > 1 serving/day of sugar-sweetened beverages was significantly associated with increased systolic and diastolic blood pressure, ($p < 0.001$ and $p < 0.05$ after adjusting for weight and height, respectively) (Brown *et al.* 2011). The effect of increased blood pressure was not observed with diet soda consumption. Another study ($n = 810$ adults) showed that decreasing sugar-sweetened beverage intake was associated with a reduction in systolic and diastolic blood pressure (Chen *et al.* 2010). The mechanism of hypertension in response to fructose is complex but may be mediated by increased sodium absorption in the intestine, by inhibition of systemic endothelial function, and by stimulation of the sympathetic nervous system (Madero *et al.* 2011). In addition, some of the effects of fructose to increase blood pressure may be the consequence of fructose-induced increases in intracellular and serum uric acid. First, fructose-induced hypertension in rats is largely ameliorated by lowering uric acid levels (Sanchez-Lozada *et al.* 2007). Second, in a study of overweight men, the rise in ambulatory blood pressure in response to 200 g of oral fructose per day for two weeks was blocked in those subjects who were concomitantly administered allopurinol (Perez-Pozo *et al.* 2010).

42.2.7   Fructose and the Metabolic Syndrome

As previously discussed, fructose has been associated with obesity, insulin resistance, DM, hypertriglycidemia, low HDL cholesterol, and hypertension. Thus, it is conceivable that fructose may be a risk factor for metabolic syndrome (Table 42.1). Metabolic syndrome is defined by three or more of the following: high blood pressure, waist circumference > 35 inches (females) or 40 inches (males), high fasting plasma glucose, high plasma triglyceride, and low HDL cholesterol. Recent data from the Framingham Heart Study ($n = 6,039$ participants) showed that consumption of more than one can of soft drink per day was significantly associated with the prevalence of the metabolic syndrome. Furthermore, upon prospective follow-up of individuals without the metabolic syndrome at inclusion, consumption of more than one soft drink per day was associated with an increased risk of developing the metabolic syndrome (Dhingra *et al.* 2007).

Table 42.1 Features of metabolic syndrome – need 3 or more for diagnosis.

Risk Factor	Defining Level
Abdominal obesity, given as waist circumference	
Men	$>102\,\text{cm}$ ($>40\,\text{in}$)
Women	$>88\,\text{cm}$ ($>35\,\text{in}$)
Triglycerides	$\geq 150\,\text{mg/dL}$
HDL cholesterol	
Men	$<40\,\text{mg/dL}$
Women	$<50\,\text{mg/dL}$
Blood pressure	$\geq 130/\geq 85\,\text{mm Hg}$
Fasting glucose	$\geq 110\,\text{mg/dL}$

42.3 Caveats

Presently, epidemiologic studies provide an incomplete, and at times, discordant appraisal of the effects of fructose on human disease conditions. Part of the inability of existing studies to clearly define the health implications of fructose may be explained by the fact that intake of sucrose, fructose, fruit juices, and/or sweetened beverages was not recorded separately, thus precluding an accurate assessment of total fructose intake. In addition, fructose is essentially consumed as either sucrose or high-fructose corn syrup, with the consequence that glucose intake varies with fructose intake. Confounding factors (age, total calorie intake, intake of other dietary nutrients, associated physical activity and life-style factors) were not controlled in existing studies. Thus, further experimental and clinical studies are warranted to accurately assess the implications of fructose consumption on health.

42.4 Conclusion

The potential danger of fructose consumption and its links to various metabolic disorders have been widely documented. Deleterious effects of high fructose intake on body weight, insulin sensitivity/glucose homeostasis, dyslipidemia, and hypertension have been identified both in animal studies as well as small clinical trials and/or epidemiologic studies. The effects were often documented at high levels of fructose intake. Nevertheless, there is solid evidence that even at moderate doses fructose can cause alterations in metabolism. Although data are scarcer, there is evidence that fructose increases intrahepatic lipids and leads to insulin resistance. At present, short-term intervention studies suggest that a high-fructose intake can increase the risk of metabolic syndrome. However, there is no evidence that moderate intake of natural fructose as would be consumed with fruits, vegetable or honey is unsafe. Studies aimed at delineating the dose threshold at which fructose starts to chronically exert deleterious effects remain to be performed. High fructose consumption is associated with other risk factors (*e.g.* hypercaloric diet, diets risk in saturated fat, and/or low physical activity) which may also contribute to all metabolic disorders discussed herein. Given the number of confounding variables which must (and presently are not) taken into consideration when assessing the effects of fructose on health outcomes, well-designed prospective interventional studies are clearly needed to define role of increased fructose consumption on human health.

Summary Points

- This chapter focused on fructose function and the metabolic syndrome.
- Fructose consumption has substantially increased over the past 30 years.
- The increased consumption of fructose has paralleled the rise in obesity and diabetes.
- Fructose plays a direct role in *de novo* lipogenesis and insulin resistance contributing to complications of the metabolic syndrome.
- High fructose intake alters lipid homeostasis and increase triglycerides.
- High fructose intake can increase serum uric acid.
- High fructose intake increases blood pressure which may in turn increase the risk for chronic kidney disease and/or cardiovascular disease.
- Current epidemiologic and experimental studies have not controlled well for potential confounding factors which may be associated with fructose consumption and metabolic syndrome.
- Well-designed prospective clinical and epidemiologic studies are necessary to clearly define the effect(s) of fructose consumption at varying doses on human health and disease.

Key Facts

- Metabolic syndrome is a combination of factors that multiply a person's risk for heart disease, diabetes and stroke.
- Almost 35% of American adults have metabolic syndrome.
- Many of the factors for metabolic syndrome have no symptoms until severe damage has been done and/or complications are advanced enough to become clinically apparent.
- The exact mechanism(s) for the complex pathways underlying metabolic syndrome are not completely known.
- The term 'metabolic syndrome' dates back to at least the late 1950s, but came into common usage in the late 1970s to describe various associations of risk factors with diabetes.
- It is unknown whether obesity and insulin resistance are the cause or consequence of metabolic syndrome.
- Lifestyle modification (decrease calorie consumption and increased physical activity) is the first-line therapy for metabolic syndrome.

Definitions of Words and Terms

Diabetes mellitus: a complication of insulin resistance where the cells of the body fail to use insulin properly resulting in high blood sugars.

Disaccharide: a carbohydrate consisting of two sugar molecules bond together.

Dyslipidemia: a condition characterized by increased cholesterol, triglycerides, and low-density (*i.e.* 'bad' fat) lipoproteins and low high (*i.e.* 'good' fat) density lipoproteins in the blood.

Enzymatic isomerization: the enzymatic reaction used to convert D-glucose to D-fructose for the purpose of making high fructose corn syrup.

Fructose: a simple water soluable sugar found in honey, fruits and some vegatables.

Hepatic steatosis: the increased accumulation of triglycerides in the liver.

High fructose corn syrup: a common sweetener that is used in soft drinks, breakfast cereals, cookies, snacks and many other baked goods, which consist of a combination of glucose and fructose.

Hypertension: a chronic medical condition where the blood pressure in the arteries is elevated. Normal blood pressure is at or below 120/80 mmHg. High blood pressure is present if it is at or above 140/90 mmHg.

Hyperuricemia: a condition where the level of uric acid in the blood is elevated above normal. High uric acid may cause gout.

Insulin resistance: a physiologic state where the cells in the body responsible for the metabolism of glucose become resistant to the hormone insulin.

Lipogenesis: the process in which acetyl-CoA, an intermediary product of sugar metabolism, is converted to fats.

Lipoproteins: a biochemical assembly of lipids bonded to proteins that allow fats to move through the body and within cells. Decreased high density ('good') lipoprotein and increased low density ('bad') lipoproteins are risk factors for cardiovascular disease.

Metabolic syndrome: a poorly understood constellation of medical conditions (increased waist circumference, elevated blood lipids, elevated blood glucose and/or high blood pressure) when occurring together increase the risk for diabetes and heart disease.

Monosaccharide: the simplest form of a carbohydrate consisting of only one sugar molecule.

Obesity: a medical condition defined as body mass exceeds $30 \, \text{kg/m}^2$ or a condition in which excess body fat has accumulated and is associated with adverse effects on health.

Soft drinks: a non-alcoholic beverage that contains water, possible carbonation, sugar, high-fructose corn syrup, coloring and/or a flavoring agent. Diet soft-drinks contain a sugar substitute.

Steatohepatits: inflammation in the liver as a consequence of increased fat deposition in liver.

Triglycerides: an ester derived from glycerol and three fatty acids. In humans, triglycerides are the body's mechanism for storing unused calories.

List of Abbreviations

BMI   body mass index
JNK   c-Jun terminal kinase
DM   diabetes mellitus
HDL   high density lipoprotein
HFCS   high fructose corn syrup
RR   relative risk
USDA   United States Department of Agriculture
CDC   Center for Disease Control

References

Bazzano, L.A., Li, T.Y., Joshipura, K.J., and Hu, F.B., 2008. Intake of fruit, vegetables, and fruit juices and risk of diabetes in women. *Diabetes Care*. 31: 1311–1317.

Berkey, C.S., Rockett, H.R., Field, A.E., Gillman, M.W., and Colditz, G.A., 2004. Sugar-added beverages and adolescent weight change. *Obesity Research*. 12: 778–788.

Bocarsly, M.E., Powell, E.S., Avena, N.M., and Hoebel, B.G., 2010. High-fructose corn syrup causes characteristics of obesity in rats: increased body weight, body fat and triglyceride levels. *Pharmacology, Biochemistry and Behavior*. 97: 101–106.

Bolton-Smith, C., and Woodward, M., 1994. Dietary composition and fat to sugar ratios in relation to obesity. *International Journal of Obesity and Related Metabolic Disorders*. 18: 820–828.

Bomback, A.S., Derebail, V.K., Shoham, D.A., Anderson, C.A., Steffen, L.M., Rosamond, W.D., and Kshirsagar, A.V., 2010. Sugar-sweetened soda consumption, hyperuricemia, and kidney disease. *Kidney International*. 77: 609–616.

Bray, G.A., 2007. How bad is fructose? *American Journal of Clinical Nutrition*. 86: 895–896.

Brown, I.J., Stamler, J., Van Horn, L., Robertson, C.E., Chan, Q., Dyer, A.R., Huang, C.C., Rodriguez, B.L., Zhao, L., Daviglus, M.L., Ueshima, H., Elliott, P., International Study of Macro/Micronutrients and Blood Pressure Research Group. 2011. Sugar-sweetened beverage, sugar intake of individuals, and their blood pressure: international study of macro/micronutrients and blood pressure. *Hypertension*. 57: 695–701.

Busserolles, J., Gueux, E., Rock, E., Mazur, A., Rayssiguier, Y.. 2002. Substituting honey for refined carbohydrates protects rats from hypertriglyceridemic and prooxidative effects of fructose. *Journal of Nutrition*.132: 3379–3382.

Busserolles, J., Gueux, E., Rock, E., Mazur, A., Rayssiguier, Y., 2003. High fructose feeding of magnesium deficient rats is associated with increased plasma triglyceride concentration and increased oxidative stress. *Magnesium Research*. 16: 7–12.

Center for Disease Control (CDC). Overweight and Obesity (OO). Available at: http://www.cdc.gov/obesity/data/index.html. Accessed 26 March 2011.

Chen, L., Caballero, B., Mitchell, D.C., Loria, C., Lin, P.H., Champagne, C.M., Elmer, P.J., Ard, J.D., Batch, B.C., Anderson, C.A., and Appel, L.J., 2010. Reducing consumption of sugar-sweetened beverages is associated with reduced blood pressure: a prospective study among United States adults. *Circulation.* 121: 2398–406.

Choi, H.K., Willett, W., and Curhan, G., 2010. Fructose-rich beverages and risk of gout in women. *Journal of the American Medical Association.* 304: 2270–2278.

Cirillo, P., Sato, W., Reungjui, S., Heinig, M., Gersch, M., Sautin, Y., Nakagawa, T., and Johnson, R.J., 2006. Uric acid, the metabolic syndrome, and renal disease. *Journal of American Society of Nephrology.* 17: S165–168.

Dhingra, R., Sullivan, L., Jacques, P.F., Wang, T.J., Fox, C.S., Meigs, J.B., D'Agostino, R.B., Gaziano, J.M., and Vasan, R.S., 2007. Soft drink consumption and risk of developing cardiometabolic risk factors and the metabolic syndrome in middle-aged adults in the community. *Circulation.* 116(5): 480–8.

Ditschuneit, H.H., Flechtner-Mors, M., Johnson, T.D., and Adler, G., 1999. Metabolic and weight-loss effects of a long-term dietary intervention in obese patients. *American Journal of Clinical Nutrition.* 69: 198–204.

Ebbeling, C.B., Feldman, H.A., Osganian, S.K., Chomitz, V.R., Ellenbogen, S.J., and Ludwig, D.S., 2006. Effects of decreasing sugar-sweetened beverage consumption on body weight in adolescents: a randomized, controlled pilot study. *Pediatrics.* 117: 673–680.

Faeh, D., Minehira, K., Schwarz, J.M., Periasamy, R., Park, S., and Tappy, L., 2005. Effect of fructose overfeeding and fish oil administration on hepatic de novo lipogenesis and insulin sensitivity in healthy men. *Diabetes.* 54: 1907–1913.

Feig, D.I., Kang, D.H., Nakagawa, T., Mazzali, M., and Johnson, R.J., 2006. Uric acid and hypertension. *Current Hypertension Reports.* 8: 111–115.

Fiaschi, E., Baggio, B., Favaro, S., Antonello, A., Camerin, E., Todesco, S., and Borsatti, A., 1977. Fructose-induced hyperuricemia in essential hypertension. *Metabolism.* 26: 1219–1223.

French, S.A., Lin, B.H., and Guthrie, J.F., 2003. National trends in soft drink consumption among children and adolescents age 6 to 17 years: prevalence, amounts, and sources, 1977/1978 to 1994/1998. *Journal of the American Dietetic Association.* 103: 1326–1331.

Giammattei, J., Blix, G., Marshak, H.H., Wollitzer, A.O., and Pettitt, D.J., 2003. Television watching and soft drink consumption: associations with obesity in 11- to 13-year-old schoolchildren. *Archives of Pediatric & Adolescent Medicine.* 157: 82–886.

Havel, P.J., 2005. Dietary fructose: implications for dysregulation of energy homeostasis and lipid/carbohydrate metabolism. *Nutrition Reviews.* 63: 133–157.

Hawkins, M., Gabriely, I., Wozniak, R., Vilcu, C., Shamoon, H., and Rossetti, L., 2002. Fructose improves the ability of hyperglycemia per se to regulate glucose production in type 2 diabetes. *Diabetes.* 51: 606–614.

Hwang, I.S., Ho, H., Hoffman, B.B., and Reaven, G.M., 1987. Fructose-induced insulin resistance and hypertension in rats. *Hypertension*. 10: 512–516.

Janket, S.J., Manson, J.E., Sesso, H., Buring, J.E., and Liu, S., 2003. A prospective study of sugar intake and risk of type 2 diabetes in women. *Diabetes Care*. 26: 1008–1015.

Lê, K.A., Ith, M., Kreis, R., Faeh, D., Bortolotti, M., Tran, C., Boesch, C., and Tappy, L., 2009. Fructose overconsumption causes dyslipidemia and ectopic lipid deposition in healthy subjects with and without a family history of type 2 diabetes. *American Journal of Clinical Nutrition*. 89: 1760–1765.

Ludwig, D.S., Peterson, K.E., and Gortmaker, S.L., 2001. Relation between consumption of sugar-sweetened drinks and childhood obesity: a prospective, observational analysis. *Lancet*. 357: 505–508.

Madero, M., Perez-Pozo, S.E., Jalal, D., Johnson, R.J., and Sánchez-Lozada, L.G., 2011. Dietary fructose and hypertension. *Current Hypertension Reports*. 13: 29–35.

Maillard, G., Charles, M.A., Lafay, L., Thibult, N., Vray, M., Borys, J.M., Basdevant, A., Eschwège, E., and Romon, M., 2000. Macronutrient energy intake and adiposity in non obese prepubertal children aged 5-11 y (the Fleurbaix Laventie Ville Santé Study). *International Journal of Obesity and Related Metabolic Disorders*. 24: 1608–1617.

Marshall, R.O., and Kooi, E.R., 1957. Enzymatic conversion of D-glucose to D-fructose. *Science*. 125: 648-649.

Montonen, J., Jarvinen, R., Knekt, P., Heliovaara, M., and Reunanen, A., 2007. Consumption of sweetened beverages and intakes of fructose and glucose predict type 2 diabetes occurrence. *The Journal of Nutrition*. 137: 1447–1454.

Nakagawa, T., Tuttle, K.R., Short, R.A., and Johnson, R.J., 2005. Hypothesis: fructose-induced hyperuricemia as a causal mechanism for the epidemic of the metabolic syndrome. *Nature Clinical Practice Nephrology*. 1: 80–86.

Pagliassotti, M.J., and Prach, P.A., 1995. Quantity of sucrose alters the tissue pattern and time course of insulin resistance in young rats. *American Journal of Physiology*. 269: R641–646.

Pagliassotti, M.J., Prach, P.A., Koppenhafer, T.A., and Pan, D.A., 1996. Changes in insulin action, triglycerides, and lipid composition during sucrose feeding in rats. *American Journal of Physiology*. 271: R1319–1326.

Pagliassotti, M.J., Shahrokhi, K.A., and Moscarello, M., 1994. Involvement of liver and skeletal muscle in sucrose-induced insulin resistance: dose-response studies. *American Journal of Physiology*. 266: R1637–1644.

Palmer, J.R., Boggs, D.A., Krishnan, S., Hu, F.B., Singer, M., and Rosenberg, L., 2008. Sugar-sweetened beverages and incidence of type 2 diabetes mellitus in African American women. *Archives of Internal Medicine*. 168: 1487–1492.

Perez-Pozo, S.E., Schold, J., Nakagawa, T., Sanchez-Lozada, L.G., Johnson, R.J., and Lillo, J.L., 2010. Excessive fructose intake induces the features of metabolic syndrome in healthy adult men: role of uric acid in the hypertensive response. *International Journal of Obesity (London)*. 34: 454–461.

Putnam, J., and Alshouse, J.E., 1999. *Food Consumption, Prices and Expenditures. 1970-1997.* Washington, DC: Food and Rural Economics Division, Economics Research Service, US Dept of Agriculture. Statistical Bulletin No. 965.

Raben, A., Vasilaras, T.H., Møller, A.C., and Astrup A. 2002. Sucrose compared with artificial sweeteners: different effects on ad libitum food intake and body weight after 10 wk of supplementation in overweight subjects. *American Journal of Clinical Nutrition.* 76: 721–729.

Raivio, K.O., Becker, A., Meyer, L.J., Greene, M.L., Nuki, G., and Seegmiller, J.E., 1975. Stimulation of human purine synthesis de novo by fructose infusion. *Metabolism.* 24: 861–869.

Reiser, S., Powell, A.S., Scholfield, D.J., Panda, P., Ellwood, K.C., and Canary, J.J., 1989. Blood lipids, lipoproteins, apoproteins, and uric acid in men fed diets containing fructose or high-amylose cornstarch. *American Journal of Clinical Nutrition.* 49: 832–839.

Sánchez-Lozada, L.G., Tapia, E., Jiménez, A., Bautista, P., Cristóbal, M., Nepomuceno, T., Soto, V., Avila-Casado, C., Nakagawa, T., Johnson, R.J., Herrera-Acosta, J., and Franco, M., 2007. Fructose-induced metabolic syndrome is associated with glomerular hypertension and renal microvascular damage in rats. *American Journal of Physiology Renal Physiology.* 292: F423–429.

Schulze, M.B., Manson, J.E., Ludwig, D.S., Colditz, G.A., Stampfer, M.J., Willett, W.C., and Hu, F.B., 2004. Sugar-sweetened beverages, weight gain, and incidence of type 2 diabetes in young and middle-aged women. *Journal of the American Medical Association.* 292: 927–934.

Sestoft L., 1985. An evaluation of biochemical aspects of intravenous fructose, sorbitol and xylitol administration in man. *Acta Anaesthesiologica Scandinavica Supplement.* 82: 19–29.

Shapiro, A., Mu, W., Roncal, C., Cheng, K.Y., Johnson, R.J., and Scarpace, P.J., 2008. Fructose-induced leptin resistance exacerbates weight gain in response to subsequent high-fat feeding. *American Journal of Physiology Regulatory, Integrative and Comparative Physiology.* 295: R1370–1375.

Silbernagel, G., Machann, J., Unmuth, S., Schick, F., Stefan, N., Häring, H.U., and Fritsche, A., 2011. Effects of 4-week very-high-fructose/glucose diets on insulin sensitivity, visceral fat and intrahepatic lipids: an exploratory trial. *The British Journal of Nutrition.* 106: 79–86.

Stanhope, K.L., Griffen, S.C., Bair, B.R., Swarbrick, M.M., Keim, N.L., and Havel, P.J., 2008. Twenty-four-hour endocrine and metabolic profiles following consumption of high-fructose corn syrup-, sucrose-, fructose-, and glucose-sweetened beverages with meals. *American Journal of Clinical Nutrition.* 87: 1194–1203.

Teff, K.L., Elliott, S.S., Tschöp, M., Kieffer, T.J., Rader, D., Heiman, M., Townsend, R.R., Keim, N.L., D'Alessio, D., and Havel, P.J., 2004, Dietary fructose reduces circulating insulin and leptin, attenuates postprandial suppression of ghrelin, and increases triglycerides in women. *Journal of Clinical Endocrinology and Metabolism.* 89(6): 2963–2972.

Tetri, L.H., Basaranoglu, M., Brunt, E.M., Yerian, L.M., and Neuschwander-Tetri, B.A., 2008 Severe NAFLD with hepatic necroinflammatory changes in mice fed trans fats and a high-fructose corn syrup equivalent. *American Journal of Physiology: Gastrointestinal and Liver Physiology.* 295: G987–995.

Tordoff, M.G., and Alleva, A.M., 1990. Effect of drinking soda sweetened with aspartame or high-fructose corn syrup on food intake and body weight. *American Journal of Clinical Nutrition.* 51: 963–969.

United States Department of Agriculture (USDA). Sugar and Sweeteners: Recommended Data (SSRD). Available at: http://www.ers.usda.gov/Briefing/Sugar/Data.htm; Accessed 26 March 2011.

Vartanian, L.R., Schwartz, M.B., and Brownell, K.D., 2007. Effects of soft drink consumption on nutrition and health: a systematic review and meta-analysis. *American Journal of Public Health.* 97: 667–675.

Vos, M.B., Kimmons, J.E., Gillespie, C., Welsh, J., and Blanck, H.M., 2008. Dietary fructose consumption among US children and adults: the Third National Health and Nutrition Examination Survey. *The Medscape Journal of Medicine.* 10: 160.

Wei, Y., Wang, D., and Pagliassotti, M.J., 2005. Fructose selectively modulates c-jun N-terminal kinase activity and insulin signaling in rat primary hepatocytes. *Journal of Nutrition.* 135: 1642–1646.

Wei, Y., Wang, D., Topczewski, F., and Pagliassotti, M.J., 2007. Fructose-mediated stress signaling in the liver: implications for hepatic insulin resistance. *Journal of Nutritional Biochemistry.* 18: 1–9.

White, J.S., 2008. Straight talk about high-fructose corn syrup: what it is and what it ain't. *American Journal Clinical Nutrition.* 88: 1716S–1721S.

CHAPTER 43

Fructose and Nonalcoholic Fatty Liver Disease

FLAVIO FRANCINI, MARÍA L MASSA AND
JUAN J. GAGLIARDINO*

CENEXA – Centro de Endocrinología Experimental y Aplicada (UNLP –
CONICET LA PLATA, Centro Colaborador OPS/OMS en Diabetes),
Facultad de Ciencias Médicas UNLP, Calles 60 y 120 s/n, 1900 La Plata,
Argentina
*Email: cenexaar@yahoo.com.ar

43.1 Introduction

Nonalcoholic fatty liver disease (NAFLD) is a common clinical entity characterized by insulin resistance and fat accumulation in the liver, in the absence of other apparent causes of fat deposit such as alcohol abuse, viral or autoimmune hepatitis, and alpha 1 antitrypsin deficiency (Ali and Cusi 2009). NAFLD comprises a continuum of disease ranging from steatosis, the earliest relatively benign and indolent state of liver injury, to steatohepatitis (NASH) with necroinflammation induced by severe lipotoxicity. This latter condition is the most current common cause of cirrhosis and indication for liver transplant, thus representing an important emerging public health problem frequently associated with overweight/obesity and type 2 diabetes (T2DM). Atherogenic dyslipidemia (high triglyceride, low cholesterol-HDL and increased cholesterol-LDL levels) is another common clinical feature present in patients with NAFLD. What could apparently be seen as independent entities (obesity, T2DM, cardiovascular disease and NAFLD) are in fact closely linked by a

Food and Nutritional Components in Focus No. 3
Dietary Sugars: Chemistry, Analysis, Function and Effects
Edited by Victor R Preedy
© The Royal Society of Chemistry 2012
Published by the Royal Society of Chemistry, www.rsc.org

common underlying metabolic abnormality: the metabolic syndrome (Ali and Cusi 2009).

The progression of all these entities can be effectively prevented or even delayed by appropriate lifestyle changes. However, sedentary habits and consumption of high calorie diets with an unbalanced composition of nutrients are common characteristics of modern societies. In fact, the annual *per capita* consumption of fructose in the US has increased from ~0.2 kg in 1970 to ~28 kg in 1997 (Wei *et al.* 2006). Consumption of soft and fruit drinks, which are the major source of fructose in the form of high fructose corn syrups (HFCS), has increased from 3.9% in 1977 to 9.2% of the total energy intake in 2001 (Nielsen and Popkin 2004).

With these data, and evidence that the conversion rate to diabetes increases as a function of the amount of soft drink consumption (Hu and Malik 2010), several authors have suggested that the increased use of HFCS and refined carbohydrates registered in the last decades has contributed to the current epidemics of obesity and T2DM (Samuel 2011). The increased intake of HFCS recorded in the last two decades may also affect insulin sensitivity, carbohydrate and lipid metabolism, and hepatic steatosis. In this regard, Toshimitsu *et al.* (2007) reported that patients with fatty liver consume more simple carbohydrates than the general population, and Assy *et al.* (2008) showed that consumption of soft drink beverages (rich in HFCS) is the most common risk factor for hepatic steatosis in the absence of other classic risk factors. Population-based studies also demonstrated that subjects with NAFLD have consumed significantly greater amounts of sweetened beverages (2-fold) than the mean intake in controls (Ouyang *et al.* 2008); further, a pilot study in humans has shown that reduction of fructose intake may exert beneficial effects on the progression of NAFLD (Zelber-Sagi *et al.* 2007).

43.2   Fructose Metabolism in the Liver

The liver is the primary site for fructose extraction and metabolism; consequently, it has been postulated that an increase in fructose flow and availability will impair the hepatic metabolism of carbohydrates and lipids (Pagliassotti and Horton 2004).

Although glucose and fructose share some properties, they are chemically different, and such difference is responsible for their differential intestinal absorption into the portal vein and the higher fructose splanchnic extraction (Samuel 2011). Indeed, all the fructose absorbed goes initially very fast into the liver, which has an active system ("fructose pathway") for its metabolism: the enzymes fructokinase, aldolase B and triokinase (Mayes 1993). Due to the activity of this enzymatic system, 55 to 71% of fructose remains in the liver (according to the feeding or starved state), while a small amount is offered to peripheral tissues through a very low blood concentration (Mayes 1993). In hepatocytes, fructose is transported into the cytoplasm by the glucose transporter 2 (GLUT2) and rapidly phosphorylated to fructose-1-phosphate by fructokinase and thereafter converted to glyceraldehyde and dihydroxyacetone-P (DHA-P) by aldolase B. Through this pathway, fructose bypasses the main glycolysis regulatory point: phosphorylation by phosphofructokinase-1 (PFK-1) (Figure 43.1). Unlike PFK-1, fructokinase is not inhibited by ATP and consequently is not under the tight energetic control of glycolysis. It must be stressed that the unique metabolic effects of fructose in the liver depend on these two properties: (a) its very fast uptake by the liver, and (b) its channelling into the glycolytic and gluconeogenetic pathways at the triose phosphate level, bypassing the PFK-1 regulatory step (Mayes 1993). Additionally and unlike glucokinase, fructokinase activity is not affected by insulin.

As mentioned before, fructose-1-phosphate is split by aldolase B to yield glyceraldehyde and DHA-P. While triose phosphate isomerase converts DHA-P into glyceraldehyde-3-phosphate (G-3-P), triokinase phosphorylates glyceraldehyde and also yields G-3-P. At this step, G-3-P goes into the glycolysis generating acetyl-coenzyme A (acetyl-CoA), that can be either oxidized in the mitochondria or channeled to fatty acid synthesis. Trioses also contribute to lipid metabolism providing glycerol-3-phosphate, which in turn serves as backbone for triglyceride synthesis (Fried and Rao 2003) (Figure 43.2). Importantly, fructose conversion to triose-phosphate is a fast insulin-

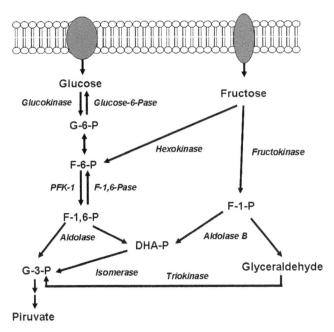

Figure 43.1 "Fructose pathway" in the liver. The activity of fructokinase, aldolase B and triokinase is responsible for the deposit of fructose in the liver. While GLUT2 is responsible for the cellular uptake of fructose, it is phosphorylated by fructokinase and thereafter converted to glyceraldehyde and DHA-P by aldolase B, bypassing the glycolysis regulatory point: phosphorylation by PFK-1.

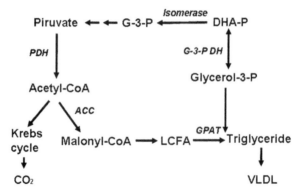

Figure 43.2   Fructose provides the substrates for lipid biosynthesis after its breakdown to trioses. DHA-P generated by aldolase B activity is converted
into G-3-P that goes into the glycolysis pathway generating acetyl-CoA;
this compound can be thereafter either oxidized in the mitochondria or
channeled to fatty acid synthesis. Trioses also contribute to lipid metabolism providing glycerol-3-phosphate, which in turn serves as backbone
for triglyceride synthesis.

independent pathway because of the low Km of fructokinase for its substrate
and the absence of a negative feedback by ATP or citrate. Due to all these
characteristics, fructose increases the provision of substrates at the triose
phosphate level and consequently fuels, in an unregulated way, all the metabolic pathways starting at this point (glycolysis, glyconeogenesis and lipogenesis) (Figure 43.3).

43.3   The "Lipid Connection"

The main proportion of triose-phosphate produced by fructose metabolism is
geared to glucose and glycogen synthesis through gluconeogenesis (Koo *et al.*
2008), but as mentioned before, fructose-derived carbons can be converted into
fatty acid precursors – acetyl-CoA – or provide the glycerol backbone for triglyceride synthesis; in fact, several studies have demonstrated that dietary
fructose enhances lipid synthesis (Figure 43.2).

   The enhanced lipogenic pathways increase the deposit of triglycerides not
only in adipose tissue but also in the liver, leading to steatosis and eventually to
insulin resistance and dyslipidemia. Since enhanced lipid metabolism is tightly
linked to the onset of obesity and insulin resistance, it has been suggested that
the mechanism by which fructose promotes the development of the metabolic
syndrome is related to its lipogenic properties (Dekker *et al.* 2010). Supporting
this assumption, Stanhope *et al.* (2009) demonstrated that the enhanced intake
of fructose-rich beverages during 8 weeks increases postprandial plasma triglyceride levels and fat accumulation in the abdomen. This *de novo* lipogenesis
effect induced by high fructose intake was also recorded after one week,
demonstrating that it is a fast process (Le *et al.* 2009).

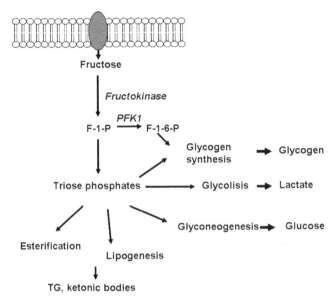

Figure 43.3 Fructose provision of substrates. Fructose increases the provision of substrates at the triose phosphate level and consequently fuels, in an unregulated manner, all the metabolic pathways starting at this point, namely glycolysis, glyconeogenesis and lipogenesis.

The liver "fructose pathway" promotes an increase in DHA-P and, in turn, in pyruvate and glycerol-3-phosphate concentration. In patients with NAFLD, the activity, protein and gene expression of fructokinase – the first enzyme of the fructose pathway – was higher than in controls, which is consistent with the known up-regulatory effect of fructose upon this liver enzyme in the rat (Ouyang *et al.* 2008). While glycerol-3-phosphate acts as a backbone for trygliceride synthesis, pyruvate could have two different metabolic fates, either be converted to lactate or enter into the mitochondria to form acetyl-CoA by the pyruvate dehidrogenase complex (PDH). Carmona and Freedland (1989) reported earlier that high fructose diets enhanced ^3H_2O incorporation into fat, thus suggesting the activation of PDH. Later, Park *et al.* (1992) demonstrated that the chronic intake of fructose increased the dephosphorylated active form of PDH. This change was explained by a parallel decrease in the activity of mitochondrial PDH kinase (PDK), one regulator of PDH activity. Although these changes could explain the mechanisms by which fructose induces hypertriglyceridaemia in rats (Park *et al.* 1992), there is no evidence that this mechanism also occurs in humans.

Interestingly, while an infusion of fructose, together with nonesterified fatty acids (NEFA) increased liver fatty acid esterification and VLDL secretion, it also decreased fatty acid oxidation. The rate-limiting enzyme carnitin-palmitoyltransferase I (CPT1) is responsible for the co-regulation of both processes, fatty acid synthesis and oxidation. CPT1 catalyzes fatty acid transport into the

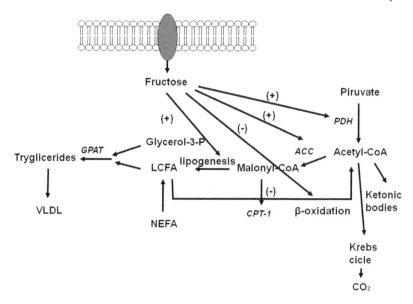

Figure 43.4   Fructose co-regulates lipid synthesis and oxidation. Fructose and non esterified fatty acids (NEFA) increase liver fatty acid esterification and VLDL secretion, and decrease fatty acid oxidation. The rate-limiting enzyme carnitin-palmitoyltransferase I (CPT1) is responsible for the co-regulation of fatty acid synthesis and oxidation. This enzyme is inhibited by the acetil CoA carboxilase (ACC) product malonyl-CoA. Since fructose metabolism increases malonyl-CoA production, this effect can explain the switch from lipid oxidation to its synthesis that follows the increase of fructose intake.

mitochondrial matrix to its further oxidation, while fatty acids that cannot enter into this pathway are esterified into triglycerides. This enzyme is inhibited by the acetil CoA carboxilase (ACC) product malonyl-CoA. Since fructose metabolism increases malonyl-CoA, a CPT1 inhibitor, this effect can explain the switch from lipid oxidation to lipid synthesis induced by the increased fructose intake (Figure 43.4).

43.4   Fructose and the Master Transcriptional Regulators of Lipogenesis

As mentioned before, high fructose diets up-regulate *de novo* lipogenesis; therefore, great efforts were directed to the identification of master transcriptional regulators of such a process. One of these regulators is hepatic sterol regulatory element binding protein (SREBP), a key transcription factor responsible for the regulation of fatty acid and cholesterol biosynthesis. SREBP binds to sterol responsive elements (SRE) in the DNA, inducing the expression of lipogenesis-related genes (Bennett *et al.* 1995). A high fructose diet increases significantly the liver expression of SREBP-1c, which in turn enhances the expression of several target genes related to lipid synthesis, *e.g.*, fatty acid synthase (FAS), ACC and stearoyl-CoA desaturase (SCD) (Miyazaki *et al.* 2004). Although the stimulation of SREBP involves the activation of the downstream insulin signaling pathway (MAP kinases and ERK1/2), when sufficient carbohydrates are available, a clear nutritional regulation of SREBP-1c and lipogenic genes occurs in a completely independent insulin pathway (Matsuzaka *et al.* 2004). In fact, high fructose feeding causes insulin resistance with increased hepatic SREBP-1c mRNA levels (Nagai *et al.* 2002). Such increase is accompanied by a parallel increase in protein-tyrosine phosphatase 1B (PTP-1B), a major negative regulator of insulin signal transduction (Shimizu *et al.* 2003). In the liver of rats with fructose-induced insulin resistance, the enhanced expression of PTP-1B leads to an increased mRNA and promoter activity of SREBP-1c, thus suggesting that this phosphatase might be the link among high carbohydrate feeding, insulin resistance and lipogenesis (Shimizu *et al.* 2003). Since the regulation of PTP-1B upon SREBP-1 expression (probably up-regulating the Sp1 transcriptional activity) occurs *via* an increase in protein phosphatase 2A (PP2A) activity, the authors have postulated the activation of the PTP1B-PP2A axis as a novel regulatory mechanism of gene expression in the triglyceride biosynthesis pathway (Shimizu *et al.* 2003) (Figure 43.5).

As already mentioned, fructose enhances SCD gene expression with the consequent increase in the *de novo* synthesis of oleate, which activates SREBP-1c. In fact, the fructose overload failed to induce SREBP-1 or lipogenic genes in SCD1-/- mice, but that effect was partially restored supplementing the diet with oleate. Since long-term administration of fructose induces the expression of SCD1 and other liver lipogenic genes in SREBP-1c-/- mice, it is assumed that the fructose-induced oleate production could activate lipogenic genes through either a SREBP-1c-dependent or -independent mechanism (Miyazaki *et al.* 2004). In the first pathway, SREBP-1c can induce SCD1 gene expression in a positive feedback loop, resulting in a further increase of oleate production.

Besides this transcriptional regulation of lipogenic genes *via* SREBP activation, it has also been shown that fructose induces the post-transcriptional regulation of rat liver FAS by increasing the stability of its mRNA (Katsurada *et al.* 1990). Another possible pathway involves a transcriptional co-activator, peroxisome proliferator activated receptor (PPAR)-γ coactivator 1β (PGC1β). PGC1β is a nuclear receptor co-activator that increases the activity of several key transcription factors such as PPARγ, PPARα, estrogen-related receptors and liver X receptor (LXR) (Lin *et al.* 2005). The role of PGC1β in fructose-induced lipogenesis (associated to its ability to bind to and transactivate SREBP-1c) was clarified using antisense oligonucleotides (ASOs) to silence its expression (Nagai *et al.* 2002). Fructose administration to ASO animals decreased the expansion of white adipose tissue and also prevented the increase of fasting glucose, insulin and plasma triglyceride levels; these changes were accompanied by a large decrease in SREBP1c expression.

Based on chromatin immunoprecipitation assays, fructose increased the binding of both SREBP1c and LXR to the SREBP1c promoter, and this

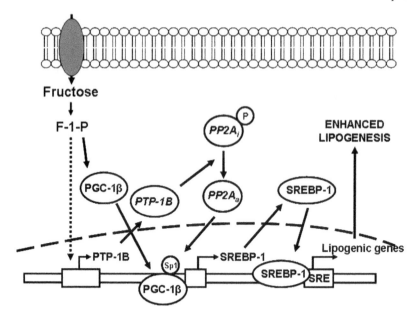

Figure 43.5 Fructose and the master transcriptional regulators of lipogenesis. Fructose increases liver expression of sterol regulatory element binding protein (SREBP), a key transcription factor responsible for the regulation of fatty acid and cholesterol biosynthesis. SREBP binds to sterol responsive elements (SRE) at the DNA level, inducing the expression of lipogenesis-related genes, which in turn enhance the expression of several target genes related to lipid synthesis. Such increase is accompanied by a parallel increase in protein-tyrosine phosphatase 1B (PTP-1B), leading to an increase of mRNA concentration and promoter activity of SREBP-1c; this regulation occurs *via* an increase in the activity of protein phosphatase 2A (PP2A).

interaction was lower in PGC1β ASO animals. This decreased expression of SREBP-1c leads to a decreased induction of lipogenic enzymes, with the consequent fall in acylylglycerol deposits. In brief, even when the mechanism by which a fructose increases PGC1β activity is not completely understood, this co-activator plays a key role in liver lipid metabolism by regulating the fructose-induced hepatic *de novo* lipogenesis (Figure 43.5).

Transcription factor X-box binding protein (XBP) 1, a key regulator of the unfolded protein response (or endoplasmic reticulum [ER] stress response), has recently emerged as a novel regulator of the fructose-mediated lipogenic pathways (Lee *et al.* 2008). The stimulated ER stress activates the inositol-requiring transmembrane kinase/endonuclease (IRE1) which, together with PERK (PKR-like ER-associated kinase or protein kinase RNA-like ER kinase) activate transcription factor 6 (ATF6), promoting the unfolded protein response (Calfon *et al.* 2002). IRE1 induces the XBP1 mRNA splicing to generate mature XBP1 mRNA, encoding a transcriptional factor (XBP1s) that binds to several ER chaperone genes involved in the expression of many proteins related to ER membrane expansion, including lipogenic enzymes (Glimcher 2010). Hepatic XBP1s protein was markedly increased in mice fed with a fructose-rich diet. This effect, however, was not associated to changes in other ER stress markers, suggesting a novel function of XBP1s in the lipogenic pathway independent of (or unrelated to) the ER stress response. Interestingly, XBP1-deficient mice exhibited a lower expression of critical lipogenic genes (such as those encoding FAS, SCD and ACC1), and these animals are characterized by an impaired liver triglyceride secretion and an impaired *de novo* fatty acid and sterol synthesis (Lee *et al.* 2008).

Carbohydrate response element binding protein (ChREBP) is a transcription factor of the bHLHLZ family. In addition to SRE, carbohydrate response element (ChRE) is present in the promoter of several lipogenic genes such as ACC, FAS and possibly SCD1 and glycerol-3-phosphate acyl transferase 1 (GPAT1). In the presence of a high glucose concentration, the increased ChREBP binding to ChRE contributes to the induction of lipogenic genes. The fact that ChREBP-null mice are intolerant to high sucrose/fructose diets suggests a critical role of ChREBP in the mechanism of fructose-induced lipogenesis (Iizuka *et al.* 2004). In this regard, Koo *et al.* (2008) demonstrated that the expression of the above mentioned lipogenic genes and ChREBP itself were highly increased in rats fed with fructose. Additionally, fructose-induced ChREBP activity can also be regulated by a post-translational mechanism: translocation from the cytosol to the nucleus increases its binding activity to a target DNA sequence (Koo *et al.* 2009). The underlying mechanism by which fructose activates this transcription factor remains unknown.

It has been postulated that xylulose 5-phosphate (X-5-P), an intermediate metabolite of the pentose phosphate pathway, could be the missing link between fructose metabolism and ChREBP activation, through an indirect pathway that involves activation of PPA2, with the consequent dephosphorylation/activation of ChREBP (Kabashima *et al.* 2003; Uyeda and Repa 2006) (Figure 43.6). Fructose increased the gene expression and the activity of glucose-6-phosphate dehydrogenase, the first and rate-limiting enzyme of the pentose pathway (Koo *et al.* 2008), with the consequent increase in the postulated mediator X-5-P.

43.5 Fructose, Hepatic Leptin Resistance and Impaired β-oxidation of Fatty Acids

Although the increased fat deposit in the liver mainly results from the enhanced lipogenic activity induced by fructose, the reduced fatty acid utilization *via* β-oxidation also contributes to such effect; this is due to an increase in ACC activity that increments the amount of malonyl-CoA that inhibits the activity of CPT-1, with the consequent decrease of fatty acid transport into the mitochondria for their oxidation.

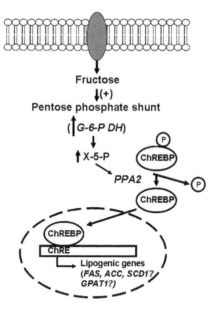

Figure 43.6   Xylulose 5-phosphate (X-5-P) is the link between fructose metabolism and gene activation. Fructose increases the gene expression and the activity of glucose-6-phosphate dehydrogenase, the first and rate-limiting enzyme of the pentose pathway, with the consequent increase in X-5-P. X-5-P could be the link between fructose metabolism and ChREBP activation, through a pathway that involves PPA2.

According to some authors, the reduction of liver fatty acid oxidation induced by fructose could be due to an impairment of the leptin-signal transduction pathway (leptin resistance) secondary to a hyperleptinemic state (Vilà *et al.* 2008). The link between these two processes comprises the inactivation of two proteins related to fatty acid catabolism: adenosine monophosphate-activated protein kinase (AMPK) and PPARα. Such inactivation is due to a deficient phosphorylation of the Ser/Thr residues in several proteins involved in the intracellular leptin signal: the lack of AMPK activation is mediated by inactivation of the mitogen-activated protein kinase (MAPK) pathway, while trans-repression of PPARα accounts for the defective leptin signaling through the Janus-activated kinase-2 (JAK-2)/Akt pathway. Additionally, binding of PPARα to the activated, unphosphorilated form of FoxO1 decreases its activity and fatty acid oxidation (Vilà *et al.* 2008) (Figure 43.7). The fructose-induced increase in PP2A activity (probably through X-5-P production; see the previous section) could be the main responsible for the above mentioned deficient phosphorylation of the Ser/Thr residues. Closing the vicious circle, fructose would reinforce leptin resistance by activating the signal transducer and activator of transcription-3 (STAT-3) pathway, which increases the protein expression of the suppressor of cytokine signaling-3 (SOC-3), which on time blocks the JAK-2 and MAPK pathways (Vilà *et al.* 2008).

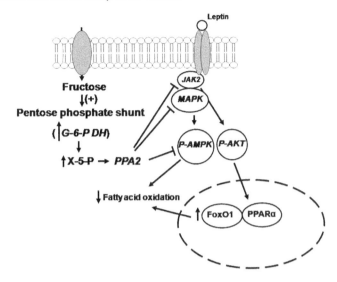

Figure 43.7 Fructose, hepatic leptin resistance and impaired fatty acid β-oxidation. The reduction of liver fatty acid oxidation induced by fructose could involve an impairment of the leptin-signal transduction pathway. The link between these two processes comprises the inactivation of proteins related to fatty acid catabolism due to a deficient phosphorylation process, namely, adenosine monophosphate-activated protein kinase (AMPK) and PPARα. Additionally, binding of PPARα to the activated, unphosphorylated form of FoxO1, decreases its activity and fatty acid oxidation. The fructose-induced increase in PP2A activity could be the main responsible for the above mentioned deficient phosphorylation rate.

43.6 The "Inflammatory Connection"

The role of tumor necrosis factor α (TNFα) in the pathogenesis of NAFLD, largely suggested by evidence obtained in different animal models, has been recently causally involved in the onset of fructose-induced NAFLD (Kanuri *et al.* 2011). It has been shown that fructose simultaneously increases the intestinal translocation of bacterial endotoxin and TNFα and subsequently induces liver steatosis in mice. These changes were attenuated by either antibiotic administration or the loss of the endotoxin toll-like receptor (TLR)-4 (Bergheim *et al.* 2008; Spruss *et al.* 2009). The knock-out (KO) of the TNFα receptor (TNFR1$-/-$) has also demonstrated a causal role of TNFα in the onset of fructose-induced NAFLD: in TNFR1-/- mice, both inflammatory markers and hepatic fat accumulation were attenuated after chronic fructose administration.

It has been suggested that the induction of TNFα may suppress the activation of AMPK in the liver, leading to the induction of SREBP1 and subsequently FAS, thereby contributing to the development of fatty liver (Kanuri *et al.* 2011). However, the exact underlying mechanisms and signaling pathways involved are not completely clear and should be further studied.

On the other hand, TNFα would also induce the hepatic protein expression of the plasminogen activator inhibitor-1 (PAI-1) (Kanuri *et al.* 2011) and the KO of this protein prevents the onset of fructose-induced steatosis. These data reinforce the concept of the causative role of TNFα, mediating the early phase of liver damage. In a pilot study, it was also demonstrated that patients with NAFLD consumed significantly more fructose than controls, and had significantly higher plasma levels of endotoxin and PAI-1 as well as of hepatic TLR-4 and PAI-1 mRNA expression; these data suggest that these changes may contribute to the development of NAFLD also in humans (Thuy *et al.* 2008).

43.7   Epigenetic Induction of Fructose-related Lipogenesis

In the previous section, we showed that environmental factors – in our case the intake of fructose – significantly affect the expression of genes involved in the regulation of carbohydrate and lipid metabolism. It is not clear however, the underlying mechanism by which fructose induced such changes.

In the last years, epigenetics have provided important clues to explain the mechanisms by which environmental changes affect gene activity (Holness *et al.* 2010). In this context, the epigenetic reprogramming of a subset of genes involved in lipid and carbohydrate metabolism could be the potential "missing" link between the increased fructose supply and the impaired liver metabolic response that leads to liver steatosis. Recently, Caton *et al.* (2011) demonstrated that incubation of rat hepatocytes with fructose induced an increase in sirtuin 1 (SIRT1) protein content and activity. SIRT1 belongs to the SIRT family of oxidized nicotinamide adenine dinucleotide-dependent protein deacetylases that regulate several transcription factors and/or co-regulators (Holness *et al.* 2010). These changes in SIRT1 were accompanied by a parallel induction of genes related to lipogenesis, an effect that was abolished by addition of an SIRT1 inhibitor to the culture media. Even when the mechanism was demonstrated in an *in vitro* assay, epigenomic modification *via* SIRT1 could play a role in the enhanced lipogenesis induced by fructose observed in the liver of rodents and humans.

43.8   Conclusion

NAFLD is a common clinical entity that in many cases produces a serious liver injury that requires its replacement. There is clear evidence that unhealthy diets, such as those rich in fructose, play a significant role in NAFLD development and progression, and that it can be effectively prevented by appropriate lifestyle changes. However, many people develop this pathological entity, with the consequent increase in costs of care and a decrease in their quality of life. Despite the multiple research activities devoted to understand the mechanisms by which unhealthy diets trigger NAFLD, there are still many unanswered questions. Hopefully, future research using innovative approaches would provide key answers and opportunities to develop effective drugs to prevent the development and progression of NAFLD.

Summary Points

- Nonalcoholic fatty liver disease (NAFLD) is an important emerging public health problem which is expanding worldwide.
- NAFLD is the liver component of the metabolic syndrome, frequently associated with obesity, type 2 diabetes (T2DM) and cardiovascular disease.
- The high intake of high fructose corn syrups (HFCS) and refined carbohydrates registered in the last decades could be partly responsible for the current epidemics of obesity and T2DM.
- Due to the existence of a "fructose pathway", high fructose intake provides an uncontrolled amount of substrates for *de novo* lipogenesis.
- The mechanism by which fructose activates lipogenesis is complex and involves several transcription regulators.
- Other mechanisms involved in the process by which the high fructose intake induces NAFLD are the inflammatory cascade, the hepatic leptin resistance and the reprogramming of genes involved in lipid and carbohydrate metabolism through epigenetic mechanisms.

Key Facts of Fructose and Nonalcoholic Fatty Liver Disease

- Fructose, or fruit sugar, is a simple monosaccharide found in plants that is absorbed directly from the gut to the bloodstream during digestion.
- Commercial fructose is usually derived from sugar cane and corn, and one of its common forms is high-fructose corn syrup (HFCS), that contains a mixture of glucose and fructose as monosaccharides.
- The liver is the primary site for fructose extraction and metabolism. An increase in fructose flow and availability can impair the hepatic metabolism of lipids and carbohydrates.
- The increased intake of HFCS may also affect insulin sensitivity and fat liver deposits.
- It has been suggested that the increased use of HFCS has contributed to the current epidemics of obesity and type 2 diabetes (T2DM).
- Nonalcoholic fatty liver disease (NAFLD) is a common clinical entity characterized by insulin resistance and fat accumulation into the liver, in the absence of other apparent causes of fat deposits.
- NAFLD comprises a continuum of disease ranging from steatosis, the earliest relatively benign and indolent state of liver injury, to steatohepatitis (NASH) with necroinflammation induced by severe lipotoxicity.

- Obesity, T2DM, cardiovascular disease and NAFLD are closely linked by a common underlying clinical metabolic abnormality: the metabolic syndrome.

Definitions of Words and Terms

Insulin resistance. Abnormal state present in several diseases, characterized by a lower response to insulin stimulus.

Autoimmune hepatitis. Hepatic lesion secondary to a self-autoimmune attack to liver cells.

Lipotoxicity. Cell toxicity induced by high lipid levels.

Atherogenic dyslipidemia. Vascular lesion characterized by abnormal deposits of lipids in the vessel lumen, with the consequent decrease in vascular inner diameter.

T2DM. Type 2 diabetes mellitus, the most common clinical presentation of diabetes.

Cardiovascular disease. Abnormal state characterized by impairment of cardiac and vascular organs.

Metabolic syndrome. Human disease characterized by the simultaneous presence of obesity, impaired glucose tolerance or type 2 diabetes, dyslipidemia and hypertension.

Fatty liver. Accumulation of an abnormal amount of fat in the gland.

Toll-like receptor. Proteins located at the cell membrane level that participate in immunoreactions.

Epigenetic. Related to changes in gene expression induced by mechanisms other than changes in DNA sequence, such as DNA methylation or histone deacetylation.

Sirtuins. Proteins with histone deacetylase activity that have been related to delayed aging, the decrease of apoptosis and stress resistance, and also as part of the alert reaction triggered by a low-calorie intake.

Glycolytic pathway or glycolysis. it is involved in the metabolic pathway that generates pyruvate from glucose.

Gluconeogenetic pathway. Metabolic pathway that involves the generation of glucose from non-carbohydrate carbon sources such as lactate, glycerol, and glucogenic amino acids.

Lipogenic pathway. Metabolic pathway leading to the synthesis of lipids.

List of Abbreviations

ACC	Acetil CoA carboxilase
Akt	AK strain trasforming
AMPK	Adenosine monophosphate-activated protein kinase
ASO	Antisense oligonucleotide
ATF6	Activating transcription factor 6
ChRE	Carbohydrate response element
ChREBP	Carbohydrate regulatory element binding protein

CoA	Coenzyme A
CPT1	Carnitin-palmitoyltransferase I
DHA-P	Dihydroxyacetone-phosphate
ER	Endoplasmic reticulum
ERK1/2	Extracellular signal-regulated kinase 1/2
FAS	Fatty acid synthase
G-3-P	Glyceraldehyde-3-phosphate
GLUT2	Glucose transporter 2
GPAT1	Glycerol-3-phosphate acyl transferase 1
HDL	High density lipoprotein
HFCS	High fructose corn syrups
IRE1	Inositol-requiring transmembrane kinase/endonuclease
JAK-2	Janus-activated kinase-2
LDL	Low-density lipoprotein
LXR	Liver X receptor
MAPK	Mitogen-activated protein kinase
NAFLD	Nonalcoholic fatty liver disease
NASH	Nonalcoholic steatohepatitis
NEFA	Non esterified fatty acids
PAI-1	Plasminogen activator inhibitor-1
PDH	Pyruvate dehidrogenase complex
PDK	Pyruvate dehidrogenase kinase
PERK	Protein kinase RNA-like ER kinase
PFK-1	Phosphofructokinase-1
PGC1β	Peroxisome proliferator activated receptor (PPAR)-γ coactivator 1β
PPA2	Protein phosphatase 2A
PTP-1B	Protein-tyrosine phosphatase 1B
SCD	Stearoyl-CoA desaturase
SIRT	Sirtuin
SRE	Sterol responsive elements
SREBP	Sterol regulatory element binding protein
STAT-3	Signal transducer and activator of transcription-3
T2DM	Type 2 diabetes mellitus
TLR	Toll-like receptor
TNFα	Tumor necrosis factor α
TNFR	Tumor necrosis factor α-receptor
VLDL	Very-low-density lipoprotein
X-5-P	Xylulose 5-phosphate
XBP	X-box binding protein

Acknowledgements

This review was supported by grants from CONICET and Universidad Nacional de La Plata, Argentina. The authors are grateful to Adriana Di Maggio for careful manuscript editing.

References

Ali, R., and Cusi, K., 2009. New diagnostic and treatment approaches in nonalcoholic fatty liver disease (NAFLD). *Annals of Medicine*. 41: 265–278.

Assy, N., Nasser, G., Kamayse, I., Nseir, W., Beniashvili, Z., Djibre, A., and Grosovski, M., 2008. Soft drink consumption linked with fatty liver in the absence of traditional risk factors. *Canadian Journal of Gastroenterology*. 22: 811–816.

Bennett, M.K., Lopez, J.M., Sanchez, H.B., and Osborne, T.F., 1995. Sterol regulation of fatty acid synthase promoter. Coordinate feedback regulation of two major lipid pathways. *The Journal of Biological Chemistry*. 270: 25578–25583.

Bergheim, I., Weber, S., Vos, M., Kramer, S., Volynets, V., Kaserouni, S., McClain, C.J., and Bischoff, S.C., 2008. Antibiotics protect against fructose-induced hepatic lipid accumulation in mice: role of endotoxin. *Journal of Hepatology*. 48: 983–992.

Calfon, M., Zeng, H., Uramo, F., Till, J.H., Hubbard, S.R., Harding, H.P., Clark, S.G., and Ron, D., 2002. IRE1 couples endoplasmic reticulum load to secretory capacity by processing the XBP1 mRNA. *Nature*. 415: 92–96.

Carmona, A., and Freedland, R.A., 1989. Comparison among the lipogenic potential of various substrates in rat hepatocytes: the differential effects of fructose-containing diets on hepatic lipogenesis. *Journal of Nutrition*. 119: 1304–1310.

Caton, P.W., Nayuni, N.K., Khan, N.Q., Wood, E.G., and Corder, R., 2011. Fructose induces gluconeogenesis and lipogenesis through a SIRT1-dependent mechanism. *Journal of Endocrinology*. 208: 273–783.

Dekker, M.J., Su, Q., Baker, C., Rutledge, A.C., and Adeli, K., 2010. Fructose: a highly lipogenic nutrient implicated in insulin resistance, hepatic steatosis, and the metabolic syndrome. *American Journal of Physiology – Endocrinology and Metabolism*. 299: E685–E694.

Fried, S.K., and Rao, S.P., 2003. Sugars, hypertriglyceridemia, and cardio-vascular disease. *American Journal of Clinical Nutrition*. 78: 873S–880S.

Glimcher, L.H., 2010. XBP-1: the last two decades. *Annals of the Rheumatic Diseases*. 69Suppl.1:i67–71.

Holnes, M.J., Caton, P.W., and Sugden, M.C., 2010. Acute and long-term nutrient-led modifications of gene expression: potential role of SIRT1 as a central co-ordinator of short and longer-term programming of tissue function. *Nutrition*. 26: 491–501.

Hu, F.B., and Malik, V.S., 2010. Sugar-sweetened beverages and risk of obesity and type 2 diabetes: epidemiologic evidence. *Physiology & Behavior*. 100: 47–54.

Iizuka, K., Bruick, R.K., Liang, G., Horton, J.D., and Uyeda, K., 2004. Deficiency of carbohydrate response element-binding protein (ChREBP) reduces lipogenesis as well as glycolysis. *Proceedings of the National Academy of Sciences of the United States of America*. 101: 7281–7286.

Kabashima, T., Kawaguchi, T., Wadzinski, B.E., and Uyeda, K., 2003. Xylulose 5-phosphate mediates glucose-induced lipogenesis by xylulose 5-phosphate-activated protein phosphatase in rat liver. *Proceedings of the National Academy of Sciences of the United States of America*. 100: 5107–5112.

Kanuri, G., Spruss, A., Wagnerberger, S., Bischoff, S.C., and Bergheim, I., 2011. Role of tumor necrosis factor alfa (TNF alfa) in the onset of fructose-induced nonalcoholic fatty liver disease in mice. *The Journal of Nutritional Biochemistry*. 22: 527–534.

Katsurada, A., Iritani, N., Fukuda, H., Matsumura, Y., Nishimoto, N., Noguchi, T., and Tanaka, T., 1990. Effect of nutrients and hormones on transcriptional and post-transcriptional regulation of fatty acid synthase in rat liver. *European Journal of Biochemistry*. 190: 427–433.

Koo, H.-Y., Wallig, M.A., Chung, B.H., Nara, T.Y., Cho, B.H.S., and Nakamura, M.T., 2008. Dietary fructose induces a wide range of genes with distinct shift in carbohydrate and lipid metabolism in fed and fasted rat liver. *Biochimica et Biophysica Acta*. 1782: 341–348.

Koo, H.-Y., Miyashita, M., Cho, B.H.S., and Nakamura, M.T., 2009. Replacing dietary glucose with fructose increases ChREBP activity and SREBP-1 protein in rat liver nucleus. *Biochemical and Biophysical Research Communications*. 390: 285–289.

Le, K.A., Ith, M., Kreis, R., Faeh, D., Bortolotti, M., Tran, C., Boesch, C., and Tappy, L., 2009. Fructose overconsumption causes dyslipidemia and ectopic deposition in healthy subjects with and without a family history of type 2 diabetes. *The American Journal of Clinical Nutrition*. 89: 1760–1765.

Lee, A.-H., Scapa, E.F., Cohen, D.E., and Glimcher, L.H., 2008. Regulation of hepatic lipogenesis by the trascriptional factor XBP1. *Science*. 320: 1492–1496.

Lin, J., Handschin, C., and Spiegelman, B.M., 2005. Metabolic control through the PGC-1 family of transcription coactivators. *Cell Metabolism*. 1: 361–370.

Matsuzaka, T., Shimano, H., Yahagi, N., Amemiya-Kudo, M., Okazaki, H., Tamura, Y., Iizuka, Y., Ohashi, K., Tomita, S., Sekiya, M., Hasty, A., Nakagawa, Y., Sone, H., Toyoshima, H., Ishibashi, S., Osuga, J., and Yamada, N., 2004. Insulin-independent induction of sterol regulatory element-binding protein-1c expression in the livers of streptozotocin-treated mice. *Diabetes*. 53: 560–569.

Mayes, P.A., 1993. Intermediary metabolism of fructose. *The American Journal of Clinical Nutrition*. 58(5 Suppl): 754S–765S.

Miyazaki, M., Dobrzyn, A., Man, W.C., Chu, K., Sampath, H., Kim, H.J., and Ntambi, J.M., 2004. Stearoyl-CoA desaturase 1 gene expression is necessary for fructose-mediated induction of lipogenic gene expression by sterol regulatory element-binding protein-1c-dependent and -independent mechanisms. *The Journal of Biological Chemistry*. 279: 25164–25171.

Nagai, Y., Nishio, Y., Nakamura, T., Maegawa, H., Kikkawa, R., and Kashiwagi, A., 2002. Amelioration of high fructose-induced metabolic derangements by activation of PPARalpha. *American Journal of Physiology – Endocrinology and Metabolism*. 282: E1180–1190.

Nielsen, S.J., and Popkin, B.M., 2004. Changes in beverage intake between 1977 and 2001. *American Journal of Preventive Medicine.* 27: 205–209.

Ouyang, X., Cirillo, P., Sautin, Y., McCall, S., Bruchette, J.L., Mae Diehl, A., Johnson, R.J., and Abdelmalek, M.F., 2008. Fructose consumption as a risk factor for non-alcoholic fatty liver disease. *Journal of Hepatology.* 48: 993–999.

Pagliassotti, M., and Horton, T., 2004. Sucrose, insulin action and biologic complexity. *Recent Research Developments in Physiology.* 2: 337–353.

Park, O.-J., Cesar, D., Faix, D., Wu, K., Shackleton, C.H.L., and Hellerstein, M.K., 1992. Mechanism of fructose-induced hypertriglyceridaemia in the rat. *Biochemical Journal.* 282: 753–757.

Samuel, V.T., 2011. Fructose induced lipogenesis: from sugar to fat to induced insulin resistance. *Trends in Endocrinology and Metabolism.* 22: 60–65.

Shimizu, S., Ugi, S., Maegawa, H., Egawa, K., Nishio, Y., Yoshizaki, T., Shi, K., Nagai, Y., Morino, K., Nemoto, K., Nakamura, T., Bryer-Ash, M., and Kashiwagi, A., 2003. Protein-tyrosine phosphatase 1B as new activator for hepatic lipogenesis via sterol regulatory element-binding protein-1 gene expression. *The Journal of Biological Chemistry.* 278: 43095–43101.

Spruss, A., Kanuri, G., Wagnerberger, S., Haub, S., Bischoff, S.C., and Berheim, I., 2009. Toll-like receptor 4 is involved in the development of fructose-induced hepatic steatosis in mice. *Hepatology.* 50: 1094–1104.

Stanhope, K.L., Schwarz, J.M., Keim, N.L., Griffen, S.C., Bremer, A.A., Graham, J.L., Hatcher, B., Cox, C.L., Dyachenko, A., Zhang, W., McGahan, J.P., Seibert, A., Krauss, R.M., Chiu, S., Schaefer, E.J., Ai, M., Otokozawa, S., Nakajima, K., Nakano, T., Beysen, C., Hellerstein, M.K., Berglund, L., and Havel, P.J., 2009. Consuming fructose-sweetened, not glucose-sweetened beverages increases visceral adiposity and lipids and decreases insulin sensitivity in overweight/obese humans. *The Journal of Clinical Investigation.* 119: 1322–1334.

Thuy, S., Ladurner, R., Volynets, V., Wagner, S., Strahl, S., Königsrainer, A., Maier, K.P., Bischoff, S.C., and Bergheim, I., 2008. Nonalcoholic fatty liver disease in humans is associated with increased plasma endotoxin and plasminogen activator inhibitor 1 concentrations and with fructose intake. *The Journal of Nutrition.* 138: 1452–1455.

Toshimitsu, K., Matsuura, B., Ohkubo, I., Niiya, T., Furukawa, S., Hiasa, Y., Kawamura, M., Ebihara, K., and Onji, M., 2007. Dietary habits and nutrient intake in non-alcoholic statohepatitis. *Nutrition.* 23: 46–52.

Uyeda, K., and Repa, J.J., 2006. Carbohydrate response element binding protein, ChREBP, a transcriptional factor coupling hepatic glucose utilization and lipid synthesis. *Cell Metabolism.* 4: 107–110.

Vilà, L., Roglans, N., Alegret, M., Sánchez, R.M., Vázquez-Carrera, M., and Laguna, J.C., 2008. Suppressor of cytokine signaling-3 (SOCS-3) and a deficit of serine/threonine (Ser/Thr) phosphoproteins involved in leptin transduction mediate the effect of fructose on rat liver lipid metabolism. *Hepatology.* 48: 1506–1516.

Wei, Y., Wang, D., Topczewski, F., and Pagliassotti, M., 2006. Fructose-mediated stress signaling in the liver: implications for hepatic insulin resistance. *The Journal of Nutritional Biochemistry*. 18: 1–9.

Zelber-Sagi, S., Nitzan-Kaluski, D., Goldsmith, R., Webb, M., Blendis, L., Halpern, Z., and Oren, R., 2007. Long term nutritional intake and the risk for non-alcoholic fatty liver disease [NAFLD]: a population based study. *Journal of Hepatology*. 47: 711–717.

CHAPTER 44

High Sucrose Diet and Antioxidant Defense

KAMAL A. AMIN,*[a] G. M. SAFWAT[b] AND
RAJAVENTHAN SRIRAJASKANTHAN[c]

[a] Biochemistry Department, Faculty of Veterinary Medicine, Beni-Suef
University, Beni-Suef, Egypt 62511; [b] Department of Biochemistry,
Faculty of Veterinary Medicine, Beni Suef, Egypt 62511; [c] University
Hospital Lewisham, London, SE13 6LH, UK
*Email: Kaamin10@yahoo.com; kamalamin2001@yahoo.com

44.1 Introduction

44.1.1 Overview of Sucrose

44.1.1.1 Prevalence of Sucrose and its Use

Pure sucrose is well known and widely used as a sweetener in the modern, industrialized world in cooked, uncooked and baked preparation of food and beverages. Sucrose is most commonly associated with use in food, drink juice and chocolate.

Generally, sucrose is extracted from sugar cane or sugar beet and then purified and crystallized as refined white sugar (Figure 44.1). Other minor commercial sources are sorghum, pineapple, carrot roots and sugar maples.

A dramatic rise in the prevalence of insulin resistance and type 2 diabetes mellitus has been paralleled by increasing dietary consumption of sugar.

Food and Nutritional Components in Focus No. 3
Dietary Sugars: Chemistry, Analysis, Function and Effects
Edited by Victor R Preedy
© The Royal Society of Chemistry 2012
Published by the Royal Society of Chemistry, www.rsc.org

Figure 44.1 A photo of refined, white sugar and its consumption.
Photo illustration from New York time, prop stylist: Nell Tivnan.
Source: USDA 2009 estimates.

The use of added sweeteners containing sucrose and high-fructose corn syrup has increased by 25% over the past three decades (Figure 44.1).

44.1.2 Sucrose Function, Oxidant/Antioxidant Qualities and its Body Health Effects

Sucrose has several functional roles in foods which expand beyond its sweetness, including preservative, texture and flavor modifying qualities. It is broadly used as a main ingredient in many confectionery ingredients and desserts. Sucrose provides humans with a sudden increase of energy once consumed; however, this effect is not sustained.

44.1.2.1 Effect of Sucrose on Food Intake

High Sucrose Diets (HSD) intake can occur due to hyperphagia which is induced by sucrose which affects the appetite and centers in the hypothalamus. Neuropeptide Y (NPY) and proopinne melanocortin (POMC) both receive information about nutritional status and level of energy storage through insulin and leptin signaling mediated by specific receptor located POMC and NPY neurons. Sucrose consumption increases calorie intake through up regulation of hypothalamic CB1 mRNA and down regulation of NPY mRNA. Also consumption of sucrose solution results in body weight gain through activation of hunger signals and depression of satiety signals. Conversely, sucrase deficiency – the enzyme that metabolises sucrose – leads to the development of malabsorption with diarrhea and flatulence.

44.1.2.2 Effect of Sucrose on Body Weight and Mood

Data from Amin *et al.* (2011) showed significant increase in body weight and BMI especially in the 8th wk in a high sucrose diet. Increased BW gain,

Table 44.1 Demonstrates the body weight gain, serum glucose and TG in different diet feeding groups. This table shows significant elevation in weight gain, serum TG and TC with HSD and HFD, which is supported by other studies.

	Body Weight gain (g/day)	Glucose (mg/dl)	Serum Triglyceride (mg/dl)	Total Cholesterol (mg/dl)	Reference
-Control	40	59.0 ± 1.6 a	85 ± 2.3[a]	81 ± 1.4 a	Amin et al.
-HSD	120	71.0 ± 3.2 a	350 ± 3.9[b]	320 ± 3.3 b	(2011)
-HFD	190	111.0 ± 4.1b	384.01 ± 11b	351.01 ± 4.6[b]	(Feeding for 12 wks)
-Control	185.1 ± 3.62	—	51.50 ± 0.92	—	Sweazea
-HSD	208.5 ± 4.10*		95.00 ± 8.11*		et al.
-HFD	220.8 ± 4.37*#		52.11 ± 2.11#		(2010). (Feeding for 6 Wks)
	—	—	(mmol/l)	(mmol/l)	Robert et al.
-Control			1.51 ± 0.13a	2.29 ± 0.11a	(2008).
-HSD			1.76 ± 0.17a	2.22 ± 0.12a	(Feeding
-Potato			1.25 ± 0.16a	1.95 ± 0.05b	for 3 Wks)
-Control	85.4 ± 5.1[a]	92.4 ± 3.9[a]	98.69 ± 6.24[a]	103.19 ± 2.70[a]	Amin et al.
-HSD	108.2 ± 3.23[c]	113.32 ± 5.9[b]	118.10 ± 3.40[c]	105.84 ± 4.20[a]	(2004).
-SFO	127.5 ± 7.08[b]	93.14 ± 2.0[a]	69.52 ± 6.99[b]	89.82 ± 2.52[b]	(Feeding
-Beef Tallow	113.2 ± 4.03[bc]	89.79 ± 3.67[a]	134.97 ± 8.86[c]	121.13 ± 2.59[c]	for 7 wks)

HSD; high sucrose diet. HFD; high fat diet. SFO; Sunflower oil.
Data expressed as means ± SEM; *p < 0.05 from control; #p < 0.05 from HSD (Sweazea et al. (2010)).

(see Table 44.1) and BMI may be due to increased caloric intake resulting in more adipose tissue deposition than starch diet. These results vary between sex, age and ethnicity. Furthermore, HSD appear to induce mitochondrial dysfunction in adipose tissue, which may be related to greater weight gain and metabolic impairment (Lomba et al. 2009).

Overweight persons who consumed quite large amounts of sucrose (28% of energy), mostly as beverages, had increased energy intake, body weight, fat mass, and blood pressure after 10 wk. These effects were not noticed in a similar group of subjects who consumed synthetic sweeteners (Raben et al. 2002).

The long-term effect of sucrose on appetite and mood remain uncertain. Normal weight subjects compensate for sucrose added blind to the diet. Overweight women do not respond adversely to sucrose added blind to the diet, but compensate for it by reducing voluntary energy intake (Reid et al. 2010).

Sucrose and fructose intake is synchronized differentially by the dopamine system and reasonable interaction between diet composition and dopamine control of carbohydrate intake were verified in diet-induced obese rats. HSD may raise the level of neurotransmitter dopamine that have role in behavior, cognition and mood.

Mice fed diets rich in sucrose became obese, sucrose in the diet also counteracted the well-known anti-inflammatory effect of fish oil in adipose tissue

(Ma *et al.* 2011). *De novo* lipogenesis plays a significant role in increasing adiposity in response to carbohydrate feeding in rodents. In the liver, *de novo* lipogenesis was reduced by polyunsaturated, but not saturated fats. The high-carbohydrate diet as HSD in the present study produced a high glycemic response, promoting carbohydrate oxidation at the expense of fat oxidation, leading to weight gain and visceral adiposity (Janette *et al.* 2002).

44.1.2.3 Effect of Sucrose on Lipid Profiles

Many reports have indicated that feeding a high-sucrose diet induces an increase in plasma TG in human and animals (Preuffer *et al.* 1992). Moreover, HSD induces hepatic synthesis of TG from glucose and transport it to blood through VLDL and stored in adipose tissue.

Diets high in sucrose used in artificial foods and beverages produce large amounts of fructose and glucose which is transported to the liver via the hepatic portal vein. The structure of sucrose is shown in Figure 44.2. Fructose undergoes faster hepatic glycolysis than does glucose because it bypasses the regulatory reaction catalyzed by phosphofructokinase (Figure 44.3). This allows fructose to overflow the pathways in the liver, leading to enhanced fatty acid biosynthesis, increased fatty acids esterification and increased VLDL secretion and raise serum triacylglycerols and ultimately elevate LDL cholesterol.

Sucrose and fructose have a greater effect in raising blood lipids, particularly triacylglycerols, than do other carbohydrates.

Feeding rats HSD induced significant hypertriglyceridemia and hypercholesterolemia in comparison with control (Amin *et al.* 2011) (Table 44.1), which may be due to consumption of HSD, increases long-chain acyl CoA levels in muscle, an early indication of the disturbance of lipid metabolism (Chicco *et al.* 2003).

44.1.2.4 Effect of Sucrose on Blood Glucose Level

Fukuchi *et al.* (2004) demonstrated in the short term HSD has no significant effect on blood glucose level. They showed that rats fed a sucrose diet for 4 wks had significantly larger visceral fat pads and hypertriglyceridemia, however,

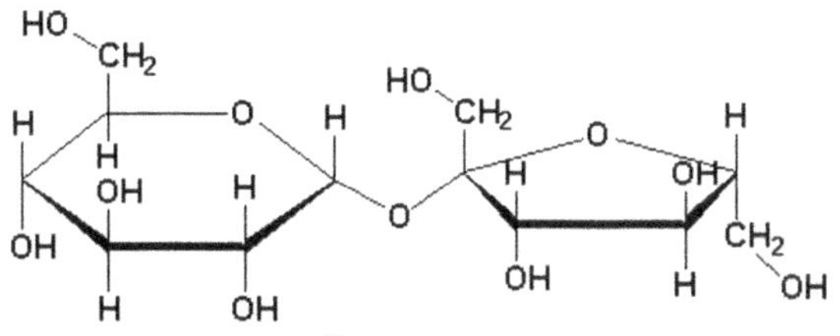

Sucrose

Figure 44.2 The chemical structure of sucrose. The diagram shows the sucrose structure which consists of 2 molecules as follow *O*-α-D-Glucopyranosyl-(1 → 2)-β-D-fructofuranoside.

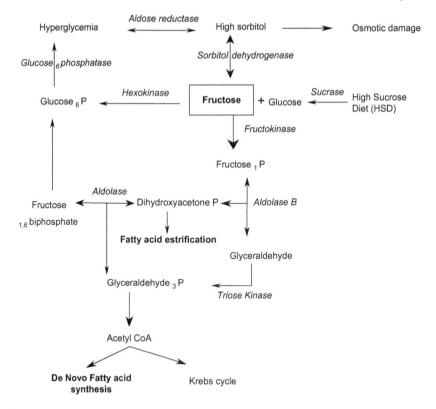

Figure 44.3 Present several metabolic pathway of sucrose and fructose (Modified from *Harper Review of Biochemistry*). This demonstrates the rapid conversion of sucrose to fatty acid through fructokinase. The conversion of fructose into water-attracting sorbitol, and the pathway of hexokinase for glucose formation.

neither plasma glucose nor insulin levels were significantly higher, while hyperglycemia and insulin resistance occur after 20 wks of feeding HSD. In contrast, a study by Amin *et al.* (2004) reported significant hyperglycemia in HSD compared with other diets (Table 44.1).

There was significant increase in G6PD activities in HSD which may be attributed to metabolic alteration produced from a sucrose rich diet, among these changes was increased level of G6pD, which is the first enzyme of hexose monophosphate pathway and induction of G6pD in paranchymal cells that support the escalating needs for NADPH for synthesis of fatty acids (Amin *et al.* 2011).

44.1.2.5   Sucrose and Metabolic Syndrome

Sugar-sweetened soft drinks may increase the risk of diabetes due to their large amounts of high-fructose corn syrup, rapidly raising blood glucose.

Diet, thereby contributing to a high glycemic index of the diet and promoting the development of obesity and diabetes. Higher consumption of sugar-sweetened beverages was associated with both greater weight gain and an increased risk of type 2 diabetes, independent of known risk factors. Sucrose-sweetened soft drinks and food might also increase risk of type 2 diabetes due to their readily absorbable carbohydrates.

44.1.2.6   High Sucrose Diet and Risk of Cancer

Recent papers link the high sucrose diet and pancreatic cancer. Rather than being causal, the short-term increase in pancreatic cancer risk associated with high available carbohydrate and low fat intake may be capturing dietary changes associated with subclinical disease (Meinhold *et al.* 2010).

A diet high in sucrose and dextrin acts as a promoter of cancer development and has been demonstrated to cause progression towards malignancy of tumors in the colon of rats (Poulsen *et al.* 2001).

Due to the large amounts of high-fructose, consumption of sugar-sweetened soft drinks might therefore contribute to a high glycemic load of the overall diet, which is a risk factor for pancreatic cancer. In addition, cola-type soft drinks contain caramel coloring, which is rich in advanced glycation end-products that might increase insulin resistance (Schernhammer *et al.* 2005).

Forthcoming data on the effects of different types of dietary sugars on cancer incidence have been limited. Added sugars were positively associated with risk of esophageal adenocarcinoma: supplementary fructose was associated with risk of small intestinal cancer. No association between dietary sugars and risk of colorectal or any other major cancer. Statistically significant associations observed for the rare cancers are of interest and warrant further investigation (Tasevska *et al.* 2011).

44.1.3   High Sucrose Diet, Antioxidant and Oxidative Stress Marker

44.1.3.1   High Sucrose Diet and Different Tissue Antioxidant

Several studies in human and animal models have shown that consumption of high sucrose from 2–8 weeks induces oxidative damage in heart and kidney. The mechanisms of oxidative damage involved may be through increased MDA, GSSG while decreasing GSH and altering antioxidant enzymes (SOD, catalase and GPx). When fed HSD, levels of ascorbic acid were significantly reduced (Table 44.4).

Reactive oxygen species, such as superoxide radicals, are thought to trigger the pathogenesis of various diseases. About 3–10% of the oxygen utilized by tissues is changed to its reactive intermediates, which damage the functions of cells and tissues (Noor *et al.* 2002). Free radicals are mainly generated in the mitochondrial membrane, owing to electron leakage from the electron transport chain as a direct result of the increased electron flow needed to meet the higher energetic requirements.

Lipid peroxidation is among the main effects resulting from free radical release in biological membranes. Many studies support the notion that the degree of fatty acid unsaturation of the membrane phospholipids determines susceptibility to peroxidation (Halliwell and Chirico 1993). In addition, it has been demonstrated that, after the period of adaptation, the fatty acid composition of a membrane is affected by the lipid profile of the diet.

The reduced form of glutathione (GSH) is a multifunctional tripeptide that directly or indirectly regulates a number of biological processes, such as DNA synthesis, ion transport, enzyme activity, transcription, signal transduction, and antioxidant defenses. Under physiological conditions, most protein sulfhydryls are maintained in the reduced state because of a high intracellular concentration of GSH relative to oxidized glutathione (GSSG), which is necessary for normal cell function. This condition is largely due to the activities of two major pathways that supply GSH; γ-glutamylcysteine synthetase, the ATP-dependent, rate limiting step in GSH synthesis and glutathione reductase, which catalyzes the reduction of GSSG to GSH using NADPH derived from the hexose monophosphate pathway (Deneke 2000).

Heart GSH levels are not affected while blood GSH decreased significantly in HSD fed rats than controls. This decrement in GSH levels consequently increasing its oxidized form (GSSG) following the initiation of oxidative stress. The change in GSH status is due to an increased cellular demand for GSH which was utilized for inactivation of hydroperoxides through the activity of glutathione peroxidase, which generates GSSG as a by-product. Depletion of intracellular GSH and accumulation of GSSG, dramatically impair cell function (Li *et al.* 2003). Importantly, GSH protects proteins from oxidation by a direct thiol-disulfide exchange reaction or by acting as a cofactor for oxidoreductase-mediated reduction of sulfhydryls.

The brain appears to be especially involved in generation and detoxification of ROS. The brain cells utilize 20% of the oxygen consumed by the body; this indicates generation of a large quantity of ROS during oxidative phosphorylation in brain. In addition, its high iron content (Gerlach *et al.* 1994) is able to catalyze the generation of ROS.

TBARs were significantly elevated in the liver and heart of rats fed sunflower oil and HSD than Beef tallow and control groups (Table 44.2). A significant positive correlation between liver TBARs and PUFA content and recommended that feeding oils rich in PUFA raises tissue level of these fatty acids, therefore increasing tissue lipid peroxidation and reducing the antioxidative status. Furthermore, Busserolles *et al.* (2002) reported that pro-oxidant and a great cardiac susceptibility to peroxidation occurred in HSD fed rats.

There was a significant increase in serum triacylglycerol plus liver and heart TBARs levels on a HSD intake. Conversely, serum vitamin E and C levels and blood GSH were significantly decreased in rats fed HSD (Table 44.3). These results indicate that a short-term consumption of HSD negatively affects the balance of free radical production and antioxidant defence in rats; leading to

Table 44.2 Comparison of oxidative stress marker, MDA levels in plasma and hepatic, cardiac and renal tissues in different diet fed groups. Serum, hepatic and cardiac MDA oxidative stress markers increased significantly in HSD compared with control and HFD (with some authors).

	Control	HSD	HFD	References
-Serum MDA (nmol/ml)	2.6 ± 0.22^a	7.3 ± 0.37^c ↑	4.9 ± 0.31^b	Amin *et al.* (2011)
-Renal MDA (nmol/g)	$1.2 \pm 0.15a$	$4.8. \pm 0.5c$ ↑	$6.6 \pm 0.27b$	
Plasma MDA (nmol/L)	21.29 ± 2.20	$32.54 \pm 2.56^*$ ↑	$29.88 \pm 1.59^*$	Sweazea *et al.* 2010
-Hepatic MDA (nmol/g)	$495.0 \pm 14.6a$	$520.4 \pm 22.9a$	—	Robert *et al.* (2008) (Feeding for 3 Wks)
-Cardiac MDA (nmol/g)	$285.2 \pm 45.8a$	$264.9 \pm 64.7a$		
-Plasma MDA (nmol/ml)	$6.96 \pm 0.841^{(a)}$	$10.11 \pm 0.67^{(bc)}$ ↑	$9.52 \pm 0.19^{(c)}$	Amin *et al.* (2004)
-Hepatic MDA	$21.85 \pm 1.95^{(a)}$	$46.13 \pm 11.13^{(bc)}$ ↑	$29.59 \pm 2.99^{(ac)}$	
-Cardiac MDA	$20.84 \pm 1.38^{(a)}$		$20.35 \pm 1.6^{(a)}$	
-Renal MDA (nmol/gm)	$23.21 \pm 1.10^{(a)}$	$30.8 \pm 4.33^{(b)}$ ↑ $28.21 \pm 3.77^{(ab)}$	$25.53 \pm 0.67^{(a)}$	

Different litters indicate significant difference ($p < 0.05$). Similar litter indicates no significant difference. HSD; high sucrose diet. HFD; high fat diet. SFO; Sunflower oil.
Data expressed as means $\pm$ SEM; $^*p < 0.05$ from control; $\#p < 0.05$ from HSD.

increased lipid susceptibility to peroxidation (Amin *et al.* 2004; Amin *et al.* 2011; Busserolles 2002).

The harmful oxidative stress of a high sucrose diet could be attributed to its fructose content, where reducing sugars are known to produce ROS mainly through the glycation reaction. Fructose is recognized as having a stronger reducing capacity than glucose and the glycation reaction being easily induced by fructose. Also, fructose-induced oxidative stress might modulate transcription factors that are sensitive to changes in the redox state of the cell (Allen and Tresini 2000). Furthermore, HSD alters cellular metabolism which in turn accelerates oxidative stress, leading to lipid peroxidation.

Fructose bypasses the main regulatory step in glycolysis, catalyzed by phosphor-fructokinase, and quickly stimulates fatty acid synthesis and hepatic triacylglycerol secretion (Figure 44.3).

Elevated intracellular glucose concentrations in HSD and an adequate supply of NADPH cause aldose reductase to produce a significant increase in the amount of sorbitol, which cannot pass efficiently through cell membranes and, therefore, remains trapped inside the cell (see Figure 44.3). This is exacerbated when sorbitol dehydrogenase is low or absent, for example, in retina, lens, kidney, and nerve cells. As a result, sorbitol accumulates in these cells, causing strong osmotic effects and, therefore, cell swell as a result of water retention. Some of the pathologic alterations associated with diabetes can be attributed,

Table 44.3 List of antioxidant GSH levels and catalase activity in plasma,
hepatic, cardiac and renal tissues in different diet feeding groups.
This table lists the antioxidant levels and catalase activity in different tissues.

	Control	*HSD*	*HFD*	*References*
-Hepatic GSH (mg/g tissue)	124.24 ± 1.39	—	93.32 ± 9.62 ↓	Noeman *et al.* (2011)
-Cardiac GSH (mg/g tissue)	98.2 ± 3.67	—	90.54 ± 5.59 ↓	
-Renal GSH	101 ± 2.38	—	89.89 ± 1.9	
-Blood GSH (mg/dl)	$5.30 \pm 0.55^{(a)}$	$2.2 \pm 0.48^{(b)}$ ↓	$4.25 \pm 0.37^{(a)}$	Amin *et al.* (2004)
-Hepatic GSH (μmol/gm)	$9.29 \pm 0.44^{(a)}$	$7.68 \pm 0.64^{(a)}$	$7.71 \pm 0.50^{(a)}$	
-Cardiac GSH	$7.14 \pm 0.52^{(a)}$	$6.52 \pm 0.99^{(ab)}$	$5.97 \pm 0.1^{(ab)}$	
-Renal GSH (μmol/gm)	$5.7 \pm 0.51^{(a)}$	$4.43 \pm 0.43^{(a)}$	$4.98 \pm 0.72^{(a)}$	
Renal Catalase (U/gm)	$4.4 \pm 0.29a$	14.5 ± 0.34 b ↑	$14.2 \pm 1.3b$ ↑	Amin *et al.* (2011)
-Hepatic Catalase (μmol H_2O_2 consumed/mg protein/min)	665.80 ± 6.77	—	664.26 ± 6.66 (NS)	Noeman *et al.* (2011)
-Cardiac catalase	586.5 ± 9.5	—	$591.8 \pm 4.75*$ ↑	
-Renal catalase	585.9 ± 3.7	—	$577.37 \pm 7.69**$	

Different litters indicate significant difference (p<0.05). Similar litter indicate no significant difference. HSD; high sucrose diet. HFD; high fat diet. SFO; Sunflower oil.
Data expressed as means $\pm$ SEM; *p<0.05 from control.

in part, to this phenomenon, including cataract development, peripheral neuropathy, and vascular troubles leading to nephropathy and retinopathy.

Lipid peroxidation products were relatively higher in the liver than in the
kidney and heart in the different diet groups (Beef tallow, Sunflower oil and
HSD), which may be attributable at least partly to the fact that liver is the
major organ for processing and storing nutrients and xenobiotics absorbed by
the body.

Recent findings show that a sucrose-based diet, in addition to its hyperli-
pemic effect, has a pro-oxidant effect in rats when compared with a starch-
based diet.

Increased intake of HSD was associated with increased susceptibility of
membranes to oxidation and an increased requirement for vitamin E which is
the major lipid-soluble free radical chain-breaking antioxidant found in
plasma, red cells, and tissues decreasing its blood level.

When providing dietary sucrose, an adequate intake of antioxidants is
advised for preventing oxidative stress and may increase tissue oxidative
stability.

Protection against lipid peroxidation differs from tissue to tissue. It is
important to estimate the risk of the presence of dietary factors inducing higher

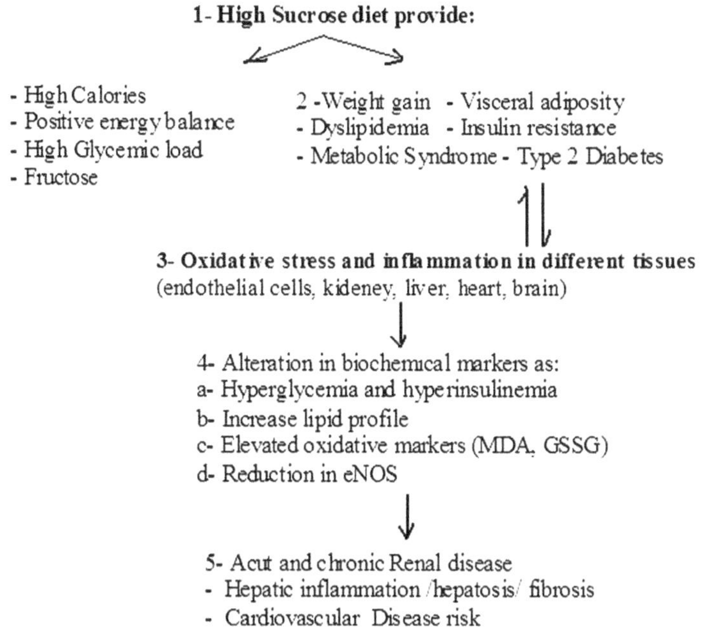

1- High Sucrose diet provide:

- High Calories
- Positive energy balance
- High Glycemic load
- Fructose

2 - Weight gain - Visceral adiposity
- Dyslipidemia - Insulin resistance
- Metabolic Syndrome - Type 2 Diabetes

3- Oxidative stress and inflammation in different tissues
(endothelial cells, kidney, liver, heart, brain)

4- Alteration in biochemical markers as:
a- Hyperglycemia and hyperinsulinemia
b- Increase lipid profile
c- Elevated oxidative markers (MDA, GSSG)
d- Reduction in eNOS

5- Acut and chronic Renal disease
- Hepatic inflammation /hepatosis/ fibrosis
- Cardiovascular Disease risk

Figure 44.4 The effect of HSD on biochemical and oxidative markers in different tissues. The figure summarizes that HSD provides weight gain and adiposity, which result in harmful oxidative stress in different tissues *via* changes in biochemical markers.

susceptibility to peroxidation, which may compromise the antioxidant capacity of blood and organs with lower defense systems. Summary about the effect of HSD on biochemical and oxidative markers in different tissues have been demonstrated in (Figure 44.4).

44.1.3.2 Effect of High Sucrose Diet on Cardiac Tissue Antioxidant

HSD could change cellular metabolism through several pathways and thereby increase oxidative stress. Dietary sucrose, as opposed to other complex carbohydrates as potato, may have a differential effect on net oxidative stress and that the difference is reflected in the accumulation of advanced glycation product (Robert *et al.* 2008).

A number of studies considering HSD and cardiac dysfunction have been carried out. Cardiac contractile dysfunction has been reported in sucrose fed animal at the whole heart and single cardiomyocyte level. This dysfunction was associated with slower uptake of calcium into sarcoplasmic reticulum during relaxation (Wold *et al.* 2005).

Two weeks of high sucrose feeding to rats produced pro-oxidant effect *via* increasing TBARS and lower Cu-Zn-SOD activity in heart (Busserolles *et al.*

2002). Therefore, the increased oxidative stress and the associated lipid peroxidation could be due to oxygen free radical production and/or decreased protection from nonenzymatic or enzymatic antioxidants.

Cardiac insulin resistance has been reported to be associated with elevated ROS by generating enzymes NADPH oxidase. Six weeks of 30% sucrose in drinking water reduced levels of the myocardial antioxidant SOD and GPx, in rats. Moreover, HSD decreases the myocardial ratio of GSH/GSSG indicating oxidative stress (Mellor *et al.* 2010).

MDA concentrations were reported to be significantly greater in the heart and liver of obese rats compared to the lean, supporting the opinion that obesity predisposes the myocardium and liver to oxidative stress. In addition HSD (68% total energy) decreased the myocardial ratio of reduced glutathione/oxidized glutathione indicative of oxidative stress.

Sucrose rich-diet increased plasma insulin levels, in rats, and increased cardiac β-oxidation and coronary flow-rate, but reduced glycolytic flux and contractility during normoxic baseline function of isolated perfused hearts. Sucrose rich-diet impaired early post-ischemic recovery of isolated heart cardiac mechanical function and further augmented cardiac beta-oxidation but reduced glycolytic and lactate flux (Gonsolin *et al.* 2007). Systolic and diastolic blood pressure increased in the sucrose group by 3.8 and 4.1 mm Hg, respectively and decreased in the sweetener group by 3.1 and 1.2 mm Hg, respectively.

The consumption of the palatable diet enriched with sucrose for 4 months induced overall corporal and metabolic changes, and decreased nitric oxide metabolites and ectonucleotidase activity, thereby promoting an appropriate environment for the development of cardiovascular and renal diseases, without apparent changes in insulin levels.

Rats fed HSD had higher susceptibility to lipid peroxidation in heart and liver tissues than those fed the control or potato-based diets. Hyperlipidemia and hyperglycemia have been associated with increased oxidative damage, affecting both lipoproteins and the antioxidant condition. Post-prandial raises in lipid and carbohydrate levels lead to increased oxidative stress, which has been associated with an increased risk for atherosclerosis, hepatic and related disorders.

44.1.3.3 Effect of High Sucrose Diet on Renal Function and Antioxidant Marker

Lifestyle factors and diet play a role in the development of kidney disease during several stages including development of obesity and the metabolic syndrome and occurrence of obesity-related glomerulopathy. Over-consumption of soft drink containing sucrose is currently considered to be a main public health concern with implications for renal diseases.

Metabolic syndrome (MS) resulting from high sucrose diet is a risk factor for proteinuria and chronic kidney disease. In part this may be mediated secondary to hypertension and diabetes mellitus, which are aspects of metabolic syndrome.

Rats fed a diet with HSD for 12 weeks showed hypertriglyceridemia, increased LDL production, increased oxidative stress and renal alteration.

Moreover, this suggests an association between lipid peroxidation, obesity and nephropathy (Amin *et al.* 2011) (Tables 44.2 and 44.3).

Renal catalase activity was significantly increased in HSD. These findings occur in response to oxidative stress and important to balance the elevated ROS resulting from the activation of biochemical pathway, increased lipid peroxidation and producing oxidative stress. Increased catalase activity was indicative of elevated ROS and consequently higher oxidative stress (Table 44.3).

Feeding HSD raises LDL, TC, TG and renal lipid peroxidation (MDA) levels, which can have effects on the kidney because MDA act as tissue toxic agent. These changes can be monitored by an increased level of urea and creatinine and estimated glomerular filtration rates estimations.

HSD significantly decreased serum and renal nitric oxide (NO) compared with control group. NO plays an important role in the regulation of renal blood flow to the renal medulla and in the tubular regulation of sodium excretion. The decrement in NO induces vasoconstriction that affects the kidney.

In study by Amin *et al.* (2011) collectively, feeding HFD and HSD to rats resulted in significant elevating in BW, BMI, feed intake, adipose tissue, blood levels of glucose, TG, TC, LDL, MDA and catalase activity while, significantly, decreased serum and renal NO levels. These markers are implicated in renal disorders (Figure 44.4).

44.1.3.4 The Effect of High Sucrose Diet on Antioxidant Vitamins Sufficiency

Increased intake of HSD was associated with increased susceptibility of membranes to oxidation and an increased requirement for vitamin E which is the major lipid-soluble free radical chain-breaking antioxidant found in plasma, red cells, and tissues produce decreasing in its blood level (Table 44.4). Increased requirement of vitamin E is suggested to counteract the pro-oxidant effect of high sucrose consumption.

In the males, the plasma triglyceride elevation secondary to sucrose was accompanied by significantly lowered plasma alpha-tocopherol and a significantly lowered alpha-tocopherol/TG ratio (30%), suggesting that vitamin E depletion may predispose lipoproteins to subsequent oxidative stress. Unlike

Table 44.4 Demonstrates the concentration of serum antioxidant vitamin E and C of rat fed different diets. Illustrates the exhaustion of vitamin E and C by HSD administration.

	Vitamin E (mg/dl)	*Vitamin C (mg/dl)*	*References*
Control	$0.95 \pm 0.06^{[a]}$	$1.1 \pm 0.07^{[a]}$	Amin *et al.* (2004)
HSD	$0.46 \pm 0.07^{[b]}$ ↓	$0.63 \pm 0.07^{[b]}$ ↓	
Sunflower oil	$0.60 \pm 0.03^{[b]}$ ↓	$0.74 \pm 0.09^{[b]}$ ↓	
Beef tallow	$0.52 \pm 0.02^{[b]}$ ↓	$0.79 \pm 0.04^{[b]}$ ↓	

Different litters indicate significant difference ($p < 0.05$).
Similar litter indicate no significant difference.

other fat-soluble vitamins, vitamin E has no specific transport protein, but transported in plasma lipoproteins and tocopherol is secreted in the liver in VLDL and protects lipoproteins from oxidation. When lipoproteins are depleted of antioxidants, unsaturated fatty acids are quickly oxidised and the vitamin E exhaustion in sucrose-fed rats may predispose lipoproteins to subsequent oxidative stress (Busserolles *et al.* 2002). In the same manner, vitamin C deficiency was shown to augment LDL oxidation susceptibility.

44.1.3.5   The Effect of High Sucrose Diet on Blood Antioxidant Ability

Fructose-fed rats showed hyperglycemia, hyperinsulinemia and hyper-trigly-ceridemia at the end of four weeks. Enhanced plasma lipid peroxidation and inadequate cellular antioxidant defense system were observed in them (Srividhya *et al.* 2002).

Plasma MDA level significantly increased in rats fed SFO, B.T and HSD in comparison with control, also plasma level of TBARs increased significantly with SFO than BT and control (Table 44.2). These results indicated that the elevation of PUFA in blood leads to increase an intracellular oxidative stress, resulting in the activation of oxidative stress-responsive transcription factors, indicating that a high SFO diet potentiates susceptibility of rats to lipid peroxidation and augments requirement for antioxidants to provide adequate protection (Amin *et al.* 2004; Busserolles *et al.* 2002).

Higher amount of sugar resulted in significant increasing of the erythrocyte glutathione reductase as well as the rate of lipid peroxidation in blood hence, changes the antioxidant status.

GSH plays a central defensive role against oxidative insults as an endogenous scavenger of free radicals. Its blood level is a sensitive indicator of antioxidant status in circulation. Significant lowering in blood GSH in HSD fed rats was recorded.

The recorded exhaustion in blood GSH indicating oxidative stress where the integral RBCs membrane was disturbed due to intake HSD, hyperlipaemia which lead to oxidative damage due to lipid peroxidation from generated peroxide. As the GSH is the main antioxidant constituent of RBCs, it reacts with oxygen free radicals species protecting lipid membrane.

The ferric reducing ability of plasma (FRAP) value, which reflects the antioxidative capacity of plasma, was decreased in sucrose-fed rats, compared to control rats (-17%, $P < 0.05$, control *vs.* sucrose), whereas potato consumption tended to slightly increased the FRAP value (Robert *et al.* 2008).

44.1.3.6   The Effect of High Sucrose Diet on Vascular Reactivity and its Mechanism

Recent researches have been studying the relation of HSD and blood capillaries' healthiness. The data highlight the importance of the type of diet as it can produce divergent effects on vascular reactivity pathways despite both groups developing increased body mass, adiposity and oxidative stress. Although rats in the HSD-fed group develop similar levels of oxidative stress as observed in the HFD rats, the impaired vasodilation is not as severe and the mechanisms of impaired vasodilation are divergent.

In the HFD group, the impaired vasodilation appears to be mediated by scavenging of NO. In contrast, H_2O_2 is implicated in the impaired vasodilatory responses in vessels from HSD rats. In conclusion, the impaired vasodilatory responses to acetylcholine in rats fed either HSD or HFD are mediated by ROS scavenging of NO, impaired smooth muscle sensitivity to NO as well as by inflammatory factors.

Further studies demonstrated that arteries from HSD rats have harmed vascular smooth muscle sensitivity to NO contributing to the impaired acetylcholine-mediated vasodilation. Therefore, it is evident that the residual response to acetylcholine following high calorie feeding is NO-independent and likely involves other endothelium-dependent vasodilatory pathways.

HSD containing fructose induce endothelial dysfunction because high-fructose diet leads to increased NO inactivation, secondary to enhanced formation of ROS, and decreased vascular relaxation through impaired eNOS activity caused by relative deficiency of tetrahydrobiopterin in endothelial cells. Since sucrose contains high amounts of fructose, the ROS-induced tetrahydrobiopterin depletion may contribute to endothelial dysfunction (Roberts *et al.* 2005).

Feeding rats either a high fat or high sucrose diet results in the development of oxidative stress as well as impaired vasodilation (Sweazea *et al.* 2010). Oxidative stress diminishes vasodilatory responsiveness in HSD and HFD rats through ROS-mediated scavenging of NO and decreased smooth muscle sensitivity to NO. Inflammation also plays a significant role in the impaired vasodilation (Sweazea *et al.* 2010).

Endothelial nitric oxide synthase is important in linking metabolic and vascular disease and indicate the ability of a Westernized diet (HSD and HFD) to induce endothelial dysfunction; which is the early hallmark of atherogenesis and to alter metabolic and vascular homeostasis (Bourgoin *et al.* 2008).

Adipose tissue has a role in secreting factor that impairs endothelial dependent dilatation *via* inhibition of NO synthase thereby mediating NO production. NO decreased in HSD, initiating vasoconstriction that affects the kidney vasculature. The data of Amin *et al.* (2011) suggested potential mechanisms of renal dysfunction, oxidative stress and NO in obesity induced by HSD.

44.2   Conclusion

Studies of sucrose feeding have provided valuable insight to understand how genes and nutritional environment interact in the genesis of cardiomyopathy, nephropathy and hepatic disorders. Further work is required to determine the underlying pathogenesis and mediators of this organ damage.

Adaption of the diet to decrease the consumption of simple sugars, such as sucrose and promote intake of complex carbohydrates from whole grain bread, cereals, legumes and potatoes is required to reduce the increasing health burden. Public health measures need to be increased and policies introduced to promote healthy eating and avoidance of high sucrose intake.

The consumption of sucrose may be "hidden" in juices and drinks or other manufactured foods. The importance of this relates to the desire to reduce obesity, and associated disorders, as well as curtailing oxidative stress in different organs such as the liver, cardiovascular, kidney, pancreas and brain tissues. Efforts should be made to ensure the public is aware of the risks of high fructose and sucrose containing juices.

Although increasing dietary sucrose enhances the energy supply for the body, it may adversely affect lipid peroxidation and the antioxidant ability. When consuming dietary sucrose, an adequate intake of antioxidants is advised for preventing oxidative stress and may improve tissue antioxidant stability.

There appears to be different levels of protection against lipid peroxidation from tissue to tissue. It is important to estimate the risk of certain dietary factors inducing higher susceptibility to peroxidation, which may compromise the antioxidant capacity of blood and organs with lower defense systems.

Summary Points

- This chapter focuses on the effect of HSD on food intake, body weight, lipid profiles, blood glucose and risk of cancer.
- The use of sweeteners containing sucrose and high-sucrose food, drink juice has increased over the past 3 decades.
- HSD affects appetite and increases food intake, induces synthesis of TG, increasing adiposity and leading to weight gain and visceral adiposity.
- Higher consumption of sucrose was associated with increased risk of type 2 diabetes and recently may increase pancreatic and intestinal cancer risk.
- Consumption of high sucrose range from 2–8 weeks facilitates oxidative damage.
- Reactive oxygen species, such as superoxide radicals, are thought to underlie the pathogenesis of various diseases.
- A short-term consumption of HSD negatively affects the balance of ROS production (indicated by elevating MDA and GSSG levels) and antioxidant defense (indicated by lowering GSH, Vitamin E and C levels and changes in catalase activity), leading to increased lipid susceptibility to peroxidation. Therefore, using HSD has the potential for production of oxidative stress.
- Pro-oxidant and a great cardiac susceptibility to peroxidation occurred in HSD fed rats.

- Soft drink containing sucrose overconsumption is now considered to be a major public health concern with proposition for renal diseases via renal lipid peroxidation, lipid profile and nitric oxide.
- Improved requirement of vitamin E and C is suggested to counteract the prooxidant effect of high sucrose consumption.

Key Facts

- High sucrose is the main ingredient in drinking juice and foods, produce changes in the adiposity and metabolic markers.
- **Figure 44.3**: this figure clearly reveals that sucrose metabolism through fructokinase pathway rapidly metabolised to fatty acid synthesis.
- **Figure 44.4**: HSD provides higher calories, dyslipidemia, visceral adiposity, resulting in oxidative stress and down regulation of biochemical and oxidative markers leading to susceptibility to tissues and organ injury.
- **Table 44.2**: this table presents comparison of oxidative stress marker, MDA levels in plasma and hepatic, cardiac and renal tissues. Plasma and renal oxidative indicator MDA was elevated in HSD indicating decreased the antioxidant ability of these tissues and its susceptibility for renopathy and cardiopathy.
- HSD intake is associated with numerous oxidative health conditions related to cardiovascular, renal and blood illness.

Definitions of Words and Terms

Insulin resistance: Reduced response of the tissues (liver, muscles and adipose tissues) to normal action of insulin, where tissues are unable to respond effectively to circulating insulin, resulting in hyperinsulinemia, hyperglycemia and type 2 diabetes.

Oxidative stress: A profound imbalance between oxidants and antioxidants and disturbance in this balance can results in toxic effect by ROS that can damage cell components resulting in nephropathy and cardiomyopathy.

Antioxidant marker: These are indicators that measure in the blood and tissues for interpretation of the extent of oxidation antioxidant in the body.

Enzymatic antioxidants: these are enzymes that catalyse the scavenging of the reactive oxygen species, present in the blood and different tissue.

Vascular reactivity: It is the response of the blood vessels either through vasodilatation or vasoconstriction to metabolites as NO and acetylcholine which mediated by ROS.

Cardiac antioxidant: It is the ability of the cardiac tissue to remove oxidant production *via* its antioxidant defense of SOD, GSH.

Metabolic syndrome: Disorders characterized by visceral obesity associated with the clustering of metabolic disturbances and pathophysiological cardiovascular, hepatic and renal risk factors. It is characterized by impaired glucose tolerance, dyslipidemia, and hypertension.

Diabetes Mellitus: A condition associated with metabolic syndrome, more common in the Western world and diagnosed by having raised fasting blood glucose. The overall condition is due to insulin resistance or deficiency and is associated with numerous co-morbidity.

Free radicals: Are generated as the body uses oxygen, these by-products cause oxidative damage to the cells of the body.

List of Abbreviations

HSD	high sucrose diet
LDL	low denisity lipoproteins
VLDL	very Low density lipoproteins
TC	total cholesterol
TG	triacylglycerol
BW	body weight
BMI	body mass index
MDA	malodialdehyde
ROS	reactive oxygen species
GSH	reduced glutathione
GPx	glutathione peroxidase
SOD	super oxide dismutase
GSSG	oxidized glutathione
G6pD	glucose 6-phospate dehydrogenase
TBARs	thiobarbituric acid reactive substances
eNOS	endothelial nitric oxide synthase

References

Allen, R.G., and Tresini, M., 2000. Oxidative stress and gene regulation. *Free Radical Biology and Medicine*. 28: 463–499.

Amin, K.A., Kamel, H.H., and Abd Eltawab, M.A., 2011. Protective effect of Garcinia against renal oxidative stress and biomarkers induced by high fat and sucrose diet. *Lipids in Health and Disease*. 10: 6.

Amin, K.A., Ali, K.M., and Yousria, A., El-Arnaooty., 2004. Effect of induced obesity by different diets on lipid profiles, antioxidant defense, oxidative stress and associated disorders in rat. Bulletin Egyptian Society. *Physiological Science*. 24(1): 177–195.

Busserolles, J., Zimowska, W., Rock, E., Rayssiguier, Y., and Mazur, A., 2002. Rats fed a high sucrose diet have altered heart antioxidant enzyme activity and gene expression. *Life Science*. 71: 1303–1312.

Bourgoin, F., Bachelard, H., Badeau, M., Mélançon, S., Pitre, M., Larivière, R., and Nadeau, A., 2008. Endothelial and vascular dysfunctions and insulin resistance in rats fed a high-fat, high-sucrose diet. *American Journal Physiology Heart Circ Physiol*. 295(3): H1044–H1055.

Chicco, A., D'Alessandro, M.E., Karabatas, L., Pastorale, C., Basabe, J.C., and Lombardo, Y.B., 2003. Muscle lipid metabolism and insulin secretion are altered in insulin-resistant rats fed a high sucrose diet. *Journal Nutrition.* 133: 127–133.

Deneke, S.M., 2000. Thiol-based antioxidants. *Current Topics in Cellular Regulation.* 36: 151–180.

Fukuchi, S., hamaguch, K., Seike, M., Himeno, k., Skata, T., and Yoshimatsu, H., 2004. Role of fatty acid composition in development of metabolic disorders in sucrose induces obese rats. *Experimental Boilogy Medicine.* 229(6): 486–493.

Gerlach, M., Ben-Shachar, D., Riederer, P., and Youdim, M.H., 1994. Altered brain metabolism of iron as a cause of neurodegenerative diseases. *Journal Neurochemistry.* 63: 793–807.

Gonsolin, D., Couturier, K., Garait, B., Rondel, S., Novel-Chaté, V., Peltier, S., Faure, P., Gachon, P., Boirie, Y., Keriel, C., Favier, R., Pepe, S., Demaison, L., and Leverve, X., 2007. High dietary sucrose triggers hyper-insulinemia, increases myocardial beta-oxidation, reduces glycolytic flux and delays post-ischemic contractile recovery. *Molecular Cell Biochemistry.* 295(1–2): 217–228.

Halliwell, B., and Chirico, S., 1993. Lipid peroxidation: Its mechanism, measurement and significance. *American Journal Clinical Nutrition.* 57(S): 715S–722S.

Janette, C.B., Susanna, H.A., Dorota, B.P., and McMillan, J., 2002. Glycemic index and obesity. *American Journal Clinical Nutrition.* 76(1): 281S–285S.

Li, S., Li, X., and Rozanski, G.J., 2003. Regulation of glutathione in cardiac myocytes. *Journal of Molecular and Cellular Cardiology.* 35: 145–1152.

Lomba, A., Milagro, F.I., García-Díaz, D.F., Campión, J., Marzo, F., and Martínez, J.A., 2009. A high-sucrose isocaloric pair-fed model induces obesity and impairs NDUFB6 gene function in rat adipose tissue. *Journal Nutrigenet Nutrigenomics.* 2(6): 267–272.

Ma, T., Liaset, B., Hao, Q., Petersen, R.K., Fjære, E., Ngo, H.T., Lillefosse, H.H., Ringholm, S., Sonne, S.B., Treebak, J.T., Pilegaard, H., Frøyland, L., Kristiansen, K., and Madsen, L., 2011. Sucrose counteracts the anti-inflammatory effect of fish oil in adipose tissue and increases obesity development in mice. *PLoS One.* 6(6): e21647.

Meinhold, C.L., Dodd, K.W., Jiao L., Flood, A., Shikany, J.M., Genkinger, J.M., Hayes, R.B., and Stolzenberg-Solomon, R.Z., 2010. Available carbohydrates, glycemic load, and pancreatic cancer: is there a link? *American Journal Epidemiology.* 171(11): 1174–1182.

Mellor, K.M., Ritchie R.H., and Delbridge L.M., 2010. Reactive oxygen species and insulin-resistant cardiomyopathy. *Clinical Experimental Pharmacology Physiology.* 37(2): 222–228.

Noeman, S.A., Hamooda, H.E., and Baalash, A.A., 2011. Biochemical study of oxidative stress markers in the liver, kidney and heart of high fat diet induced obesity in rats. *Diabetology and Metabolic Syndrome.* 3(1): 17.

Noor, R., Mittal., S., and Iqbal, J., 2002. Superoxide dismutase–applications and relevance to human diseases. *Medical Science Monitor.* 8(9): 210–215.

Poulsen, M., Mølck, A.M., Thorup, I. Breinholt, V., and Meyer, O., 2001. The influence of simple sugars and starch given during pre- or post-initiation on aberrant crypt foci in rat colon. *Cancer Letters.* 167(2): 135–143.

Preuffer, M., and Barth, C.A., 1992. Dietary sucrose but not starch promotes protein-induced differences in rates of VLDL secretion and plasma lipid concentrations in rats. *Journal Nutrition.* 122: 1582–1586.

Raben, A., Vasilaras, T.H., Møller, A.C., and Astrup, A., 2002. Sucrose compared with artificial sweeteners: different effects on ad libitum food intake and body weight after 10 wk of supplementation in overweight subjects. *American Journal Clinical Nutrition.* 76(4): 721–729.

Reid, M., Hammersley, R., and Duffy, M., 2010. Effects of sucrose drinks on macronutrient intake, body weight, and mood state in overweight women over 4 weeks. *Appetite.* 55(1): 130–136.

Robert, L., Narcy, A., Rayssiguier, Y., Mazur, A., and Rémésy, C., 2008. Lipid metabolism and antioxidant status in sucrose vs. potato-fed rats. *Journal of the American College of Nutrition.* 27(1): 109–116.

Roberts, C.K., Barnard, R.J., Sindhu, R.K., Jurczak, M., Ehdaie, A., and Vaziri, N.D., 2005. A high-fat, refined-carbohydrate diet induces endothelial dysfunction and oxidant/antioxidant imbalance and depresses NOS protein expression. *Journal Applied Physiology.* 98(1): 203–210.

Schernhammer, E.S., Hu, F.B., Giovannucci, E., Michaud, D.S., Colditz, G.A., Stampfer, M.J., and Fuchs, C.S., 2005. Sugar-sweetened soft drink consumption and risk of pancreatic cancer in two prospective cohorts. *Cancer Epidemiological Biomarkers Preview.* 14(9): 2098–2105.

Srividhya, S., Ravichandran, M.K., and Anuradha, C.V., 2002. Metformin attenuates blood lipid peroxidation and potentiates antioxidant defense in high fructose-fed rats. *Journal Biochemical Molecular Biology Biophysics.* 6(6): 379–385.

Sweazea, K.L., Lekic, M., and Walker, B.R., 2010. Comparison of mechanisms involved in impaired vascular reactivity between high sucrose and high fat diets in rats. *Nutrition and Metabolism (Lond).* 4(7): 48.

Tasevska, N., Jiao, L., Cross, A.J., Kipnis, V., Subar, A.F., Hollenbeck, A., Schatzkin, A., and Potischman, N., 2012. Sugars in diet and risk of cancer in the NIH-AARP Diet and Health Study. *International Journal of Cancer.* 130(1): 159–169.

Wold, L.E., Dutta, K., Mason, M.M., Ren, J., Cala, S.E., Schwanke, M.L., and Davidoff, A.J., 2005. Impaired SERCA function contributes to cardiomyocyte dysfunction in insulin resistant rats. *Journal Molecular Cell Cardiology.* 39(2): 297–307.

CHAPTER 45
Sugars in the Diet of Young Children

ERKKOLA MAIJALIISA,*[a] RUOTTINEN SOILE[b] AND
VIRTANEN SUVI M[c]

[a] Division of Nutrition, Department of Food and Environmental Sciences,
PO BOX 66, FI-00014 University of Helsinki, Finland; [b] Research Centre, of
Applied and Preventive Cardiovascular Medicine (CAPC), University of
Turku, Kiinamyllynkatu 10, FI-20520, Turku, Finland, PO Box 30,
FI-00271, Helsinki, Finland, Email: soile.ruottinen@utu.fi; [c] Nutrition Unit,
Department of Lifestyle and Participation, National Institute for Health and
Welfare, Other insitutions: School of Health Sciences, University of
Tampere, Tampere, Finland and Science Center of Pirkanmaa Hospital
District, Tampere, Finland
*Email: maijaliisa.erkkola@helsinki.fi

45.1   Introduction

Newborn infants have an innate preference for sweet tastes (Ventura and
Mennella 2011). The positive hedonic response to sweet taste, and the preference for a greater intensity of sweetness among children as compared to
adults, is universal. Hence, as a population group, children have the highest
sugar intake in relation to energy (e.g. Alexy et al. 2003; Buyken et al. 2008;
Gibson and Boyd 2009; Kyttälä et al. 2010; Patterson et al. 2010; Reedy and
Krebs-Smith 2010). In addition to nutrient dilution of the diet (Erkkola et al.
2009; Farris et al. 1998; Gibson 2007; Gibson and Boyd 2009; Kranz et al. 2005;
Ruottinen et al. 2008), high sugar intake may result in various short- and

Food and Nutritional Components in Focus No. 3
Dietary Sugars: Chemistry, Analysis, Function and Effects
Edited by Victor R Preedy
© The Royal Society of Chemistry 2012
Published by the Royal Society of Chemistry, www.rsc.org

long-term health impacts, such as caries, obesity, and increased risk of chronic diseases (World Health Organization 2003).

The differences in terms used in publications describing sugar intakes make between-country comparisons regarding the actual intake of different sugars difficult and partly inaccurate. In most studies, naturally occurring sugars are not distinguished from sugars added to foods during manufacturing. The majority of sugars added to foods either by manufacturers or by consumers comprise sucrose. In the present review, the term sugar refers to both added and naturally occurring sugar, unless stated otherwise. It should be kept in mind that the sugar composition of many food products, especially that of those industrially produced, is not clear. Sucrose composition estimates often include other mono- and disaccharides and invert sugar (a mixture of glucose and fructose). The present overview starts by briefly describing dietary guidelines, and assessment methods in children, followed by a description of sugar sources and intake in children. Finally, we present the current evidence of the association between sugar intake and health in children.

45.2   Dietary Guidelines and Assessment of Intake

45.2.1   Dietary Guidelines for Sugar Intake

In the World Health Organization's report "Diet, nutrition and the prevention of chronic diseases" (2003), the population goal for *free sugars* is set to be less than 10 E% (percentage of total daily energy intake). The report suggests that in order to reduce chronic diseases, children's consumption of sugar-sweetened beverages should be limited. In the US Dietary reference intake (DRI) report, a maximal intake level suggested for added sugars is 25% from total energy intake (Institute of Medicine of the National Academies, IoM 2005), based on trends indicating that people with diets at or above this level are more likely to have poorer intakes of important essential nutrients (Report of the Dietary Guidelines Advisory Committee on the Dietary Guidelines for Americans 2010). The term *added sugars* is commonly used in the United States and is defined by the Dietary Guidelines for Americans (2005) as follows: "Sugars and syrups that are added to foods during processing or preparation. Added sugars do not include naturally occurring sugars such as those that occur in milk and fruits." It has been highlighted that reducing the current high consumption of hard fats and added sugars may be the most critically needed change for better energy balance in childhood in the US (Williams 2010).

According to the European Food Safety Authority's (EFSA) NDA Panel (EFSA 2010), the available data from Europe are not sufficient to set an upper limit for sugar intake. The British Nutrition Foundation (2009) has set a Dietary Reference Value of 47% for total carbohydrates, of which non-milk extrinsic sugars distribution is 10% of total energy intake. According to the Nordic Nutrition Recommendations (NNR) (Nordic Council of Ministers 2004) the intake of refined sugars in infants over 6 months and in older children should not exceed 10 E%. Refined sugars include sucrose, fructose, glucose, starch hydrolysates (glucose syrup, high-fructose syrup) and other isolated sugar preparations such as food components used as such or added during food preparation and manufacturing. According to the NNR, a limitation for the intake of refined sugars is necessary to ensure an adequate intake of essential nutrients and dietary fibre, especially in children and adults with low energy intake.

45.2.2 Dietary Assessment of Sugar Intake in Children

Choosing an appropriate dietary assessment method is a complex decision that is based on the objective of the data collection, the level of accuracy sought and the amount of available resources. In nutritional studies, it is usually essential to know the length of time the diet should be recorded in order to obtain results that are sufficiently accurate to assess the individual's usual intake, and to rank individuals within the group (Willett 1998). This is especially challenging in studies among children whose dietary day-to-day variability changes with age (Black *et al.* 1983; Erkkola *et al.* 2011). Food frequency questionnaires capture quantitative data retrospectively on usual, long-term intake, whereas food records and 24-hour recalls provide highly detailed information on short-term intake. The number of days required to measure nutrient intake in groups of individuals is a function of the ratio of the variation within individuals to the variation between individuals (Black *et al.* 1983). The smaller the ratio, the fewer the number of days needed for recording. In young children, the diversity of foods eaten increases rapidly after the child starts partaking in family meals. Consequently, the ratio of within-subject to between-subject variance increases with increasing age, and is slightly higher in girls than in boys (Erkkola *et al.* 2011).

The optimal number of replicates depends on both the degree of accuracy that is needed and the variability of the nutrient in question. In a Finnish study, 2 days of diet record would have been needed to achieve $r \geq 0.8$ between observed and true sucrose intake in 1-year-olds, whereas 5 days would be needed for 6-year-olds (Erkkola *et al.* 2011). Similarly, to achieve $r \geq 0.9$ between observed and true sucrose intake, 4 and 13 days of diet record would have been needed in 1- and 6-year-olds, respectively. In many countries, 'sweet day' traditions contributes to daily variation in sucrose intake, and, consequently, the child's average sucrose intake could differ according to the type of days recorded (Erkkola *et al.* 2009; Sepp *et al.* 2001).

When assessing diet in infancy and early childhood, all information has to, by necessity, be obtained by surrogate reporters. Factors that lessen the time parents spend with the child, such as employment outside the home and larger family size, can possibly lead to less accurate reporting of the child's diet (Willett 1998). The person who spends the most time with the child is thus the best surrogate reporter. The process of food registration neither affects the appetite nor the food preferences of the child. However, it is possible that parents want to pad out what they actually give to their child or report having given them. Parents may also simplify the child's diet due to the burden of recording. The food items commonly underreported are sugar-rich foods such as cakes, soft drinks and sweets, while those overreported tend to be healthier foods such as bread, fruits and potatoes (Andersen *et al.* 2004). When assessing sugar intake, this could imply that subjects in the top quarter of the sugar distribution are more likely to underreport their consumption of sugar-rich foods than subjects with lower intake. The true association could then be obscured and, for example, the influence of the intake of sugar on the intake of nutrients would be stronger than that observed. However, systematic misreporting associated with unhealthy foods (including sugary soft drinks, chocolate and sweets) could not be detected in a Norwegian study in which 9-year-old children completed a food diary under parental auspices (Lillegaard and Andersen 2005). Moreover, added sugar intake as a percentage of energy intake did not significantly differ between the underreporters and the acceptable reporters.

45.3 Sugar Sources and Intake in Children

45.3.1 Dietary Sources of Sugars

Sugars added by manufacturers account for the major proportion, around 80%, of the total sugar intake in children's diet, with naturally occurring sugar and sugar added by the consumers accounting for 15%, and 5%, respectively (Erkkola *et al.* 2009; Linseisen *et al.* 1998). The majority of added sugars (sugars and syrups that are added to foods by manufacturers or consumers, excluding naturally occurring sucrose) in the diet comprised sucrose, although fructose, glucose and different syrups were also found.

Beverages (juice and soft drinks), sweets and chocolate, and sweetened milk products are the key contributors to the intake of added sugars (Figure 45.1); and fresh fruits and berries and fruit and berry juices to the intake of naturally

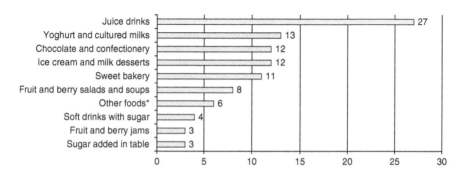

Figure 45.1   The average contribution (%) of food groups to the intake of added sucrose among the Finnish children (n = 471), (Erkkola *et al.* 2008).
*The food groups are based on quantity and quality of sucrose. Other foods include all other foods which are not mentioned in the Table, such as meat, fish, etc.

occurring sucrose (Alexy *et al.* 2003; Erkkola *et al.* 2009; Farris *et al.* 1998; Joyce *et al.* 2008; Øverby *et al.* 2004; Patterson *et al.* 2010; Ruottinen *et al.* 2004). In most countries, more than half of the added sugars in children's diet are derived from soft drinks and sweets (Lyhne and Ovensen 1999; Øverby *et al.* 2004, Patterson *et al.* 2010; Reedy and Krebs-Smith 2010). Sweetened milk products and desserts are also significant contributors to the intake of sugar, the former mainly in countries which have a strong national tradition favoring the use of milk products (Erkkola *et al.* 2009; Farris *et al.* 1998; Kranz *et al.* 2005; Popkin and Nielsen 2003; Reedy *et al.* 2010; Ruottinen *et al.* 2004). Repeated studies indicate that in time the sources of sugar have changed as soft drinks have become the main source of added sugars in many countries (Moynihan *et al.* 2005; Popkin and Nielsen 2003).

45.3.2 Sugar Intake in Children

Increased consumption of caloric sweeteners is one element in the world's dietary changes, represented by a 74 kcal daily increase between 1962 and 2000 (Popkin and Nielsen 2003). If both urbanization and the processing of the food supply continue unabated, this trend in the worldwide diet will persist. However, in terms of the actual intake of sugars between countries, comparisons are difficult to make and are also partly inaccurate due to discrepancies in the calculation or mode of expression of sugar. The proportion of sucrose from total energy intake among children in the Western world has almost constantly been above the recommended upper limit of 10 E % as reported in studies from different Western countries (Table 45.1).

In children, the percentage of energy intake from sugars increases with increasing age (*e.g.* Buyken *et al.* 2008; Emmett *et al.* 2002; Kyttälä *et al.* 2010; Øverby *et al.* 2004; Reedy and Krebs-Smith 2010). In a Finnish cohort study, it was found that consumption of foods that contain high amounts of added sugars was already common among the 1-year-olds and increased thereafter (Kyttälä *et al.* 2010), (Figure 45.2). Most of the children aged 2–6 years consumed all the main sugar-containing food groups at least once during the three recording days. Consequently, the average proportion of sucrose from total energy intake increased from 5.5 E% in 1-year-olds to 13.4% in 6-year-olds.

The sugar intake of a child is strongly determined by the availability of foods rich in sugar. Children who are cared for outside their home have a lower total sugar intake during the week than during the weekend (Lehtisalo *et al.* 2010; Sepp *et al.* 2001). This difference may be due to differences in meal habits at home compared to day-cares, or to different meal habits during weekends as compared with weekdays. A snack-dominated meal pattern that is associated with higher sugar intake (Ovaskainen *et al.* 2006) may be more common among children cared for at home and hence also during weekends. As the frequency of eating occasions of added sugar intakes increases, the percentage of energy from added sugar intakes also increases (Joyce *et al.* 2008). In a UK cohort study among 3-year olds, the total sugar intake was slightly higher when

Table 45.1 Sugar intake as the proportion of total energy intake in Western children.

Author, Year, Data Source, Country	Method	Subjects	Sugar Definition	Estimates of Intake
Ruottinen et al. 2011, The STRIP Project, Finland	3–4-day food record	n = 543 10-y follow-up from 13-mo of age	Sucrose	5.4 E% in 13-mo-olds 9.4 E% in 3-y-olds 9.8 E% in 5-y-olds 9.4 E% in 7-y-olds 9.6 E% in 9-y-olds
Kyttälä et al. 2010, Erkkola et al. 2008, The DIPP Study, Finland	3-day food record	n = 2 535 1–7-y-olds	Sucrose, accordingly in 3-y-olds: added sucrose, naturally occurring sucrose	5.5. E% in 1-y-olds, 12.2 E% in 2-y-olds, 13.3 E% in 3-y-olds, of which: added sucrose 11.3 E% naturally occurring sucrose 2 E% 13.7 E% in 4-y-olds, 13.4 E% in 6-y-olds
Patterson et al. 2010, The European Youth Heart Study, Sweden	24-hour recall	n = 551 10-y-olds	Sucrose	10.6 E%
Reedy & Krebs-Smith 2010, The NHANES survey, US	24-hour recall	n = 3 778 2–18-y-olds	Added sugars	18 E%
Gibson & Boyd 2009, The National Diet and Nutrition Survey, Great Britain	7-day weighed dietary records	n = 1688 4–18-y-olds	Added sugars	15.4 E%

Study	Method	n / age	Sugar type	E%
Buyken *et al.* 2008, The DONALD study, Germany	3-day weighed food record	n = 380 2- and 7-y-olds	Added sugar	9.5 E% in 2-y-olds 14 E% in 7-y-olds
Joyce *et al.* 2008, The National Children's Food Survey, Ireland	7-day food record	n = 594 5–12-y-olds	Added sugar	14.6 E%
Garemo *et al.* 2007, Sweden	7-day food record	n = 132 4-y-olds	Sucrose	12 E%
Villa *et al.* 2007, Estonia	24-hour recall	n = 444 9-y-olds	Energy from sugar, sweets and drinks	14 E%
Kranz *et al.* 2005, the CSFII-study, US	2-day dietary intake	n = 5437 2–5-y-olds	Added sugars	15 E% in 2–3-y-olds 17 E% in 4–5-y-olds
Øverby *et al.* 2004, Norway	4-day pre-coded food record	n = 391 4-y-olds, n = 810 9-y-olds	Added sugar	15 E% in 4-y-olds 17 E% in 9-y-olds
Alexy *et al.* 2002; 2003, The DONALD study, Germany	3-day weighed dietary record	n = 849 2–18-y-olds	Added sugar	11 E% in 2–3-y-olds 13 E% in 4–8-y-olds
Emmett *et al.* 2002, The ALSPAC study, Great Britain	3-day food records	n = 1026 18-mo-olds n = 863 43-mo-olds	Non-milk extrinsic (NME) sugar	12 E% in 18-mo-olds 23 E% in 43-mo-olds
Lyhne and Ovensen 1999, Denmark	7-day food record	n = 649 4-6-y-olds and 7–10-y-olds	Added sugars	13 E% and 14 E% in 4-6-y-old boys and girls, respectively, 14 E% in 7-10-y-olds
Farris *et al.* 1998, The Bogalusa Heart Study, US	24-hour recall	n = 568 10-y-olds	Sugar	15 E%

Figure 45.2 Consumption of foods rich in sugars among children in a Finnish birth
cohort study by age; proportion of consumers during the 3 study days (%).

someone else other than the main carer provided the child's meal (Emmett *et al.*
2002). In a large US cohort, limiting availability of soft drinks at school was
associated with a decreased overall consumption (Fernandes 2008).

Having a milk-restricted diet seems to endow children with greater risk for
high added sugar intake (Alexy *et al.* 2002; Erkkola *et al.* 2009; Fiorito *et al.*
2010). A substantial proportion of changes in the added sugar intake of children can be attributed to changes in their beverage consumption patterns. The
inverse association found between added sugar intake and riboflavin and calcium intake points to a reduction in milk consumption when sugar intake
increases. It is an enormous challenge to replace milk in a child's diet with
another drink that would provide similar nutrients without any increase in
added sugar intake. In a US longitudinal study, early intake of sweetened soft
drinks predicted different patterns of beverage and nutrient intake across
childhood and adolescence (Fiorito *et al.* 2010). The diets of soft drink consumers were higher in added sugars and lower in many nutrients.

45.3.3 Socioeconomic Determinants of Sugar Intake in Children

During the first year of life, patterns of food intake change dramatically. The
infant moves from a completely milk-based diet to the consumption of a wide
range of foods and there is potential for significant variation in how this change
is achieved. This early learning of eating is constrained by the child's genetic

predispositions, which includes the unlearned preference for sweet and salty tastes, and the rejection of sour and bitter tastes (Birch 1998; Birch and Fisher 1998). Children usually learn to prefer energy-dense foods and foods offered in positive contexts (Birch 1998). Repeated exposure to new foods, especially those that are not sweet, is critical for acceptance (Birch and Fisher 1998). Food preferences are strongly influenced by social, demographic and lifestyle factors related to the family, particularly of the mother (Birch 1998). Early feeding may also affect food consumption later in life.

High maternal age, high educational level of parents and small family size have all been associated with a lower proportion of sugars in the total energy intake of children (Erkkola *et al.* 2009; Kyttälä *et al.* 2010; Øverby *et al.* 2004; Ruottinen *et al.* 2004). It is important to note that naturally occurring and added sugars relate differently to the background variables. An overall, high intake of naturally occurring sugars seems to be more related to a better overall dietary quality than high intake of added sugars, a phenomenon clearly explained by the differences in their main dietary sources.

45.4 Sugar Intake and Quality of Diet

Several observational studies among children suggest that diets containing a high proportion of sugars, especially added ones, are lower in vitamins and minerals than diets containing a moderate proportion of added sugars (Alexy *et al.* 2003; Erkkola *et al.* 2009; Farris *et al.* 1998; Gibson 2007; Gibson and Boyd 2009; Kranz *et al.* 2005; Ruottinen *et al.* 2008). Consequently, a high proportion of added sugars in the diet has a mostly unfavorable impact on the intake of recommended foods and key nutrients in children. A nutrient dilution effect of high sugar intake can be seen in decreased amounts of several essential nutrients as sugar intake increases. However, intakes of most nutrients appear adequate regardless of the level of sucrose intake (Erkkola *et al.* 2009; Gibson 2007) or the intake has been lower than recommendations only in the highest sugar intake group (Øverby *et al.* 2004). As concluded by Gibson and Boyd (2009), the impact of added sugars on micronutrient intakes appears to be modest but may have relevance for those children who have inadequate micronutrient intakes coupled with a diet high in added sugars. The most critical nutritional consequence of increased added sugar intake in children is their association with decreased consumption of vegetables and decreased intake of critical nutrients such as vitamin D, iron, and calcium. Studies have shown that vegetable consumption can decrease as much as 40% (at its highest), between low to high added sucrose E% quarters (Erkkola *et al.* 2009; Gibson and Boyd 2009; Øverby *et al.* 2004; Ruottinen *et al.* 2008).

The association between added sugar intake and fat intake has turned out to be inverse (sugar:fat see-saw phenomenon) in children, which indicates that added sugar intake at least partially replaces fat intake (Alexy *et al.* 2003; Erkkola *et al.* 2009; Farris *et al.* 1998; Garemo *et al.* 2007; Gibson 2007). However, there is also evidence that a high-sugar-diet can also be high in fat (Emmett and Heaton 1995). Regardless of the quantity, the quality of dietary fat is likely to be better in the diet of children with low sugar intake than those with high sugar intake (Erkkola *et al.* 2009; Ruottinen *et al.* 2008).

The form in which sugars are consumed seems to be an important modifier of the impact of dilution; soft drinks, sugar and sweets are more likely to have an adverse impact on diet quality whereas dairy foods, milk drinks and pre-sweetened cereals may have a positive impact (Gibson 2007). Foods simultaneously fortified with vitamins and minerals and those sweetened with added sugars can mask nutrient dilution (Alexy *et al.* 2002). Forshee and Storey (2004) have debated that the use of a percentage of daily energy from added sugars as the key variable in the analysis makes it impossible to separate the association of added sugars on micronutrient intake from that of total energy or energy from other sources, and could, therefore, create serious statistical analysis and interpretation problems. In their review, Rennie and Livingstone (2007) concluded that there are currently insufficient data to draw firm conclusions on associations between dietary added sugar intake and micronutrient intake. However, the current evidence base does not show any advantages in the high consumption of added sugars, in relation to micronutrient intake either. Further work is warranted to distinguish between sources of added sugar with regards to the impact on micronutrient intakes, and to overcome the statistical challenges.

45.5   Associations Between Sugar Intake and Health in Children

Sugar is unique among carbohydrates for its effect on appetite because its high hedonic properties seem to overrule the regulatory control (Anderson 1995). The high requirement of growing children for energy drives their preference for sugary foods (Ventura and Mennella 2011). Intake of added sugars as percentage of energy intake is positively associated with energy intake (Alexy *et al.* 2002). Children with a high sugar intake are at risk of high energy intake (Alexy *et al.* 2003; Farris *et al.* 1998; Kranz *et al.* 2005; Linseisen *et al.* 1998; Lyhne and Ovesen 1999). High intake of sugars or increased consumption of foods with high sugar content may affect cardiovascular and metabolic disease risk factors in children (Kavey 2010). The positive association between consumption of sugar-sweetened beverages and obesity development is well documented, even in children (Lim *et al.* 2009; Ludwig *et al.* 2001; WHO 2003). However, findings regarding the effects of children's sugar intake on the development of obesity are inconsistent (*e.g.*, Hill and Prentice 1995; O'Neil *et al.* 2011; Ruottinen *et al.* 2008; Ruxton *et al.* 1999; Villa *et al.* 2007) as there are also studies showing an inverse association between high sugar intake and BMI in children (Ruxton *et al.* 1999; Villa *et al.* 2007).

Consumption of sugar-sweetened beverages has been associated with the development of chronic diseases such as metabolic syndrome, type 2 diabetes and cardiometabolic outcomes (Duffey *et al.* 2010; Hu and Malik 2010). Especially high-fructose diets have been linked to insulin resistance, non-alcoholic fatty liver disease, deposition of visceral fat and hypertension (Hu and

Malik 2010; Spruss and Bergheim 2009; Tappy *et al.* 2010). High sugar intake may cause adverse changes in lipoproteins (Howard and Wylie-Rosett 2000). Sugar intake in children has been associated with increased plasma triglyceride (Morrison *et al.* 1980; Ruottinen *et al.* 2009) and decreased HDL cholesterol concentration (Kouvalainen *et al.* 1982; Welsh *et al.* 2011). Findings on the association between sugar intake and total/LDL cholesterol concentrations in children are inconsistent (Cowin *et al.* 2001; Welsh *et al.* 2011). Welsh and co-workers (2011) concluded that consumption of added sugars among US adolescents is positively associated with varied cardiovascular disease risk factors.

From the 1970s to the 1990s the dental health of children improved in most industrialised countries (Burt and Szpunar 1994). At the same time, there was no similar concomitant decrease in sucrose consumption. The basic concept of sugar consumption as a risk factor of dental caries has been seriously questioned, although a number of epidemiological and experimental studies have established the role of sugar in dental caries development for a long time. There are studies in children providing support for the positive association between sucrose intake and caries in cross-sectional (Gibson and Williams 1999) and longitudinal study settings (Burt and Szpunar 1994; Meurman and Pieni-häkkinen 2010; Ruottinen *et al.* 2004). During the past 20 years caries has become common again, and dental caries remains a major public health problem in the EU and in other developed countries (Moynihan and Petersen 2004; Sheiman 2001). Restriction of children's sugar consumption still has a role to play in the prevention of caries, and it is important to notice that the habit of excessive daily sucrose intake may start early in childhood. The dental risk of dietary sugars is dependent on the frequency and the amount of sugar intake, but the prevalence of caries is also modified by other dietary, dental hygiene, social, genetic, and behavioural factors (König and Navia 1995).

Summary Points

The number of recording days needed to accurately measure sugar intake in children increases with age. The foods rich in sugar are often underreported.

Soft drinks, sweets, chocolate and sweetened milk products are the key contributors to the intake of sugars among children.

A child's sugar intake is strongly determined by the availability of drinks rich in sugar.

The proportion of sugars intake in the total energy intake among children in the Western countries, is commonly above the recommended upper limit of 10 E%. Sugar intake increases rapidly after the child starts partaking in family meals.

Being cared for at home, low maternal age, having less educated parents, big family size, and having a milk-restricted diet have all been associated with a higher proportion of sugars in the total energy intake of children.

A high proportion of added sugar in the diet has amostly unfavorable impact on the intake of recommended foods and key nutrients in children. Further work is warranted to distinguish between sources of added sugar with regards to the impact on micronutrient intakes.

High intake of sugars may increase the risk of obesity and chronic diseases such as metabolic syndrome, type 2 diabetes and cardiovascular diseases. The dental risk of dietary sugars is dependent on the frequency and amount of sugar intake.

Key Facts of Sugars in the Diet of Young Children

- It is recommended that the intake of sugars should not exceed 10% of the energy in children's diets. The recommendation is seldom met.
- In most countries, more than a half of the sugars in children's diets derive from soft drinks and sweets.
- Diets that are rich in sugar are usually rather low in vitamins and minerals.
- Consumption of sugar-sweetened beverages should be limited in children to reduce obesity and chronic diseases.
- Frequent sugar consumption is a risk factor of dental caries.

Definitions of Words and Terms

Added sugars: Sugars and syrups that are added to foods during processing or preparation. Added sugars do not include naturally occurring sugars such as those that occur in milk and fruits.

Extrinsic sugars: Sugars that are not located within the cellular structure of food, and are mainly found in fruit juice and added to processed foods.

Free sugars: All monosaccharides and disaccharides added to foods by manufacturer, cook, or consumer, and sugars which are naturally present in honey, syrups, and fruit juices.

Intrinsic (naturally occurring) sugars: Naturally occurring sugars that are an integral part of certain unprocessed foodstuffs enclosed in the cell.

Invert sugar: Mixture of glucose and fructose.

Non-milk extrinsic sugars: The extrinsic sugars which are not from milk, that is excluding lactose, nor contained within plant cell wall.

Refined sugars: Refined sugars include sucrose, fructose, glucose, starch hydrolysates (glucose syrup, high-fructose syrup) and other isolated sugar preparations such as food components used as such or added during food preparation and manufacturing.

Sugar: Any free monosaccharide or disaccharide present in foods. Sometimes refers strictly to sucrose (table sugar).

List of Abbreviations

BMI   body mass index
E%   percentage of total daily energy intake
DRI   dietary reference intake

EFSA   European Food Safety Authority
IoM   Institute of Medicine of the National Academies
kcal   kilocalorie
NDA   EFSA Panel on Dietetic Products, Nutrition and Allergies
NNR   Nordic Nutrition Recommendations
WHO   World Health Organization

References

Alexy, U., Sichert-Hellert, W., and Kersting, M., 2002. Fortification masks nutrient dilution due to added sugars in the diet of children and adolescents. *Journal of Nutrition*. 132: 2785–2791.

Alexy, U., Sichert-Hellert, W., and Kersting, M., 2003. Associations between intake of added sugars and intakes of nutrients and food groups in the diets of German children and adolescents. *British Journal of Nutrition*. 90: 441–447.

Andersen, L.F, Lande, B., Trygg, K., and Hay, G., 2004. Validation of a semi-quantitative food-frequency questionnaire used among 2-year-old Norwegian children. *Public Health Nutrition*. 7: 757–764.

Anderson, G.H., 1995.Sugars, sweetness, and food intake. *American Journal of Clinical Nutrition*. 62(Suppl): 195S–202S.

Birsch, L.L., 1998. Development of food acceptance patterns in the first years of life. *Proceedings of the Nutrient Society*. 57: 617–624.

Birch, L.L., and Fisher, J.O., 1998. Development of eating behaviours among children and adolescents. *Pediatrics*. 101: 539–549.

Black, A.E, Cole, T.J, Wiles, S.J., and White, F., 1983. Daily variation in food intake of infants from 2 to 18 months. *Human Nutrition. Applied Nutrition*. 37: 448–458.

Burt, B.A., and Szpunar, S.M., 1994. The Michigan study: the relationship between sugars intake and dental caries over three years. *International Dental Journal*. 44: 230–240.

Buyken, A.E, Cheng, G., Günther, A.L, Liese, A.D, Remer, T., and Karaolis-Danckert, N., 2008. Relation of dietary glycemic index, glycemic load, added sugar intake, or fiber intake to the development of body composition between ages 2 and 7 y. *American Journal of Clinical Nutrition*. 88: 755–762.

Cowin, I.S., Emmett, P.M; ALSPAC Study Team. 2001. Avon Longitudinal Study of Pregnancy and Childhood. Associations between dietary intakes and blood cholesterol concentrations at 31 months. *European Journal of Clinical Nutrition*. 55: 39–49.

Duffey, K.J., Gordon-Larsen, P., Steffen, L.M., Jacobs, D.Rjr., and Popkin, B.M., 2010. Drinking caloric beverages increases the risk of adverse-cardiometabolic outcomes in the Coronary Artery Risk Development in Young Adults (CARDIA) Study. *American Journal of Clinical Nutrition*. 92: 954–959.

Emmett, P.M., and Heaton, K.W., 1995. Is extrinsic sugar a vehicle for dietary fat? *Lancet.* 345: 1537–1540.

Emmett, P., Rogers, I., Symes, C., and ALSPAC Study Team., 2002. Avon Longitudinal Study of Pregnancy and Childhood. Food and nutrient intakes of a population sample of 3-year-old children in the south west of England in 1996. *Public Health Nutrition.* 5: 55–64.

Erkkola, M., Kronberg-Kippilä, C., Kyttälä, P., Lehtisalo, J., Reinivuo, H., Tapanainen, H., Veijola, R., Knip, M., Ovaskainen, M-L., and Virtanen, S.M., 2009. Sucrose in the diet of 3-year-old Finnish children: sources, determinants and impact on food and nutrient intake. *British Journal of Nutrition.* 101: 1209–1217.

Erkkola, M., Kyttälä, P., Takkinen, H-M., Kronberg-Kippilä, C., Nevalainen, J., Simell, O., Ilonen, J., Veijola, R., Knip, M., and Virtanen, S.M., 2011. Nutrient intake variability and number of days needed to assess intake in preschool children. *British Journal of Nutrition.* 106: 130–140.

European food safety authority (EFSA), 2010. Scientific opinion on dietary reference values for carbohydrates and dietary fibre. EFSA Panel on dietetic products, nutrition, and allergies (NDA). Scientific opinion., Parma, Italy. *EFSA Journal.* 8:1462: 1–77.

Farris, R.P, Nicklas, T.A, Myers, L., and Berenson, G.S., 1998. Nutrient intake and food group consumption of 10-year-olds by sugar intake level: the Bogalusa Heart Study. *Journal of the American College of Nutrition.* 17: 579–585.

Fernandes, M.M., 2008. The effect of soft drink availability in elementary schools on consumption. *Journal of American Dietetic Association.* 108: 1445–1452.

Forshee, R.A., and Storey, M.L., 2004. Controversy and statistical issues in the use of nutrient densities in assessing diet quality. *Journal of Nutrition.* 134: 2733–2737.

Fiorito, L.M., Marini, M., Mitchell, D.C., Smiciklas-Wright, H., and Birch, L.L., 2010. Girls' early sweetened carbonated beverage intake predicts different patterns of beverage and nutrient intake across childhood and adolescence. *Journal of American Dietetic Association.* 110: 543–550.

Garemo, M., Lenner, R.A., and Strandvik, B., 2007. Swedish pre-school children eat too much junk food and sucrose. *Acta Paediatrica.* 96: 266–272.

Gibson, S.A. 2007. Dietary sugars intake and micronutrient adequacy: a systematic review of the evidence. *Nutrition Research Reviews.* 20: 121–131.

Gibson, S., and Boyd, A., 2009. Associations between added sugars and micronutrient intakes and status: further analysis of data from the National Diet and Nutrition Survey of Young People aged 4 to 18 years. *British Journal of Nutrition.* 101: 100-107.

Gibson, S., and Williams, S., 1999. Dental caries in pre-school children: Associations with social class, toothbrushing habit and consumption of sugars and sugar-containing foods. *Caries Research.* 33: 101–113.

Hill, J.O., and Prentice, A.M., 1995. Sugar and body weight regulation. *American Journal of Clinical Nutrition.* 62(Suppl): 264S-274S; discussion 273S–274S.

Howard, B.V., and Wylie-Rosett, J., 2000. Sugar and cardiovascular disease a statement for healthcare professionals from the committee on nutrition of the council on nutrition, physical activity, and metabolism of the American heart association. *Circulation.* 106: 523–527.

Hu, F.B., and Malik, V. S, 2010. Sugar-sweetened beverages and risk of obesity and type 2 diabetes. *Epidemiologic Evidence Physiology & Behavior.* 100: 47–54.

Institute of Medicine of the National Academies, Food and Nutrition Board, (2005). *Dietary references intakes (DRI) for energy, carbohydrate, fiber, fat, fatty acids, cholesterol, protein, and amino acids.* Washington, DC: The National Academies Press.

Joyce, T., McCarthy, S.N., and Gibney, M.J., 2008. Relationship between energy from added sugars and frequency of added sugars intake in Irish children, teenagers and adults. *British Journal of Nutrition.* 99: 1117–1126.

Kavey, R.-E.W., 2010. How sweet it is: sugar-sweetened beverage consumption, obesity, and cardiovascular risk in childhood. *Journal of American Dietetic Association.* 110: 1456–1460.

Kouvalainen, K., Uhari, M., Akerblom, H.K., Viikari, J., Räsänen, L., Ahola, M., Suoninen, P., Pietikäinen, M., Pesonen, E., Lähde, P.L., Dahl, M., Nikkari, T., Seppänen, A., and Vuori, I. 1982. Nutrient intake and blood lipids in children. *Klinische Padiatrie.* 194: 307–309.

Kranz, S., Smiciklas-Wright, H., Siega-Riz, A.M., and Mitchell, D., 2005. Adverse effect of high added sugar consumption on dietary intake in American preschoolers. *Journal of Pediatrics.* 146: 105–111.

Kyttälä, P., Erkkola, M., Kronberg-Kippilä, C., Tapanainen, H., Veijola, R., Simell, O., Knip, M., and Virtanen, S.M., 2010. Food consumption and nutrient intake in Finnish 1- to 6-year-old children. *Public Health Nutrition.* 13(Suppl): S947–S956.

König, K.G., and Navia, J.M, 1995. Nutritional role of sugars in oral health. *American Journal of Clinican Nutrition.* 62(Suppl): S275–S282; discussion S282S–S283.

Lehtisalo, J., Erkkola, M., Tapanainen, H., Kronberg-Kippilä, C., Veijola, R., Knip, M., and Virtanen, S.M., 2010. Food consumption and nutrient intake in day care and at home in 3-year-old Finnish children. *Public Health Nutrition.* 13(Suppl): S957–S964.

Lillegaard, I.T., and Andersen, L.F., 2005. Validation of a pre-coded food diary with energy expenditure, comparison of under-reporters v. acceptable reporters. *British Journal of Nutrition.* 94: 998–1003.

Linseisen, J., Gedrich, K., Karg, G., and Wolfram, G., 1998. Sucrose intake in Germany. *Z Ernahrungswiss.* 37: 303–314.

Lim, S., Zoellner, JM., Lee, MJ., Burt, BA., Sandretto, AM., Sohn, W., Ismail, AI., and Lepkowski, JM., 2009. Obesity and sugar-sweetened beverages in African-american preschool children: a longitudinal study. *Obesity.* 17: 1262–1268.

Ludwig, DS., Peterson, KE., and Gortmaker, SL.., 2001. Relation between consumption of sugar-sweetened drinks and childhood obesity: a prospective, observational analysis. *Lancet.* 357: 505–508.

Lyhne, N., and Ovesen, L., 1999. Added sugars and nutrient density in the diet of Danish children. *Scandinavian Journal of Nutrition*. 43: 4–7.

Meurman, PK., and Pienihäkkinen, K., 2010. Factors associated with caries increment: a longitudinal study from 18 months to 5 years of age. *Caries Research*. 44: 519–524.

Morrison, J.A., Larsen, R., Glatfelter, L., Boggs, D., Burton, K., Smith, C., Kelly, K., Mellies, M.J., Khoury, P., and Glueck, C.J., 1980. Interrelationships between nutrient intake and plasma lipids and lipoproteins in school-children aged 6 to 19: the Princeton School District Study. *Pediatrics*. 65: 727–734.

Moynihan, P., 2005. The interrelationship between diet and oral health. *The Proceedings of the Nutrition Society*. 64: 571–580.

Moynihan, P., and Petersen, PE., 2004. Diet, nutrition and the prevention of dental diseases. *Public Health Nutrition*. 7: 201–226.

Nordic Council of Ministers (NNR) 2004. *Nordic Nutrition Recommendations 2004, Integrating nutrition and physical activity*. Copenhagen. Nord, 13.

O'Neil, CE., Fulgoni 3rd, VL., and Nicklas, TA., 2011. Association of candy consumption with body weight measures, other health risk factors for cardiovascular disease, and diet quality in US children and adolescents: NHANES 1999-2004. *Food & Nutrition Research*. 55.

Ovaskainen, M-L., Reinivuo, H., Tapanainen, H., Hannila, M.L, Korhonen, T., and Pakkala, H., 2006. Snacks as an element of energy intake and food consumption. *European Journal of Clinical Nutrition*. 60: 494–501.

Øverby, N.C., Lillegaard, I.T., Johansson, L., and Andersen, L.F., 2004. High intake of added sugar among Norwegian children and adolescents. *Public Health Nutrition*. 7: 285–293.

Patterson, E., Wärnberg, J., Kearney, J., and Sjöström, M., 2010. Sources of saturated fat and sucrose in the diets of Swedish children and adolescents in the European Youth Heart Study: strategies for improving intakes. *Public Health Nutrition*. 13: 1955–1964.

Popkin, B.M., and Nielsen, S.J., 2003. The sweetening of the world's diet. *Obesity Research*. 11: 1325–1332.

Reedy, J., and Krebs-Smith, S.M., 2010. Dietary sources of energy, solid fats, and added sugars among children and adolescents in the United States. *Journal of American Dietetic Association*. 110: 1477–1484.

Rennie, K.L., and Livingstone, M.B., 2007. Associations between dietary added sugar intake and micronutrient intake: a systematic review. *British Journal of Nutrition*. 97: 832–841.

Ruottinen, S., Karjalainen, S., Pienihäkkinen, K., Lagström, H., Niinikoski, H., Salminen, M., Rönnemaa, T., and Simell, O., 2004. Sucrose intake since infancy and dental health in 10-year-old children. *Caries Research*. 38: 142–148.

Ruottinen, S., Niinikoski, H., Lagström, H., Rönnemaa, T., Hakanen, M., Viikari, J., Jokinen, E., and Simell, O., 2008. High sucrose intake is associated with poor quality of diet and growth between 13 months and 9 years of age: the special Turku Coronary Risk Factor Intervention Project. *Pediatrics*. 121: e1676–1685.

Ruottinen, S., 2011. Carbohydrate intake in children – association with dietary intakes, growth, serum lipids, and dental health. The STRIP Project. Doctoral theses. Turku: University of Turku, Department of Pediatrics and Department of Medicine. Ser. D, Tom. 947.

Ruxton, C.H., Garceau, F.J., and Cottrell, R.C., 1999. Guidelines for sugar consumption in Europe: is a quantitative approach justified? *European Journal of Clinical Nutrition.* 53: 503–513.

Sepp, H., Lennernäs, M., Pettersson, R., and Abrahamsson, L., 2001. Children's nutrient intake at preschool and at home. *Acta Paediatrica.* 90: 483–491.

Sheiham, A., 2001. Dietary effects on dental diseases. *Public Health Nutrition.* 4: 569–591.

Spruss, A., and Bergheim, I., 2009. Reviews: Current topics. Dietary fructose and intestinal barrier: potential risk factor in the pathogenesis of nonalcoholic fatty liver disease. *Journal of Nutritional Biochemistry.* 20: 657–662.

Tappy, L.,F., Le, K.,A., Tran, C., and Paquot, N., 2010. Fructose and metabolic diseases: New findings, new questions. *Nutrition.* 26: 1044–1049.

U.S. Department of Health and Human Services, U.S. Department of Agriculture. Dietary guidelines for Americans 2005. Available at: http://www.health.gov/dietaryguidelines/ Accessed Nov. 28, 2009.

U.S. Department of Health and Human Services, U.S. Department of Agriculture. Dietary guidelines for Americans, 2010. Available at: http://www.health.gov/dietaryguidelines/Accessed Jun. 15, 2011.

Ventura, A.K., and Mennella, J.A., 2011. Innate and learned preferences for sweet taste during childhood. *Current Opinion in Clinical Nutrition and Metabolic Care.* 14: 379–384.

Villa, I., Yngve, A., Poortvliet, E., Grjibovski, A., Liiv, K., Sjöström, M., and Harro, M. 2007. Dietary intake among under-, normal- and overweight 9- and 15-year-old Estonian and Swedish schoolchildren. *Public Health Nutrition.* 10: 311–322.

Welsh, J.A., Sharma, A., Cunningham, S.A., and Vos, M.B. 2011. Consumption of added sugars and indicators of cardiovascular disease risk among US adolescents. *Circulation.* 123: 249–257.

Williams, C. 2010. Resource 1: Children's dietary intakes. Supplemental information related to the Report of the Dietary Guidelines Advisory Committee on the Dietary Guidelines for Americans, 2010. Available at: http://www.cnpp.usda.gov/Publications/DietaryGuidelines/2010/DGAC/Report/Resources.pdf#xml=http://65.216.150.153/texis/search/pdfhi.txt?query=children+and+sugar+and+williams&pr=CNPP&prox=page&rorder=500&rprox=500&rdfreq=500&rwfreq=500&rlead=500&rdepth=0&sufs=2&order=r&mode=&opts=&cq=&sr=&id=4e1e3f647. Accessed 4 July 2011 .Willett, W. 1998. Nutritional epidemiology, 2nd ed. Oxford University Press, New York, USA.

World Health Organization (WHO) 2003. Diet, nutrition and the prevention of chronic diseases. Report of a Joint WHO/FAO Expert Consultation. Geneva, World Health Organization, 2002. World Health Organization Technical Report Series, 916.

CHAPTER 46

Lactose: Uses in Industry and Biomedical Importance
Lactose a Functional Disaccharide

BYONG LEE[a] AND ANDREW SZILAGYI[*b]

[a] Professor (AAFC Secondment) in Department of Microbiology/
Immunology, McGill University, Duff Medical Building, Room 511,
3557 University Street, Montreal, Quebec, Canada H3A 2B4; [b] Assistant
Professor Medicine, Division of Gastroenterology, Department of Medicine,
Jewish General Hospital 3755 Cote st Catherine Rd Room G325, McGill
University School of Medicine, Montreal, Quebec, Canada H3T 1E2
*Email: aszilagy@jgh.mcgill.ca

46.1 Introduction

Lactose, (glucose and galactose), is the principal carbohydrate in most mammalians' milks [about 4.8% in cow's milk, (Adam *et al.* 2004)] adding proper solubility and viscosity. In neonates it represents about 30% nutritional value and is split into its components by lactase phlorizin hydrolase (LPH). While glucose is a basic nutrient, galactose is involved in the development of brain (Adam *et al.* 2004). Lactose is a most intriguing and controversial component of diet and is readily manufactured for industrial uses.

There are two relevant aspects, both because of wide use of dairy foods. The first is the application to the food and pharmaceutical industry. The second is

Food and Nutritional Components in Focus No. 3
Dietary Sugars: Chemistry, Analysis, Function and Effects
Edited by Victor R Preedy
© The Royal Society of Chemistry 2012
Published by the Royal Society of Chemistry, www.rsc.org

the function of lactose in biology and medicine. We review its biochemistry, and industrial uses. We briefly describe division of adults into those who can (lactase persistent, LP) or cannot digest lactose (lactase non-persistent, LNP). The relationship of lactose to intolerance is mentioned but details are left for another chapter. Possible relationships of LP/LNP interactions modifying some disease risks are discussed.

46.2 Industrial Uses of Lactose

In 2000, in the USA, production of lactose from whey was 520 million pounds, of which 42% were exported. Increase in production, derived from whey, a byproduct of cheese making, has focused on development of new uses. Lactose can be applied in diverse food products, including bakery goods, confections, dry mixes, dairy foods, dried vegetables, snacks and infant formula (US Dairy Export Council 2003). Because lactose has no color, with less sweetness than sugar, it is commonly used in dried vegetable processes to reduce sweetness and prevent discoloration. Lactose serves as a building block for other compounds in food products including: lactulose, lactitol, lactobionic acid, and galacto-oligosaccharides (Park 2009), which act as prebiotics (Gibson and Roberfroid 1995). Markets are expanding for use of whey and byproducts of lactose, in sports and nutrition drinks (Baek and Lee 2009; Lee 2008).

In the pharmaceutical industry, lactose serves as an efficient filler for tablets because of its physical properties (*i.e.*, compressibility) and low price as well as a stabilizer of aroma and pharmaceutical products (Chambers and Ferretti 1979).

46.2.1 Biochemical Description of Lactose

Lactose consists of two aldohexoses-β-D-galactose and glucose-linked so that the aldehydo group at the anomeric carbon of glucose is free to react (Figure 46.1).

Two isomeric forms exist, α- and β-lactose, differing in steric configuration of the hydroxyl group of C-1 moiety of glucose. Different crystal polymorphs exist, depending on the crystallization conditions. The α- and β-lactose differ in

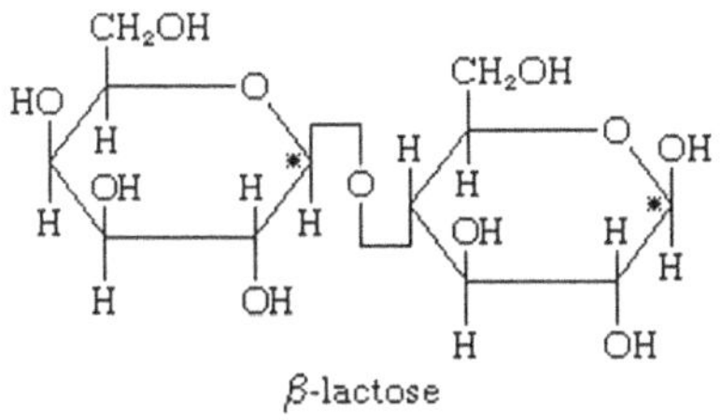

β-lactose

Figure 46.1 Lactose consists of two aldohexoses β-D-galactose and glucose. The aldehydo group at the anomeric carbon of glucose is free to react as a reducing sugar (on the carbon of glucose allows reaction).

solubility and are temperature dependent. If α-lactose is brought into water, less dissolves initially than later, because of mutarotation: α-lactose is converted to β, hence the α-concentration diminishes, and more α can dissolve. If β-lactose is brought into water, more dissolves initially than later (at least below 70 °C): on mutarotation, more α-lactose forms than can stay dissolved, and α-lactose starts to crystallize.

Manufacture of lactose consists of a five-step process: concentration, coagulation and removal of whey, further concentration, crystallization, and recovery of crystals (usually by centrifugation) (Ganzle *et al.* 2008). These are usually washed and may be dissolved and recrystallized to yield a high purity product. The normal form of commercial lactose is α-monohydrate, with a limited market for β-lactose, produced by crystallizing above 93.5 °C.

46.2.2   Analysis of Lactose

Lactose is a common excipient in the pharmaceutical and dairy industry, requiring reliable automated analytical methods assessing quality and quantity of input batches. Several procedures are available, including polarimetry, gravimetry, infrared, colorimetry, enzymatic, gas-liquid chromatography (GLC), and high pressure liquid chromatography (HPLC). Most of these are time consuming, insensitive, or nonspecific. The most common method HPLC, equipped with refractive index detection suffers from poor resolution, reproducibility and selectivity (Altria *et al.* 1999). The criteria for selecting an ideal analytical method follows that for enzymatic methods. Lactose is first hydrolyzed to D-glucose and D-galactose by β-galactosidase. β-D-galactose is oxidized by nicotinamide-adenine dinucleotide (NAD^+) to galactonic acid in the presence of β-galactose dehydrogenase. The amount of reduced NADH formed is stoichiometric with the amount of lactose and is measured at 340 nm in a spectrophotometer possessing a slit width of ≤ 10 nm. Determinations by enzymatic method, average 49% lower than by the Association of Official Analytical Chemists gravimetric method (Lynch and Barbano 2007).

Gas liquid chromatography (GLC) uses sugar derivatization with trimethylsilyl or others, on a fused silica capillary column. A new, less time consuming, capillary electrophoresis method has also been validated for analysis. Separation is achieved by an on-capillary chelation of lactose with borate ion, that can be detected at 195 nm. This method is now used in routine quality control in pharmaceutical industries (Altira *et al.* 1999). Other infrared spectroscopy techniques such as Fourier transform (FTIR) (Lefier *et al.* 1996) and mid-infrared (mid-IR) (Barbano *et al.* 2010) has also been tried for the analyses of milk components and pharmaceutical ingredients.

46.2.3   Food Uses of Lactose

Lactose is used as an additive to baby food, simulating the composition of human milk. Due to its flavor-enhancing properties and low sweetness, lactose is used in numerous other food products such as cakes, biscuits, chocolate, confectionery, soups, and sauces (Schaafsma 2008). Overall characteristics, such as flavor, mouth feel, texture, color, and stability are affected. Typically, 10% (occasionally 20%) lactose can be used in most systems as a part-substitute for sucrose or skimmed milk powder without altering product texture. The concentration of lactose in condensed milk is critical in protecting the texture of the milk product. Condensed milk has 60% water removed and sucrose added to produce a total carbohydrate concentration of 56%. Too much lactose results in a grainy product, and too little creates a slimy texture. These characteristics limit lactose as a confection and a substitute for sucrose.

Lactose is an additive in powdered foods, as a free-flowing agent (humectants). Lactose glass is used in increasing the dispensability of certain foods. Alternatively, powdered lactose glass is added to the powdered food, which is then wetted slightly, causing lactose to crystallize into agglomerates that entrap other components.

In baking and confectionary industries, lactose is used to: react with protein amino groups to give color and flavor (Maillard browning), emulsify and cream properties of shortenings, improve product quality, facilitate baking operations, give increased loaf volume, external appearance scores, and extend shelf-life freshness by 50–100%.

The Maillard browning reaction, however, is a problem in the manufacture and storage of dried milk and whey powder. Stock products should be stored under dry, cool conditions and rotated, to avoid browning during storage.

Hydrolyzed lactose syrup is currently used in nonalcoholic beverages, ice cream, yogurt, salad dressings, jams, jellies, toffees, fudge, and boiled sweets. Its sweetness is increased with partial hydrolysis to form a solution of galactose, glucose, and lactose. These three sugars have synergistic effects on sweetness and glucose, galactose increase solubility.

46.2.4 Bioactive Lactose Derived Products

46.2.4.1 Lactulose and Lactitol

Lactulose (β-D-galactosyl-D-fructofuranose) and lactitol (β-D-galactosyl-D-glucitol) are synthetic derivatives by alkali isomerization or reduction, respectively. Lactulose is also formed in small amounts during sterilization of milk. Neither lactulose, nor lactitol are digested in the small intestine. Their energy content is $2\,\text{kcal}\,\text{g}^{-1}$ (about 50% of digestible sugars) and are metabolized by colonic bacteria into short chain fatty acids (SCFA; acetate, propionate, butyrate).

Lactulose and lactitol do not cause tooth decay and their sweetness makes them suitable for application in a large variety of products, including chewing gum, confectionery and ice cream.

46.2.4.2 Lactobionic Acid

Lactobionic acid (β-D-galactosyl-gluconic acid) does not exist in nature, and is produced by oxidation of lactose. It combines a sweet taste with nutritional energy value approximately $2\,\mathrm{kcal\,g^{-1}}$, a pH-reducing effect and strong mineral-complexing properties, allowing it to be used as a food ingredient. The physiological effects of lactobionic acid have not been investigated in detail. However, it can be considered a resistant carbohydrate, bypassing intestinal hydrolysis and will be fermented by intestinal flora, exerting prebiotic effects (Park 2009). Lactobionic acid interferes with lactose digestion, by competing with lactose for mucosal β-galactosidase. However, before its application as a food ingredient, more knowledge is required about tolerance and effects on the intestinal flora in humans.

46.2.4.3   *Lactosucrose and Tagatose*

The trisaccharide lactosucrose is produced by transfructosylation of lactose by bacterial or fungal fructosyltransferases (levansucrases or inulosucrases). Using sucrose or raffinose, it acts as a fructosyl donor. Its sweetness relative to sucrose is 0.3–0.6. Lactosucrose behaves as a prebiotic, stimulating bifidobacteria, decreasing fecal pH and inhibiting growth of clostridia (Ogata *et al.* 1993). The synthesis of lactosucrose is a general property of bacterial levansucrase activity, frequently found in food fermenting lactic acid bacteria, (particularly *L. reuteri*, *L. pontis* and *L. acidophilus*). These organisms have been successfully used to generate high levels of oligosaccharides (Tieking *et al.* 2005).

D-Tagatose is a hexoketose monosaccharide sweetener (1.5 kcal/g), which is an isomer of D-galactose prepared from lactose hydrolysis of whey in (Kim 2004) (Figure 46.2). Recently, there has been industrial interest in

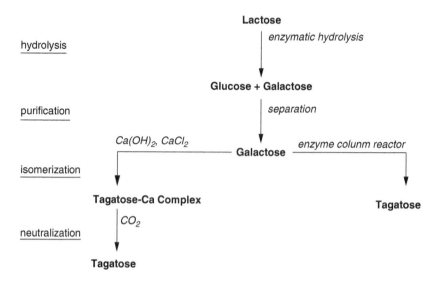

Figure 46.2   Schematic diagram of chemical and enzymatic tagatose production. Calcium and L-arabinose isomerase (AI) mediate the isomerization of galactose. The process requires an additional neutralization step for purification (Kim 2004).

D-tagatose as a low-calorie sugar-substituting sweetener used for treating type 2 diabetes, as well as obesity (Espinosa and Fogelfeld 2010; Lu *et al.* 2008).

46.2.4.4 Galacto-oligosaccharides

Lactose can be used as substrate for the synthesis of oligosaccharides, like *trans*-galactosyl-oligosaccharide (TOS). The general formula of galactosyl-oligosaccharides is Gal-(Gal)n-Glu. The galactosyl-oligosaccharides (GOS) are resistant to intestinal hydrolysis and have gained popularity, because of their sweet taste, low energy value (approx. $2\,kcal\,g^{-1}$), prebiotic properties and enhancement of mineral absorption. GOS are increasingly added to other foods (*e.g.* chewing gum, dairy products, and baked goods). Industrial production processes can extract oligosaccharides from natural sources, by hydrolyzing polysaccharides, or by enzymatic and chemical synthesis from disaccharide substrates. Worldwide, there are 13 classes of food-grade oligosaccharides currently produced commercially (Table 46.1) (Baek and Lee 2009).

46.2.4.5 Synbiotics

A wide range of prebiotics have been isolated from plant materials, including β-glucans from oats, inulin from chicory root, oligosaccharides from beans,

Table 46.1 Non-digestible oligosaccharides with bifidogenic functions, commercially available.*

Compound	Molecular Structure[a]	Production Method
Cyclodextrins	(Gu)$_n$	Starch by cyclodextrin glycosyl-transferase (CGTase) and alpha-amylase
Fructooligosaccharides	(Fr)$_n$–Gu	Transfructosylation from sucrose by beta-fructofuranosidase or inulin hydrolysate
Galactooligosaccharides	(Ga)$_n$–Gu	Transgalactosylation from lactose by beta-galactosidase
Gentiooligosaccharides	(Gu)$_n$	Transgalactosylation of starch by beta-glucosidase
Glycosylsucrose	(Gu)$_n$–Fr	Transglucosylation of sucrose and lactose by cyclomaltodextrin gluca-notransferase
Malto/ Isomaltooligosaccharides	(Gu)$_n$	Transgalactosylation of starch by pullanase, isoamylase, alpha-amylases/ transglucosidase
Isomaltulose (or palatinose)	(Gu–Fr)$_n$	Transglucosidase of sucrose
Lactosucrose	Ga–Gu–Fr	Transglycosylation of lactose and sucrose by beta-fructofuranosidase
Lactulose	Ga–Fr	Isomerisation of lactose by alkali
Raffinose	Ga–Gu–Fr	Extraction of plant materials by water, or alcohol
Soybean oligosaccharides	(Ga)$_n$–Gu–Fr	Extraction of soy whey
Xylooligosaccharides	(Xy)$_n$	Xylan hydrolysis by xylanase or acids

[a]Ga, galactose; Gu, glucose; Fr, fructose; Xy, xylose.
*Adapted from Baek and Lee (2009).

onions and leek are a few examples. The combination of probiotics and pre-biotics (called synbiotics) increase effectiveness of probiotic preparations for therapeutic use. Exact effects are not fully elucidated and can vary by geo-graphic host distribution of resident flora, as well as by diet, aging and disease states (McFarland 2006). There are several mechanisms by which affected bacteria exert benefits. These include: interference with pathogens, production of antibacterial agents, modification of mucosal permeability, alteration of the innate immune reaction and metabolism of carbohydrates to SCFA. These result in anti-inflammatory (Broekaert and Walker 2007; Tannock 2005), antineoplastic and other disease-modifying effects.

46.2.5   Pharmaceutical Uses of Lactose

In the pharmaceutical industry, lactose is used to fill or bind drugs, acting as a carrier. For example, recrystallised lactose has been recently studied as a dry powder delivery system for inhalers (Kaialy *et al.* 2010). Lactose can be used in conjunction with other additives to modify coating and flow properties of pharmaceutical powders using an intensive mechanical dry processing (Zhou *et al.* 2010). Lactulose and lactitol can be used to treat constipation. They are also used in the treatment of patients with hepatic encephalopathy (intoxication of the brain caused by failure of the liver to convert ammonia to urea). These prebiotics reduce ammonia formation by intestinal flora. Like lactose, both lactulose and lactitol will promote mineral absorption. Overdose of lactulose and lactitol cause intestinal discomfort and diarrhea.

46.3   Biomedical Significance of Lactose

Lactose is relevant as a nutrient in all mammalian neonates as outlined above. All (with rare exceptions) have the ability to digest lactose for a variable time after birth. Inheritance of LP is dominant and LNP is recessive. In humans, decline in intestinal lactase is largely irreversible and cannot be reconstituted by regular lactose consumption. Genetic polymorphisms determining LP pheno-type are complex and have interesting geographic distributions (Ingram *et al.* 2007). Interplay between lactose ingestion and digestion status has several effects on medical problems. The unusual geographic population LP/LNP distributions may affect disease risks.

A more detailed discussion of the genetics of lactase is provided in another chapter of this book. However, it is important to emphasize that about 65% of the world's population are LNP while the rest are LP (Swallow 2003). The pattern of LP/LNP distributions will be mentioned further below.

46.3.1   Digestion of Lactose

Digestion of lactose occurs at the intestinal brush border. The highest con-centration of LPH occurs in the proximal intestinal epithelium, (mainly jejunum), where the optimal pH 5.5–6 for activity is attained (Arola and Tamm 1994). LPH splits lactose into its components, which are transported into apical enterocytes through sodium-glucose transporter-1 (SGLT1) (reviewed by Drozdowski and Thomson 2006). While glucose is utilized directly, galactose in large amounts is toxic (to eye and liver) and has to be converted to glucose by epimerization via the Leloir pathway. Both components serve as energy sources, but milk provides large amounts of galactosylceramide and othergalactolipids, required for the rapidly developing infant brain (Adam *et al.* 2004; Lebea and Pretorius 2005).

Lactose stimulates vitamin D, independently of intestinal calcium transport. This could be due to decreased luminal pH and increased calcium solubility by fermentation of undigested lactose in the intestinal lumen (Schaafsma 2008). It has a lower glycemic index than sucrose or glucose and is less cariogenic than other sugars.

The inability or ability to digest lactose can be assessed clinically by several methods, the details of which are outlined elsewhere in this book. The two most common tests measure outcome as a response to a lactose load. In the case of the lactose tolerance test, ability of the host to digest lactose is measured in rise of absorbed blood glucose. In the case of the lactose breath hydrogen test, metabolism of undigested lactose by intestinal bacteria is measured as a response with exhaled hydrogen gas. This physiological distinction is important when one interprets results since both tests have distinct individual discrepancies. However, they also individually have good predictability for genetic lactase trait (Babu *et al.* 2010). To date, this correlation has been carried out only for the north European genetic polymorphism.

46.3.2 Lactose Intolerance

The main clinical relevance of this disaccharide has been the problem of gastrointestinal symptoms putatively induced by undigested lactose. Who gets these symptoms and why and how best to manage them continues to elicit much clinical research and public interest. There is also confusion about terms such as lactase persistence/non-persistence, lactose digestion, lactose intolerance and lactose sensitivity. Detailed analyses of these aspects are outlined elsewhere in this book. A possible impact of LP/LNP distributions on disease risks will be described in detail below.

46.3.3 Impact of Lactose and Digestion Status on Different Diseases

The notion that some disease risks may be linked with digestion status of lactose originated in the early 1970s. A European group suggested that lactose maldigestion may protect against colorectal cancer. In Europe it was postulated that an apparent south-north gradient of increasing lactose digestion may have been linked in evolution to decreasing sunshine at northern latitudes (discussed in Gerbault *et al.* 2009). As such, LP status developed to compensate for rickets and bone disease allowing LP populations consumption of dairy foods (calcium assimilation hypothesis). Other diseases seemingly correlated with LP/LNP distributions included ovarian cancer and the inflammatory bowel diseases. In 2008 our group evaluated 8 diseases, including 6 cancers (breast, prostate colorectal, ovary, lung and stomach) and the inflammatory bowel diseases (Crohn's disease and ulcerative colitis) (Shrier *et al.* 2008). Rates of disease were evaluated in relationship to national LP/LNP frequencies that were available at the time in the literature. Also, relationships between these diseases and yearly dairy food consumption were evaluated. Finally, outcomes from these data were compared with available meta-analyses of the effects of dairy foods on some of the cancers evaluated.The results suggested three effects depending on LP or LNP status. In the case of prostate and ovarian cancer dairy foods seemed to exert a deleterious effect and this was more or less supported by then available meta-analyses. The effect was more pronounced in LP dominant countries. In the case of colorectal cancer the expected increased risk with LP dominance was not found, and patient level data suggested protection by dairy foods. This latter finding raised a hypothesis that lactose may behave as a prebiotic in LNP populations selectively, thereby exerting a protective effect. The third intreaction was the finding that increasing LNP portions of the national population were related to altered risk of diseases, independent of dairy foods.

46.3.4   Gene Food Interaction

In interaction of diseases with dairy food consumption, two different possible effects could constitute a genetic food interaction. In the first scenario; dairy foods increase risk for some diseases (prostate, ovarian cancer) reduction of intake by LNP subjects may be one facet of reduced risk. The LNP status is associated in general with decreased dairy food intake.

In the second scenario, exemplified by colorectal cancer, a genetic food interaction is possible through the selective prebiotic effect of lactose in LNP populations. The hypothesis was based on the observation that protection against CRC occurred in both high dairy consuming, high LP populations and low dairy consuming, low LP populations (*e.g.* Asians). In LNP subjects, regular consumption of lactose could lead to colonic adaptation (Hertzler and Savaiano 1996; Szilagyi 2002).

In colonic adaptation alterations in the fecal flora occur such that bifido-bacteria (Szilagyi *et al.* 2010) and, lactobacilli increase while bacteroides and clostridia decrease (Ito and Kimura 1993). These bacterial changes may be associated with benefits to the host (Gibson and Roberfroid 1995). Some of these occur putatively through bacterial metabolism of lactose, which lead to the formation of SCFA through the Leloir pathway (Figure 46.3).

Although both acetate and propionate are thought to provide nutritional and anti-inflammatory benefits, butyric acid has been traditionally attributed to

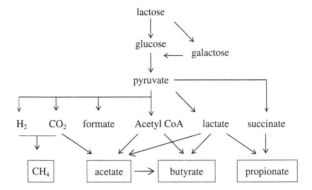

Figure 46.3   Metabolism of lactose by intestinal bacteria. The enzyme β-galactosidase splits lactose. The Leloir pathway converts galactose to glucose. Gases are formed and used by bacteria for production of methane (some 5–8% of the general population). Short chain fatty acids (SCFA) especially butyrate are produced by second tier stimulation of butyrogenic bacteria. Each SCFA is in turn used by the host for a variety of purposes. Reprinted with permission from He *et al.* (2008).

provide primary colonocyte nutrition and anti-inflammatory and antineo-plastic effects (Al-Laham *et al.* 2010; Wong *et al.* 2006).

It is of note that while lactic acid bacteria (bifidobacteria and lactobacilli) do not produce butyrate, lactic acid and acetate stimulate second tier bacteria which can produce butyrate (Duncan *et al.* 2004; Morrison *et al.* 2006). In summary, the effects of lactose containing foods on disease risk constitute a direct gene–food interaction which depends on LP/LNP status.

46.3.5   Prediction of Disease Risk by Population Proportion of LP/LNP Status

In the above paradigm on disease LP/LNP status interactions, the apparent protection against several diseases as national LNP proportion increase, affords a third interaction. The explanation of this observation is not immediately apparent. The 8 diseases evaluated for outcome with LP/LNP status reveal a similar relationship in risk with that previously published for latitudinal effects (Grant 2007). The putative risk modifier is UVB and skin synthesis of vitamin D (Zitterman 2003). However, apparent decrease in risk with increase in LNP proportions show the effects in an eastern direction on a global scale, as well. Anthropological geneticists have shed light on this curiosity.

As noted above, lactase distributions in Europe are affected by latitudinal decrease toward the equator. However, on a global scale there is strong evidence that LP dominance is rather a consequence of ancestral cattle herding which then led to the emergence of LP genotype with several different polymorphisms accounting for the trait.

This is labeled the gene-culture co evolution hypothesis (discussed in Itan *et al.* 2009). While in non-European locales latitudinal relationship with LP/LNP are less apparent, in Europe they strongly correlate (Itan *et al.* 2009). Since about 50% of the diseases evaluated for LP/LNP related risk originated from Europe, this contribution biased overall outcome. It seems that similar disease risks are strongly correlated with 2 different variables. The observation suggests that both variables exert impact on such diseases. Putative mechanisms whereby LP/LNP status could exert influence is *via* co-evolution of genes predisposing to a variety of diseases.

Summary Points

- Lactose is a key carbohydrate in mammalian milk and is important for both nutrition and development of neonates.
- Lactose is used in the food and pharmaceutical industry, and is an important building block for prebiotic compounds.
- The genetics of lactase divide the world's population into lactose digesters (dominant trait, 35%) and maldigesters (recessive trait, 65%).
- Lactose maldigestion may lead to symptoms.
- However, the distribution of LP/LNP populations may modify risks of some diseases.
- These observations should lead to further research in this field.

Key Facts: Lactose a Functional Disaccharide: Uses in Industry and Biomedical Importance

- Lactose is synthesized in mammary glands from hexoses; glucose and galactose.
- Industrial lactose is mainly derived from whey during cheese production.
- Uses include additive to baby food, cakes, biscuits, chocolate, chocolate products, sugar confectionery, soups, sauces where it affects flavor, mouth feel, texture, color, and stability.
- Derivatives of lactose serve as colonic floral modifiers, possibly producing health benefits.
- In pharmaceuticals, physical characteristics of lactose make it ideal as a filler and vector for different medications.

Key Facts: Biomedical Importance of Lactose

- With rare exceptions, all neonates digest lactose.
- Two-thirds of adults lose lactose digestion ability.
- Lactose intolerance leads to symptoms, but the cause may be multifactorial.
- There are geographic patterns of lactase status in population distributions, which may influence different disease risks.

Definitions of Words and Terms

Whey: byproduct of cheese-making, high lactose content; concentrated for the production of biofuels and bioproducts www.ncbioconsortium.org/index.asp.

Lactase Phlorizin Hydrolase: enzyme in the superficial layer of the small intestine (duodenum and jejunum), splits lactose into monosaccharides.

Lactase Persistent: adults retain ability to digest lactose.

Lactase Non-persistent: adults lose ability to digest lactose.

Lactose maldigestion: reduction in LPH, reduced ability to digest lactose.

Lactose intolerance: symptoms due to ingestion of lactose.

Adaptation to lactose in LNP people: altered colonic bacteria in response to regular lactose ingestion may improve symptoms and offer other health benefits.

Probiotics: ingested live non-pathogenic microorganisms, traverse gastric acidity and are embedded in the lower intestine of the host. In adequate numbers they confer health benefits.

Prebiotics: undigested carbohydrates which bypass digestion, selectively stimulate growth and/or metabolism of bacteria, exerting health benefits.

Synbiotics: synergistic or additive effects of pre and probiotics.

Genetic polymorphism: genetics of two (or more) phenotypes which exist in the same population, like the dichotomy of lactose digestion.

Genetic promoter: a segment of DNA usually occurring upstream from a gene coding region and acting as a controlling element in the expression of that gene.

Intron: a gene segment between sequences of DNA, that is removed before protein formation and itself does not contain coding for protein synthesis.

List of Abbreviations

CRC	Colorectal cancer
GLC	Gas-liquid chromatography
GOS	Galactosyl-oligosaccharides
HPLC	High pressure liquid chromatography
LABS	Lactic acid producing bacteria
LI	Lactose intolerance
LM	Lactose maldigester
LNP	Lactase non-persistent adult
LP	Lactase persistent adult
LPH	Lactase phlorizin hydrolase (lactase)
PCR	Polymerase chain reaction test
SCFA	Short chain fatty acids
TOS	Trans-galacto-oligosaccharides

References

Adam, A.C., Rubio-Texeira, M., and Polaina, J., 2004. Lactose: The milk sugar from a biotechnological perspective. *Critical Reviews in Food and Nutrition.* 44: 553–557.

Al-Laham, S.H., Peppelenbosch, M.P., Roelfsen, H., Vonk, R.J., and Venema, K., 2010. Biological effects of propionic acid in humans; metabolism, potential applications and underlying mechanisms. *Biochimica et Biophysica Acta.* 1801: 1175–1183.

Altria, K., Ennis, K., and Sadler, R., 1999. Quantitative and selective analysis of lactose by capillary electrophoresis. *Chromatographic.* 49: 406–410.

Arola, H., and Tamm, A., 1994. Metabolism of lactose in the human body. *Scandanavian Journal of Gastroenterology Supplement.* 202: 21–25.

Babu, J., Kumar, S., Babu, P., Prasad, J.H., and Ghoshal, U.C., 2010. Frequency of lactose malabsorption among healthy southern and northern Indian populations by genetic analysis and lactose hydrogen breath and tolerance tests. *American Journal of Clinical Nutrition.* 91: 140–146.

Baek, Y.J., and Lee, B.H., 2009. Chapter 13. Probiotics and prebiotics as bioactive components in dairy products. In *Bioactive Components in Milk and Dairy Products* (Ed: Y.W. Park), pp. 449, Wiley-Blackwell Publ., New York.

Barbano, D.M., Wojciechowski., K.L., and Lynch, J.M., 2010. Effect of preservatives on the accuracy of mid-infrared milk component testing. *Journal of Dairy Science.* 93: 6000–6011.

Chambers, J.V., and Ferretti, A., 1979. Industrial application of whey/lactose. *Journal of Dairy Science.* 62: 112–116.

Drozdowski, L.A., and Thomson, A.B.R., 2006. Intestinal sugar transport. *World Journal of Gastroenterology.* 12: 1857–1670.

Duncan, S.H., Louis, P., and Flint, H.J., 2004. Lactate-utilizing bacteria, isolated from human feces, that produce butyrate as a major fermentation product. *Applied and Environmental Microbiology.* 70: 5810–5817.

Espinosa, I., and Fogelfeld, L., 2010. Tagatose: from a sweetener to a new diabetic medication? *Expert Opinion on Investigational Drugs.* 19: 285–294.

Gangle, M.G., Haase, G., and Jelen, P., 2008. Lactose: Crystallization, hydrolysis and value added derivatives. *International Dairy Journal.* 18: 685–694.

Gerbault, P., Moret, C., Currat, M., and Sanchez-Mazas, A., 2009. Impact of selection and demography on the diffusion of lactase persistence. *PLoS ONE* 4: e6369 doi:10.1371/journal.pone.0006369.

Gibson, G.R., and Roberfroid, M.B., 1995. Dietary modulation of the human colonic microbiota: introducing the concept of prebiotics. *Journal of Nutrition.* 125: 1401–1412.

Grant, W.B., 2007. A meta-analysis of second cancers after a diagnosis of non-melanoma skin cancer: Additional evidence that solar ultraviolet–B irradiance reduces the risk of internal cancers. *Journal of Steroid Biochemistry and Molecular Biology.* 103: 668–674.

He, T., Venema, K., Priebe, M.G., Welling, G.W., Brummer, R.-J., and Vonk, R.J., 2008. The role of colonic metabolism in lactose intolerance. *European Journal of Clinical Investigation*. 38: 541–547.

Hertzler, S.R., and Savaiano, D.A., 1996. Colonic adaptation to daily lactose feeding in lactose maldigesters reduces lactose intolerance. *American Journal of Clinical Nutrition*. 64: 232–236.

Ingram, C.J.E., Elamin, M.F., Mulcare, C.A., Weale, M.E., Tarekegen, A., Raga, T.O., Bekele, E., Elamin, F.M., Thomas, M.G., Bradman, N., and Swallow, D.M., 2007. A novel polymorphism associated with lactose tolerance in Africa: multiple causes for lactase persistence? *Human Genetics*. 120: 779–788.

Itan, Y., Powell, A., Beaumont, M.A., Burger, J., and Thomas, M.G., 2009. The origins of lactase persistence in Europe. *PloS One Comput Biol*. 5: e1000491.doi:10.1371/journal.pcbi.1000491.

Ito, M., and Kimura, M., 1993. Influence of lactose on faecal microflora in lactose maldigesters. *Microbial Ecology in Health and Disease*. 6: 73–76.

Kaialy, W., Martin, G.P., Ticehurst, M.D., Royall, P., Mohammad, M.A., Murphy, J., and Nokhodchi, A., 2010. Characterization and deposition studies of recrystallised lactose from binary mixtures of ethanol/butanol for improved drug delivery from dry powder inhalers. *American Association of Pharmaceutical Scientists Journal*. 13: 30–43.

Kim, P., 2004. Current studies on biological tagatose production using l-arabinose isomerase: a review and future perspective. *Applied Microbiology and Biotechnology*. 65: 243–249.

Lebea, P.J., and Pretorius, P.J., 2005. The molecular relationship between deficient UDP-galactose uridyl transferase (GALT) and ceramide galactosyltransferase (CGT) enzyme function: A possible cause for poor long-term prognosis in classic galactosemia. *Medical Hypotheses*. 65: 1051–1057.

Lee, B.H., 2008. Structure, function and applications of microbial beta-galactosidase (lactase), In: *Carbohydrate-active Enzymes: Structure, Function and Applications*. (Ed: K.H. Park). p. 326, Woodhead Publisher, UK.

Lefier, D., Grappin, R., and Pochet, S., 1996. Determination of fat, protein, and lactose in raw milk by Fourier transform infrared spectroscopy and by analysis with a conventional filter-based milk analyzer. *Journal of Association of Official Analytical Chemists International*. 79: 711–717.

Lu, Y., Levin G.V., and Donner, T.W., 2008. Tagatose, a new antidiabetic and obesity control drug. *Diabetes Obesity and Metabolism*. 10: 109–134.

Lynch, J.M., Barbano, D. M., and Fleming, J.M., 2007. Determination of the lactose content of fluid milk by spectrophotometric enzymatic analysis using weight additions and path length adjustment: Collaborative study. *Journal of Association of Official Analytical Chemists International*. 90: 196–216.

McFarland, L.V., 2006. Meta-analysis of probiotics for the prevention of antibiotic associated diarrhea and the treatment of *Clostridium difficile* disease. *American Journal of Gastroenterology*. 101: 812–822.

Morrison, D.J., Mackay, W.G., Edwards, C.A., Preston, T., Dodson B., and Weaver L.T., 2006. Butyrate production from oligofructose fermentation by the human faecal flora: what is the contribution of extracellular acetate and lactate. *British Journal of Nutrition.* 96: 570–577.

Ogata, Y., Fujita, H., Ishigami, K., Hara, A., Terada, H., Hara, I., Fujimori, I., and Misuoka, T., 1993. Effect of a small amount of 4G-beta-D-galacto-sylsucrose (lactosucrose) on fecal flora and fecal properties. *Journal of Japanese Society of Nutrition and Food Science.* 46: 317–323.

Park, Y.W., 2009. Bioactive components in goat milk, In: *Bioactive Components in Milk and Dairy Products* (Ed: Y.W. Park), p. 449, Wiley-Blackwell Publ., New York.

Schaafsma, G., 2008. Lactose and lactose derivatives as bioactive ingredients in human nutrition. *International Dairy Journal.* 18: 458–465.

Shrier, I., Szilagyi, A., and Correa, J.A., 2008. Impact of lactose containing foods and the genetics of lactase on diseases: An analytical review of population data. *Nutrition and Cancer.* 60: 292–300.

Swallow, D.M., 2003. Genetics of lactase persistence and lactose intolerance. *Annals of Human Genetics.* 37: 197–219.

Szilagyi, A., 2002. Lactose: A potential prebiotic. *Alimentary Pharmacology and Therapeutics.* 16: 1591–1602.

Szilagyi, A., Shrier, I., Heilpern, D., Je, J., Park, S., Chong, G., Lalonde, C., Cote, L-F., and Lee, B., 2010. Differential impact of lactose/lactase phenotype on colonic microflora. *Canadian Journal of Gastroenterology.* 24: 373–379.

Tannock, G.W., 1998. Studies of the intestinal microflora: A prerequisite for the development of probiotics. *International Dairy Journal.* 8: 527–533.

Wong, J.M.W., de Souza, R., Kendall, C.W.C., Emam, A., and Jenkins, D.J.A., 2006. Colonic health: fermentation and short chain fatty acids. *Journal of Clinical Gastroenterology.* 40: 235–243.

Zhou, Q., Armstrong, B., Larson, I., Stewart, P.J., and Morton D.A., 2010. Improving powder flow properties of a cohesive lactose monohydrate powder by intensive mechanical dry coating. *Journal of Pharmacological Sciences.* 99: 969–981.

Zitterman, A., 2003. Vitamin D in preventive medicine: are we ignoring the evidence? *British Journal of Nutrition.* 89: 552–572.

CHAPTER 47

Technology and Biotechnology of Lactose Contained in Raw Food Materials

MAGDALINI SOUPIONI,*[a] MARIA KANELLAKI[b] AND LOULOUDA A. BOSNEA[c]

[a] Food Biotechnology Group, Department of Chemistry, University of Patras, 26500 Patras, Greece; [b] Food Biotechnology Group, Department of Chemistry, University of Patras, 26500 Patras, Greece; [c] Food Biotechnology, Group Department of Chemistry, University of Patras, 26500 Patras, Greece
*Email: m.soupioni@chemistry.upatras.gr

47.1 Introduction

Lactose is the main sugar present in the milk of most mammals; non-mammalian sources are very rare. The lactose content of milk varies between 3.6 and 5.5 % and its fermentation by lactose-fermenting micro-organisms, is responsible for the production of many important dairy products. This microbial activity may produce desirable organoleptic characteristics and increase the product's shelf life mainly due to the high organic acid production.

Dairy products are believed to have originated from the Middle East and have evolved over thousands of years (Tamime and Robinson 1999). Gradually, these products spread worldwide and became very popular because of their potential health and dietary benefits. Methods for fermented dairy product production were subsequently improved, using not only endogenous lactic

Food and Nutritional Components in Focus No. 3
Dietary Sugars: Chemistry, Analysis, Function and Effects
Edited by Victor R Preedy
© The Royal Society of Chemistry 2012
Published by the Royal Society of Chemistry, www.rsc.org

acid bacteria but also other starter cultures of micro-organisms, *e.g.*, yeasts, or mixed cultures like kefir. Today, cheeses, yogurt, buttermilk, sour cream, kefir and kumiss alcoholic beverage are probably the most popular and widely consumed fermented milk and whey products.

In cheese manufacture, a large quantity of the lactose from the milk is removed in the whey. In Greece in particular, a part of whey is used for whey-cheese production (myzithra, manouri, anthotyros), while in other countries (*e.g.* Caucasus, Turkestan) is used for alcoholic beverage preparation. However, in many countries, whey accumulates in considerable amounts as a waste product. So, whey represents an important environmental problem due to its bulk capacity and its high organic load of 40–70 g/L biological oxygen demand (BOD) and 60–80 g/L chemical oxygen demand (COD) (Athanasiadis *et al.* 2004). Lactose is largely responsible for the high BOD and COD. Protein recovery reduces the COD of whey only by about $10 \text{ g} \cdot \text{L}^{-1}$ (Domingues *et al.* 1999; Siso 1996). At the same time, whey is of great nutritious value as it contains approximately 0.8–1.6% proteins, 0.3–1.3% fats and 4.8–5.3% lactose. Generally, around 9 l of whey are generated from the production of 1 kg of cheese (Kosikowski 1979). The world whey production is over 160 million tons per year (estimated as 9-fold the cheese production), showing a 1–2% annual growth rate (OECD-FAO 2008; Smithers 2008).

Therefore, further whey processing aims to reduce its polluting load and possible production of useful fermentation products such as ethanol. So, except for whey lactose fermentation, the development of recovery processes was essential. Lactose, hydrolyzed whey syrups, whey powder, demineralised or not, and partially desugared whey protein concentrates were produced. Whey and whey products can be used in animal feed, dietetic foods, bread, confectionery, candies and beverages.

In recent years, improvements in process technology and biotechnology of lactose have been accompanied by massive changes in the scale of milk-whey/dairy processing operations, and the manufacture of a wide range of dairy and other novel related products. The increasing economic value of these products around the world reflects consumers' acceptability and increases basic research into starter cultures of micro-organisms, immobilized cell technology, manufacturing methods and mechanisation over the past couple of decades.

47.2   Chemical Properties of Lactose

Lactose (4-O-β-D-galactopyranosyl-(1-4)-β-D-glucose, $C_{12}H_{22}O_{11}$) (Adam *et al.* 2004; Gänzle *et al.* 2008), is the principal carbohydrate in milk and cheese whey. Lactose is a disaccharide derived from the condensation of D-galactose and D-glucose, which form a β-1 → 4 glycosidic linkage (Figure 47.1). Its systematic name is β-D-galactopyranosyl-(1 → 4)-D-glucose. The glucose can be in either the α-pyranose form or the β-pyranose form, whereas the galactose can

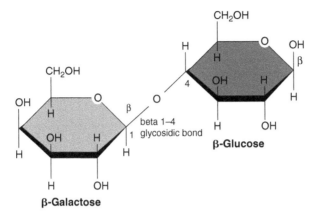

Figure 47.1   Chemical structure of β-lactose (Modified from http://academic. brooklyn.cuny.edu/biology/bio4fv/page/disaccharide.html). The chemical structure of disaccharide β-lactose is presented. Also, the glycosidic bond which joins the monosaccharide glucose to monosaccharide galactose is shown.

only have the β-pyranose form: hence α-lactose and β-lactose refer to anomeric form of the glucopyranose ring alone.

Two isomers of lactose exist α-lactose and β-lactose, which differ in their specific rotation to polarised light, ($+89.4°$ or $+35.0°$, respectively, in water at 20 °C), in their melting point and their solubility in water *etc.* The most stable form is α-lactose monohydrate, $C_{12}H_{22}O_{11} \cdot H_2O$ and the ratio of isomers is temperature dependent. The solubility and sweetness of lactose is low compared to other sugars (1/6 in relation to sucrose) (Gänzle *et al.* 2008).

In general, there are considerable differences in composition and properties of milk from the four main milk-producing species, such as cows, sheep, goats and buffalos (Table 47.1). Indeed, bovine, ovine and buffalo milk contain ~ 4.8 g lactose 100 g^{-1}, whereas caprine milk generally contains a lower level of lactose (Harper 1992). In the milk of some mammals, such as sea lions, some seals and opossums, lactose is absent, or only present in very low concentrations (Jenness and Holt 1987). Lactose is responsible for $\sim 50\%$ of the osmotic pressure of milk, which is equal to that of blood. In bovine milk, the concentration of lactose decreases progressively and significantly with lactation stage, and with increasing somatic cell count of the milk (Walstra and Jenness 1984) – in both cases due to the influx of NaCl from the blood and the resultant need to maintain the osmotic equilibrium (Fox 2003).

It must be mentioned that composition of milk and consequently whey, depends on collection period of the year, type of operations to which milk is subjected after recollection (maximal heating temperature, centrifugal velocity, homogenization process, *etc.*) and to which the whey is subjected after rennet separation (pasteurization, pre-concentration, *etc.*).

Table 47.1 Composition (%) of the milk of selected species (Modified from Fox and McSweeney 1998). The accurate chemical composition of the milk of selected species (%) is shown and it is obvious that the precise components of raw milk vary by species.

Species	Total Solids	Fat	Protein	Lactose	Ash
Human	12.2	3.8	1.0	7.0	0.2
Cow	12.7	3.7	3.4	4.8	0.7
Buffalo	16.8	7.4	3.8	4.8	0.8
Goat	12.3	4.5	2.9	4.1	0.8
Sheep	19.3	7.4	4.5	4.8	1.0
Pig	18.8	6.8	4.8	5.5	–
Horse	11.2	1.9	2.5	6.2	0.5
Donkey	11.7	1.4	2.0	7.4	0.5
Reindeer	33.1	16.9	11.5	2.8	–

Lactose has a low solubility in water and also a low sweetening power that can be advantageous in certain food applications. A significant disadvantage of lactose is the high intolerance of human organisms to it. It is estimated that approximately 70% of the world population is lactose intolerant. However, hydrolysation of lactose with the help of β-*galactosidase* facilitates its consumption.

47.2.1 Biosynthesis of Lactose

Lactose is essentially unique to mammary secretions. It is synthesized from glucose absorbed from blood. One molecule of glucose is isomerised to UDP-galactose *via* the Leloir pathway and then is linked to another molecule by lactose synthetase, a two-component enzyme leading to the synthesis of lactose.

Mammals have a control mechanism that enables termination of the lactose synthesis when necessary, mainly to regulate and control osmotic pressure when there is an influx of NaCl, especially during mastitis or in late lactation. The ability to control osmotic pressure is sufficiently important to justify an elaborate control mechanism.

47.3 Recovery of Lactose – Lactose Products and Derivatives

Lactose is a raw material for a set of products with applications in food and pharmaceutical industries. The major use of powder lactose occurs in the food industry as ingredient for various foods such as milk formulas for children and as in medicine production for the pharmaceutical industry.

Most of the lactose powder is produced from whey or whey permeate recovery by a process involving crystallization (Gänzle *et al.* 2008). More specifically, lactose production involves whey concentration or ultra-filtration permeate by vacuum concentration, lactose crystallization from the concentrate, recovery of the crystals by centrifugation and then drying. Lactose may also be recovered by precipitation with $Ca(OH)_2$, usually in the presence of ethanol, methanol or acetone. Lactose has low hydroscopicity when properly crystallized, which makes it an attractive sugar for use in icings for confectionary products.

Generally speaking, recovered lactose may exist in three categories of products: (a) as powder lactose or lactose syrup; (b) as raw material for pharmaceutical and food products – lactose derivatives; (c) as substrate for fermentation processes – fermented lactose.

Recovered lactose has several other applications in food products. The major uses for lactose include food ingredient, ingredient in infant formula, filler or a coating agent for tablets in the pharmaceutical industry and raw material for the production of added-value lactose derivatives. Specifically, lactose has been used as starting point for lactose derivatives such as lactulose (product of the "healthy food"), lactilol (sweetener for diabetic patients), lactobionic acid, lactosyl urea, galacto-oligosaccharides (GOS) (probiotic foods) and lactosucrose (Audic *et al.* 2003; Gänzle *et al.* 2008). The production of this type of derivatives results from the use of high purity lactose powder that require economically prohibitive production processes in order to reach high-value products.

Although the demand for lactose has increased recently, it is unlikely that a profitable market exists for all the lactose potentially available. Since the disposal of whey or UF permeate by dumping into waterways is no longer permitted, profitable, or at least inexpensive, ways of utilizing lactose have been sought for several years. For many years, the most promising of these was considered to be hydrolysis to glucose and galactose, but other modifications are attracting increasing attention.

47.4   Fermentation of Lactose

Besides lactose's application in food and pharmaceutical products formulation, whey lactose is used as raw material for countless products and as substrate in various fermentation processes, in particular, in some fermentation processes to obtain fuels (Borzacconia and Etchebehereb 2009; Davila-Vazquez *et al.* 2008; Guimarães *et al.* 2008; Ozmihci and Kargi 2009; Venetsaneas *et al.* 2009). As fermentation substrate, lactose has been used in fermentation processes of petrochemical and food products.

Lactose is an interesting carbon source for the production of various bioproducts, through biotechnological means, primarily because it is the major component of cheese whey, the main waste of dairy activities. Therefore, whey streams could be used as an abundant and renewable potential raw material for microbial fermentations (Panesar *et al.* 2007).

The main limitation of this type of application is the reduced number of micro-organisms able to use lactose as substrate. Despite this, various processes have been proposed and a considerable number of them have been already implemented at an industrial level. Among the various substances produced, the

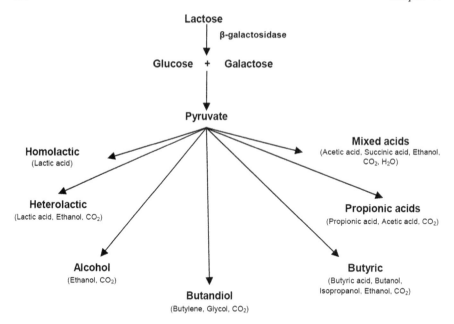

Figure 47.2 Lactose Fermentation. A simplified scheme of lactose fermentation is shown. Many biochemical changes caused during processes include action of microorganisms or enzymes on lactose substrate and significant products produced.

most relevant are ethanol, biogas (methane), bio-hydrogen, single cell protein (SCP), beverages, lactic acid, acetic acid, bio-polymers, bio-surfactants, *etc.*

Depending on the micro-organisms involved, lactose can be fermented to lactic acid, ethanol, butyric acid, propionic acid and some aromatic compounds, since is hydrolyzed first to its components glucose and galactose by the enzyme β-galactosidase (or lactase) (Figure 47.2). Enzymatic hydrolysis presents problems, as lactase is not widely found, *e.g.* only some rare yeasts do contain it (*Kluyveromyces fragilis*). Also, galactose must be transformed into glucose in order to enter the biological cycles (during anaerobic glycolysis and the pentose-phosphate cycle).

The number of micro-organisms that can use lactose as a source of carbon and energy is limited, yet include bacteria, yeasts and filamentous fungi. Bacteria have evolved different strategies for the uptake and hydrolysis of lactose (Domingues *et al.* 2010). The most effective implies the simultaneous phosphorylation and translocation of the sugar across the cell membrane, existing at least two alternative mechanisms for uptake (a lactose-proton symporter and a lactose-galactose antiporter). Once inside the bacterial cell, the phosphorylated lactose is hydrolysed by a phospho-β-galactosidase (an enzyme that recognises phosphorylated lactose). When the uptake mechanism does not involve phosphorylation, lactose is cleaved intracellularly by a β-galactosidase (Adam *et al.* 2004).

The yeasts that assimilate lactose aerobically are widespread, but those that ferment lactose are rather rare, including *e.g.*, *Kluyveromyces lactis, K. marxianus* and *Candida pseudotropicalis* (Fukuhara 2006).

47.4.1 Lactic Acid Fermentation

Lactic acid bacteria (LAB) are among the most important lactose-consuming micro-organisms, due to their occurrence in milk and dairy products. They are a cluster of Gram-positive bacteria, belong to the phylum of *Firmicutes* and count the genera *Lactobacillus, Lactococcus, Leuconostoc Pediococcus* and *Streptococcus*. The most important species are presented in Table 47.2 (Belitz *et al.* 2004). LAB may be either of the mesophilic or thermophilic type. Besides their food-related significance, the importance of LAB in biotechnology is extended to the production of lactic acid, *e.g.*, from lactose metabolism during whey fermentation (Panesar *et al.* 2007). There are two major pathways for sugar fermentation by lactic acid bacteria: homolactic (homofermentation) and heterolactic (heterofermentation) fermentation. Through homofermentation

Table 47.2 Lactic acid bacteria (LAB) (Modified from Belitz *et al.* 2004). The most important species of lactic acid bacteria fermenting milk and the proportion of produced L-Lactic acid are presented. The quantity of lactic acid depends on the bacterial strain and the culture conditions (Belitz *et al.* 2004).

Micro-organisms	L-lactic Acid (%)[a]	Remarks
Lactobacillus bulgaricus	0, 6–4	thermophilic,
L. lactis	0	homofermentative
L. delbrueckii	70	D-, L- or
L. helveticus	60	D,L-Lactic acid
L. acidophilus		
L. casei		mesophilic, homofermentative D-, L- or D,L-Lactic acid
L. brevii		heterofermentative
L. fermentum		D,L-Lactic acid
Lactococcus lactis subsp. lactis	92–99	mesophilic, homofermentative
Lactococcus lactis subsp. cremoris	99	
Leuconostoc cremoris		heterofermentative D-Lactic acid
Pediococcus acidilactici		thermophilic, homofermentative D,L-Lactic acid
Streptococcus thermophilus	99	thermophilic, homofermentative

[a]Orientation values.

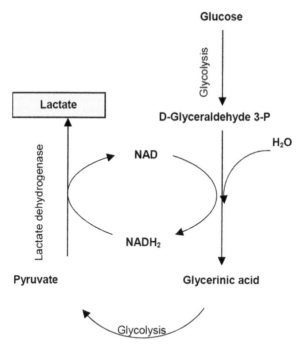

Figure 47.3 Homolactic fermentation. A simplified scheme of homolactic fermentation is illustrated, which is a major pathway for sugar fermentation in
lactic acid bacteria, where glucose is converted mainly to lactic acid.

glucose is converted mainly to lactic acid via the Embden–Meyerhof glycolytic
pathway (Figure 47.3). Through heterofermentation, glucose is converted to
lactic acid, ethanol, acetic acid and possibly CO_2 *via* a phosphoketolase-
dependent pathway (Figure 47.4). Apart from the type of fermentation both
enantiomers as D- and L-Lactic acid are formed in varying amounts depended
on the micro-organisms involved.

Also, galactose is metabolised either through the tagatose 6-phosphate
pathway or *via* the Leloir pathway by some LAB species. First it is converted to
glucose-1- phosphate which can enter the glycolytic pathway and converted to
lactic acid as well.

47.4.2 Alcoholic Fermentation

Lactose can be fermented by some species of yeast, *e.g. Kluyveromyces* spp.,
(*Kluyveromyces fragilis, K. marxianus*), to ethanol in small quantities by the
following reaction (1):

$$C_{12}H_{22}O_{11} + H_2O \rightarrow 4C_2H_5OH + 4CO_2 \qquad (1)$$

Similarly to other micro-organisms present in milk, *K. lactis* is adapted for
the efficient utilization of lactose. The ability of this yeast to metabolise lactose
results from the presence of a membrane protein named lactose permease and
β-galactosidase (Rubio-Texeira 2006). The lactose uptake in *K. lactis* is

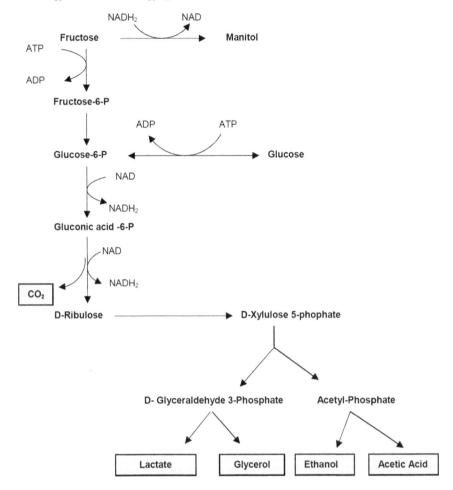

Figure 47.4   Heterolactic fermentation. A simplified scheme of homolactic fermentation is illustrated, which is a major pathway for sugar fermentation in lactic acid bacteria, where glucose is converted mainly to lactic acid, ethanol and acetic acid.

mediated by a transport system inducible by lactose and galactose (the inducer is intracellular galactose) (Dickson and Barr 1983). Uptake is mediated by a carrier and is saturated at high substrate concentrations.

47.4.3   Propionic Acid Fermentation

Lactose is metabolized by propionibacteria like *Propionibacterium acidipropionici* or *P. shermanii etc.* (Sheng-Tsiung Hsu and Shang-Tian Yang 1991) producing propionic acid by the following reaction (2):

$$1,5C_6H_{12}O_6 \rightarrow 2CH_3CH_2COOH + CH_3COOH + CO_2 + H_2O + 6ATP \quad (2)$$

Propionibacteria are gram-positive, non-sporeforming, rod-shaped, facultative anaerobes. Like most organic acid fermentations, the propionic acid fermentation is inhibited by acidic pH and the fermentation product, propionic acid.

47.4.4 Butyric Acid Fermentation

Lactose is metabolized by bacteria of the genus *Butyrivibrio* into butyric acid by the following reaction (3), since is hydrolyzed first to glucose and galactose.

$$C_6H_{12}O_6 \rightarrow C_4H_8O_2 + 2CO_2 + 2H_2 \qquad (3)$$

Mainly a type species, *Butyrivibrio fibrisolvens* ferments glucose with the production of carbon dioxide, hydrogen, and butyric, formic, and lactic acids (Figure 47.5).

The cells of bacteria of the genus *Butyrivibrio* are universally described as being gram negative, and they produce an unequivocal gram-negative reaction in the standard staining procedure.

47.4.5 Production of Aroma Substances

Apart from the main products mentioned before, during lactose fermentation, some indigenous milk micro-organisms also produce various aroma substances, which contribute to the good aroma or to the aroma defects of milk and dairy products. For example, diaketyl, ethanal, dimethylsulfide, acetic acid and various aldehydes, ketones and esters with characteristic aroma are formed as metabolic products of LAB. Methyl thioacetate has also been found in cheeses which undergo propionic acid fermentation, ethyl esters produced by *Pseudomonas fragii*, 2- or 3-methylbutanal are formed by *Strept. lactis var. maltigenes* etc. (Belitz *et al.* 2004).

47.5 Products from Fermentation-based Modifications of Lactose

Since whey utilization methods can be treated as waste management, the advantages of designing and implementing such methods are of great importance. In the last two decades, many research efforts have been undertaken and been also proposed for whey valorisation (Dominguez *et al.* 2000; 2001; 2003; Ferrari *et al.* 2001; Gomes *et al.* 2003; Panesar *et al.* 2007; Zohri 2000).

The earliest ways to dispose of whey included piping it into rivers, lakes or the ocean, funneling into caves, spreading over fields and feeding into ruminants. Whey disposal by these means provides no valuable product and also introduces direct costs of whey handling and transport.

Another major application for the lactose in whey or permeate involves its use as a substrate for the production of valuable compounds by fermentation. The classical examples are ethanol and single cell protein (SCP) production and other alternative bio-products such as biogas, organic acids, amino acids,

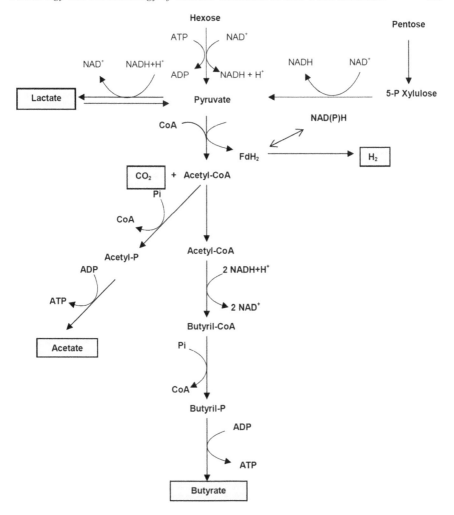

Figure 47.5 Butyric fermentation. A simplified scheme of butyric fermentation is presented, where proper micro-organisms ferment glucose with the production of carbon dioxide, hydrogen, and butyric, formic, and lactic acids.

vitamins, polysaccharides, oils, enzymes and other compounds. (Audic *et al.* 2003; Pesta *et al.* 2007; Siso 1996).

For example. lactic acid produced from whey may be used as a food acidulant, as a component in the manufacture of plastics, or converted to ammonium lactate as a source of nitrogen for animal nutrition. It can be converted to propionic acid by *Propionibacterium* spp, which has many food applications. Also, potable ethanol is being produced commercially from lactose in whey or UF permeate. The ethanol may also be used for industrial purposes or as a fuel but is probably not cost-competitive with ethanol produced by fermentation of sucrose or chemically. The ethanol may also be oxidized to acetic acid.

The mother liquor remaining from the production of lactic acid or ethanol may be subjected to anaerobic digestion with the production of methane (CH_4) for use as a fuel; several such plants are in commercial use. Lactose can also be used as a substrate for *Xanthomonas campestris* in the production of xanthan gum, which has several food and industrial applications.

47.5.1   Fermentation of Lactose to Ethanol

The first step in most procedures for cheese whey valorization consists of the recovery of the protein fraction. Alcoholic fermentation is an interesting alternative for the bioremediation of the polluting permeate that remains after separation of the whey proteins. The fermentation of lactose to ethanol using yeasts has been frequently referred in the literature since the 1940s (Rogosa *et al.* 1947; Webb and Whittier 1948; Whittier, 1944). Although the yeasts that assimilate lactose aerobically are widespread, those that ferment lactose are rather rare (Fukuhara 2006), including *e.g. Kluyveromyces lactis, K. marxianus*, and *Candida pseudotropicalis*.

Even though whey ethanol is potable and therefore can be used in foods and beverages, pharmaceutical and cosmetic industries, is hardly economically competitive with the currently established processes, using cane sugar and cornstarch as substrates, or with emerging second-generation technologies using lignocellulosic biomass as raw material.

Direct fermentation of whey or whey permeate to ethanol is generally not economically feasible because the low lactose content results in low ethanol concentrations (2–3% v/v), making the distillation process too expensive. Thus, it is important to start the fermentation with high concentration of lactose, which can be achieved by concentrating the whey. Alternatively, the sugar concentration can be increased by mixing the native whey with high-sugar condensed materials such as molasses, although in such sugar mixtures yeasts may exhibit catabolite repression and not be able to consume lactose (Oda and Nakamura 2009).

During the last 50 years, many authors have addressed the production of ethanol from lactose, mostly referring the yeasts *Kluyveromyces fragilis, K. marxianus* and *Candida pseudotropicalis*. There are a few cases of industrial plants that produce ethanol from whey or permeate, mostly using *Kluyveromyces* yeasts. Also, the use of *S. cerevisiae* for lactose fermentation has attracted much attention. The initial strategies involved the fermentation of pre-hydrolyzed lactose solutions to mixtures of glucose and galactose. Furthermore, the designing of lactose-consuming *S. cerevisiae* strains has been attempted by several strategies, such as protoplast fusion, expression of heterologous ß-galactosidases secreted to the extracellular medium or simultaneous expression of the permease and ß-galactosidase of *K. lactis* (Rubio-Texeira 2005).

Furthermore, many researchers today work whey utilization as a low-cost raw material for the production of different fermented products of high additive value by implementing modern biotechnological methods such as immobilization.

47.6   Fermentation of Lactose by Kefir

Kefir is a well-known natural mixed culture (Figure 47.6) consisting of various yeasts species (*Kluyveromyces*, *Candida*, *Saccharomyces*, and *Pichia*), various lactic acid bacteria of the genera *Lactobacillus, Lactococcus, Leuconostoc* and acetic acid bacteria, being in a unique symbiotic formation (Garrote *et al.* 1997; Witthuhn *et al.* 2005).

As was previously mentioned, most of these species convert lactose and therefore kefir has the potential for producing almost all the aforementioned fermentation products from whey as a substrate. So, many efforts have been made in the last two decades to use kefir co-culture for whey fermentation and the production of various bio-products.

In this respect, lactose converting by kefir has been evaluated for the production of potable and fuel-grade alcohol (Athanasiadis *et al.* 2002; Kourkoutas *et al.* 2002; Petsas *et al.* 2002), kefir-like whey drinks (Paraskevopoulou *et al.* 2003), lactic acid (Elezi *et al.* 2003; Kourkoutas *et al.* 2005), baker's yeast (Harta *et al.* 2004; Plessas *et al.* 2004), single cell protein (SCP) as livestock feed (Plessas *et al.* 2008), probiotic starter cultures for fermented milk products (Kourkoutas *et al.* 2005) and cheese ripening (Dimitrellou *et al.* 2007; Papavasiliou *et al.* 2008).

The effect of various conditions on lactose uptake rate in whey fermentation using kefir were previously investigated using as radiotracer ^{14}C-labelled lactose (Golfinopoulos *et al.* 2009; Golfinopoulos *et al.* 2011), in order to approach new ways for the fermentation rate increment. So, it was found that lactose

Figure 47.6   Kefir grains (Photo by M. Soupioni). The photo indicates the resilient polysaccharide matrix (kefiran) which forms the kefir grains resembling cauliflower. In these grains a community of lactic acid bacteria and yeasts are embedded in a symbiotic association.

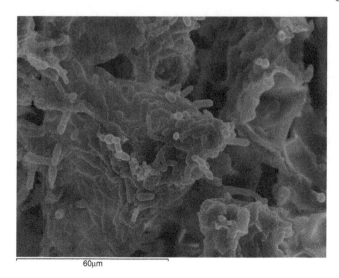

60μm

Figure 47.7   Micrographs of a scanning electron microscope showing kefir cells
immobilized on brewery spent grains. The characteristic shapes of bacteria ranging from spheres to rods are shown.

uptake rate was strongly correlated to fermentation rate and increased during
whey fermentation by free kefir cells, as temperature was increased up to 30 °C.
Moreover, the highest lactose uptake rate was recorded at 5.5 pH value, while
high cell concentration didn't play any role in the fermentation of whey and
ethanol didn't affect whey lactose uptake.

Also, studies in order to find alternative and more effective processes for
whey fermentation illustrated that whey fermentation by kefir cells immobilized
on delignified cellulosic materials (DCM) and brewery spent grains (BSG)
was faster compared to cells immobilized on gluten pellets (GP) or free cells
(Figure 47.7). So, such processes could involve cell immobilization on nano and
microtubular biopolymers, the use of promoters and the removal of ions and
mainly Ca^{++} from whey. The latter signify the fact of much higher uptake of
ions such as Ca^{++} by the cell wall, as it was observed in the case of glucose
(Akrida-Demertzi *et al.* 1991), which reduces lactose uptake by kefir.

47.7   Bio-surfactants

Surfactants are tensoactive substances of microbial origin and are usually
byproducts of microbial, fungal and yeast metabolism (Rodrigues *et al.* 2006).
They are made of amphipathic molecules constituted of a hydrophobic and a
hydrophilic portion (Rahman and Gakpe 2008).

Surfactant properties make them appropriate for a wide range of
applications in the cleaning products industry, soap and detergents production, petroleum, bio-remediation, cosmetics and hygiene products. Since,

biosurfactants present various advantages compared to synthetic surfactants they can find multiple applications even though they remain low due to high production costs. The use of cheese whey as alternative substrate which is easily available, inexpensive and major pollutant could reduce the overall bio- surfactant production cost. Several scientific attempts have been made so far, with encouraging results for potential industrial application.

47.8   Biogas

Biogas is a gaseous mixture used as an alternative energy source made mainly by methane and carbon dioxide produced by anaerobic digestion of the organic material present in a substrate by the action of bacteria (Frigon *et al.* 2009; Venetsaneas *et al.* 2009). The usual biogas composition is about 65 to 70% (v/v) methane, and 25 to 30% (v/v) carbon dioxide. However, depending on the microbiological process involved, other gases, such as hydrogen may be produced.

Biological processes for residual treatment incorporate a variety of microbial species and, therefore, present a sufficiently wide metabolic versatility. For example, some processes involve bacterial species able to degrade complex compounds and synthesized artificially, while others include bacteria that only degrade simple organic molecules, such as acetic acid, producing fuels, such as methane (CH_4) or hydrogen (H_2). Normally, substrates used for these fer- mentation reactions are simple sugars or starch that are economically impractical due to its high costs. Residual or residual waters with high con- centrations in carbohydrates can be used for economical reasons.

The anaerobic decomposition of cheese includes prior pretreatment opera- tions of filtration and fat separation. Whey is then skimmed and introduced in the bioreactor for biogas production. Usually, the produced slurries can be recovered and further used as fertilizers. The remaining liquid could be intro- duced again in the effluent treatment system, after aeration to stop the anaerobic decomposition process, since it has a low organic load (Castelló *et al.* 2009).

Summary Points

- This chapter focuses on promotional aspects for biotechnological and technological processing of lactose.
- The lactose contained in milk can be fermented by proper species of bacteria and yeasts in order to produce many innovative products.
- Kefir is a mixed culture consisting of yeasts and lactic acid bacteria, able to ferment lactose.
- Therefore, using whey as raw material, kefir has the potential for pro- duction of kefir-like whey-based drinks, potable and fuel alcohol, as well as kefir-yeast biomass for use as baker's yeast.
- Recently, the research on lactose fermentation by kefir was increased:
  o Various fermentation promoters and immobilized cell technologies were effectively used.

   ○ Starter cultures of microorganisms and manufacturing methods were strongly improved.
- Approaching new ways for the lactose fermentation rate increment lactose uptake rate by kefir during lactose fermentation was recorded by using ^{14}C-labelled lactose.

Key Facts

Table 2 This table lists the most important species of some lactic acid bacteria genera. The first lactic acid bacterium was isolated by Joseph Lister in 1873 and was the organism that we now refer to as *Lactococcus lactis*. This is a species of great significance in the fermentation of lactose for making milk products.

Figure 6 This figure clearly presents the cauliflower shape of kefir grains. The grain is a symbiotic association of microorganisms embedded in a resilient polysaccharide matrix (kefiran) which is composed of equal amounts of glucose and galactose. East Europeans produce the refreshing fermented beverage *Kefir* by inoculating milk with kefir grains. This alcoholic drink is originated at region of mountain Caucasus where its consumption is a tradition.

Definitions of Words and Terms

Cheese whey. Whey or *milk serum* is the dairy liquid waste generated in great volumes annually. It is obtained mainly after removal of cheese from curdled milk and curd ripening. Whey represents 85–90% of the milk volume and retains 55% of milk nutrients. It consists of water (93–94%) and lactose (4.8–5.5%) but also contains proteins (0.8–1.0%), fats (0.3–1.3%) and minerals ($\sim 0.6\%$).

Lactic acid bacteria. Lactic acid bacteria are micro-organisms naturally present in the intestine and very important in the fermentation of many foodstuffs. There are 16 genera of lactic acid bacteria, some 12 of which are active in a food context and produce acids and other substances.

Yeasts. Yeasts are heterotrophic organisms plentiful on the surfaces of plant tissues (including flowers and fruit) involved in many fermentation processes. Normally, the word yeast is synonymous with brewer's yeast or baker's yeast, namely, *Saccharomyces cerevisiae*, which ferments glucose to ethanol and carbon dioxide.

Kefir. Kefir yeast is a mixed culture used in the production of a traditional Russian (Caucasus region) soft alcoholic drink *Kefir* from milk. Kefir microflora is consisting of a diverse spectrum of species and genera including (83–90%) lactic acid bacteria (*Lactobacilli, Lactococci, Leuconostoc*), (10–17%) yeasts (*Kluyeveromyces, Candida, Saccharomyces, Pichia*) and sometimes acetic acid bacteria (*Acetobacter*). The various micro-organisms, sharing symbiotic relationships, are embedded in a polysaccharide matrix (kefiran).

BOD. BOD is an abbreviation for Biological Oxygen Demand, a parameter of organic pollution applied to waters that involves the measurement of the dissolved oxygen used by microorganisms in the oxidation of organic matter.

COD. COD is an abbreviation for Chemical Oxygen Demand, a test used to measure the content of organic pollution of waters that measures the oxygen equivalent of the organic matter that can be oxidized by using a strong chemical oxidizing agent (*e.g.* dichromate, $Cr_2O_7^{-2}$).

Myzithra, manouri, anthotyros. The myzithra, manouri and anthotyros are traditional Greek semi-soft unpasteurized fresh white cheeses made from milk or/and the drained whey from feta production (sheep or goat milk). They are similar to feta cheese but creamier and less salty or salt-free.

Homolactic and heterolactic fermentation. Homolactic and heterolactic fermentation are two major pathways for sugar fermentation in lactic acid bacteria. Through homolactic fermentation glucose is converted mainly to lactic acid via the Embden–Meyerhof glycolytic pathway and through heterolactic fermentation glucose is converted to lactic acid, ethanol and acetic acid via a phosphoketolase-dependent pathway.

Xanthan gum. Xanthan gum is a polysaccharide which can be regarded as a cellulose derivative. Its main chain consists of 1,4 linked β-glucopyranose residues, glucose and mannose. The xanthan gum molecular weight is $> 10^6$ but is quite soluble in water. It is useful in food canning, in instant puddings, in the production of salad dressings and improves the freeze-thaw stability of starch gels.

Microorganisms cell immobilization. Micro-organisms cell immobilization is the localization of the cells on surfaces of different kinds of support materials with preservation of some desired catalytic activity. Numerous biotechnological processes are advantaged by immobilization techniques and therefore several such techniques and support materials have been proposed.

SCP. SCP is an abbreviation for "Single Cell Protein", a term used to describe microbial biomass derived from either uni-and multicellular organisms (bacteria, yeasts, filamentous fungi or algae) cultivated specifically as food-stuffs (food or feed additives).

Starter cultures. Starter cultures consist of the relevant micro-organisms that are inoculated directly into food materials to overwhelm the existing flora and bring about desired changes in the finished fermented material.

UF permeate. UF permeate is an abbreviation for Ultra Filtration permeate, which means the output from an ultra filtration unit.

Lactulose. Lactulose is a synthetic disaccharide obtained from lactose through an isomerization reaction. Lactulose is widely used in the pharmaceutical industry as a laxative and anti-hyperammonaemic against diseases like acute and chronic constipation.

Lactilol. Lactilol is a disaccharide made of galactose and sorbitol that is used as osmotic laxative. Normally, reaches colon without any change and then interacts with the intestinal flora producing carboxylic acids such as lactic, butyric, propionic, and acetic, thus reducing the absorption of ammonia and increases osmolarity.

Lactobionic acid. Lactobionic acid is an organic acid resulting from the chemical or microbial oxidation of lactose that finds application in cosmetics due to its anti-aging, antioxidant and healing properties. It is also used in the preservation of organs to be transplanted.

Galacto-oligossaccharides (GOS). Galacto-oligossaccharides are known as prebiotic additives that increase in population of probiotic bacteria in gastrointestinal tract, has antagonistic effect and suppress the activity of putrefaction bacteria and formation of toxic products by fermentation. They are obtained from enzymatic transgalactosilation of lactose.

WPC. WPC is an abbreviation for Whey Protein Concentrates

List of Abbreviations

BOD	Biological Oxygen Demand
COD	Chemical Oxygen Demand
LAB	Lactic acid bacteria
SCP	Single Cell protein
GOS	Galacto-oligossaccharides
UF permeate	Ultra Filtration permeate
WPC	Whey Protein Concentrates
UDP-Gal	Uridine Diphosphate Galactose
DCM	Delignified Cellulosic Materials
BSG	Brewery Spent Grains
GP	Gluten Pellets

References

Adam, A.C., Rubio-Texeira, M., and Polaina, J., 2004. Lactose: the milk sugar from a biotechnological perspective. *Critical Reviews in Food Science and Nutrition.* 44: 553–557.

Akrida-Demertzi, K., and Koutinas, A.A., 1991. Optimization of sucrose ethanol fermentation for K, Na, Ca and Cu metal contents. *Applied Biochemistry and Biotechnology.* 30: 1–7.

Athanasiadis, I., Paraskevopoulou A., Blekas G., and Kiosseoglou V., 2004. Development of a novel whey beverage by fermentation with kefir granules. Effect of various treatments. *Biotechnology Progress.* 20: 1091–1095.

Athanasiadis, I., Boskou, D., Kanellaki, M., Kiosseoglou, V., and Koutinas, A.A., 2002. Whey liquid waste of the dairy industry as raw material for potable alcohol production by kefir granules. *Journal of Agricultural and Food Chemistry.* 50: 7231–7234.

Audic, J.L., Chaufer, B., and Daufin, G., 2003. Non-food applications of milk components and dairy co-products: a review. *Lait.* 83: 417–438.

Belitz, H.-D., Grosch, W., and Schieberle P., 2004. *Food chemistry.* Springer-Verlag, Berlin, Germany.

Castelló, E., Santosa C., Iglesias T., Paolino G., Wenzelb J., Borzacconia, L. and Etchebehereb C., 2009. Feasibility of biohydrogen production from cheese whey using a UASB reactor: Links between microbial community and reactor performance. *International Journal of Hydrogen Energy*. 34: 5674–5682.

Davila-Vazquez, G., Arriaga, S., Alatriste-Mondragón, F., León-Rodríguez, A., Rosales-Colunga, L., and Razo-Flores, E., 2008. Fermentative bio-hydrogen production: trends and perspectives. *Reviews in Environmental Science and Biotechnology*. 7: 27–45.

Dickson, R.C., and Barr, K., 1983. Characterization of lactose transport in *Kluyveromyces lactis. Journal of Bacteriology*. 154: 1245–51.

Dimitrellou, D. Kourkoutas, Y., Banat, I.M., Marchant, R., and Koutinas, A.A., 2007. Whey cheese production using freeze-dried kefir co-culture as a starter. *Journal of Applied Microbiology*. 103: 1170–1183.

Domingues, L., Dantas, M.M., Lima,N., and Teixeira, J.A., 1999. Continuous ethanol fermentation of lactose by a recombinant flocculating *Saccharomyces cerevisiae* strain. *Biotechnology and Bioengineering*. 64: 692–697.

Domingues, L., Guimaraes, P.M.R., and Oliveira, C. Metabolic engineering of *Saccharomyces cerevisiae* for lactose/whey fermentation. *Bioengineered Bugs*. 2010.

Elezi, O., Kourkoutas, Y., Koutinas, A.A., Kanellaki, M., Bezirtzoglou, E., Barnett, Y.A., and Nigam, P., 2003. Food additive lactic acid production by immobilized cells of *Lactobacillus brevis* on delignified cellulosic material. *Journal of Agricultural and Food Chemistry*. 51: 5285–5289.

Ferrari, D.M., Bianco, R., Froche, C., and Loperena, L.M., 2001. Baker's yeast production from molasses/cheese whey mixtures. *Biotechnology Letters*. 23: 1–4.

Fox, P., 2003. Milk Proteins: General and Historical Aspects in P.F. Fox and P.L.H., McSweeny, ed, *Advanced Dairy Chemistry*. 3rd ed. Kluwer Academic, New York.

Fox, P.F., and McSweeney, P.L.H., 1998. *Dairy chemistry and biochemistry*. Glasgow, U.K.: Blackie Academic & Professional.

Frigon, J., Breton, J., Bruneau, T., Moletta, R., and Guiot, S., 2009. The treatment of cheese whey wastewater by sequential anaerobic and aerobic steps in a single digester at pilot scale. *Bioresource Technology*. 100: 4156–4163.

Fukuhara, H., 2006. *Kluyveromyces lactis*—a retrospective. *FEMS Yeast Research*. 6: 323–4.

Gänzle, M.G., Haase, G., and Jelen, P., 2008. Lactose: crystallization, hydrolysis and value-added derivatives. *International Dairy Journal*. 18: 685–694.

Garrote, G.L., Abraham, A.G., and De Antoni, G.L., 1997. Preservation of kefir grains, a comparative study. *Lebensmittel-Wissenschaft Und-Technologie*. 30: 77–84.

Golfinopoulos, A., Kopsahelis, N., Tsaousi, K., Koutinas, A.A., and Soupioni, M. 2011. Research perspectives and role of lactose uptake rate revealed by its study using 14C-labelled lactose in whey fermentation. *Bioresource Technology*. 102: 4204–4209.

Golfinopoulos, A., Papaioannou, L., Soupioni, M., and Koutinas, A.A., 2009. Lactose uptake rate by kefir yeast using 14C- labelled lactose to explain kinetic aspects in its fermentation. *Bioresource Technology*. 100: 5210–5213.

Gomes, M., Mota, M.J.C., and Teixeira, J.A., 2003. Cheese whey treatment and valorization process with continuous ethanolic fermentation. European patent, EP1041153, 09/24/2003.

Guimarães, P., Teixeira, J., and Domingues, L., 2008. Fermentation of high concentrations of lactose to ethanol by engineered flocculent *Saccharomyces cerevisiae*. *Biotechnol Letters*. 30: 1953–1958.

Harper, W.J., 1992. Functional properties of whey protein concentrates and their relationship to ultrafiltration. In *New Applications of Membrane Processes*. pp. 77–108. IDF Special Issue, 9201.

Harta, O., Iconomopoulou, M., Bekatorou, A., Nigam, P., and Kontominas, M. 2004. Effect of various carbohydrate substrates on the production of kefir grains for use as a novel baking starter. *Food Chemistry*. 88: 237–242.

Jenness, R., and Holt, C., 1987. Casein and Lactose concentrations in milk of 31 species are negatively correlated. *Experientia*. 43: 1015–1018.

Kosikowski, F.V., 1979. Whey utilization and whey products. *Journal of Dairy Science*. 62: 1149–1160.

Kourkoutas, Y. Kanellaki, M., Koutinas, A.A., and Tzia, C., 2005. Effect of fermentation conditions and immobilization supports on the wine-making. *Journal of Food Engineering*. 69:115–123.

Kourkoutas, Y. Psarianos, C., Koutinas, A.A., Kanellaki, M., Banat, I.M., and Marchant, R., 2002. Continuous whey fermentation using kefir yeast immobilized on delignified cellulosic material. *Journal of Agriculture and Food Chemistry*. 50: 2543–2547.

Oda, Y., and Nakamura, K., 2009. Production of ethanol from the mixture of beet molasses and cheese whey by a 2-deoxyglucose-resistant mutant of *Kluyveromyces marxianus*. *FEMS Yeast Research*. 9: 742–748.

OECD-FAO Agricultural Outlook 2008–2017 Highlights. Paris: Organisation for Economic Co-operation and Development—Food and Agriculture Organization of the United Nations, 2008.

Ozmihci, S., and Kargi, F., 2009. Fermentation of cheese whey powder solution to ethanol in a packed-column bioreactor: effects of feed sugar concentration. *Chemical Technology and Biotechnology*. 84: 106–111.

Panesar, P.S., Kennedy, J.F., Gandhi, D.N., and Bunko, K., 2007 Bioutilisation of whey for lactic acid production. *Food Chemistry*. 105:1–14.

Papavasiliou, G., Kourkoutas, Y., Rapti, A., Sipsas, V., Koutinas, A.A., and Soupioni, M., 2008. Freeze-dried kefir co-culture using whey. *International Dairy Journal*. 13: 247–254.

Paraskevopoulou, A., Athanasiadis, I., Blekas,G., Koutinas, A.A., Kanellaki, M., and Kiosseoglou, V., 2003. Influence of polysaccharide addition on stability of a cheese whey kefir-milk mixture. *Food Hydrocolloids*. 17: 615–620.

Pesta, G., Meyer-Pittroff, R., and Russ, W., 2007. Utilization of whey. In: V. Oreopoulou and W. Russ, Editors, *Utilization of By-products and Treatment of Waste in the Food Industry*. Springer.

Petsas, I., Psarianos, K., Bekatorou, A., Koutinas, A.A., Banat, I.M., and Marchant, R., 2002. Improvement of kefir yeast by mutation with N-methyl-N-nitrosoguanidine. *Biotechnology Letters*. 24: 557–560.

Plessas, S., Bosnea, L.A., Psarianos, C., Koutinas, A.A., Marchant, R., and Banat, I.M., 2008. Lactic acid production by mixed cultures of *Kluyveromyces marxianus, Lactobacillus delbrueckii ssp. bulgaricus* and *L. helveticus*. *Bioresource Technology*. 99: 5951–5955.

Plessas, S., Pherson, L., Bekatorou, A., Nigam, P., and Koutinas, A.A., 2004. Bread making using kefir grains as baker's yeast. *Food Chemistry*. 93: 585–589.

Rahman, P., and Gakpe E., 2008. Production, Characterization and applications of biosurfactants – review. *Biotechnology*. 7: 360–370.

Rodrigues, L., Moldes, A., Teixeira, J., and Oliveira, R., 2006. Kinetic study of fermentative biosurfactant production by Lactobacillus strains. *Biochemical Engineering Journal*. 28: 109–116.

Rogosa, M., Browne, H.H., and Whittier, E.O., 1947. Ethyl alcohol from whey. *Journal of Dairy Science*. 30: 263–269.

Rubio-Texeira, M., 2006. Endless versatility in the biotechnological applications of Kluyveromyces *LAC* genes. *Biotechnology Advances*. 24: 212–25.

Sheng-Tsiung Hsu and Shang-Tian Yang. 1991. Propionic acid fermentation of lactose by *Propionibacterium acidipropionici*: Effects of pH. *Biotechnology and Bioengineering*. 38: 571–578.

Siso, M.I.G., 1996. The biotechnological utilization of cheese whey: a review. *Bioresource Technology*. 57: 1–11.

Smithers, G.W., 2008. Whey and whey proteins—from gutter-to-gold. *International Dairy Journal*. 18: 695–704.

Tamime, A.Y., and Robinson, R.K. 1999. *Yoghurt science and technology* (2nd ed). CRC Press, New York.

Venetsaneas, N., Antonopoulou, G., Stamatelatou, K., Kornaros, M., and Lyberato, G., 2009. Using cheese whey for hydrogen and methane generation in a two-stage continuous process with alternative pH controlling approaches. *Bioresource Technology*. 100: 3713–3717.

Walstra, P., and Jenness, R., In: P. Walstra and R. Jenness, Editors, *Dairy Chemistry and Physics*, Wiley, New York (1984), pp. 186–197.

Webb, B.H., and Whittier, E.O., 1948. The utilization of whey: a review. *Journal of Dairy Science*. 31: 139–164.

Whittier, E.O., 1944. Lactose and its utilization: a review. *Journal of Dairy Science*. 7: 505–537.

Witthuhn, R., Schoeman, T., and Britz, T., 2005. Characterization of the microbial population at different stages of Kefir production and Kefir grain mass cultivation. *International Dairy Journal*. 15: 383–389.

Zohri, A.A., 2000. Glycerol production from cheese whey by selected fungal cultures. *Journal of Food Science and Technology*. 37: 533–538.

CHAPTER 48

Lactose Intolerance and the Consumption of Dairy Foods

JEANETTE N. KEITH*[a] AND RAVI CHHATRALA[b]

[a] Director, Gastroenterology and Bariatric Medicine, Cooper Green Mercy Hospital, 1515 6th Avenue South, Birmingham, AL 35233, US; [b] Internal Medicine Residency Program, Department of Internal Medicine, University at Buffalo, The State University of New York, 462 Grider Street, Buffalo, NY 14215, US, Email: ravichhatrala@hotmail.com
*Email: jeanettenkeith@gmail.com

48.1 Introduction

Lactose intolerance (LI) is a clinical syndrome associated with preventable health consequences. The b-galactosidase, lactase-phlorizin hydrolase, or "lactase" (LPH), is required for digestion of lactose. Inadequate or low levels of the enzyme (aka: lactase non-persistence, LNP) is associated with impaired lactose digestion in adults as complete absence of LPH is rare. Under cultural and societal influences, individuals perceive themselves to be lactose intolerant (LI). Often, objective testing does not support the diagnosis of LNP despite symptoms. Interestingly, many dairy consumers have LNP and are asymptomatic. Misperceptions of tolerance negatively impact the consumption of dairy foods, particularly in populations with a high prevalence of LNP. Recent data also challenge our understanding of the prevalence of true lactose intolerance, suggesting that it is not as prevalent as once reported. Further, during the 2010 National Institutes of Health (NIH) Lactose Intolerance Consensus Development Conference, it was noted that confusion in the literature regarding the

Food and Nutritional Components in Focus No. 3
Dietary Sugars: Chemistry, Analysis, Function and Effects
Edited by Victor R Preedy
© The Royal Society of Chemistry 2012
Published by the Royal Society of Chemistry, www.rsc.org

synonymous use of distinctly different terms related to lactose digestion negatively impacts clinical care. Scientific advances show adverse health consequences associated with dairy avoidance and improvement in health outcome parameters with a dairy-rich diet. This chapter will review the clinical significance of lactase non-persistence, define the clinical syndrome of lactose intolerance, identify methods of diagnosing lactose intolerance used in clinical practice, outline the health implications of dairy food avoidance, and describe specific strategies to manage lactose intolerance in the clinical setting.

48.2   Clinical Significance of Lactase Non-persistence

48.2.1   Lactose Digestion

Lactase is found primarily on the intestinal brush border expressed in absorptive enterocytes on the villi of the small intestine as noted in Figure 48.1. The expression of lactase is quantified graphically along the intestinal axis, producing a bell-shaped image with the highest concentration in the jejunum and gradual decline on either side, proximally towards the duodenum and distally towards the terminal ileum. *In utero*, the expression of lactase is minimal prior to 24 weeks gestation but slowly increases during the third trimester and peaks at term (Robayo-Torres and Nichols 2007). However, in some populations, LNP or hypolactasia occurs, resulting in dose-dependent malabsorption of lactose (aka: lactose maldigestion, LM). Several different forms of hypolactasia or lactase deficiency exist.

48.2.2   Lactase Deficiency

The various forms of lactase deficiency have been previously defined in the literature. Early recognition, confirmation of diagnosis and evidence-based

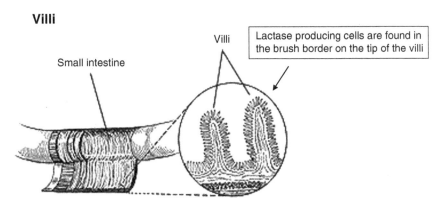

Figure 48.1   Small intestine villi.
Source: http://www.freeclipartnow.com/science/medicine/anatomy/. Accessed 14 January 2012.

Table 48.1 Clinical management of lactase deficiency by form of deficiency.

Form	Defect	Dairy Food Tolerance	Clinical Management
Congenital lactase deficiency	Complete absence of lactase from birth	Lifelong intolerance	Dairy food substitutes, enzyme replacement
Familial lactase deficiency	Production of a non-functional lactase enzyme despite normal levels	Lifelong intolerance	Dairy food substitutes, enzyme replacement
Primary lactase deficiency	Normal, physiologic age-related decline in lactase activity	Majority Dairy Tolerant	Lactose intolerant individuals respond well to clinical interventions including dietary modification and colonic adaptation
Secondary lactase deficiency	Temporary decline in lactase activity due to mucosal injury	Self-limited intolerance	Responds to dairy food avoidance during the acute injury; tolerance returns within 3–4 weeks of mucosal healing
Developmental lactase deficiency	Low lactase activity due to lack of intestinal development in premature infants	Self-limited intolerance	Tolerance is based on gut maturity

Used with permission: J. Keith, M.D.

interventions are essential for optimal treatment outcomes. Clinical management strategies for treating lactase deficiency based on the specific defects are outlined in Table 48.1.

48.2.2.1 Congenital Lactase Deficiency

True congenital lactase deficiency (CLD) is a rare life-threatening condition and only two autosomal recessive forms have been reported. The first genetic form of CLD was described in breast-fed siblings in 1959 by Holzel *et al.* (Holzel 1959) and is due to defective lactase production resulting in little or no active lactase in the small intestine. The second form of CLD is "familial lactase deficiency" (FLD). In contrast to CLD, these infants have normal enzyme levels but produce a defective lactase enzyme. In both cases, the infants typically present early in life with foul-smelling diarrhea, colicky abdominal pain, weight loss and malnutrition (Holzel 1964). Treatment typically requires lactose avoidance with nutritional supplementation and lactase enzyme replacement.

48.2.2.2 Primary Lactase Deficiency

This age-related decline in LPH is a normal physiologic change. Complete loss of LPH is rare; thus, maintenance of residual LPH activity in adults allows for a person-specific amount of lactose to be consumed with minimal or no symptoms. Individuals present with non-specific symptoms that typically include diarrhea, abdominal bloating, gaseous distension, nausea and pain. The physical exam is often normal. Treatment includes lactase supplementation and strategies that promote adequate consumption of dairy foods to meet the recommended daily intake for calcium and vitamin D (see section 48.5).

48.2.2.3   Secondary Lactase Deficiency

Lactase-producing cells are particularly prone to damage and require a longer time to recover following injury (Holzel 1968). When injury occurs, there is a self-limited reduction in lactase activity known as secondary hypolactasia or secondary lactase deficiency (SLD). Any cause of intestinal mucosal injury such as gastroenteritis, celiac disease, chemotherapy, radiation, nonsteroidal anti-inflammatory drugs (NSAIDs), *etc.* can give rise to SLD. The hypolactasia usually reverses within 3–4 weeks of recovery from the inciting event. Therefore, true lactose intolerance does not typically occur. In the case of preterm babies, there is a specialized form of SLD, developmental lactase deficiency (DLD). Intestinal lactase activity in premature infants varies from nearly absent at 24 weeks to 70% by weeks 34–35 (Robayo-Torres and Nichols 2007). There is a transient intolerance to milk with the severity determined by gestation at birth. While lactase supplementation or lactose avoidance is required during early development, lactose consumption is generally well tolerated once the GI tract matures.

48.2.3   Lactose Maldigestion

48.2.3.1   Clinical Measures of LM

In the setting of LNP, ingested but unabsorbed lactose reaches the colon, where colonic bacteria digest the lactose generating methane gas (CH_4), hydrogen gas (H_2) and short-chain fatty acids (SCFA). This intra-luminal phenomenon is known as LM and is usually asymptomatic. In the office setting, LM may be identified by objective, but indirect measures such as the lactose tolerance test (LTT), lactose hydrogen breath test (HBT), and the stool acidity test (SAT). For infants and young children, the LTT, HBT, SAT and the lactose challenge test (LChT) are performed. Occasionally, small bowel biopsies are obtained (Heyman 2006). For the HBT to be effective, the person being tested must be old enough to blow up a balloon. The SAT measures fecal pH to identify reducing substances in feces such as SCFA from malabsorbed carbohydrates. However, the test is not specific for lactose malabsorption and is less sensitive; thus, careful clinical correlation is needed for accurate interpretation of test results.

In comparison studies, the HBT has been shown to be superior to LTT as a measure of LM (Arvanitakis *et al.* 1977; Hovde and Farup 2009; Newcomer *et al.* 1975). In the adult population, HBT is most commonly used for diagnostic purposes. However, the HBT is time sensitive and a little cumbersome to perform, limiting its clinical use. It should be noted that a test dose of 50 grams of lactose was historically used as the gold standard. This amount of lactose in the test dose is equivalent to 4 glasses of milk and when placed in water, is the least absorbable form for lactose. In the more recent literature, a test dose of 25 grams or a 2 cup equivalent, while still supra-physiologic, is the current gold standard for testing, which affects the reported prevalence rates for LM. The sensitivity and specificity of the test is 69–100% and 89–100%, respectively (Arola 1994).

False positive results are rare unless there is bacterial overgrowth in the small intestine or surgically altered anatomy as in bariatric surgery. The false negativity of the HBT is 5–15%. False negative results can be due to an absence of H_2 producing bacteria in the colon, recent antibiotic use, *etc.* Sometimes, colonic bacteria convert H_2 to CH_4, giving rise to false negative test results. It can be avoided by simultaneously measuring CH_4 along with H_2. The test is considered positive if CH_4 increases by at least 12 parts per million from baseline or combined H_2 and CH_4 increase by 15 parts per million.

48.2.3.2   Prevalence of LM

Schrimshaw *et al.* reviewed 97 studies from 45 countries and found that more than two-thirds of the adult population worldwide has LM based on LTT or HBT, using the 50 gram test dose (Schrimshaw and Murray 1988). When compared to European populations, certain ethnic groups were found to have a higher prevalence of LM. Specifically, in Africans and African Americans, the prevalence of LM was 65–100%; in Latin American populations, LM ranged from 45–94% and was noted to be as high as 100% in Asian populations. Yet, the presence of LM does not necessarily mean that person has LI. Therefore, the identification of LM requires correlation with symptoms to avoid overestimating the prevalence of true LI.

48.3   Lactose Intolerance

48.3.1   Clinical Definition of LI

In February 2010, a consensus development panel was assembled by the National Institutes of Health to address the growing concerns regarding the health impact of LI. LM and LI are clinically distinct conditions even though both share the same pathophysiology, which is hypolactasia. As part of the proceedings, LI was defined as the symptomatic lactose ingestion requiring the presence of clinical symptoms and documentation of LM using an objective measure (Brannon *et al.* 2010). Therefore, LM and LI should not be used interchangeably.

48.3.2   Prevalence of LI

The prevalence of LI based on measures of LM had been reported in earlier literature to be as high as 80–90% in adults with LNP (Bayless and Rosensweig 1966). Other investigators noted a lower prevalence of LI (Savaiano *et al.* 2006). Notably, two recent studies document that the prevalence of self-reported LI is much lower than previously reported. Keith *et al.* surveyed 2016 African Americans regarding perceptions of tolerance to dairy foods. The authors found that 24% of the adults had self-described LI as opposed to 80–90% anticipated based on the earlier literature, and of those, only 19% had been diagnosed by a healthcare professional. Interestingly enough, 87% of the respondents failed to consume the recommended daily amount of dairy foods, citing LI as the primary reason (Keith 2011). In a national survey by Nicklas *et al.* only 12.3% of the 3452 adult participants reported LI. Racial differences in prevalence of self-described LI were noted, as 7.72% of Caucasian, 10.05% of Hispanics and 19.5% of African Americans reported symptoms (Nicklas *et al.* 2009). Thus, the true prevalence of LI is unknown but is less than previously reported.

48.3.3 Clinical Assessment

The symptoms of LI are non-specific. The development of symptoms depend on multiple factors, such as amount of lactose ingested at the time, gastric emptying time, the lactase activity level, foods co-ingested, degree of colonic adaption, perception of tolerance, and the degree of fermentation of the lactose containing food. Among these factors, the most important one appears to be the amount of lactose ingested (Wilt *et al.* 2010).

Thorough evaluation of symptomatic patients is recommended as the physical exam is often unrevealing. The differential diagnosis for LI as outlined in Table 48.2 includes common diseases such as celiac disease and food allergies (Gracey and Burke 1973). Food allergy is the result of an abnormal immune response triggered in the host after an ingestion of a particular food (Holzel 1968).The reaction is to the milk protein as opposed to the milk sugar but may mimic symptoms of LI. Because of the potential for severe reactions and anaphylaxis, early diagnosis of food allergy is essential as dairy avoidance along with nutrient supplementation would be indicated.

Table 48.2 Differential diagnosis for lactose intolerance.

Antibiotic associated diarrhea	Gallbladder disease
Anxiety disorders	Gastroesophageal reflux disease (GERD)
Bacterial overgrowth	Gastrointestinal dysmotility disorder
Carcinoid syndrome	Giardiasis
Celiac disease	Infectious diarrhea (viral, parasitic, bacterial)
Depression	Inflammatory bowel disease
Diabetic diarrhea	Irritable bowel disease
Diabetic gastroparesis	Malignancy
Disaccharidase deficiency (fructose, etc.)	Mesenteric ischemia
Diverticular disease (small bowel, colonic)	Mixed connective tissue disease
	Non-diabetic gastroparesis
Eosinophilic gastroenteritis	Scleroderma
Fiber intolerance	Soy intolerance
Food allergies	Zollinger Ellison Syndrome

Used with permission: J. Keith, M.D.

In clinical care, some clinicians use the "lactose challenge test" (LChT) to rapidly identify individuals with LI. Individuals consume a dairy free diet until symptoms abate. Next, the person is asked to consume a large quantity of lactose in a single serving on an empty stomach. The challenge is considered positive if symptoms of gastrointestinal distress are induced following the lactose ingestion. During the challenge, the amount of lactose consumed typically exceeds the lactase activity in a potentially dairy tolerant individual with modest residual LPH activity. It often induces symptoms with such severity that symptomatic individuals elect to become dairy avoiders. The primary clinical limitations of the LChT is that it has never been validated as a diagnostic test for LI; nor does it distinguish LI from secondary causes such as celiac disease and may promote misperceptions of tolerance. There is a modified version of the LChT that incorporates the use of lactose free products or lactase supplementation with gradual reintroduction of dairy foods to determine if associated GI symptoms are related to lactose ingestion. Although this avoids the negative nutritional implications of dairy food avoidance and is better tolerated by individuals, this approach has also not been validated in a clinical trial but may be useful, especially in pediatric populations (Heyman 2006). If the LChT approach is used, it should be done on a limited basis followed by re-introduction of dairy foods to the level of tolerance to avoid the health consequences of dairy avoidance. The next section will review four of the major health consequences associated with dairy avoidance.

48.4   Clinical Implications of LI

48.4.1   Health Consequences of Dairy Avoidance

48.4.1.1   Diet Quality

In a 2006 study, the 7-day diet diaries of 272 healthy pre-menopausal women were reviewed to assess their diet quality. The diets were scored by giving one point for each of the nine key nutrients consumed at 70% or more of the recommended daily allowance. When diet quality relative to calcium intake was assessed, 151 women had low calcium intakes along with low dairy food intake based on a diet quality score below 5 or poor. Of the remaining participants, 121 women had high diet quality scores that were associated with high calcium intake from dairy foods. Only 10% of participants had low diet quality scores when consuming the recommended servings of dairy foods while over 50% of the dairy avoiders had poor diet quality scores. When calcium supplements were added to the 151 women with low calcium (*i.e.* low dairy) intakes, 56 participants still had low diet quality scores. This study highlights the importance of adequate dairy food intake for improvement of diet quality and emphasizes the superiority of whole foods to supplementation alone relative to total diet quality (Weaver and Heaney 2006).

48.4.1.2   Osteoporosis

Decreased bone mineral density that results in osteopenia and osteoporosis affects all races and ethnic groups including African Americans, Native Americans and Hispanics (Siris *et al.* 2001). Low intakes of calcium and vitamin D have been identified as risk factors for its development (NAMS 2010). Low dairy food intake in individuals with LI has been associated with decreased bone density, increasing the risk for osteoporosis (Heaney *et al.* 2011). The National Osteoporosis Foundation estimates that over 10 million Americans of all races have osteoporosis and in 2005, osteoporosis-related hip fractures accounted for $19 billion dollars in health care costs. Adequate intake of calcium and vitamin D, especially during peak bone building years is preventative (NOF 2012).

48.4.1.3   Metabolic Syndrome and Cardiovascular Disease

In a recent prospective study of 3417 participants, dietary parameters with a focus on calcium and dairy foods were correlated with metabolic parameters and their changes over the subsequent 9 years. Foods were divided into two food groups: (1) cheese and, (2) milk and milk products except cheese. Total dairy product consumption was defined as the combination of the two food groups. The investigators found that total dairy product consumption, dairy (minus cheese) and dietary calcium were inversely associated with the development of metabolic syndrome, and impaired glucose tolerance. Cheese consumption was inversely associated with metabolic syndrome but not with impaired glucose tolerance or type 2 diabetes. Higher total dairy and cheese consumption were associated with lower insulin levels and a lower increase in waist circumference. Total dairy food products, cheese, milk and milk products were associated with lower diastolic blood pressure and triglycerides with a lower gain in BMI. These are additional data that demonstrate a role for dairy foods as protective against metabolic syndrome, diabetes, hypertension, dyslipidemia and obesity. These findings are further supported by Nicklas *et al.* who recently found that individuals with LI had lower levels of intake of calcium and were more likely to have physician diagnosed diabetes and hypertension. The odds of these diagnoses were decreased by one-third for each 1000 mg per day increase in calcium intake highlighting the need to effectively manage LI (Nicklas *et al.* 2011).

48.4.1.4   Vitamin D Deficiency

Vitamin D is a metabolically active nutrient that is involved in maintaining bone health by regulating the uptake of calcium from the GI tract. Deficiency of vitamin D has been linked to osteoporosis, rickets, metabolic syndrome, certain cancers, hypertension, diabetes and multiple sclerosis (NIH 2011). African Americans who are LI are 383 times more likely to be vitamin D deficient increasing their risk for these diet-preventable conditions (Wooten and Price 2004). In November of 2010, the Institute of Medicine addressed increasing concerns by the United States and Canadian governments regarding calcium and vitamin D nutriture. The report outlines the increased need for both calcium and vitamin D in the diet. The primary dietary sources of calcium and vitamin D in the Unites States and Canada are dairy foods (IOM 2011). In July 2011, the Endocrine Society published updated practice guidelines encouraging increased vitamin D consumption from supplementation, sun exposure and adequate consumption of dietary sources such as dairy foods (Holick *et al.* 2011). Therefore, dairy avoidance for any reason, including LI, increases the risk for these preventable conditions.

48.4.2 Impact on Public Health

After a review of the existing data, the 2010 Lactose Intolerance Consensus Development Panel reported "The public health burden from deficiencies attributable to lactose intolerance has not been established. However, many adults and children who avoid dairy products—which constitute a readily accessible source of calcium, vitamin D, and other nutrients—are not ingesting adequate amounts of these essential nutrients. For example, most African American adolescents consume inadequate amounts of calcium and vitamin D because they avoid dairy products. Deficient intakes of calcium and vitamin D are risk factors for decreased bone mineral density. This may increase the risk of fracture throughout the life cycle, especially in postmenopausal women. Very low intake of vitamin D can lead to the development of rickets, especially in children of African descent and other highly pigmented individuals. Although reduced-lactose dairy and non-dairy alternative products are typically fortified with calcium, vitamin D, and other nutrients, they may be more expensive and less widely available than conventional dairy products. The bioequivalence of these and other calcium supplements is uncertain."

This statement from the NIH Lactose Intolerance Consensus Statement panel is particularly worrisome when considering dairy food consumption data in the United States. As a group, African Americans consume 1.1 servings per day of dairy foods and Hispanics consume 1.5–1.9 servings per day depending on nationality and acculturation (Fulgoni *et al.* 2007; Sharma *et al.* 2004). Notably, the prevalence of diet preventable diseases associated with inadequate calcium and vitamin D are more common in these two populations.

Perceived LI has a transgenerational effect because parents with perceived LI may restrict asymptomatic children from consuming a lactose-containing diet. Keith *et al.* (2011) found that 43% of their respondents avoided milk and perceived themselves to be LI due to word of mouth or because a family member had been diagnosed. Of concern, 79% of those with LI reduced or eliminated dairy foods without supplementation, whereas only 21% added lactose-free dairy foods to the diet. Therefore, evidence-based management strategies should be incorporated in medical practices.

48.5   Clinical Management of LI

48.5.1   From Bench to Bedside

Myths about LI, insufficient knowledge, limited understanding of symptoms as well as social and cultural factors affect the individual's perception of LI and consumption of dairy foods. A four-step intervention is offered as a suggested bedside strategy for management of lactose intolerance is outlined in Table 48.3. It incorporates the basic principles of inquiry, identification, information and implementation. In addition to confirming the diagnosis with objective testing, all bedside interventions should be accompanied to a regularly scheduled re-assessment of the patient's nutritional status to identify potential nutrient deficiencies and prevent unnecessary disease.

48.5.2   Specific Dietary Strategies

Dairy avoidance increases the risk for certain adverse health outcomes, whereas consumption of a dairy-rich diet reduces risks due to the superior quality of dairy food nutrients (Weaver 2006). Simple measures such as consuming lactose-containing products in smaller, divided doses and mixing with other foods in a meal (notably protein and fat) improves the tolerance. The regular consumption of lactose in foods also results in resolution of symptoms in LNP over

Table 48.3   Bedside approach to managing lactose intolerance.

The Four "I's"	Bedside Approach to the Management of Lactose Intolerance
Inquire	Inquire about perception of tolerance, symptom onset and past experience Inquire about previous attempts to consume dairy foods Inquire about testing method and confirm diagnosis Inquire about personal fears or concerns
Identify	Identify the association between the dairy food and the symptoms Identify physical exam findings and objective test results Identify potential clinical masqueraders of disease Identify applicable race or culture specific considerations Identify clinical barriers to dairy food consumption
Inform	Inform patients of clinical definitions of LM and LI in simple terms Inform the patient about the 2010 consensus development conference Inform individuals of the health consequences of dairy avoidance Inform clients which food groups to encourage
Implement	Implement evidence based, culturally appropriate recommendations Implement dietary strategies that include three servings per day of dairy Implement strategies that promote adequate calcium and vitamin D intake Implement tips for tolerance for symptomatic individuals
Clinical caveat	*2010 NIH Lactose Intolerance Consensus Development Conference criteria to define lactose intolerance requires both confirmation of symptoms that are temporally associated with dairy food ingestion and evidence of lactose maldigestion by objective testing*

Used with permission: J. Keith, M.D.

a period of time. Moreover, a progressive increase in the lactose dose does not cause any progressive worsening of symptoms and leads to improved tolerance (Hertzler and Savaiano 1996). Studies have demonstrated that ingestion of lactose in LNP alters the ability of colonic bacteria to metabolize lactose and is a clinically useful strategy to improve tolerance and consumption. The time for adaptation varies ranging from weeks to months, but is usually 3–4 weeks. Gradual reintroduction is recommended following periods of cessation of intake to avoid the recurrence of symptoms.

Changes in dietary habits such as adding yogurt and aged cheese are well tolerated due to low lactose content. It is noteworthy that chocolate-flavored milk has a beneficial effect on suppressing symptoms of LI. The chocolate flavoring, which contains a small amount of fat, is thought to slow digestion long enough to allow for increased nutrient absorption with minimal or no symptoms. Addition of cocoa or chocolate flavoring not only improved subjective symptoms of bloating and cramping significantly, but also improved the digestion of lactose confirmed objectively by HBT as noted in Figure 48.2 (Lee and Hardy 1989). The degree of benefit obtained from probiotics in LI individuals depends on multiple factors such as microbial strain used as the probiotic, the amount of micro-organisms ingested, the amount of lactase produced in the host, simultaneous ingestion of yogurt, *etc*. The lactobacillus bulgaricus and *Streptococcus thermophillus* are the most effective strains. Other strains of lactobacillus are not as effective as these two, when compared to the changes in HBT and symptomatic improvement (Onwulata *et al.* 1989).

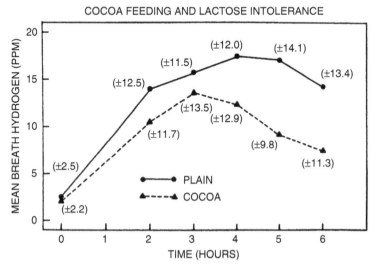

Figure 48.2 Changes in breath hydrogen value with time after administration of plain and cocoa milk formulas (Lee and Hardy 1989). ●, milk without added cocoa; ▲, milk with added cocoa. Each point is the mean ± standard deviation of the values.

Table 48.4 Tips for tolerance for the lactose intolerant individual.

Tips for Tolerance	*Examples*
Lactose free dairy foods are now widely available in the stores	Lactaid® lactose free milk, cottage cheese and ice cream
Lactase enzyme supplements taken with the first bite or sip of dairy foods	Lactaid® enzymes with first bite or sip of food in an acidic stomach, repeat supplement if the meal lasts longer than 30–45 minutes
Dairy foods are best tolerated with protein and fat in the meal which slows digestion long enough for the healthy nutrients to be absorbed	Have milk as part of lunch and dinner, not on an empty stomach
Slowly introduce dairy foods and consume with a meal	Begin with $\frac{1}{4}$–1/2 glass of flavored milks or hot chocolate made with milk then slowly increase
Add dairy to your favorite food	Try adding a small amount of milk to tomato soup, try yogurt-based fruit smoothies, or salad dressing
Choose aged cheeses that naturally have less lactose	Try Cheddar, Colby, Swiss or Parmesan cheese; top a slice of grilled turkey burger with a slice of cheese
Top healthy foods with shredded cheese as a dip or dressing	Salads or vegetables topped with shredded aged cheese
Try different flavors and types of yogurts on your favorite fruit, vegetable or cereal	Yogurt with fruit, yogurt with granola, yogurt based salad dressing, yogurt dips
It takes about 3–4 weeks for the gut to adapt to a new food	Start with $\frac{1}{4}$–1/2 cup of milk or $\frac{1}{2}$ serving of yogurt with a meal then gradually increase

Modified From: National Dairy Council Lactose Intolerance Health Education Toolkit. http://www.nationaldairycouncil.org/EducationMaterials/HealthProfessionalsEducationKits/Pages/LactoseIntoleranceHealthEducationKit.aspx. Used with permission: J. Keith, M.D.

Although fluid milk substitutes such as rice drink and soy beverages are available, the 2010 NIH Consensus Development Panel recommended reintroduction of dairy foods as the primary intervention when managing LNP individuals. Table 48.4 outlines specific tips for tolerance that are patient-friendly.

Summary Points

- There are three main types of lactase deficiency, but it is rare for the complete absence of lactase to occur in adults.
- True lactose intolerance requires the documentation of lactose maldigestion by an objective measure and the temporal association of symptoms with lactose ingestion.
- Lactose intolerance does occur but is not as prevalent as the older literature suggests.
- Dairy avoidance is associated with adverse health consequences and a dairy-enriched diet can reduce health risks.

- There are specific dietary strategies available that allow for dairy consumption in lactose-intolerant individuals, such that lactose intolerance does not require dairy avoidance.
- Adaptation to the re-introduction of dairy foods into the diet of a person with LNP requires at least 3–4 weeks of constant dietary modification.

Key Facts

Key Facts for Healthcare Professionals

- The prevalence of LI is not as widespread as was previously reported.
- The diagnosis of LI requires both documentation of LM and clinical symptoms temporally related to lactose ingestion.
- Due to the negative health consequences of dairy avoidance, LI should be managed with effective interventions that promote high diet quality, adequate nutrient consumption and improved dairy food tolerance.
- Based on the 2010 NIH Lactose Intolerance Consensus Conference report, re-introduction of dairy foods should be the primary intervention when treating individuals with LI.

Definitions of Words and Terms

Lactase non-persistence: The loss or reduction in lactase activity that is regulated by the lactase gene on chromosome 2q21-22.

Lactose maldigestion: The intraluminal process that occurs when unabsorbed lactose passes into the colon and is digested by colonic flora producing hydrogen gas, methane gas and short chain fatty acids.

Lactose intolerance: The clinical syndrome of symptomatic lactose maldigestion in humans that requires documentation of lactose maldigestion and the temporal association of symptoms to the lactose ingestion.

Primary lactase deficiency: The normal age-related decline in lactase enzyme activity following weaning with maintenance of residual lactase activity in adults which allows for digestion of a person-specific amount of lactose.

Secondary lactase deficiency: a self-limited and reversible reduction in lactase activity due to non-specific injury to the lactase-producing brush border enterocytes on the villi in the small intestine.

List of Abbreviations

CH_4	Methane gas
CLD	Congenital lactase deficiency
DLD	Developmental lactase deficiency
FLD	Familial lactase deficiency
H_2	Hydrogen gas
HBT	Hydrogen breath test

LChT   Lactose challenge test
LI   Lactose intolerance
LM   Lactose malabsorption
LNP   Lactase non-persistent
LP   Lactase persistent
LPH   Lactase-phlorizin hydrolase
LTT   Lactose tolerance test
NIH   National Institutes of Health
PLD   Primary lactase deficiency
SAT   Stool acidity test
SCFA   Short chain fatty acids
SLD   Secondary lactase deficiency

References

Arola, H., 1994. Diagnosis of Hypolactasia and Lactose Malabsorption, *Scandinavian Journal of Gastroenterology* 29(s202): 26–35.

Arvanitakis, C., Chen, G.H., Folscroft, J., and Klotz, A.P., 1977. Lactase deficiency--a comparative study of diagnostic methods, *The American Journal of Clinical Nutrition* 30(10): 1597–1602.

Bayless, T.M., and Rosensweig, N.S., 1966. A Racial Difference in Incidence of Lactase Deficiency, *JAMA: The Journal of the American Medical Association* 197(12): 968–972.

Brannon, P.M., Carpenter, T.O., Fernandez, J.R., Gilsanz V., Gould, J.B., Hall, K.E., Hui, S.L., Lupton, J.R., Mennella, J., Miller, N.J., Osganian, S.K., Sellmeyer, D.E., Suchy, F.J., and Wolf, M.A., 2010. NIH Consensus Development Conference Statement: Lactose Intolerance and Health, *NIH Consensus and State-of-the-science Statements* 27(2): Bethesda, MD U.S. Dept. of Health and Human Services, Public Health Service, National Institutes of Health, Office of Medical Applications Research.

Fulgoni, V., Nicholls, J., Reed, A., Buckley, R., Kafer, K., Huth, P., DiRenzo and Miller, G.D., 2007. Dairy Consumption and Related Nutrient Intake in African American Adults and Children in the United States: Continuing Survey of Food Intakes by Individuals 1994–1996, 1998 and the National Health and Nutrition Examination Survey 1999–2000, *J Am Diet Assoc* 107, 256–264.

Gracey, M., and Burke, V., 1973. Sugar-induced diarrhea in children, *Arch Dis Child* 48, 331–336.

Heaney, R.P., Keith, J.N., and Duyff, R.L., 2011. Unintended Consequences of Dairy Avoidance, *White Paper* http://www.nationaldairycouncil.org/SiteCollectionDocuments/health_wellness/lactose_intolerance/Unintended%20Consequences%20of%20Dairy%20Avoidance%20-%202011.pdf accessed 12 January 2012.

Hertzler, S.R., and Savaiano, D.A., 1996. Colonic adaptation to daily lactose feeding in lactose maldigesters reduces lactose intolerance, *The American Journal of Clinical Nutrition* 64(2): 232–236.

Heyman, and Melvin, B., 2006. Lactose Intolerance in Infants, Children and Adolescents, *Pediatrics* 118(3): 1279–1286.

Holick, M.F., Binkley, N.C., Biscchoff-Ferrari, H.A., Gordon, C.M., Hanley, D.A., Heaney, R.P., Murrad, M.H., and Weaver, C.M., 2011. Evaluation, Treatment and Prevention of Vitamin D Deficiency: an Endocrine Society Clinical Practice Guideline, *J Clin Endocrinol Metab* 96, 1911–1930.

Holzel, A., 1964. Nutritional consequences of altered carbohydrate absorption in infancy and childhood, *Proc Nutr Soc* 23, 123–129.

Holzel, A., 1968. Defects of sugar absorption, *Proc Roy Soc Med* 61, 1095–1099.

Holzel, A., Schwartz, V., and Sutcliffe, K.W., 1959. Defective lactose absorption causing malnutrition in infancy, *Lancet* 1, 1126–1128.

Hovde, O., and Farup, P., 2009. A comparison of diagnostic tests for lactose malabsorption - which one is the best?, *BMC Gastroenterology* 9(1): 1–7.

IOM, 2011. Dietary Reference Intakes for Calcium and Vitamin D., Washington DC: The National Academies Press. Institute of Medicine, Food and Nutrition Board.

Keith, J.N., Nicholls, J., Reed, A., Kafer, K., and Miller, G.D., 2011. The Prevalence of Self-reported Lactose Intolerance and the Consumption of Dairy Foods Among African American Adults Are Less Than Expected, *J Natl Med Assoc* 103, 36–45.

Lee, C.M., and Hardy, C.M., 1989. Cocoa feeding and human lactose intolerance, *The American Journal of Clinical Nutrition* 49(5): 840–844.

NAMS 2010. Management of osteoporosis in postmenopausal women: 2010 position statement of The North American Menopause Society, *Menopause* 17(1): 25–54.

Newcomer, A.D., McGill, D.B., Thomas, P.J., and Hofmann, A.F., 1975. Prospective Comparison of Indirect Methods for Detecting Lactase Deficiency, *New England Journal of Medicine* 293(24): 1232–1236.

Nicklas, T.A., Qu H., Hughes, S.O., He M., Wagner, S.E., Foushee, H.R., and Shewchuk R.M., 2011. Self-perceived lactose intolerance results in lower intakes of calcium and dairy foods and is assicated with hypertension and diabetes in adults, *Am J Clin Nutr* 94(1): 191–198.

Nicklas, T.A., Qu, H., Hughes, S.O., Wagner, S.E., Foushee, H.R., and Shewchuk, R.M., 2009. Prevalence of self-reported lactose intolerance in multiethnic sample of adults, *Nutrition Today* 44(5): 222–227.

NIH Dietary Supplement Fact Sheet: Vitamin D, http://ods.od.nih.gov/factsheets/vitamind, accessed January 30, 2012.

NOF Osteoporosis Fast Facts, http://www.nof.org/node/40, accessed January 30, 2012.

Onwulata, C.I., Rao, D.R., and Vankineni, P., 1989. Relative efficiency of yogurt, sweet acidophilus milk, hydrolyzed-lactose milk, and a commercial lactase tablet in alleviating lactose maldigestion, *The American Journal of Clinical Nutrition* 49(6): 1233–1237.

Robayo-Torres, C.C., and Nichols, B.L., 2007. Molecular Differentiation of Congenital Lactase Deficiency from Adult-Type Hypolactasia, *Nutrition Reviews* 65(2): 95–98.

Savaiano, D.A., Boushey, C.J., and McCabe, G.P., 2006. Lactose Intolerance Symptoms Assessed by Meta-Analysis: A Grain of Truth That Leads to Exaggeration, *The Journal of Nutrition* 136(4): 1107–1113.

Schrimshaw, N.S., and Murray, E.B., 1988. Chapter 2 Prevalence of lactose maldigestion, *The American Journal of Clinical Nutrition* 48(4): 1086–1098.

Sharma, S., Murphy, S.P., Wilkens, L.R., Shen, L., Hankin, J.H., Monroe, K.R., Henderson, B., and Koleonel, L.N., 2004. Adherence to the Food Guide Pyramid Recommendations Among African Americans and Latinos: Results from the Multiethnc Cohort, *J Am Diet Assoc* 104, 1873–1877.

Siris, E.S., Miller, P.D., Barrett-Connor, E, Faulkner, K.G., Wehren, L.E., Abbott, T.A., Berger, M.L., Santora, A.C., and Sherwood, L.M., 2001. Identification and Fracture Outcomes of Undiagnosed Low Bone Mineral Density in Postmenopausal Women, *JAMA: The Journal of the American Medical Association* 286(22): 2815–2822.

Weaver, C.M., and Heaney, R.P., 2006. *Food Sources, Supplements and Bio-availability.*, ed. Weaver, C.M., and Heaney, R.P. (Eds). (Calcium in Human Health.; Totowa, NJ: Humana Press).

Wilt, T.J., Shaukat, A., Shamliyan, T., Taylor, B.C., MacDonald, R., Tacklind, J., Rutks, I., Schwarzenberg, S.J., Kane, R.L., and Levitt, M., 2010. *Evid. Rep. Technol. Assess (Full Rep.)* 192, 1–410.

Wooten, W.J., and Price, W., 2004. Consensus Report of the National Medical Association: The Role of Dairy and Dairy Nutrients in the Diet of African Americans, *J Natl Med Assoc* 96, 1S–31S.

Subject Index

Illustrations and figures are in **bold**. Tables are in *italics*.